*Elements
of Chemical
Reaction
Engineering*

Third Edition

ISBN 0-13-531708-8

90000

9 780135 317082

Elements of Chemical Reaction Engineering

Third Edition

H. SCOTT FOGLER

Ame and Catherine Vennema Professor
of Chemical Engineering
The University of Michigan, Ann Arbor

Prentice Hall PTR
Upper Saddle River, New Jersey 07458
http://www.phptr.com

Library of Congress Cataloging-in-Publication Data

Fogler, H. Scott.
    Elements of chemical reaction engineering / H. Scott Fogler. --
3rd ed.
      p. cm. -- (Prentice-Hall international series in the physical
and chemical engineering sciences)
    Includes bibliographical references and index.
    ISBN 0-13-531708-8 (cloth)
    1. Chemical reactors. I. Title. II. Series.
TP157.F65 1999   96-1052
660'28--DC20   CIP

Acquisitions editor: Bernard M. Goodwin
Cover design director: Jerry Votta
Manufacturing manager: Alexis R. Heydt
Marketing manager: Kaylie Smith
Text composition: Preparé/Emilcomp

Prentice Hall books are widely used by corporations and
government agencies for training, marketing, and resale.

The publisher offers discounts on this book when ordered
in bulk quantities. For more information contact:

    Corporate Sales Department
    Phone: 800-382-3419
    Fax: 201-236-7141
    E-mail: corpsales@prenhall.com

    Or write:

    Prentice Hall PTR
    Corp. Sales Dept.
    One Lake Street
    Upper Saddle River, New Jersey 07458

Printed in the United States of America

10 9 8 7 6 5 4 3

ISBN 0-13-531708-8

Dedicated to the memory of

Professors

Giuseppe Parravano
Joseph J. Martin
Donald L. Katz

of the University of Michigan
whose standards and lifelong achievements
serve to inspire us

Contents

12   *DIFFUSION AND REACTION IN POROUS CATALYSTS*   **738**

13   *DISTRIBUTIONS OF RESIDENCE TIMES FOR CHEMICAL REACTORS*   **809**

Preface

A. The Audience

This book is intended for use as both an undergraduate- and graduate-level text in chemical reaction engineering. The level of difficulty will depend on the choice of chapters to be covered and the type and degree of difficulty of problems assigned. Most problems requiring significant numerical computations can be solved with a personal computer using either POLYMATH or MATLAB.

B. The Goals

B.1. To Develop a Fundamental Understanding of Reaction Engineering

The first goal of this book is to enable the reader to develop a clear understanding of the fundamentals of chemical reaction engineering. This goal will be achieved by presenting a structure that allows the reader to solve reaction engineering problems through reasoning rather than through memorization and recall of numerous equations and the restrictions and conditions under which each equation applies. To accomplish this, we use (1) conventional problems that reinforce the student's understanding of the basic concepts and principles (included at the end of each chapter); (2) problems whose solution requires reading the literature, handbooks, or other textbooks on chemical engineering kinetics; and (3) problems that give students practice in problem definition and alternative pathways to solutions. The algorithms presented in the text for reactor design provide a framework through which one can develop confidence through reasoning rather than memorization.

To give a reference point as to the level of understanding required in the profession, a number of reaction engineering problems from the *California Board of Registration for Civil and Professional Engineers—Chemical Engineering Examinations* (PECEE) are included. Typically, each problem should require approximately one-half hour to solve. Hints on how to work the California exam problems can be found in the *Summary Notes* and in the *Thoughts on Problem Solving* on the CD-ROM.

The second and third goals of this book are to increase the student's critical thinking skills and creative thinking skills by presenting heuristics and problems that encourage the student to practice these skills.

B.2. To Develop Critical Thinking Skills

Due to the rapid addition of new information and the advancement of science and technology that occur almost daily, an engineer must constantly expand his or her horizons beyond simply gathering information and relying on the basic engineering principles.

A number of homework problems have been included that are designed to enhance critical thinking skills. Socratic questioning is at the heart of critical thinking and a number of homework problems draw from R. W. Paul's six types of Socratic questions:[1]

(1) *Questions for clarification*: Why do you say that? How does this relate to our discussion?

(2) *Questions that probe assumptions*: What could we assume instead? How can you verify or disprove that assumption?

(3) *Questions that probe reasons and evidence*: What would be an example?

(4) *Questions about viewpoints and perspectives*: What would be an alternative?

(5) *Questions that probe implications and consequences*: What generalizations can you make? What are the consequences of that assumption?

(6) *Questions about the question*: What was the point of this question? Why do you think I asked this question?

Practice in critical thinking can be achieved by assigning additional parts to the problems at the end of each chapter that utilize R. W. Paul's approach. Most of these problems have more than one part to them. The instructor may wish to assign all or some of the parts. In addition, the instructor could add the following parts to any of the problems:

- Describe how you went about solving this problem.
- How reasonable is each assumption you made in solving this problem?

---

[1] Paul, R. W., *Critical Thinking* (Published by the Foundation for Critical Thinking, Santa Rosa, CA, 1992).

- Ask another question or suggest another calculation that can be made for this problem.
- Write a few sentences about what you learned from working this homework problem and what you think the point of the problem is.

Another important exercise in this text that fosters critical thinking is the critiquing of journal articles. For the last 20 years, students in the graduate reaction engineering class at the University of Michigan have been required to carry out an in-depth critique of a journal article on chemical engineering kinetics. Although the students were told that choosing an article with erroneous data or reasoning was not necessary for a successful critique, finding an error made the whole assignment much more fun and interesting. Consequently, a select number of problems at the end of chapters involve the critique of journal articles on reaction engineering which may or may not have major or minor inconsistencies. In some cases, a small hint is given to guide the student in his or her analysis.

B.3. To Develop Creative Thinking Skills

To help develop creative thinking skills, a number of problems are open-ended to various degrees. Beginning with Chapter 4, the first problem in each chapter provides students the opportunity to practice their creative skills by making up and solving an original problem. Problem 4-1 gives some guidelines for developing original problems. A number of techniques that can aid the students in practicing their creativity (e.g., lateral thinking and brainstorming) can be found in Fogler and LeBlanc.[2]

"What if..." problems can serve to develop both critical and creative thinking skills. The second problem of each chapter (e.g., 4-2) contains *"What if..."* questions that encourage the student to think beyond a single answer or operating condition. These problems can be used in conjunction with the living example problems on the CD to explore the problem. Here, questioning can be carried out by varying the parameters in the problems.

One of the major goals at the undergraduate level is to bring the students to the point where they can solve complex reaction systems, such as multiple reactions with heat effects, and then ask **"What if..."** questions and look for optimum operating conditions. One problem whose solution exemplifies this goal is the Manufacture of Styrene, **Problem 8-30**.

(1) Ethylbenzene → Styrene + Hydrogen: *Endothermic*
(2) Ethylbenzene → Benzene + Ethylene: *Endothermic*
(3) Ethylbenzene + Hydrogen → Toluene + Methane: *Exothermic*

In this problem, the students can find a number of operating conditions which maximize the yield and selectivity.

The parameters can also be easily varied in the example problems by loading the POLYMATH or MATLAB programs from the CD onto a computer to explore and answer *"What if..."* questions.

---

[2] Fogler, H. S. and S. E. LeBlanc, *Strategies for Creative Problem Solving* (Upper Saddle River, NJ: Prentice Hall, 1995).

C. The Structure

The strategy behind the presentation of material is to continually build on a few basic ideas in chemical reaction engineering to solve a wide variety of problems. These ideas are referred to as the *Pillars of Chemical Reaction Engineering*, on which different applications rest. The pillars holding up the application of chemical reaction engineering are shown in Figure P-1.

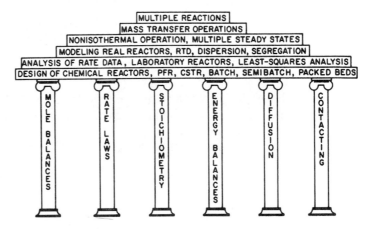

Figure P-1   Pillars of Chemical Reaction Engineering.

The architecture and construction of the structure shown in Figure P-1 had many participants, most notably Professors Amundson, Aris, Smith, Levenspiel, and Denbigh. The contents of this book may be studied in virtually any order after the first four chapters, with few restrictions. A flow diagram showing possible paths is shown in Figure P-2.

In a three-hour undergraduate course at the University of Michigan, approximately eight chapters are covered in the following order: Chapters 1, 2, 3, 4, and 6, Sections 5.1–5.3, and Chapters 8, 10, and parts of either 7 or 13. Complete sample syllabi for a 3-credit-hour course and a 4-credit-hour course can be found on the CD-ROM.

The reader will observe that although metric units are used primarily in this text (e.g., $kmol/m^3$, J/mol), a variety of other units are also employed (e.g., lb/ft^3). This is intentional. It is our feeling that whereas most papers published in the future will use the metric system, today's engineers as well as those graduating over the next ten years will be caught in the transition between English, SI, and metric units. As a result, engineers will be faced with extracting information and reaction rate data from older literature which uses English units as well as the current literature using metric units, and they should be equally at ease with both.

The notes in the margins are meant to serve two purposes. First, they act as guides or as commentary as one reads through the material. Second, they identify key equations and relationships that are used to solve chemical reaction engineering problems.

Finally, in addition to developing the intellectual skills discussed above, this is a book for the professional bookshelf. It is a "how to" book with numerous

| Margin Notes |

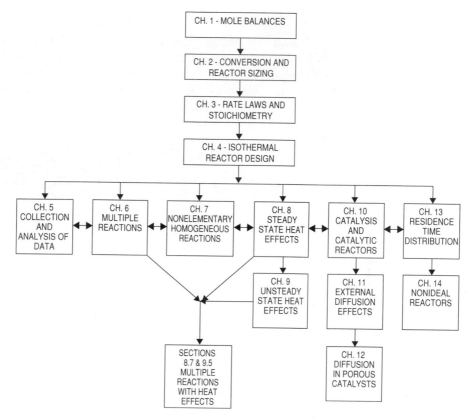

Figure P-2   Sequences for Studying the Text.

examples and clear explanations, rather than an outline of the principles and the philosophy of chemical reaction engineering. There are many other applications described in the text.

D. The Applications

Important applications of chemical reaction engineering (CRE) of all kinds can be found both inside and outside the chemical process industries (CPI). In this text, examples from the chemical process industries include the manufacture of ethylene oxide, phthalic anhydride, ethylene glycol, metaxylene, styrene, sulfur trioxide, propylene glycol, ketene, and *i*-butane just to name a few. Also, plant safety in the CPI is addressed in both example problems and homework problems. These are real industrial reactions with actual data and reaction rate law parameters.

Because of the wide versatility of the principles of CRE, a number of examples outside the CPI are included, such as the use of wetlands to degrade toxic chemicals, smog formation, longevity of motor oils, oil recovery, and pharmacokinetics (cobra bites, SADD-MADD, drug delivery). A sampling of the applications is shown graphically in the following figures.

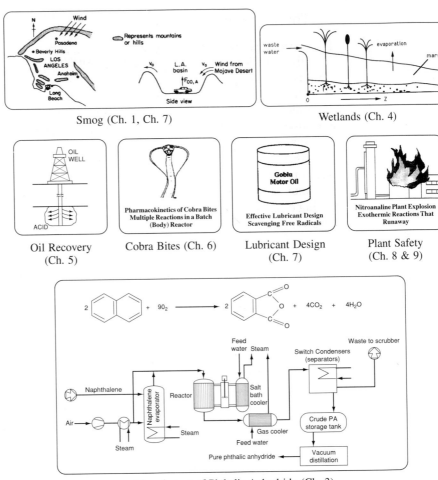

Smog (Ch. 1, Ch. 7)

Wetlands (Ch. 4)

Oil Recovery
(Ch. 5)

Cobra Bites (Ch. 6)

Lubricant Design
(Ch. 7)

Plant Safety
(Ch. 8 & 9)

Manufacture of Phthalic Anhydride (Ch. 3)

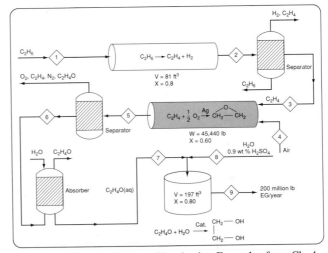

Chemical Plant for Ethylene Glycol using Examples from Ch. 4

E. The Components of the CD-ROM

The primary purpose of the CD-ROM is to serve as an enrichment resource. Its objectives are fourfold: (1) To provide the option/opportunity for further study or clarification on a particular concept or topic through Summary Notes, additional examples, interactive computing modules and web modules, (2) To provide the opportunity to practice critical thinking skills, creative thinking skills, and problem solving skills through the use of **"What if..."** questions and "living example problems," (3) To provide additional technical material for the professional reference shelf, (4) To provide other tutorial information, such as additional homework problems, thoughts on problem solving, how to use computational software in chemical reaction engineering, and representative course structures. The following components are listed at the end of most chapters and can be accessed, by chapter, on the CD.

- **Learning Resources**

  These resources give an overview of the material in each chapter and provide extra explanations, examples, and applications to reinforce the basic concepts of chemical reaction engineering. The learning resources on the CD-ROM include:

  1. *Summary Notes*

     These are Summary Notes that will give an overview of each chapter, and are taken from lecture notes from an undergraduate class at Michigan.

  2. *Web Modules*

     These modules which apply key concepts to both standard and non-standard reaction engineering problems (e.g., the use of wetlands to degrade toxic chemicals, cobra bites) can be loaded directly from the CD-ROM. Additional Web Modules are expected to be added over the next several years. (http://www.engin.umich.edu/~cre)

  3. *Interactive Computer Modules*

     Students can use the corresponding Interactive Computer Modules to review the important chapter concepts and then apply them to real problems in a unique and entertaining fashion. The Murder Mystery module has long been a favorite with students across the nation.

  4. *Solved Problems*

     A number of solved problems are presented along with problem-solving heuristics. Problem-solving strategies and additional worked example problems are available in the **Thoughts on Problem Solving** section of the CD-ROM.

- **Living Example Problems**

  A copy of POLYMATH is provided on the CD-ROM for the students to use to solve the homework problems. The example problems that use an ODE solver (e.g., POLYMATH) are referred to as "living example problems" because the students can load the POLYMATH program directly onto their own computer in order to study the problem. Students are encouraged to change parameter values and to "play with" the key variables and assumptions. Using the living example problems to explore the problem and asking *"What if…"* questions provides the opportunity to practice critical and creative thinking skills.

- **Professional Reference Shelf**
  This section of the CD-ROM contains:
  1. material that is important to the practicing engineer, although it is typically not included in the majority of chemical reaction engineering courses.
  2. material that gives a more detailed explanation of derivations that were abbreviated in the text. The intermediate steps to these derivations are given on the CD-ROM.

- **Additional Homework Problems**
  New problems were developed for this edition that provide a greater opportunity to use today's computing power to solve realistic problems.

- **Other CD-ROM Material**
  In addition to the components listed at the end of each chapter the following components are included on the CD-ROM:
  1. *Software ToolBox*
     Instructions on how to use the different software packages (POLYMATH, MATLAB, and ASPEN PLUS) to solve examples.
  2. *Representative Syllabi for a 3- and a 4-Credit Course*
     The syllabi give a sample pace at which the course could be taught as well as suggested homework problems.
  3. FAQ
     These are Frequently Asked Questions (FAQ's) from undergraduate students taking reation engineering.

- **Virtual Reality Module (WWW)**
  This module provides an opportunity to move inside a catalyst pellet to observe surface reactions and coking. It can be found at http://www.engin.umich.edu/labs/vrichel.

F. The Integration of the Text and the CD-ROM

There are a number of ways one can use the CD in conjunction with the text. The CD provides *enrichment resources* for the reader in the form of interactive tutorials. Pathways on how to use the materials to learn chemical reaction engineering are shown in Figure P-3 and P-4. The keys to the CRE learning flowsheets are

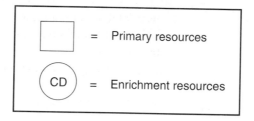

F.1. For the University Student

In developing a fundamental understanding of the material, the student may wish to use only the primary resources without using the CD-ROM, (i.e., using only the boxes shown in Figure P-3) or the student may use a few or all of the interactive tutorials in the CD-ROM (i.e., the circles shown in Figure P-3). However, to practice the skills that enhance critical and creative thinking, the students are strongly encouraged to use the *Living Example Problems* and vary the model parameters to ask and answer **"What if..."** questions.

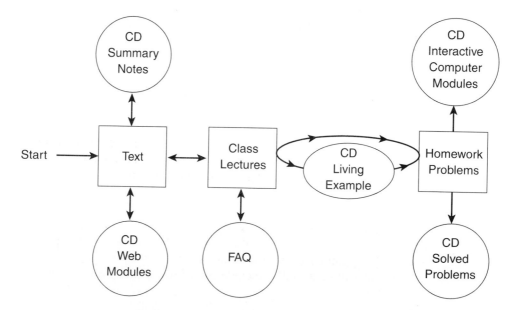

Figure P-3 A Student Pathway to Integrate the Class Text and CD.

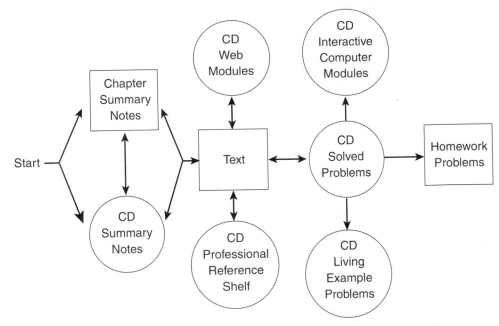

Figure P-4 A Problem-Solving Pathway to Integrate the text and the CD.

One notes that while the author recommends studying the living examples before working home problems, they may be bypassed, as is the case with all the enrichment resources if time is not available. However, class testing of the enrichment resources reveals that they not only greatly aid in learning the material but they may also serve to motivate students through the novel use of CRE principles.

F.2.   For the Practicing Engineer

Practicing engineers may want to first review the CD summary notes or the summaries at the end of each chapter to refresh their memories as to what they have previously studied. They can then focus on the topics that they want to study in the text using the web modules, solved problems, and interactive computer modules as tutorials. They can also learn more about specialty topics by using the CD reference shelf. The flow diagram is shown in Figure P-4.

G. The Web

The Web site (http://www.engin.umich.edu/~cre) will be used to update the text and the CD-ROM. It will identify typographical and other errors in the 1st and 2nd printings of the 3rd edition of the text. In the near future, additional material will be added to include more solved problems as well as additional Web Modules.

H. What's New

The main thrust of the new edition is to enable the student to solve *Digital Age*[3] reaction engineering problems. Consequently the content, example problems, and homework problems focus on achieving this goal. These problems provide the students an opportunity to practice their critical and creative thinking skills by "playing with" the problems through parameter variations. Consequently, some of the text material, e.g., control of chemical reactors and safety, was added because it provides opportunities to formulate and solve problems. For example, in the *Case Study* on safety, the student can use the CD-ROM to carry out a post-mortem on the nitroanaline explosion in Example 9-2 to find out what would have happened if the cooling had failed for 5 minutes instead of 10 minutes. Significant effort has been devoted to developing example and homework problems that foster critical and creative thinking.

The use of mole balances in terms of concentrations and flow rates rather than conversions is introduced early in the text so they can be easily applied to membrane reactors and multiple reactions. The 3rd edition contains more industrial chemistry with real reactors and real reactions and extends the wide range of applications to which chemical reaction engineering principles can be applied (i.e., cobra bites, drug medication, ecological engineering). New material includes spherical reactors, recycle reactors, trickle bed reactors, fluidized bed reactors, regression of rate data, etching of semiconductors, multiple reactions in RTD models, the application of process control to CSTRs, safety, collision theory, transition state theory, and an example using computational chemistry to calculate an activation energy. The material that has been greatly expanded includes polymerization, heat effects in batch reactors and in multiple reactions, catalysts and catalytic reactions, experimental design, and reactor staging. The living example problems on the CD-ROM are in both POLYMATH and MATLAB.

A large number of enrichment resources are provided on the CD-ROM that can help the student over difficult spots. However, if there is a time constraint, or the reader's computer breaks down, the reader need only read the text and proceed along the pathway of the boxes shown in Figures P-3 and P-4.

I. Acknowledgments

Many of the problems at the end of the various chapters were selected from the *California Board of Registration for Civil and Professional Engineers—Chemical Engineering Examinations (PECEE)* in past years. The permission for use of these problems, which, incidentally, may be obtained from the Documents Section, California Board of Registration for Civil and Professional Engineers—Chemical Engineering, 1004 6th Street, Sacramento, CA 95814, is gratefully acknowledged. (Note: These problems have been copyrighted by the California Board of Registration and may not be reproduced without their permission.)

---

[3] Fogler, H. S., "*Teaching Critical Thinking, Creative Thinking, and Problem Solving in the Digital Age*" (Phillips Lecture, Oklahoma State University Press, April 25, 1997).

However, all intensive laws tend often to have exceptions. Very interesting concepts take orderly, responsible statements. Virtually all laws intrinsically are natural thoughts. General observations become laws under experimentation.

There are so many colleagues and students who contributed to this book that it would require another chapter to thank them all in an appropriate manner.

I would like to again acknowledge all my friends and colleagues for their contributions to the 1st and 2nd editions (See Introduction, CD-ROM). For the 3rd edition, I would like to give special recognition to the students who contributed so much to the CD-ROM: In particular, Dieter Schweiss, Anuj Hasija, Jim Piana, and Susan Fugett, with thanks also to Anurag Murial, Gavin Sy, Scott Conaway, Mayur Valanju, Matt Robinson, Tim Mashue, Lisa Ingalls, Sean Conners, Gustavo Bolaños, and Ellyne Buckingham. Further, Tim Hubbard, Jessica Hamman, David Johnson, Kylas Subramanian, Sumate Charoenchaidet, Lisa Ingalls, Probjot Singh, Abe Sendijarevic, and Nicholas R. Abu-Absi worked on the solution manual. Jason Ferns, Rob Drewitt, and Probjot Singh contributed to the problems, while Professor Andy Hrymak, Probjot Singh, Marty Johnson, Sumate Charoenchaidet, N. Vijay, and K. Subramanian helped with proofreading the galleys. Thanks to my graduate students Venkat Ramachandran, Chris Fredd, Dong Kim, Barry Wolf, Probjot Singh, Vaibhav Nalwaya, and Ann Wattana for their patience and understanding. Barbara Zieder (copy-editing), Lisa Garboski (production), and Yvette Raven (CD-ROM) did an excellent job in bringing the project to a successful completion. Bernard Goodwin of Prentice Hall was extremely helpful and supportive throughout. The stimulating discussions with Professors John Falconer, D. B. Battacharia, Richard Braatz, Kristi Anseth, and Al Weimer are greatly appreciated. I also appreciate the friendship and insights provided by Dr. Lee Brown, who contributed to chapters 8, 12, 13, and 14. Professor Mike Cutlip gave not only suggestions and a critical reading of many sections, but most important provided continuous support and encouragement throughout the course of this project. Laura Bracken is so much a part of this manuscript through her excellent deciphering of equations and scribbles, and typing, her organization, and always present wonderful disposition. Thanks *Radar*!! Finally, to my wife Janet, love and thanks. Without her enormous help and support the project would not have been possible.

HSF
Ann Arbor

For updates on the CD and typographical errors for this printing see the web site:

http://www.engin.umich.edu/~cre

*Elements
of Chemical
Reaction
Engineering*

Third Edition

Mole Balances   **1**

The first step to knowledge
is to know that we are ignorant.

Socrates (470–399 B.C.)

Chemical kinetics and reactor design are at the heart of producing almost all industrial chemicals. It is primarily a knowledge of chemical kinetics and reactor design that distinguishes the chemical engineer from other engineers. The selection of a reaction system that operates in the safest and most efficient manner can be the key to the economic success or failure of a chemical plant. For example, if a reaction system produced a large amount of undesirable product, subsequent purification and separation of the desired product could make the entire process economically unfeasible. The chemical kinetic principles learned here, in addition to the production of chemicals, can be applied in areas such as living systems, waste treatment, and air and water pollution. Some of the examples and problems used to illustrate the principles of chemical reaction engineering are: the use of wetlands to remove toxic chemicals from rivers, increasing the octane number of gasoline, the production of anti-freeze starting from ethane, the manufacture of computer chips, and the application of enzyme kinetics to improve an artificial kidney.

This book focuses on a variety of chemical reaction engineering topics. It is concerned with the rate at which chemical reactions take place, together with the mechanism and rate-limiting steps that control the reaction process. The sizing of chemical reactors to achieve production goals is an important segment. How materials behave within reactors, both chemically and physically, is significant to the designer of a chemical process, as is how the data from chemical reactors should be recorded, processed, and interpreted.

Before entering into discussions of the conditions that affect chemical reaction rates and reactor design, it is necessary to account for the various chemical species entering and leaving a reaction system. This accounting process is achieved through overall mole balances on individual species in the reacting system. In this chapter we develop a general mole balance that can be applied to any species (usually a chemical compound) entering, leaving, and/or remaining within the reaction system volume. After defining the rate of reaction, $-r_A$, and discussing the earlier difficulties of properly defining the chemical reaction rate, in this chapter we show how the general balance equation may be used to develop a preliminary form of the design equations of the most common industrial reactors: batch, continuous-stirred tank (CSTR), and tubular. In developing these equations, the assumptions pertaining to the modeling of each type of reactor are delineated. Finally, a brief summary and series of short review questions are given at the end of the chapter.

1.1 Definition of the Rate of Reaction, $-r_A$

nicotine

We begin our study by performing mole balances on each chemical species in the system. Here, the term *chemical species* refers to any chemical compound or element with a given identity. The identity of a chemical species is determined by the *kind, number,* and *configuration* of that species' atoms. For example, the species nicotine (a bad tobacco alkaloid) is made up of a fixed number of specific elements in a definite molecular arrangement or configuration. The structure shown illustrates the kind, number, and configuration of the species nicotine (responsible for "nicotine fits") on a molecular level.

Even though two chemical compounds have exactly the same number of atoms of each element, they could still be different species because of different configurations. For example, 2-butene has four carbon atoms and eight hydrogen atoms; however, the atoms in this compound can form two different arrangements.

cis-2-butene   *trans*-2-butene

As a consequence of the different configurations, these two isomers display different chemical and physical properties. Therefore, we consider them as two different species even though each has the same number of atoms of each element.

When has a chemical reaction taken place?

We say that a *chemical reaction* has taken place when a detectable number of molecules of one or more species have lost their identity and assumed a new form by a change in the kind or number of atoms in the compound and/or by a change in structure or configuration of these atoms. In this classical approach to chemical change, it is assumed that the total mass is neither created nor destroyed when a chemical reaction occurs. The mass referred to is the total collective mass of all the different species in the system. However, when considering the individual species involved in a particular reaction, we do speak of the rate of disappearance of mass of a particular species. The rate of disappearance of a species, say species A, is the number of A molecules that lose their chemical identity per unit time per unit volume through the breaking and subsequent re-forming of chemical bonds during the course of the reaction. In order for a particular species to "appear" in the system, some prescribed fraction of another species must lose its chemical identity.

There are three basic ways a species may lose its chemical identity. One way is by *decomposition,* in which a molecule is broken down into smaller molecules, atoms, or atom fragments. For example, if benzene and propylene are formed from a cumene molecule,

$$CH(CH_3)_2$$

cumene   benzene   propylene

the cumene molecule has lost its identity (i.e., disappeared) by breaking its bonds to form these molecules. A second way that a molecule may lose its species identity is through *combination* with another molecule or atom. In the example above, the propylene molecule would lose its species identity if the reaction were carried out in the reverse direction so that it combined with benzene to form cumene.

The third way a species may lose its identity is through *isomerization,* such as the reaction

$$\underset{\text{CH}_2=\overset{\overset{\displaystyle CH_3}{\displaystyle |}}{C}-CH_2CH_3}{} \longrightarrow \underset{CH_3\overset{\overset{\displaystyle CH_3}{\displaystyle |}}{C}=CHCH_3}{}$$

A species can lose its identity by decomposition, combination, or isomerization

Here, although the molecule neither adds other molecules to itself nor breaks into smaller molecules, it still loses its identity through a change in configuration.

To summarize this point, we say that a given number of molecules (e.g., mole) of a particular chemical species have reacted or disappeared when the molecules have lost their chemical identity.

The rate at which a given chemical reaction proceeds can be expressed in several ways. It can be expressed either as the rate of disappearance of reactants *or* the rate of formation of products. For example, the insecticide DDT (dichlorodiphenyltrichloroethane) is produced from chlorobenzene and chloral in the presence of fuming sulfuric acid.

$$2C_6H_5Cl + CCl_3CHO \longrightarrow (C_6H_4Cl)_2CHCCl_3 + H_2O$$

What is $-r_A$? r_A?

Letting the symbol A represent the chemical chloral, the numerical value of the **rate of reaction**, $-r_A$, is defined as *the number of moles of chloral reacting (disappearing) per unit time per unit volume* (mol/dm$^3 \cdot$ s). In the next chapter we delineate the prescribed relationship between the rate of formation of one species, r_j (e.g., DDT), and the rate of disappearance of another species, $-r_i$ (e.g., chlorobenzene), in a chemical reaction.

In heterogeneous reaction systems, the rate of reaction is usually expressed in measures other than volume, such as reaction surface area or catalyst weight. Thus for a gas–solid catalytic reaction, the dimensions of this rate, $-r_A'$, *are the number of moles of A reacted per unit time per unit mass of catalyst* (mol/s·g catalyst). Most of the introductory discussions on chemical reaction engineering in this book focus on homogeneous systems.

The mathematical definition of a chemical reaction rate has been a source of confusion in chemical and chemical engineering literature for many years. The origin of this confusion stems from laboratory bench-scale experiments that were carried out to obtain chemical reaction rate data. These early experiments were batch-type, in which the reaction vessel was closed and rigid; consequently, the ensuing reaction took place at constant volume. The reactants were mixed together at time $t = 0$ and the concentration of one of the reactants, C_A, was measured at various times t. The rate of reaction was determined from the slope of a plot of C_A as a function of time. Letting r_A be the rate of formation of A per unit volume (e.g., g mol/s·dm^3), the investigators then defined and reported the chemical reaction rate as

$$r_A = \frac{dC_A}{dt} \tag{1-1}$$

However, this definition was for a *constant-volume batch reactor.*

As a result of the limitations and restrictions given, Equation (1-1) is a rather limited and confusing definition of the chemical reaction rate. For amplification of this point, consider the following steady-flow system in which the saponification of ethyl acetate is carried out.

Example 1–1   Is Sodium Hydroxide Reacting?

Sodium hydroxide and ethyl acetate are continuously fed to a rapidly stirred tank in which they react to form sodium acetate and ethanol:

$$NaOH + CH_3COOC_2H_5 \longrightarrow CH_3COONa + C_2H_5OH$$

(Figure E1-1.1). The product stream, containing sodium acetate and ethanol, together with the unreacted sodium hydroxide and ethyl acetate, is continuously withdrawn from the tank at a rate equal to the total feed rate. The contents of the tank in which this reaction is taking place may be considered to be perfectly mixed. Because the system is operated at steady state, if we were to withdraw liquid samples at some location in the tank at various times and analyze them chemically, we would find that the concentrations of the individual species in the different samples were identical. That is, the concentration of the sample taken at 1 P.M. is the same as that of the sample taken at 3 P.M. Because the species concentrations are constant and therefore do not change with time,

$$\frac{dC_A}{dt} = 0 \tag{E1-1.1}$$

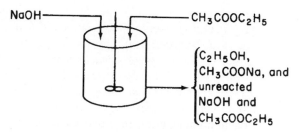

Figure E1-1.1   Well mixed reaction vessel.

where $A \equiv NaOH$. Substitution of Equation (E1-1.1) into Equation (1-1) leads to

$$r_A = 0 \qquad \qquad \text{(E1-1.2)}$$

which is incorrect because C_2H_5OH and CH_3COONa *are* being formed from NaOH and $CH_3COOC_2H_5$ at a finite rate. Consequently, the rate of reaction as defined by Equation (1-1) cannot apply to a flow system and is incorrect if it is *defined* in this manner.

Definition of r_j

By now you should be convinced that Equation (1-1) is not the definition of the chemical reaction rate. We shall simply say that r_j *is the rate of formation of species j per unit volume.* It is the number of moles of species j generated per unit volume per unit time. The rate equation for r_j is solely a function of the properties of the reacting materials [e.g., species concentration (i.e. activities), temperature, pressure, or type of catalyst, if any] at a point in the system and is independent of the type of system (i.e., batch or continuous flow) in which the reaction is carried out. However, since the properties of the reacting materials can vary with position in a chemical reactor, r_j can in turn be a function of position and can vary from point to point in the system.

What is $-r_A$ a function of?

The chemical reaction rate is an intensive quantity and depends on temperature and concentration. The reaction rate equation (i.e., the rate law) is essentially an algebraic equation involving concentration, not a differential equation.[1] For example, the algebraic form of the rate law $-r_A$ for the reaction

$$A \longrightarrow \text{products}$$

may be a linear function of concentration,

$$-r_A = kC_A$$

or it may be some other algebraic function of concentration, such as

---

[1] For further elaboration on this point, see *Chem. Eng. Sci., 25,* 337 (1970); B. L. Crynes and H. S. Fogler, eds., *AIChE Modular Instruction Series E: Kinetics,* Vol. 1 (New York: AIChE, 1981), p. 1; and R. L. Kabel, "Rates," *Chem. Eng. Commun., 9,* 15 (1981).

$$-r_A = kC_A^2 \qquad (1\text{-}2)$$

or

$$-r_A = \frac{k_1 C_A}{1 + k_2 C_A}$$

For a given reaction, the particular concentration dependence that the rate law follows (i.e., $-r_A = kC_A$ or $-r_A = kC_A^2$ or ...) must be determined from experimental observation. Equation (1-2) states that the rate of disappearance of A is equal to a rate constant k times the square of the concentration of A. By convention, r_A is the rate of formation of A; consequently, $-r_A$ is the rate of disappearance of A. Throughout this book the phrase *rate of generation* means exactly the same as the phrase *rate of formation*, and these phrases are used interchangeably.

1.2 The General Mole Balance Equation

To perform a mole balance on any system, the system boundaries must first be specified. The volume enclosed by these boundaries will be referred to as the *system volume*. We shall perform a mole balance on species j in a system volume, where species j represents the particular chemical species of interest, such as water or NaOH (Figure 1-1).

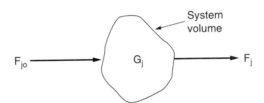

Figure 1-1 Balance on system volume.

A mole balance on species j at any instant in time, t, yields the following equation:

$$\begin{bmatrix} \text{rate of flow} \\ \text{of } j \textit{ into} \\ \text{the system} \\ (\text{moles/time}) \end{bmatrix} + \begin{bmatrix} \text{rate of } \textit{generation} \\ \text{of } j \text{ by chemical} \\ \text{reaction within} \\ \text{the system} \\ (\text{moles/time}) \end{bmatrix} - \begin{bmatrix} \text{rate of flow} \\ \text{of } j \textit{ out of} \\ \text{the system} \\ (\text{moles/time}) \end{bmatrix} = \begin{bmatrix} \text{rate of} \\ \textit{accumulation} \\ \text{of } j \text{ within} \\ \text{the system} \\ (\text{moles/time}) \end{bmatrix}$$

$$\text{in} \quad + \quad \text{generation} \quad - \quad \text{out} \quad = \quad \text{accumulation}$$

$$F_{j0} \quad + \quad G_j \quad - \quad F_j \quad = \quad \frac{dN_j}{dt} \qquad (1\text{-}3)$$

where N_j represents the number of moles of species j in the system at time t. If all the system variables (e.g., temperature, catalytic activity, concentration of the chemical species) are spatially uniform throughout the system volume, the rate of generation of species j, G_j, is just the product of the reaction volume, V, and the rate of formation of species j, r_j.

$$G_j = r_j \cdot V$$

$$\frac{\text{moles}}{\text{time}} = \frac{\text{moles}}{\text{time} \cdot \text{volume}} \cdot \text{volume}$$

Suppose now that the rate of formation of species j for the reaction varies with the position in the system volume. That is, it has a value r_{j1} at location 1, which is surrounded by a small volume, ΔV_1, within which the rate is uniform: similarly, the reaction rate has a value r_{j2} at location 2 and an associated volume, ΔV_2 (Figure 1-2). The rate of generation, ΔG_{j1}, in terms of r_{j1} and

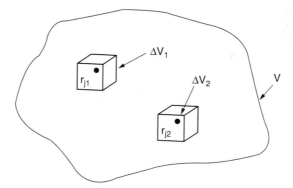

Figure 1-2   Dividing up the system volume V.

subvolume ΔV_1 is

$$\Delta G_{j1} = r_{j1} \, \Delta V_1$$

Similar expressions can be written for ΔG_{j2} and the other system subvolumes ΔV_i. The total rate of generation within the system volume is the sum of all the rates of generation in each of the subvolumes. If the total system volume is divided into M subvolumes, the total rate of generation is

$$G_j = \sum_{i=1}^{M} \Delta G_{ji} = \sum_{i=1}^{M} r_{ji} \, \Delta V_i$$

By taking the appropriate limits (i.e., let $M \to \infty$ and $\Delta V \to 0$) and making use of the definition of an integral, we can rewrite the foregoing equation in the form

$$G_j = \int^{V} r_j \, dV$$

From this equation we see that r_j will be an indirect function of position, since the properties of the reacting materials (e.g., concentration, temperature) can have different values at different locations in the reactor.

We now replace G_j in Equation (1-3),

$$F_{j0} - F_j + G_j = \frac{dN_j}{dt}$$

by its integral form to yield a form of the general mole balance equation for any chemical species j that is entering, leaving, reacting, and/or accumulating within any system volume V.

This is a basic equation for chemical reaction engineering

$$F_{j0} - F_j + \int^V r_j \, dV = \frac{dN_j}{dt} \tag{1-4}$$

From this general mole balance equation we can develop the design equations for the various types of industrial reactors: batch, semibatch, and continuous-flow. Upon evaluation of these equations we can determine the time (batch) or reactor volume (continuous-flow) necessary to convert a specified amount of the reactants to products.

1.3 Batch Reactors

A batch reactor has neither inflow nor outflow of reactants or products while the reaction is being carried out; $F_{j0} = F_j = 0$. The resulting general mole balance on species j is

$$\frac{dN_j}{dt} = \int^V r_j \, dV$$

$$\frac{-dN_A}{dt} = -r_A V$$

If the reaction mixture is perfectly mixed so that there is no variation in the rate of reaction throughout the reactor volume, we can take r_j out of the integral and write the mole balance in the form

$$\frac{dN_j}{dt} = r_j V \tag{1-5}$$

Figure 1-3 shows two different types of batch reactors used for gas-phase reactions. Reactor A is a constant-volume (variable-pressure) reactor and Reactor B is a constant-pressure (variable-volume) reactor. At time $t = 0$, the reactants are injected into the reactor and the reaction is initiated. To see clearly the different forms the mole balance will take for each type of reactor, consider the following examples, in which the gas-phase decomposition of dimethyl ether is taking place to form methane, hydrogen, and carbon monoxide:

$$(CH_3)_2O \longrightarrow CH_4 + H_2 + CO$$

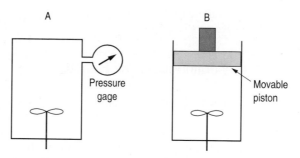

Figure 1-3 Batch reactors for gas-phase reactions.

Example 1–2   Constant Volume or Constant Pressure:
Does It Make a Difference?

Write the mole balance for dimethyl ether in terms of the reactor volume, concentration, and rate of formation of dimethyl ether for both a constant-pressure and a constant-volume batch reactor.

Solution

To reduce the number of subscripts, we write the reaction symbolically as

$$A \longrightarrow M + H + C$$

where A is dimethyl ether, M is methane, H is hydrogen, and C is carbon monoxide. For both batch reactors, the mole balance on A is

$$\frac{1}{V}\frac{dN_A}{dt} = r_A \tag{1-5}$$

In writing the mole balance for dimethyl ether for a batch reactor, the only assumption made is that there are no spatial variations in r_A.

Constant-volume batch reactor. The reactor is perfectly mixed so that the concentration of the reacting species is spatially uniform. Because the volume is constant we can take V inside the differential and write the mole balance in terms of the concentration of A:

$$\frac{1}{V}\frac{dN_A}{dt} = \frac{d(N_A/V)}{dt} = \frac{dC_A}{dt} = r_A \tag{E1-2.1}$$

Constant-pressure batch reactor. To write the mole balance for this reactor in terms of concentration, we again use the fact that

$$N_A = C_A V$$

$$\frac{1}{V}\frac{dN_A}{dt} = \frac{1}{V}\frac{d(C_A V)}{dt} = \frac{dC_A}{dt} + \frac{C_A}{V}\frac{dV}{dt} = r_A \tag{E1-2.2}$$

$$r_A = \frac{dC_A}{dt} + \frac{C_A d \ln V}{dt} \tag{E1-2.3}$$

The difference between equations (E1-2.1) and (E1-2.3) for the two different types of reactors is apparent.

1.4 Continuous-Flow Reactors

1.4.1 Continuous-Stirred Tank Reactor

A type of reactor used very commonly in industrial processing is a stirred tank operated continuously (Figure 1-4). It is referred to as the *continuous-stirred tank reactor* (CSTR) or *backmix reactor.* The CSTR is normally run at steady state and is usually operated so as to be quite well mixed. As a result of the latter quality, the CSTR is generally modeled as having no spatial variations in concentration, temperature, or reaction rate throughout the vessel. Since the temperature and concentration are identical everywhere within the reaction vessel, they are the same at the exit point as they are elsewhere in the tank. Thus the temperature and concentration in the exit stream are modeled as being the same as those inside the reactor. In systems where mixing is highly nonideal, the well-mixed model is inadequate and we must resort to other modeling techniques, such as residence-time distributions, to obtain meaningful results. This topic is discussed in Chapters 13 and 14.

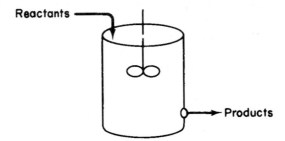

Figure 1-4 Continuous-stirred tank reactor.

When the general mole balance equation

$$F_{j0} - F_j + \int^V r_j \, dV = \frac{dN_j}{dt} \tag{1-4}$$

is applied to a CSTR operated at steady state (i.e., conditions do not change with time),

$$\frac{dN_j}{dt} = 0$$

in which there are no spatial variations in the rate of reaction,

$$\int^V r_j \, dV = V r_j$$

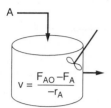

$$v = \frac{F_{AO} - F_A}{-r_A}$$

it takes the familiar form known as the *design equation* for a CSTR:

$$V = \frac{F_{j0} - F_j}{-r_j} \qquad (1\text{-}6)$$

The CSTR design equation gives the reactor volume necessary to reduce the entering flow rate of species, j, F_{j0}, to the exit flow rate F_j. We note that the CSTR is modeled such that the conditions in the exit stream (e.g., concentration, temperature) are identical to those in the tank. The molar flow rate F_j is just the product of the concentration of species j and the volumetric flow rate v:

$$F_j = C_j \cdot v$$

$$\frac{moles}{time} = \frac{moles}{volume} \cdot \frac{volume}{time} \qquad (1\text{-}7)$$

1.4.2 Tubular Reactor

In addition to the CSTR and batch reactors, another type of reactor commonly used in industry is the *tubular reactor*. It consists of a cylindrical pipe and is normally operated at steady state, as is the CSTR. For the purposes of the material presented here, we consider systems in which the flow is highly turbulent and the flow field may be modeled by that of plug flow. That is, there is no radial variation in concentration and the reactor is referred to as a plug-flow reactor (PFR). (The laminar flow reactor is discussed in Chapter 13.)

In the tubular reactor, the reactants are continually consumed as they flow down the length of the reactor. In modeling the tubular reactor, we assume that the concentration varies continuously in the axial direction through the reactor. Consequently, the reaction rate, which is a function of concentration for all but zero-order reactions, will also vary axially. The general mole balance equation is given by Equation (1-4):

$$F_{j0} - F_j + \int^{V} r_j \, dV = \frac{dN_j}{dt} \qquad (1\text{-}4)$$

To develop the PFR design equation we shall divide (conceptually) the reactor into a number of subvolumes so that within each subvolume ΔV, the reaction rate may be considered spatially uniform (Figure 1-5). We now focus our attention on the subvolume that is located a distance y from the entrance of the reactor. We let $F_j(y)$ represent the molar flow rate of species j into volume ΔV at y and $F_j(y + \Delta y)$ the molar flow of species j out of the volume at the location $(y + \Delta y)$. In a spatially uniform subvolume ΔV,

$$\int^{\Delta V} r_j \, dV = r_j \, \Delta V$$

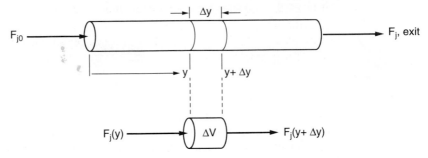

Figure 1-5 Tubular reactor.

For a tubular reactor operated at steady state,

$$\frac{dN_j}{dt} = 0$$

Equation (1-4) becomes

$$F_j(y) - F_j(y + \Delta y) + r_j \, \Delta V = 0 \qquad (1\text{-}8)$$

In this expression r_j is an indirect function of y. That is, r_j is a function of reactant concentration, which is a function of the position y down the reactor. The volume ΔV is the product of the cross-sectional area A of the reactor and the reactor length Δy.

$$\Delta V = A \, \Delta y$$

We now substitute in Equation (1-8) for ΔV and then divide by Δy to obtain

$$-\left[\frac{F_j(y + \Delta y) - F_j(y)}{\Delta y}\right] = -Ar_j$$

The term in brackets resembles the definition of the derivative

$$\lim_{\Delta x \to 0}\left[\frac{f(x + \Delta x) - f(x)}{\Delta x}\right] = \frac{df}{dx}$$

Taking the limit as Δy approaches zero, we obtain

$$-\frac{dF_j}{dy} = -Ar_j$$

or dividing by -1, we have

$$\frac{dF_j}{dy} = Ar_j \qquad (1\text{-}9)$$

It is usually most convenient to have the reactor volume V rather than the reactor length y as the independent variable. Accordingly, we shall change variables using the relation $dV = A\,dy$ to obtain one form of the design equation for a tubular reactor:

$$\frac{dF_j}{dV} = r_j \qquad (1\text{-}10)$$

We also note that for a reactor in which the cross-sectional area A varies along the length of the reactor, the design equation remains unchanged. This equation can be generalized for the reactor shown in Figure 1-6, in a manner simi-

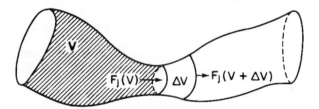

Figure 1-6

lar to that presented above, by utilizing the volume coordinate V rather than a linear coordinate y. After passing through volume V, species j enters subvolume ΔV at volume V at a molar flow rate $F_j(V)$. Species j leaves subvolume ΔV at volume $(V + \Delta V)$, at a molar flow rate $F_j(V + \Delta V)$. As before, ΔV is chosen small enough so that there is no spatial variation of reaction rate within the subvolume:

$$G_j = \int^{\Delta V} r_j\,dV = r_j\,\Delta V \qquad (1\text{-}11)$$

After accounting for steady-state operation in Equation (1-4), it is combined with Equation (1-11) to yield

$$F_j(V) - F_j(V + \Delta V) + r_j\,\Delta V = 0$$

Rearranging gives

$$\frac{F_j(V + \Delta V) - F_j(V)}{\Delta V} = r_j$$

and taking the limit as $\Delta V \to 0$, we again obtain Equation (1-10):

$$\boxed{\frac{dF_j}{dV} = r_j} \qquad (1\text{-}10)$$

Consequently, we see that Equation (1-10) applies equally well to our model of tubular reactors of variable and constant cross-sectional area, although it is doubtful that one would find a reactor of the shape shown in Figure 1-6, unless designed by Pablo Picasso. The conclusion drawn from the application of the design equation is an important one: The extent of reaction achieved in a plug-flow tubular reactor (PFR) does not depend on its shape, only on its total volume.

1.4.3 Packed-Bed Reactor

The principal difference between reactor design calculations involving homogeneous reactions and those involving fluid–solid heterogeneous reactions is that for the latter, the reaction rate is based on mass of solid catalyst, W, rather than on reactor volume, V. For a fluid–solid heterogeneous system, the rate of reaction of a substance A is defined as

$$-r'_A = \text{g mol A reacted/s} \cdot \text{g catalyst}$$

The mass of solid is used because the amount of the catalyst is what is important to the rate of reaction. The reactor volume that contains the catalyst is of secondary significance.

In the three idealized types of reactors just discussed [the perfectly mixed batch reactor, the plug-flow tubular reactor, and the perfectly mixed continuous-stirred tank reactor (CSTR)], the design equations (i.e., mole balances) were developed based on reactor volume. The derivation of the design equation for a packed-bed catalytic reactor will be carried out in a manner analogous to the development of the tubular design equation. To accomplish this derivation, we simply replace the volume coordinate in Equation (1-8) with the catalyst weight coordinate W (Figure 1-7). As with the PFR, the PBR is assumed to have

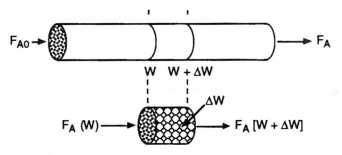

Figure 1-7   Packed-bed reactor schematic.

no radial gradients in concentration, temperature, or reaction rate. The generalized mole balance on species A over catalyst weight ΔW results in the equation

$$
\begin{array}{ccccccc}
\text{in} & - & \text{out} & + & \text{generation} & = & \text{accumulation} \\
F_A(W) & - & F_A(W + \Delta W) & + & r'_A \, \Delta W & = & 0
\end{array}
\qquad (1\text{-}12)
$$

The dimensions of the generation term in Equation (1-12) are

$$(r'_A)\,\Delta W \equiv \frac{\text{moles A}}{(\text{time})(\text{mass of catalyst})} \cdot (\text{mass of catalyst}) \equiv \frac{\text{moles A}}{\text{time}}$$

which are, as expected, the same dimension of the molar flow rate F_A. After dividing by ΔW and taking the limit as $\Delta W \to 0$, we arrive at the differential form of the mole balance for a packed-bed reactor:

Use differential form of design equation for catalyst decay and pressure drop

$$\frac{dF_A}{dW} = r'_A \qquad (1\text{-}13)$$

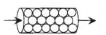

When pressure drop through the reactor (see Section 4.4) and catalyst decay (see Section 10.7) are neglected, the integral form of the packed-catalyst-bed design equation can be used to calculate the catalyst weight.

$$W = \int_{F_{A0}}^{F_A} \frac{dF_A}{r'_A} \qquad (1\text{-}14)$$

To obtain some insight into things to come, consider the following example of how one can use the tubular reactor design equation (1-10).

Example 1–3 How Large Is It?

The first-order reaction

$$A \longrightarrow B$$

is carried out in a tubular reactor in which the volumetric flow rate, v, is constant. Derive an equation relating the reactor volume to the entering and exiting concentrations of A, the rate constant k, and the volumetric flow rate v. Determine the reactor volume necessary to reduce the exiting concentration to 10% of the entering concentration when the volumetric flow rate is 10 dm³/min (i.e., liters/min) and the specific reaction rate, k, is 0.23 min^{-1}.

Solution

For a tubular reactor, the mole balance on species A ($j = A$) was shown to be

$$\frac{dF_A}{dV} = r_A \qquad (1\text{-}10)$$

For a first-order reaction, the rate law (discussed in Chapter 3) is

$$-r_A = kC_A \qquad (E1\text{-}3.1)$$

Since the volumetric flow rate, v_0, is constant,

Reactor sizing

$$\frac{dF_A}{dV} = \frac{d(C_A v_0)}{dV} = v_0 \frac{dC_A}{dV} = r_A \qquad (E1\text{-}3.2)$$

Substituting for r_A in Equation (E1-3.1) yields

$$-\frac{v_0 dC_A}{dV} = -r_A = kC_A \qquad (E1\text{-}3.3)$$

Rearranging gives

$$-\frac{v_0}{k}\left(\frac{dC_A}{C_A}\right) = dV$$

Using the conditions at the entrance of the reactor that when $V = 0$, then $C_A = C_{A0}$,

$$-\frac{v_0}{k}\int_{C_{A0}}^{C_A}\frac{dC_A}{C_A} = \int_0^V dV \tag{E1-3.4}$$

This equation gives

$$V = \frac{v_0}{k}\ln\frac{C_{A0}}{C_A} \tag{E1-3.5}$$

Substituting C_{A0}, C_A, v_0, and k in Equation (E1-3.5), we have

$$V = \frac{10\ \text{dm}^3/\text{min}}{0.23\ \text{min}^{-1}}\ln\frac{C_{A0}}{0.1C_{A0}} = \frac{10\ \text{dm}^3}{0.23}\ln 10 = 100\ \text{dm}^3 \ (\text{i.e., } 100\ \text{L; } 0.1\ \text{m}^3)$$

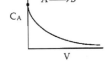

We see that a reactor volume of 0.1 m³ is necessary to convert 90% of species A entering into product B.

In the remainder of this chapter we look at slightly more detailed drawings of some typical industrial reactors and point out a few of the advantages and disadvantages of each.[2]

1.5 Industrial Reactors

When is a batch reactor used?

A batch reactor is used for small-scale operation, for testing new processes that have not been fully developed, for the manufacture of expensive products, and for processes that are difficult to convert to continuous operations. The reactor can be charged (i.e., filled) through the holes at the top (Figure 1-8). The batch reactor has the advantage of high conversions that can be obtained by leaving the reactant in the reactor for long periods of time, but it also has the disadvantages of high labor costs per batch and the difficulty of large-scale production.

Liquid-Phase Reactions. Although a semibatch reactor (Figure 1-9) has essentially the same disadvantages as the batch reactor, it has the advantages of good temperature control and the capability of minimizing unwanted side reactions through the maintenance of a low concentration of one of the reactants. The semibatch reactor is also used for two-phase reactions in which a gas is usually bubbled continuously through the liquid.

What are the advantages and disadvantages of a CSTR?

A continuous-stirred tank reactor (CSTR) is used when intense agitation is required. A photo showing a cutaway view of a Pfaudler CSTR/batch reactor is presented in Figure 1-10. Table 1-1 gives the typical sizes (along with that of

[2] *Chem. Eng.*, 63(10), 211 (1956). See also *AIChE Modular Instruction Series E,* Vol. 5 (1984).

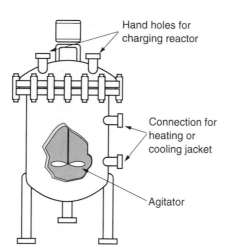

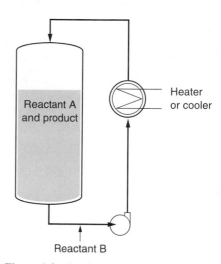

Figure 1-8 Simple batch homogeneous reactor. [Excerpted by special permission from *Chem. Eng., 63*(10), 211 (Oct. 1956). Copyright 1956 by McGraw-Hill, Inc., New York, NY 10020.]

Figure 1-9 Semibatch reactor. [Excerpted by special permission from *Chem. Eng., 63*(10), 211 (Oct. 1956). Copyright 1956 by McGraw-Hill, Inc., New York, NY 10020.]

Figure 1-10   CSTR/batch reactor. (Courtesy of Pfaudler, Inc.)

TABLE 1-1.   REPRESENTATIVE PFAUDLER CSTR/BATCH REACTOR
SIZES AND 1996 PRICES

Volume	Price	Volume	Price
5 Gallons (wastebasket)	$27,000	1000 Gallons (2 Jacuzzis)	$80,000
50 Gallons (garbage can)	$35,000	4000 Gallons (8 Jacuzzis)	$143,000
500 Gallons (Jacuzzi)	$67,000	8000 Gallons (gasoline tanker)	$253,000

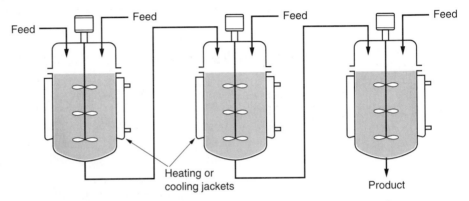

Figure 1-11   Battery of stirred tanks. [Excerpted by special permission from
Chem. Eng., 63(10), 211 (Oct. 1956). Copyright 1956 by McGraw-Hill, Inc., New
York, NY 10020.]

the comparable size of a familiar object) and costs for batch and CSTR reactors. All reactors are glass lined and the prices include heating/cooling jacket,
motor, mixer, and baffles. The reactors can be operated at temperatures between
20 and 450°F and at pressures up to 100 psi.

The CSTR can either be used by itself or, in the manner shown in Figure
1-11, as part of a series or battery of CSTRs. It is relatively easy to maintain
good temperature control with a CSTR. There is, however, the disadvantage
that the conversion of reactant per volume of reactor is the smallest of the flow
reactors. Consequently, very large reactors are necessary to obtain high
conversions.

If you are not able to afford to purchase a new reactor, it may be possible
to find a used reactor that may fit your needs. Previously owned reactors are
much less expensive and can be purchased from equipment clearinghouses
such as Universal Process Equipment or Loeb Equipment Supply.

Example 1–4 Liquid-Phase Industrial Process Flowsheet

A battery of four CSTRs similar to those in Figure 1-10 are shown in the plant flow-sheet (Figure E1-4.1) for the commercial production of nitrobenzene. In 1995, 1.65 billion pounds of nitrobenzene were produced.

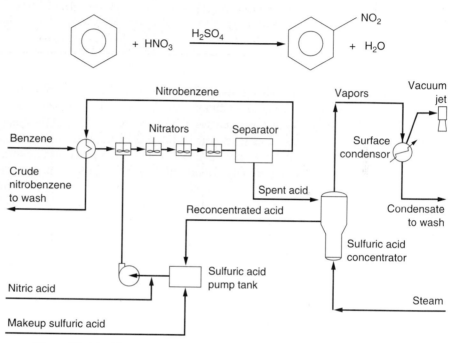

Note: Heat Exchange between Benzene feed and Nitrobenzene product

Figure E1-4.1 Flowsheet for the production of nitrobenzene. [Adapted from *Process Technology and Flowsheet*, Vol. II, reprints from *Chemical Engineering* (New York: McGraw-Hill, 1983), p. 125.]

In 1980 the operating requirements (per ton of nitrobenzene) were as follows (utilities and feedstock requirements have been minimized by recycling sulfuric acid):

Raw materials
Benzene	0.64 ton
Nitric acid (100%)	0.515 ton
Sulfuric acid (100%)	0.0033 ton
Caustic soda	0.004 ton

Utilities
Cooling water	14,200 gal
Steam	800 lb
Electricity	20 kWh
Compressed air	180 Scf/m

The feed consists of 3 to 7% HNO_3, 59 to 67% H_2SO_4, and 28 to 37% water. Sulfuric acid is necessary to adsorb the water and energy generated by the heat of reaction. The plant, which produces 15,000 lb nitrobenzene/h, requires one or two operators per shift together with a plant supervisor and part-time foreman. This exothermic reaction is carried out essentially adiabatically, so that the temperature of the feed stream rises from 90°C to 135°C at the exit. One observes that the nitrobenzene stream from the separator is used to heat the benzene feed. However, care must be taken so that the temperature never exceeds 190°C, where secondary reactions could result in an explosion. One of the safety precautions is the installation of relief valves that will rupture before the temperature approaches 190°C, thereby allowing a boil-off of water and benzene, which would drop the reactor temperature.

What are the advantages and disadvantages of a PFR?

CSTR: liquids
PFR: gases

Gas-Phase Reactions.   The tubular reactor [i.e., plug-flow reactor (PFR)] is relatively easy to maintain (no moving parts), and it usually produces the highest conversion per reactor volume of any of the flow reactors. The disadvantage of the tubular reactor is that it is difficult to control temperature within the reactor, and hot spots can occur when the reaction is exothermic. The tubular reactor is commonly found either in the form of one long tube or as one of a number of shorter reactors arranged in a tube bank as shown in Figure 1-12. Most homogeneous liquid-phase flow reactors are CSTRs, whereas most homogeneous gas-phase flow reactors are tubular.

The costs of PFR and PBR (without catalyst) are similar to the costs of heat exchangers and thus can be found in *Plant Design and Economics for*

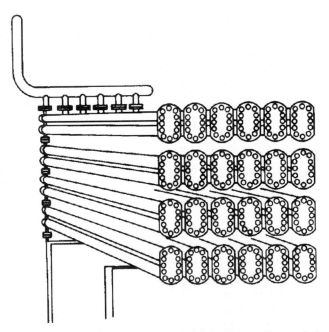

Figure 1-12   Longitudinal tubular reactor. [Excerpted by special permission from *Chem. Eng., 63*(10), 211 (Oct. 1956). Copyright 1956 by McGraw-Hill, Inc., New York, NY 10020.]

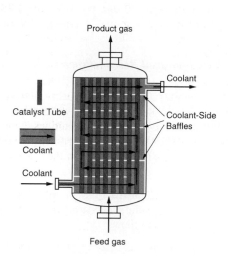

Figure 1-13 Longitudinal catalytic packed-bed reactor. [From Cropley, American Institute of Chemical Engineers, *86*(2), 34 (1990). Reproduced with permission of the American Institute of Chemical Engineers, Copyright © 1990 AIChE. All rights reserved.]

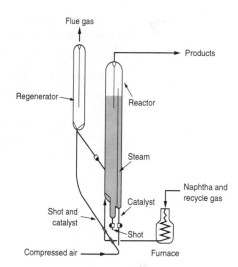

Figure 1-14 Fluidized-bed catalytic reactor. [Excerpted by special permission from *Chem. Eng.*, *63*(10), 211 (Oct. 1956). Copyright 1956 by McGraw-Hill, Inc., New York, NY 10020.]

Chemical Engineers, 4th ed., by M. S. Peters and K. D. Timmerhaus (New York: McGraw-Hill, 1991). From Figure 15-12 of this book, one can get an estimate of the purchase cost per foot of $1 for a 1-in. pipe and $2 per foot for a 2-in. pipe for single tubes and approximately $20 to $50 per square foot of surface area for fixed-tube sheet exchangers.

A packed-bed (also called a fixed-bed) reactor is essentially a tubular reactor that is packed with solid catalyst particles (Figure 1-13). This heterogeneous reaction system is used most frequently to catalyze gas reactions. This reactor has the same difficulties with temperature control as other tubular reactors, and in addition, the catalyst is usually troublesome to replace. On occasion, channeling of the gas flow occurs, resulting in ineffective use of parts of the reactor bed. The advantage of the packed-bed reactor is that for most reactions it gives the highest conversion per weight of catalyst of any catalytic reactor.

Another type of catalytic reactor in common use is the fluidized-bed (Figure 1-14). The fluidized-bed reactor is analogous to the CSTR in that its contents, though heterogeneous, are well mixed, resulting in an even temperature distribution throughout the bed. The fluidized-bed reactor cannot be modeled as either a CSTR or a tubular reactor (PFR), but requires a model of its own. The temperature is relatively uniform throughout, thus avoiding hot spots. This type of reactor can handle large amounts of feed and solids and has good temperature control; consequently, it is used in a large number of applications. The advantages of the ease of catalyst replacement or regeneration are sometimes offset by the high cost of the reactor and catalyst regeneration equipment.

Example 1–5 Gas-Phase Industrial Reactor/Process

Synthesis gas contains a mixture of carbon monoxide and hydrogen and can be obtained from the combustion of coal or natural gas. This gas can be used to produce synthetic crude by the Fischer–Tropsch reaction. Describe two industrial reactors used to convert synthesis gas to a mixture of hydrocarbons by the Fischer–Tropsch process.

Solution

Reactions. The Fischer–Tropsch reaction converts synthesis gas into a mixture of alkanes and alkenes over a solid catalyst usually containing iron. The basic reaction for paraffin formation is as follows

$$nCO + (2n+1)H_2 \longrightarrow C_nH_{2n+2} + nH_2O \tag{E1-5.1}$$

Making Gasoline

For example, when octane, a component of gasoline, is formed, Equation (E1-5.1) becomes

$$8CO + 17H_2 \longrightarrow C_8H_{18} + 8H_2O \tag{E1-5.2}$$

Similarly, for the formation of olefins,

$$nCO + 2nH_2 \longrightarrow C_nH_{2n} + nH_2O \tag{E1-5.3}$$

For ethylene formation, Equation (E1-5.3) becomes

$$2CO + 4H_2 \longrightarrow C_2H_4 + 2H_2O \tag{E1-5.4}$$

The other type of main reaction that occurs in this process is the water-gas-shift reaction

$$H_2O + CO \rightleftharpoons CO_2 + H_2 \tag{E1-5.5}$$

In addition to the simultaneous formation of paraffins and olefins, side reactions also take place to produce small quantities of acids and nonacids (e.g., ethanol).

Reactors. Two types of reactors will be discussed, a *straight-through transport reactor*, which is also referred to as a *riser* or *circulating fluidized bed*, and a *packed-bed reactor* (PBR), which is also referred to as a *fixed-bed reactor*.

Riser. Because the catalyst used in the process decays rapidly at high temperatures (e.g., 350°C), a *straight-through transport reactor* (STTR) (Chapter 10) is used. This type of reactor is also called a *riser* and/or a *circulating bed*. A schematic diagram is shown in Figure E1-5.1. Here the catalyst particles are fed to the bottom of the reactor and are shot up through the reactor together with the entering reactant gas mixture and then separated from the gas in a settling hopper. The volumetric gas feed rate of 3×10^5 m³/h is roughly equivalent to feeding the volume of gas contained in the University of Michigan football stadium to the reactor each hour.

A schematic and photo of an industrial *straight-through transport reactor* used at Sasol are shown in Figure E1-5.2 together with the composition of the feed and product streams. The products that are condensed out of the product stream

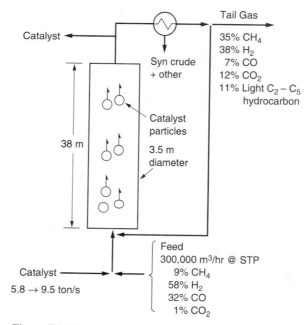

Figure E1-5.1   Schematic of Sasol Fischer–Tropsch process.

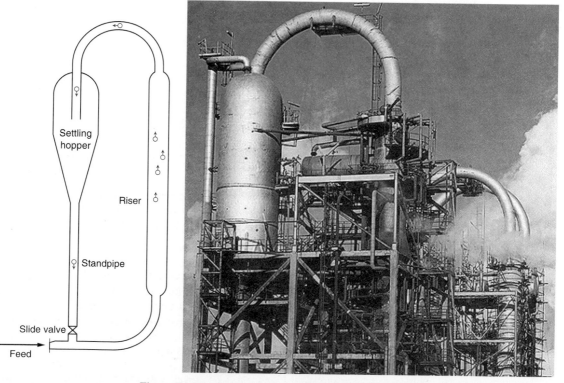

Figure E1-5.2   The reactor is 3.5 m in diameter and 38 m tall. (Schematic and photo courtesy of Sasol/Sastech PT Limited.)

before the stream is recycled include Synoil (a synthetic crude), water, methyl ethyl ketone (MEK), alcohols, acids, and aldehydes. The reactor is operated at 25 atm and 350°C and at any one time contains 150 tons of catalyst. The catalyst feed rate is 6 to 9.5 tons/s, and the gas recycle ratio is 2:1.

Packed Bed. The packed-bed reactor used at the Sasol plant to carry out Fischer–Tropsch synthesis reaction is shown in Figure E1-5.3. Synthesis gas is fed at a rate of 30,000 m³/h (STP) at 240°C and 27 atm to the packed-bed reactor. The reactor contains 2050 tubes, each of which is 5.0 cm in diameter and 12 m in length. The iron-based catalyst that fills these tubes usually contains K_2O and SiO_2 and has a specific area on the order of 200 m²/g. The reaction products are light hydrocarbons along with a wax that is used in candles and printing inks. Approximately 50% conversion of the reactant is achieved in the reactor.

Use to produce wax for candles and printing inks.

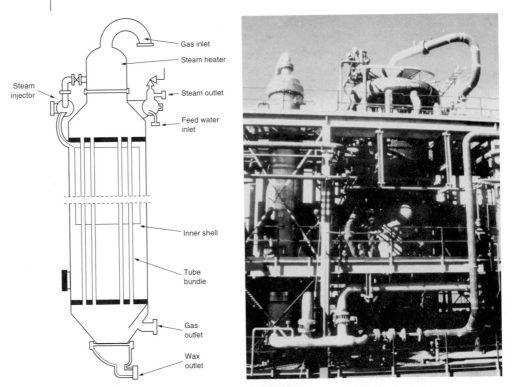

Figure E1-5.3 Packed-bed reactor. (Schematic and photograph courtesy of Sasol/Sastech PT Limited.)

The aim of the preceding discussion on commercial reactors is to give a more detailed picture of each of the major types of industrial reactors: batch, semibatch, CSTR, tubular, fixed-bed (packed-bed), and fluidized-bed. Many variations and modifications of these commercial reactors are in current use; for further elaboration, refer to the detailed discussion of industrial reactors given by Walas.[3]

---

[3] S. M. Walas, *Reaction Kinetics for Chemical Engineers* (New York: McGraw-Hill, 1959), Chap. 11.

ST Montgomery 8/94

SUMMARY

1. A mole balance on species j, which enters, leaves, reacts, and accumulates in a system volume V, is

$$F_{j0} - F_j + \int^V r_j \, dV = \frac{dN_j}{dt} \qquad (S1\text{-}1)$$

2. The kinetic rate law for r_j is:
   - Solely a function of properties of reacting materials [e.g., concentration (activities), temperature, pressure, catalyst or solvent (if any)].
   - An intensive quantity.
   - An algebraic equation, not a differential equation.

   For homogeneous catalytic systems, typical units of $-r_j$ may be gram moles per second per liter; for heterogeneous systems, typical units of r_j' may be gram moles per second per gram of catalyst. By convention, $-r_A$ is the rate of disappearance of species A and r_A is the rate of formation of species A.

3. Mole balances on four common reactors are as follows:

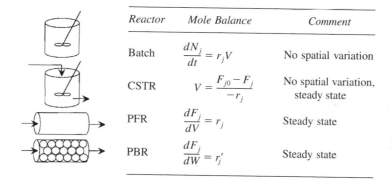

Reactor	Mole Balance	Comment
Batch	$\dfrac{dN_j}{dt} = r_j V$	No spatial variation
CSTR	$V = \dfrac{F_{j0} - F_j}{-r_j}$	No spatial variation, steady state
PFR	$\dfrac{dF_j}{dV} = r_j$	Steady state
PBR	$\dfrac{dF_j}{dW} = r_j'$	Steady state

QUESTIONS AND PROBLEMS

I wish I had an answer for that, because I'm getting tired of answering that question.

Yogi Berra, New York Yankees
Sports Illustrated, June 11, 1984

The subscript to each of the problem numbers indicates the level of difficulty: A, least difficult; D, most difficult.

A = ●   B = ■   C = ◆   D = ◆◆

In each of the questions and problems below, rather than just drawing a box around your answer, write a sentence or two describing how you solved the problem, the assumptions you made, the reasonableness of your answer, what you learned, and any other facts that you want to include. You may wish to refer to W. Strunk and E. B. White, *The Elements of Style* (New York: Macmillian, 1979) and Joseph M. Williams, *Style: Ten Lessons in Clarity & Grace* (Glenview, Ill.: Scott, Foresman, 1989) to enhance the quality of your sentences.

P1-1$_A$ After reading each page, ask yourself a question. Make a list of the most important things that you learned in this chapter.

P1-2$_A$ **What if:**
(a) the benzene feed stream in Example 1-4 were not preheated by the product stream? What would be the consequences?
(b) you needed the cost of a 6000-gallon and a 15,000-gallon Pfaudler reactor? What would they be?
(c) the exit concentration of A in Example 1-3 were specified at 0.1% of the entering concentration?
(d) the volume of the movable piston in Example 1-2 varied in a manner similar to a car cylinder, $V = V_0 + V_1 \sin \omega t$?
(e) only one operator showed up to run the nitrobenzene plant, what would be some of your first concerns?

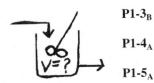

P1-3$_B$ Calculate the volume of a CSTR for the conditions used to calculate the plug-flow reactor volume in Example 1-3.

P1-4$_A$ Calculate the time to reduce the number of moles of A to 1% of its initial value in a constant-volume batch reactor for the reaction and data in Example 1-3.

P1-5$_A$ What assumptions were made in the derivation of the design equation for:
(a) the batch reactor?
(b) the CSTR?
(c) the plug-flow reactor (PFR)?
(d) the packed-bed reactor (PBR)?
(e) State in words the meanings of $-r_A$, $-r'_A$, and r'_A. Is the reaction rate $-r_A$ an extensive quantity? Explain.

P1-6$_A$ What is the difference between the rate of reaction for a homogeneous system, $-r_A$, and the rate of reaction for a heterogeneous system, $-r'_A$? Use the mole balance to derive an equation analogous to Equation (1-6) for a fluidized CSTR containing catalyst particles in terms of the catalyst weight, W, and other appropriate terms.

P1-7$_A$ How can you convert the general mole balance equation for a given species, Equation (1-4), to a general mass balance equation for that species?

P1-8$_A$ The United States produces 24% of the world's chemical products. According to the yearly "Facts and Figures" issue of *Chemical and Engineering News* (*C&E News*, June 24, 1996), the following were the 10 most produced chemicals in 1995:

Chemical	Billions of Pounds	Chemical	Billions of Pounds
1. H_2SO_4	95.36	6. NH_3	35.60
2. N_2	68.04	7. H_3PO_4	26.19
3. O_2	53.48	8. NaOH	26.19
4. C_2H_4	46.97	9. C_3H_6	25.69
5. CaO	41.23	10. Cl_2	25.09

(a) What were the 10 most produced chemicals for the year that just ended? Were there any significant changes from the 1995 statistics?

The same issue of *C&E News* gives the following chemical companies as the top 10 in total sales in 1995. (Also see http://www.chemweek.com)

Company	Sales (billions of dollars)
1. Dow	19.73
2. Dupont	18.43
3. Exxon	11.73
4. Hoechst Celanese	7.39
5. Monsanto	7.25
6. General Electric	6.63
7. Mobil	6.15
8. Union Carbide	5.89
9. Amoco	5.66
10. Occidental Petroleum	5.41

(b) What 10 companies were tops in sales for the year just ended? Did any significant changes occur compared to the 1995 statistics?

(c) Why do you think H_2SO_4 is the most produced chemical? What are some of its uses?

(d) What is the current annual production rate (lb/yr) of ethylene, ethylene oxide, and benzene?

(e) Why do you suspect there are so few organic chemicals in the top 10?

P1-9$_A$ Referring to the text material and the additional references on commercial reactors given at the end of this chapter, fill in the following table:

Type of Reactor	Characteristics	Kinds of Phases Present	Use	Advantages	Disadvantages
___	___	___	___	___	___
___	___	___	___	___	___
___	___	___	___	___	___
___	___	___	___	___	___

P1-10$_B$ Schematic diagrams of the Los Angeles basin are shown in Figure P1-10. The basin floor covers approximately 700 square miles (2×10^{10} ft^2) and is almost completely surrounded by mountain ranges. If one assumes an inversion height in the basin of 2000 ft, the corresponding volume of air in the basin is 4×10^{13} ft^3. We shall use this system volume to model the accumulation and depletion of air pollutants. As a very rough first approximation, we shall treat the Los Angeles basin as a well-mixed container (analogous to a CSTR) in which there are no spatial variations in pollutant concentrations. Consider only the pollutant carbon monoxide and assume that the source of CO is from automobile exhaust and that, on the average, there are 400,000 cars operating in the basin at any one time. Each car gives off roughly 3000 standard cubic feet of exhaust each hour containing 2 mol % carbon monoxide.

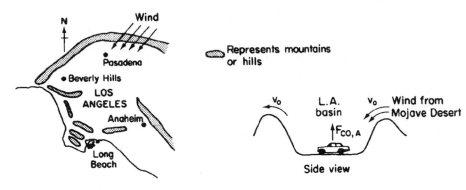

Figure P1-10

We shall perform an unsteady-state mole balance on CO as it is depleted from the basin area by a Santa Ana wind. Santa Ana winds are high-velocity winds that originate in the Mojave Desert just to the northeast of Los Angeles. This clean desert air flows into the basin through a corridor assumed to be 20 miles wide and 2000 ft high (inversion height) replacing the polluted air, which flows out to sea or toward the south. The concentration of CO in the Santa Ana wind entering the basin is 0.08 ppm (2.04×10^{-10} lb mol/ft³).

(a) How many pound moles of gas are in the system volume we have chosen for the Los Angeles basin if the temperature is 75°F and the pressure is 1 atm? (Values of the ideal gas constant may be found in Appendix B.)

(b) What is the rate, $F_{CO,A}$, at which all autos emit carbon monoxide into the basin (lb mol CO/h)?

(c) What is the volumetric flow rate (ft³/h) of a 15-mph wind through the corridor 20 miles wide and 2000 ft high? (*Ans.*: 1.67×10^{13} ft³/h.)

(d) At what rate, $F_{CO,S}$, does the Santa Ana wind bring carbon monoxide into the basin (lb mol/h)?

(e) Assuming that the volumetric flow rates entering and leaving the basin are identical, $v = v_0$, show that the unsteady mole balance on CO within the basin becomes

$$F_{CO,A} + F_{CO,S} - v_0 C_{CO} = V \frac{dC_{CO}}{dt} \qquad \text{(P1-10.1)}$$

(f) Verify that the solution to Equation (P1-10.1) is

$$t = \frac{V}{v_0} \ln \frac{F_{CO,A} + F_{CO,S} - v_0 C_{CO,0}}{F_{CO,A} + F_{CO,S} - v_0 C_{CO}} \qquad \text{(P1-10.2)}$$

(g) If the initial concentration of carbon monoxide in the basin before the Santa Ana wind starts to blow is 8 ppm (2.04×10^{-8} lb mol/ft³), calculate the time required for the carbon monoxide to reach a level of 2 ppm.

(h) Repeat parts (b) through (g) for another pollutant, NO. The concentration of NO in the auto exhaust is 1500 ppm (3.84×10^{-6} lb mol/ft³), and the initial NO concentration in the basin is 0.5 ppm. If there is no NO in the Santa Ana wind, calculate the time for the NO concentration to reach 0.1 ppm. What is the lowest concentration of NO that could be reached?

P1-11$_A$ The reaction

$$A \longrightarrow B$$

is to be carried out isothermally in a continuous-flow reactor. Calculate both the CSTR and PFR reactor volumes necessary to consume 99% of A (i.e., $C_A = 0.01C_{A0}$) when the entering molar flow rate is 5 mol/h, assuming the reaction rate $-r_A$ is:

(a) $-r_A = k$ with $k = 0.05 \dfrac{\text{mol}}{\text{h} \cdot \text{dm}^3}$ (*Ans.*: $V = 99$ dm^3)

(b) $-r_A = kC_A$ with $k = 0.0001$ s^{-1}

(c) $-r_A = kC_A^2$ with $k = 3 \dfrac{\text{dm}^3}{\text{mol} \cdot \text{h}}$ (*Ans.*: $V_{CSTR} = 66{,}000$ dm^3)

The entering volumetric flow rate is 10 dm^3/h. [*Note*: $F_A = C_A v$. For a constant volumetric flow rate $v = v_0$, then $F_A = C_A v_0$. Also, $C_{A0} = F_{A0}/v_0 = (5 \text{ mol/h})/(10 \text{ dm}^3/\text{h}) = 0.5 \text{ mol/dm}^3$.]

P1-12$_C$ The gas-phase reaction

$$A \longrightarrow B + C$$

is carried out isothermally in a 20-dm^3 constant-volume batch reactor. Twenty moles of pure A is initially placed in the reactor. The reactor is well mixed.

(a) If the reaction is first order:

$$-r_A = kC_A \quad \text{with } k = 0.865 \text{ min}^{-1}$$

calculate the time necessary to reduce the number of moles of A in the reactor to 0.2 mol. (*Note*: $N_A = C_A V$.) (*Ans.*: $t = 5.3$ min)

(b) If the reaction is second order:

$$-r_A = kC_A^2 \quad \text{with } k = \frac{2 \text{ dm}^3}{\text{mol} \cdot \text{min}}$$

calculate the time necessary to consume 19.0 mol of A.

(c) If the temperature is 127°C, what is the initial total pressure? What is the final total pressure assuming the reaction goes to completion?

P1-13$_A$ **(a)** How many cubic feet (at STP) enter the packed-bed reactor described in Example 1-5 every second? How long does a molecule spend, on the average, in the reactor? [*Hint*: What is the gas velocity in each tube assuming a 30% porosity (volume of gas/volume of reactor) for the packed bed?]

(b) Estimate the time that a catalyst particle and a gas-phase molecule spend in the Sasol straight-through transport reactor (STTR). What is the bulk density of the catalyst (kg cat/m^3) in the STTR?

P1-14$_A$ Write a one-paragraph summary of a journal article on chemical kinetics or reaction engineering. The article must have been published within the last five years.[4] What did you learn from this article? Why is the article important?

P1-15$_A$ **(a)** What journals, books, or papers give you costs of industrial (*not* laboratory, e.g., Fisher catalog) chemicals and catalysts?

(b) List various journals, books, or other sources where you will find details about the construction and safety of industrial reactors.

---

[4] See the Supplementary Reading list at the end of the chapter, particularly item 4.

P1-16$_C$ What are typical operating conditions (temperature, pressure) of a catalytic cracking reactor used in petroleum refining?

P1-17$_A$ View the photos and schematics on the CD-ROM under Elements of Chemical Reaction Engineering–Chapter 1. Look at the quicktime videos. Write a paragraph describing two or more of the reactors. What similarities and differences do you observe between the reactors on the Web and in the text?

P1-18$_A$ (a) There are initially 500 rabbits (x) and 200 foxes (y) on Farmer Oat's property. Use POLYMATH or MATLAB to plot the concentration of foxes and rabbits as a function of time for a period of up to 500 days. The predator–prey relationships are given by the following set of coupled ordinary differential equations:

$$\frac{dx}{dt} = k_1 x - k_2 x \cdot y$$

$$\frac{dy}{dt} = k_3 x \cdot y - k_4 y$$

Constant for growth of rabbits $k_1 = 0.02$ day^{-1}
Constant for death of rabbits $k_2 = 0.00004/($day $\times$ no. of foxes$)$
Constant for growth of foxes after eating rabbits $k_3 = 0.0004/($day $\times$ no. of rabbits$)$
Constant for death of foxes $k_4 = 0.04$ day^{-1}
What do your results look like for the case of $k_3 = 0.00004/($day $\times$ no. of rabbits$)$ and $t_{final} = 800$ days? Also plot the number of foxes versus the number of rabbits. Explain why the curves look the way they do.
Vary the parameters k_1, k_2, k_3, and k_4. Discuss which parameters can or cannot be larger than others. Write a paragraph describing what you find.

(b) Use POLYMATH or MATLAB to solve the following set of nonlinear algebraic equations:

$$x^3 y - 4y^2 + 3x = 1$$

$$6y^2 - 9xy = 5$$

with initial guesses of $x = 2$, $y = 2$. Try to become familiar with the edit keys in POLYMATH MatLab. See CD-ROM for instructions.

P1-19$_A$ (a) Surf the World Wide Web and make a list of the links that are relevant to chemical reaction engineering. Pick the five most relevant links and write two or three sentences about each.

(b) Check the reaction engineering 3rd ed. web site (http://www.engin.umich. edu./~cre) to learn what material has been added and any typographical errors that have been found in the first printing.

P1-20$_A$ Surf the CD-ROM included with the text.
(a) Approximately how many additional solved example problems are there?
(b) List at least one video clip.
(c) In what lectures are activation energy discussed?
(d) What photos are in the Wetlands Module?

After Reading Each Page in
This Book, Ask Yourself a Question
About What You Read

CD-ROM MATERIAL

CD-ROM
○
LINKS

- **Learning Resources**
  1. Summary Notes for Lectures 1 and 2
  2. Web Modules
     A. Problem Solving Algorithm for Closed-Ended Problems
     B. Hints for Getting Unstuck on a Problem
  3. Interactive Computer Modules
     A. Quiz Show I
  4. Solved Problems
     A. CDP1-A$_B$ Batch Reactor Calculations: A Hint of Things to Come
- **Professional Reference Shelf**
  1. Photographs of Real Reactors
- **FAQ [Frequently Asked Questions]**– In Updates/FAQ icon section
- **Additional Homework Problems**

CDP1-A$_A$ Calculate the time to consume 80% of species A in a constant-volume batch reactor for a first- and a second-order reaction. **(Includes Solution)**

CDP1-B$_A$ Derive the differential mole balance equation for a foam reactor. [2nd Ed. P1-10$_B$]

SUPPLEMENTARY READING

1. For further elaboration of the development of the general balance equation, see

   DIXON, D. C., *Chem. Eng. Sci., 25*, 337 (1970).

   FELDER, R. M., and R. W. ROUSSEAU, *Elementary Principles of Chemical Processes*, 2nd ed. New York: Wiley, 1986, Chap. 4.

   HIMMELBLAU, D. M., *Basic Principles and Calculations in Chemical Engineering*, 6th ed. Upper Saddle River, N.J.: Prentice Hall, 1996, Chaps. 2 and 6.

   HOLLAND, C. D., and R. G. ANTHONY, *Fundamentals of Chemical Reaction Engineering*, 2nd ed. Upper Saddle River, N.J.: Prentice Hall, 1989, Chap. 1.

2. A detailed explanation of a number of topics in this chapter can be found in

   CRYNES, B. L., and H. S. FOGLER, eds., *AIChE Modular Instruction Series E: Kinetics*, Vols. 1 and 2. New York: AIChE, 1981.

3. An excellent description of the various types of commercial reactors used in industry is found in Chapter 11 of

   WALAS, S. M., *Reaction Kinetics for Chemical Engineers*. New York: McGraw-Hill, 1959.

   A somewhat different discussion of the usage, advantages, and limitations of various reactor types can be found in

   DENBIGH, K. G., and J. C. R. TURNER, *Chemical Reactor Theory*, 2nd ed. Cambridge: Cambridge University Press, 1971, pp. 1–10.

4. A discussion of some of the most important industrial processes is presented by

> MEYERS, R.A., *Handbook of Chemical Production Processes*. New York: McGraw-Hill, 1986.

See also

> MCKETTA, J. J., ed., *Encyclopedia of Chemical Processes and Design*. New York: Marcel Dekker, 1976.

A similar book, which describes a larger number of processes, is

> SHREVE, R. N., and J. A. BRINK, JR., *Chemical Process Industries*, 4th ed. New York: McGraw-Hill, 1977.

5. The following journals may be useful in obtaining information on chemical reaction engineering: *International Journal of Chemical Kinetics, Journal of Catalysis, Journal of Applied Catalysis, AIChE Journal, Chemical Engineering Science, Canadian Journal of Chemical Engineering, Chemical Engineering Communications, Journal of Physical Chemistry,* and *Industrial and Engineering Chemistry Research.*

6. The price of chemicals can be found in such journals as the *Chemical Marketing Reporter, Chemical Weekly,* and *Chemical Engineering News.*

Conversion 2
and Reactor Sizing

> Be more concerned with your character than with your
> reputation, because character is what you really are
> while reputation is merely what others think you are.
>
> John Wooden, coach, UCLA Bruins

The first chapter focused on the general mole balance equation; the balance was applied to the four most common types of industrial reactors, and a design equation was developed for each reactor type. In Chapter 2 we first define *conversion* and then rewrite the design equations in terms of conversion. After carrying out this operation, we show how one may *size* a reactor (i.e., determine the reactor volume necessary to achieve a specified conversion) once the relationship between reaction rate, r_A, and conversion is known.

2.1 Definition of Conversion

In defining conversion, we choose one of the reactants as the basis of calculation and then relate the other species involved in the reaction to this basis. In most instances it is best to choose the limiting reactant as the basis of calculation. We develop the stoichiometric relationships and design equations by considering the general reaction

$$aA + bB \longrightarrow cC + dD \qquad (2\text{-}1)$$

The uppercase letters represent chemical species and the lowercase letters represent stoichiometric coefficients. Taking species A as our *basis of calculation*, we divide the reaction expression through by the stoichiometric coefficient of species A, in order to arrange the reaction expression in the form

$$A + \frac{b}{a}\,B \longrightarrow \frac{c}{a}\,C + \frac{d}{a}\,D \tag{2-2}$$

to put every quantity on a "per mole of A" basis.

Now we ask such questions as "How can we quantify how far a reaction [e.g., Equation (2-2)] has progressed?" or "How many moles of C are formed for every mole A consumed?" A convenient way to answer these questions is to define a parameter called *conversion*. The conversion X_A is the number of moles of A that have reacted per mole of A fed to the system:

Definition of X

$$X_A = \frac{\text{moles of A reacted}}{\text{moles of A fed}}$$

Because we are defining conversion with respect to our basis of calculation [A in Equation (2-2)], we eliminate the subscript A for the sake of brevity and let $X \equiv X_A$.

2.2 Design Equations

2.2.1 Batch Systems

In most batch reactors, the longer a reactant is in the reactor, the more reactant is converted to product until either equilibrium is reached or the reactant is exhausted. Consequently, in batch systems the conversion X is a function of the time the reactants spend in the reactor. If N_{A0} is the number of moles of A initially, then the total number of moles of A that have reacted after a time t is $[N_{A0}X]$

$$\begin{bmatrix} \text{moles of A} \\ \text{consumed} \end{bmatrix} = \begin{bmatrix} \text{moles of A} \\ \text{fed} \end{bmatrix} \cdot \begin{bmatrix} \text{moles of A reacted} \\ \text{mole of A fed} \end{bmatrix}$$

$$\begin{bmatrix} \text{moles of A} \\ \text{reacted} \\ \text{(consumed)} \end{bmatrix} = [N_{A0}] \quad \cdot \quad [X] \tag{2-3}$$

Now, the number of moles of A that remain in the reactor after a time t, N_A, can be expressed in terms of N_{A0} and X:

$$\begin{bmatrix} \text{moles of A} \\ \text{in reactor} \\ \text{at time } t \end{bmatrix} = \begin{bmatrix} \text{moles of A} \\ \text{initially fed} \\ \text{to reactor at} \\ t = 0 \end{bmatrix} - \begin{bmatrix} \text{moles of A that} \\ \text{have been con-} \\ \text{sumed by chemical} \\ \text{reaction} \end{bmatrix}$$

$$[N_A] \quad = \quad [N_{A0}] \quad - \quad [N_{A0}X]$$

The number of moles of A in the reactor after a conversion X has been achieved is

$$\boxed{N_A = N_{A0} - N_{A0}X = N_{A0}(1-X)}$$ (2-4)

When no spatial variations in reaction rate exist, the mole balance on species A for a batch system reduces to the following equation:

$$\frac{dN_A}{dt} = r_A V$$ (2-5)

This equation is valid whether or not the reactor volume is constant. In the general reaction

$$A + \frac{b}{a}B \longrightarrow \frac{c}{a}C + \frac{d}{a}D$$ (2-2)

reactant A is disappearing; therefore, we multiply both sides of Equation (2-5) by -1 to obtain the mole balance for the batch reactor in the form

$$-\frac{dN_A}{dt} = (-r_A)V$$

The rate of disappearance of A, $-r_A$, in this reaction might be given by a rate law similar to Equation (1-2), such as $-r_A = kC_A C_B$.

For batch reactors we are interested in determining how long to leave the reactants in the reactor to achieve a certain conversion X. To determine this length of time, we transform the mole balance, Equation (2-5), in terms of conversion by differentiating Equation (2-4),

$$N_A = N_{A0} - N_{A0}X$$ (2-4)

with respect to time, while remembering that N_{A0} is the number of moles of A initially present and is therefore a constant with respect to time.

$$\frac{dN_A}{dt} = 0 - N_{A0}\frac{dX}{dt}$$

Combining the above with Equation (2-5) yields

$$-N_{A0}\frac{dX}{dt} = r_A V$$

For a batch reactor, the design equation in differential form is

Batch reactor
design equation

$$\boxed{N_{A0}\frac{dX}{dt} = -r_A V}$$ (2-6)

The differential forms of the design equations often appear in reactor analysis and are particularly useful in the interpretation of reaction rate data.

Constant-volume batch reactors are found very frequently in industry. In particular, the laboratory bomb reactor for gas-phase reactions is widely used for obtaining reaction rate information on a small scale. Liquid-phase reactions in which the volume change during reaction is insignificant are frequently carried out in batch reactors when small-scale production is desired or operating difficulties rule out the use of continuous systems. For a *constant-volume batch reactor*, Equation (2-5) can be arranged into the form

$$-\frac{1}{V}\frac{dN_A}{dt} = -\frac{d(N_A/V)}{dt} = -\frac{dC_A}{dt} = -r_A \tag{2-7}$$

For batch-reactor systems in which the volume varies while the reaction is proceeding, the volume may usually be expressed either as a function of time alone or of conversion alone, for either adiabatic or isothermal reactors. Consequently, the variables of the differential equation (2-6) can be separated in one of the following ways:

$$V\,dt = N_{A0}\frac{dX}{-r_A}$$

or

$$dt = N_{A0}\frac{dX}{-r_A V}$$

These equations are integrated with the limits that the reaction begins at time zero (i.e., $t = 0$, $X = 0$). When the volume is varied by some external source in a specific manner (such as a car cylinder piston compressing the reacting gas according to the equation $V = V_1 + V_2 \sin \omega t$), the equation relating time and conversion that one would use is

$$\int_0^t V\,dt = N_{A0}\int_0^X \frac{dX}{-r_A} \tag{2-8}$$

However, for the more common batch reactors in which volume is not a predetermined function of time, the time t necessary to achieve a conversion X is

$$\boxed{t = N_{A0}\int_0^{X(t)} \frac{dX}{-r_A V}} \tag{2-9}$$

Design
Equation

Equation (2-6) is the differential form of the design equation, and Equations (2-8) and (2-9) are the integral forms for a batch reactor. The differential form is generally used in the interpretation of laboratory rate data.

2.2.2 Flow Systems

Normally, conversion increases with the time the reactants spend in the reactor. For continuous-flow systems, this time usually increases with increasing reactor volume; consequently, the conversion X is a function of reactor volume V. If F_{A0} is the molar flow rate of species A fed to a system operated at steady state, the molar rate at which species A is reacting within the entire system will be $F_{A0}X$.

$$[F_{A0}] \cdot [X] = \frac{\text{moles of A fed}}{\text{time}} \cdot \frac{\text{moles of A reacted}}{\text{mole of A fed}}$$

$$[F_{A0} \cdot X] = \frac{\text{moles of A reacted}}{\text{time}}$$

The molar feed rate of A *to* the system *minus* the rate of reaction of A within the system *equals* the molar flow rate of A leaving the system F_A. The preceding sentence can be written in the form of the following mathematical statement:

$$\begin{bmatrix} \text{molar flow rate} \\ \text{at which A is} \\ \text{fed to the system} \end{bmatrix} - \begin{bmatrix} \text{molar rate at} \\ \text{which A is} \\ \text{consumed within} \\ \text{the system} \end{bmatrix} = \begin{bmatrix} \text{molar flow rate} \\ \text{at which A leaves} \\ \text{the system} \end{bmatrix}$$

$$[F_{A0}] \qquad - \qquad [F_{A0}X] \qquad = \qquad [F_A]$$

Rearranging gives

$$\boxed{F_A = F_{A0}(1 - X)} \qquad (2\text{-}10)$$

The entering molar flow rate, F_{A0} (mol/s), is just the product of the entering concentration, C_{A0} (mol/dm^3), and the entering volumetric flow rate, v_0 (dm^3/s):

$$F_{A0} = C_{A0}v_0$$

For liquid systems, C_{A0} is commonly given in terms of molarity, for example, $C_{A0} = 2$ mol/dm^3. For gas systems, C_{A0} can be calculated from the entering temperature and pressure using the ideal gas law or some other gas law. For an ideal gas (see Appendix B):

$$C_{A0} = \frac{P_{A0}}{RT_0} = \frac{y_{A0}P_0}{RT_0}$$

where C_{A0} = entering concentration, mol/dm³

y_{A0} = entering mole fraction of A

P_0 = entering total pressure, kPa

T_0 = entering temperature, K

P_{A0} = entering partial pressure, kPa

$$R = \text{ideal gas constant} \left(\text{e.g., } R = 8.314 \frac{\text{kPa} \cdot \text{dm}^3}{\text{mol} \cdot K}; \text{ see Appendix B} \right)$$

Example 2–1   Using the Ideal Gas Law to Calculate C_{A0}

A gas mixture consists of 50% A and 50% inerts at 10 atm (1013 kPa) and enters the reactor with a flow rate of 6 dm³/s at 300°F (422.2 K). Calculate the entering concentration of A, C_{A0}, and the entering molar flow rate, F_{A0}. The ideal gas constant is

$$R = 0.082 \text{ dm}^3 \cdot \text{atm/mol} \cdot K \quad \text{(Appendix B)}$$

Solution

We recall that for an ideal gas:

$$C_{A0} = \frac{P_{A0}}{RT_0} = \frac{y_{A0}P_0}{RT_0} \tag{E2-1.1}$$

where P_0 = 10 atm

y_{A0} = 0.5

P_{A0} = initial partial pressure = $y_{A0}P_0$ = (0.5)(10 atm) = 5 atm

T_0 = initial temperature = 300°F = 149°C = 422.2 K

$$R = \frac{0.82 \text{ dm}^3 \cdot \text{atm}}{\text{mol} \cdot K}$$

We could also solve for the partial pressure in terms of the concentration

$$P_{A0} = C_{A0}RT_0 \tag{E2-1.2}$$

Substituting values in Equation (E2-1.1) yields

$$C_{A0} = \frac{0.5(10 \text{ atm})}{0.082 \text{ dm}^3 \cdot \text{atm/mol} \cdot K(422.2 \text{ K})} = 0.14442 \frac{\text{mol}}{\text{dm}^3}$$

Keeping only the significant figures gives us

$$C_{A0} = 0.144 \text{ mol/dm}^3 = 0.144 \text{ kmol/m}^3 = 0.144 \text{ mol/L}$$

The entering molar flow rate, F_{A0}, is just the product of the entering concentration, C_{A0}, and the entering volumetric flow rate, v_0:

$$F_{A0} = C_{A0} v_0 = (0.14442 \text{ mol/dm}^3)(6.0 \text{ dm}^3/\text{s}) = 0.867 \text{ mol/s}$$

We will use this value of F_{A0} together with either Table 2-2 or Figure 2-1 to size a number of reactor schemes in Examples 2-2 through 2-5.

Now that we have a relationship [Equation (2-10)] between the molar flow rate and conversion, it is possible to express the design equations (i.e., mole balances) in terms of conversion for the *flow* reactors examined in Chapter 1.

CSTR or Backmix Reactor.   The equation resulting from a mole balance on species A for the reaction

$$A + \frac{b}{a} B \longrightarrow \frac{c}{a} C + \frac{d}{a} D \qquad (2\text{-}2)$$

occuring in a CSTR was given by Equation (1-6), which can be arranged to

$$F_{A0} - F_A = -r_A V \qquad (2\text{-}11)$$

We now substitute for the exiting molar flow rate of A, F_A, in terms of the conversion X and the entering molar flow rate, F_{A0}, by using Equation (2-10) in the form

$$F_{A0} - F_A = F_{A0} X$$

and combining it with Equation (2-11) to give

$$F_{A0} X = -r_A V \qquad (2\text{-}12)$$

We can rearrange Equation (2-12) to determine the CSTR volume necessary to achieve a specified conversion X.

$$\boxed{V = \frac{F_{A0} X}{(-r_A)_{\text{exit}}}} \qquad (2\text{-}13)$$

Since the exit composition from the reactor is identical to the composition inside the reactor, the rate of reaction is evaluated at the exit conditions.

Tubular Flow Reactor (PFR).   After multiplying both sides of the tubular reactor design equation (1-10) by -1, we express the mole balance equation for species A in the reaction given by Equation (2-2) as

$$\frac{-dF_A}{dV} = -r_A \qquad (2\text{-}14)$$

For a flow system, F_A has previously been given in terms of the entering molar flow rate F_{A0} and the conversion X:

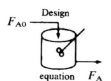

Design
equation

$$F_A = F_{A0} - F_{A0}X \tag{2-10}$$

Substituting Equation (2-10) into (2-14) gives the differential form of the design equation for a plug-flow reactor:

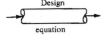

Design equation

$$F_{A0}\frac{dX}{dV} = -r_A \tag{2-15}$$

We now separate the variables and integrate with the limit $V = 0$ when $X = 0$ to obtain the plug-flow reactor volume necessary to achieve a specified conversion X:

$$V = F_{A0}\int_0^X \frac{dX}{-r_A} \tag{2-16}$$

To carry out the integrations in the batch and plug-flow reactor design equations (2-9) and (2-16), as well as to evaluate the CSTR design equation (2-13), we need to know how the reaction rate $-r_A$ varies with the concentration (hence conversion) of the reacting species. This relationship between reaction rate and concentration is developed in Chapter 3.

Packed-Bed Reactor.   The derivation of the differential and integral forms of the design equations for a packed-bed reactor are analogous to those for a PFR [cf. Equations (2-15) and (2-16)]. That is, substituting for F_A in Equation (1-11) gives

PBR design equation

$$F_{A0}\frac{dX}{dW} = -r'_A \tag{2-17}$$

The differential form of the design equation [i.e., Equation (2-17)] *must* be used when analyzing reactors that have a pressure drop along the length of the reactor. We discuss pressure drop in packed-bed reactors in Chapter 4.

Integrating with the limits $W = 0$ at $X = 0$ gives

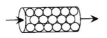

$$W = F_{A0}\int_0^X \frac{dX}{-r'_A} \tag{2-18}$$

Equation (2-18) can be used to determine the catalyst weight W necessary to achieve a conversion X when the total pressure remains constant.

2.3 Applications of the Design Equations for Continuous-Flow Reactors

The rate of disappearance of A, $-r_A$, is almost always a function of the concentrations of the various species present. When a single reaction is occurring, each of the concentrations can be expressed as a function of the conversion X (see Chapter 3); consequently, $-r_A$ can be expressed as a function of X.

A particularly simple functional dependence, yet one that occurs on many occasions, is $-r_A = kC_{A0}(1 - X)$. For this dependence, a plot of the reciprocal rate of reaction $(-1/r_A)$ as a function of conversion yields a curve similar to the one shown in Figure 2-1, where

$$\frac{1}{-r_A} = \frac{1}{kC_{A0}}\left(\frac{1}{1-X}\right)$$

To illustrate the design of a series of reactors, we consider the isothermal gas-phase decomposition reaction

$$A \longrightarrow B + C$$

The laboratory measurements given in Table 2-1 show the chemical reaction rate as a function of conversion. The temperature was 300°F (422.2 K), the total pressure 10 atm (1013 kPa), and the initial charge an equimolar mixture of A and inerts.

TABLE 2-1   RAW DATA

X	$-r_A$ (mol/dm$^3 \cdot$ s)
0.0	0.0053
0.1	0.0052
0.2	0.0050
0.3	0.0045
0.4	0.0040
0.5	0.0033
0.6	0.0025
0.7	0.0018
0.8	0.00125
0.85	0.00100

If we know $-r_A$ as a function of X, we can size any isothermal reaction system.

The rate data in Table 2-1 have been converted to reciprocal rates, $1/-r_A$ in Table 2-2, which are now used to arrive at the desired plot of $1/-r_A$ as a function of X, shown in Figure 2-1. We will use this figure to illustrate how one can size each of the reactors in a number of different reactor sequences. The volumetric feed to each reactor sequence will be 6.0 dm^3/s. First, though, some initial conditions should be evaluated. If a reaction is carried out isothermally, the rate is usually greatest at the start of the reaction when the concentration of reactant is greatest [i.e., when there is negligible conversion $(X = 0)$]. Hence $(1/-r_A)$ will be small. Near the end of the reaction, when the reactant concentration is small (i.e., the conversion is large), the reaction rate will be small. Consequently, $(1/-r_A)$ is large. For irreversible reactions of greater than zero-order,

TABLE 2-2   PROCESSED DATA

X	0.0	0.1	0.2	0.3	0.4	0.5	0.6	0.7	0.8	0.85
$-r_A$	0.0053	0.0052	0.0050	0.0045	0.0040	0.0033	0.0025	0.0018	0.00125	0.001
$1/-r_A$	189	192	200	222	250	303	400	556	800	1000

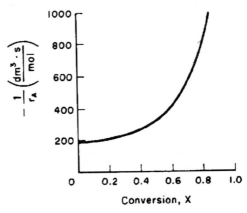

Figure 2-1 Processed Data.

$$A \longrightarrow B + C$$

$$-\frac{1}{r_A} \rightarrow \infty \quad \text{as} \quad X \rightarrow 1$$

For reversible reactions in which the equilibrium conversion is X_e,

$$A \rightleftharpoons B + C$$

$$-\frac{1}{r_A} \rightarrow \infty \quad \text{as} \quad X \rightarrow X_e$$

These characteristics are illustrated in Figure 2-1. The majority of reactions exhibit qualitatively similar curves for isothermal operation.

Example 2–2 Sizing a CSTR

(a) Using the data in either Table 2-2 or Figure 2-1, calculate the volume necessary to achieve 80% conversion in a CSTR. (b) Shade the area in Figure 2-1 which when multiplied by F_{A0} would give the volume of a CSTR necessary to achieve 80% conversion (i.e., $X = 0.8$).

Solution

From Example 2-1, knowing the entering conditions $v_0 = 6$ dm³/s, $P_0 = 10$ atm, $y_{A0} = 0.5$, $T_0 = 422.2$ K, we can use the ideal gas law to calculate the entering molar flow rate of A, i.e.,

$$F_{A0} = C_{A0}v_0 = \frac{y_{A0}P_0}{RT_0}v_0 = \frac{(0.5)(1013 \text{ kPa}) \cdot 6 \text{ dm}^3/\text{s}}{[8.314 \text{ kPa dm}^3/(\text{mol})(\text{K})](422.2 \text{ K})} = 0.867 \text{ mol/s}$$

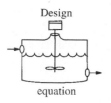

Design

equation

(a) Equation (2-13) gives the volume of a CSTR as a function of F_{A0}, X, and $-r_A$:

$$V = \frac{F_{A0} X}{-r_A} \qquad (2\text{-}13)$$

In a CSTR, the composition, temperature, and conversion of the effluent stream are identical to that of the fluid within the reactor, since perfect mixing is assumed. Therefore, we need to find the value of $-r_A$ (or reciprocal thereof) at $X = 0.8$. From either Table 2-2 or Figure 2-1 we see that when $X = 0.8$, then

$$1/-r_A = 800 \frac{dm^3 \cdot s}{mol}$$

Substitution into Equation (2-13) gives

$$V = 0.867 \frac{mol}{s} \left(\frac{800\ dm^3 \cdot s}{mol} \right) (0.8) \qquad (E2\text{-}2.1)$$

$$= 554.9\ dm^3 = 554.9\ L$$

(b) Shade the area in Figure 2-1 which when multiplied by F_{A0} yields the CSTR volume. Rearranging Equation (2-13) gives

$$V = F_{A0} \left(\frac{1}{-r_A} \right) X \qquad (2\text{-}13)$$

$$\frac{V}{F_{A0}} = \left[\frac{1}{-r_A} \right]_{x=0.8} (0.8) \qquad (E2\text{-}2.2)$$

Plots of $1/-r_A$ vs. X are sometimes referred to as Levenspiel plots (after Octave Levenspiel)

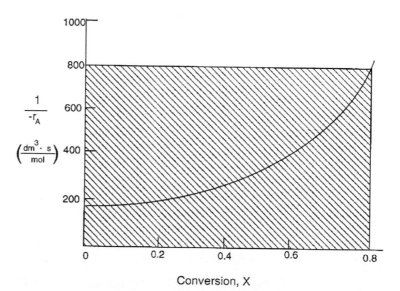

Figure E2-2.1   Levenspiel CSTR plot.

In Figure E2-2.1 the value of V/F_{A0} is equal to the area of a rectangle with a height $1/-r_A = 800\ dm^3 \cdot s/mol$ and a base $X = 0.8$. This rectangle is shaded in the figure. To calculate the reactor volume, we multiply the area of the rectangle by F_{A0}.

$$V = 0.867 \frac{\text{mol}}{\text{s}} \left[800 \frac{\text{dm} \cdot \text{s}}{\text{mol}} (0.8) \right] = 554.9 \text{ dm}^3$$

The CSTR volume necessary to achieve 80% conversion at the specified temperature and pressure is 555 dm³.

Example 2–3 Sizing a PFR

The reaction described by the data in Tables 2-1 and 2-2 is to be carried out in a PFR. The entering molar flow rate is 5 mol/s. Calculate the reactor volume necessary to achieve 80% conversion in a PFR. **(a)** First, use one of the integration formulas given in Appendix A.4 to determine the PFR reactor volume. **(b)** Next, shade the area in Figure 2-1 which when multiplied by F_{A0} would give the PFR volume. **(c)** Make a qualitative sketch of the conversion, X, and the rate of reaction, $-r_A$, down the length (volume) of the reactor.

Solution

(a) For the PFR, the differential form of the mole balance is

$$F_{A0} \frac{dX}{dV} = -r_A \tag{2-15}$$

Rearranging and integrating gives

$$V = F_{A0} \int_0^X \frac{dX}{-r_A} \tag{2-16}$$

For 80% conversion, we will use the five-point quadratic formula with $\Delta X = 0.2$.

$$V = F_{A0} \int_0^X \frac{dX}{-r_A}$$

$$= F_{A0} \frac{\Delta X}{3} \left[\frac{1}{-r_A(X=0)} + \frac{4}{-r_A(0.2)} + \frac{2}{-r_A(0.4)} + \frac{4}{-r_A(0.6)} + \frac{1}{-r_A(0.8)} \right]$$

Using values of $1/-r_A$ in Table 2-2 yields

$$V = (0.867 \text{ mol/s})(0.2/3)[189 + 4(200) + 2(250) + 4(400) + (800)] \frac{\text{s} \cdot \text{dm}^3}{\text{mol}}$$

$$= (0.867 \text{ mol/s})(259.3 \text{ s} \cdot \text{dm}^3/\text{mol})$$

$$= 225 \text{ dm}^3$$

(b) The integral in Equation (2-16) can be evaluated for the area under the curve of a plot of $(1/-r_A)$ versus X.

$$\frac{V}{F_{A0}} = \int_0^{0.8} \frac{dX}{-r_A} = \text{area under the curve between } X = 0 \text{ and } X = 0.8$$
$$\text{(see appropriate shaded area in Figure E2-3.1)}$$

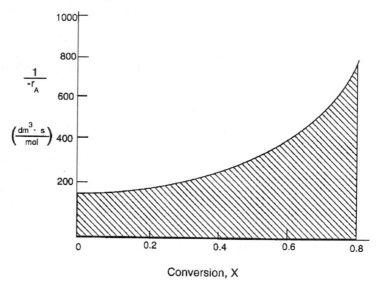

Figure E2-3.1 Levenspiel PFR plot.

The product of this area and F_{A0} will give the tubular reactor volume necessary to achieve the specified conversion of A. For 80% conversion, the shaded area is roughly equal to 260 dm³·(s/mol). The tubular reactor volume can be determined by multiplying this area [in dm³·(s/mol)] by F_{A0} (mol/s). Consequently, for an entering molar flow rate of 0.867 mol/s the PFR volume necessary to achieve 80% conversion is 225 dm³.

(c) Sketch $-r_A$ and X down the length of the reactor. We know that as we proceed down the reactor and more and more of the reactant is consumed, the concentration of reactant decreases, as does the rate of disappearance of A. However, the conversion increases as more and more reactant is converted to product. For $X = 0.2$ we calculate the corresponding reactor volume using Simpson's rule with $\Delta X = 0.1$.

$$V = F_{A0} \int_0^{0.2} \frac{dX}{-r_A} = F_{A0} \frac{\Delta X}{3} \left[\frac{1}{-r_A(X=0)} + \frac{4}{-r_A(X=0.1)} + \frac{1}{-r_A(X=0.2)} \right]$$

$$= 0.867 \frac{\text{mol}}{\text{s}} \left[\frac{0.1}{3} [189 + 4(192) + 200] \right] \frac{\text{dm}^3}{\text{mol} \cdot \text{s}}$$

$$= 33.4 \text{ dm}^3$$

For $X = 0.4$, we can again use Simpson's rule with $\Delta X = 0.2$:

$$V = F_{A0} \frac{\Delta X}{3} \left[\frac{1}{-r_A(X=0)} + \frac{4}{-r_A(X=0.2)} + \frac{1}{-r_A(X=0.4)} \right]$$

$$= 0.867 \frac{\text{mol}}{\text{s}} \left[\frac{0.2}{3} [189 + 4(200) + 250] \right] \frac{\text{dm}^3}{\text{mol} \cdot \text{s}}$$

$$= 71.6 \text{ dm}^3$$

We can continue in this manner to arrive at Table E2-3.1.

TABLE E2-3.1.   CONVERSION PROFILE

$V(\text{dm}^3)$	0	33.4	71.6	126	225
X	0	0.2	0.4	0.6	0.8
$-r_A\left(\dfrac{\text{mol}}{\text{dm}^3 \cdot s}\right)$	0.0053	0.005	0.004	0.0025	0.00125

which is shown in Figure E2-3.2.

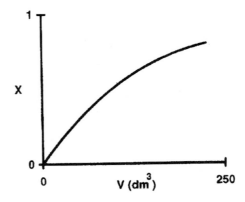

Figure E2-3.2 Conversion profile.

Rather than using Simpson's rule we could have used the data in Table 2-2 to fit $-r_A(X)$ to a polynomial and then used POLYMATH to integrate the design equation to obtain the conversion profile.

Example 2–4 Comparing CSTR and PFR Sizes

It is interesting to compare the volumes of a CSTR and a plug-flow reactor (PFR) required for the same job. To do this we shall use the data in Figure 2-1 to learn which reactor would require the smaller volume to achieve a conversion of 60%: a CSTR or a PFR. The feed conditions are the same in both cases. The entering molar flow rate is 5 mol/s.

Solution

For the CSTR:

$$\frac{V}{F_{A0}} = \left(\frac{1}{-r_A}\right)X = (400)\,(0.6) = 240\,\frac{\text{dm}^3 \cdot s}{\text{mol}}$$

This is also the area of the rectangle with vertices $(X, 1/-r_A)$ of $(0, 0)$, $(0, 400)$, $(0.6, 400)$, and $(0.6, 0)$. The CSTR volume necessary to achieve 60% conversion is

$$V = \left(\frac{5 \text{ mol}}{s}\right)\left(\frac{240 \text{ dm}^3 \cdot s}{\text{mol}}\right) = 1200 \text{ dm}^3$$

For the plug-flow (tubular) reactor:

$$F_{A0}\frac{dX}{dV} = -r_A$$

Integrating and rearranging Equation (2-15) yields

$$\frac{V}{F_{A0}} = \int_0^{0.6} \frac{dX}{-r_A} = \frac{\Delta X}{3}\left[\frac{1}{-r_A(0.0)} + \frac{4}{-r_A(0.3)} + \frac{1}{-r_A(0.6)}\right]$$

$$= \frac{0.3}{3} \times [189 + 4(222) + 400]$$

$$= 148 \frac{\text{dm}^3 \cdot s}{\text{mol}}$$

The PFR volume necessary to achieve 60% conversion is

$$V = \left(\frac{5 \text{ mol}}{s}\right)\left(148 \frac{\text{dm}^3 \cdot s}{\text{mol}}\right) = 740 \text{ dm}^3$$

For the same flow rate F_{A0} the plug-flow reactor requires a smaller volume than the CSTR to achieve a conversion of 60%. This comparison can be seen in Figure E2-4.1. For isothermal reactions of greater than zero order, the PFR will always require a smaller volume than the CSTR to achieve the same conversion.

Generally, the isothermal tubular reactor volume is smaller than the CSTR for the same conversion

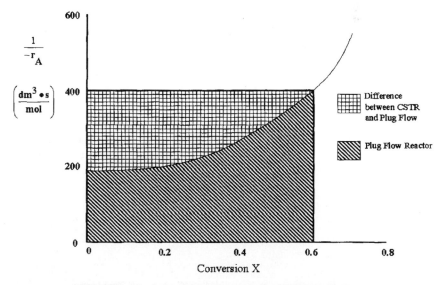

Figure E2-4.1   Levenspiel plot comparing CSTR and PFR size.

2.4 Reactors in Series

Many times reactors are connected in series so that the exit stream of one reactor is the feed stream for another reactor. When this arrangement is used it is often possible to speed calculations by defining conversion in terms of location at a point downstream rather than with respect to any single reactor. That is, the conversion X is the *total number of moles* of A that have reacted up to that point per mole of A fed to the *first* reactor. However, this definition can only be used provided that there are no side streams withdrawn and the feed stream enters only the first reactor in the series.

As an example, the relationships between conversion and molar flow rates for the reactor sequence shown in Figure 2-2 are given by the following equations:

$$F_{A1} = F_{A0} - F_{A0} X_1$$

$$F_{A2} = F_{A0} - F_{A0} X_2$$

$$F_{A3} = F_{A0} - F_{A0} X_3$$

where

$$\boxed{X_2 = \frac{\text{total moles of A reacted up to point 2}}{\text{mole of A fed to first reactor}}}$$

Similar definitions exist for X_1 and X_3.

The volume V_1 is given by Equation (2-16):

Reactor 1
$$V_1 = F_{A0} \int_0^{X_1} \frac{dX}{-r_A}$$

A mole balance on species A for the CSTR in the middle gives

$$\text{in} - \text{out} + \text{generation} = 0$$

$$F_{A1} - F_{A2} + r_{A2} V_2 = 0$$

Rearranging gives us

$-r_{A2}$ is evaluated
at X_2 for the CSTR
in this series
arrangement

$$V_2 = \frac{F_{A1} - F_{A2}}{-r_{A2}}$$

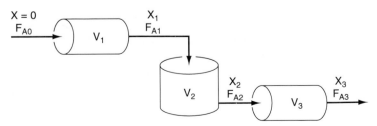

Figure 2-2   PFR-CSTR-PFR in series.

The corresponding rate of reaction $-r_{A2}$ is evaluated at the conversion X_2. Substituting for F_{A1} and F_{A2} yields

Reactor 2

$$V_2 = \frac{F_{A0}(X_2 - X_1)}{-r_{A2}} \qquad (2\text{-}19)$$

The volume for the third reactor, V_3, is found by integrating Equation (2-15) between the limits X_2 and X_3:

Reactor 3

$$V_3 = F_{A0} \int_{X_2}^{X_3} \frac{dX}{-r_A}$$

To demonstrate these ideas, let us consider three different schemes of reactors in series: two CSTRs, two PFRs, and a PFR connected to a CSTR. To size these reactors we shall use laboratory data that give the reaction rate at different conversions. The reactors will operate at the same temperature and pressure as were used in obtaining the laboratory data.

We will now use the value of F_{A0} calculated in Example 2-1 together with Figure 2-1 to size each of the reactors for the three reactor schemes. The first scheme to be considered is the two CSTRs in series shown in Figure 2-3. For the first reactor in which the rate of disappearance of A is $-r_{A1}$ at conversion X_1, the volume necessary to achieve the conversion X_1 is

$$V_1 = F_{A0} \left(\frac{1}{-r_{A1}} \right) X_1$$

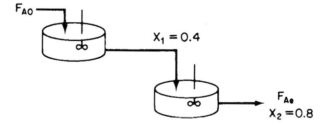

Figure 2-3   Two CSTRs in series.

In the second reactor, the rate of disappearance of A, $-r_{A2}$, is evaluated and the conversion is that of the exit stream of reactor 2, X_2. The volume necessary to increase the conversion in reactor 2 from X_1 to X_2 was derived previously and is given by Equation (2-19):

$$V_2 = \frac{F_{A0}(X_2 - X_1)}{-r_{A2}} \qquad (2\text{-}19)$$

Example 2–5 Comparing Volumes for CSTRs in Series

For the two CSTRs in series, 40% conversion is achieved in the first reactor. What is the total volume of the two reactors necessary for 80% overall conversion of the species A entering reactor 1? (If F_{A2} is the molar flow rate of A exiting from the last reactor in the sequence, $F_{A2} = 0.2F_{A0}$.)

Solution

$$F_{A0} = 0.867 \text{ mol/s}$$

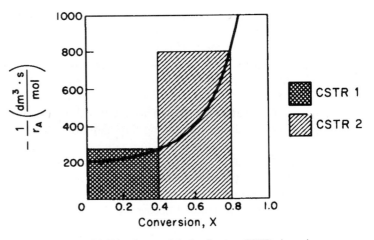

Figure E2-5.1   Levenspiel plot for two CSTRs in series.

For reactor 1 we observe from either Table 2-2 or Figure E2-5.1 that when $X = 0.4$, then

$$\frac{1}{-r_{A1}} = 250 \text{ dm}^3 \cdot \text{s/mol}$$

Then

$$V_1 = F_{A0}\left(\frac{1}{-r_{A1}}\right)X = 0.867\left(\frac{1}{-r_{A1}}\right)(0.4) = (0.867)(250)(0.4)$$

$$V_1 = 86.7 \text{ dm}^3 \text{ (liters)}$$

For reactor 2, when $X_2 = 0.8$, then $(1/-r_A) = 800 \text{ dm}^3 \cdot \text{s/mol}$, and

$$V_2 = F_{A0}\left(\frac{1}{-r_{A2}}\right)(X_2 - X_1)$$

$$= \left(0.867 \frac{\text{mol}}{\text{s}}\right)\left(800 \frac{\text{dm}^3 \cdot \text{s}}{\text{mol}}\right)(0.8 - 0.4)$$

$$= 277.4 \text{ dm}^3 \text{ (liters)}$$

To achieve the
same overall
conversion, the
total volume for
two CSTRs in
series is less than
that required for
one CSTR

Note again that for CSTRs in series the rate $-r_{A1}$ is evaluated at a conversion of 0.4 and rate $-r_{A2}$ is evaluated at a conversion of 0.8. The total volume is

$$V = V_1 + V_2 = 364 \text{ dm}^3 \text{ (liters)}$$

The volume necessary to achieve 80% conversion in one CSTR is

$$V = F_{A0}\left(\frac{1}{-r_A}\right)X = (0.867)(800)(0.8) = 555 \text{ dm}^3 \text{ (liters)}$$

Notice in Example 2-5 that the sum of the two CSTR reactor volumes (364 L) in series is less than the volume of one CSTR (555 L) to achieve the same conversion. This case does not hold true for two plug-flow reactors connected in series as shown in Figure 2-4. We can see from Figure 2-5 and from the equation

$$\int_0^{X_2} \frac{dX}{-r_A} \equiv \int_0^{X_1} \frac{dX}{-r_A} + \int_{X_1}^{X_2} \frac{dX}{-r_A}$$

that it is immaterial whether you place two plug-flow reactors in series or have one continuous plug-flow reactor; the total reactor volume required to achieve the same conversion is identical.

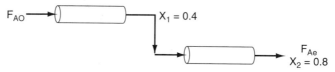

Figure 2-4   Two PFRs in series.

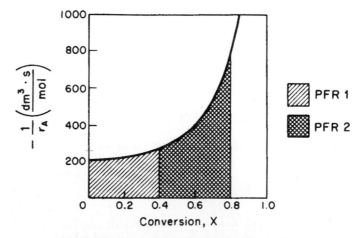

Figure 2-5   Levenspiel plot for two PFRs in series.

Example 2–6   Sizing Plug-Flow Reactors in Series

Using either the data in Table 2-2 or Figure 2-5, calculate the reactor volumes V_1 and V_2 for the plug-flow sequence shown in Figure 2-4 when the intermediate conversion is 40% and the final conversion is 80%. The entering molar flow rate is the same as in the previous examples, 0.867 mol/s.

Solution

In addition to graphical integration we could have used numerical methods to size the plug-flow reactors. In this example, we shall use Simpson's rule (see Appendix A.4) to evaluate the integrals.

The overall conversion of two PRFs in series is the same as one PRF with the same total volume.

$$\int_{X_0}^{X_2} f(X)\, dX = \frac{\Delta X}{3}\, [f(X_0) + 4f(X_1) + f(X_2)] \tag{A-23}$$

For the first reactor, $X_0 = 0$, $X_1 = 0.2$, $X_2 = 0.4$, and $\Delta X = 0.2$,

$$V_1 = F_{A0} \int_0^{0.4} \frac{dX}{-r_A} = F_{A0} \frac{\Delta X}{3} \left[\frac{1}{-r_A(0)} + 4\,\frac{1}{-r_A(0.2)} + \frac{1}{-r_A(0.4)} \right] \tag{E2-6.1}$$

Selecting the appropriate values from Table 2-2, we have

$$V_1 = (0.867 \text{ mol/s}) \left(\frac{0.2}{3} \right) [189 + 4(200) + 250] \text{ L} \cdot \text{s/mol}$$

$$= 71.6 \text{ L} = 71.6 \text{ dm}^3$$

For the second reactor,

$$V_2 = F_{A0} \int_{0.4}^{0.8} \frac{dX}{-r_A}$$

$$= F_{A0} \frac{\Delta X}{3} \left[\frac{1}{-r_A(0.4)} + 4\,\frac{1}{-r_A(0.6)} + \frac{1}{-r_A(0.8)} \right] \tag{E2-6.2}$$

$$= (0.867 \text{ mol/s}) \left(\frac{0.2}{3} \right) [250 + 4(400) + 800] \text{ L} \cdot \text{s/mol}$$

$$= 153 \text{ L} = 153 \text{ dm}^3$$

The total volume is then

$$V = V_1 + V_2 = 225 \text{ L} = 225 \text{ dm}^3$$

The final sequence we shall consider is a CSTR and plug-flow reactor in series. There are two ways in which this sequence can be arranged (Figure

2-6). If the size of each reactor is fixed, a different final conversion, X_2, will be achieved, depending on whether the CSTR, or the plug-flow reactor is placed first. If the intermediate and exit conversions are specified, the reactor volumes as well as their sums can be different for different sequencing. Figure 2-7 shows an actual system of two CSTRs and a PFR in series.

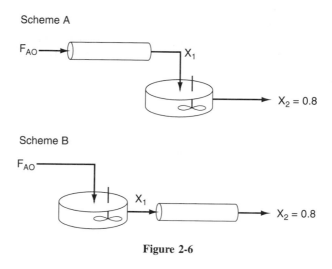

Scheme A

F_{AO} ———→ ———→ X_1

$X_2 = 0.8$

Scheme B

F_{AO} ———→

X_1

$X_2 = 0.8$

Figure 2-6

Figure 2-7   Dimersol G (an organometallic catalyst) unit (two CSTRs and one tubular reactor in series) to dimerize propylene into isohexanes. Institut Français du Petróle process. [Photo courtesy of Editions Technip (Institut Français du Petróle).]

Example 2–7 Comparing the Order of Sequencing Reactors

Calculate the individual reactor volume as well as the total reactor volume for each scheme in Figure 2-6 for the reaction data given in Table 2-2 when the intermediate conversion is 50% and $F_{A0} = 0.867$ mol/s.

Solution

We again use Figure 2-1 to arrive at Figure E2-7.1 and evaluate the design integrals.

Scheme A

Plug flow: $F_{A0} \dfrac{dX}{dV} = -r_A$

Integrating between $X = 0$ and $X = 0.5$ yields

$$V_1 = F_{A0} \int_0^{0.5} \frac{dX}{-r_A} = F_{A0} \left[\frac{\Delta X}{3} \left[f(X_0) + 4 f(X_1) + f(X_2) \right] \right]$$

$$= F_{A0} \frac{\Delta X}{3} \left[\frac{1}{-r_A(0)} + \frac{4}{-r_A(0.25)} + \frac{1}{-r_A(0.5)} \right]$$

$$= F_{A0} \frac{0.25}{3} \left[189 + 4 \times 211 + 303 \right]$$

$$= (0.867)(111)$$

$$= 97 \text{ dm}^3 \text{ (liters)}$$

CSTR: $V_2 = F_{A0} \dfrac{X_2 - X_1}{-r_{A2}} = 0.867(0.8 - 0.5)(800) = 208 \text{ dm}^3$

$$V_{\text{total}} = V_1 + V_2 = 305 \text{ dm}^3$$

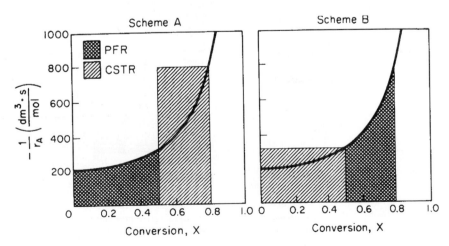

Figure E2-7.1

Scheme B

CSTR: $V_1 = \dfrac{F_{A0} X_1}{-r_{A1}} = 0.867(0.5)(303) = 131.4$ dm^3

PFR: $V_2 = F_{A0} \displaystyle\int_{0.5}^{0.8} \dfrac{dX}{-r_A} = 0.867(151) = 130.9$ dm^3

$V_{total} = 262$ dm^3 (liters)

Scheme B will give the smaller total reactor volume for an intermediate conversion of 50%. This result is shown in Figure E2-7.1. However, as is seen in Problem P2-3, the relative sizes of the reactors depend on the intermediate conversion. Compare your results in Example 2-7 with those in Problem P2-3.

The previous examples show that *if* we know the molar flow rate to the reactor and the reaction rate as a function of conversion, *then* we can calculate the reactor volume necessary to achieve a specified conversion. The reaction rate does not depend on conversion alone, however. It is also affected by the initial concentrations of the reactants, the temperature, and the pressure. Consequently, the experimental data obtained in the laboratory and presented in Table 2-1 as $-r_A$ for given values of X are useful only in the design of full-scale reactors that are to be operated at the same conditions as the laboratory experiments (temperature, pressure, initial reactant concentrations). This conditional relationship is generally true; i.e., to use laboratory data directly for sizing reactors, the laboratory and full-scale operating conditions must be identical. Usually, such circumstances are seldom encountered and we must revert to the methods described in Chapter 3 to obtain $-r_A$ as a function of X.

However, it is important for the reader to realize that if the rate of reaction is available solely as a function of conversion, $-r_A = f(X)$, or if it can be generated by some intermediate calculation, one can design a variety of reactors or combination of reactors.

Finally, let's consider approximating a PFR with a number of small, equal-volume CSTRs of V_i in series (Figure 2-8). We want to compare the total

We need only $-r_A = f(X)$ and F_{A0} to size reactors

Who's on first? Who is.

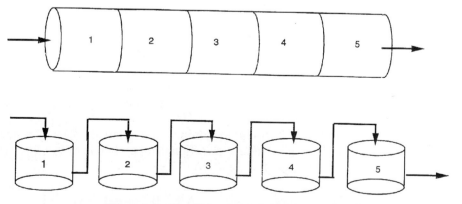

Figure 2-8 Modeling a PFR with CSTRs in series.

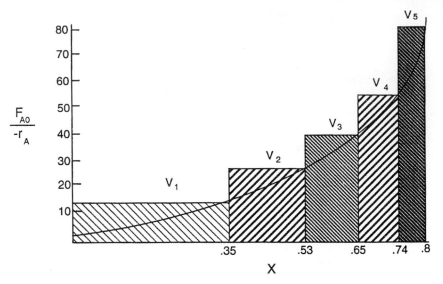

Figure 2-9 Levenspiel plot showing comparison of CSTRs in series with one PFR.

volume of all the CSTRs with the volume of one plug-flow reactor for the same conversion, say 80%. From Figure 2-9 we note a very important observation! The total volume to achieve 80% conversion for five CSTRs of equal volume in series is roughly the same as the volume of a PFR. As we make the volume of each CSTR smaller and increase the number of CSTRs, the total volume of the CSTRs and the PFR will become identical. *That is, we can model a PFR as a number of CSTRs in series.* This concept will be used later in a number of situations, such as modeling catalyst decay in packed-bed reactors or transit heat effects in PFRs.

Ordinarily, laboratory data are used to formulate a rate law, and then the reaction rate–conversion functional dependence is determined using the rate law. Preceding sections show that with the reaction rate–conversion relationship, different reactor schemes can readily be sized. In Chapter 3 we show how we obtain this relationship between reaction rate and conversion from rate law and reaction stoichiometry.

2.5 Some Further Definitions

Before proceeding to Chapter 3, some terms and equations commonly used in reaction engineering need to be defined. We also consider the special case of the plug-flow design equation when the volumetric flow rate is constant.

Relative Rates of Reaction.   The relative rates of reaction of the various species involved in a reaction can be obtained from the ratio of stoichiometric coefficients. For Reaction (2-2),

$$A + \frac{b}{a} B \longrightarrow \frac{c}{a} C + \frac{d}{a} D \qquad (2\text{-}2)$$

we see that for every mole of A that is consumed, c/a moles of C appear. In other words,

$$\text{rate of formation of C} = \frac{c}{a} \text{ (rate of disappearance of A)}$$

$$r_C = \frac{c}{a}(-r_A) = -\frac{c}{a}\, r_A$$

Similarly, the relationship between the rate of formation of C and D is

$$r_C = \frac{c}{d}\, r_D$$

The relationship can be expressed directly from the stoichiometry of the reaction,

$$aA + bB \longrightarrow cC + dD \tag{2-1}$$

for which

Remember this very important relationship for the relative rates of reaction.

$$\boxed{\frac{-r_A}{a} = \frac{-r_B}{b} = \frac{r_C}{c} = \frac{r_D}{d}} \tag{2-20}$$

Space Time. The space time, τ, is obtained by dividing reactor volume by the volumetric flow rate entering the reactor:

$$\tau \equiv \frac{V}{v_0} \tag{2-21}$$

The space time is the time necessary to process one reactor volume of fluid based on entrance conditions. For example, consider the tubular reactor shown in Figure 2-10, which is 20 m long and 0.2 m³ in volume. The dashed line in Figure 2-10 represents 0.2 m³ of fluid directly upstream of the reactor. The time it takes for this fluid to enter the reactor completely is the space time. It is also called the *holding time* or *mean residence time*.

Space time or mean residence time,
$\tau = V/v_0$

Figure 2-10

If both sides of the plug-flow reactor design equation (2-16) are divided by the entering volumetric flow rate and then the left-hand side is put in terms of space time, the equation takes the form

$$\tau = C_{A0} \int_0^X \frac{dX}{-r_A}$$

The space velocity (SV), which is defined as

$$SV \equiv \frac{v_0}{V} \qquad SV = \frac{1}{\tau} \tag{2-22}$$

might at first sight be regarded as the reciprocal of the space time. However, there is a difference in the two quantities' definitions. For the space time, the entering volumetric flow rate is measured at the entrance condition, while for the space velocity other conditions are often used. The two space velocities commonly used in industry are the liquid hourly and gas hourly space velocities, LHSV and GHSV, respectively. The v_0 in the LHSV is frequently measured as that of a liquid at 60 or 75°F, even though the feed to the reactor may be a vapor at some higher temperature. The v_0 in the GHSV is normally measured at standard temperature and pressure (STP).

For reactions in which the rate depends only on the concentration of one species [i.e., $-r_A = f(C_A)$], it is usually convenient to report $-r_A$ as a function of concentration rather than conversion. We can rewrite the design equation for a plug-flow reactor [Equation (2-16)] in terms of the concentration, C_A, rather than in terms of conversion for the special case when $v = v_0$.

$$V = F_{A0} \int_0^X \frac{dX}{-r_A} \tag{2-16}$$

$$F_{A0} = v_0 C_{A0} \tag{2-23}$$

Rearranging Equation (2-10) gives us

$$X = \frac{F_{A0} - F_A}{F_{A0}} \tag{2-24}$$

For the *special case* when $v = v_0$,

$$X = \frac{F_{A0} - F_A}{F_{A0}} = \frac{C_{A0} v_0 - C_A v}{C_{A0} v_0} = \frac{C_{A0} - C_A}{C_{A0}}$$

when $X = 0$, $C_A = C_{A0}$

when $X = X$, $C_A = C_A$

Differentiating yields

$$dX = \frac{-dC_A}{C_{A0}}$$

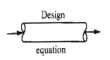

Design

equation

$$V = v_0 \int_{C_A}^{C_{A0}} \frac{dC_A}{-r_A} \tag{2-25}$$

$$\boxed{\tau = \int_{C_A}^{C_{A0}} \frac{dC_A}{-r_A}} \tag{2-26}$$

Valid only if $v = v_0$

Equation (2-26) is a form of the design equation for constant volumetric flow rate v_0 that may prove more useful in determining the space time or reactor volume for reaction rates that depend only on the concentration of one species.

Figure 2-11 shows a typical curve of the reciprocal reaction rate as a function of concentration for an isothermal reaction carried out at constant volume. For reaction orders greater than zero, the rate decreases as concentration decreases. The area under the curve gives the space time necessary to reduce the concentration of A from C_{A0} to C_{A1}.

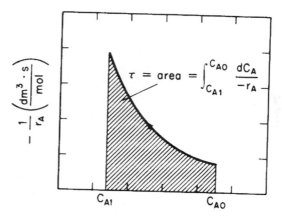

Figure 2-11   Determining the space time, τ.

To summarize these last examples, we have seen that in the design of reactors that are to be operated at conditions (e.g., temperature, initial concentration) identical to those at which the reaction rate data were obtained, detailed knowledge of the kinetic rate law $-r_A$ is not always necessary. In some instances it may be possible to scale up a laboratory-bench or pilot-plant reaction system solely from knowledge of $-r_A$ as a function of X or C_A. Unfortunately for most reactor systems, a scale-up process cannot be achieved simply from a knowledge of $-r_A$ as a function of X. In Chapter 3 we present elementary forms of the kinetic rate law from which the design equations can be evaluated, either by graphical or numerical integration or with the aid of a table of integrals.

SUMMARY

1. The points of this chapter are threefold:
    a. To define the parameter *conversion* and to rewrite the mole balances in terms of conversion.
    b. To show that by expressing $-r_A$ as a function of conversion, a number of reactors and reaction systems can be sized or a conversion be calculated from a given reactor size.
    c. To relate the relative rates of reaction of reactants and products.

2. For the reaction

$$A + \frac{b}{a} B \Longrightarrow \frac{c}{a} C + \frac{d}{a} D$$

The relative rates of reaction can be written either as

$$\boxed{\begin{array}{c} \dfrac{-r_A}{a} = \dfrac{-r_B}{b} = \dfrac{r_C}{c} = \dfrac{r_D}{d} \\[2mm] \text{or} \\[2mm] \dfrac{r_A}{-a} = \dfrac{r_B}{-b} = \dfrac{r_C}{c} = \dfrac{r_D}{d} \end{array}} \qquad \text{(S2-1)}$$

3. The conversion X is the moles of A reacted per mole of A fed.

$$\text{For batch systems:} \qquad X = \frac{N_{A0} - N_A}{N_{A0}} \qquad \text{(S2-2)}$$

$$\text{For flow systems:} \qquad X = \frac{F_{A0} - F_A}{F_{A0}} \qquad \text{(S2-3)}$$

4. For reactors in series with no side streams or multiple feeds, the evaluation of the design equations may be simplified by letting the conversion represent the total moles reacted up to a particular point in the series of reactors.

5. In terms of the conversion, the differential and integral forms of the reactor design equations become:

	Differential Form	Algebraic Form	Integral Form
Batch	$N_{A0} \dfrac{dX}{dt} = -r_A V$		$t = N_{A0} \displaystyle\int_0^X \dfrac{dX}{-r_A V}$ and
			$\displaystyle\int_0^t V(t)\, dt = N_{A0} \int_0^X \dfrac{dX}{-r_A}$
CSTR		$V = \dfrac{F_{A0}(X_{\text{out}} - X_{\text{in}})}{(-r_A)_{\text{out}}}$	
PFR	$F_{A0} \dfrac{dX}{dV} = -r_A$		$V = F_{A0} \displaystyle\int_{X_{\text{in}}}^{X_{\text{out}}} \dfrac{dX}{-r_A}$
PBR	$F_{A0} \dfrac{dX}{dW} = -r_A'$		$W = F_{A0} \displaystyle\int_{X_{\text{in}}}^{X_{\text{out}}} \dfrac{dX}{-r_A'}$

6. If the rate of disappearance is given as a function of conversion, the following graphical techniques can be used to size a CSTR and a plug-flow reactor.

Levenspiel plots

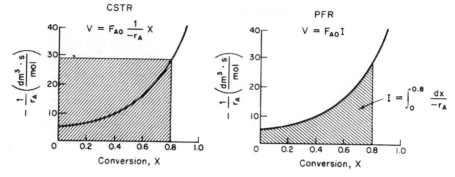

The PFR integral could also be evaluated by

$$V = \int_0^{0.8} \frac{F_{A0}}{-r_A}$$

$$= \frac{0.2}{3}\left[\frac{F_{A0}}{-r_A(0)} + \frac{4F_{A0}}{-r_A(0.2)} + \frac{2F_{A0}}{-r_A(0.4)} + \frac{4F_{A0}}{-r_A(0.6)} + \frac{F_{A0}}{-r_A(0.8)}\right]$$

(S2-4)

[see Equation (A-22) in Appendix A.4]. For the case of reactors in series, for which there are no side streams, the conversion is based on the total conversion up to a specified point. For the reaction sequence

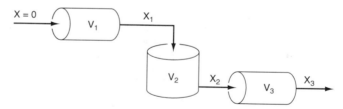

the reactor volumes can be determined from the areas under the curve of a Levenspiel plot as shown below.

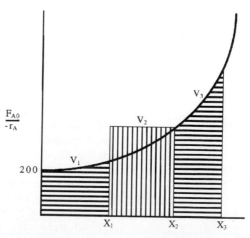

7. Space time, τ, and space velocity, SV, are given by

$$\tau = \frac{V}{v_0} \qquad\qquad (S2\text{-}5)$$

$$SV = \frac{v_0}{V} \qquad\qquad (S2\text{-}6)$$

In evaluating space velocity, the entering volumetric flow rate is usually referred to standard temperature and pressure.

8. Other definitions:

$$LHSV = \text{liquid hourly space velocity, } h^{-1}$$

$$GHSV = \text{gas hourly space velocity, } h^{-1} \text{ at STP.}$$

QUESTIONS AND PROBLEMS

The subscript to each of the problem numbers indicates the level of difficulty: A, least difficult; D, most difficult.

$$A = \bullet \quad B = \blacksquare \quad C = \blacklozenge \quad D = \blacklozenge\blacklozenge$$

In each of the questions and problems below, rather than just drawing a box around your answer, write a sentence or two describing how you solved the problem, the assumptions you made, the reasonableness of your answer, what you learned, and any other facts you want to include. You may wish to refer to W. Strunk and E. B. White, *The Elements of Style* (New York: Macmillan, 1979) and Joseph M. Williams, *Style: Ten Lessons in Clarity & Grace* (Glenview, Ill.: Scott, Foresman 1989) to enhance the quality of your sentences.

P2-1$_A$ Without referring back, make a list of the most important items you learned in this chapter. What do you believe was the overall purpose of the chapter?

P2-2$_A$ What if:
   (a) you needed to estimate the reactor volume necessary to achieve 98% conversion using the data in Table 2-1?
   (b) you were asked to use the data in Table 2-1 to calculate the CSTR reactor volume at a temperature 100°C higher than the temperature at which the data were taken?

P2-3$_A$ Redo Example 2-7 for the cases when the intermediate conversions are (a) 30%, and (b) 70%. The molar flow rate is 52 mol/min.

P2-4$_A$ The space time necessary to achieve 80% conversion in a CSTR is 5 h. Determine (if possible) the reactor volume required to process 2 ft³/min. What is the space velocity for this system?

P2-5$_A$ There are two reactors of equal volume available for your use: one a CSTR, the other a PFR. The reaction is second order $(-r_A = kC_A^2 = kC_{A0}^2(1-X)^2)$, irreversible, and is carried out isothermally.

$$A \longrightarrow B$$

There are three ways you can arrange your system:

(1) Reactors in series: CSTR followed by PFR
(2) Reactors in series: PFR followed by CSTR
(3) Reactors in parallel with half the feed rate going to each reactor after which the exit streams are mixed

(a) If possible, state which system will give the highest overall conversion.

(b) If possible, state which system will give the lowest overall conversion.

(c) If in one or more of the cases above it is not possible to obtain an answer, explain why. *(final exam, winter 1996)*

(d) Comment on whether or not this is a reasonable final exam problem.

P2-6$_B$ The exothermic reaction

$$A \longrightarrow B + C$$

was carried out adiabatically and the following data recorded:

X	0	0.2	0.4	0.5	0.6	0.8	0.9
$-r_A$ (mol/dm^3·min)	10	16.67	50	50	50	12.5	9.09

The entering molar flow rate of A was 300 mol/min.

(a) What are the PFR and CSTR volumes necessary to achieve 40% conversion? ($V_{PFR} = 7.2$ dm^3, $V_{CSTR} = 2.4$ dm^3)

(b) Over what range of conversions would the CSTR and PFR reactor volumes be identical?

(c) What is the maximum conversion that can be achieved in a 10.5-dm^3 CSTR?

(d) What conversion can be achieved if a 7.2-dm^3 PFR is followed in series by a 2.4-dm^3 CSTR?

(e) What conversion can be achieved if a 2.4-dm^3 CSTR is followed in a series by a 7.2-dm^3 PFR?

(f) Plot the conversion and rate of reaction as a function of PFR reactor volume up to a volume of 10 dm^3.

MEMBER
PROBLEM
HALL OF
FAME

P2-7$_C$ *Sgt. Nigel Ambercromby.* Worthless Chemical has been making tirene (B) from butalane (A) (both dark liquids) using a 8.0 ft^3 CSTR followed by a 3.1 ft^3 PFR. The entering flow rate is 1 ft^3/min. A conversion of approximately 81% is achieved using this arrangement. The rate is shown as a function of conversion in Figure P2-8(a). The CSTR is identical to the one of the battery of CSTRs shown in Figure 1-11. There is a preheater upstream of the CSTR that heats the feed to 60°C. One morning the plant manager, Dr. Pakbed, arrived and found that the conversion had dropped to approximately 24%. After inspecting the reactors, the PFR was found to be working perfectly, but a dent was found in the CSTR that may have been caused by something like a fork lift truck. He also notes the CSTR, which normally makes a "woosh" sound is not as noisy as it was yesterday. The manager suspects foul play and calls in Sgt. Nigel Ambercromby from Scotland Yard. What are the first four questions Sgt. Ambercromby asks? Make a list of all the things that could cause the drop in conversion. Quantify the possible explanations with numerical calculations where possible. Dr. Pakbed tells Sgt. Ambercromby that he must achieve a conversion greater than 50% to meet production schedules downstream. Sgt. Ambercromby says, "I think I know how you could do this immediately." What does Ambercromby have in mind? [with Dan Dixon, ChE 344 W'97]

P2-8$_B$ Figure P2-8a shows $C_{A0}/-r_A$ versus X_A for a nonisothermal, nonelementary, multiple-reaction liquid-phase decomposition of reactant A.

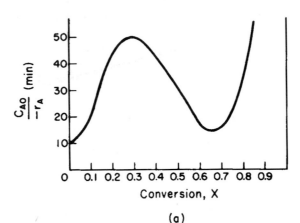

(a)

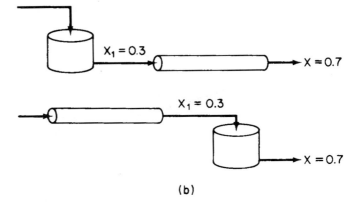

(b)

Figure P2-8

(a) Consider the two systems shown in Figure P2-8b in which a CSTR and plug-flow reactor are connected in series. The intermediate conversion is 0.3 and the final conversion is 0.7. How should the reactors be arranged to obtain the minimum total reactor volume? Explain.

(b) If the volumetric flow rate is 50 L/min, what is the minimum total reactor volume? (*Ans. V* = 750 dm³)

(c) Is there a better means (i.e., smallest total volume achieving 70% conversion other than either of the systems proposed above? (*Ans.*: 512 L)

(d) At what conversion(s) would the required reactor volume be identical for either a CSTR or a tubular PFR? (*Ans.*: X = 0.45, and X = ?)

(e) Using the information in Figure P2-8a together with the CSTR design equation, make a plot of τ versus X. If the reactor volume is 700 L and the volumetric flow rate 50 L/min, what are the possible outlet conversions (i.e., multiple steady states) for this reactor?

P2-9$_B$ The irreversible gas-phase nonelementary reaction

$$A + 2B \longrightarrow C$$

is to be carried out isothermally in a constant-pressure batch reactor. The feed is at a temperature of 227°C, a pressure of 1013 kPa, and its composition is 33.3% A and 66.7% B. Laboratory data taken under identical conditions are as follows (note that at $X = 0$, $-r_A = 0.00001$):

$-r_A$ (mol/dm$^3 \cdot$s) $\times 10^3$	0.010	0.005	0.002	0.001
X	0.0	0.2	0.4	0.6

(a) Estimate the volume of a plug-flow reactor required to achieve 30% conversion of A for an entering volumetric flow rate of 2 m^3/min.

(b) Estimate the volume of a CSTR required to take the effluent from the plug-flow reactor (PFR) above and achieve 50% total conversion (based on species A fed to the PFR).

(c) What is the total volume of the two reactors?

(d) What is the volume of a single plug-flow reactor necessary to achieve 60% conversion? 80% conversion?

(e) What is the volume of a single CSTR necessary to achieve 50% conversion?

(f) What is the volume of a second CSTR to raise the conversion from 50% to 60%?

(g) Plot the rate of reaction and conversion as a function of PFR volume.

(h) Give a critique of the answers to this problem.

P2-10$_A$ Estimate the reactor volumes of the two CSTRs and the PFR shown in Figure 2-7.

P2-11$_D$ Don't calculate anything. Just go home and relax.

P2-12$_B$ The curve shown in Figure 2-1 is typical of a reaction carried out isothermally, while the curve shown in Figure P2-12 is typical of an exothermic reaction carried out adiabatically.

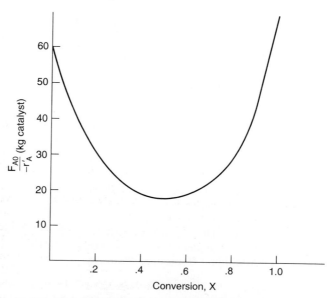

Figure P2-12   Exothermic reaction.

(a) Assuming that you have a CSTR and a PBR containing equal weights of catalyst, how should they be arranged for an isothermal reaction and for adiabatic reaction? In each case use the smallest amount of catalyst weight and still achieve 80% conversion.

(b) What is the catalyst weight necessary to achieve 80% conversion in a well-mixed reactor with catalyst particles (e.g., CSTR)?

(c) What "CSTR" weight is necessary to achieve 40% conversion?

(d) What PBR weight is necessary to achieve 80% conversion?

(e) What PBR weight is necessary to achieve 40% conversion?

(f) Plot the rate of reaction and conversion as a function of PBR volume.

(g) Write a paragraph describing how you would arrange reactors for different $-r_A'$ versus X curves.

Additional information: $F_{A0} = 2$ mol/s.

P2-13 Using POLYMATH, MatLab, Mathmatica, or some other software package, first fit the $-r_A$ versus X to a polynominal (i.e., $-r_A = a_0 + a_1 X + a_2 X^2$). Next use this polynomial and an ODE solver to plot the conversion down the length (i.e., volume) of a PFR and find the CSTR volume for 80% converison for an entering molar flow rate of 5 mol/s.

(a) Use the data in Table 2-2.

(b) Use the data in Problem 2-6.

(c) Use the data in Problem 2-9.

(d) Use the data in Figure P2-12.

P2-14$_D$ What is a typical reactor size for:[1]

(a) cracking furnaces?

(b) packed beds?

(c) fluidized beds?

(d) pilot-plant-scale CSTR?

(e) industrial-scale CSTR?

For parts (b) and (c) specify the catalyst dimensions in addition to the dimensions of the reactor.

P2-15$_B$ Review the reactor volumes calculated in each of the example problems in this chapter. Using the methanol reactor described in *Chem. Eng. Prog.*, 79(7), 64 (1983) as a basis of comparison, classify each of the reactor sizes and flow rates in the example problems as industrial, pilot plant, or laboratory scale.

CD-ROM MATERIAL

- **Learning Resources**
  1. *Summary Notes for Lectures 1 and 2*
  2. *Web Module*
     A. Hippopotamus Digestive System
  3. *Interactive Computer Modules*
     A. Reactor Staging
  4. *Solved Problems*
     A. CDP2-A$_B$ More CSTR and PFR Calculations - No Memorization
- **FAQ [Frequently Asked Questions]–** In Updates/FAQ icon section
- **Additional Homework Problems**

---

[1] See the Supplementary Reading lists for Chapters 1 and 2.

CDP2-A   Use Levenspiel plots to calculate PFR and CSTR reactor volumes given $-r_A = f(X)$. **(Includes Solution)** [2nd Ed. P2-12$_B$]

CDP2-B$_A$ An ethical dilemma as to how to determine the reactor size in a competitor's chemical plant. [2nd Ed. P2-18$_B$]

CDP2-C$_A$ Use Levenspiel plots to calculate PFR and CSTR volumes.

CDP2-D$_A$ Use Levenspiel plots to calculate CSTR and PFR volumes for the reaction

$$A + B \longrightarrow C$$

SUPPLEMENTARY READING

1. Further discussion of stoichiometry may be found in

HIMMELBLAU, D. M., *Basic Principles and Calculations in Chemical Engineering*, 6th ed. Upper Saddle River, N.J.: Prentice Hall, 1996, Chap. 2.

FELDER, R. M., and R. W. ROUSSEAU, *Elementary Principles of Chemical Processes*, 2nd ed. New York: Wiley, 1986, Chap. 4.

2. Further discussion of the proper staging of reactors in series for various rate laws, in which a plot of $-1/r_A$ versus X is given, is presented in

LEVENSPIEL, O., *Chemical Reaction Engineering*, 2nd ed. New York: Wiley, 1972, Chap. 6 (especially pp. 139–156).

HILL, C. G., *An Introduction to Chemical Engineering Kinetics and Reactor Design*. New York: Wiley, 1977, Chap. 8.

Rate Laws 3
and Stoichiometry

Kinetics is nature's way of preventing everything
from happening all at once.

–S. E. LeBlanc

We have shown that in order to calculate the time necessary to achieve a given
conversion X in a batch system, or to calculate the reactor volume needed to
achieve a conversion X in a flow system, we need to know the reaction rate as
a function of conversion. In this chapter we show how this functional dependence is obtained. First there is a brief discussion of *chemical kinetics*, empha-
sizing definitions, which illustrates how the reaction rate depends on the
concentrations of the reacting species. This discussion is followed by instructions on how to convert the reaction rate law from the concentration dependence to a dependence on conversion. Once this dependence is achieved, we
can design a number of isothermal reaction systems.

3.1 Basic Definitions

A *homogeneous reaction* is one that involves only one phase. A *heterogeneous
reaction* involves more than one phase, and reaction usually occurs at or very
near the interface between the phases. An *irreversible reaction* is one that proceeds in only one direction and continues in that direction until the reactants

Types of reactions   are exhausted. A *reversible reaction*, on the other hand, can proceed in either
direction, depending on the concentrations of reactants and products relative to
the corresponding equilibrium concentrations. An irreversible reaction behaves
as if no equilibrium condition exists. Strictly speaking, no chemical reaction is
completely irreversible, but in very many reactions the equilibrium point lies
so far to the right that they are treated as irreversible reactions.

3.1.1 The Reaction Rate Constant

In the chemical reactions considered in the following paragraphs, we take as the basis of calculation a species A, which is one of the reactants that is disappearing as a result of the reaction. The limiting reactant is usually chosen as our basis for calculation. The rate of disappearance of A, $-r_A$, depends on temperature and composition. For many reactions it can be written as the product of a *reaction rate constant k* and a function of the concentrations (activities) of the various species involved in the reaction:

The rate law gives the relationship between reaction rate and concentration

$$-r_A = [k_A(T)][fn(C_A, C_B, \ldots)] \qquad (3\text{-}1)$$

The algebraic equation that relates $-r_A$ to the species concentrations is called the kinetic expression or **rate law**. The specific rate of reaction, k_A, like the reaction rate $-r_A$, is always referred to a particular species in the reactions and normally should be subscripted with respect to that species. However, for reactions in which the stoichiometric coefficient is 1 for all species involved in the reaction, for example,

$$1\,NaOH + 1\,HCl \longrightarrow 1\,NaCl + 1\,H_2O$$

we shall delete the subscript on the specific reaction rate:

$$k = k_{NaOH} = k_{HCl} = k_{NaCl} = k_{H_2O}$$

The reaction rate constant k is not truly a constant, but is merely independent of the concentrations of the species involved in the reaction. The quantity k is also referred to as the **specific reaction rate (constant).** It is almost always strongly dependent on temperature. In gas-phase reactions, it depends on the catalyst and may be a function of total pressure. In liquid systems it can also be a function of total pressure, and in addition can depend on other parameters, such as ionic strength and choice of solvent. These other variables normally exhibit much less effect on the specific reaction rate than does temperature, so for the purposes of the material presented here it will be assumed that k_A depends only on temperature. This assumption is valid in most laboratory and industrial reactions and seems to work quite well.

It was the great Swedish chemist Arrhenius who first suggested that the temperature dependence of the specific reaction rate, k_A, could be correlated by an equation of the type

Arrhenius equation

$$k_A(T) = A e^{-E/RT} \qquad (3\text{-}2)$$

where A = preexponential factor or frequency factor

   E = activation energy, J/mol or cal/mol

   R = gas constant = 8.314 J/mol·K = 1.987 cal/mol·K

   T = absolute temperature, K

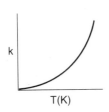

T(K)

Equation (3-2), known as the *Arrhenius equation*, has been verified empirically to give the temperature behavior of most reaction rate constants within experimental accuracy over fairly large temperature ranges.

The activation energy E has been equated with a minimum energy that must be possessed by reacting molecules before the reaction will occur. From the kinetic theory of gases, the factor $e^{-E/RT}$ gives the fraction of the collisions between molecules that together have this minimum energy E. Although this might be an acceptable elementary explanation, some suggest that E is nothing more than an empirical parameter correlating the specific reaction rate to temperature.[1] **(See Appendix G)** Other authors take exception to this interpretation; for example, Tolman's[2] interpretation of activation energy is that it is the difference between *the average energy of those molecules that do react* and *the average energy of all reactant molecules*. Nevertheless, postulation of the Arrhenius equation remains the greatest single step in chemical kinetics, and retains its usefulness today, nearly a century later.

The activation energy is determined experimentally by carrying out the reaction at several different temperatures. After taking the natural logarithm of Equation (3-2),

Calculation of the
activation energy

$$\ln k_A = \ln A - \frac{E}{R}\left(\frac{1}{T}\right) \tag{3-3}$$

it can be seen that a plot of $\ln k_A$ versus $1/T$ should be a straight line whose slope is proportional to the activation energy.

Example 3–1   Determination of the Activation Energy

Calculate the activation energy for the decomposition of benzene diazonium chloride to give chlorobenzene and nitrogen:

using the following information for this first-order reaction:

k (s^{-1})	0.00043	0.00103	0.00180	0.00355	0.00717
T (K)	313.0	319.0	323.0	328.0	333.0

---

[1] M. Karplus, R. N. Porter, and R. D. Sharma, *J. Chem. Phys.*, *43*, 3259 (1965); D. G. Truhlar, *J. Chem. Educ.*, *55*(5), 310 (1978).

[2] R. C. Tolman, *Statistical Mechanics with Applications to Physics and Chemistry* (New York: Chemical Catalog Company, 1927), pp. 260–270.

Solution

By converting Equation (3-3) to log base 10,

$$\log k = \log A - \frac{E}{2.3R}\left(\frac{1}{T}\right)$$ (E3-1.1)

we can use semilog paper to determine E quite readily by first forming the following table from the data above:

k (s^{-1})	0.00043	0.00103	0.00180	0.00355	0.00717
$1000/T$ (K^{-1})	3.20	3.14	3.10	3.05	3.0

Then we plot the data directly on semilog paper as shown in Figure E3-1.1.

Finding the
activation energy.
Plot (ln k) vs. (1/T)

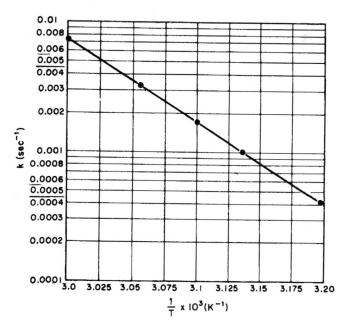

Figure E3-1.1

Although the slope can be determined by a number of methods as described in Appendix D, the decade method is chosen here. For two different points on Figure E3-1.1, we have

$$\log k_1 = \log A - \frac{E}{2.3R}\left(\frac{1}{T_1}\right)$$

$$\log k_2 = \log A - \frac{E}{2.3R}\left(\frac{1}{T_2}\right)$$

Subtracting yields

$$\log \frac{k_2}{k_1} = \frac{-E}{2.3R}\left(\frac{1}{T_2} - \frac{1}{T_1}\right)$$

(E3-1.2)

$$E = -\frac{(2.3)(R)\log(k_2/k_1)}{1/T_2 - 1/T_1}$$

To use the decade method, choose $1/T_1$ and $1/T_2$ so that $k_2 = 0.1k_1$. Then, $\log(k_1/k_2) = 1$.

When $k_1 = 0.005$: $\dfrac{1}{T_1} = 0.00303$

When $k_2 = 0.0005$: $\dfrac{1}{T_2} = 0.00319$

Therefore,

$$E = \frac{2.303R}{1/T_2 - 1/T_1} = \frac{(2.303)(8.314 \text{ J/mol} \cdot \text{K})}{(0.00319 - 0.00303)/\text{K}}$$

$$= 120 \frac{\text{kJ}}{\text{mol}} \text{ or } 28.7 \text{ kcal/mol}$$

The rate does not always double for a temperature increase of 10°C

There is a rule of thumb that states that the rate of reaction doubles for every 10°C increase in temperature. However, this is true only for a specific combination of activation energy and temperature. For example, if the activation energy is 53.6 kJ/mol, the rate will double only if the temperature is raised from 300 K to 310 K. If the activation energy is 147 kJ/mol, the rule will be valid only if the temperature is raised from 500 K to 510 K. (See Problem P3-5 for the derivation of this relationship.)

The larger the activation energy, the more temperature-sensitive is the rate of reaction. While there are no typical values of the frequency factor and activation energy for a first-order gas-phase reaction, if one were forced to make a guess, values of A and E might be 10^{13} s^{-1} and 300 kJ/mol. However, for families of reactions (e.g., halogenation), a number of correlations can be used to estimate the activation energy. One such correlation is the Polanyi-Semenov equation, which relates activation energy to the heat of reaction (see Problem P3-20). Another correlation relates activation energy to differences in bond strengths between products and reactants.[3] While activation energy cannot be currently predicted a priori, significant research efforts are under way to calculate activation energies from first principles.[4] (Also see Appendix J)

---

[3] M. Boudart, *Kinetics of Chemical Processes* (Upper Saddle River, N.J.: Prentice Hall, 1968), p. 168. J. W. Moore and R. G. Pearson, *Kinetics and Mechanics,* 3rd ed. (New York: Wiley, 1981), p. 199. S. W. Benson, *Thermochemical Kinetics,* 2nd ed. (New York: Wiley, 1976).

[4] S. M. Senkan, *Detailed Chemical Kinetic Modeling: Chemical Reaction Engineering of the Future,* Advances in Chemical Engineering, Vol. 18 (San Diego: Academic Press, 1992), pp. 95–96.

Other expressions similar to the Arrhenius equation exist. One such expression is the temperature dependence derived from transition-state theory, which takes a form similar to Equation (3-2):

$$k(T) = A'T^n e^{-E'/RT} \tag{3-4}$$

in which $0 \leq n \leq 1$.

If Equations (3-2) and (3-4) are used to describe the temperature dependence for the same reaction data, it will be found that the activation energies E and E' will differ slightly.

3.1.2 The Reaction Order and Rate Law

The dependence of the reaction rate $-r_A$ on the concentrations of the species present, $\mathrm{fn}(C_j)$, is almost without exception determined by experimental observation. Although the functional dependence may be postulated from theory, experiments are necessary to confirm the proposed form. One of the most common general forms of this dependence is the product of concentrations of the individual reacting species, each of which is raised to a power, for example,

$$\boxed{-r_A = k_A C_A^\alpha C_B^\beta} \tag{3-5}$$

The exponents of the concentrations in Equation (3-5) lead to the concept of *reaction order*. The **order of a reaction** refers to the powers to which the concentrations are raised in the kinetic rate law.[†] In Equation (3-5), the reaction is α *order with respect to reactant* A, and β *order with respect to reactant* B. The overall order of the reaction, n, is

Overall reaction order

$$n = \alpha + \beta$$

For example, in the gas-phase reaction

$$2NO + O_2 \longrightarrow 2NO_2$$

Strictly Speaking   [†] Strictly speaking, the reaction rates should be written in terms of the activities, a_i ($a_i = \gamma_i C_i$, where γ_i is the activity coefficient):

$$-r_A = k_A' a_A^\alpha a_B^\beta$$

However, for many reacting systems, the activity coefficients, γ_i, do not change appreciably during the course of the reaction and they are adsorbed in the specific reaction rate:

$$-r_A = k_A' a_A^\alpha a_B^\beta = k_A'(\gamma_A C_A)^\alpha (\gamma_B C_B)^\alpha = (k_A' \gamma_A^\alpha \gamma_B^\beta) C_A^\alpha C_B^\beta = k_A C_A^\alpha C_B^\beta$$

the kinetic *rate law* is

$$-r_{NO} = k_{NO} C_{NO}^2 C_{O_2}$$

This reaction is second-order with respect to nitric oxide, first-order with respect to oxygen, and overall is a third-order reaction. In general, first- and second-order reactions are more commonly observed than zero- and third-order reactions.

The overall order of a reaction does not have to be an integer, nor does the order have to be an integer with respect to any individual component. As an example, consider the gas-phase synthesis of phosgene:

$$CO + Cl_2 \longrightarrow COCl_2$$

in which the kinetic *rate law* is

$$-r_{CO} = k C_{CO} C_{Cl_2}^{3/2}$$

This reaction is first-order with respect to carbon monoxide, three-halves order with respect to chlorine, and five-halves order overall.

Sometimes reactions have complex rate expressions that cannot be separated into solely temperature-dependent and concentration-dependent portions. In the decomposition of nitrous oxide over platinum,

$$2N_2O \xrightarrow{\ Pt\ } 2N_2 + O_2$$

the kinetic *rate law* is

$$-r_{N_2O} = \frac{k_{N_2O} C_{N_2O}}{1 + k' C_{O_2}}$$

Apparent reaction
orders

Both k_{N_2O} and k' are strongly temperature-dependent. When a rate expression such as the one described above occurs, we can speak of reaction orders only under certain limiting conditions. For example, at very low concentrations of oxygen, the second term in the denominator would be negligible ($1 \gg k' C_{O_2}$) and the reaction would be "apparent" first-order with respect to nitrous oxide and first-order overall. However, if the concentration of oxygen were large enough so that the number 1 in the denominator were insignificant in comparison with the second term, $k' C_{O_2} (k' C_{O_2} \gg 1)$, the *apparent* reaction order would be -1 with respect to oxygen and 1 with respect to nitrous oxide. Rate expressions of this type are very common for liquid and gaseous reactions promoted by solid catalysts (see Chapter 10). They also occur occasionally in homogeneous reaction systems (see Chapter 7).

The units of the specific reaction rate, k_A, vary with the order of the reaction. Consider a reaction involving only one reactant, such as

$$A \longrightarrow products$$

For this type of reaction, the rate laws corresponding to a zero-, first-, second-, third-order reaction, together with typical units for the corresponding rate constants, are:

$$\text{Zero-order:} \qquad -r_A = k_A: \qquad\qquad\qquad\qquad\qquad (3\text{-}6)$$
$$\{k\} = \text{mol/(dm)}^3 \cdot \text{s}$$

$$\text{First-order:} \qquad -r_A = k_A C_A: \qquad\qquad\qquad\qquad (3\text{-}7)$$
$$\{k\} = \text{s}^{-1}$$

$$\text{Second-order:} \qquad -r_A = k_A C_A^2: \qquad\qquad\qquad\qquad (3\text{-}8)$$
$$\{k\} = \text{(dm)}^3/\text{mol} \cdot \text{s}$$

$$\text{Third-order:} \qquad -r_A = k_A C_A^3: \qquad\qquad\qquad\qquad (3\text{-}9)$$
$$\{k\} = \text{(dm}^3/\text{mol)}^2 \cdot \text{s}^{-1}$$

Where do you find rate laws? The activation energy, frequency factor, and reaction orders for a large number of gas- and liquid-phase reactions can be found in the National Bureau of Standards' circulars and supplements.[5] Also consult the journals listed at the end of Chapter 1.

3.1.3 Elementary Rate Laws and Molecularity

A reaction has an *elementary rate law* if the reaction order of each species is identical with the stoichiometric coefficient of that species for the reaction *as written*. For example, the oxidation of nitric oxide presented above has an elementary rate law under this definition, while the phosgene synthesis reaction does not. Another example of this type of reaction with an elementary rate law is the gas-phase reaction between hydrogen and iodine to form hydrogen iodide:

$$H_2 + I_2 \longrightarrow 2HI$$

for which the rate law is

$$-r_{H_2} = k C_{H_2} C_{I_2}$$

In some circles when a reaction has an elementary rate law it is referred to as an *elementary reaction*. A more restrictive definition of an elementary

Very important references, but you should also look in the other literature before going to the lab

---

[5] *Tables of Chemical Kinetics: Homogeneous Reactions*, National Bureau of Standards Circular 510 (Sept. 28, 1951); Suppl. 1 (Nov. 14, 1956); Suppl. 2 (Aug. 5, 1960); Suppl. 3 (Sept. 15, 1961) (Washington, D.C.: U.S. Government Printing Office). *Chemical Kinetics and Photochemical Data for Use in Stratospheric Modeling*, Evaluate No. 10, JPL Publication 92–20, Aug. 15, 1992, Jet Propulsion Laboratories, Pasadena, Calif.

reaction is sometimes encountered, and it involves the mechanism or molecular path of the reaction. This definition is discussed in Chapter 7.[†]

In the study of reaction orders and kinetic mechanisms, reference is sometimes made to the *molecularity* of a reaction. The molecularity is the number of atoms, ions, or molecules involved (colliding) in the rate-limiting step of the reaction. The terms *unimolecular, bimolecular,* and *termolecular* refer to reactions involving, respectively, one, two, or three atoms (or molecules) interacting or colliding in any one reaction step.

The most common example of a unimolecular reaction is radioactive decay, such as the spontaneous emission of an alpha particle from uranium 238 to give thorium and helium:

$$_{92}U^{238} \longrightarrow \; _{90}Th^{234} + \; _2He^4$$

If the hydrogen–iodine and the nitric oxide oxidation reactions did indeed result simply from the collision of the molecular species named in the overall stoichiometric equations, they would be bimolecular and termolecular reactions, respectively. There is considerable doubt, though, about whether this actually occurs.

The reaction between methyl bromide and sodium hydroxide is classified as a nucleophilic aliphatic substitution:[6]

$$NaOH + CH_3Br \longrightarrow CH_3OH + NaBr$$

This irreversible reaction has an elementary rate law and is carried out in aqueous ethanol. Therefore, like almost all liquid-phase reactions, the density remains almost constant throughout the reaction. It is a general principle that for most liquid-phase reactions, the volume V for a batch reaction system and the volumetric flow rate v for a continuous-flow system will not change appreciably during the course of a chemical reaction.

V (batch) and *v* (flow) are constant for most liquids

We want to write the rate of disappearance of methyl bromide, $-r_{MB}$, in terms of the appropriate concentrations. Because this reaction is elementary the reaction orders agree with the stoichiometric coefficients.

$$1NaOH + 1CH_3Br \longrightarrow 1CH_3OH + 1NaBr$$

$\alpha = 1$, first-order with respect to sodium hydroxide

$\beta = 1$, first-order with respect to methyl bromide (MB)

$$-r_{MB} = kC_{NaOH}C_{CH_3Br}$$

Overall, this reaction is second-order.

Strictly Speaking

[†] Strictly speaking, elementary reactions involve only single steps such as one iodide molecule colliding and reacting with one hydrogen molecule. However, most reactions involve multiple steps and pathways. For many of these reactions, the powers in the rate laws surprisingly agree with the stoichiometric coefficients. Consequently, to facilitate describing this class of reactions, reactions where the rate law powers and stoichiometric coefficients are identical may also be referred to as elementary reactions.

[6] R. T. Morrison and R. N. Boyd, *Organic Chemistry*, 4th ed. (Needham Heights, Mass.: Allyn and Bacon, 1983).

Example 3–2 Describing a Reaction

Another nucleophilic aliphatic substitution is the reaction between sodium hydroxide and *tert*-butyl bromide (TBB):

$$
NaOH + CH_3\!-\!\underset{\underset{Br}{|}}{\overset{\overset{CH_3}{|}}{C}}\!-\!CH_3 \longrightarrow CH_3\!-\!\underset{\underset{OH}{|}}{\overset{\overset{CH_3}{|}}{C}}\!-\!CH_3 + NaBr
$$

State the reaction order with respect to each species as well as the overall reaction order and generally describe this reaction.

Solution

Just because this reaction is similar to the previous nucleophilic aliphatic substitution, one should not jump to the conclusion that the rate law and kinetics will be similar. *The rate law is determined from experimental observation. It relates the rate of reaction at a particular point to the species concentrations at that same point.* In this case if one consults an organic chemistry text,[7] one will find that the rate law is

$$-r_{TBB} = kC_{TBB} \tag{E3-2.1}$$

Using the definitions above, the reaction of sodium hydroxide with *tert*-butyl bromide (TBB) *can be described* as an irreversible, homogeneous, liquid-phase reaction which is first-order with respect to *tert*-butyl bromide, zero-order with respect to sodium hydroxide, overall first-order, and nonelementary.

3.1.4 Reversible Reactions

All rate laws for reversible reactions *must* reduce to the thermodynamic relationship relating the reacting species concentrations at equilibrium. At equilibrium, the rate of reaction is identically zero for all species (i.e., $-r_A \equiv 0$). That is, for the general reaction

$$aA + bB \rightleftharpoons cC + dD \tag{2-1}$$

the concentrations at equilibrium are related by the thermodynamic relationship (see Appendix C).

Thermodynamic
Equilibrium
Relationship

$$K_C = \frac{C_{Ce}^c C_{De}^d}{C_{Ae}^a C_{Be}^b} \tag{3-10}$$

The units of K_C are $(\text{mol/dm}^3)^{d+c-b-a}$.

To illustrate how to write rate laws for reversible reactions we will use the combination of two benzene molecules to form one molecule of hydrogen

---

[7] Ibid.

and one of diphenyl. In this discussion we shall consider this gas-phase reaction to be elementary and reversible:

$$2C_6H_6 \underset{k_{-B}}{\overset{k_B}{\rightleftharpoons}} C_{12}H_{10} + H_2$$

or symbolically,

$$2B \underset{k_{-B}}{\overset{k_B}{\rightleftharpoons}} D + H_2$$

The forward and reverse specific reaction rate constants, k_B and k_{-B}, respectively, will *be defined with respect to benzene.*

Benzene (B) is being depleted by the forward reaction

$$2C_6H_6 \xrightarrow{k_B} C_{12}H_{10} + H_2$$

in which the rate of disappearance of benzene is

$$-r_{B, forward} = k_B C_B^2$$

If we multiply both sides of this equation by -1, we obtain the expression for the rate of formation of benzene for the forward reaction:

$$r_{B, forward} = -k_B C_B^2 \tag{3-11}$$

For the reverse reaction between diphenyl (D) and hydrogen (H_2),

$$C_{12}H_{10} + H_2 \xrightarrow{k_{-B}} 2C_6H_6$$

the rate of formation of benzene is given as

$$r_{B, reverse} = k_{-B} C_D C_{H_2} \tag{3-12}$$

The net rate of formation of benzene is the sum of the rates of formation from the forward reaction [i.e., Equation (3-11)] and the reverse reaction [i.e., Equation (3-12)]:

$$r_B \equiv r_{B, net} = r_{B, forward} + r_{B, reverse}$$

$$r_B = -k_B C_B^2 + k_{-B} C_D C_{H_2} \tag{3-13}$$

Multiplying both sides of Equation (3-13) by -1, we obtain the rate law for the rate of disappearance of benzene, $-r_B$:

Elementary reversible
$A \rightleftharpoons B$

$$-r_A = k\left(C_A - \frac{C_B}{K_c}\right)$$

$$-r_B = k_B C_B^2 - k_{-B} C_D C_{H_2} = k_B\left(C_B^2 - \frac{k_{-B}}{k_B} C_D C_{H_2}\right)$$

$$-r_B = k_B\left(C_B^2 - \frac{C_D C_{H_2}}{K_C}\right) \tag{3-14}$$

where

$$\frac{k_B}{k_{-B}} = K_C = \text{concentration equilibrium constant}$$

The equilibrium constant decreases with increasing temperature for exothermic reactions and increases with increasing temperature for endothermic reactions.

We need to check to see if the rate law given by Equation (3-14) is thermodynamically consistent at equilibrium. Using Equation (3-10) and substituting the appropriate species concentration and exponents, thermodynamics tells us that

$$K_C = \frac{C_{De}C_{H_2e}}{C_{Be}^2} \tag{3-15}$$

At equilibrium, $-r_B \equiv 0$, and the rate law given by Equation (3-14) becomes

At equilibrium the rate law must reduce to an equation consistent with thermodynamic equilibrium

$$-r_B \equiv 0 = k_B\left[C_{Be}^2 - \frac{C_{De}C_{H_2e}}{K_C} \right]$$

Rearranging, we obtain

$$K_C = \frac{C_{De}C_{H_2e}}{C_{Be}^2}$$

which is identical to Equation (3-15).

A further discussion of the equilibrium constant and its thermodynamic relationship is given in **Appendix C**.

Finally, we want to rewrite the rate of formation of diphenyl and hydrogen in terms of concentration. The rate of formation of these species must have the same functional dependence on concentrations as does the rate of disappearance of benzene. The rate of formation of diphenyl is

$$r_D = k_D\left[C_B^2 - \frac{C_D C_{H_2}}{K_C} \right] \tag{3-16}$$

Using the relationship given by Equation (2-20) for the general reaction

$$\boxed{\frac{r_A}{-a} = \frac{r_B}{-b} = \frac{r_C}{c} = \frac{r_D}{d}} \tag{2-20}$$

we can obtain the relationship between the various specific reaction rates, k_B, k_D:

$$\frac{r_D}{1} = \frac{r_B}{-2} = \frac{k_B}{2}\left[C_B^2 - \frac{C_D C_{H_2}}{K_C} \right] \tag{3-17}$$

Comparing Equations (3-16) and (3-17), we see the relationship between the specific reaction rate with respect to diphenyl and the specific reaction rate with respect to benzene is

$$k_D = \frac{k_B}{2}$$

Example 3–3 Formulating a Reversible Rate Law

The exothermic reaction

$$A + 2B \longrightarrow 2D \tag{E3-3.1}$$

is virtually irreversible at low temperatures and the rate law is

$$-r_A = k_A C_A^{1/2} C_B \tag{E3-3.2}$$

Suggest a rate law that is valid at high temperatures, where the reaction is reversible:

$$A + 2B \rightleftharpoons 2D \tag{E3-3.3}$$

Solution

These criteria must
be satisfied

The rate law for the reversible reaction *must*

1. satisfy thermodynamic relationships at equilibrium, and
2. reduce to the irreversible rate law when the concentration of one or more of the reaction products is zero.

We know from thermodynamics that the equilibrium relationship for Reaction (E3-3.1) as written is

$$K_C = \frac{C_{De}^2}{C_{Ae} C_{Be}^2} \quad \text{with units} \quad [K_C] = \frac{dm^3}{mol} \tag{E3-3.4}$$

Rearranging Equation (E3-3.4) in the form

$$C_{Ae} C_{Be}^2 - \frac{C_{De}^2}{K_C} = 0$$

suggests that we try a reversible rate law of the form

$$-r_A = k_A \left[C_A C_B^2 - \frac{C_D^2}{K_C} \right] \tag{E3-3.5}$$

Equation (E3-3.5) satisfies the equilibrium conditions but does not simplify to the initial, irreversible rate when $C_D = 0$. Substituting $C_D = 0$ into the equation being tested yields

$$-r_{A0} = k_A C_{A0} C_{B0}^2 \tag{E3-3.6}$$

Equation (E3-3.6) does not agree with Equation (E3-3.2) and therefore the rate law given by Equation (E3-3.5) is not valid.

The one-half power in the rate law suggests that we might take the square root of Equation (E3-3.4):

$$\sqrt{K_C} = \frac{C_{De}}{C_{Ae}^{1/2} C_{Be}} = K_{C2} \qquad [K_{C2}] \text{ is in units of } \left(\frac{dm^3}{mol}\right)^{1/2} \tag{E3-3.7}$$

Rearranging gives

$$C_{Ae}^{1/2} C_{Be} - \frac{C_{De}}{K_{C2}} = 0 \tag{E3-3.8}$$

Using this new equilibrium constant, K_{C2}, we can formulate another suggestion for the reaction rate expression:

$$\boxed{-r_A = k_A \left[C_A^{1/2} C_B - \frac{C_D}{K_{C2}} \right]} \tag{E3-3.9}$$

Note that this expression satisfies both the thermodynamic relationship (see the definition of K_{C2}) and reduces to the irreversible rate law when $C_D = 0$. The form of the irreversible rate law provides a big clue as to the form of the reversible reaction rate expression.

3.1.5 Nonelementary Rate Laws and Reactions

It is interesting to note that although the reaction orders correspond to the stoichiometric coefficients for the reaction between hydrogen and iodine, the rate expression for the reaction between hydrogen and another halogen, bromine, is quite complex. This nonelementary reaction

$$H_2 + Br_2 \longrightarrow 2HBr$$

proceeds by a free-radical mechanism, and its reaction rate law is

$$r_{HBr} = \frac{k_1 C_{H_2} C_{Br_2}^{1/2}}{k_2 + C_{HBr}/C_{Br_2}} \tag{3-18}$$

Another reaction involving free radicals is the vapor-phase decomposition of acetaldehyde:

$$CH_3CHO \longrightarrow CH_4 + CO$$

At a temperature of about 500°C, the order of the reaction is three-halves with respect to acetaldehyde.

$$-r_{CH_3CHO} = kC_{CH_3CHO}^{3/2} \qquad (3\text{-}19)$$

In many gas–solid catalyzed reactions it is sometimes preferable to write the rate law in terms of partial pressures rather than concentrations. One such example is the reversible catalytic decomposition of cumene, C, to form benzene, B, and propylene, P:

$$C_6H_5CH(CH_3)_2 \;\rightleftharpoons\; C_6H_6 + C_3H_6$$

The reaction can be written symbolically as

$$C \;\rightleftharpoons\; B + P$$

It was found experimentally that the reaction follows Langmuir–Hinshelwood kinetics and the rate law is (see Chapter 10)

$$-r_C' = \frac{k(P_C - P_B P_P / K_P)}{1 + K_C P_C + K_B P_B} \qquad (3\text{-}20)$$

where K_P is the pressure equilibrium constant with units of atm (or kPa); K_C and K_B are the adsorption constants with units of atm^{-1} (or kPa); and the specific reaction rate, k, has units of

$$[k] = \frac{\text{mol cumene}}{\text{kg cat} \cdot \text{s} \cdot \text{atm}}$$

We see that at equilibrium ($-r_C' = 0$) the rate law for the reversible reaction is indeed thermodynamically consistent:

$$-r_C' = 0 = k\,\frac{P_{Ce} - P_{Be} P_{Pe} / K_P}{1 + K_C P_{Ce} + K_B P_{Be}}$$

Solving for K_P yields

$$K_P = \frac{P_{Be} P_{Pe}}{P_{Ce}}$$

which is identical to the expression obtained from thermodynamics.

To express the rate of decomposition of cumene $-r_C'$ as a function of conversion, replace the partial pressure with concentration, using the ideal gas law:

$$P_C = C_C RT \qquad (3\text{-}21)$$

and then express concentration in terms of conversion.

The rate of reaction per unit weight of catalyst, $-r_A'$, and the rate of reaction per unit volume, $-r_A$, are related through the bulk density ρ_b of the *catalyst particles* in the fluid media:

$$-r_A = \rho_b(-r_A')$$

$$\frac{\text{moles}}{\text{time} \cdot \text{volume}} = \left(\frac{\text{mass}}{\text{volume}}\right)\left(\frac{\text{moles}}{\text{time} \cdot \text{mass}}\right)$$

(3-22)

In fluidized catalytic beds the bulk density is normally a function of the flow rate through the bed.

3.2 Present Status of Our Approach to Reactor Sizing and Design

In Chapter 2 we showed how it was possible to size CSTRs, PFRs, and PBRs using the design equations in Table 3-1 *if* the rate of disappearance of A is known as a function of conversion, X:

$$-r_A = g(X)$$

TABLE 3-1.   DESIGN EQUATIONS

	Differential Form		*Algebraic Form*		*Integral Form*	
Batch	$N_{A0}\dfrac{dX}{dt} = -r_A V$	(2-6)			$t = N_{A0}\displaystyle\int_0^X \dfrac{dX}{-r_A V}$	(2-9)
Backmix (CSTR)			$V = \dfrac{F_{A0}X}{-r_A}$	(2-13)		
Tubular (PFR)	$F_{A0}\dfrac{dX}{dV} = -r_A$	(2-15)			$V = F_{A0}\displaystyle\int_0^X \dfrac{dX}{-r_A}$	(2-16)
Packed bed (PBR)	$F_{A0}\dfrac{dX}{dW} = -r_A'$	(2-17)			$W = F_{A0}\displaystyle\int_0^X \dfrac{dX}{-r_A'}$	(2-18)

The design equations

In general, information in the form $-r_A = g(X)$ is not available. However, we have seen in Section 3.1 that the rate of disappearance of A, $-r_A$, is normally expressed in terms of the concentration of the reacting species. This functionality,

$$-r_A = k[\text{fn}(C_A, C_B, \dots)]$$

(3-1)

$-r_A = f(C_j)$

$+$

$C_j = h_j(X)$

$\downarrow$

$-r_A = g(X)$

and then we can design isothermal reactors

is called a *rate law*. In Section 3.3 we show how the concentration of the reacting species may be written in terms of the conversion X,

$$C_j = h_j(X)$$

With these additional relationships, one observes that if the rate law is given and the concentrations can be expressed as a function of conversion, *then in fact we have* $-r_A$ *as a function of X and this is all that is needed to evaluate the design equations.* One can use either the numerical techniques described in Chapter 2, or, as we shall see in Chapter 4, a table of integrals.

3.3 Stoichiometric Table

Now that we have shown how the rate law can be expressed as a function of concentrations, we need only express concentration as a function of conversion in order to carry out calculations similar to those presented in Chapter 2 to size reactors. If the rate law depends on more than one species, we must relate the concentrations of the different species to each other. This relationship is most easily established with the aid of a stoichiometric table. This table presents the stoichiometric relationships between reacting molecules for a single reaction. That is, it tells us how many molecules of one species will be formed during a chemical reaction when a given number of molecules of another species disappears. These relationships will be developed for the general reaction

$$aA + bB \; \xrightleftharpoons \; cC + dD \tag{2-1}$$

Recall that we have already used stoichiometry to relate the relative rates of reaction for Equation (2-1):

$$\frac{r_A}{-a} = \frac{r_B}{-b} = \frac{r_C}{c} = \frac{r_D}{d} \tag{2-20}$$

In formulating our stoichiometric table we shall take species A as our basis of calculation (i.e., limiting reactant) and then divide through by the stoichiometric coefficient of A,

$$A + \frac{b}{a} B \; \longrightarrow \; \frac{c}{a} C + \frac{d}{a} D \tag{2-2}$$

in order to put everything on a basis of "per mole of A."

Next, we develop the stoichiometric relationships for reacting species that give the change in the number of moles of each species (i.e., A, B, C, and D).

3.3.1 Batch Systems

Figure 3-1 shows a batch system in which we will carry out the reaction given by Equation (2-2). At time $t = 0$ we will open the reactor and place a number of moles of species A, B, C, D, and I (N_{A0}, N_{B0}, N_{C0}, N_{D0}, and N_I, respectively) into the reactor.

Species A is our basis of calculation and N_{A0} is the number of moles of A initially present in the reactor. Of these, $N_{A0}X$ moles of A are consumed in the system as a result of the chemical reaction, leaving ($N_{A0} - N_{A0}X$) moles of A in the system. That is, the number of moles of A remaining in the reactor after conversion X has been achieved is

$$N_A = N_{A0} - N_{A0}X = N_{A0}(1 - X)$$

The complete stoichiometric table for the reaction shown in Equation (2-2) taking place in a batch reactor is presented in Table 3-2.

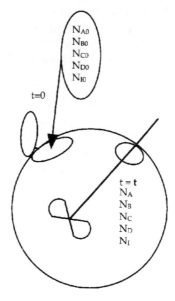

$$t=0$$

$$N_{A0}$$
$$N_{B0}$$
$$N_{C0}$$
$$N_{D0}$$
$$N_{I0}$$

$$t = t$$
$$N_A$$
$$N_B$$
$$N_C$$
$$N_D$$
$$N_I$$

Figure 3-1 Batch reactor.

To determine the number of moles of each species remaining after $N_{A0}X$ moles of A have reacted, we form the stoichiometric table (Table 3-2). This stoichiometric table presents the following information:

Column 1: the particular species
Column 2: the number of moles of each species initially present
Column 3: the change in the number of moles brought about by reaction
Column 4: the number of moles remaining in the system at time t

Components of the stoichiometric table

TABLE 3-2.   STOICHIOMETRIC TABLE FOR A BATCH SYSTEM

Species	Initially (mol)	Change (mol)	Remaining (mol)
A	N_{A0}	$-(N_{A0}X)$	$N_A = N_{A0} - N_{A0}X$
B	N_{B0}	$-\dfrac{b}{a}(N_{A0}X)$	$N_B = N_{B0} - \dfrac{b}{a}N_{A0}X$
C	N_{C0}	$\dfrac{c}{a}(N_{A0}X)$	$N_C = N_{C0} + \dfrac{c}{a}N_{A0}X$
D	N_{D0}	$\dfrac{d}{a}(N_{A0}X)$	$N_D = N_{D0} + \dfrac{d}{a}N_{A0}X$
I (inerts)	N_{I0}	—	$N_I = N_{I0}$
Totals	N_{T0}		$N_T = N_{T0} + \left(\dfrac{d}{a} + \dfrac{c}{a} - \dfrac{b}{a} - 1\right)N_{A0}X$

To calculate the number of moles of species B remaining at time t we recall that at time t the number of moles of A that have reacted is $N_{A0}X$. For every mole of A that reacts, b/a moles of B must react; therefore, the total number of moles of B that have reacted is

$$\text{moles B reacted} = \frac{\text{moles B reacted}}{\text{moles A reacted}} \cdot \text{moles A reacted}$$

$$= \frac{b}{a}(N_{A0}X)$$

Because B is disappearing from the system, the sign of the "change" is negative. N_{B0} is the number of moles initially in the system. Therefore, the number of moles of B remaining in the system, N_B, is given in the last column of Table 3-2 as

$$N_B = N_{B0} - \frac{b}{a}N_{A0}X$$

The complete stoichiometric table delineated in Table 3-2 is for all species in the reaction

$$A + \frac{b}{a}B \longrightarrow \frac{c}{a}C + \frac{d}{a}D \qquad\qquad (2\text{-}2)$$

The stoichiometric coefficients in parentheses $(d/a + c/a - b/a - 1)$ represent the increase in the total number of moles per mole of A reacted. Because this term occurs often in our calculations it is given the symbol δ:

$$\boxed{\delta = \frac{d}{a} + \frac{c}{a} - \frac{b}{a} - 1} \qquad\qquad (3\text{-}23)$$

The total number of moles can now be calculated from the equation

$$N_T = N_{T0} + \delta N_{A0}X$$

We recall from Chapter 1 that the kinetic rate law (e.g., $-r_A = kC_A^2$) is a function solely of the intensive properties of the reacting materials (e.g., temperature, pressure, concentration, and catalysts, if any). The reaction rate, $-r_A$, usually depends on the concentration of the reacting species raised to some power. Consequently, to determine the reaction rate as a function of conversion X, we need to know the concentrations of the reacting species as a function of conversion.

We want $C_j = h_j(X)$

The concentration of A is the number of moles of A per unit volume:

Batch concentration

$$C_A = \frac{N_A}{V}$$

After writing similar equations for B, C, and D, we use the stoichiometric table to express the concentration of each component in terms of the conversion X:

$$C_A = \frac{N_A}{V} = \frac{N_{A0}(1-X)}{V}$$

$$C_B = \frac{N_B}{V} = \frac{N_{B0} - (b/a)N_{A0}X}{V}$$

$$C_C = \frac{N_C}{V} = \frac{N_{C0} + (c/a)N_{A0}X}{V}$$

$$C_D = \frac{N_D}{V} = \frac{N_{D0} + (d/a)N_{A0}X}{V}$$

(3-24)

We further simplify these equations by defining the parameter Θ_i, which allows us to factor N_{A0} in each of the expressions for concentration:

$$\Theta_i = \frac{N_{i0}}{N_{A0}} = \frac{C_{i0}}{C_{A0}} = \frac{y_{i0}}{y_{A0}}$$

$$C_B = \frac{N_{A0}[N_{B0}/N_{A0} - (b/a)X]}{V} = \frac{N_{A0}[\Theta_B - (b/a)X]}{V} \qquad \Theta_B = \frac{N_{B0}}{N_{A0}}$$

(3-25)

$$C_C = \frac{N_{A0}[\Theta_C + (c/a)X]}{V}$$

$$C_D = \frac{N_{A0}[\Theta_D + (d/a)X]}{V}$$

We need $V(X)$ to obtain $C_j = h_j(X)$

We now need only to find volume as a function of conversion to obtain the species concentration as a function of conversion.

3.3.2 Constant-Volume Reaction Systems

Some significant simplifications in the reactor design equations are possible when the reacting system undergoes no change in volume as the reaction progresses. These systems are called constant-volume, or constant-density, because of the invariance of either volume or density during the reaction process. This situation may arise from several causes. In gas-phase batch systems, the reactor is usually a sealed vessel with appropriate instruments to measure pressure and temperature within the reactor. The volume within this vessel is fixed and will not change, and is therefore a constant-volume system. The laboratory bomb reactor is a typical example of this type of reactor.

Another example of a constant-volume gas-phase isothermal reaction occurs when the number of moles of product equals the number of moles of reactant. The water-gas shift reaction, important in coal gasification and many other processes, is one of these:

$$CO + H_2O \rightleftharpoons CO_2 + H_2$$

In this reaction, 2 mol of reactant forms 2 mol of product. When the number of reactant molecules forms an equal number of product molecules at the *same* temperature and pressure, the volume of the reacting mixture will not change if the conditions are such that the ideal gas law is applicable, or if the compressibility factors of the products and reactants are approximately equal.

For liquid-phase reactions taking place in solution, the solvent usually dominates the situation. As a result, changes in the density of the solute do not affect the overall density of the solution significantly and therefore it is essentially a constant-volume reaction process. Most liquid-phase organic reactions do not change density during the reaction, and represent still another case to which the constant-volume simplifications apply. An important exception to this general rule exists for polymerization processes.

For the constant-volume systems described above, Equation (3-25) can be simplified to give the following expressions relating concentration and conversion:

$$V = V_0$$

$$C_A = \frac{N_{A0}(1-X)}{V_0} = C_{A0}(1-X)$$

Concentration as a function of conversion when no volume change occurs with reaction

$$C_B = \frac{N_{A0}[\Theta_B - (b/a)X]}{V_0} = C_{A0}\left(\Theta_B - \frac{b}{a}X\right) \qquad (3\text{-}26)$$

$$C_C = C_{A0}\left(\Theta_C + \frac{c}{a}X\right)$$

$$C_D = C_{A0}\left(\Theta_D + \frac{d}{a}X\right)$$

Example 3–4 Expressing $C_j = h_j(X)$ for a Liquid-Phase Reaction

Soap consists of the sodium and potassium salts of various fatty acids such as oleic, stearic, palmitic, lauric, and myristic acids. The saponification for the formation of soap from aqueous caustic soda and glyceryl stearate is

$$3NaOH(aq) + (C_{17}H_{35}COO)_3C_3H_5 \longrightarrow 3C_{17}H_{35}COONa + C_3H_5(OH)_3$$

Letting X represent the conversion of sodium hydroxide (the moles of sodium hydroxide reacted per mole of sodium hydroxide initially present), set up a stoichiometric table expressing the concentration of each species in terms of its initial concentration and the conversion X.

Solution

Because we are taking sodium hydroxide as our basis, we divide through by the stoichiometric coefficient of sodium hydroxide to put the reaction expression in the form

Choosing a basis of calculation

$$\text{NaOH} + \tfrac{1}{3}(C_{17}H_{35}COO)_3C_3H_5 \longrightarrow C_{17}H_{35}COONa + \tfrac{1}{3}C_3H_5(OH)_3$$

$$\text{A} \quad + \quad \tfrac{1}{3}\text{B} \qquad \longrightarrow \qquad \text{C} \quad + \quad \tfrac{1}{3}\text{D}$$

We may then perform the calculations shown in Table E3-4.1. Because this is a liquid-phase reaction, the density ρ is considered to be constant; therefore, $V = V_0$.

$$C_A = \frac{N_A}{V} = \frac{N_A}{V_0} = \frac{N_{A0}(1-X)}{V_0} = C_{A0}(1-X)$$

$$\Theta_B = \frac{C_{B0}}{C_{A0}} \qquad \Theta_C = \frac{C_{C0}}{C_{A0}} \qquad \Theta_D = \frac{C_{D0}}{C_{A0}}$$

TABLE E3-4.1.   STOICHIOMETRIC TABLE FOR LIQUID-PHASE SOAP REACTION

Stoichiometric table (batch)

Species	Symbol	Initially	Change	Remaining	Concentration
NaOH	A	N_{A0}	$-N_{A0}X$	$N_{A0}(1-X)$	$C_{A0}(1-X)$
$(C_{17}H_{35}COO)_3C_3H_5$	B	N_{B0}	$-\tfrac{1}{3}N_{A0}X$	$N_{A0}\left(\Theta_B - \dfrac{X}{3}\right)$	$C_{A0}\left(\Theta_B - \dfrac{X}{3}\right)$
$C_{17}H_{35}COONa$	C	N_{C0}	$N_{A0}X$	$N_{A0}(\Theta_C + X)$	$C_{A0}(\Theta_C + X)$
$C_3H_5(OH)_3$	D	N_{D0}	$\tfrac{1}{3}N_{A0}X$	$N_{A0}\left(\Theta_D + \dfrac{X}{3}\right)$	$C_{A0}\left(\Theta_D + \dfrac{X}{3}\right)$
Water (inert)	I	$\dfrac{N_{I0}}{N_{T0}}$	$\overline{\overline{0}}$	$\dfrac{N_{I0}}{N_T = N_{T0}}$	C_{I0}

Example 3–5   What Is the Limiting Reactant?

Having set up the stoichiometric table in Example 3-4, one can now readily use it to calculate the concentrations at a given conversion. If the initial mixture consists solely of sodium hydroxide at a concentration of 10 mol/L (i.e., 10 mol/dm³ or 10 kmol/m³) and of glyceryl stearate at a concentration of 2 g mol/L, what is the concentration of glycerine when the conversion of sodium hydroxide is **(a)** 20% and **(b)** 90%?

Solution

Only the reactants NaOH and $(C_{17}H_{35}COO)_3C_3H_5$ are initially present; therefore, $\Theta_C = \Theta_D = 0$.

(a) For 20% conversion:

$$C_D = C_{A0}\left(\frac{X}{3}\right) = (10)\left(\frac{0.2}{3}\right) = 0.67 \text{ g mol/L} = 0.67 \text{ mol/dm}^3$$

$$C_B = C_{A0}\left(\Theta_B - \frac{X}{3}\right) = 10\left(\frac{2}{10} - \frac{0.2}{3}\right) = 10\,(0.133) = 1.33 \text{ mol/dm}^3$$

(b) For 90% conversion:

$$C_D = C_{A0}\left(\frac{X}{3}\right) = 10\left(\frac{0.9}{3}\right) = 3 \text{ mol/dm}^3$$

Let us find C_B:

$$C_B = 10\left(\frac{2}{10} - \frac{0.9}{3}\right) = 10(0.2 - 0.3) = -1 \text{ mol/dm}^3$$

Negative concentration—impossible!

Ninety percent conversion of NaOH is not possible, because glyceryl stearate is the limiting reactant. Consequently, all the glyceryl stearate is used up before 90% of the NaOH could be reacted. It is important to choose the limiting reactant as the basis of calculation.

The basis of calculation should be the limiting reactant

3.3.3 Flow Systems

The form of the stoichiometric table for a continuous-flow system (see Figure 3-2) is virtually identical to that for a batch system (Table 3-2) except that we replace N_{j0} by F_{j0} and N_j by F_j (Table 3-3). Taking A as the basis, divide Equation (2-1) through by the stoichiometric coefficient of A to obtain

$$A + \frac{b}{a}B \longrightarrow \frac{c}{a}C + \frac{d}{a}D \tag{2-2}$$

For a flow system, the concentration C_A at a given point can be determined from F_A and the volumetric flow rate v at that point:

Definition of concentration for flow system

$$C_A = \frac{F_A}{v} = \frac{\text{moles/time}}{\text{liters/time}} = \frac{\text{moles}}{\text{liter}} \tag{3-27}$$

Units of v are typically given in terms of liters per second, cubic decimeters per second, or cubic feet per minute. We now can write the concentrations of A, B, C, and D for the general reaction given by Equation (2-2) in terms of the entering molar flow rate (F_{A0}, F_{B0}, F_{C0}, F_{D0}), the conversion X, and the volumetric flow rate, v.

$$C_A = \frac{F_A}{v} = \frac{F_{A0}}{v}(1-X) \qquad C_B = \frac{F_B}{v} = \frac{F_{B0} - (b/a)F_{A0}X}{v}$$

$$C_C = \frac{F_C}{v} = \frac{F_{C0} + (c/a)F_{A0}X}{v} \qquad C_D = \frac{F_D}{v} = \frac{F_{D0} + (d/a)F_{A0}X}{v} \tag{3-28}$$

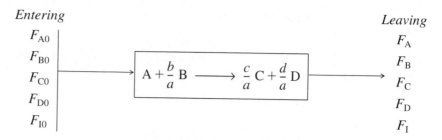

Figure 3-2 Flow reactor.

where

$$\Theta_B = \frac{F_{B0}}{F_{A0}} = \frac{C_{B0} v_0}{C_{A0} v_0} = \frac{C_{B0}}{C_{A0}} = \frac{y_{B0}}{y_{A0}}$$

and Θ_C, Θ_D, and Θ_I are defined similarly.

TABLE 3-3.   STOICHIOMETRIC TABLE FOR A FLOW SYSTEM

Stoichiometric table (flow)

Species	Feed Rate to Reactor (mol/time)	Change within Reactor (mol/time)	Effluent Rate from Reactor (mol/time)
A	F_{A0}	$-F_{A0}X$	$F_A = F_{A0}(1-X)$
B	$F_{B0} = \Theta_B F_{A0}$	$-\dfrac{b}{a} F_{A0}X$	$F_B = F_{A0}\left(\Theta_B - \dfrac{b}{a}X\right)$
C	$F_{C0} = \Theta_C F_{A0}$	$\dfrac{c}{a} F_{A0}X$	$F_C = F_{A0}\left(\Theta_C + \dfrac{c}{a}X\right)$
D	$F_{D0} = \Theta_D F_{A0}$	$\dfrac{d}{a} F_{A0}X$	$F_D = F_{A0}\left(\Theta_D + \dfrac{d}{a}X\right)$
I	$F_{I0} = \Theta_I F_{A0}$	—	$F_I = F_{A0}\Theta_I$
	F_{T0}		$F_T = F_{T0} + \left(\dfrac{d}{a} + \dfrac{c}{a} - \dfrac{b}{a} - 1\right)F_{A0}X$
			$F_T = F_{T0} + \delta F_{A0}X$

For liquids, volume change with reaction is negligible when no phase changes are taking place. Consequently, we can take

$$v = v_0$$

For liquids

$$C_A = C_{A0}(1 - X)$$

$$C_B = C_{A0}\left(\Theta_B - \frac{b}{a}X\right)$$

Therefore, for a given
rate law we have
$$-r_A = g(X)$$

Then

$$C_A = \frac{F_{A0}}{v_0}(1 - X) = C_{A0}(1 - X)$$

$$C_B = C_{A0}\left(\Theta_B - \frac{b}{a}X\right) \quad \text{etc.}$$

(3-29)

However, for gas-phase reactions the volumetric flow rate most often changes during the course of the reaction due to a change in the total number of moles or in temperature or pressure. One cannot always use Equation (3-29) to express concentration as a function of conversion for gas-phase reactions.

3.3.4 Volume Change with Reaction

In our previous discussions, we considered primarily systems in which the reaction volume or volumetric flow rate did not vary as the reaction progressed. Most batch and liquid-phase and some gas-phase systems fall into this category. There are other systems, though, in which either V or v do vary, and these will now be considered.

A situation in which a varying flow rate occurs quite frequently is in gas-phase reactions that do not have an equal number of product and reactant moles. For example, in the synthesis of ammonia,

$$N_2 + 3H_2 \underset{\longleftarrow}{\overset{\longrightarrow}{\rule{0pt}{0pt}}} 2NH_3$$

4 mol of reactants gives 2 mol of product. In flow systems where this type of reaction occurs, the molar flow rate will be changing as the reaction progresses. Because only equal numbers of moles occupy equal volumes in the gas phase at the same temperature and pressure, the volumetric flow rate will also change.

Another variable-volume situation, which occurs much less frequently, is in batch reactors where volume changes with time. Examples of this situation are the combustion chamber of the internal-combustion engine and the expanding gases within the breech and barrel of a firearm as it is fired.

In the stoichiometric tables presented on the preceding pages, it was not necessary to make assumptions concerning a volume change in the first four columns of the table (i.e., the species, initial number of moles or molar feed rate, change within the reactor, and the remaining number of moles or the molar effluent rate). All of these columns of the stoichiometric table are independent of the volume or density and they are *identical* for constant-volume (constant-density) and varying-volume (varying-density) situations. Only when concentration is expressed as a function of conversion does variable density enter the picture.

Individual concentrations can be determined by expressing the volume V for a batch system (or volumetric flow rate v for a flow system) as a function of conversion using the following equation of state:

Equation of state

$$PV = ZN_T RT \qquad (3\text{-}30)$$

in which V and N_T are defined as before and

T = temperature, K

P = total pressure, atm (kPa; 1 atm = 101.3 kPa)

Z = compressibility factor

R = gas constant = 0.08206 $dm^3 \cdot atm/g \ mol \cdot K$

This equation is valid at any point in the system at any time t. At time $t = 0$ (i.e., when the reaction is initiated), Equation (3-30) becomes

$$P_0 V_0 = Z_0 N_{T0} RT_0 \qquad (3\text{-}31)$$

Dividing Equation (3-30) by Equation (3-31) and rearranging yields

$$V = V_0 \left(\frac{P_0}{P} \right) \frac{T}{T_0} \left(\frac{Z}{Z_0} \right) \frac{N_T}{N_{T0}} \qquad (3\text{-}32)$$

We now want to express the volume V as a function of the conversion X. Recalling the equation for the total number of moles in Table 3-2,

$$N_T = N_{T0} + \delta N_{A0} X \qquad (3\text{-}33)$$

we divide through by N_{T0}:

$$\frac{N_T}{N_{T0}} = 1 + \frac{N_{A0}}{N_{T0}} \delta X = 1 + \delta y_{A0} X \qquad (3\text{-}34)$$

where y_{A0} is the mole fraction of A initially present. If all the species in the generalized reaction are in the gas phase, then

$$\boxed{\delta = \frac{d}{a} + \frac{c}{a} - \frac{b}{a} - 1} \qquad (3\text{-}23)$$

Equation (3-34) is further simplified by letting

$$\varepsilon = \frac{\text{change in total number of moles for complete conversion}}{\text{total number of moles fed to the reactor}}$$

Definitions of δ and ε In symbols,

$$\varepsilon = \left(\frac{d}{a} + \frac{c}{a} - \frac{b}{a} - 1 \right) \frac{N_{A0}}{N_{T0}} = y_{A0} \delta \qquad (3\text{-}35)$$

$$\boxed{\varepsilon = y_{A0} \delta} \qquad (3\text{-}36)$$

Equation (3-32) now becomes

$$V = V_0 \left(\frac{P_0}{P}\right) \frac{T}{T_0} \left(\frac{Z}{Z_0}\right) (1 + \varepsilon X) \tag{3-37}$$

In gas-phase systems that we shall be studying, the temperatures and pressures are such that the compressibility factor will not change significantly during the course of the reaction; hence $Z_0 \cong Z$. For a batch system the volume of gas at any time t is

Volume of gas for
a variable volume
batch reaction

$$\boxed{V = V_0 \left(\frac{P_0}{P}\right)(1 + \varepsilon X) \frac{T}{T_0}} \tag{3-38}$$

Equation (3-38) applies only to a *variable-volume* batch reactor. If the reactor is a rigid steel container of constant volume, then of course $V = V_0$. For a constant-volume container, $V = V_0$, and Equation (3-38) can be used to calculate the pressure inside the reactor as a function of temperature and conversion.

An expression similar to Equation (3-38) for a variable-volume batch reactor exists for a variable-volume flow system. To derive the concentrations of the species in terms of conversion for a variable-volume flow system, we shall use the relationships for the total concentration. The total concentration at any point in the reactor is

$$C_T = \frac{F_T}{v} = \frac{P}{ZRT} \tag{3-39}$$

At the entrance to the reactor,

$$C_{T0} = \frac{F_{T0}}{v_0} = \frac{P_0}{Z_0 R T_0} \tag{3-40}$$

Taking the ratio of Equation (3-40) to Equation (3-39) and assuming negligible changes in the compressibility factor, we have upon rearrangement

$$\boxed{v = v_0 \left(\frac{F_T}{F_{T0}}\right) \frac{P_0}{P} \left(\frac{T}{T_0}\right)} \tag{3-41}$$

From Table 3-3, the total molar flow rate is

$$F_T = F_{T0} + F_{A0}\, \delta X \tag{3-42}$$

Substituting for F_T in Equation (3-41) gives

$$v = v_0 \frac{F_{T0} + F_{A0}\, \delta X}{F_{T0}} \left(\frac{P_0}{P}\right) \frac{T}{T_0}$$

$$= v_0 \left(1 + \frac{F_{A0}}{F_{T0}} \delta X\right) \frac{P_0}{P} \left(\frac{T}{T_0}\right) = v_0 (1 + y_{A0}\, \delta X) \frac{P_0}{P} \left(\frac{T}{T_0}\right) \tag{3-43}$$

Gas-phase
volumetric flow
rate

$$v = v_0(1 + \varepsilon X) \frac{P_0}{P}\left(\frac{T}{T_0}\right) \qquad (3\text{-}44)$$

We can now express the concentration of species j for a flow system in terms of conversion:

$$C_j = \frac{F_j}{v} = \frac{F_j}{v_0\left(\dfrac{F_T}{F_{T0}}\dfrac{P_0}{P}\dfrac{T}{T_0}\right)} = \left(\frac{F_{T0}}{v_0}\right)\left(\frac{F_j}{F_T}\right)\left(\frac{P}{P_0}\right)\left(\frac{T_0}{T}\right)$$

For multiple
reactions
(Chapter 6)

$$C_j = C_{T0}\left(\frac{F_j}{F_T}\right)\left(\frac{P}{P_0}\right)\left(\frac{T_0}{T}\right) \qquad (3\text{-}45)$$

We will use this form of the concentration equation for multiple gas-phase reactions and for membrane reactors. Substituting for F_j and F_T in terms of conversion in Equation (3-45) yields

$$C_j = C_{T0}\frac{F_{A0}(\Theta_j + v_j X)}{F_{T0} + F_{A0}\delta X}\left(\frac{P}{P_0}\right)\left(\frac{T_0}{T}\right)$$

Dividing numerator and denominator by F_{T0}, we have

$$C_j = C_{T0}\left(\frac{F_{A0}}{F_{T0}}\right)\frac{\Theta_j + v_i X}{1 + (F_{A0}/F_{T0})\,\delta X}\left(\frac{P}{P_0}\right)\left(\frac{T_0}{T}\right)$$

Recalling $y_{A0} = F_{A0}/F_{T0}$ and $C_{A0} = y_{A0}C_{T0}$, then

Gas-phase
concentration as a
function of
conversion

$$C_j = \frac{C_{A0}(\Theta_j + v_j X)}{1 + \varepsilon X}\left(\frac{P}{P_0}\right)\frac{T_0}{T} \qquad (3\text{-}46)$$

where v_i is the stoichiometric coefficient, which is negative for reactants and positive for products. For example, for the reaction

$$A + \frac{b}{a}B \longrightarrow \frac{c}{a}C + \frac{d}{a}D \qquad (2\text{-}2)$$

$v_A = -1$, $v_B = -b/a$, $v_C = c/a$, and $v_D = d/a$.
    The stoichiometry table for the gas-phase reaction (2-2) is given in Table 3-4.

TABLE 3-4.   CONCENTRATIONS IN A VARIABLE-VOLUME GAS FLOW SYSTEM

$$C_A = \frac{F_A}{v} = \frac{F_{A0}(1-X)}{v} \quad = \frac{F_{A0}(1-X)}{v_0(1+\varepsilon X)}\left(\frac{T_0}{T}\right)\frac{P}{P_0} \quad = C_{A0}\left(\frac{1-X}{1+\varepsilon X}\right)\frac{T_0}{T}\left(\frac{P}{P_0}\right)$$

$$C_B = \frac{F_B}{v} = \frac{F_{A0}[\Theta_B - (b/a)X]}{v} \quad = \frac{F_{A0}[\Theta_B - (b/a)X]}{v_0(1+\varepsilon X)}\left(\frac{T_0}{T}\right)\frac{P}{P_0} \quad = C_{A0}\left(\frac{\Theta_B - (b/a)X}{1+\varepsilon X}\right)\frac{T_0}{T}\left(\frac{P}{P_0}\right)$$

$$C_C = \frac{F_C}{v} = \frac{F_{A0}[\Theta_C + (c/a)X]}{v} \quad = \frac{F_{A0}[\Theta_C + (c/a)X]}{v_0(1+\varepsilon X)}\left(\frac{T_0}{T}\right)\frac{P}{P_0} \quad = C_{A0}\left(\frac{\Theta_C + (c/a)X}{1+\varepsilon X}\right)\frac{T_0}{T}\left(\frac{P}{P_0}\right)$$

$$C_D = \frac{F_D}{v} = \frac{F_{A0}[\Theta_D + (d/a)X]}{v} \quad = \frac{F_{A0}[\Theta_D + (d/a)X]}{v_0(1+\varepsilon X)}\left(\frac{T_0}{T}\right)\frac{P}{P_0} \quad = C_{A0}\left(\frac{\Theta_D + (d/a)X}{1+\varepsilon X}\right)\frac{T_0}{T}\left(\frac{P}{P_0}\right)$$

$$C_I = \frac{F_I}{v} = \frac{F_{A0}\Theta_I}{v} \quad = \frac{F_{A0}\Theta_I}{v_0(1+\varepsilon X)}\left(\frac{T_0}{T}\right)\frac{P}{P_0} \quad = \frac{C_{A0}\Theta_I}{1+\varepsilon X}\left(\frac{T_0}{T}\right)\frac{P}{P_0}$$

At last!
We now have
$C_j = h_j(X)$
and
$-r_A = g(X)$
for variable-volume
gas-phase reactions

Example 3–6 Manipulation of the Equation for $C_j = h_j(X)$

Show under what conditions and manipulation the expression for C_B for a gas flow system reduces to that given in Table 3-4.

Solution

For a flow system the concentration is *defined* as

$$C_B = \frac{F_B}{v} \tag{E3-6.1}$$

From Table 3-3, the molar flow rate and conversion are related by

$$F_B = F_{A0}\left(\Theta_B - \frac{b}{a}X\right) \tag{E3-6.2}$$

Combining Equations (E3-6.1) and (E3-6.2) yields

$$C_B = \frac{F_{A0}[\Theta_B - (b/a)X]}{v} \tag{E3-6.3}$$

This equation for v
is only for a gas-
phase reaction

Using Equation (3-44) gives us

$$v = v_0(1+\varepsilon X)\frac{P_0}{P}\left(\frac{T}{T_0}\right) \tag{3-44}$$

to substitute for the volumetric flow rate gives

$$C_B = \frac{F_B}{v} = \frac{F_{A0}[\Theta_B - (b/a)X]}{v_0(1+\varepsilon X)}\left(\frac{P}{P_0}\right)\frac{T_0}{T} \tag{E3-6.4}$$

Recalling $\dfrac{F_{A0}}{v_0} = C_{A0}$, we obtain

$$C_B = C_{A0} \left[\frac{\Theta_B - (b/a)X}{1 + \varepsilon X} \right] \frac{P}{P_0} \left(\frac{T_0}{T} \right)$$

$\frac{F}{Q}$

which is identical to the concentration expression for a variable-volume batch reactor.

Similarly, substituting ε and the appropriate Θ's into different concentration expressions for a flow system gives the same concentration expressions as those in Table 3-4 for a variable-volume batch reaction in the gas phase.

One of the major objectives of this chapter is to learn how to express any given rate law $-r_A$ as a function of conversion. The schematic diagram in Figure 3-3 helps to summarize our discussion on this point. The concentration of the key reactant, A (the basis of our calculations), is expressed as a function of conversion in both flow and batch systems, for various conditions of temperature, pressure, and volume.

Example 3–7 Determining $C_j = h_j(X)$ for a Gas-Phase Reaction

A mixture of 28% SO_2 and 72% air is charged to a flow reactor in which SO_2 is oxidized.

$$2SO_2 + O_2 \longrightarrow 2SO_3$$

First, set up a stoichiometric table using only the symbols (i.e., Θ_i, F_i) and then prepare a second stoichiometric table evaluating numerically as many symbols as possible for the case when the total pressure is 1485 kPa and the temperature is constant at 227°C.

Solution

Taking SO_2 as the basis of calculation, we divide the reaction through by the stoichiometric coefficient of our chosen basis of calculation:

$$SO_2 + \tfrac{1}{2}O_2 \longrightarrow SO_3$$

The initial stoichiometric table is given as Table E3-7.1. Initially, 72% of the total number of moles is air containing 21% O_2 and 79% N_2.

$$F_{A0} = (0.28)(F_{T0})$$

$$F_{B0} = (0.72)(0.21)(F_{T0})$$

$$\Theta_B = \frac{F_{B0}}{F_{A0}} = \frac{(0.72)(0.21)}{0.28} = 0.54$$

$$\Theta_I = \frac{F_{I0}}{F_{A0}} = \frac{(0.72)(0.79)}{0.28} = 2.03$$

To write concentration in terms of conversion, we must express the volumetric flow rate as a function of conversion.

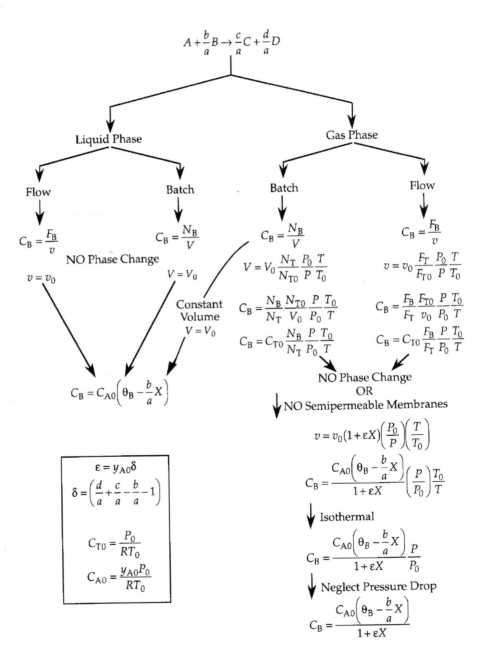

Figure 3-3 Expressing concentration as a function of conversion.

TABLE E3-7.1.   STOICHIOMETRIC TABLE FOR $SO_2 + \frac{1}{2}O_2 \longrightarrow SO_3$

Species	Symbol	Initially	Change	Remaining
SO_2	A	F_{A0}	$-F_{A0}X$	$F_A = F_{A0}(1-X)$
O_2	B	$F_{B0} = \Theta_B F_{A0}$	$-\dfrac{F_{A0}X}{2}$	$F_B = F_{A0}\left(\Theta_B - \dfrac{1}{2}X\right)$
SO_3	C	0	$+F_{A0}X$	$F_C = F_{A0}X$
N_2	I	$F_{I0} = \Theta_I F_{A0}$	—	$F_I = F_{I0} = \Theta_I F_{A0}$
		$\overline{F_{T0}}$		$F_T = F_{T0} - \dfrac{F_{A0}X}{2}$

$$C_A = \frac{F_A}{v} = \frac{F_{A0}(1-X)}{v}$$

Recalling Equation (3-44), we have

$$v = v_0(1 + \varepsilon X)\frac{P_0}{P}\left(\frac{T}{T_0}\right)$$

Neglecting
pressure
drop, $P = P_0$

Neglecting pressure drop in the reaction, $P = P_0$, yields

$$v = v_0(1 + \varepsilon X)\frac{T}{T_0}$$

If the reaction is also carried out isothermally, $T = T_0$, we obtain

$$v = v_0(1 + \varepsilon X)$$

Isothermal
operation, $T = T_0$

$$C_A = \frac{F_{A0}(1-X)}{v_0(1 + \varepsilon X)} = C_{A0}\left(\frac{1-X}{1 + \varepsilon X}\right)$$

The concentration of A initially is equal to the mole fraction of A initially multiplied by the total concentration. The total concentration can be calculated from an equation of state such as the ideal gas law:

$$C_{A0} = y_{A0}C_{T0} = y_{A0}\left(\frac{P_0}{RT_0}\right)$$

$$= 0.28\left(\frac{1485 \text{ kPa}}{8.314 \text{ kPa} \cdot \text{dm}^3/\text{mol} \cdot \text{K} \times 500 \text{ K}}\right)$$

$$= 0.1 \text{ mol/dm}^3$$

The total concentration is

$$C_{\mathrm{T}} = \frac{F_{\mathrm{T}}}{v} = \frac{F_{\mathrm{T0}} + y_{\mathrm{A0}}\delta X F_{\mathrm{T0}}}{v_0(1+\varepsilon X)} = \frac{F_{\mathrm{T0}}(1+\varepsilon X)}{v_0(1+\varepsilon X)} = \frac{F_{\mathrm{T0}}}{v_0} = C_{\mathrm{T0}} = \frac{P_0}{RT_0}$$

$$= \frac{1485\ \mathrm{kPa}}{(8.314\ \mathrm{kPa}\cdot\mathrm{dm^3/mol\cdot K})(500\ \mathrm{K})} = 0.357\,\frac{\mathrm{mol}}{\mathrm{dm^3}}$$

We now evaluate ε.

$$\varepsilon = y_{\mathrm{A0}}\,\delta = (0.28)(1 - 1 - \tfrac{1}{2}) = -0.14$$

$$C_{\mathrm{A}} = C_{\mathrm{A0}}\left(\frac{1-X}{1+\varepsilon X}\right) = 0.1\left(\frac{1-X}{1-0.14X}\right)\ \mathrm{mol/dm^3}$$

$$C_{\mathrm{B}} = C_{\mathrm{A0}}\left(\frac{\Theta_{\mathrm{B}} - \tfrac{1}{2}X}{1+\varepsilon X}\right) = \frac{0.1\,(0.54 - 0.5X)}{1 - 0.14X}\ \mathrm{mol/dm^3}$$

$$C_{\mathrm{C}} = \frac{C_{\mathrm{A0}}X}{1+\varepsilon X} = \frac{0.1X}{1 - 0.14X}\ \mathrm{mol/dm^3}$$

$$C_{\mathrm{I}} = \frac{C_{\mathrm{A0}}\Theta_{\mathrm{I}}}{1+\varepsilon X} = \frac{(0.1)(2.03)}{1 - 0.14X}\ \mathrm{mol/dm^3}$$

The concentrations of different species at various conversions are calculated in Table E3-7.2 and plotted in Figure E3-7.1. *Note* that the concentration of N_2 is changing even though it is an inert species in this reaction.

TABLE E3-7.2.   CONCENTRATION AS A FUNCTION OF CONVERSION

		C_i (g mol/dm³)				
Species		$X = 0.0$	$X = 0.25$	$X = 0.5$	$X = 0.75$	$X = 1.0$
SO$_2$	$C_{\mathrm{A}} =$	0.100	0.078	0.054	0.028	0.000
O$_2$	$C_{\mathrm{B}} =$	0.054	0.043	0.031	0.018	0.005
SO$_3$	$C_{\mathrm{C}} =$	0.000	0.026	0.054	0.084	0.116
N$_2$	$C_{\mathrm{I}} =$	0.203	0.210	0.218	0.227	0.236
	$C_T =$	0.357	0.357	0.357	0.357	0.357

The concentration of the inert is not constant

We are now in a position to express $-r_{\mathrm{A}}$ as a function of X. For example, *if* the rate law for this reaction *were* first order in SO$_2$ (i.e., A) and in O$_2$ (i.e., B), with $k = 200\ \mathrm{dm^3/mol\cdot s}$, then the rate law becomes

Use Eq. (E3–7.2) to obtain

$$-r_{\mathrm{A}} = kC_{\mathrm{A}}C_{\mathrm{B}} = kC_{\mathrm{A0}}^2\,\frac{(1-X)(\Theta_{\mathrm{B}} - 0.5X)}{(1+\varepsilon X)^2} = \frac{2(1-X)(0.54 - 0.5X)}{(1 - 0.14X)^2} \qquad \text{(E3-7.1)}$$

Taking the reciprocal of $-r_{\mathrm{A}}$ yields

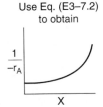

$$\frac{1}{-r_{\mathrm{A}}} = \frac{0.5\,(1 - 0.14X)^2}{(1-X)(0.54 - 0.5X)} \qquad \text{(E3-7.2)}$$

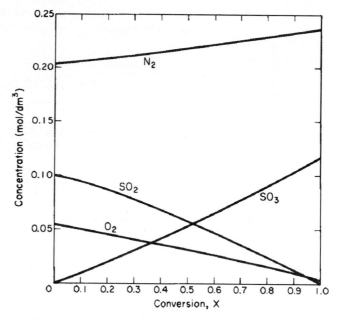

Figure E3-7.1   Concentration as a function of conversion.

We see that we could size a variety of combinations of *isothermal reactors* using the techniques discussed in Chapter 2.

All the reactions used thus far in this chapter have been irreversible reactions. The procedure one uses for the isothermal reactor design of reversible reactions is virtually the same as that for irreversible reactions, with one notable exception. First calculate the maximum conversion that can be achieved at the isothermal reaction temperature. This value is the equilibrium conversion. In the following example it will be shown how our algorithm for reactor design is easily extended to reversible reactions.

Example 3–8   Calculating the Equilibrium Conversion

The reversible gas-phase decomposition of nitrogen tetroxide, N_2O_4, to nitrogen dioxide, NO_2,

$$N_2O_4 \rightleftharpoons 2NO_2$$

is to be carried out at constant temperature and pressure. The feed consists of pure N_2O_4 at 340 K and 2 atm. The concentration equilibrium constant at 340 K is 0.1 mol/dm³.

(a)   Calculate the equilibrium conversion of N_2O_4 in a constant-volume batch reactor.

(b)   Calculate the equilibrium conversion of N_2O_4 in a flow reactor.

(c)   Assuming the reaction is elementary, express the rate of reaction solely as a function of conversion for a flow system and for a batch system.

Solution

$$N_2O_4 \xrightleftharpoons{} 2NO_2$$

$$A \xrightleftharpoons{} 2B$$

At equilibrium the concentrations of the reacting species are related by the relationship dictated by thermodynamics [see Equation (3-10) and Appendix C]

$$K_C = \frac{C_{Be}^2}{C_{Ae}} \tag{E3-8.1}$$

(a) Batch system—constant volume, $V = V_0$. See Table E3-8.1.

TABLE E3-8.1.   STOICHIOMETRIC TABLE

Species	Symbol	Initial	Change	Remaining
N_2O_4	A	N_{A0}	$-N_{A0}X$	$N_A = N_{A0}(1-X)$
NO_2	B	0	$+2N_{A0}X$	$N_B = 2N_{A0}X$
		$N_{T0} = N_{A0}$		$N_T = N_{T0} + N_{A0}X$

For batch systems $C_i = N_i / V$,

$$C_A = \frac{N_A}{V} = \frac{N_A}{V_0} = \frac{N_{A0}(1-X)}{V_0} = C_{A0}(1-X) \tag{E3-8.2}$$

$$C_B = \frac{N_B}{V} = \frac{N_B}{V_0} = \frac{2N_{A0}X}{V_0} = 2C_{A0}X \tag{E3-8.3}$$

$$C_{A0} = \frac{y_{A0}P_0}{RT_0} = \frac{(1)(2 \text{ atm})}{(0.082 \text{ atm} \cdot \text{dm}^3/\text{gmol} \cdot \text{K})(340 \text{ K})}$$

$$= 0.07174 \text{ mol/dm}^3$$

At equilibrium $X = X_e$ and we substitute Equations (E3-8.2) and (E3-8.3) into Equation (E3-8.1),

$$K_C = \frac{C_{Be}^2}{C_{Ae}} = \frac{4C_{A0}^2 X_e^2}{C_{A0}(1-X_e)} = \frac{4C_{A0}X_e^2}{1-X_e}$$

$$X_e = \sqrt{\frac{K_C(1-X_e)}{4C_{A0}}} \tag{E3-8.4}$$

We will use POLYMATH to solve for the equilibrium conversion and let xeb represent the equilibrium conversion in a constant-volume batch reactor. Equation (E3-8.4) written in POLYMATH format becomes

$$f(xeb) = xeb - [kc*(1-xeb)/(4*cao)] ** 0.5$$

The POLYMATH program and solution are given in Tables E3-8.2 and E3-8.3. The equilibrium conversion in a constant-volume batch reactor is

$$\boxed{X_{eb} = 0.44}$$

TABLE E3-8.2.   POLYMATH PROGRAM

Equations:	Initial Values:
f(Xeb)=Xeb-(Kc*(1-Xeb)/(4*Cao))**0.5	0.5
f(Xef)=Xef-(Kc*(1-Xef)*(1+eps*Xef)/(4*Cao))**0.5	0.5
Kc=0.1	
Cao=0.07174	
eps=1	

TABLE E3-8.3.   POLYMATH SOLUTION

N2O4 EQUILIBRIUM CONVERSION FOR BATCH AND FLOW SYSTEMS

	Solution	
Variable	Value	f()
Xeb	0.44126	3.661e-16
Xef	0.508355	-3.274e-17
Kc	0.1	
Cao	0.07174	
eps	1	

(b) Flow system. The stoichiometric table is the same as that for a batch system except that the number of moles of each species, N_i, is replaced by the molar flow rate of that species, F_i. For constant temperature and pressure the volumetric flow rate is $v = v_0(1 + \varepsilon X)$ and the resulting concentrations of species A and B are

$$C_A = \frac{F_A}{v} = \frac{F_{A0}(1-X)}{v} = \frac{F_{A0}(1-X)}{v_0(1+\varepsilon X)} = \frac{C_{A0}(1-X)}{1+\varepsilon X} \qquad \text{(E3-8.5)}$$

$$C_B = \frac{F_B}{v} = \frac{2F_{A0}X}{v_0(1+\varepsilon X)} = \frac{2C_{A0}X}{1+\varepsilon X} \qquad \text{(E3-8.6)}$$

At equilibrium $X = X_e$ and we can substitute Equations (E3-8.5) and (E3-8.6) into Equation (E3-8.1) to obtain the expression

$$K_C = \frac{C_{Be}^2}{C_{Ae}} = \frac{[2C_{A0}X_e/(1+\varepsilon X_e)]^2}{C_{A0}(1-X_e)/(1+\varepsilon X_e)}$$

Simplifying gives

$$K_C = \frac{4C_{A0}X_e^2}{(1-X_e)(1+\varepsilon X_e)} \qquad \text{(E3-8.7)}$$

Rearranging to use POLYMATH yields

$$X_e = \sqrt{\frac{K_C(1 - X_e)(1 + \varepsilon X_e)}{4C_{A0}}} \qquad (E3\text{-}8.8)$$

For a pure N_2O_4 feed, $\varepsilon = y_{A0}\,\delta = 1(2-1) = 1$.

We shall let xef represent the equilibrium conversion in a flow system. Equation (E3-8.8) written in the POLYMATH format becomes

$$f(xef) = xef - [kc*(1 - xef)*(1 + eps*xef)/4/cao] ** 0.5$$

This solution is also shown in Tables E3-8.2 and E3-8.3.

Note that the equilibrium conversion in a flow reactor (i.e., $X_{ef} = 0.51$), with negligible pressure drop, is greater than the equilibrium conversion in a constant-volume batch reactor ($X_{eb} = 0.44$). Recalling Le Châtelier's principle, can you suggest an explanation for this difference in X_e?

(c) Rate laws. Assuming that the reaction follows an elementary rate law, then

$$-r_A = k_A\left[C_A - \frac{C_B^2}{K_C}\right] \qquad (E3\text{-}8.9)$$

1. For a flow system, $C_A = F_A/v$ and $C_B = F_B/v$ with $v = v_0(1 + \varepsilon X)$. Consequently, we can substitute Equations (E3-8.5) and (E3-8.6) into Equation (E3-8.9) to obtain

$-r_A = f(X)$ for a flow reactor

$$-r_A = k_A\left[\frac{C_{A0}(1 - X)}{1 + \varepsilon X} - \frac{4C_{A0}^2 X^2}{K_C(1 + \varepsilon X)^2}\right] \qquad (E3\text{-}8.10)$$

Let's check to see if at equilibrium this equation reduces to the same equation as that obtained from thermodynamics. At equilibrium $-r_A \equiv 0$:

$$0 = \frac{k_A C_{A0}}{1 + \varepsilon X_e}\left[1 - X_e - \frac{4C_{A0}X_e^2}{K_C(1 + \varepsilon X_e)}\right]$$

Rearranging gives us

$$X_e = \sqrt{\frac{K_C(1 - X_e)(1 + \varepsilon X_e)}{4C_{A0}}} \qquad (E3\text{-}8.8)$$

It must agree with the value calculated from thermodynamic value and it does!

2. For a constant volume ($V = V_0$) batch system, $C_A = N_A/V_0$ and $C_B = N_B/V_0$. Substituting Equations (E3-8.2) and (E3-8.3) into the rate law, we obtain the rate of disappearance of A as a function of conversion:

$-r_A = f(X)$ for a batch reactor with $V = V_0$

$$-r_A = k_A\left[C_A - \frac{C_B^2}{K_C}\right] = k_A\left[C_{A0}(1 - X) - \frac{4C_{A0}^2 X^2}{K_C}\right] \qquad (E3\text{-}8.11)$$

As expected, the dependence of reaction rate on conversion for a constant-volume batch system [i.e., Equation (E3-8.11)] is different than that for a flow system [Equation (E3-8.10)] for gas-phase reactions.

3.4 Expressing Concentrations in Terms Other Than Conversion

As we shall see later in the book, there are some instances in which it is much more convenient to work in terms of the number of moles (N_A, N_B) or molar flow rates $(F_A, F_B,$ etc.) rather than conversion. Membrane reactors and gas-phase multiple reactions are two such cases where molar flow rates rather than conversion are preferred. Consequently, the concentrations in the rate laws need to be expressed in terms of the molar flow rates. We start by recalling and combining Equations (3-40) and (3-41):

$$C_{T0} = \frac{F_{T0}}{v_0} = \frac{P_0}{Z_0 R T_0}$$

Used for:
• Multiple rxns
• Membranes
• Unsteady state

$$v = \left(\frac{v_0}{F_{T0}}\right) F_T \frac{P_0}{P} \frac{T}{T_0}$$

to give

$$v = \frac{F_T}{C_{T0}} \frac{P_0}{P} \frac{T}{T_0} \tag{3-47}$$

For the case of an ideal gas $(Z = 1)$, the concentration is

$$C_A = \frac{F_A}{v}$$

Substituting for v gives

$$\boxed{C_A = C_{T0} \frac{F_A}{F_T} \frac{P}{P_0} \frac{T_0}{T}} \tag{3-48}$$

In general $(j = A, B, C, D, I)$

$$\boxed{C_j = C_{T0} \frac{F_j}{F_T} \frac{P}{P_0} \frac{T_0}{T}} \tag{3-45}$$

with the total molar flow rate given as the sum of the flow rates of the individual species:

$$F_T = F_A + F_B + F_C + F_D + F_I \tag{3-49}$$

The molar flow rate of each species F_j is obtained from a mole balance on each species.

Example 3–9   PFR Mole Balances in Terms of Molar Flow Rates

Reconsider the elementary gas reaction discussed in Example 3-8.

$$N_2O_4 \xrightleftharpoons{} 2NO_2$$
$$A \xrightleftharpoons{} 2B$$

The reaction is to be carried out isothermally ($T = T_0$) and isobarically ($P = P_0$) in a PFR. Express the rate law and mole balances in terms of the molar flow rates.

Solution

Mole balance:
$$\frac{dF_A}{dV} = r_A \tag{E3-9.1}$$

$$\frac{dF_B}{dV} = r_B \tag{E3-9.2}$$

Rate law:
$$-r_A = k_A \left[C_A - \frac{C_B^2}{K_C} \right] \tag{E3-9.3}$$

Stoichiometry:
$$\frac{r_A}{-1} = \frac{r_B}{2}$$

Then

$$r_B = -2r_A \tag{E3-9.4}$$

Combine:

Using Equation (3-45) to substitute for the concentrations of A and B when $T = T_0$ and $P = P_0$, Equation (E3-9.3) becomes

$$-r_A = k_A \left[C_{T0} \left(\frac{F_A}{F_T} \right) - \frac{C_{T0}^2}{K_C} \left(\frac{F_B}{F_T} \right)^2 \right] \tag{E3-9.5}$$

where the total molar flow rate is just the sum of the flow rates of A and B:

$$F_T = F_A + F_B \tag{E3-9.6}$$

and the total concentration of the entrance to the reactor (P_0, T_0) is calculated from the equation

$$C_{T0} = \frac{P_0}{RT_0} \tag{E3-9.7}$$

Combining Equations (E3-9.5) and (E3-9.6), we obtain

$$\frac{dF_A}{dV} = r_A = -k_A C_{T0} \left[\frac{F_A}{F_A + F_B} - \frac{C_{T0}}{K_C} \left(\frac{F_B}{F_A + F_B} \right)^2 \right] \tag{E3-9.8}$$

and combining Equations (E3-9.2), (E3-9.4), and (E3-9.8) gives

$$\frac{dF_B}{dV} = r_B = -2r_A = 2k_A C_{T0} \left[\frac{F_A}{F_A + F_B} - \frac{C_{T0}}{K_C} \left(\frac{F_B}{F_A + F_B} \right)^2 \right] \qquad (E3\text{-}9.9)$$

Equations (E3-9.8) and (E3-9.9) can now be solved numerically, preferably by a software package such as POLYMATH or MATLAB. (See Chapter 4)

3.5 Reactions with Phase Change

When Equation (3-36) is used to evaluate ε, it should be remembered from the derivation of this equation that δ represents the change in the number of moles in the gas phase per mole of A reacted. As the last example in this chapter, we consider a gas-phase reaction in which condensation occurs. An example of this class of reactions is

$$C_2H_6(g) + 2Br_2(g) \longrightarrow C_2H_4Br_2(g,l) + 2HBr(g)$$

Another example of phase change during reaction is chemical vapor deposition (CVD), a process used to manufacture microelectronic materials. Here, gas-phase reactants are deposited (analogous to condensation) as thin films on solid surfaces (see Problem P3-25). One such reaction is the production of gallium arsenide, which is used in computer chips.

$$GaCl_2(g) + \tfrac{1}{2}As_2(g) + H_2(g) \longrightarrow GaAs(s) + 2HCl(g)$$

The development of continuous-flow CVD reactors where solid wafers and gases continuously pass through the reactor is currently under way (see Sections 10.8 and 12.10).

We now will develop our stoichiometric table for reactions with phase change. When one of the products condenses during the course of a reaction, calculation of the change in volume or volumetric flow rate must be undertaken in a slightly different manner. Consider another isothermal reaction:

$$A(g) + 2B(g) \longrightarrow C(g) + D(g,l)$$

The vapor pressure of species D at temperature T is P_v.

The gas-phase concentration of the product D will increase until the corresponding mole fraction at which condensation begins is reached:

At $P_D = P_v$,
$y_D = y_{D,e}$ and
condensation starts

$$\boxed{y_{D,e} = \frac{P_v}{P_T}} \qquad (3\text{-}50)$$

Once saturation is reached in the gas phase, every mole of D produced condenses. To account for the effects of condensation on the concentrations of the reacting species, we now write two columns for the number of moles (or molar flow rates) in our stoichiometric table (Table 3-5). One column gives the molar flow rates of each species before condensation has begun and the other column gives these quantities after condensation has begun. We use X_c to refer to the conversion of A at which the condensation of D begins. Note that we must rearrange the equation for the total molar flow rate to write it explicitly in terms of F_{A0}, X, and $y_{D,e}$. We use the equations for the mole fraction of species D to calculate the conversion at which condensation begins.

TABLE 3-5.   STOICHIOMETRIC TABLE FOR REACTION WITH CONDENSATION
$$A(g) + 2B(g) \longrightarrow C(g) + D(g, l)$$

Species	Entering	Change	Before Condensation $P_D < P_v$ Leaving	After Condensation $P_D = P_v$ Leaving
A(g)	F_{A0}	$-F_{A0}X$	$F_A = F_{A0}(1 - X)$	$F_{A0}(1 - X)$
B(g)	$F_{B0} = 2F_{A0}$	$-2F_{A0}X$	$F_B = F_{A0}(2 - 2X)$	$F_{A0}(2 - 2X)$
C(g)	—	$F_{A0}X$	$F_C = F_{A0}X$	$F_{A0}X$
D(g)	$\overline{F_{T0} = 3F_{A0}}$	$F_{A0}X$	$\dfrac{F_D = F_{A0}X}{F_T = F_{A0}(3 - X)}$	$\dfrac{F_D = y_{D,e}F_T}{F_T = y_{D,e}F_T + 3F_{A0} - 2F_{A0}X}$

An extra column is added to the table for phase changes

Solve for F_T:
$$F_T = 2F_{A0}(1.5 - X)/(1 - y_{D,e})$$

Example 3–10 Expressing $-r_A = g(X)$ for Reactions with Phase Change

For the reaction just discussed, calculate the conversion at which condensation begins and express the concentration of the reacting species and the rate of reaction as a function of conversion. The reaction is first-order in both species A and species B. The feed contains only A and B in stoichiometric amounts and the reaction is carried out isothermally. The total pressure is 101.3 kPa (1 atm) and species D has a vapor pressure of 16 kPa (120 mmHg) at the isothermal reaction temperature of 300 K.

Solution

At the point where condensation begins,

$$X = X_c$$

From the stoichiometric table,

$$y_{D,e} = \frac{F_D}{F_T} = \frac{F_{A0}X_c}{F_{A0}(3 - X_c)} = \frac{X_c}{3 - X_c} \qquad \text{(E3-10.1)}$$

At saturation,

$$y_{D,e} = \frac{P_{vD}}{P_T} = \frac{16}{101.3} = 0.158 \qquad \text{(E3-10.2)}$$

Equating Equations (E3-10.1) and (E3-10.2) gives

$$0.158 = \frac{X_c}{3 - X_c} \tag{E3-10.3}$$

Solving for X_c yields

$$\boxed{X_c = 0.41}$$

Before condensation begins: For $X < X_c$ there is no condensation and one can use the basic equations for δ and ε to calculate the concentrations; i.e.,

$$C_A = \frac{C_{A0}(1 - X)}{1 + \varepsilon X} \tag{E3-10.4}$$

$$\varepsilon = y_{A0}\delta = 0.33\,(1 + 1 - 2 - 1) = -0.33$$

$$C_A = C_{A0}\left(\frac{1 - X}{1 - 0.33X}\right) \tag{E3-10.5}$$

$$C_B = C_{A0}\left(\frac{2 - 2X}{1 - 0.33X}\right) = \frac{2C_{A0}(1 - X)}{(1 - 0.33X)} \tag{E3-10.6}$$

Because the temperature and pressure are constant, the total concentration is constant.

$$C_T = \frac{P}{ZRT} = \frac{P_0}{Z_0 R T_0} = C_{T0} \tag{E3-10.7}$$

The reaction rate is first-order in A and in B for $X < X_c$:

$$\boxed{-r_A = 2kC_{A0}^2\left[\frac{(1 - X)^2}{(1 - 0.33X)^2}\right]} \tag{E3-10.8}$$

After condensation begins: For $X > X_c$ the partial pressure of D is equal to the vapor pressure $(P_D = P_v)$. The volumetric flow rate is related to the total molar flow rate through the ideal gas equation of state:

$-r_A$ will be a different function of conversion before and after condensation

$$F_T = C_T v \tag{E3-10.9}$$

$$F_{T0} = C_{T0} v_0 \tag{E3-10.10}$$

Then, taking the ratio of Equation (E3-10.9) to Equation (E3-10.10) and rearranging, we have

$$v = v_0\left(\frac{F_T}{F_{T0}}\right) = v_0\left[\frac{2F_{A0}(1.5 - X)}{3F_{A0}(1 - y_{D,e})}\right] = \frac{v_0}{1 - y_{D,e}}\frac{1.5 - X}{1.5} \tag{E3-10.11}$$

We must use the column in the stoichiometric table labeled "after condensation" in conjunction with Equation (E3-10.11) to determine C_A and C_B.

$$C_A = \frac{F_A}{v} = \frac{1.5\,F_{A0}(1-X)}{v_0(1.5-X)/(1-y_{D,e})} = 1.5\,C_{A0}(1-y_{D,e})\frac{1-X}{1.5-X} \quad \text{(E3-10.12)}$$

$$C_B = \frac{F_B}{v} = \frac{1.5\,F_{A0}(2-2X)}{v_0(1.5-X)/(1-y_{D,e})} = 3\,C_{A0}(1-y_{D,e})\frac{1-X}{1.5-X} \quad \text{(E3-10.13)}$$

The rate law for $X > X_c$ is

$$-r_A = 4.5\,kC_{A0}^2(1-y_{D,e})^2\,\frac{(1-X)^2}{(1.5-X)^2} \quad \text{(E3-10.14)}$$

Before condensation, for $X < X_c$, the gas-phase molar flow rate of D is $F_D = F_{A0}X$. After condensation begins (i.e., $X > X_c$), the molar flow rate of D in the gas phase is

$$F_D(g) = y_{D,e}F_T = \frac{y_{D,e}}{1-y_{D,e}}\,2F_{A0}(1.5-X) = 0.375\,F_{A0}(1.5-X) \quad \text{(E3-10.15)}$$

The liquid molar flow rate of D is

$$F_D(l) = F_{A0}X - F_D(g) = F_{A0}(1.375X - 0.563)$$

Plots of the molar flow rates of species D and the total, together with the concentration of A, are shown in Figure E3-10.1 as a function of conversion.

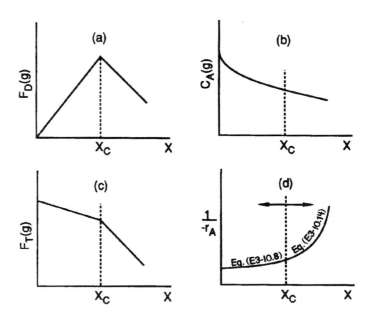

Figure E3-10.1

If we know specific values of v_0, C_{A0}, and k, we can use Figure E3-10.1(d) to size a variety of combinations of CSTRs and PFRs.

SUMMARY

1. *Reaction order* is determined from experimental observation:

$$A + B \longrightarrow C \qquad \text{(S3-1)}$$

$$-r_A = k C_A^\alpha C_B^\beta$$

The reaction in Equation (S3-1) is α order with respect to species A and β order with respect to species B, whereas the overall order is $\alpha + \beta$. Reaction order is determined from experimental observation. If $\alpha = 1$ and $\beta = 2$, we would say that the reaction is first-order with respect to A, second-order with respect to B, and overall third-order.

2. In addition to the reaction order, the following terms were defined:
   a. Elementary reaction
   b. Reversible and irreversible reactions
   c. Homogeneous and heterogeneous reactions

3. The temperature dependence of a specific reaction rate is given by the *Arrhenius equation*,

$$k = A e^{-E/RT} \qquad \text{(S3-2)}$$

where A is the frequency factor and E the activation energy.

4. The *stoichiometric table* for the reaction

$$A + \frac{b}{a} B \longrightarrow \frac{c}{a} C + \frac{d}{a} D \qquad \text{(S2-2)}$$

being carried out in a flow system is:

Species	Entering	Change	Leaving
A	F_{A0}	$-F_{A0}X$	$F_{A0}(1 - X)$
B	F_{B0}	$-\left(\dfrac{b}{a}\right) F_{A0}X$	$F_{A0}\left(\Theta_B - \dfrac{b}{a} X\right)$
C	F_{C0}	$\left(\dfrac{c}{a}\right) F_{A0}X$	$F_{A0}\left(\Theta_C + \dfrac{c}{a} X\right)$
D	F_{D0}	$\left(\dfrac{d}{a}\right) F_{A0}X$	$F_{A0}\left(\Theta_D + \dfrac{d}{a} X\right)$
I	$\dfrac{F_{I0}}{F_{T0}}$	—	$\dfrac{F_{I0}}{F_T = F_{T0} + \delta F_{A0}X}$

$$\text{where} \qquad \delta = \frac{d}{a} + \frac{c}{a} - \frac{b}{a} - 1$$

The relative rates of reaction are

$$\frac{r_A}{-a} = \frac{r_B}{-b} = \frac{r_C}{c} = \frac{r_D}{d} \qquad (2\text{-}20)$$

5. In the case of ideal gases, Equations (S3-3) through (S3-6) relate volume and volumetric flow rate to conversion.

Batch constant volume: $V = V_0$ $\qquad$ (S3-3)

Batch variable volume: $V = V_0 \left(\dfrac{P_0}{P}\right)(1 + \varepsilon X)\dfrac{T}{T_0}$ $\qquad$ (S3-4)

Flow systems: $v = v_0 \left(\dfrac{P_0}{P}\right)(1 + \varepsilon X)\dfrac{T}{T_0}$ $\qquad$ (S3-5)

where the change in the number of moles per mole of A fed is

$$\varepsilon = y_{A0}\delta \qquad (S3\text{-}6)$$

and the change in the number of moles per mole of A reacted is

$$\delta = \frac{d}{a} + \frac{c}{a} - \frac{b}{a} - 1 \qquad (S3\text{-}7)$$

6. For the ideal gas-phase reaction

$$A + \frac{b}{a} B \longrightarrow \frac{c}{a} C + \frac{d}{a} D \qquad (2\text{-}2)$$

the volumetric flow rate is

$$v = v_0 \left(\frac{F_T}{F_{T0}}\right)\frac{P_0}{P}\left(\frac{T}{T_0}\right) \qquad (S3\text{-}8)$$

Using the stoichiometric table along with the definitions of concentration (e.g., $C_A = F_A/v$), the concentrations of A and C are:

$$C_A = \frac{F_A}{v} = \frac{F_{A0}(1 - X)}{v} = C_{A0}\left[\frac{1 - X}{1 + \varepsilon X}\right]\frac{P}{P_0}\left(\frac{T_0}{T}\right) \qquad (S3\text{-}9)$$

$$C_C = \frac{F_C}{v} = C_{A0}\left[\frac{\Theta_C + (c/a)X}{1 + \varepsilon X}\right]\frac{P}{P_0}\left(\frac{T_0}{T}\right) \qquad (S3\text{-}10)$$

7. When the reactants and products are incompressible liquids, the concentrations of species A and C in the reaction given by Equation (2-2) can be written as

$$C_A = C_{A0}(1 - X) \tag{S3-11}$$

$$C_C = C_{A0}\left(\Theta_C + \frac{c}{a}X\right) \tag{S3-12}$$

Equations (S3-11) and (S3-12) also hold for gas-phase reactions carried out at constant volume in batch systems.

8. When using measures other than conversion for reactor design, the mole balances are written for each species in the reacting mixture:

$$\frac{dF_A}{dV} = r_A, \quad \frac{dF_B}{dV} = r_B, \quad \frac{dF_C}{dV} = r_C, \quad \frac{dF_D}{dV} = r_D$$

The mole balances are then coupled through their relative rates of reaction. If

$$-r_A = k\, C_A^\alpha C_B^\beta$$

then

$$r_B = \frac{b}{a}r_A \quad r_C = -\frac{c}{a}r_A \quad r_D = -\frac{d}{a}r_A$$

Concentration can also be expressed in terms of the number of moles (batch) in molar flow rates (flow).

Gas: $\quad C_A = C_{T0}\dfrac{F_A}{F_T}\dfrac{P}{P_0}\dfrac{T_0}{T}$ $\qquad\qquad$ (S3-13)

$$F_T = F_A + F_B + F_C + F_D + F_I$$

Liquid: $\quad C_A = \dfrac{F_A}{v_0}$ $\qquad\qquad\qquad\qquad$ (S3-14)

9. For reactions in which condensation occurs, e.g.,

$$A(g) + B(g) \longrightarrow C(g,l)$$

before condensation, with $P = P_0$, $T = T_0$, $\Theta_B = 1$,

$$v = v_0(1 + \varepsilon X) = v_0(1 - 0.5X)$$

$$\tag{S3-15}$$

$$C_C = \frac{C_{A0}X}{1 - 0.5X}$$

and **after condensation** $(X > X_c)$,

$$v = \frac{v_0(1 - X)}{1 - y_{C,e}} \quad F_C = y_C F_T \quad y_{C,e} = \frac{P_{vC}}{P_T} \quad C_C = \frac{y_C}{y_{A0}}C_{A0} \tag{S3-16}$$

where $y_{C,e} = P_{ve}/P_0$ and is the mole fraction of C at which condensation begins.

QUESTIONS AND PROBLEMS

The subscript to each of the problem numbers indicates the level of difficulty: A, least difficult; D, most difficult.

$$A = \bullet \quad B = \blacksquare \quad C = \blacklozenge \quad D = \blacklozenge\blacklozenge$$

In each of the questions and problems below, rather than just drawing a box around your answer, write a sentence or two describing how you solved the problem, the assumptions you made, the reasonableness of your answer, what you learned, and any other facts that you want to include. You may wish to refer to W. Strunk and E. B. White, *The Elements of Style* (New York: Macmillian, 1979) and Joseph M. Williams, *Style: Ten Lessons in Clarity & Grace* (Glenview, Ill.: Scott, Foresman, 1989) to enhance the quality of your sentences.

P3-1$_C$ (a) List the important concepts that you learned from this chapter. What concepts are you not clear about?

(b) Explain the strategy to evaluate reactor design equations and how this chapter expands on Chapter 2.

(c) Read through all the problems at the end of this chapter. Make up and solve an original problem based on the material in this chapter. (1) Use real data and reactions. (2) Make up your data. Identify what concepts the problem is trying to enforce and why the problem is important. Novel applications (e.g., environmental, food processing) are encouraged. At the end of the problem and solution, describe the process used to generate the idea of the problem. (*Hint*: The journals listed at the end of Chapter 1 may be helpful in obtaining real data.)

P3-2$_A$ What if:

(a) you were asked to give an example of the material discussed in this chapter that applies to things you observe every day; what would you describe? (*Hint*: See Problem 3-3$_A$.)

(b) a catalyst were added to increase the reaction rate by a factor of 10 in Example 3-8? How would your answers change?

(c) very, very little NaOH were used in Example 3-2 compared to the amount of TBB? Would the rate of reaction be affected? What might be the rate law with respect to TBB?

(d) a plot of ln k vs. $(1/T)$ were not linear, but a curve whose slope was shallow at high T and steep at low T or vice versa. How would you explain such curves? (*Hint*: one example A → B and A → C)

(e) someone suggested that you bake a 9-in.-diameter cake for 15 minutes at 400°F instead of the cookbook's recommendation of 30 minutes at 325°F? How would you develop a plot of cooking time versus oven temperature?

P3-3$_A$ The frequency of flashing of fireflies and the frequency of chirping of crickets as a function of temperature are given below [*J. Chem. Educ.*, *5*, 343 (1972) Reprinted by permission.].

For fireflies:

T (°C)	21.0	25.00	30.0
Flashes/min	9.0	12.16	16.2

For crickets:

T (°C)	14.2	20.3	27.0
Chirps/min	80	126	200

The running speed of ants and the flight speed of honeybees as a funtion of temperature are given below [Source: B. Heinrich, "The Hot-Blooded Insects" (Harvard University Press, Cambridge, MA, 1993)].

For ants:

T (°C)	10	20	30	38
V (cm/s)	0.5	2	3.4	6.5

For honeybees:

T (°C)	25	30	35	40
V (cm/s)	0.7	1.8	3	?

(a) What do the firefly and cricket have in common?

(b) What is the velocity of the honeybee at 40°C? At –5°C

(c) Do the bees, ants, crickets, and fireflies have anything in common? If so, what is it? You may also do a pairwise comparison.

(d) Would more data help clarify the relationships among frequency, speed, and temperature? If so, in what temperature should the data be obtained? Pick an insect and explain how you would carry out the experiment to obtain more data.

P3-4_B Corrosion of high-nickel stainless steel plates was found to occur in a distillation column used at DuPont to separate HCN and water. Sulfuric acid is always added at the top of the column to prevent polymerization of HCN. Water collects at the bottom of the column and HCN at the top. The amount of corrosion on each tray is shown in Figure P3-4 as a function of plate location in the column.

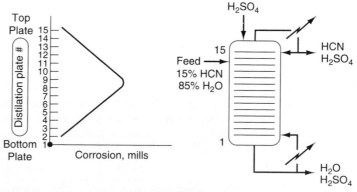

Figure P3-4

The bottom-most temperature of the column is approximately 125°C and the topmost is 100°C. The corrosion rate is a function of temperature and the concentration of a HCN–H_2SO_4 complex. Suggest an explanation for the observed corrosion plate profile in the column. What effect would the column operating conditions have on the corrosion profile?

P3-5$_B$ The rule of thumb that the rate of reaction doubles for a 10°C increase in temperature occurs only at a specific temperature for a given activation energy.

(a) Develop a relationship between the temperature and activation energy for which the rule of thumb holds. Neglect any variation of concentration with temperature.

(b) Determine the activation energy and frequency factor from the following data:

k (min^{-1})	0.001	0.050
T (°C)	00.0	100.0

P3-6$_A$ In each of the following reactions determine the specific reaction rate constant for each of the other species in the reaction. Assume that k_A in each case has a value of 25 with the appropriate combination of units of mol, dm^3, g cat, and s.

(a) For the reaction

$$2A + B \longrightarrow C$$

the rate law is $-r_A = k_A C_A^2 C_B$. (Partial ans.: For $k_A = 25$ dm^6/mol$^2 \cdot$s, $k_C = k_A/2 = 12.5$ dm^6/mol$^2 \cdot$ s and $r_C = 12.5\ C_A^2 C_B$.)

(b) For the reaction

$$\tfrac{1}{2}A + \tfrac{3}{2}B \longrightarrow C$$

the rate law is $-r_A = k_A C_A C_B$.

(c) For the solid catalyzed reaction

$$4A + 5B \longrightarrow 4C + 6D$$

the rate law is $-r_A' = k_A C_A^2 C_B$ (see Problem 3-13) [$k_D = ?,\ k_B = ?$].

(d) In the homogeneous gas-phase reaction

$$CH_4 + \tfrac{3}{2}O_2 \longrightarrow HCOOH + H_2O$$

What is the relationship between r_{CH_4} and r_{O_2}?

(1) $r_{CH_4} = r_{O_2}$

(2) Cannot tell without the data

(3) $r_{CH_4} = \tfrac{2}{3}r_{O_2}$

(4) $r_{CH_4} = \tfrac{3}{2}r_{O_2}$

(5) None of the above

P3-7$_A$ Set up a stoichiometric table for each of the following reactions and express the concentration of each species in the reaction as a function of conversion evaluating all constants (e.g., ε, Θ).

(a) The liquid-phase reaction

$$\underset{\displaystyle CH_2{-}CH_2}{\overset{\displaystyle O}{\diagup\ \diagdown}} + H_2O \ \xrightarrow{H_2SO_4}\ \underset{\displaystyle CH_2{-}OH}{\overset{\displaystyle CH_2{-}OH}{\vert}}$$

The initial concentrations of ethylene oxide and water are 1 lb mol/ft³ and 3.47 lb-mol/ft³ (62.41 lb/ft³ ÷ 18), respectively.

(b) The isothermal, isobaric gas-phase pyrolysis

$$C_2H_6 \longrightarrow C_2H_4 + H_2$$

Pure ethane enters the flow reactor at 6 atm and 1100 K.
How would your equation for the concentration change if the reaction were to be carried out in a constant-volume batch reactor?

(c) The isothermal, isobaric, catalytic gas-phase oxidation

$$C_2H_4 + \tfrac{1}{2}O_2 \longrightarrow \overset{\displaystyle O}{\overset{\displaystyle \triangle}{CH_2-CH_2}}$$

The feed enters a PBR at 6 atm and 260°C and is a stoichiometric mixture of oxygen and ethylene.

P3-8   There were 5430 million pounds of ethylene oxide produced in the United States in 1995. The flowsheet for the commercial production of ethylene oxide (EO) by oxidation of ethylene is shown below. We note that the process essentially consists of two systems, a reaction system and a separation system. Describe how your answers to P3-7 **(c)** would change if air is used as a feed? This reaction is studied further in *Example 4-6*.

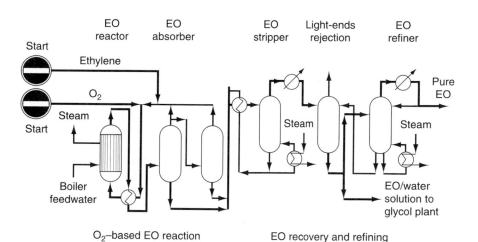

Figure P3-8   EO plant flowsheet. [Adapted from R. A. Meyers, ed., *Handbook of Chemical Production Processes, Chemical Process Technology Handbook Series*, McGraw-Hill, 1983, p. 1.5-5. ISBN 0-67-041-765-2.]

P3-9   Rework Problem 3-7 to write the combined mole balance rate law along the lines discussed in Section 3.4. Assume each reaction is elementary.

(a) Write the CSTR mole balance and rate law for each species solely in terms of concentration and rate law parameters for P3-7(a).

(b) For Problem 3-7(a) write the combined PFR mole balance on each species and rate law solely in terms of the molar flow rates and rate law parameters.

(c) For Problem 3-7(b) write the combined PFR mole balance on each species and rate law solely in terms of the molar flow rates and rate law parameters.

(d) For Problem 3-7(c), write the combined PFR mole balance and rate law solely in terms of the molar flow rates for a PFR.

P3-10$_B$ For each of the following reactions and rate laws at low temperatures, suggest a rate law at high temperatures. The reactions are highly exothermic and therefore reversible at high temperatures.

(a) The reaction

$$A \longrightarrow B$$

is irreversible at low temperatures, and the rate law is

$$-r_A = kC_A$$

(b) The reaction

$$A + 2B \longrightarrow 2D$$

is irreversible at low temperatures and the rate law is

$$-r_A = kC_A^{1/2} C_B$$

(c) The catalytic reaction

$$A + B \longrightarrow C + D$$

is irreversible at low temperatures and the rate law is

$$-r_A' = \frac{kP_A P_B}{1 + K_A P_A + K_B P_B}$$

In each case, make sure that the rate laws at high temperatures are thermodynamically consistent at equilibrium (cf. Appendix C).

P3-11$_B$ There were 820 million pounds of phthalic anhydride produced in the United States in 1995. One of the end uses of phthalic anhydride is in the fiberglass of sailboat hulls. Phthalic anhydride can be produced by the partial oxidation of naphthalene in either a fixed or a fluidized catalytic bed. A flowsheet for the commercial process is shown in Figure P3-11. Here the reaction is carried out in a fixed-bed reactor with a vanadium pentoxide catalyst packed in 25-mm-diameter tubes. A production rate of 31,000 tons per year would require 15,000 tubes.

Set up a stoichiometric table for this reaction for an initial mixture of 3.5% naphthalene and 96.5% air (mol %), and use this table to develop the relations listed below. $P_0 = 10$ atm and $T_0 = 500$ K.

(a) For an isothermal flow reactor in which there is no pressure drop, determine each of the following as a function of the conversion of naphthalene, X_N.

    (1) The partial pressures of O_2 and CO_2 (*Ans.:* $P_{CO_2} = 0.345$ [5.8 − 9/2 X]/(1 − 0.0175 X))

    (2) The concentrations of O_2 and naphthalene (*Ans.:* $C_N = 0.084$ (1 − X)/(1 − 0.0175 X))

    (3) The volumetric flow rate v

(b) Repeat part (a) when a pressure drop occurs in the reactor.

(c) If the reaction just happened to be first order in oxygen and second order in naphthalene with a value of k_N of 0.01 mol^2/dm^6 · s, write an equation for $-r_N$ solely as a function of conversion for parts (a) and (b).

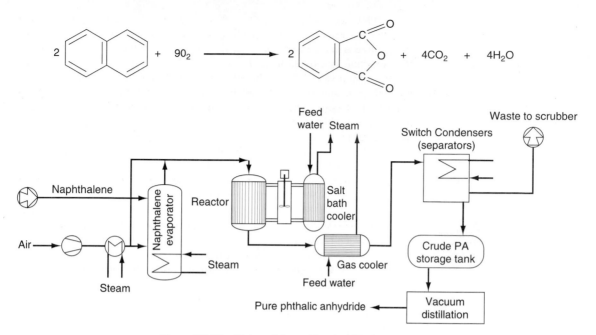

Figure P3-11   [Adapted from *Chemical Engineering, Process Technology and Flowsheet*, Vol. IIK pp. 111 and 125.]

(d) Rework part (c) for stoichiometric feed of pure oxygen. What would be the advantages and disadvantages of using pure oxygen rather than air?

(e) What safety features are or should be included in this reaction system? (*Hint*: See the flowsheet reference.)

[For explosive limits of this reaction, see *Chem. Eng. Prog.*, 66, 49 (1970).]

P3-12$_A$ **(a)** Taking H_2 as your basis of calculation, construct a complete stoichiometric table for the reaction

$$\tfrac{1}{2}N_2 + \tfrac{3}{2}H_2 \longrightarrow NH_3$$

for an isobaric, isothermal flow system with equimolar feeds of N_2 and H_2.

(b) If the entering total pressure is 16.4 atm and the entering temperature is 1727°C, calculate the concentrations of ammonia and hydrogen when the conversion of H_2 is 60%. (*Ans.*: $C_{H_2} = 0.025$ g mol/L, $C_{NH_3} = 0.025$ g mol/L.)

(c) If you took N_2 as your basis of calculation, could 60% conversion of N_2 be achieved?

P3-13$_A$ Nitric acid is made commercially from nitric oxide. Nitric oxide is produced by the gas-phase oxidation of ammonia:

$$4NH_3 + 5O_2 \longrightarrow 4NO + 6H_2O$$

The feed consists of 15 mol % ammonia in air at 8.2 atm and 227°C.

(a) What is the total entering concentration?

(b) What is the entering concentration of ammonia?

(c) Set up a stoichiometric table with ammonia as your basis of calculation.

Then

    (1) Express P_i and C_i for all species as functions of conversion for a constant-pressure batch reactor operated isothermally. Express volume as a function of X.

    (2) Express P_i and C_i for all species as functions of conversion for a constant-volume reactor. Express P_T as a function of X.

    (3) Express P_i and C_i for all species as functions of conversion for a flow reactor.

  (d) Referring to Section 3.4, write the combined mole balance and rate law [cf. Equations (E3-9.8 and E3-9.9)] solely in terms of the molar flow rates and rate law parameters. Assume elementary reaction.

P3-14$_B$ Reconsider the decomposition of nitrogen tetroxide discussed in Example 3-8. The reaction is to be carried out in PFR and also in a constant-volume batch reactor at 2 atm and 340 K. Only N_2O_4 *and* an inert I are to be fed to the reactors. Plot the equilibrium conversion as a function of inert mole fraction in the feed for both a constant-volume batch reactor and a plug flow reactor. Why is the equilibrium conversion lower for the batch system than the flow system in Example 3-8? Will this lower equilibrium conversion result always be the case for batch systems?

P3-15$_A$ **(a)** Express the rate of formation of hydrogen bromide in terms of the constants k_1 and k_2 and the conversion of bromine, X. Evaluate numerically all other quantities. The feed consists of 25% hydrogen, 25% bromine, and 50% inerts at a pressure of 10 atm and a temperature of 400°C.

  (b) Write the rate of decomposition of cumene, $-r_C'$, in terms of conversion, initial concentration of cumene, and the specific rate and equilibrium constants. The initial mixture consists of 75% cumene and 25% inerts.

P3-16$_A$ The gas-phase reaction

$$2A + 4B \longrightarrow 2C$$

which is first-order in A and first-order in B is to be carried out *isothermally in a plug-flow reactor*. The entering volumetric flow rate is 2.5 dm³/min, and the feed is equimolar in A and B. The entering temperature and pressure are 727°C and 10 atm, respectively. The specific reaction rate at this temperature is 4 dm³/g mol·min and the activation energy is 15,000 cal/g mol.

  (a) What is the volumetric flow rate when the conversion of A is 25%? (*Ans.:* $v = 1.88$ dm³/min.)

  (b) What is the rate of reaction at the entrance to the reactor (i.e., $X = 0$)? (*Ans.:* $-r_A = 1.49 \times 10^{-2}$ g mol/dm³·min.)

  (c) What is the rate of reaction when the conversion of A is 40%? (*Hint:* First express $-r_A$ as a function of X alone.) (*Ans.:* $-r_A = 4.95 \times 10^{-3}$ g mol/dm³·min.)

  (d) What is the concentration of A at the entrance to the reactor? (*Ans.:* $C_{A0} = 6.09 \times 10^{-2}$ g mol/dm³.)

  (e) What is the concentration of A at 40% conversion of A? (*Ans.:* $C_A = 6.09 \times 10^{-2}$ g mol/dm³.)

  (f) What is the value of the specific reaction rate at 1227°C? (*Ans.:* $k = 49.6$ dm³/g mol·min.)

P3-17$_B$ Calculate the equilibrium conversion and concentrations for each of the following reactions.

  (a) The liquid-phase reaction

$$A + B \rightleftharpoons C$$

with $C_{A0} = C_{B0} = 2$ mol/dm^3 and $K_C = 10$ dm^3/mol.

(b) The gas-phase reaction

$$A \rightleftharpoons 3C$$

carried out in a flow reactor with no pressure drop. Pure A enters at a temperature of 400 K and 10 atm. At this temperature, $K_C = 0.25$ dm^3/mol^2.

(c) The gas-phase reaction in part (b) carried out in a constant-volume batch reaction.

(d) The gas-phase reaction in part (b) carried out in a constant-pressure batch reaction.

P3-18$_B$ Consider a *cylindrical batch reactor* that has one end fitted with a frictionless piston attached to a spring (Figure P3-18). The reaction

$$A + B \longrightarrow 8C$$

with the rate expression

$$-r_A = k_1 C_A^2 C_B$$

is taking place in this type of reactor.

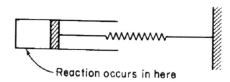

Reaction occurs in here

Figure P3-18

(a) Write the rate law solely as a function of conversion, numerically evaluating all possible symbols. (*Ans.:* $-r_A = 5.03 \times 10^{-9} [(1 - X)^3/(1 + 3X)^{3/2}]$ lb mol/ft^3·s.)

(b) What is the conversion and rate of reaction when $V = 0.2$ ft^3? (*Ans.:* $X = 0.259$, $-r_A = 8.63 \times 10^{-10}$ lb mol/ft^3·s.)

Additional information:

Equal moles of A and B are present at $t = 0$

Initial volume: 0.15 ft^3

Value of k_1: 1.0 (ft^3/lb mol)^2·s^{-1}

The relationship between the volume of the reactor and pressure within the reactor is

$$V = (0.1)(P) \qquad (V \text{ in ft}^3, P \text{ in atm})$$

Temperature of system (considered constant): 140°F

Gas constant: 0.73 ft^3·atm/lb mol·°R

P3-19$_C$ Find the reaction rate parameters (i.e., reaction order, specific reaction rate at one temperature, and the activation energy) for:

(a) Three industrial reactions

(b) Three laboratory reactions

(c) Three reactions discussed in the literature during the last year

P3-20$_C$ For families of reactions, the Polanyi–Semenov equation can be used to estimate activation energies from the heats of reaction, ΔH_R according to the equation

$$E = C - \alpha(-\Delta H_R) \qquad\qquad (P3\text{-}20.1)$$

where α and C are constants. For exothermic reactions $\alpha = -0.25$ and $C = 48$ kJ/mol, while for endothermic reactions $\alpha = -0.75$ and $C = 48$ kJ/mol. However, these values may vary somewhat from reaction family to reaction family [K. J. Laidler, *Theories of Chemical Reaction Rates* (New York, R. E. Krieger, 1979), p. 38]. (Also see Appendix J)

(a) Why is this a reasonable correlation?

Consider the following family of reactions:

	E (kcal/mol)	$-\Delta H_R$ (kcal/mol)
H + RBr $\longrightarrow$ HBr + R	6.8	17.5
H + R'Br $\longrightarrow$ HBr + R'	6.0	20.0

(b) Estimate the activation energy for the reaction

$$CH_3\cdot + RBr \longrightarrow CH_3Br + R\cdot$$

which has an exothermic heat of reaction of 6 kcal/mol (i.e., $\Delta H_R = -6$ kcal/mol).

P3-21$_B$ The gas-phase reaction between chlorine and methane to form carbon tetrachloride and hydrochloric acid is to be carried out at 75°C and at 950 kPa in a continuous-flow reactor. The vapor pressure of carbon tetrachloride at 75°C is approximately 95 kPa. Set up a stoichiometric table for this reaction with phase change. Calculate the conversion of methane at which *condensation* begins. Plot the concentrations and molar flow rates of each species as well as the total molar flow rate as a function of conversion for a stoichiometric feed. The volumetric flow rate is 0.4 dm³/s.

P3-22$_B$ The reaction

$$C_2H_6(g) + 2Br_2(g) \longrightarrow C_2H_4Br_2(g,l) + 2HBr(g)$$

is to be carried out at 200°C and 2500 kPa. The vapor pressure of 1,2-dibromoethane at 200°C is 506.5 kPa. With $k = 0.01$ dm⁶/mol²·min. The reaction is first order in C_2H_5 and second order in Br_2. Calculate the conversion of ethane at which *condensation* begins. Plot the concentration and molar flow rates of each species as well as the total molar flow rate as a function of conversion for a stoichiometric feed. The volumetric flow rate is 0.5 dm³/s. (*Ans.:* $X_{cond} = 0.609$.) Are there a set of feed conditions (e.g., equal molar) such that the concentration of $C_2H_6(g)$ will be constant after condensation begins?

P3-23$_B$ **Chemical vapor deposition (CVD)** is a process used in the microelectronics industry to deposit thin films of constant thickness on silicon wafers. This process is of particular importance in the manufacturing of very large scale integrated circuits. One of the common coatings is Si_3N_4, which is produced according to the reaction

$$3SiH_4(g) + 4NH_3(g) \longrightarrow Si_3N_4(s) + 12H_2(g)$$

This dielectric is typically more resistant to oxidation than other coatings. Set up a stoichiometric table for this reaction and plot the concentration of each species as a function of conversion. The entering pressure is 1 Pa and the temperature is constant at 700°C. The feed is equimolar in NH_3 and SiH_4.

P3-24$_B$ It is proposed to produce ethanol by one of two reactions:

$$C_2H_5Cl + OH^- \longleftrightarrow C_2H_5OH + Cl^- \qquad (1)$$

$$C_2H_5Br + OH^- \longrightarrow C_2H_5OH + Br^- \qquad (2)$$

Use SPARTAN (see Appendix J) or some other software package to answer the following:
  (a) What is the ratio of the rates of reaction at 25°C? 100°C? 500°C?
  (b) Which reaction scheme would you choose to make ethanol? (*Hint*: Consult *Chemical Marketing Reporter* or www.chemweek.com for chemical prices).

[Professor R. Baldwin, Colorado School of Mines]

CD-ROM MATERIAL

- **Learning Resources**
  1. *Summary Notes for Lectures 3 and 4*
  3. *Interactive Computer Modules*
      A. Quiz Show II
  4. *Solved Problems*
      A. CDP3-A$_B$ Activation Energy for a Beetle Pushing a Ball of Dung.
      B. CDP3-B$_B$ Microelectronics Industry and the Stoichiometric Table.
- **FAQ [Frequently Asked Questions]**– In Updates/FAQ icon section
- **Additional Homework Problems**

CDP3-A$_B$ Estimate how fast a Tenebrionid Beetle can push a ball of dung at 41.5°C. **(Solution included.)**

CDP3-B$_B$ Silicon is used in the manufacture of microelectronics devices. Set up a stochiometric table for the reaction **(Solution included.)**

$$Si\,HCl_3(g) + H_2(g) \longrightarrow Si(s) + HCl(g) + Si_xH_gCl_z(g)$$

[2nd Ed. P3-16$_B$]

CDP3-C$_B$ The elementary reaction $A(g) + B(g) \longrightarrow C(g)$ takes place in a square duct containing liquid B, which evaporates into the gas to react with A. [2nd Ed. P3-20$_B$]

CDP3-D$_B$ Condensation occurs in the gas phase reaction

$$CH_4(g) + 2Cl_2(g) \xrightarrow{\ h\nu\ } CH_2Cl_2(g,l) + 2HCl(g)$$

[2nd Ed. P3-17$_B$]

CDP3-E$_B$ Set up a stoichiometric Table for the reaction

$$C_6H_5COCH + 2NH_3 \longrightarrow C_6H_5ONH_2 + NH_2Cl$$

[2nd Ed. P3-10$_B$]

SUPPLEMENTARY READING

1. Two references relating to the discussion of activation energy have already been cited in this chapter. Activation energy is usually discussed in terms of either collision theory or transition-state theory. A concise and readable account of these two theories can be found in

> LAIDLER, K. J. *Chemical Kinetics*. New York: Harper & Row, 1987, Chap. 3.

An expanded but still elementary presentation can be found in

> GARDINER, W. C., *Rates and Mechanism of Chemical Reactions*. New York: W. A. Benjamin, 1969, Chaps. 4 and 5.
>
> MOORE, J. W., and R. G. PEARSON, *Kinetics and Mechanism*, 3rd ed. New York: Wiley, 1981, Chaps. 4 and 5.

A more advanced treatise of activation energies and collision and transition-state theories is

> BENSON, S. W., *The Foundations of Chemical Kinetics*. New York: McGraw-Hill, 1960.
>
> J. I. STEINFELD, J. S. FRANCISCO, W. L. HASE, *Chemical Kinetics and Dynamics*, Prentice Hall, New Jersey: 1989.

2. The books listed above also give the rate laws and activation energies for a number of reactions; in addition, as mentioned earlier in this chapter, an extensive listing of rate laws and activation energies can be found in NBS circulars:

> NATIONAL BUREAU OF STANDARDS, *Tables of Chemical Kinetics: Homogeneous Reactions*. Circular 510, Sept. 28, 1951; Supplement 1, Nov. 14, 1956; Supplement 2, Aug. 5, 1960; Supplement 3, Sept. 15, 1961. Washington, D.C.: U.S. Government Printing Office.

3. Also consult the current chemistry literature for the appropriate algebraic form of the rate law for a given reaction. For example, check the *Journal of Physical Chemistry* in addition to the journals listed in Section 4 of the Supplementary Reading section in Chapter 4.

Isothermal Reactor Design $\qquad$ **4**

Why, a four-year-old child could understand this.
Someone get me a four-year-old child.

Groucho Marx

Tying everything together

In this chapter we bring all the material in the preceding three chapters together to arrive at a logical structure for the design of various types of reactors. By using this structure, one should be able to solve reactor engineering problems through reasoning rather than memorization of numerous equations together with the various restrictions and conditions under which each equation applies (i.e., whether there is a change in the total number of moles, etc.). In perhaps no other area of engineering is mere formula plugging more hazardous; the number of physical situations that can arise appears infinite, and the chances of a simple formula being sufficient for the adequate design of a real reactor are vanishingly small.

This chapter focuses attention on reactors that are operated isothermally. We begin by studying a liquid-phase batch reactor to determine the specific reaction rate constant needed for the design of a CSTR. After illustrating the design of a CSTR from batch reaction rate data, we carry out the design of a tubular reactor for a gas-phase pyrolysis reaction. This is followed by a discussion of pressure drop in packed-bed reactors, equilibrium conversion, and finally, the principles of unsteady operation and semibatch reactors.

4.1 Design Structure for Isothermal Reactors

The following procedure is presented as a pathway for one to follow in the design of isothermal (and in some cases nonisothermal) reactors. It is the author's experience that following this structure, shown in Figure 4-1, will lead to a greater understanding of isothermal reactor design. We begin by applying

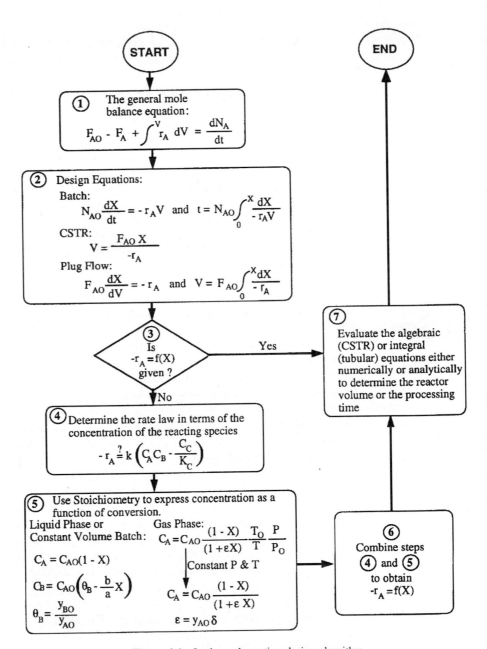

Figure 4-1 Isothermal reaction design algorithm.

our general mole balance equation (level 1) to a specific reactor to arrive at the design equation for that reactor (level 2). If the feed conditions are specified (e.g., N_{A0} or F_{A0}), all that is required to evaluate the design equation is the rate of reaction as a function of conversion at the same conditions as those at which the reactor is to be operated (e.g., temperature and pressure). When $-r_A = f(X)$ is given, one can go directly from level 3 to level 7 to determine either the time or reactor volume necessary to achieve the specified conversion.

Use the algorithm
rather than
memorizing
equations

If the rate of reaction is not given explicitly as a function of conversion, the rate law must be determined (level 4) by either finding it in books or journals or by determining it experimentally in the laboratory. Techniques for obtaining and analyzing rate data to determine the reaction order and rate constant are presented in Chapter 5. After the rate law has been established, one has only to use stoichiometry (level 5) together with the conditions of the system (e.g., constant volume, temperature) to express concentration as a function of conversion. By combining the information in levels 4 and 5, one can express the rate of reaction as a function of conversion and arrive at level 6. It is now possible to determine either the time or reactor volume necessary to achieve the desired conversion by substituting the relationship relating conversion and rate of reaction into the appropriate design equation. The design equation is then evaluated in the appropriate manner (i.e., analytically using a table of integrals, or numerically using an ODE solver). Although this structure emphasizes the determination of a reaction time or volume for a specified conversion, it can also readily be used for other types of reactor calculations, such as determining the conversion for a specified volume. Different manipulations can be performed in level 7 to answer the types of questions mentioned here.

The structure shown in Figure 4-1 allows one to develop a few basic concepts and then to arrange the parameters (equations) associated with each concept in a variety of ways. Without such a structure, one is faced with the possibility of choosing or perhaps memorizing the correct equation from a *multitude of equations* that can arise for a variety of different reactions, reactors, and sets of conditions. The challenge is to put everything together in an orderly and logical fashion so that we can proceed to arrive at the correct equation for a given situation.

Fortunately, by using an algorithm to formulate CRE problems, which happens to be analogous to ordering dinner from a fixed-price menu in a fine French restaurant, we can eliminate virtually all memorization. In both of these algorithms we must make choices in each category. For example, in ordering from a French menu, we begin by choosing one dish from the *appetizers* listed. Step 1 in the analog in CRE is to begin by choosing the mole balance for one of the three types of reactors shown. In step 2 we choose the rate law (*entrée*), and in step 3 we specify whether the reaction is gas *or* liquid phase (*cheese* or *dessert*). Finally, in step 4 we combine steps 1, 2, and 3 and obtain an analytical solution or solve the equations using an ordinary differential equation (ODE) solver. (See complete French menu on the CD-ROM)

We now will apply this algorithm to a specific situation. The first step is to derive or apply the mole balance equation for the system at hand. Suppose that we have, as shown in Figure 4-2, mole balances for three reactors, three

French
Menu
Analogy

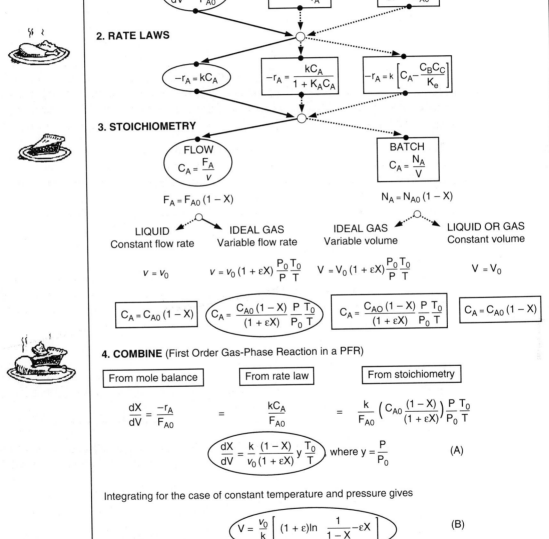

Figure 4-2 Algorithm for isothermal reactors.

rate laws, and the equations for concentrations for both liquid and gas phases. In Figure 4-2 the algorithm is used to formulate the equation to calculate the *PFR reactor volume for a first-order gas-phase reaction* and the pathway to arrive at this equation is shown by the ovals connected to the dark lines through the algorithm. The dashed lines and the boxes represent other pathways for solutions to other situations. For the reactor and reaction specified we will choose

1. the **mole balance** on species A for a PFR,
2. the **rate law** for an irreversible first-order reaction,
3. the equation for the concentration of species A in the gas phase (**stoichiometry**), and then
4. **combine** to calculate the volume necessary to achieve a given conversion or calculate the conversion that can be achieved in a specified reaction volume.

For the case of isothermal operation with no pressure drop, we were able to obtain an analytical solution, given by equation B, which gives the reactor volume necessary to achieve a conversion X for a gas-phase reaction carried out isothermally in a PFR. However, in the majority of situations, analytical solutions to the ordinary differential equations appearing in the combine step are not possible. Consequently, we include POLYMATH, or some other ODE solver such as MATLAB, in our menu in that it makes obtaining solutions to the differential equations much more palatable.

4.2 Scale-Up of Liquid-Phase Batch Reactor Data to the Design of a CSTR

One of the jobs in which chemical engineers are involved is the scale-up of laboratory experiments to pilot-plant operation or to full-scale production. In the past, a pilot plant would be designed based on laboratory data. However, owing to the high cost of a pilot-plant study, this step is beginning to be surpassed in many instances by designing a full-scale plant from the operation of a laboratory-bench-scale unit called a *microplant*. To make this jump successfully requires a thorough understanding of the chemical kinetics and transport limitations. In this section we show how to analyze a laboratory-scale batch reactor in which a liquid-phase reaction of known order is being carried out. After determining the specific reaction rate, k, from a batch experiment, we use it in the design of a full-scale flow reactor.

4.2.1 Batch Operation

In modeling a batch reactor, we have assumed that there is no inflow or outflow of material and that the reactor is well mixed. For most liquid-phase reactions, the density change with reaction is usually small and can be neglected (i.e., $V = V_0$). In addition, for *gas phases* in which the batch reactor volume remains constant, we also have $V = V_0$. Consequently, for constant-volume ($V = V_0$) (e.g., closed metal vessels) batch reactors the mole balance

$$\frac{1}{V}\left(\frac{dN_A}{dt}\right) = r_A \tag{4-1}$$

can be written in terms of concentration.

$$\frac{1}{V}\frac{dN_A}{dt} = \frac{1}{V_0}\frac{dN_A}{dt} = \frac{d(N_A/V_0)}{dt} = \frac{dC_A}{dt} = r_A \tag{4-2}$$

Generally, when analyzing laboratory experiments it is best to process the data in terms of the measured variable. Since concentration is the measured variable for most liquid-phase reactions, the general mole balance equation applied to reactions in which there is no volume change becomes

Mole balance
$$-\frac{dC_A}{dt} = -r_A$$

We consider the reaction

$$A \longrightarrow B$$

which is irreversible and second order in A. The rate at which A is being consumed is given by the rate law

Rate law
$$-r_A = kC_A^2 \tag{4-3}$$

We **combine** the **rate law** and the **mole balance** to obtain

$$-\frac{dC_A}{dt} = kC_A^2$$

$$\hspace{10cm} (4-4)$$

$$-\frac{dC_A}{kC_A^2} = dt$$

Initially, $C_A = C_{A0}$ at $t = 0$. If the reaction is carried out isothermally, we can integrate this equation to obtain the reactant concentration at any time t:

$$\frac{-1}{k}\int_{C_{A0}}^{C_A}\frac{dC_A}{C_A^2} = \int_0^t dt$$

Second-order,
isothermal,
liquid-phase
batch reaction

$$\boxed{\frac{1}{k}\left(\frac{1}{C_A} - \frac{1}{C_{A0}}\right) = t} \tag{4-5}$$

This time is the time t needed to reduce the reactant concentration in a batch reactor from an initial value C_{A0} to some specified value C_A.

The total cycle time in any batch operation is considerably longer than the reaction time, t_R, as one must account for the time necessary to fill (t_f) and empty (t_e) the reactor together with the time necessary to clean the reactor between batches, t_c. In some cases the reaction time calculated from Equation (4-5) may be only a small fraction of the total cycle time, t_t.

$$t_t = t_f + t_e + t_c + t_R$$

Typical cycle times for a batch polymerization process are shown in Table 4-1. Batch polymerization reaction times may vary between 5 and 60 h. Clearly, decreasing the reaction time with a 60-h reaction is a critical problem. As the reaction time is reduced, it becomes important to use large lines and pumps to achieve rapid transfers and to utilize efficient sequencing to minimize the cycle time.

TABLE 4-1.   TYPICAL CYCLE TIMES FOR A BATCH
POLYMERIZATION PROCESS

Activity	Time (h)
1. Charge feed to the reactor and agitate, t_f	1.5–3.0
2. Heat to reaction temperature, t_e	1.0–2.0
3. Carry out reaction, t_R	(varies)
4. Empty and clean reactor, t_c	0.5–1.0
Total time excluding reaction	3.0–6.0

Batch operation times

It is important to have a grasp of the order of magnitude of batch reaction times, t_R, in Table 4-1 to achieve a given conversion, say 90%, for the different values of the specific reaction rate, k. We can obtain these estimates by considering the irreversible reaction

$$A \longrightarrow B$$

carried out in a constant-volume batch reactor for a first- and a second-order reaction. We start with a mole balance and then follow our algorithm as shown in Table 4-2.

TABLE 4-2.   ALGORITHM TO ESTIMATE REACTION TIMES

Mole balance		$\dfrac{dX}{dt} = \dfrac{-r_A}{N_{A0}} V$	
Rate law	First-order		Second-order
	$-r_A = kC_A$		$-r_A = kC_A^2$
Stoichiometry $(V = V_0)$		$C_A = \dfrac{N_A}{V_0} = C_{A0}(1-X)$	
Combine	$\dfrac{dX}{dt} = k(1-X)$		$\dfrac{dX}{dt} = kC_{A0}(1-X)^2$
Integrate	$t = \dfrac{1}{k} \ln \dfrac{1}{1-X}$		$t = \dfrac{X}{kC_{A0}(1-X)}$

For first-order reactions the reaction time to reach 90% conversion (i.e., $X = 0.9$) in a constant-volume batch reactor scales as

$$t_R = \frac{1}{k} \ln \frac{1}{1-X} = \frac{1}{k} \ln \frac{1}{1-0.9} = \frac{2.3}{k}$$

If $k = 10^{-4}\ s^{-1}$,

$$t_R = \frac{2.3}{10^{-4}\ s^{-1}} = 23,000\ s = 6.4\ h$$

The time necessary to achieve 90% conversion in a batch reactor for an irreversible first-order reaction in which the specific reaction rate is $10^{-4}\ s^{-1}$ is 6.4 h.
    For second-order reactions, we have

$$t_R = \frac{1}{kC_{A0}} \frac{X}{1-X} = \frac{0.9}{kC_{A0}(1-0.9)} = \frac{9}{kC_{A0}}$$

If $kC_{A0} = 10^{-3}\ s^{-1}$,

$$t_R = \frac{9}{10^{-3}\ s^{-1}} = 9000\ s = 2.5\ h$$

Table 4-3 gives the *order of magnitude* of time to achieve 90% conversion for first- and second-order irreversible batch reactions.

Estimating Reaction
Times

TABLE 4-3.   BATCH REACTION TIMES

Reaction Time t_R	First-Order k (s^{-1})	Second-Order kC_{A0} (s^{-1})
Hours	10^{-4}	10^{-3}
Minutes	10^{-2}	10^{-1}
Seconds	1	10
Milliseconds	1000	10,000

Example 4–1 Determining k from Batch Data

It is desired to design a CSTR to produce 200 million pounds of ethylene glycol per year by hydrolyzing ethylene oxide. However, before the design can be carried out, it is necessary to perform and analyze a batch reactor experiment to determine the specific reaction rate constant. Since the reaction will be carried out isothermally, the specific reaction rate will need to be determined only at the reaction temperature of the CSTR. At high temperatures there is a significant by-product formation, while at temperatures below 40°C the reaction does not proceed at a significant rate; consequently, a temperature of 55°C has been chosen. Since the water is usually present in excess, its concentration may be considered constant during the course of the reaction. The reaction is first-order in ethylene oxide.

$$\underset{A}{CH_2\!-\!CH_2} + \underset{+\ B}{H_2O} \xrightarrow[\text{catalyst}]{H_2SO_4} \underset{C}{\overset{CH_2\!-\!OH}{\underset{CH_2\!-\!OH}{|}}}$$

In the laboratory experiment, 500 mL of a 2 M solution (2 kmol/m^3) of eth- ylene oxide in water was mixed with 500 mL of water containing 0.9 wt % sulfuric acid, which is a catalyst. The temperature was maintained at 55°C. The concentra- tion of ethylene glycol was recorded as a function of time (Table E4-1.1). From these data, determine the specific reaction rate at 55°C.

TABLE E4-1.1.   CONCENTRATION-TIME DATA

Time (min)	Concentration of Ethylene Glycol (kmol/m^3)[a]
0.0	0.000
0.5	0.145
1.0	0.270
1.5	0.376
2.0	0.467
3.0	0.610
4.0	0.715
6.0	0.848
10.0	0.957

[a]1 kmol/m^3 = 1 mol/dm^3 = 1 mol/L.

Check 10 types of homework problems on the CD-ROM for more solved examples using this algorithm.

In this example we use the problem-solving algorithm (A through G) that is given in the CD-ROM and on the web "http://www.engin.umich.edu/~problemsolving". You may wish to follow this algorithm in solving the other examples in this chapter and the problems given at the end of the chapter. However, to conserve space it will not be repeated for other example problems.

A. *Problem statement.* Determine the specific reaction rate, k_A.

B. *Sketch*:

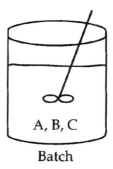

A, B, C

Batch

C. *Identify*:

　　C1. Relevant theories

　　　　Rate law: $-r_A = k_A C_A$

　　　　Mole balance: $\dfrac{dN_A}{dt} = r_A V$

C2. Variables

   Dependent: concentrations

   Independent: time

C3. Knowns and unknowns

   Knowns: concentration of ethylene glycol as a function of time

   Unknowns:

   1. Concentration of ethylene oxide as a function of time

   2. Specific reaction rate

   3. Reactor volume

C4. Inputs and outputs: reactant fed all at once to a batch reactor

C5. Missing information: None; does not appear that other sources need to be sought.

D. *Assumptions and approximations*:

Assumptions

   1. Well mixed

   2. All reactants enter at the same time

   3. No side reactions

   4. Negligible filling and emptying time

   5. Isothermal operation

Approximations

   1. Water in excess so that its concentration is essentially constant.

E. *Specification.* The problem is neither overspecified nor underspecified.

F. *Related material.* This problem uses the mole balances developed in Chapter 1 for a batch reactor and the stoichiometry and rate laws developed in Chapter 3.

G. *Use an algorithm.* For an isothermal reaction, use the chemical reaction engineering algorithm shown in Figure 4-2.

Following the
algorithm

Solution

1. A **mole balance** on a batch reactor that is well mixed is

$$\frac{1}{V}\frac{dN_A}{dt} = r_A \qquad (E4\text{-}1.1)$$

2. The **rate law** is

$$-r_A = kC_A \qquad (E4\text{-}1.2)$$

Since water is present in such excess, the concentration of water at any time t is virtually the same as the initial concentration and the rate law is independent of the concentration of H_2O. ($C_B \cong C_{B0}$.)

Rate Law

3. **Stoichiometry.** Liquid phase, no volume change, $V = V_0$ (Table E4-1.2):

TABLE E4-1.2.   STOICHIOMETRIC TABLE

Species	Symbol	Initial	Change	Remaining	Concentration
CH_2CH_2O	A	N_{A0}	$-N_{A0}X$	$N_A = N_{A0}(1-X)$	$C_A = C_{A0}(1-X)$
H_2O	B	$\Theta_B N_{A0}$	$-N_{A0}X$	$N_B = N_{A0}(\Theta_B - X)$	$C_B = C_{A0}(\Theta_B - X)$
					$C_B \approx C_{A0}\Theta_B = C_{B0}$
$(CH_2OH)_2$	C	0	$N_{A0}X$	$N_C = N_{A0}X$	$C_C = C_{A0}X$
		N_{T0}		$N_T = N_{T0} - N_{A0}X$	

Stoichiometric table for constant volume

$$C_A = \frac{N_A}{V} = \frac{N_A}{V_0}$$

$$\frac{1}{V_0}\left(\frac{dN_A}{dt}\right) = \frac{d(N_A/V_0)}{dt} = \frac{dC_A}{dt}$$

4. **Combining** the rate law and the mole balance, we have

$$-\frac{dC_A}{dt} = kC_A \tag{E4-1.3}$$

For isothermal operation we can integrate this equation,

$$-\int_{C_{A0}}^{C_A}\frac{dC_A}{C_A} = \int_0^t k\,dt$$

Combining mole balance, rate law, and stoichiometry

using the initial condition that when $t = 0$, then $C_A = C_{A0}$. The initial concentration of A after mixing the two volumes together is 1.0 kmol/m³ (1 mol/L).

5. **Integrating** yields

$$\ln\frac{C_{A0}}{C_A} = kt \tag{E4-1.4}$$

The concentration of ethylene oxide at any time t is

$$C_A = C_{A0}e^{-kt} \tag{E4-1.5}$$

The concentration of ethylene glycol at any time t can be obtained from the reaction stoichiometry:

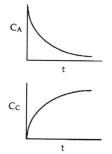

$$A + B \longrightarrow C$$
$$N_C = N_{A0}X = N_{A0} - N_A$$

For liquid-phase reactions $V = V_0$,

$$C_C = \frac{N_C}{V} = \frac{N_C}{V_0} = C_{A0} - C_A = C_{A0}(1 - e^{-kt}) \tag{E4-1.6}$$

Rearranging and taking the logarithm of both sides yields

$$\ln \frac{C_{A0} - C_C}{C_{A0}} = -kt \qquad \text{(E4-1.7)}$$

We see that a plot of $\ln[(C_{A0} - C_C)/C_{A0}]$ as a function of t will be a straight line with a slope $-k$. Calculating the quantity $(C_{A0} - C_C)/C_{A0}$ (Table E4-1.3) and then plot-

TABLE E4-1.3

t (min)	C_C (kmol/m³)	$\dfrac{C_{A0} - C_C}{C_{A0}}$
0.0	0.000	1.000
0.5	0.145	0.855
1.0	0.270	0.730
1.5	0.376	0.624
2.0	0.467	0.533
3.0	0.610	0.390
4.0	0.715	0.285
6.0	0.848	0.152
10.0	0.957	0.043

Evaluating the specific reaction rate from *batch reactor* concentration–time data

ting $(C_{A0} - C_C)/C_{A0}$ versus t on semilogarithmic paper is shown in Figure E4-1.1. The slope of this plot is also equal to $-k$. Using the decade method (see Appendix D) between $(C_{A0} - C_C)/C_{A0} = 0.6$ ($t = 1.55$ min) and $(C_{A0} - C_C)/C_{A0} = 0.06$ ($t = 8.95$ min) to evaluate the slope

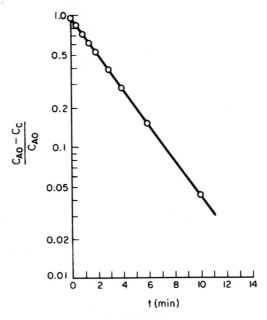

Figure E4-1.1

$$k = \frac{\ln 10}{t_2 - t_1} = \frac{2.3}{8.95 - 1.55} = 0.311 \text{ min}^{-1} \qquad \text{(E4-1.8)}$$

the rate law becomes

$$\boxed{-r_A = (0.311 \text{ min}^{-1})C_A} \qquad \text{(E4-1.9)}$$

This rate law can now be used in the design of an industrial CSTR.

4.2.2 Design of CSTRs

In Chapter 2 we derived the following design equation for a CSTR:

Mole balance

$$V = \frac{F_{A0}X}{(-r_A)_{\text{exit}}} \qquad \text{(2-13)}$$

which gives the volume V necessary to achieve a conversion X. When the volumetric flow rate does not change with reaction, (i.e., $v = v_0$) we can write

$$V = v_0\left(\frac{C_{A0} - C_A}{-r_A}\right) \qquad \text{(4-6)}$$

or in terms of the space time,

$$\tau = \frac{V}{v_0} = \frac{C_{A0} - C_A}{-r_A} \qquad \text{(4-7)}$$

For a first-order irreversible reaction, the rate law is

Rate law

$$-r_A = kC_A$$

We can combine the rate law and mole balance to give

Combine

$$\tau = \frac{C_{A0} - C_A}{kC_A}$$

Solving for the effluent concentration of A, C_A, we obtain

$$C_A = \frac{C_{A0}}{1 + \tau k} \qquad \text{(4-8)}$$

For the case we are considering, there is no volume change during the course of the reaction, so we can use Equation (3-29),

$$C_A = C_{A0}(1 - X) \qquad \text{(3-29)}$$

Relationship between space time and conversion for a first-order liquid-phase reaction

to combine with Equation (4-8) to give

$$X = \frac{\tau k}{1 + \tau k} \qquad \text{(4-9)}$$

$$Da = \frac{-r_{A0}V}{F_{A0}}$$

For a first-order reaction the product τk is often referred to as the reaction **Damköhler number**.

The Damköhler is a dimensionless number that can give us a quick estimate of the degree of conversion that can be achieved in continuous-flow reactions. The Damköhler number is the ratio of the rate of reaction of A to the rate of convective transport of A at the entrance to the reactor. For first- and second-order irreversible reactions the Damköhler numbers are

$$Da = \frac{-r_{A0}V}{F_{A0}} = \frac{kC_{A0}V}{v_0 C_{A0}} = \tau k$$

and

$$Da = \frac{kC_{A0}^2 V}{v_0 C_{A0}} = \tau k C_{A0}$$

respectively. It is important to know what values of the Damköhler number, Da, give high and low conversion in continuous-flow reactors. A value of Da = 0.1 or less will usually give less than 10% conversion and a value of

$$0.1 < Da < 10$$

Da = 10.0 or greater will usually give greater than 90% conversion.

CSTRs in Series.   A first-order reaction with no volume change ($v = v_0$) is to be carried out in two CSTRs placed in series (Figure 4-3). The effluent con-

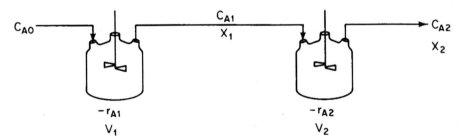

Figure 4-3   Two CSTRs in series.

centration of A from reactor 1 is

$$C_{A1} = \frac{C_{A0}}{1 + \tau_1 k_1}$$

From a mole balance on reactor 2,

$$V_2 = \frac{F_{A1} - F_{A2}}{-r_{A2}} = \frac{v_0(C_{A1} - C_{A2})}{k_2 C_{A2}}$$

Solving for C_{A2}, the concentration exiting the second reactor, we get

1st order reaction
$$C_{A2} = \frac{C_{A1}}{1 + \tau_2 k_2} = \frac{C_{A0}}{(1 + \tau_2 k_2)(1 + \tau_1 k_1)}$$

If instead of two CSTRs in series we had n equal-sized CSTRs connected in series ($\tau_1 = \tau_2 = \cdots = \tau_n = \tau$) operating at the same temperature ($k_1 = k_2 = \cdots = k_n = k$), the concentration leaving the last reactor would be

$$C_{A_n} = \frac{C_{A0}}{(1 + \tau k)^n} = \frac{C_{A0}}{(1 + Da)^n} \qquad (4\text{-}10)$$

The conversion for these n tank reactors in series would be

Conversion as a function of the number of tanks in series

$$\boxed{X = 1 - \frac{1}{(1 + \tau k)^n}} \qquad (4\text{-}11)$$

A plot of the conversion as a function of the number of reactors in series for a first-order reaction is shown in Figure 4-4 for various values of the Damköhler

CSTRs in series

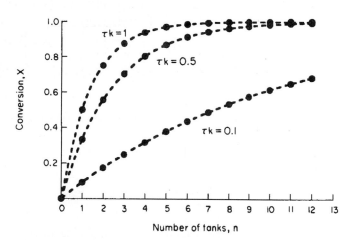

Figure 4-4 Conversion as a function of the number of tanks in series for different Damköhler numbers for a first-order reaction.

number τk. Observe from Figure 4-4 that when the product of the space time and the specific reaction rate is relatively large, say, Da ≥ 1, approximately 90% conversion is achieved in two or three reactors; thus the cost of adding subsequent reactors might not be justified. When the product τk is small, Da ~ 0.1, the conversion continues to increase significantly with each reactor added.

The rate of disappearance of A in the nth reactor is

$$-r_{A_n} = kC_{A_n} = k\frac{C_{A0}}{(1 + \tau k)^n}$$

CSTRs in Parallel. We now consider the case in which equal-sized reactors are placed in parallel rather than in series, and the feed is distributed equally among each of the reactors (Figure 4-5). A balance on any reactor, say i, gives

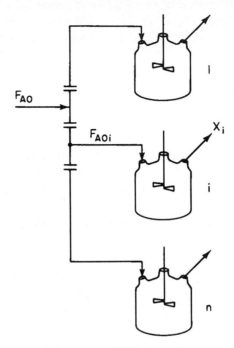

Figure 4-5 CSTRs in parallel.

the individual reactor volume

$$V_i = F_{A0i}\left(\frac{X_i}{-r_{Ai}}\right) \qquad (4\text{-}12)$$

Since the reactors are of equal size, operate at the same temperature, and have identical feed rates, the conversion will be the same for each reaction:

$$X_1 = X_2 = \cdots = X_n = X$$

as will be the rate of reaction in each reactor

$$-r_{A1} = -r_{A2} = \cdots = -r_{An} = -r_A$$

The volume of each individual reactor, V_i, is related to the total volume, V, of all the reactors by the equation

$$V_i = \frac{V}{n}$$

A similar relationship exists for the total molar flow rate:

$$F_{A0i} = \frac{F_{A0}}{n}$$

Substituting these values into Equation (4-12) yields

$$\frac{V}{n} = \frac{F_{A0}}{n}\left(\frac{X_i}{-r_{A_i}}\right)$$

or

Conversion for
tanks in parallel

$$V = \frac{F_{A0}X_i}{-r_{Ai}} = \frac{F_{A0}X}{-r_A} \qquad (4\text{-}13)$$

This result shows that the conversion achieved in any one of the reactors in parallel is identical to what would be achieved if the reactant were fed in one stream to one large reactor of volume V!

A Second-Order Reaction in a CSTR.   For a second-order liquid-phase reaction being carried out in a CSTR, the **combination** of the **rate law** and the **design equation** yields

$$V = \frac{F_{A0}X}{kC_A^2} \qquad (4\text{-}14)$$

For constant density $v = v_0$, $F_{A0}X = v_0(C_{A0} - C_A)$, then

$$\tau = \frac{V}{v_0} = \frac{C_{A0} - C_A}{kC_A^2}$$

Using our definition of conversion, we have

$$\tau = \frac{X}{kC_{A0}(1 - X)^2} \qquad (4\text{-}15)$$

We solve Equation (4-15) for the conversion X:

Conversion for
a second-order
liquid-phase
reaction
in a CSTR

$$X = \frac{(1 + 2\tau k C_{A0}) - \sqrt{(1 + 2\tau k C_{A0})^2 - (2\tau k C_{A0})^2}}{2\tau k C_{A0}}$$

$$= \frac{(1 + 2\tau k C_{A0}) - \sqrt{1 + 4\tau k C_{A0}}}{2\tau k C_{A0}} \qquad (4\text{-}16)$$

$$= \frac{(1 + 2\mathrm{Da}) - \sqrt{1 + 4\mathrm{Da}}}{2\mathrm{Da}}$$

The minus sign must be chosen in the quadratic equation because X cannot be greater than 1. Conversion is plotted as a function of the Damköhler parameter, $\tau k C_{A0}$, in Figure 4-6. Observe from this figure that at high conver-

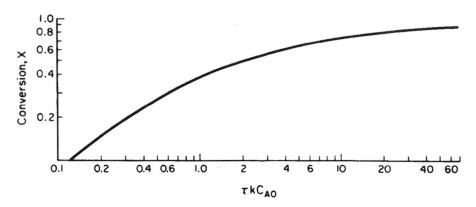

Figure 4-6   Conversion as a function of the Damköhler number ($\tau k C_{A0}$) for a second order reaction in a CSTR.

sions (say 70%) a 10-fold increase in the reactor volume (or increase in the specific reaction rate by raising the temperature) will increase the conversion only to 85%. This is a consequence of the fact that the CSTR operates under the condition of the lowest value of the reactant concentration (i.e., the exit concentration), and consequently the smallest value of the rate of reaction.

Example 4–2   Producing 200 Million Pounds per Year in a CSTR

Close to 5.2 billion pounds of ethylene glycol were produced in 1995, which ranked it the twenty-sixth most produced chemical in the nation that year on a total pound basis. About one-half of the ethylene glycol is used for *antifreeze* while the other half is used in the manufacture of polyesters. In the polyester category, 88% was used for fibers and 12% for the manufacture of bottles and films. The 1997 selling price for ethylene glycol was $0.38 per pound.

It is desired to produce 200 million pounds per year of ethylene glycol. The reactor is to be operated isothermally. A 1 lb mol/ft³ solution of ethylene oxide in water is fed to the reactor together with an equal volumetric solution of water containing 0.9 wt % of the catalyst H_2SO_4. If 80% conversion is to be achieved, determine the necessary reactor volume. How many 800-gal reactors would be required if they are arranged in parallel? What is the corresponding conversion? How many 800-gal reactors would be required if they are arranged in series? What is the corresponding conversion? The specific reaction rate constant is 0.311 min^{-1}, as determined in Example 4-1.

Solution

Assumption: Ethylene glycol is the only reaction product formed.

$$\underset{A}{\underset{\text{CH}_2\text{—CH}_2}{\overset{\text{O}}{\triangle}}} + \underset{B}{H_2O} \xrightarrow[\text{catalyst}]{H_2SO_4} \underset{C}{\overset{\text{CH}_2\text{—OH}}{\underset{\text{CH}_2\text{—OH}}{|}}}$$

The specified production rate in lb mol/min is

$$F_C = 2 \times 10^8 \frac{\text{lb}}{\text{yr}} \times \frac{1 \text{ yr}}{365 \text{ days}} \times \frac{1 \text{ day}}{24 \text{ h}} \times \frac{1 \text{ h}}{60 \text{ min}} \times \frac{1 \text{lb mol}}{62 \text{ lb}} = 6.137 \frac{\text{lb mol}}{\text{min}}$$

From the reaction stoichiometry

$$F_C = F_{A0} X$$

we find the required molar flow rate of ethylene oxide to be

$$F_{A0} = \frac{F_C}{X} = \frac{6.137}{0.8} = 7.67 \frac{\text{lb mol}}{\text{min}} \quad (58.0 \text{ g mol/s})$$

We can now calculate the reactor volume using the following equations:

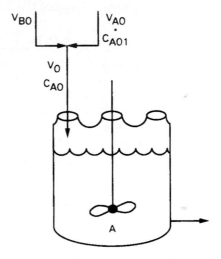

Figure E4-2.1   Single CSTR.

Following the algorithm

1. **Design equation**:

$$V = \frac{F_{A0} X}{-r_A} \tag{E4-2.1}$$

2. **Rate law**:

$$-r_A = kC_A \tag{E4-2.2}$$

3. **Stoichiometry**. Liquid phase $(v = v_0)$:

$$C_A = C_{A0}(1 - X) \tag{E4-2.3}$$

4. **Combining**:

$$V = \frac{F_{A0} X}{kC_{A0}(1 - X)} = \frac{v_0 X}{k(1 - X)} \tag{E4-2.4}$$

The entering volumetric flow rate of stream A, with $C_{A01} = 1 \cdot \dfrac{mol}{dm^3}$ before mixing, is

$$v_{A0} = \frac{F_{A0}}{C_{A01}} = \frac{7.67 \text{ lb mol/min}}{1 \text{ lb mol/ft}^3} = 7.67 \frac{ft^3}{min}$$

From the problem statement

$$v_{B0} = v_{A0}$$

The total entering volumetric flow rate of liquid is

$$v_0 = v_{A0} + v_{B0} = 15.34 \frac{ft^3}{min} \quad (7.24 \text{ dm}^3/\text{s})$$

5. **Substituting** in Equation (E4-2.4), recalling that $k = 0.311 \text{ min}^{-1}$, yields

$$V = \frac{v_0 X}{k(1-X)} = 15.34 \frac{ft^3}{min} \frac{0.8}{(0.311 \text{ min}^{-1})(1-0.8)} = 197.3 \text{ ft}^3$$
$$= 1480 \text{ gal } (5.6 \text{ m}^3)$$

A tank 5 ft in diameter and approximately 10 ft tall is necessary to achieve 80% conversion.

6. **CSTRs in parallel.** For two 800-gal CSTRs arranged in parallel with 7.67 ft³/min ($v_0/2$) fed to each reactor, the conversion achieved can be calculated from

$$X = \frac{\tau k}{1 + \tau k} \qquad\qquad \text{(E4-2.5)}$$

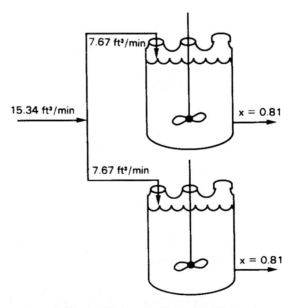

Figure E4-2.2 CSTRs in parallel.

where

$$\tau = \frac{V}{v_0/2} = \left(800 \text{ gal} \times \frac{1 \text{ ft}^3}{7.48 \text{ gal}}\right) \times \frac{1}{7.67 \text{ ft}^3/\text{min}} = 13.94 \text{ min}$$

The **Damköhler number** is

$$\text{Da} = \tau k = 13.94 \text{ min} \times \frac{0.311}{\text{min}} = 4.34$$

Substituting into Equation (E4-2.5) gives us

$$X = \frac{4.34}{1 + 4.34} = 0.81$$

7. **CSTRs in series.** If the 800-gal reactors are arranged in series, the conversion in the first reactor is

$$X_1 = \frac{\tau_1 k}{1 + \tau_1 k} \tag{E4-2.6}$$

where

$$\tau = \frac{V_1}{v_{01}} = \left(800 \text{ gal} \times \frac{1 \text{ ft}^3}{7.48 \text{ gal}}\right) \times \frac{1}{15.34 \text{ ft}^3/\text{min}} = 6.97 \text{ min}$$

The Damköhler number is

$$\text{Da}_1 = \tau_1 k = 6.97 \text{ min} \times \frac{0.311}{\text{min}} = 2.167$$

$$X_1 = \frac{2.167}{3.167} = 0.684$$

To calculate the conversion exiting the second reactor, we recall that $V_1 = V_2 = V$ and $v_{01} = v_{02} = v_0$; then

$$\tau_1 = \tau_2 = \tau$$

Combining the mole balance on the second reactor with the rate law, we obtain

$$V = \frac{F_{A0}(X_2 - X_1)}{-r_{A2}} = \frac{C_{A0}v_0(X_2 - X_1)}{kC_{A0}(1 - X_2)} = \frac{v_0}{k}\left(\frac{X_2 - X_1}{1 - X_2}\right) \tag{E4-2.7}$$

Solving for the conversion exiting the second reactor yields

$$X_2 = \frac{X_1 + \tau k}{1 + \tau k} = \frac{0.684 + 2.167}{1 + 2.167} = 0.90$$

The same result could have been obtained from Equation (4-11):

$$X = 1 - \frac{1}{(1 + \tau k)^n} = 1 - \frac{1}{(1 + 2.167)^2} = 0.90$$

Two million pounds of ethylene glycol per year can be produced using two 800-gal (3.0-m³) reactors in series.

Conversion in the series arrangement is greater than in parallel for CSTRs. From our discussion of reactor staging in Chapter 2, we could have predicted that the series arrangement would have given the higher conversion.

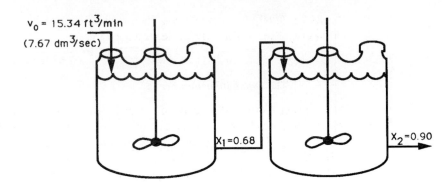

Figure E4-2.3 CSTRs in series.

Two equal-sized CSTRs in series will give a higher conversion than two CSTRs in parallel of the same size when the reaction order is greater than zero.

We can find information about the safety of ethylene glycol and other chemicals from the World Wide Web (WWW) (Table 4-4). One source is the Vermont Safety information on the Internet (Vermont SIRI). For example, we can learn from the *Control Measures* that we should use neoprene gloves when handling the material, and that we should avoid breathing the vapors. If we

TABLE 4-4.   ACCESSING SAFETY INFORMATION

Safety Information
MSDS

1. Type in
    http://www.siri.org/
2. When the first screen appears, click on "Material Safety Data Sheets." ("MSDS")
3. When the next page appears, type in the chemical you want to find.
    Example: Find ethylene glycol
    Then click on Enter.
4. The next page will show a list of a number of companies that provide the data on ethylene glycol.
    EXCITON CHEMICAL
    KALAMA INTERNATIONAL
    DOW CHEMICAL USA
    etc.
    Let's click on "EXCITON." The materials safety data sheet provided by EXCITON will appear.
5. Scroll "ethylene glycol" for information you desire.
    General Information
    Ingredients Information
    Physical/Chemical Characteristics
    Fire and Explosion Hazard Data
    Reactivity Data
    Health Hazard Data
    Control Measures
    Transportation Data
    Disposal Data
    Label Data

click on "Dow Chemical USA" and scroll the *Reactivity Data*, we would find that ethylene glycol will ignite in air at 413°C.

4.3 Tubular Reactors

Gas-phase reactions are carried out primarily in tubular reactors where the flow is generally turbulent. By assuming that there is no dispersion and there are no radial gradients in either temperature, velocity, or concentration, we can model the flow in the reactor as plug-flow. Laminar reactors are discussed in Chapter 13 and dispersion effects in Chapter 14. The differential form of the design equation

PFR mole
balance

$$F_{A0} \frac{dX}{dV} = -r_A$$

must be used when there is a pressure drop in the reactor or heat exchange between the PFR and the surroundings. In the absence of pressure drop or heat exchange the integral form of the *plug flow design* equation is used,

PFR design
equation

$$V = F_{A0} \int_0^X \frac{dX}{-r_A}$$

Substituting the rate law for the special case of a second-order reaction gives us

Rate law

$$V = F_{A0} \int_0^X \frac{dX}{kC_A^2}$$

For constant-temperature and constant-pressure gas-phase reactions, the concentration is expressed as a function of conversion:

Stoichiometry
(gas-phase)

$$C_A = \frac{F_A}{v} = \frac{F_A}{v_0(1+\varepsilon X)} = \frac{F_{A0}\,(1-X)}{v_0\,(1+\varepsilon X)} = C_{A0}\frac{(1-X)}{(1+\varepsilon X)}$$

and then substituted into the design equation:

Combine

$$V = F_{A0} \int_0^X \frac{(1+\varepsilon X)^2}{kC_{A0}^2(1-X)^2}\,dX$$

The entering concentration C_{A0} can be taken outside the integral sign since it is not a function of conversion. Since the reaction is carried out isothermally, the specific reaction rate constant, k, can also be taken outside the integral sign.

For an isothermal
reaction, k is
constant

$$V = \frac{F_{A0}}{kC_{A0}^2} \int_0^X \frac{(1+\varepsilon X)^2}{(1-X)^2}\,dX$$

From the integral equations in Appendix A.1, we find that

Reactor volume
for a second-order
gas-phase reaction

$$V = \frac{v_0}{kC_{A0}} \left[2\varepsilon(1+\varepsilon)\ln(1-X) + \varepsilon^2 X + \frac{(1+\varepsilon)^2 X}{1-X} \right] \qquad (4\text{-}17)$$

If we divide both sides of Equation (4-17) by the cross-sectional area of the reactor, A_c, we obtain the following equation relating reactor length to conversion:

$$L = \frac{v_0}{kC_{A0}A_c} \left[2\varepsilon(1+\varepsilon)\ln(1-X) + \varepsilon^2 X + \frac{(1+\varepsilon)^2 X}{1-X} \right]$$

A plot of conversion along the length of the reactor is shown for four different reactions and values of ε in Figure 4-7 to illustrate the effects of volume

The importance of
changes in
volumetric flow rate
(i.e., $\varepsilon \neq 0$) with
reaction

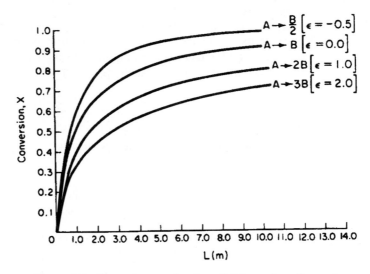

Figure 4-7   Conversion as a function of distance down the reactor.

change on reaction parameters. The following typical parameter values were chosen to arrive at these curves:

$$k = 5.0 \text{ dm}^3/\text{mol} \cdot \text{s} \qquad A_c = 1.0 \text{ dm}^2$$

$$C_{A0} = 0.2 \text{ mol/dm}^3 \qquad v_0 = 1 \text{ dm}^3/\text{s}$$

$$L = 1.0 \left[2\varepsilon(1+\varepsilon)\ln(1-X) + \varepsilon^2 X + \frac{(1+\varepsilon)^2 X}{1-X} \right] \text{ meters}$$

We observe from this figure that for **identical rate-law parameters**, the reaction that has a decrease in the total number of moles (i.e., $\varepsilon = -0.5$) will have the highest conversion for a fixed reactor length. This relationship should be expected for fixed temperature and pressure because the volumetric flow rate,

$$v = (1 - 0.5X)v_0$$

decreases with increasing conversion, and the reactant spends more time in the reactor than reactants that produce no net change in the total number of moles (e.g., A → B and $\varepsilon = 0$). Similarly, reactants that produce an increase in the total number of moles upon reaction (e.g., $\varepsilon = 2$) will spend less time in the reactor than reactants of reactions for which ε is zero or negative.

Example 4–3   Neglecting Volume Change with Reaction

The gas-phase cracking reaction

$$A \longrightarrow 2B + C$$

is to be carried out in a tubular reactor. The reaction is second-order and the parameter values are the same as those used to construct Figure 4-7. If 60% conversion is desired, what error will result if volume change is neglected ($\varepsilon = 0$) in sizing the reactor?

Solution

In Figure E4-3.1 (taken from Figure 4-7), we see that a reactor length of 1.5 m is required to achieve 60% conversion for $\varepsilon = 0$. However, by correctly accounting for volume change $[\varepsilon = (1)(2 + 1 - 1) = 2]$, we see that a reactor length of 5.0 m would be required. If we had used the 1.5-m-long reactor, we would have achieved only 40% conversion.

Look at the poor design that could result

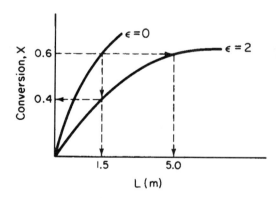

Figure E4-3.1

Example 4–4   Producing 300 Million Pounds per Year of Ethylene in a Plug-Flow Reactor: Design of a Full-Scale Tubular Reactor

The economics

Ethylene ranks fourth in the United States in total pounds of chemicals produced each year and it is the number one organic chemical produced each year. Over 35 billion pounds were produced in 1997 and sold for $0.25 per pound. Sixty-five percent of the ethylene produced is used in the manufacture of fabricated plastics, 20% for ethylene oxide and ethylene glycol, 5% for fibers, and 5% for solvents.

The uses

Determine the plug-flow reactor volume necessary to produce 300 million pounds of ethylene a year from cracking a feed stream of pure ethane. The reaction is irreversible and follows an elementary rate law. We want to achieve 80% conversion of ethane, operating the reactor isothermally at 1100 K at a pressure of 6 atm.

Solution

$$C_2H_6 \longrightarrow C_2H_4 + H_2$$

Let A = C_2H_6, B = C_2H_4, and C = H_2. In symbols,

$$A \longrightarrow B + C$$

The molar flow rate of ethylene exiting the reactor is

$$F_B = 300 \times 10^6 \frac{lb}{year} \times \frac{1 \text{ year}}{365 \text{ days}} \times \frac{1 \text{ day}}{24 \text{ h}} \times \frac{1 \text{ h}}{3600 \text{ s}} \times \frac{lb \text{ mol}}{28 \text{ lb}}$$

$$= 0.340 \frac{lb \text{ mol}}{s}$$

Next calculate the molar feed rate of ethane. To produce 0.34 lb mol/s of ethylene when 80% conversion is achieved,

$$F_B = F_{A0} X$$

$$F_{A0} = \frac{0.34}{0.8} = 0.425 \frac{lb \text{ mol}}{s}$$

1. **Plug-flow design equation**:

$$F_{A0} \frac{dX}{dV} = -r_A$$

Rearranging and integrating for the case of no pressure drop and isothermal operation yields

$$V = F_{A0} \int_0^X \frac{dX}{-r_A} \tag{E4-4.1}$$

2. **Rate law:**[1]

$$-r_A = kC_A \qquad \text{with} \quad k = 0.072 \text{ s}^{-1} \text{ at } 1000 \text{ K} \tag{E4-4.2}$$

The activation energy is 82 kcal/g mol.

3. **Stoichiometry.** For isothermal operation and negligible pressure drop, the concentration of ethane is calculated as follows:

---

[1] *Ind. Eng. Chem. Process Des. Dev., 14,* 218 (1975); *Ind. Eng. Chem.,* 59(5), 70 (1967).

Gas phase, constant T and P:

$$v = v_0 \frac{F_T}{F_{T0}} = v_0 (1 + \varepsilon X):$$

$$C_A = \frac{F_A}{v} = \frac{F_{A0}(1 - X)}{v_0(1 + \varepsilon X)} = C_{A0}\left(\frac{1 - X}{1 + \varepsilon X}\right) \tag{E4-4.3}$$

$$C_C = \frac{C_{A0} X}{(1 + \varepsilon X)} \tag{E4-4.4}$$

4. We now **combine** Equations (E4-4.1) through (E4-4.3) to obtain

Combining the
design equation,
rate law, and
stoichiometry

$$V = F_{A0} \int_0^X \frac{dX}{k C_{A0}(1 - X)/(1 + \varepsilon X)} = F_{A0} \int_0^X \frac{(1 + \varepsilon X)\, dX}{k C_{A0}(1 - X)}$$

$$= \frac{F_{A0}}{C_{A0}} \int_0^X \frac{(1 + \varepsilon X)\, dX}{k(1 - X)} \tag{E4-4.5}$$

Since the reaction is carried out isothermally, we can take k outside the integral sign and use Appendix A.1 to carry out our integration.

Analytical solution

$$V = \frac{F_{A0}}{k C_{A0}} \int_0^X \frac{(1 + \varepsilon X)\, dX}{1 - X} = \frac{F_{A0}}{k C_{A0}} \left[(1 + \varepsilon) \ln \frac{1}{1 - X} - \varepsilon X \right] \tag{E4-4.6}$$

5. **Parameter evaluation**:

$$C_{A0} = y_{A0} C_{T0} = \frac{y_{A0} P_0}{R T_0} = (1)\left(\frac{6 \text{ atm}}{0.73 \text{ ft}^3 \cdot \text{atm/lb mol} \cdot {}^\circ\text{R} \times 1980{}^\circ\text{R}} \right)$$

$$= 0.00415 \frac{\text{lb mol}}{\text{ft}^3}$$

$$\varepsilon = y_{A0} \delta = (1)(1 + 1 - 1) = 1$$

We need to calculate k at 1100 K.

$$k(T_2) = k(T_1) \exp\left[\frac{E}{R}\left(\frac{1}{T_1} - \frac{1}{T_2} \right) \right]$$

$$= k(T_1) \exp\left[\frac{E}{R}\left(\frac{T_2 - T_1}{T_1 T_2} \right) \right] \tag{E4-4.7}$$

$$= \frac{0.072}{\text{s}} \exp\left[\frac{82.000 \text{ cal/g mol}(1100 - 1000) \text{ K}}{1.987 \text{ cal/(g mol} \cdot \text{K})(1000 \text{ K})(1100 \text{ K})} \right]$$

$$= 3.07 \text{ s}^{-1}$$

Substituting into Equation (E4-4.6) yields

$$V = \frac{0.425 \text{ lb mol/s}}{(3.07/\text{s})(0.00415 \text{ lb mol/ft}^3)}\left[(1+1)\ln\frac{1}{1-X}-(1)X\right]$$

(E4-4.8)

$$= 33.36 \text{ ft}^3\left[2\ln\left(\frac{1}{1-X}\right)-X\right]$$

For $X = 0.8$,

$$V = 33.36 \text{ ft}^3\left[2\ln\left(\frac{1}{1-0.8}\right)-0.8\right]$$

$$= 80.7 \text{ ft}^3 = (2280 \text{ dm}^3 = 2.28 \text{ m}^3)$$

It was decided to use a bank of 2-in. schedule 80 pipes in parallel that are 40 ft in length. For pipe schedule 80, the cross-sectional area is 0.0205 ft². The number of pipes necessary is

The number of PFRs
in parallel

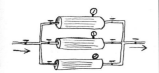

$$n = \frac{80.7 \text{ ft}^3}{(0.0205 \text{ ft}^2)(40 \text{ ft})} = 98.4$$

Equation (E4-4.7) was used along with $A_C = 0.0205$ ft² and Equations (E4-4.3) and (E4-4.4) to obtain Figure E4-4.1. Using a bank of 100 pipes will give us the reactor

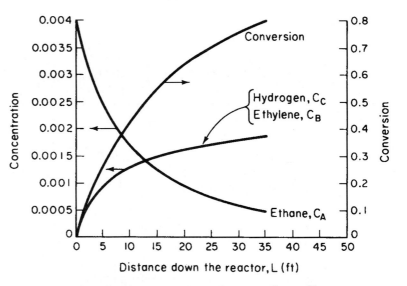

Figure E4-4.1 Conversion and concentration profiles.

volume necessary to make 300 million pounds per year of ethylene from ethane. The concentration and conversion profiles down any one of the pipes are shown in Figure E4-4.1.

4.4 Pressure Drop in Reactors

Pressure drop is ignored for liquid-phase kinetics calculations

In liquid-phase reactions, the concentration of reactants is insignificantly affected by even relatively large changes in the total pressure. Consequently, we can totally ignore the effect of pressure drop on the rate of reaction when sizing liquid-phase chemical reactors. However, in gas-phase reactions, the concentration of the reacting species is proportional to the total pressure and consequently, proper accounting for the effects of pressure drop on the reaction system can, in many instances, be a key factor in the success or failure of the reactor operation.

4.4.1 Pressure Drop and the Rate Law

We now focus our attention on accounting for the pressure drop in the rate law. For an ideal gas, the concentration of reacting species i is

For gas-phase reactions pressure drop may be very important

$$C_i = \frac{F_i}{v} = \frac{F_{A0}(\Theta_i + v_i X)}{v_0(1 + \varepsilon X)(P_0/P)(T/T_0)} \tag{3-46}$$

For isothermal operation

$$C_i = C_{A0}\left(\frac{\Theta_i + v_i X}{1 + \varepsilon X}\right)\frac{P}{P_0} \tag{4-18}$$

We now must determine the ratio P/P_0 as a function of volume V or the catalyst weight, W, to account for pressure drop. We then can combine the concentration, rate law, and design equation. However, whenever accounting for the effects of pressure drop, *the differential form of the mole balance (design equation) must be used*.

If, for example, the second-order isomerization reaction

$$A \longrightarrow B$$

When $P \neq P_0$ one must use the differential forms of the PFR/PBR design equations

is being carried out in a packed-bed reactor, the **differential form of the mole balance** equation in terms of catalyst weight is

$$F_{A0}\frac{dX}{dW} = -r'_A \qquad \left(\frac{\text{gram moles}}{\text{gram catalyst}\cdot\text{min}}\right) \tag{2-17}$$

The **rate law** is

$$-r'_A = kC_A^2 \tag{4-19}$$

From **stoichiometry** for gas-phase reactions,

$$C_A = \frac{C_{A0}(1-X)}{1 + \varepsilon X}\frac{P}{P_0}\frac{T_0}{T}$$

and the rate law can be written as

$$-r_A' = k \left[\frac{C_{A0}(1-X)}{1+\varepsilon X} \frac{P}{P_0} \frac{T_0}{T} \right]^2 \tag{4-20}$$

Note from Equation (4-20) that the larger the pressure drop (i.e., the smaller P) from frictional losses, the smaller the reaction rate!

Combining Equation (4-20) with the mole balance (2-17) and assuming isothermal operation $(T = T_0)$ gives

$$F_{A0} \frac{dX}{dW} = k \left[\frac{C_{A0}(1-X)}{1+\varepsilon X} \right]^2 \left(\frac{P}{P_0} \right)^2$$

Dividing by F_{A0} (i.e., $v_0 C_{A0}$) yields

$$\frac{dX}{dW} = \frac{kC_{A0}}{v_0} \left(\frac{1-X}{1+\varepsilon X} \right)^2 \left(\frac{P}{P_0} \right)^2$$

For isothermal operation $(T = T_0)$ the right-hand side is a function of only conversion and pressure:

Another Equation is needed. (e.g., $P = f(W)$)

$$\frac{dX}{dW} = F_1(X,P) \tag{4-21}$$

We now need to relate the pressure drop to the catalyst weight in order to determine the conversion as a function of catalyst weight.

4.4.2 Flow Through a Packed Bed

The majority of gas-phase reactions are catalyzed by passing the reactant through a packed bed of catalyst particles. The equation used most to calculate pressure drop in a packed porous bed is the **Ergun equation:**[2]

Ergun equation

$$\frac{dP}{dz} = -\frac{G}{\rho g_c D_p} \left(\frac{1-\phi}{\phi^3} \right) \left[\frac{150(1-\phi)\mu}{D_p} + 1.75G \right] \tag{4-22}$$

---

[2] R. B. Bird, W. E. Stewart, and E. N. Lightfoot, *Transport Phenomena* (New York: Wiley, 1960), p. 200.

where P = pressure, lb/ft^2

$$\phi = \text{porosity} = \frac{\text{volume of void}}{\text{total bed volume}}$$

$$1 - \phi = \frac{\text{volume of solid}}{\text{total bed volume}}$$

g_c = 32.174 lb$_m$ · ft/s^2 · lb$_f$ (conversion factor)

= 4.17 × 10^8 lb$_m$ · ft/h^2 · lb$_f$

(recall that for the metric system g_c = 1.0)

D_p = diameter of particle in the bed, ft

μ = viscosity of gas passing through the bed, lb$_m$/ft · h

z = length down the packed bed of pipe, ft

u = superficial velocity = volumetric flow ÷ cross-sectional area of pipe, ft/h

ρ = gas density, lb/ft^3

$G = \rho u$ = superficial mass velocity, (g/cm^2 · s) or (lb$_m$/ft^2 · h)

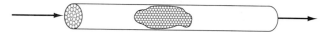

In calculating the pressure drop using the Ergun equation, the only parameter that varies with pressure on the right-hand side of Equation (4-22) is the gas density, ρ. We are now going to calculate the pressure drop through the bed.

Because the reactor is operated at steady state, the mass flow rate at any point down the reactor, $\dot{m}$ (kg/s), is equal to the entering mass flow rate, $\dot{m}_0$ (i.e., equation of continuity),

$$\dot{m}_0 = \dot{m}$$

$$\rho_0 v_0 = \rho v$$

Recalling Equation (3-41), we have

$$v = v_0 \frac{P_0}{P}\left(\frac{T}{T_0}\right)\frac{F_T}{F_{T0}} \tag{3-41}$$

$$\rho = \rho_0 \frac{v_0}{v} = \rho_0 \frac{P}{P_0}\left(\frac{T_0}{T}\right)\frac{F_{T0}}{F_T} \tag{4-23}$$

Combining Equations (4-22) and (4-23) gives

$$\frac{dP}{dz} = -\frac{G(1-\phi)}{\rho_0 g_c D_P \phi^3}\left[\frac{150(1-\phi)\mu}{D_P} + 1.752G\right]\frac{P_0}{P}\left(\frac{T}{T_0}\right)\frac{F_T}{F_{T0}}$$

Simplifying yields

$$\frac{dP}{dz} = -\beta_0 \frac{P_0}{P}\left(\frac{T}{T_0}\right)\frac{F_T}{F_{T0}} \qquad (4\text{-}24)$$

where

$$\beta_0 = \frac{G(1-\phi)}{\rho_0 g_c D_P \phi^3}\left[\frac{150(1-\phi)\mu}{D_P} + 1.75G\right] \qquad (4\text{-}25)$$

For tubular packed-bed reactors we are more interested in catalyst weight rather than the distance z down the reactor. The catalyst weight up to a distance of z down the reactor is

$$W = (1-\phi)A_c z \times \rho_c$$

Use this form for
multiple reactions
and membrane
reactors

$$\begin{bmatrix}\text{weight of}\\ \text{catalyst}\end{bmatrix} = \begin{bmatrix}\text{volume of}\\ \text{solids}\end{bmatrix} \times \begin{bmatrix}\text{density of}\\ \text{solid catalyst}\end{bmatrix} \qquad (4\text{-}26)$$

where A_c is the cross-sectional area. The bulk density of the catalyst, ρ_b (mass of catalyst per volume of reactor bed), is just the product of the solid density, ρ_c, the fraction of solids, $(1-\phi)$:

$$\rho_b = \rho_c(1-\phi)$$

Using the relationship between z and W [Equation (4-26)] we can change our variables to express the Ergun equation in terms of catalyst weight:

$$\frac{dP}{dW} = -\frac{\beta_0}{A_c(1-\phi)\rho_c}\frac{P_0}{P}\left(\frac{T}{T_0}\right)\frac{F_T}{F_{T0}} \qquad (4\text{-}27)$$

Further simplification yields

$$\frac{dP}{dW} = -\frac{\alpha}{2}\frac{T}{T_0}\frac{P_0}{P/P_0}\left(\frac{F_T}{F_{T0}}\right) \qquad (4\text{-}28)$$

where

$$\alpha = \frac{2\beta_0}{A_c \rho_c(1-\phi)P_0} \qquad (4\text{-}29)$$

Equation (4-28) will be the one we use when multiple reactions are occurring or when there is pressure drop in a membrane reactor. However, for single reactions in packed-bed reactors it is more convenient to express the Ergun equation in terms of the conversion X. Recalling Equation (3-42) for F_T,

$$F_T = F_{T0} + F_{A0}\delta X = F_{T0}\left(1 + \frac{F_{A0}}{F_{T0}}\delta X\right) \qquad (3\text{-}42)$$

and the development leading to Equation (3-43),

$$\frac{F_T}{F_{T0}} = 1 + \varepsilon X$$

where, as before,

$$\varepsilon = y_{A0}\delta = \frac{F_{A0}}{F_{T0}}\delta$$

Equation (4-28) can now be written as

Differential form of Ergun equation for the pressure drop in packed beds

$$\boxed{\frac{dP}{dW} = -\frac{\alpha}{2}\frac{T}{T_0}\frac{P_0}{P/P_0}(1 + \varepsilon X)} \qquad (4\text{-}30)$$

We note that when ε is negative the pressure drop ΔP will be less (i.e., higher pressure) than that for $\varepsilon = 0$. When ε is positive, the pressure drop ΔP will be greater than when $\varepsilon = 0$.

For isothermal operation, Equation (4-30) is only a function of conversion and pressure:

$$\frac{dP}{dW} = F_2(X, P) \qquad (4\text{-}31)$$

Recalling Equation (4-21),

Two coupled equations to be solved numerically

$$\frac{dX}{dW} = F_1(X, P) \qquad (4\text{-}21)$$

we see that we have two coupled first-order differential equations, (4-31) and (4-21), that must be solved simultaneously. A variety of software packages and numerical integration schemes are available for this purpose.

Analytical Solution.   If $\varepsilon = 0$, *or* if we can neglect (εX) with respect to 1.0 (i.e., $1 \gg \varepsilon X$), we can obtain an analytical solution to Equation (4-30) for isothermal operation (i.e., $T = T_0$). For isothermal operation with $\varepsilon = 0$, Equation (4-30) becomes

Isothermal with
$\varepsilon = 0$

$$\frac{dP}{dW} = -\frac{\alpha P_0}{2(P/P_0)} \qquad (4\text{-}32)$$

Rearranging gives us

$$\frac{2P}{P_0} \frac{d(P/P_0)}{dW} = -\alpha$$

Taking P/P_0 inside the derivative, we have

$$\frac{d(P/P_0)^2}{dW} = -\alpha$$

Integrating with $P = P_0$ at $W = 0$ yields

$$\left(\frac{P}{P_0}\right)^2 = 1 - \alpha W$$

Taking the square root of both sides gives

Pressure ratio
only for $\varepsilon = 0$

$$\boxed{\frac{P}{P_0} = (1 - \alpha W)^{1/2}} \qquad (4\text{-}33)$$

where again

$$\boxed{\alpha = \frac{2\beta_0}{A_c(1-\phi)\rho_c P_0}}$$

Equation (4-33) can be used to substitute for the pressure in the rate law, in which case the mole balance can be written solely as a function of conversion and catalyst weight. The resulting equation can readily be solved either analytically or numerically.

If we wish to express the pressure in terms of reactor length z, we can use Equation (4-26) to substitute for W in Equation (4-33). Then

$$\frac{P}{P_0} = \left(1 - \frac{2\beta_0 z}{P_0}\right)^{1/2} \qquad (4\text{-}34)$$

Example 4–5 Calculating Pressure Drop in a Packed Bed

Calculate the pressure drop in a 60 ft length of 1 1/2-in. schedule 40 pipe packed with catalyst pellets 1/4-in. in diameter when 104.4 lb/h of gas is passing through the bed. The temperature is constant along the length of pipe at 260°C. The void fraction is 45% and the properties of the gas are similar to those of air at this temperature. The entering pressure is 10 atm.

Solution

At the end of the reactor $z = L$ and Equation (4-34) becomes

$$\frac{P}{P_0} = \left(1 - \frac{2\beta_0 L}{P_0}\right)^{1/2} \tag{E4-5.1}$$

$$\beta_0 = \frac{G(1 - \phi)}{g_c \rho_0 D_p \phi^3}\left[\frac{150(1 - \phi)\mu}{D_p} + 1.75G\right] \tag{4-25}$$

Evaluating the pressure drop parameters

$$G = \frac{\dot{m}}{A_c} \tag{E4-5.2}$$

For $1\frac{1}{2}$-in. schedule 40 pipe, $A_c = 0.01414$ ft^2:

$$G = \frac{104.4 \text{ lb}_m/\text{h}}{0.01414 \text{ ft}^2} = 7383.3 \frac{\text{lb}_m}{\text{h} \cdot \text{ft}^2}$$

For air at 260°C and 10 atm,

$$\mu = 0.0673 \text{ lb}_m/\text{ft} \cdot \text{h}$$
$$\rho_0 = 0.413 \text{ lb}_m/\text{ft}^3 \tag{E4-5.3}$$

From the problem statement,

$$D_p = \tfrac{1}{4}\text{in.} = 0.0208 \text{ ft}$$

$$g_c = 4.17 \times 10^8 \frac{\text{lb}_m \cdot \text{ft}}{\text{lb}_f \cdot \text{h}^2}$$

Substituting the values above into Equation (4-25) gives us

$$\beta_0 = \left[\frac{7383.3 \text{ lb}_m/\text{ft}^2 \cdot \text{h}(1 - 0.45)}{(4.17 \times 10^8 \text{ lb}_m \cdot \text{ft}/\text{lb}_f \cdot \text{h}^2)(0.413 \text{ lb}_m/\text{ft}^3)(0.0208 \text{ ft})(0.45)^3}\right]$$
$$\times \left[\frac{150(1 - 0.45)(0.0673 \text{ lb}_m/\text{ft} \cdot \text{h})}{0.0208 \text{ ft}} + 1.75(7383.3) \frac{\text{lb}_m}{\text{ft}^2 \cdot \text{h}}\right] \tag{E4-5.3}$$

$$= 0.01244 \frac{\text{lb}_f \cdot \text{h}}{\text{ft} \cdot \text{lb}_m}(266.9 + 12,920.8) \frac{\text{lb}_m}{\text{ft}^2 \cdot \text{h}} = 164.1 \frac{\text{lb}_f}{\text{ft}^3}$$

$$= 164.1 \frac{\text{lb}_f}{\text{ft}^3} \times \frac{1 \text{ ft}^2}{144 \text{ in.}^2} \times \frac{1 \text{ atm}}{14.7 \text{ lb}_f/\text{in.}^2} \tag{E4-5.4}$$

$$= 0.0775 \frac{\text{atm}}{\text{ft}} = 25.8 \frac{\text{kPa}}{\text{m}}$$

$$\frac{P}{P_0} = \left(1 - \frac{2\beta_0 L}{P_0}\right)^{1/2} = \left(1 - \frac{2 \times 0.0775 \text{ atm/ft} \times 60 \text{ ft}}{10 \text{ atm}}\right)^{1/2} \quad \text{(E4-5.5)}$$

$$P = 0.265 P_0 = 2.65 \text{ atm}$$
$$\Delta P = P_0 - P = 10 - 2.65 = 7.35 \text{ atm} \quad \text{(E4-5.6)}$$

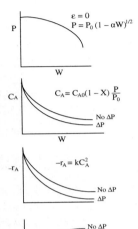

Reaction with Pressure Drop

Analytical solution: Now that we have expressed pressure as a function of catalyst weight [Equation (4-33)] we can return to the second-order isothermal reaction,

$$A \longrightarrow B$$

to relate conversion and catalyst weight. Recall our mole balance, rate law, and stoichiometry.

Mole balance:
$$F_{A0} \frac{dX}{dW} = -r'_A \quad \text{(2-17)}$$

Rate law:
$$-r'_A = k C_A^2 \quad \text{(4-19)}$$

Stoichiometry. Gas-phase isothermal reaction with $\varepsilon = 0$:

$$C_A = C_{A0}(1 - X) \frac{P}{P_0} \quad \text{(4-35)}$$

Using Equation (4-33) to substitute for P/P_0 in terms of the catalyst weight, we obtain

Only
for
$\varepsilon = 0$

$$C_A = C_{A0}(1 - X)(1 - \alpha W)^{1/2}$$

Combining:
$$\frac{dX}{dW} = \frac{k C_{A0}^2}{F_{A0}} (1 - X)^2 [(1 - \alpha W)^{1/2}]^2$$

Separating variables:
$$\frac{F_{A0}}{k C_{A0}^2} \frac{dX}{(1 - X)^2} = (1 - \alpha W) \, dW$$

Integrating with limits $X = 0$ when $W = 0$ and substituting for $F_{A0} = C_{A0} v_0$ yields

$$\frac{v_0}{k C_{A0}} \left(\frac{X}{1 - X}\right) = W \left(1 - \frac{\alpha W}{2}\right)$$

Solving for conversion gives

$$X = \cfrac{\dfrac{kC_{A0}W}{v_0}\left(1 - \dfrac{\alpha W}{2}\right)}{1 + \dfrac{kC_{A0}W}{v_0}\left(1 - \dfrac{\alpha W}{2}\right)} \qquad (4\text{-}36)$$

Solving for the catalyst weight, we have

Catalyst weight for
second-order
reaction in PFR
with ΔP

$$W = \frac{1 - \{1 - [(2v_0\alpha)/kC_{A0}][X/(1 - X)]\}^{1/2}}{\alpha} \qquad (4\text{-}37)$$

We now proceed (Example 4-6) to combine pressure drop with reaction in a packed bed for the case where we will assume that $\varepsilon X \ll 1$ in the Ergun equation but not in the rate law in order to obtain an analytical solution. Example 4-7 removes this assumption and solves Equations (4-21) and (4-31) numerically.

Example 4–6   Calculating X in a Reactor with Pressure Drop

The economics

Approximately 7 billion pounds of ethylene oxide were produced in the United States in 1997. The 1997 selling price was $0.58 a pound, amounting to a commercial value of $4.0 billion. Over 60% of the ethylene oxide produced is used to make ethylene glycol. The major end uses of ethylene oxide are antifreeze (30%), polyester (30%), surfactants (10%), and solvents (5%). We want to calculate the catalyst weight necessary to achieve 60% conversion when ethylene oxide is to be made by the vapor-phase catalytic oxidation of ethylene with air.

$$C_2H_4 + \tfrac{1}{2}O_2 \longrightarrow \overset{O}{\overset{\triangle}{CH_2-CH_2}}$$

$$A \ + \tfrac{1}{2}B \ \longrightarrow \ C$$

Ethylene and oxygen are fed in stoichiometric proportions to a packed-bed reactor operated isothermally at 260°C. Ethylene is fed at a rate of 0.30 lb mol/s at a pressure of 10 atm. It is proposed to use 10 banks of $1\tfrac{1}{2}$-in.-diameter schedule 40 tubes packed with catalyst with 100 tubes per bank. Consequently, the molar flow rate to each tube is to be 3×10^{-4} lb mol/s. The properties of the reacting fluid are to be considered identical to those of air at this temperature and pressure. The density of the $\tfrac{1}{4}$-in.-catalyst particles is 120 lb/ft³ and the bed void fraction is 0.45. The rate law is

$$-r'_A = kP_A^{1/3}P_B^{2/3} \qquad \text{lb mol/lb cat} \cdot \text{h}$$

with[3]

$$k = 0.0141 \ \frac{\text{lb mol}}{\text{atm} \cdot \text{lb cat.} \cdot \text{h}} \ \text{at } 260°C$$

---

[3] *Ind. Eng. Chem., 45*, 234 (1953).

Solution

1. **Differential mole balance**:

$$F_{A0} \frac{dX}{dW} = -r_A' \tag{E4-6.1}$$

2. **Rate law**:

$$-r_A' = kP_A^{1/3} P_B^{2/3} = k(C_A RT)^{1/3}(C_B RT)^{2/3} \tag{E4-6.2}$$

$$= kRTC_A^{1/3} C_B^{2/3} \tag{E4-6.3}$$

The algorithm

3. **Stoichiometry**. Gas-phase, isothermal $v = v_0(1 + \varepsilon X)(P_0/P)$:

$$C_A = \frac{F_A}{v} = \frac{C_{A0}(1-X)}{1+\varepsilon X}\left(\frac{P}{P_0}\right) \tag{E4-6.4}$$

$$C_B = \frac{F_B}{v} = \frac{C_{A0}(\Theta_B - X/2)}{1+\varepsilon X}\left(\frac{P}{P_0}\right) \tag{E4-6.5}$$

4. **Combining** the rate law and concentrations:

$$-r_A' = kRT_0\left[\frac{C_{A0}(1-X)}{1+\varepsilon X}\left(\frac{P}{P_0}\right)\right]^{1/3} \cdot \left[\frac{C_{A0}(\Theta_B - X/2)}{1+\varepsilon X}\left(\frac{P}{P_0}\right)\right]^{2/3} \tag{E4-6.6}$$

$$= \frac{kC_{A0}RT_0}{1+\varepsilon X}\left(\frac{P}{P_0}\right)(1-X)^{1/3}\left(\Theta_B - \frac{X}{2}\right)^{2/3} \tag{E4-6.7}$$

For stoichiometric feed, $\Theta_B = \frac{1}{2}$:

$$-r_A' = kP_{A0}\frac{(1-X)^{1/3}(1/2 - X/2)^{2/3}}{1+\varepsilon X}\left(\frac{P}{P_0}\right) \tag{E4-6.8}$$

$$= k'\left(\frac{1-X}{1+\varepsilon X}\right)\frac{P}{P_0}$$

where $k' = kP_{A0}\left(\tfrac{1}{2}\right)^{2/3} = 0.63kP_{A0}$.

5. **Developing the design equation**. For a packed-bed reactor, the relationship between P and W when $\varepsilon X \ll 1$ is

Eq. (4-33) is valid only if $\varepsilon = 0$ or $\varepsilon X \ll 1$

$$\frac{P}{P_0} = (1 - \alpha W)^{1/2} \tag{4-33}$$

(Note that we will check this assumption in Example 4-7.) Combining Equations (E4-6.8) and (4-33) gives us

$$-r_A' = k'\left(\frac{1-X}{1+\varepsilon X}\right)(1 - \alpha W)^{1/2} \tag{E4-6.9}$$

Combining Equations (E4-6.9) and (E4-6.1), we have

$$F_{A0} \frac{dX}{dW} = -r'_A = k' \left(\frac{1 - X}{1 + \varepsilon X} \right) (1 - \alpha W)^{1/2}$$

Separating variables to form the integrals yields

$$\int_0^X \frac{F_{A0}(1 + \varepsilon X)\, dX}{k'(1 - X)} = \int_0^W (1 - \alpha W)^{1/2}\, dW$$

Integrating gives us

$$\frac{F_{A0}}{k'} \left[(1 + \varepsilon)\, \ln \frac{1}{1 - X} - \varepsilon X \right] = \frac{2}{3\alpha} [1 - (1 - \alpha W)^{3/2}] \qquad \text{(E4-6.10)}$$

Solving for W, we obtain

$$W = \frac{1 - [1 - (3\alpha F_{A0}/2k')\{(1 + \varepsilon)\, \ln[1/(1 - X)] - \varepsilon X\}]^{2/3}}{\alpha} \qquad \text{(E4-6.11)}$$

6. **Parameter evaluation per tube** (i.e., divide feed rates by 1000):

Ethylene:	$F_{A0} = 3 \times 10^{-4}$ lb mol/s = 1.08 lb mol/h
Oxygen:	$F_{B0} = 1.5 \times 10^{-4}$ lb mol/s = 0.54 lb mol/h

$$I = \text{inerts} = N_2: \quad F_I = 1.5 \times 10^{-4} \text{ lb mol/s} \times \frac{0.79 \text{ mol } N_2}{0.21 \text{ mol } O_2}$$

$$F_I = 5.64 \times 10^{-4} \text{ lb mol/s} = 2.03 \text{ lb mol/h}$$

Summing: $\qquad F_{T0} = F_{A0} + F_{B0} + F_I = 3.65$ mol/h

$$y_{A0} = \frac{F_{A0}}{F_{T0}} = \frac{1.08}{3.65} = 0.30$$

$$\varepsilon = y_{A0}\delta = (0.3)(1 - \tfrac{1}{2} - 1) = -0.15$$

$$P_{A0} = y_{A0}P_0 = 3.0 \text{ atm}$$

$$k' = kP_{A0}\left(\tfrac{1}{2}\right)^{2/3} = 0.0141 \frac{\text{lb mol}}{\text{atm} \cdot \text{lb cat} \cdot \text{h}} \times 3 \text{ atm} \times 0.63 = 0.0266 \frac{\text{lb mol}}{\text{h} \cdot \text{lb cat}}$$

$$W = \frac{1 - [1 - (3\alpha F_{A0}/2k')\{(1 - 0.15)\, \ln[1/(1 - 0.6)] - (-0.15)(0.6)\}]^{2/3}}{\alpha}$$

For 60% conversion, Equation (E4-6.11) becomes

$$W = \frac{1 - (1 - 1.303\alpha F_{A0}/k')^{2/3}}{\alpha} \qquad \text{(E4-6.12)}$$

In order to calculate α,

Evaluating the
pressure drop
parameters

$$\alpha = \frac{2\beta_0}{A_c(1 - \phi)\rho_c P_0}$$

we need the superficial mass velocity, G. The mass flow rates of each entering species are:

$$\dot{m}_{A0} = 1.08 \ \frac{\text{lb mol}}{\text{h}} \times 28 \ \frac{\text{lb}}{\text{lb mol}} = 30.24 \ \text{lb/h}$$

$$\dot{m}_{B0} = 0.54 \ \frac{\text{lb mol}}{\text{h}} \times 32 \ \frac{\text{lb}}{\text{lb mol}} = 17.28 \ \text{lb/h}$$

$$\dot{m}_{I0} = 2.03 \ \frac{\text{lb mol}}{\text{h}} \times 28 \ \frac{\text{lb}}{\text{lb mol}} = 56.84 \ \text{lb/h}$$

The total mass flow rate is

$$\dot{m}_{T0} = 104.4 \ \frac{\text{lb}}{\text{h}}$$

$$G = \frac{\dot{m}_{T0}}{A_c} = \frac{104.4 \ \text{lb/h}}{0.01414 \ \text{ft}^2} = 7383.3 \ \frac{\text{lb}}{\text{h} \cdot \text{ft}^2}$$

This is essentially the same superficial mass velocity, temperature, and pressure as in Example 4-5. Consequently, we can use the value of β_0 calculated in Example 4-5.

$$\beta_0 = 0.0775 \ \frac{\text{atm}}{\text{ft}}$$

$$\alpha = \frac{2\beta_0}{A_c(1-\phi)\rho_c P_0} = \frac{(2)(0.0775) \ \text{atm/ft}}{(0.01414 \ \text{ft}^2)(0.55)(120 \ \text{lb cat/ft}^3)(10 \ \text{atm})}$$

$$= \frac{0.0166}{\text{lb cat}}$$

Substituting into Equation (E4-6.12) yields

$$W = \frac{1 - \left[1 - \dfrac{\left(1.303 \ \dfrac{0.0166}{\text{lb cat}} \right)\left(1.08 \ \dfrac{\text{lb mol}}{\text{h}} \right)}{0.0266 \ \dfrac{\text{lb mol}}{\text{lb cat} \cdot \text{h}}} \right]^{2/3}}{0.0166/\text{lb cat}}$$

$$= 45.4 \ \text{lb of catalyst per tube}$$

$$\text{or } 45{,}400 \ \text{lb of catalyst total}$$

This catalyst weight corresponds to a pressure drop of approximately 5 atm. If we had neglected pressure drop, the result would have been

$$W = \frac{F_{A0}}{k'} \left[(1+\varepsilon) \ln \frac{1}{1-X} - \varepsilon X \right]$$

$$= \frac{1.08}{0.0266} \times \left[(1-0.15) \ln \frac{1}{1-0.6} - (-0.15)(0.6) \right]$$

$$= 35.3 \ \text{lb of catalyst per tube (neglecting pressure drop)}$$

Neglecting pressure drop results in poor design (here 53% vs. 60% conversion)

and we would have had insufficient catalyst to achieve the desired conversion. Substituting this catalyst weight (i.e., 35,300 lb total) into Equation (E4-6.10) gives a conversion of only 53%.

Example 4–7 Pressure Drop with Reaction—Numerical Solution

Rework Example 4-6 for the case where volume change is *not* neglected in the Ergun equation and the two coupled differential equations describing the variation of conversion and pressure with catalyst weight are solved numerically.

Solution

Rather than rederive everything starting from the mole balance, rate law, stoichiometry, and pressure drop equations, we will use the equations developed in Example 4-6. Combining Equations (E4-6.1) and (E4-6.8) gives

$$F_{A0}\frac{dX}{dW} = k'\left(\frac{1-X}{1+\varepsilon X}\right)\frac{P}{P_0} \tag{E4-7.1}$$

Recalling Equation (4-30) and assuming isothermal operation gives

$$\frac{dP}{dW} = -\frac{\alpha}{2}\frac{P_0}{(P/P_0)}(1+\varepsilon X) \tag{E4-7.2}$$

Next, we let

$$y = \frac{P}{P_0} \quad and \quad \alpha = \frac{2\beta_0}{A_c(1-\phi)\rho_c P_0}$$

so that Equations (E4-7.1) and (E4-7.2) can be written as

$$\boxed{\frac{dX}{dW} = \frac{k'}{F_{A0}}\left(\frac{1-X}{1+\varepsilon X}\right)y} \tag{E4-7.3}$$

$$\boxed{\frac{dy}{dW} = -\frac{\alpha(1+\varepsilon X)}{2y}} \tag{E4-7.4}$$

For the reaction conditions described in Example 4-6, we have the boundary conditions $W = 0$, $X = 0$, and $y = 1.0$ and the parameter values $\alpha = 0.0166$/lb cat, $\varepsilon = -0.15$, $k' = 0.0266$ lb mol/h·lb cat, and $F_{A0} = 1.08$ lb mol/h.

A large number of ordinary differential equation solver software packages (i.e., ODE solvers) which are extremely user friendly have become available. We shall use POLYMATH[4] to solve the examples in the printed text. However, the CD-ROM contains an example that uses ASPEN, as well as all the MATLAB and POLYMATH solution programs to the example programs. With POLYMATH one simply enters Equations (E4-7.3) and (E4-7.4) and the corresponding parameter value into the computer (Table E4-7.1) with the initial (rather, boundary) conditions and they are solved and displayed as shown in Figure E4-7.1.

We note that neglecting εX in the Ergun equation in Example 4-6 ($\varepsilon X = -0.09$) to obtain an analytical solution resulted in less than a 10% error.

CD Living

Example Problems

Program examples POLYMATH, MatLab can be loaded from the CD-ROM (see the Introduction)

---

[4] Developed by Professor M. Cutlip of the University of Connecticut, and Professor M. Shacham of Ben Gurion University. Available from the CACHE Corporation, P.O. Box 7939, Austin, TX 78713.

$$f = \frac{v_0}{v}$$

$$y = \frac{P}{P_0}$$

TABLE E4-7.1.   POLYMATH SCREEN SHOWING EQUATIONS TYPED
IN AND READY TO BE SOLVED.

Equations	Initial Values
`d(y)/d(w)=-alpha*(1+eps*x)/2/y`	1
`d(x)/d(w)=rate/fa0`	0
`fa0=1.00`	
`alpha=0.0166`	
`eps=-0.15`	
`kprime=0.0266`	
`f=(1+eps*x)/y`	
`rate=kprime*((1-x)/(1+eps*x))*y`	
$w_0 = 0,\quad w_f = 60$	

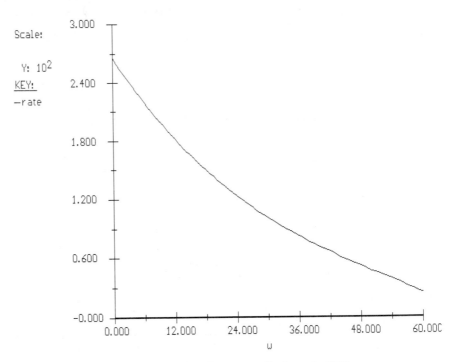

Figure E4-7.1   Reaction rate profile down the PBR.

However, larger errors will result if large values of εX are neglected! By taking into account the change in the volumetric flow rate (i.e., $\varepsilon = -0.15$) in the pressure drop term, we see that 44.0 lb of catalyst is required per tube as opposed to 45.4 lb when ε was neglected in the analytical solution, Equation (E4-7.4). Why was less catalyst required when ε was not neglected in Equation (E4-7.4)? The reason is that the numerical solution accounts for the fact that the pressure drop will be less because ε is negative.

It is also interesting to learn what happens to the volumetric flow rate along the length of the reactor. Recalling Equation (3-44),

$$v = v_0(1 + \varepsilon X) \frac{P_0}{P} \frac{T}{T_0} = \frac{v_0(1 + \varepsilon X)(T/T_0)}{P/P_0} \qquad (3\text{-}44)$$

Volumetric flow rate increases with increasing pressure drop

We let f be the ratio of the volumetric flow rate, v, to the entering volumetric flow rate, v_0, at any point down the reactor. For isothermal operation Equation (3-44) becomes

$$f = \frac{v}{v_0} = \frac{1 + \varepsilon X}{y} \qquad (E4\text{-}7.5)$$

Figure E4-7.2 shows X, y (i.e., $y = P/P_0$), and f down the length of the reactor. We see that both the conversion and the volumetric flow increase along the length of the reactor while the pressure decreases. For gas-phase reactions with orders greater than zero, this decrease in pressure will cause the reaction rate to be less than in the case of no pressure drop.

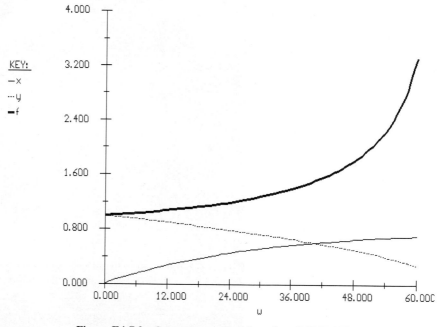

Figure E4-7.2   Output in graphical form from POLYMATH.

Effect of added catalyst on conversion

We note from Figure E4-7.2 that the catalyst weight necessary to raise the conversion the last 1% from 65% to 66% (3.5 lb) is 8.5 times more than that (0.41 lb) required to raise the conversion 1% at the reactor's entrance. Also, during the last 5% increase in conversion, the pressure decreases from 3.8 atm to 2.3 atm.

4.4.3 Spherical Packed-Bed Reactors

When small catalyst pellets are required, the pressure drop can be significant. In Example 4-6 we saw that significant design flaws can result if pressure drop is neglected or if steps are not taken to minimize pressure drop. One type of reactor that minimizes pressure drop and is also inexpensive to build is the spherical reactor, shown in Figure 4-8. In this reactor, called an ultraformer, dehydrogenation reactions such as

$$\text{paraffin} \longrightarrow \text{aromatic} + 3H_2$$

are carried out.

Figure 4-8   Spherical Ultraformer Reactor. (Courtesy of Amoco Petroleum Products.) This reactor is one in a series of six used by Amoco for reforming petroleum naphtha. Photo by K. R. Renicker, Sr.

Another advantage of spherical reactors is that they are the most economical shape for high pressures. As a first approximation we will assume that the fluid moves down through the reactor in plug flow. Consequently, because of the increase in cross-sectional area, A_c, as the fluid enters the sphere, the superficial velocity, $G = \dot{m}/A_c$, will decrease. From the Ergun equation [Equation (4-22)],

$$\frac{dP}{dz} = -\frac{G(1-\phi)}{\rho g_c D_P \phi^3}\left[\frac{150(1-\phi)\mu}{D_P} + 1.75G\right] \qquad (4\text{-}22)$$

we know that by decreasing G, the pressure drop will be reduced significantly, resulting in higher conversions.

Because the cross-sectional area of the reactor is small near the inlet and outlet, the presence of catalyst there would cause substantial pressure drop, thereby reducing the efficiency of the spherical reactor. To solve this problem, screens to hold the catalyst are placed near the reactor entrance and exit (Figures 4-9 and 4-10). Here L is the location of the screen from the center of the

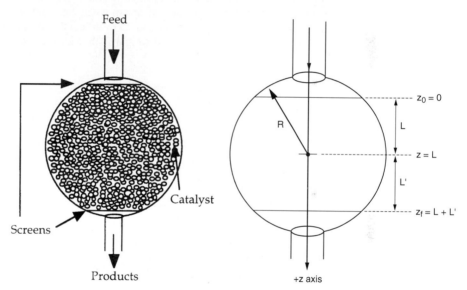

Figure 4-9   Schematic drawing of the inside of a spherical reactor.

Figure 4-10   Coordinate system and variables used with a spherical reactor. The initial and final integration values are shown as z_0 and z_f.

reactor. We can use elementary geometry and integral calculus to derive the following expressions for cross-sectional area and catalyst weight as a function of the variables defined in Figure 4-10:

$$A_c = \pi[R^2 - (z-L)^2] \qquad (4\text{-}38)$$

Spherical reactor
catalyst weight

$$W = \rho_c(1-\phi)V = \rho_c(1-\phi)\pi\left[R^2 z - \frac{1}{3}(z-L)^3 - \frac{1}{3}L^3\right] \qquad (4\text{-}39)$$

By using these formulas and the standard pressure drop algorithm, one can solve a variety of spherical reactor problems. Note that Equations (4-38) and

(4-39) make use of L and not L'. Thus, one does not need to adjust these formulas to treat spherical reactors that have different amounts of empty space at the entrance and exit (i.e., $L \neq L'$). Only the upper limit of integration needs to be changed, $z_f = L + L'$.

Example 4–8 Dehydrogenation Reactions in a Spherical Reactor

Reforming reactors are used to increase the octane number of petroleum. In a reforming process 20,000 barrels of petroleum are to be processed per day. The corresponding mass and molar feed rates are 44 kg/s and 440 mol/s, respectively. In the reformer, dehydrogenation reactions such as

$$\text{paraffin} \rightarrow \text{olefin} + H_2$$

occur. The reaction is first-order in paraffin. Assume that pure paraffin enters the reactor at a pressure of 2000 kPa and a corresponding concentration of 0.32 mol/dm^3. Compare the pressure drop and conversion when this reaction is carried out in a tubular packed bed 2.4 m in diameter and 25 m in length with that of a spherical packed bed 6 m in diameter. The catalyst weight is the same in each reactor, 173,870 kg.

$$-r'_A = k'C_A$$
$$-r'_A = \rho_B(-r'_A) = \rho_C(1-\phi)(-r'_A) = \rho_C(1-\phi)k'C_A$$

Additional information:

$$\rho_0 = 0.032 \text{ kg/dm}^3$$
$$D_P = 0.02 \text{ dm} \qquad\qquad \phi = 0.4$$
$$k' = 0.02 \text{ dm}^3/\text{kg cat} \cdot \text{s} \quad \mu = 1.5 \times 10^{-6} \text{ kg/dm} \cdot \text{s}$$
$$L = L' = 27 \text{ dm} \qquad\qquad \rho_c = 2.6 \text{ kg/dm}^3$$

Solution

We begin by performing a mole balance over the cylindrical core of thickness Δz shown in Figure E4-8.1.

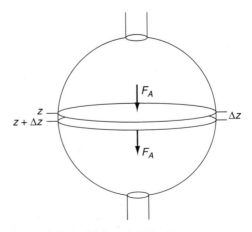

Figure E4-8.1   Spherical reactor.

Following
the algorithm

1. **Mole balance**:

$$\text{In} - \text{out} + \text{generation} = 0$$

$$F_A|_z - F_A|_{z+\Delta z} + r_A A_c \Delta z = 0$$

Dividing by Δz and taking the limit as $\Delta z \longrightarrow 0$ yields

$$\frac{dF_A}{dz} = r_A A_c$$

In terms of conversion

$$\boxed{\frac{dX}{dz} = \frac{-r_A A_c}{F_{A0}}} \tag{E4-8.1}$$

2. **Rate law**:

$$\boxed{r_A = -kC_A = -k'C_A \rho_c (1 - \phi)} \tag{E4-8.2}$$

3. **Stoichiometry**. Gas, isothermal $(T = T_0)$:

$$\boxed{C_A = C_{A0}\left(\frac{1-X}{1+\varepsilon X}\right)y} \tag{E4-8.3}$$

$$\varepsilon = y_{A0}\delta = 1 \times (1 + 1 - 1) = 1 \tag{E4-8.4}$$

where

$$y = \frac{P}{P_0} \tag{E4-8.5}$$

Note that y_{A0} (y with a subscript) represents the mole fraction and y alone represents the pressure ratio, P/P_0.

The variation in the dimensionless pressure, y, is given by incorporating the variable y in Equation (4-24):

$$\boxed{\frac{dy}{dz} = -\frac{\beta_0}{P_0 y}(1 + \varepsilon X)} \tag{E4-8.6}$$

The units of β_0 for this problem are kPa/dm³.

The equations in
boxes are the key
equations used
in the ODE solver
program

$$\beta_0 = \frac{G(1-\phi)}{\rho_0 g_c D_P \phi^3}\left[\frac{150(1-\phi)\mu}{D_P} + 1.75G\right] \tag{E4-8.7}$$

$$G = \frac{\dot{m}}{A_c} \tag{E4-8.8}$$

For a spherical reactor

$$\boxed{A_c = \pi[R^2 - (z - L)^2]} \tag{E4-8.9}$$

$$\boxed{W = \rho_c(1 - \phi)\pi\left[R^2 z - \frac{1}{3}(z - L)^3 - \frac{1}{3}L^3\right]} \tag{E4-8.10}$$

Parameter evaluation:

Recall that $g_c = 1$ for metric units.

$$\beta_0 = \left[\frac{G(1-0.4)}{(0.032 \text{ kg/dm}^3)(0.02 \text{ dm})(0.4)^3} \right]$$

$$\times \left[\frac{150(1-0.4)(1.5 \times 10^{-6} \text{ kg/dm} \cdot \text{s})}{0.02 \text{ dm}} + 1.75G \right]$$ (E4-8.11)

$$\boxed{\beta_0 = [(98.87 \text{ s}^{-1})G + (25{,}630 \text{ dm}^2/\text{kg})G^2] \times \left(0.01 \frac{\text{kPa/dm}}{\text{kg/dm}^2 \cdot \text{s}^2} \right)}$$ (E4-8.12)

The last term in brackets converts $(\text{kg}/\text{dm}^2 \cdot \text{s})$ to (kPa/dm). Recalling other parameters, $\dot{m} = 44$ kg/s, $L = 27$ dm, $R = 30$ dm, and $\rho_{\text{cat}} = 2.6$ kg/dm^3.

Table E4-8.1 shows the POLYMATH input used to solve the above equations. The MATLAB program is given as a living example problem on the CD-ROM.

TABLE E4-8.1.   POLYMATH PROGRAM

Equations	Initial Values
`d(X)/d(z)=-ra*Ac/Fao`	0
`d(y)/d(z)=-beta/Po/y*(1+X)`	1
`Fao=440`	
`Po=2000`	
`CaO=0.32`	
`R=30`	
`phi=0.4`	
`kprime=0.02`	
`L=27`	
`rhocat=2.6`	
`m=44`	
`Ca=CaO*(1-X)*y/(1+X)`	
`Ac=3.1416*(R^2-(z-L)^2)`	
`V=3.1416*(z*R^2-1.3*(z-L)^3-1/3*L^3)`	
`G=m/Ac`	
`ra=-kprime*Ca*rhocat*(1-phi)`	
`beta=(98.87*G+25630*G^2)*0.01`	
`W=rhocat*(1-phi)*V`	
$z_0 = 0, \quad z_f = 54$	

CD
Living

Example Problems

For the spherical reactor, the conversion and the pressure at the exit are

$$X = 0.81 \quad P = 1980 \text{ kPa}$$

If similar calculations are performed for the tubular packed-bed reactor (PBR), one finds that for the same catalyst weight the conversion and pressure at the exit are

A comparison between reactors

$$X = 0.71 \quad P = 308 \text{ kPa}$$

Figure E4-8.2 shows how conversion, X, and dimensionless pressure, y, vary with catalyst weight in each reactor. Here X_1 and y_1 represent the tubular reactor and X_2

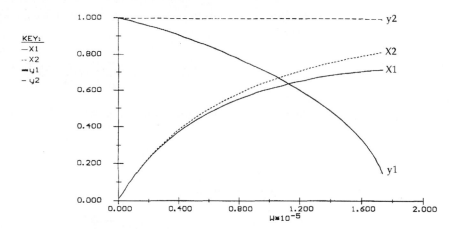

Figure E4-8.2   Pressure and conversion for: 1, tubular PBR; 2, spherical PBR.

and y_2 represent the spherical reactor. In addition to the higher conversion, the spherical reactor has the economic benefit of reducing the pumping and compression cost because of higher pressure at the exit.

Because the pressure drop in the spherical reactor is very small, one could increase the reactant flow rate significantly and still maintain adequate pressure at the exit. In fact, Amoco uses a reactor with similar specifications to process 60,000 barrels of petroleum naphtha per day.

4.4.4 Pressure Drop in Pipes

Normally, the pressure drop for gases flowing through pipes without packing can be neglected. For flow in pipes, the pressure drop along the length of the pipe is given by

$$\frac{dP}{dL} = -G\frac{du}{dL} - \frac{2fG^2}{\rho D} \tag{4-40}$$

where D = pipe diameter, cm

u = average velocity of gas, cm/s

f = Fanning friction factor

$G = \rho u$, g/cm$^2 \cdot$ s

The friction factor is a function of the Reynolds number and pipe roughness. The mass velocity G is constant along the length of the pipe. Replacing u with G/ρ, and combining with Equation (4-23) for the case of constant T and F_T, Equation (4-40) becomes

$$\rho_0 \frac{dP}{dL} - G^2\frac{dP}{P\,dL} + \frac{2fG^2}{D} = 0$$

Integrating with limits $P = P_0$ when $L = 0$, and assuming that f does not vary, we have

$$\frac{P_0^2 - P^2}{2} = G^2 \frac{P_0}{\rho_0} \left(2f \frac{L}{D} + \ln \frac{P_0}{P} \right)$$

Neglecting the second term on the right-hand side gives

$$\frac{P_0^2 - P^2}{2} = 2fG^2 \frac{P_0}{\rho_0} \frac{L}{D}$$

Rearranging, we obtain

$$\frac{P}{P_0} = \left[1 - \frac{4fG^2 V}{\rho_0 P_0 A_c D} \right]^{1/2} = (1 - \alpha_p V)^{1/2} \qquad (4\text{-}41)$$

where

$$\alpha_p = \frac{4fG^2}{A_c \rho_0 P_0 D}$$

For the flow conditions given in Example 4-4 in a 1000-ft length of $1\frac{1}{2}$-in. schedule 40 pipe ($\alpha_p = 0.0118$), the pressure drop is less than 10%.

4.5 Synthesizing a Chemical Plant

Synthesizing a chemical plant

Careful study of the various reactions, reactors, and molar flows of the reactants and products used in the example problems in this chapter reveals that they can be arranged to form a chemical plant to produce 200 million pounds of ethylene glycol from a feedstock of 402 million pounds per year of ethane. The flowsheet for the arrangement of the reactors together with the molar flow rates is shown in Figure 4-11. Here 0.425 lb mol/s of ethane is fed to 100 tubular plug-flow reactors connected in parallel; the total volume is 81 ft³ to produce 0.34 lb mol/s of ethylene (see Example 4-4). The reaction mixture is then fed to a separation unit where 0.04 lb mol/s of ethylene is lost in the separation process in the ethane and hydrogen streams that exit the separator. This process provides a molar flow rate of ethylene of 0.3 lb mol/s which enters the packed-bed catalytic reactor together with 0.15 lb mol/s of O_2 and 0.564 lb mol/s of N_2. There are 0.18 lb mol/s of ethylene oxide (see Example 4-6) produced in the 1000 pipes arranged in parallel and packed with silver-coated catalyst pellets. There is 60% conversion achieved in each pipe and the total catalyst weight in all the pipes is 45,400 lb. The effluent stream is passed to a separator where 0.03 lb mol/s of ethylene oxide is lost. The ethylene oxide stream is then contacted with water in a gas absorber to produce a 1-lb mol/ft³ solution of ethylene oxide in water. In the absorption process, 0.022 lb mol/s of ethylene oxide is lost. The ethylene oxide solution is fed to a 197-ft³ CSTR together with a stream of 0.5 wt % H_2SO_4 solution to produce ethylene glycol

Always challenge the assumptions, constraints, and boundaries of the problem

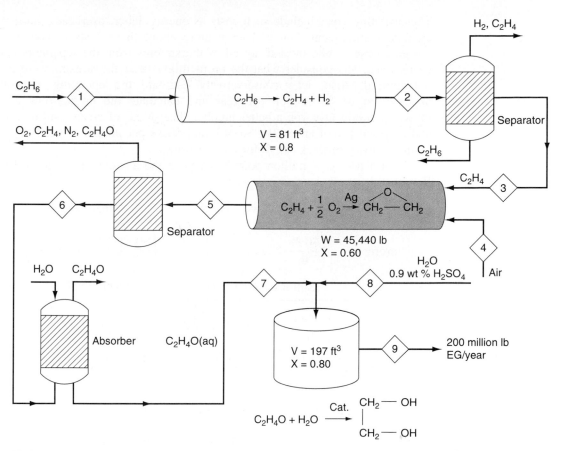

Figure 4-11   Production of ethylene glycol.

Stream	Component[a]	Flow rate (lb mol/s)	Stream	Component[a]	Flow rate (lb mol/s)
1	C_2H_6	0.425	6	EO	0.150
2	C_2H_4	0.340	7	EO	0.128
3	C_2H_4	0.300	8	H_2O	0.44
4	Air	0.714	9	EG	0.104
5	EO	0.180			

[a]EG, ethylene glycol; EO, ethylene oxide.

at a rate of 0.102 lb mol/s (see Example 4-2). This rate is equivalent to approximately 200 million pounds of ethylene glycol per year.

The profit from a chemical plant will be the difference between income from sales and the cost to produce the chemicals. An approximate formula might be

$$\text{Profit} = \text{value of products} - \text{cost of reactants}$$
$$- \text{operating costs} - \text{separation costs}$$

$$\$\$\$\$$$

The operating costs include such costs as energy, labor, overhead, and depreciation of equipment. You will learn more about these costs in your senior design course. While most if not all of the streams from the separators could be recycled, lets consider what the profit might be if the streams were to go unrecovered. Also, let's conservatively estimate the operating and other expenses to be $8 million per year and calculate the profit. Your design instructor might give you a better number. The prices of ethane, sulfuric acid, and ethylene glycol are $0.04, $0.043, and $0.38 per pound, respectively. See "http://www.chemweek.com/" for current prices.

For a feed of 4 million pounds per year and a production rate of 2 million pounds of ethylene glycol per year:

$$
\text{profit} = \left[\left(\overbrace{\frac{\$0.38}{\text{lb}} \times 2 \times 10^8 \frac{\text{lb}}{\text{year}}}^{\text{(Ethylene Glycol Cost)}} \right) - \overbrace{\left(\frac{\$0.34}{\text{lb}} \times 4 \times 10^8 \frac{\text{lb}}{\text{year}} \right)}^{\text{(Ethane Cost)}} \right.
$$

$$
\left. - \overbrace{\left(\frac{\$0.043}{\text{lb}} \times 2.26 \times 10^6 \frac{\text{lb}}{\text{year}} \right)}^{\text{(Sulfuric Acid Cost)}} - \overbrace{\$8,000,000}^{\text{Operating Cost}} \right]
$$

$$
= \$76,000,000 - \$16,000,000 - \$54,000 - \$8,000,000
$$

$$
\cong \$52 \text{ million}
$$

Using $52 million a year as a rough estimate of the profit, you can now make different approximations about the conversion, separations, recycle streams, and operating costs to learn how they affect the profit.

4.6 Using C_A (liquid) and F_A (gas) in the Mole Balances and Rate Laws

Used for:
• Multiple rxns
• Membranes
• Unsteady state

There are a number of instances when it is much more convenient to work in terms of the number of moles (N_A, N_B) or molar flow rates (F_A, F_B, etc.) rather than conversion. Membrane reactors and multiple reactions taking place in the gas phase are two such cases where molar flow rates rather than conversion are preferred. In Section 3.4 we described how we can express concentrations in terms of the molar flow rates of the reacting species rather than conversion. We will develop our algorithm using concentrations (liquids) and molar flow rates (gas) as our dependent variables. The main difference is that when conversion is used as our variable to relate one species concentration to that of another species concentration, we needed to write a mole balance on only one species, our basis of calculation. When molar flow rates and concentrations are used as our variables, we must write a mole balance on each species and then relate the mole balances to one another through the relative rates of reaction; for

$$
A + \frac{b}{a} B \longrightarrow \frac{c}{a} C + \frac{d}{a} D \tag{2-2}
$$

we have

$$\frac{r_A}{-a} = \frac{r_B}{-b} = \frac{r_C}{c} = \frac{r_D}{d} \tag{2-20}$$

Reaction (2-2) will be used together with the generic rate law

$$-r_A = k_A C_A^\alpha C_B^\beta \tag{5-2}$$

to develop the algoritms when C_i (liquids) and F_i (gases) are used as the system variables.

4.6.1 CSTRs, PFRs, PBRs, and Batch Reactors

Liquid Phase. For liquid-phase reactions in which there is no volume change, concentration is the preferred variable. The mole balances are shown in Table 4–5 in terms of concentration for the four reactor types we have been discussing. We see from Table 4-5 that we have only to specify the parameter values for the system (C_{A0}, v_0, etc.) and for the rate law (i.e., k_A, α, β) to solve the coupled ordinary differential equations for either PFR, PBR, or batch reactors or to solve the coupled algebraic equations for a CSTR.

TABLE 4-5.   MOLE BALANCES FOR LIQUID-PHASE REACTIONS

LIQUIDS

Batch	$\dfrac{dC_A}{dt} = r_A$ and	$\dfrac{dC_B}{dt} = \dfrac{b}{a} r_A$
CSTR	$V = \dfrac{v_0(C_{A0} - C_A)}{-r_A}$ and	$V = \dfrac{v_0(C_{B0} - C_B)}{-(b/a)r_A}$
PFR	$v_0 \dfrac{dC_A}{dV} = r_A$ and	$v_0 \dfrac{dC_B}{dV} = \dfrac{b}{a} r_A$
PBR	$v_0 \dfrac{dC_A}{dW} = r_A'$ and	$v_0 \dfrac{dC_B}{dW} = \dfrac{b}{a} r_A'$

Gas Phase. For gas-phase reactions, the mole balances are given identically in Table 4-6. Consequently, the concentrations in the rate laws need to be expressed in terms of the molar flow rates: for example,

$$r_j = k C_j^2$$

Rate law   We start by recalling and combining Equations (3-40) and (3-41),

$$C_{T0} = \frac{F_{T0}}{v_0} = \frac{P_0}{Z_0 R T_0} \tag{3-40}$$

Stoichiometry

$$v = \left(\frac{v_0}{F_{T0}}\right) F_T \frac{P_0}{P} \frac{T}{T_0} \qquad (3\text{-}41)$$

to give the case of an ideal gas ($Z = 1$). We now recall Equation (3-45),

$$C_j = C_{T0} \frac{F_j}{F_T} \frac{P}{P_0} \frac{T_0}{T} \qquad (3\text{-}45)$$

The total molar flow rate is given as the sum of the flow rates of the individual species:

$$F_T = \sum_{j=1}^{n} F_j$$

when species A, B, C, D, and I are the only ones present. Then

$$F_T = F_A + F_B + F_C + F_D + F_I$$

The molar flow rates for each species F_j are obtained from a mole balance on each species, as given in Table 4-6: for example,

Must write a mole balance on each species

$$\frac{dF_j}{dV} = r_j \qquad (4\text{-}42)$$

We now return to Example 3-8 to complete its solution.

TABLE 4-6.   ALGORITHM FOR GAS-PHASE REACTIONS

$$a\text{A} + b\text{B} \longrightarrow c\text{C} + d\text{D}$$

1. Mole balances:

Batch	CSTR	PFR
$\dfrac{dN_A}{dt} = r_A V$	$V = \dfrac{F_{A0} - F_A}{-r_A}$	$\dfrac{dF_A}{dV} = r_A$
$\dfrac{dN_B}{dt} = r_B V$	$V = \dfrac{F_{B0} - F_B}{-r_B}$	$\dfrac{dF_B}{dV} = r_B$
$\dfrac{dN_C}{dt} = r_C V$	$V = \dfrac{F_{C0} - F_C}{-r_C}$	$\dfrac{dF_C}{dV} = r_C$
$\dfrac{dN_D}{dt} = r_D V$	$V = \dfrac{F_{D0} - F_D}{-r_D}$	$\dfrac{dF_D}{dV} = r_D$

2. Rate Law:

$$-r_A = k_A C_A^{\alpha} C_B^{\beta}$$

3. Stoichiometry:

Relative rates of reaction:

$$\frac{r_A}{-a} = \frac{r_B}{-b} = \frac{r_C}{c} = \frac{r_D}{d}$$

TABLE 4-6.   *(continued)*

then

$$r_B = \frac{b}{a} r_A \qquad r_C = -\frac{c}{a} r_A \qquad r_D = -\frac{d}{a} r_A$$

Concentrations:

$$C_A = C_{T0} \frac{F_A}{F_T} \frac{P}{P_0} \frac{T_0}{T} \quad C_B = C_{T0} \frac{F_B}{F_T} \frac{P}{P_0} \frac{T_0}{T}$$

$$C_C = C_{T0} \frac{F_C}{F_T} \frac{P}{P_0} \frac{T_0}{T} \quad C_D = C_{T0} \frac{F_D}{F_T} \frac{P}{P_0} \frac{T_0}{T}$$

Total molar flow rate: $F_T = F_A + F_B + F_C + F_D$

4. Combine: For an isothermal operation of a PBR with no ΔP

$$\frac{dF_A}{dV} = -k_A C_{T0}^{\alpha+\beta} \left(\frac{F_A}{F_T}\right)^\alpha \left(\frac{F_B}{F_T}\right)^\beta \qquad \frac{dF_{BC}}{dV} = -\frac{b}{a} k_A C_{T0}^{\alpha+\beta} \left(\frac{F_A}{F_T}\right)^\alpha \left(\frac{F_B}{F_T}\right)^\beta$$

$$\frac{dF_C}{dV} = \frac{c}{a} k_A C_{T0}^{\alpha+\beta} \left(\frac{F_A}{F_T}\right)^\alpha \left(\frac{F_B}{F_T}\right)^\beta \qquad \frac{dF_{DC}}{dV} = \frac{d}{a} k_A C_{T0}^{\alpha+\beta} \left(\frac{F_A}{F_T}\right)^\alpha \left(\frac{F_B}{F_T}\right)^\beta$$

1. Specify parameter values: $k_A, C_{T0}, \alpha, \beta, T_0, a, b, c, d$
2. Specify entering numbers: $F_{A0}, F_{B0}, F_{C0}, F_{D0}$ and final values: V_{final}

5. Use an ODE solver.

Example 4–9 *Working in Terms of Molar Flow Rates in a PFR*

The gas-phase reaction

$$A \rightleftharpoons 2B$$

is carried out isothermally ($T = 500$ K) and isobarically ($P_0 = 4.1$ atm) in a PFR and follows an elementary rate law. Express the rate law and mole balances in terms of the molar flow rates and solve the combined equations to determine the molar flow rates along the length of a 100-dm^3 PFR.

Additional information:

$$k_A = 2.7 \text{ min}^{-1}, \ K_C = 1.2 \text{ mol/dm}^3, \ F_{A0} = 10 \text{ mol/min}$$

Solution

The algorithm | **Mole balance:**

$$\frac{dF_A}{dV} = r_A \tag{E4-9.1}$$

$$\frac{dF_B}{dV} = r_B \tag{E4-9.2}$$

Rate law:

$$-r_A = k_A \left(C_A - \frac{C_B^2}{K_C} \right) \tag{E4-9.3}$$

Stoichiometry. Using Equation (3-45) to substitute for the concentrations of A and B in terms of the molar flow rates, we have for $T = T_0$ and $P = P_0$ gives

$$C_A = C_{T0} \frac{F_A}{F_T} \qquad\qquad (E4\text{-}9.4)$$

$$C_B = C_{T0} \frac{F_B}{F_T} \qquad\qquad (E4\text{-}9.5)$$

where the total molar flow rate, F_T, is just the sum of the flow rates of A and B:

$$F_T = F_A + F_B \qquad\qquad (E4\text{-}9.6)$$

For every 1 mol of A disappearing 2 mol of B appear:

$$r_B = 2(-r_A) \qquad\qquad (E4\text{-}9.7)$$

and

$$F_B = 2(F_{A0} - F_A) \qquad\qquad (E4\text{-}9.8)$$

The total molar flow rate

$$F_T = F_A + 2F_{A0} - 2F_A = 2F_{A0} - F_A \qquad\qquad (E4\text{-}9.9)$$

and the total concentration at the entrance to the reactor (P_o, T_o) is calculated from the equation

$$C_{T0} = \frac{P_o}{RT_o} = \frac{4.1 \text{ atm}}{0.082 \dfrac{\text{dm}^3 \cdot \text{atm}}{\text{mol} \cdot \text{K}} (500 \text{ K})} = 0.1 \frac{\text{mol}}{\text{dm}^3} \qquad (E4\text{-}9.10)$$

$$C_A = C_{T0} \frac{F_A}{2F_{A0} - F_A} \qquad\qquad (E4\text{-}9.11)$$

$$C_B = C_{T0} \frac{F_B}{2F_{A0} - F_A} \qquad\qquad (E4\text{-}9.12)$$

TABLE E4-9.1.   POLYMATH PROGRAM

Equations	Initial Values
d(fa)/d(v)=ra	10
ka=2.7	
kc=1.2	
ct0=.1	
fa0=10	
fb=2*(fa0-fa)	
ft=2*fa0-fa	
ca=ct0*fa/ft	
cb=ct0*fb/ft	
ra=-ka*(ca-cb**2/kc)	
rb=-2*ra	
$v_0 = 0,$ $v_f = 100$	

CD Living
Example Problems

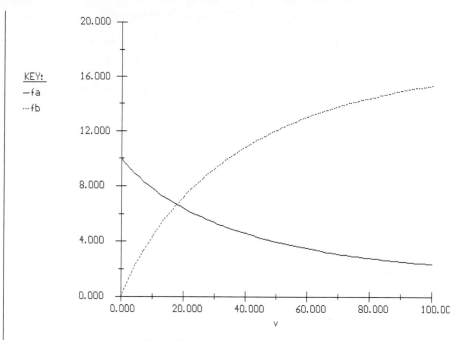

KEY:
−fa
···fb

Figure E4-9.1   Molar flow rate profiles.

The mole balance (E4-9.1) and the rate laws (E4-9.3) are combined to give

$$\frac{dF_A}{dV} = r_A = -k_A[C_A - C_B^2/K_C] \tag{E4-9.13}$$

Equations (E4-9.11) through (E4-9.13) can now be solved numerically, preferably by a software package such as POLYMATH or MATLAB. The POLYMATH program and solution are given in Table E4-9.1 for $k_A = 2.7$ min^{-1}, $K_C = 1.2$ mol/dm^3, and $F_{A0} = 10$ mol/min. The MATLAB program is included on the CD-ROM.

The molar flows of A and B at equilibrium can be found as follows:

$$K_C = \frac{C_{Be}^2}{C_{Ae}} = \frac{\left[C_{T0}\left(\dfrac{F_{Be}}{F_{Ae} + F_{Be}}\right)\right]^2}{C_{T0}\left(\dfrac{F_{Ae}}{F_{Ae} + F_{Be}}\right)} \tag{E4-9.14}$$

simplifying

$$\frac{C_{T0}F_{Be}^2}{F_{Ae}(F_{Ae} + F_{Be})} = K_C \tag{E4-9.15}$$

Substituting for F_{Be}, C_{T0}, and K_C gives

$$\frac{(0.1)(2)^2(10 - F_{Ae})^2}{F_{Ae}(2F_{A0} - F_{Ae})} = 1.2 \tag{E4-9.16}$$

Equations (E4-9.15) and (E4-9.16) solve to

$$F_{Ae} = 1.34 \quad \text{mol/min}$$
$$F_{Be} = 17.32 \text{ mol/min}$$

We note from Figure E4-9.1 that the molar flow rates begin to approach the equilibrium values near the end of the reactor.

4.6.2 Membrane Reactors

Catalytic membrane reactors can be used to increase the yield of reactions that are highly reversible over the temperature range of interest. (Some refer to this type of reaction as being thermodynamically limited.) The term *membrane reactor* describes a number of different types of reactor configurations that contain a membrane. The membrane can either provide a barrier to certain components, while being permeable to others, prevent certain components such as particulates from contacting the catalyst, or contain reactive sites and be a catalyst in itself. Like reactive distillation, the membrane reactor is another technique for driving reversible reactions to the right in order to achieve very high conversions. These high conversions can be achieved by having one of the reaction products diffuse out of a semipermeable membrane surrounding the reacting mixture.[5] As a result, the reaction will continue to proceed to the right toward completion.

By having one of the products pass throughout the membrane, we drive the reaction towards completion

Two of the main types of catalytic membrane reactors are shown in Figure 4-12. The reactor in the middle is called an *inert membrane reactor with catalyst pellets on the feed side* (IMRCF). Here the membrane is inert and serves as a barrier to the reactants and some of the products. The reactor on the bottom is a *catalytic membrane reactor* (CMR). The catalyst is deposited directly on the membrane and only specific reaction products are able to exit the permeate side. For example, in the reversible reaction

$$C_6H_{12} \xrightleftharpoons{} C_6H_6 + 3H_2$$

H$_2$ diffuses through the membrane while C$_6$H$_6$ does not

the hydrogen molecule is small enough to diffuse through the small pores of the membrane while C_6H_{12} and C_6H_6 cannot. Consequently, the reaction continues to proceed to the right even for a small value of the equilibrium constant.

Detailed modeling of the transport and reaction steps in membrane reactors is beyond the scope of this text but can be found in *Membrane Reactor Technology.*[6] The salient features, however, can be illustrated by the following example. When analyzing membrane reactors, it is much more convenient to use molar flow rates rather than conversion.

[5] R. Govind, and N. Itoh, eds., *Membrane Reactor Technology,* AIChE Symposium Series No. 268, Vol. 85 (1989). T. Sun and S. Khang, *Ind. Eng. Chem. Res., 27,* 1136 (1988).

[6] Govind and Itoh, *Membrane Reactor Technology.*

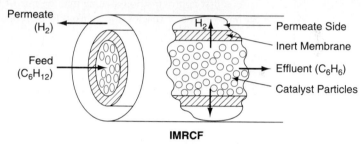

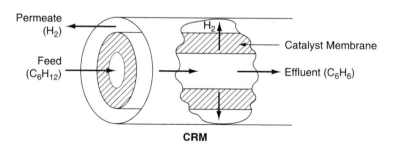

Figure 4-12   Membrane reactors. (Photo courtesy of Coors Ceramics, Golden, Colorado.)

According to the
DOE, 10 trillion
BTU/yr could be
saved by using
membrane reactors

Example 4–10 Membrane Reactor

According to The Department of Energy, an energy saving of 10 trillion Btu per year could result from the use of catalytic membrane reactors as replacements for conventional reactors for dehydrogenation reactions such as the dehydrogenation of ethylbenzene to styrene:

and of butane to butene:

$$C_4H_{10} \longrightarrow C_4H_8 + H_2$$

The dehydrogenation of propane is another reaction that has proven successful with a membrane reactor [*J. Membrane Sci.*, 77, 221 (1993)].

$$C_3H_8 \longrightarrow C_3H_6 + H_2$$

All the dehydrogenation reactions above can be represented symbolically as

$$A \underset{\longleftarrow}{\overrightarrow{\hspace{1cm}}} B + C$$

and will take place on the catalyst side of an IMRCF. The equilibrium constant for this reaction is quite small at 227°C (i.e., $K_C = 0.05$ mol/dm³). The membrane is permeable to B (e.g. H_2) but not to A and C. Pure gaseous A enters the reactor at 8.2 atm and 227°C at a rate of 10 mol/min.

As a first approximation assume that the rate of diffusion of B out of the reactor per unit volume of reactor, R_B, is taken to be proportional to the concentration of B (i.e., $R_B = k_c C_B$).

(a) Perform differential mole balances on A, B, and C to arrive at a set of coupled differential equations to solve.

(b) Plot the molar flow rates of each species as a function of space time.

Additional information: Even though this reaction is a gas–solid catalytic reaction, we will make use of the bulk catalyst density in order to write our balances in terms of reactor volume rather than catalyst weight (recall $-r_A = -r'_A \rho_b$). For the bulk catalyst density of $\rho_b = 1.5$ g/cm³ and a 2-cm inside diameter of the tube containing the catalyst pellets, the specific reaction rate, k, and the transport coefficient, k_c, are $k = 0.7$ min⁻¹ and $k_c = 0.2$ min⁻¹, respectively.

Solution

We shall choose reactor volume rather than catalyst weight as our independent variable for this example. First we shall perform mole balances on the volume element ΔV shown in Figure E4-10.1.

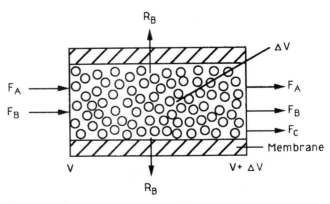

Figure E4-10.1

1. **Mole balances:**

Balance on A in the catalytic bed:

$$\underbrace{\left[\begin{array}{c} \text{In} \\ \text{by Flow} \end{array}\right]}_{F_A|_V} - \underbrace{\left[\begin{array}{c} \text{Out} \\ \text{by Flow} \end{array}\right]}_{F_A|_{V+\Delta V}} + \underbrace{\left[\begin{array}{c} \text{Generation} \end{array}\right]}_{r_A \Delta V} = \underbrace{\left[\begin{array}{c} \text{Accumulation} \end{array}\right]}_{0}$$

Dividing by ΔV and taking the limit as $\Delta V \to 0$ gives

$$\boxed{\frac{dF_A}{dV} = r_A} \qquad (E4\text{-}10.1)$$

Balance on B in the catalytic bed:

Note there are two "OUT" terms for species B

$$\underbrace{\left[\begin{array}{c} \text{In} \\ \text{by Flow} \end{array}\right]}_{F_B|_V} - \underbrace{\left[\begin{array}{c} \text{Out} \\ \text{by Flow} \end{array}\right]}_{F_B|_{V+\Delta V}} - \underbrace{\left[\begin{array}{c} \text{Out} \\ \text{by Diffusion} \end{array}\right]}_{R_B \Delta V} + \underbrace{\left[\begin{array}{c} \text{Generation} \end{array}\right]}_{r_B \Delta V} = \underbrace{\left[\begin{array}{c} \text{Accumulation} \end{array}\right]}_{0}$$

where R_B is the molar flow of B out through the membrane per unit volume of reactor. Dividing by ΔV and taking the limit as $\Delta V \to 0$ gives

$$\boxed{\frac{dF_B}{dV} = r_B - R_B} \qquad (E4\text{-}10.2)$$

The mole balance on C is carried out in an identical manner to A and the resulting equation is

$$\frac{dF_C}{dV} = r_C \qquad (E4\text{-}10.3)$$

2. **Rate law:**

$$\boxed{-r_A = k\left(C_A - \frac{C_B C_C}{K_C}\right)} \qquad (E4\text{-}10.4)$$

$$r_B = -r_A$$
$$r_C = -r_A$$

3. **Transport out the sides of the reactor.** We have assumed that

$$\boxed{R_B = k_c C_B} \qquad (E4\text{-}10.5)$$

where k_c is a transport coefficient. In general, this coefficient can be a function of the membrane and fluid properties, the fluid velocity, the tube diameter, and so on (see Chapter 11). However, in this example we assume that the main resistance to diffusion of B out of the reactor is the membrane itself and consequently, k_c is taken to be a constant.

4. **Stoichiometry.** Recalling Equation (3-45) for the case of constant temperature and pressure, we have for isothermal operation and no pressure drop ($T = T_0$, $P = P_0$),

$$C_A = C_{T0} \frac{F_A}{F_T}$$

(E4-10.6)

$$C_B = C_{T0} \frac{F_B}{F_T}$$

(E4-10.7)

$$C_C = C_{T0} \frac{F_C}{F_T}$$

(E4-10.8)

$$F_T = F_A + F_B + F_C$$

(E4-10.9)

$$-r_A = r_B = r_C$$

(E4-10.10)

5. **Combining and summarizing:**

Summary of equations describing flow and reaction in a membrane reactor

$$\frac{dF_A}{dV} = r_A$$

$$\frac{dF_B}{dV} = -r_A - k_c C_{T0} \left(\frac{F_B}{F_T} \right)$$

$$\frac{dF_C}{dV} = -r_A$$

$$-r_A = k C_{T0} \left[\left(\frac{F_A}{F_T} \right) - \frac{C_{T0}}{K_C} \left(\frac{F_B}{F_T} \right) \left(\frac{F_C}{F_T} \right) \right]$$

$$F_T = F_A + F_B + F_C$$

6. **Parameter evaluation:**

$$C_{T0} = \frac{P_0}{RT_0} = \frac{8.2 \text{ atm}}{0.082 \text{ atm} \cdot \text{dm}^3 /(\text{mol} \cdot \text{K}) \ (500 \text{ K})} = 0.2 \frac{\text{mol}}{\text{dm}^3}$$

$$k = 0.7 \text{ min}^{-1}, K_C = 0.05 \text{ mol/dm}^3, k_c = 0.2 \text{ min}^{-1}$$

$$F_{A0} = 10 \text{ mol/min}$$

$$F_{B0} = F_{C0} = 0$$

7. **Numerical solution.** Equations (E4-10.1) through (E4-10.10) were solved using POLYMATH and another ODE Solver MATLAB. The profiles of the molar flow rates are shown below. Table E4-10.1 shows the POLYMATH

programs, and Figure E4-10.2 shows the results of the numerical solution of the initial (entering) conditions.

$$A = 0 \quad F_A = F_{A0} \quad F_B = 0 \quad F_C = 0$$

TABLE E4-10.1.   POLYMATH PROGRAM

Equations	Initial Values
d(Fb)/d(V)=-ra-kc*Cto*(Fb/Ft)	0
d(Fa)/d(V)=ra	10
d(Fc)/d(V)=-ra	0
kc=0.2	
Cto=0.2	
Ft=Fa+Fb+Fc	
k=0.7	
Kc=0.05	
ra=-k*Cto*((Fa/Ft)-Cto/Kc*(Fb/Ft)*(Fc/Ft))	
$V_0 = 0, \quad V_f = 500$	

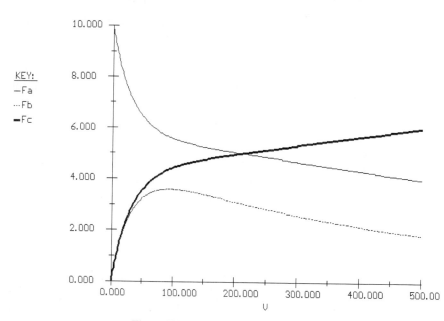

Figure E4-10.2   POLYMATH solution.

4.7 Unsteady-State Operation of Reactors

In this chapter we have already discussed the unsteady operation of one type of reactor, the batch reactor. In this section we discuss two other aspects of unsteady operation. First, the startup of a CSTR is examined to determine the time necessary to reach steady-state operation (see Figure 4-13a). Next, semi-batch reactors are discussed. In each of these cases, we are interested in predicting the concentration and conversion as a function of time. Closed-form analytical solutions to the differential equations arising from the mole balance of these reaction types can be obtained only for zero- and first-order reactions. ODE solvers must be used for other orders.

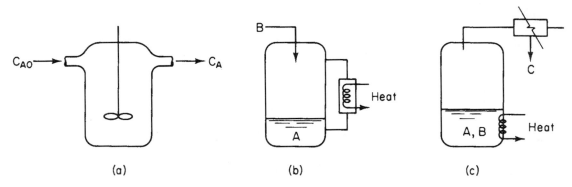

Figure 4-13   Semibatch reactors. [Excerpted by special permission from *Chem. Eng.*, *63*(10) 211 (Oct. 1956). Copyright © 1956 by McGraw-Hill, Inc., New York, NY 10020.]

There are two basic types of semibatch operations. In one type, one of the reactants in the reaction

$$A + B \rightarrow C + D$$

(e.g., B) is slowly fed to a reactor containing the other reactant (e.g., A), which has already been charged to a reactor such as that shown in Figure 4-13b. This type of reactor is generally used when unwanted side reactions occur at high concentrations of B, or the reaction is highly exothermic. In some reactions, the reactant B is a gas and is bubbled continuously through liquid reactant A. Examples of reactions used in this type of semibatch reactor operation include *ammonolysis*, *chlorination*, and *hydrolysis*. The other type of semibatch reactor is shown schematically in Figure 4-13c. Here reactants A and B are charged simultaneously and one of the products is vaporized and withdrawn continuously. Removal of one of the products in this manner (e.g., C) shifts the equilibrium towards the right, increasing the final conversion above that which would be achieved had C not been removed. In addition, removal of one of the products further concentrates the reactant, thereby producing an increased rate of reaction and decreased processing time. This type of reaction operation is called *reactive distillation*. Examples of reactions carried out in this type of reactor include *acetylation reactions* and *esterification reactions* in which water is removed.

4.7.1 Startup of a CSTR

An expanded version of this section can be found on the CD-ROM

The startup of a fixed volume CSTR under isothermal conditions is rare, but it does occur occasionally. Here we want to determine the time necessary to reach steady-state operation. We begin with the general mole balance equation applied to Figure 4-13a:

$$F_{A0} - F_A + r_A V = \frac{dN_A}{dt} \tag{4-43}$$

Conversion does not have any meaning in startup because one cannot separate the moles reacted from the moles accumulated in the CSTR. Consequently, we *must* use concentration rather than conversion as our variable in the balance equation. For liquid-phase ($v = v_0$) reactions with constant overflow ($V = V_0$), using $\tau = V_0/v_0$ we can transform Equation (4-43) to

$$C_{A0} - C_A + r_A \tau = \tau \frac{dC_A}{dt} \tag{4-44}$$

For a first-order reaction ($-r_A = kC_A$) Equation (4-44) then becomes

$$\frac{dC_A}{dt} + \frac{1 + \tau k}{\tau} C_A = \frac{C_{A0}}{\tau} \tag{4-45}$$

which solves to

$$C_A = \frac{C_{A0}}{1 + \tau k} \left\{ 1 - exp\left[-(1 + \tau k)\frac{t}{\tau} \right] \right\} \tag{4-46}$$

Letting t_s be the time necessary to reach 99% of the steady-state concentration, C_{AS}:

$$C_{AS} = \frac{C_{A0}}{1 + \tau k} \tag{4-47}$$

Rearranging Equation (4-46) for $C_A = 0.99 C_{AS}$ yields

$$t_s = 4.6 \frac{\tau}{1 + \tau k} \tag{4-48}$$

For slow reactions:

$$t_s = 4.6 \, \tau \tag{4-49}$$

For rapid reactions:

Time to reach steady state in an isothermal CSTR

$$t_s = \frac{4.6}{k} \tag{4-50}$$

For most first-order systems, steady state is achieved in three to four space times.

4.7.2 Semibatch Reactors

Of the two types of semibatch reactors described earlier, we focus attention primarily on the one with constant molar feed. A schematic diagram of this semibatch reactor is shown in Figure 4-14. We shall consider the elementary liquid-phase reaction

$$A + B \rightarrow C$$

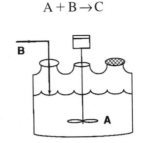

Figure 4-14   Semibatch reactor.

in which reactant B is slowly added to a vat containing reactant A. A **mole balance on species A yields**

$$
\begin{bmatrix} \text{rate} \\ \text{in} \end{bmatrix} - \begin{bmatrix} \text{rate} \\ \text{out} \end{bmatrix} + \begin{bmatrix} \text{rate of} \\ \text{generation} \end{bmatrix} = \begin{bmatrix} \text{rate of} \\ \text{accumulation} \end{bmatrix} \tag{4-51}
$$

$$
0 \quad - \quad 0 \quad + \quad r_A V(t) \quad = \quad \frac{dN_A}{dt}
$$

There are three variables that can be used to formulate and solve semibatch reactor problems: the concentration, C_j, the number of moles, N_j, and the conversion, X.

Writing the Semibatch Reactor Equations in Terms of Concentrations. Recalling that the number of moles of A is just the product of concentration of A, C_A, and the volume V, we can rewrite Equation (4-51) as

$$
r_A V = \frac{d(C_A V)}{dt} = \frac{V dC_A}{dt} + C_A \frac{dV}{dt} \tag{4-52}
$$

We note that since the reactor is being filled, the volume, V, varies with time. The reactor volume at any time t can be found from an overall **mass balance** of all species:

$$
\begin{bmatrix} \text{rate} \\ \text{in} \end{bmatrix} - \begin{bmatrix} \text{rate} \\ \text{out} \end{bmatrix} + \begin{bmatrix} \text{rate of} \\ \text{generation} \end{bmatrix} = \begin{bmatrix} \text{rate of} \\ \text{accumulation} \end{bmatrix} \tag{4-53}
$$

$$
\rho_0 v_0 \quad - \quad 0 \quad + \quad 0 \quad = \quad \frac{d(\rho V)}{dt}
$$

For a constant-density system, $\rho_0 = \rho$, and

$$\frac{dV}{dt} = v_0 \tag{4-54}$$

with the initial condition $V = V_0$ at $t = 0$, integrating for the case of constant V_0 yields

Semibatch
reactor
volume as a
function of time

$$\boxed{V = V_0 + v_0 t} \tag{4-55}$$

Substituting Equation (4-54) into the right-hand side of Equation (4-52) and rearranging gives us

$$-v_0 C_A + V r_A = \frac{V dC_A}{dt}$$

The balance of A [i.e. Equation (4-52)] can be rewritten as

Mole balance on A

$$\boxed{\frac{dC_A}{dt} = r_A - \frac{v_0}{V} C_A} \tag{4-56}$$

A mole balance of B that is fed to the reactor at a rate F_{B0} is

$$\frac{dN_B}{dt} = r_B V + F_{B0} \tag{4-57}$$

$$\frac{dVC_B}{dt} = \frac{dV}{dt} C_B + \frac{V dC_B}{dt} = r_B V + F_{B0}$$

Substituting Equation (4-55) in terms of V and differentiating, the mole balance on B becomes

Mole balance on B

$$\boxed{\frac{dC_B}{dt} = r_B + \frac{v_0(C_{B0} - C_B)}{V}} \tag{4-58}$$

If the reaction order is other than zero- or first-order, or if the reaction is nonisothermal, we must use numerical techniques to determine the conversion as a function of time. Equations (4-56) and (4-58) are easily solved with an ODE solver.

Example 4–11 Isothermal Semibatch Reactor with Second-Order Reaction

The production of methyl bromide is an irreversible liquid-phase reaction that follows an elementary rate law. The reaction

$$CNBr + CH_3NH_2 \rightarrow CH_3Br + NCNH_2$$

is carried out isothermally in a semibatch reactor. An aqueous solution of methyl amine (B) at a concentration of 0.025 g mol/dm³ is to be fed at a rate of 0.05 dm³/s to an aqueous solution of bromine cyanide (A) contained in a glass-lined reactor.

The initial volume of fluid in a vat is to be 5 dm³ with a bromine cyanide concentration of 0.05 mol/dm³. The specific reaction rate constant is

$$k = 2.2 \text{ dm}^3/\text{s} \cdot \text{mol}$$

Solve for the concentrations of bromine cyanide and methyl bromide and the rate of reaction as a function of time.

Solution

Symbolically, we write the reaction as

$$A + B \rightarrow C + D$$

The reaction is elementary; therefore, the rate law is

Rate Law

$$\boxed{-r_A = kC_A C_B} \tag{E4-11.1}$$

Substituting the rate law in Equations (4-56) and (4-58) gives

Mole balances on
A, B, C, and D

$$\boxed{\frac{dC_A}{dt} = -kC_A C_B - \frac{v_0}{V}C_A} \tag{E4-11.2}$$

$$\boxed{\frac{dC_B}{dt} = -kC_A C_B + \frac{v_0}{V}(C_{B0} - C_B)} \tag{E4-11.3}$$

$$\boxed{V = V_0 + v_0 t} \tag{E4-11.4}$$

Similarly for C and D we have

$$\frac{dN_C}{dt} = r_C V = -r_A V \tag{E4-11.5}$$

$$\frac{dN_C}{dt} = \frac{d(C_C V)}{dt} = V\frac{dC_C}{dt} + C_C\frac{dV}{dt} = V\frac{dC_C}{dt} + v_0 C_C \tag{E4-11.6}$$

Then

$$\boxed{\frac{dC_C}{dt} = kC_A C_B - \frac{v_0 C_C}{V}} \tag{E4-11.7}$$

and

$$\boxed{\frac{dC_D}{dt} = kC_A C_B - \frac{v_0 C_D}{V}} \tag{E4-11.8}$$

We could also calculate the conversion of A.

$$X = \frac{N_{A0} - N_A}{N_{A0}} \tag{E4-11.9}$$

$$X = \frac{C_{A0}V_0 - C_A V}{C_{A0}V_0} \qquad \text{(E4-11.10)}$$

The initial conditions are $t = 0$, $C_A = 0.05$, $C_B = 0$, $C_C = C_D = 0$, and $V_0 = 5$.

Equations (E4-11.2) through (E4-11.10) are easily solved with the aid of an ODE solver such as POLYMATH (Table E4-11.1).

TABLE E4-11.1.   POLYMATH PROGRAM

Equations	Initial Values
d(ca)/d(t)=-k*ca*cb-v00*ca/v	0.05
d(cb)/d(t)=-k*ca*cb+v00*(cb0-cb)/v	0
d(cc)/d(t)=k*ca*cb-v00*cc/v	0
d(cd)/d(t)=k*ca*cb-v00*cd/v	0
k=2.2	
v00=0.05	
cb0=0.025	
v0=5	
ca0=0.05	
rate=k*ca*cb	
v=v0+v00*t	
x=(ca0*v0-ca*v)/(ca0*v0)	
$t_0 = 0$, $t_f = 500$	

The concentrations of bromine cyanide (A), and methyl amine are shown as a function of time in Figure E4-11.1, and the rate is shown in Figure E4-11.2. For first- and zero-order reactions we can obtain analytical solutions for semibatch reactors operated isothermally.

Why does the concentration of CH_3 Br go through a maximum w.r.t. time?

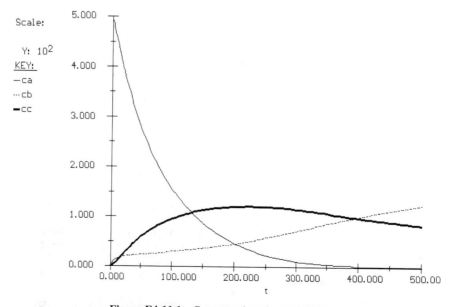

Figure E4-11.1 Concentration–time trajectories.

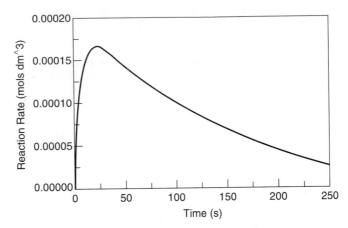

Figure E4-11.2   Reaction rate–time trajectory.

Why does the
reaction rate go
through a
maximum?

**Writing the Semibatch Reactor Equations in Terms of the Number of
Moles.**   We can also solve semibatch reactor problems by leaving the mole balance equations in terms of the number of moles of each species (i.e., N_A, N_B, N_C, and N_D).

Recalling the mole balance equations for A and B, Equations (4-51) and (4-57), respectively, along with the equation for the reactor volume, Equation (4-55), we have

$$\frac{dN_A}{dt} = r_A V(t) \tag{4-51}$$

$$\frac{dN_B}{dt} = r_A V(t) + F_{B0} \tag{4-57}$$

$$V = V_0 + v_0 t \tag{4-55}$$

One now recalls from Chapter 3 the definition of concentration for a batch system [Equation (3-24)]:

$$C_A = \frac{N_A}{V}$$

$$C_B = \frac{N_B}{V} \tag{3-24}$$

and substitutes these equations into the rate law to obtain equations that are stated solely in terms of the number of moles. For example, if $-r_A = k_A C_A C_B$, then

$$\frac{dN_A}{dt} = -k \frac{N_A N_B}{V_0 + v_0 t} \tag{4-59}$$

and

$$\frac{dN_B}{dt} = -k \frac{N_A N_B}{V_0 + v_0 t} + F_{B0} \tag{4-60}$$

One now has only to specify the parameter values (k, V_0, v_0, F_{B0}) and the initial conditions to solve these equations for N_A and N_B.

Writing the Semibatch Reactor Equations in Terms of Conversion.   Consider the reaction

$$A + B \rightleftharpoons C + D$$

in which B is fed to a vat containing only A initially. The reaction is first-order in A and first-order in B. The number of moles of A remaining at any time, t, can be found from the balance,

The limiting reactant is the one in the vat

$$\begin{bmatrix} \text{number of moles} \\ \text{of A in the vat} \\ \text{at time } t \end{bmatrix} = \begin{bmatrix} \text{number of moles} \\ \text{of A in the vat} \\ \text{initially} \end{bmatrix} - \begin{bmatrix} \text{number of moles} \\ \text{of A reacted up} \\ \text{to time } t \end{bmatrix} \quad (4\text{-}61)$$

$$N_A \qquad = \qquad N_{A0} \qquad - \qquad N_{A0}X$$

where X is the moles of A reacted per mole of A initially in the vat. Similarly, for species B.

$$\begin{bmatrix} \text{number of moles} \\ \text{of B in the vat} \\ \text{at time } t \end{bmatrix} = \begin{bmatrix} \text{number of moles} \\ \text{of B in the vat} \\ \text{initially} \end{bmatrix} + \begin{bmatrix} \text{number of moles} \\ \text{of B added to} \\ \text{the vat} \end{bmatrix} - \begin{bmatrix} \text{number of moles} \\ \text{of B reacted up} \\ \text{to time } t \end{bmatrix}$$

$$N_B \qquad = \qquad N_{Bi} \qquad + \qquad \int_0^t F_{B0} \, dt \qquad - \qquad N_{A0}X$$

$$(4\text{-}62)$$

For a constant molar feed rate

$$N_B = N_{Bi} + F_{B0}t - N_{A0}X \qquad (4\text{-}63)$$

A **mole balance** on species A gives

$$r_A V = \frac{dN_A}{dt} \qquad (4\text{-}64)$$

The number of moles of C and D can be taken directly from the stoichiometric table; for example,

$$N_C = N_{Ci} + N_{A0}X$$

For a reversible second-order reaction $A + B \rightleftharpoons C + D$ for which the **rate law** is

$$-r_A = k \left(C_A C_B - \frac{C_C C_D}{K_C} \right) \qquad (4\text{-}65)$$

Concentration
of reactants as a
function of
conversion and
time
the **concentrations of A and B are**

$$C_A = \frac{N_A}{V} = \frac{N_{A0}(1-X)}{V_0 + v_0 t} \qquad\qquad C_C = \frac{N_{A0}X}{V_0 + v_0 t}$$

$$C_B = \frac{N_B}{V} = \frac{N_{Bi} + F_{B0}t - N_{A0}X}{V_0 + v_0 t} \qquad C_D = \frac{N_{A0}X}{V_0 + v_0 t}$$

Combining equations (4-61), (4-64), and (4-65), substituting for the concentrations and dividing by N_{A0}, we obtain

$$\frac{dX}{dt} = \frac{k[(1-X)(N_{Bi} + F_{B0}t - N_{A0}X) - (N_{A0}X^2/K_C)]}{V_0 + v_0 t} \qquad (4\text{-}66)$$

Equation (4-66) needs to be solved numerically to determine the conversion as a function of time.

Equilibrium Conversion. For reversible reactions carried out in a semibatch reactor, the maximum attainable conversion (i.e., the equilibrium conversion) will change as the reaction proceeds because more reactant is continuously added to the reactor. This addition shifts the equilibrium continually to the right. Consider the reversible reaction

$$A + B \;\rightleftharpoons\; C + D$$

for which the rate law is

$$-r_A = k\left(C_A C_B - \frac{C_C C_D}{K_C}\right) \qquad (4\text{-}67)$$

If the reaction were allowed to reach equilibrium after feeding species B for a time t, the equilibrium conversion could be calculated as follows:

$$K_C = \frac{C_{Ce}C_{De}}{C_{Ae}C_{Be}} = \frac{\left(\dfrac{N_{Ce}}{V}\right)\left(\dfrac{N_{De}}{V}\right)}{\left(\dfrac{N_{Ae}}{V}\right)\left(\dfrac{N_{Be}}{V}\right)} \qquad (3\text{-}10)$$

$$= \frac{N_{Ce}N_{De}}{N_{Ae}N_{Be}}$$

The relationship between conversion and number of moles of each species is the same as shown in Table 3-1 except for species B, for which the number of moles is given by Equation (4-63). Thus

$$K_C = \frac{(N_{A0}X_e)(N_{A0}X_e)}{N_{A0}(1 - X_e)(F_{B0}t - N_{A0}X_e)}$$

$$= \frac{N_{A0}X_e^2}{(1 - X_e)(F_{B0}t - N_{A0}X_e)}$$

(4-68)

Rearranging yields

$$t = \frac{N_{A0}}{K_C F_{B0}}\left(K_C X_e + \frac{X_e^2}{1 - X_e}\right)$$

(4-69)

or

Equilibrium
conversion
in a semibatch
reactor

$$X_e = \frac{K_C\left(1 + \dfrac{F_{B0}t}{N_{A0}}\right) - \sqrt{\left[K_C\left(1 + \dfrac{F_{B0}t}{N_{A0}}\right)\right]^2 - 4(K_C - 1)K_C\dfrac{tF_{B0}}{N_{A0}}}}{2(K_C - 1)}$$

(4-70)

4.7.3 Reactive Distillation

The distillation of chemically reacting mixtures has become increasingly common in chemical industries.[7] Carrying out these two operations, reaction and distillation, simultaneously in a single unit results in significantly lower capital and operating costs. Reactive distillation is particularly attractive when one of the reaction products has a lower boiling point, resulting in its volatilization from the reacting liquid mixture. An example of reactive distillation is the production of methyl acetate:

$$CH_3COOH + CH_3OH \underset{k_2}{\overset{k_1}{\rightleftharpoons}} CH_3COOCH_3 + H_2O$$

By continually removing the volatile reaction product, methyl acetate, from the reacting liquid-phase reaction, the reverse reaction is negligible and the reaction continues to proceed towards completion in the forward direction.

Although reactive distillation will not be treated in detail, it is worthwhile to set down the governing equations. We consider the elementary reaction

$$A + B \rightleftharpoons C + D$$

in which A and B are charged in equal molar amounts and species D is continuously boiled off. A balance on species A gives

$$0 - 0 + r_A V = \frac{dN_A}{dt}$$

(4-71)

[7] H. Sawistowski and P. A. Pilavakis, *Chem. Eng. Sci.*, *43*, 355 (1988).

If we define conversion as the number of moles of A reacted per mole of A charged, then

$$N_A = N_{A0}(1 - X) \tag{4-72}$$

and

$$N_B = N_{A0}\left(\Theta_B - \frac{b}{a}X\right) = N_{A0}(1 - X) \tag{4-73}$$

Substituting Equation (4-72) into (4-71) gives

$$N_{A0}\frac{dX}{dt} = -r_A V \tag{4-74}$$

A balance on species D, which evaporates at a rate F_D after being formed, gives

$$0 - F_D + r_D V = \frac{dN_D}{dt} \tag{4-75}$$

Integrating, we have

$$\begin{bmatrix} \text{number of moles} \\ \text{of remaining D} \end{bmatrix} = \begin{bmatrix} \text{number of moles of D} \\ \text{formed by reaction} \end{bmatrix} - \begin{bmatrix} \text{number of moles of D} \\ \text{lost by vaporization} \end{bmatrix} \tag{4-76}$$

$$N_D \qquad = \qquad N_{A0}X \qquad - \qquad \int_0^t F_D \, dt$$

For the elementary reaction given,

$$-r_A = k\left(C_A C_B - \frac{C_C C_D}{K_C}\right)$$

$$C_A = \frac{N_{A0}(1 - X)}{V}$$

$$C_B = \frac{N_B}{V} = \frac{N_{A0}(1 - X)}{V}$$

$$C_C = \frac{N_C}{V} = \frac{N_{A0}X}{V}$$

$$C_D = \frac{N_D}{V} = \frac{N_{A0}X - \int_0^t F_D \, dt}{V}$$

$$-r_A V = \frac{kN_{A0}^2\left[(1 - X)^2 - \frac{X}{K_C}\left(X - \frac{1}{N_{A0}}\int_0^t F_D \, dt\right)\right]}{V} \tag{4-77}$$

We now need to determine the volume as a function of either conversion or time. An overall mass balance on all species gives

$$0 - F_D(MW_D) + 0 = \frac{d(\rho V)}{dt} \tag{4-78}$$

where MW_D is the molecular weight of D. For a constant-density system,

$$\frac{dV}{dt} = -\frac{F_D(MW_D)}{\rho} = -\alpha F_D \tag{4-79}$$

This equation must now be solved numerically and simultaneously with equations (4-74) through (4-77). However, in order to solve the set of equations above we need to specify the rate of evaporation of D, F_D.

Case 1   *Immediate Evaporation*

For the case where D evaporates immediately after being formed, we can obtain an analytical solution. For no accumulation of D in the liquid phase,

$$F_D = r_D V = -r_A V \tag{4-80}$$

$$F_D = N_{A0} \frac{dX}{dt} \tag{4-81}$$

Combining Equations (4-79) and (4-81), we have

$$\frac{dV}{dX} = -\alpha N_{A0} \tag{4-82}$$

Assuming that product D vaporizes immediately after forming

Integrating yields

$$V = V_0 - \alpha N_{A0} X$$

Rearranging, we have

$$V = V_0(1 + \varepsilon_L X) \tag{4-83}$$

where

$$\varepsilon_L = -\alpha C_{A0} = -\frac{(MW)_D C_{A0}}{\rho}$$

Combining Equation (4-77) for the case of $C_D = 0$ with (4-83), we obtain

$$\frac{dX}{dt} = kC_{A0} \left[\frac{(1-X)^2}{1 + \varepsilon_L X} \right] \tag{4-84}$$

Using the list of integrals in Appendix A.2, we can determine the true t to achieve a conversion X in the reactor:

Second-order reactive distillation

$$\boxed{t = \frac{1}{kC_{A0}} \left[\frac{(1 + \varepsilon_L)X}{1 - X} - \varepsilon_L \ln\frac{1}{1-X} \right]} \tag{4-85}$$

Case 2 *Inert gas is bubbled through the reactor*

In this case we assume the reaction product is carried off by either "boil off" or an inert gas being bubbled through the reactor. When gas is bubbled through the reactor, we assume vapor-liquid equilibrium and that the mole fractions in the gas and liquid are related by Raoult's law,

$$y_D(g) = x_D(\ell)\,\frac{P_{vD}}{P_o} = \frac{N_D}{N_A + N_B + N_C + N_D}\,\frac{P_{vD}}{P_o} \qquad (4\text{-}86)$$

where P_{vD} is the vapor pressure of D. We now set either the boil off rate by setting the heat flux to the reactor or set the molar flow rate at which an inert is bubbled through the reactor, F_I.

The molar flow rate of D in the gas phase is

$$F_D = y_D F_T$$

Assuming only D evaporates (see the CD-ROM for the situation when this is not the case) the total molar gas flow rate is the inert gas molar flow rate plus the molar flow rate of the evaporating species, in this case species D, is

$$F_T = F_I + F_D = F_I + y_D F_T$$

$$F_T = \frac{F_I}{1 - y_D}$$

The molar flow rate of D leaving the liquid is

$$F_D = \left(\frac{y_D}{1 - y_D}\right) F_I \qquad (4\text{-}87)$$

This equation is now coupled with all the mole balances, e.g.,

$$\frac{dN_A}{dt} = \frac{-k(N_A N_B - N_C N_D)}{V}$$

$$\vdots$$

$$\frac{dN_D}{dt} = \frac{k(N_A N_B - N_C N_D)}{V} - F_D$$

and solved with an ODE solver.

4.8 Recycle Reactors

Recycle reactors are used when the reaction is autocatalytic, or when it is necessary to maintain nearly isothermal operation of the reactor or to promote a certain selectivity (see Section 5.6.6). They are also used extensively in bio-chemical operations. To design recycle reactors, one simply follows the proce-dure developed in this chapter and then adds a little additional bookkeeping. A schematic diagram of the recycle reactor is shown in Figure 4-15.

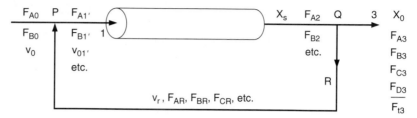

Figure 4-15   Recycle reactor.

The recycled stream is drawn off at point Q and merged with the fresh feed at point P. We shall *define* the recycle parameter R as the moles recycled per mole of product removed at point Q:

$$R = \frac{F_{tR}}{F_{t3}} = \frac{F_{AR}}{F_{A3}} = \frac{F_{iR}}{F_{i3}}$$

Two conversions are usually associated with recycle reactors: the overall conversion, X_o, and the conversion per pass, X_s:

Two conversions: X_s and X_o

$$X_s = \frac{\text{moles of A reacted in a single pass}}{\text{mole of A fed to the reactor}} \qquad (4\text{-}88)$$

$$X_o = \frac{\text{moles of A reacted overall}}{\text{mole of fresh feed}} \qquad (4\text{-}89)$$

The only new twist in calculating reactor volumes or conversions for a recycle reactor is a mole balance at the stream intersections (points P and Q) to express properly the species concentrations as a function of conversion.

As shown in the CD-ROM, along with the overall conversion and the conversion per pass are related by

$$X_s = \frac{X_o}{1 + R(1 - X_o)} \qquad (4\text{-}90)$$

The PFR design equation for a recycle reactor is also developed in the CD-ROM.

SUMMARY

1. Solution algorithm
   a. **Design equations** (Batch, CSTR, PBR):

$$N_{A0}\frac{dX}{dt} = -r_A V, \quad V = \frac{F_{A0}X}{-r_A}, \quad F_{A0}\frac{dX}{dW} = -r'_A \quad \text{(S4-1)}$$

   b. **Rate law**: for example:

$$-r_A = kC_A^2 \quad \text{(S4-2)}$$

   c. **Stoichiometry**:
      (1) *Gas phase*, constant T:

$$v = v_0\left(\frac{F_T}{F_{T0}}\right)\frac{P_o}{P} = v_0(1+\varepsilon X)\frac{P_0}{P} \quad \text{(S4-3)}$$

$$C_A = \frac{F_A}{v} = \frac{F_{A0}(1-X)}{v} = \frac{F_{A0}(1-X)}{v_0(1+\varepsilon X)}\left(\frac{P}{P_0}\right)$$

$$= C_{A0}\left(\frac{1-X}{1+\varepsilon X}\right)\left(\frac{P}{P_0}\right) \quad \text{(S4-4)}$$

      (2) *Liquid phase*:

$$v = v_0$$

$$C_A = C_{A0}(1-X)$$

   d. **Combining**, we have the following for the gas-phase reaction in a CSTR with no ΔP:

$$V = \frac{F_{A0}X(1+\varepsilon X)^2}{kC_{A0}^2(1-X)^2} \quad \text{(S4-5)}$$

   e. **Parameter evaluation**:

$$\varepsilon = y_{A0}\delta \quad \text{(S4-6)}$$

$$C_{A0} = y_{A0}\left(\frac{P_0}{RT_0}\right) \quad \text{(S4-7)}$$

   f. Solution techniques:
      (1) Numerical integration—Simpson's rule
      (2) Table of integrals
      (3) Software packages
         (a) POLYMATH
         (b) MATLAB

2. Pressure drop in isothermal reactors
   a. Variable density with $\varepsilon \neq 0$:

$$y = \frac{P}{P_0}$$

$$\frac{dy}{dW} = \frac{-\alpha}{2y} \left(\frac{F_T}{F_{T0}} \right) \left(\frac{T}{T_0} \right) \tag{S4-8}$$

$$\frac{dy}{dW} = -\frac{\alpha(1 + \varepsilon X)}{2y} \left(\frac{T}{T_0} \right) \tag{S4-9}$$

$$\alpha = \frac{2\beta_0}{A_c(1 - \phi)\rho_c P_0}$$

$$\beta_0 = \frac{G(1 - \phi)}{\rho_0 g_c D_p \phi^3} \left[\frac{150(1 - \phi)\mu}{D_p} + 1.75G \right]$$

   b. Variable density with $\varepsilon = 0$ or $\varepsilon X \ll 1$ and isothermal:

$$\frac{P}{P_0} = (1 - \alpha W)^{1/2} \tag{S4-10}$$

3. For semibatch reactors, reactant B is fed continuously to a vat initially containing only A:

$$A + B \longrightarrow C$$

   a. In terms of conversion:

$$N_{A0} \frac{dX}{dt} = -r_A V \tag{S4-11}$$

   b. Rate law:

$$-r_A = k \left(C_A C_B - \frac{C_C}{k_C} \right) \tag{S4-12}$$

$$V = V_0 + v_0 t \tag{S4-13}$$

$$C_A = \frac{N_{A0}(1 - X)}{V_0 + v_0 t} \tag{S4-14}$$

$$C_B = \frac{N_{Bi} + F_{B0} t - N_{A0} X}{V_0 + v_0 t} \tag{S4-15}$$

Equations (S4-11) through (S4-15) are combined and the resulting equation is solved numerically for conversion as a function of time.
   c. In terms of concentrations,

$$\frac{dC_A}{dt} = r_A - \frac{C_A v_0}{V} \tag{S4-16}$$

$$\frac{dC_B}{dt} = r_A + \frac{(C_{B0} - C_B)v_0}{V} \tag{S4-17}$$

$$\frac{dC_C}{dt} = -r_A - \frac{C_A v_0}{V} \tag{S4-18}$$

Equations (S4-12), (S4-13), and (S4-16) to (S4-18) are solved simultaneously.

4. Simultaneous reaction and separation
  a. Membrane reactors:

$$R_B = k_c C_B \tag{S4-19}$$

  b. Reactive distillation:

$$V = V_0(1 + \varepsilon_L X) \tag{S4-20}$$

ODE SOLVER ALGORITHM

When using an ordinary differential equation (ODE) solver such as POLYMATH or MATLAB, it is usually easier to leave the mole balances, rate laws, and concentrations as separate equations rather than combining them into a single equation as we did to obtain an analytical solution. Writing the equations separately leaves it to the computer to combine them and produce a solution. The formulations for a packed-bed reactor with pressure drop and a semibatch reactor are given below for two elementary reactions.

Gas Phase	Liquid Phase
$A + B \rightarrow 3C$	$A + B \rightarrow 2C$
Packed-Bed Reactor	*Semibatch Reactor*
$\dfrac{dX}{dW} = \dfrac{-r_A}{F_{A0}}$	$\dfrac{dN_A}{dt} = r_A V$
$r_A = -kC_A C_B$	$\dfrac{dN_B}{dt} = r_A V + F_{B0}$
$C_A = C_{A0}\dfrac{1-X}{1+\varepsilon X}y$	
	$\dfrac{dN_C}{dt} = -2r_A V$
$C_B = C_{A0}\dfrac{\theta_B - X}{1+\varepsilon X}y$	$V = V_0 + v_0 t$
$\dfrac{dy}{dW} = -\dfrac{\alpha(1+\varepsilon X)}{2y}$	$r_A = -k\left[\left(\dfrac{N_A}{V}\right)\left(\dfrac{N_B}{V}\right) - \left(\dfrac{N_C}{V}\right)^2 \middle/ K_C\right]$

(where $y = P/P_0$)

$k = 10.0$	$k = 0.15$
$\alpha = 0.01$	$K_C = 4.0$
$\varepsilon = 0.33$	$V_0 = 10.0$
$\theta_B = 2.0$	$v_0 = 0.1$
$C_{A0} = 0.01$	$F_{B0} = 0.02$
$F_{A0} = 15.0$	$N_{A0} = 0.02$
$W_{final} = 80$	$t_{final} = 200$

QUESTIONS AND PROBLEMS

The subscript to each of the problem numbers indicates the level of difficulty: A, least difficult; D, most difficult.

$$A = \bullet \qquad B = \blacksquare \qquad C = \blacklozenge \qquad D = \blacklozenge\blacklozenge$$

In each of the questions and problems below, rather than just drawing a box around your answer, write a sentence or two describing how you solved the problem, the assumptions you made, the reasonableness of your answer, what you learned, and any other facts that you want to include. You may wish to refer to W. Strunk and E. B. White, *The Elements of Style* (New York: Macmillian, 1979) and Joseph M. Williams, *Style: Ten Lessons in Clarity & Grace* (Glenview, Ill.: Scott, Foresman, 1989) to enhance the quality of your sentences. See the Preface for additional generic parts (x), (y), (z) to the home problems.

P4-1$_A$ Read through all the problems at the end of this chapter. Make up and solve an *original* problem based on the material in this chapter. **(a)** Use real data and reactions. **(b)** Make up a reaction and data. **(c)** Use an example from everyday life (e.g., making toast or cooking spaghetti). In preparing your original problem, first list the principles you want to get across and why the problem is important. Ask yourself how your example will be different from those in the text or lecture. Other things for you to consider when choosing a problem are relevance, interest, impact of the solution, time required to obtain a solution, and degree of difficulty. Look through some of the journals for data or to get some ideas for industrially important reactions or for novel applications of reaction engineering principles (the environment, food processing, etc.). The journals listed at the end of Chapter 1 may be useful for part (a). At the end of the problem and solution describe the creative process used to generate the idea for the problem.

P4-2$_B$ **What if...** you were asked to explore the example problems in this chapter to learn the effects of varying the different parameters? This sensitivity analysis can be carried out by either downloading the examples from the WWW or by loading the programs from the CD-ROM supplied with the text. For each of the example problems you investigate, write a paragraph describing your findings.

  (a) **What if** you were asked to give examples of the material in this book that are found in everyday life? What would you say?

(b) It has come time to replace the catalyst and the supplier inadvertently manufactures and ships you a catalyst whose particle size is one-fourth that of the one you are using. On realizing his mistake, he says that he will sell you the catalyst at his cost, which is roughly half price. Should you accept his offer? Prepare arguments why you should accept the catalyst along with arguments why you should not accept the catalyst.

(c) After plotting the exit conversion as a function of the pressure drop parameter α in Example 4-7, what generalization can you make? How would your answers change for an equal molar feed? What if the catalyst particles in Example 4-7 were placed in 100 tubes placed in series rather than in 100 tubes placed in parallel?

(d) Reconsider Example 4-8. Plot the conversion profile for the case when the entering pressure is increased by a factor of 5 and the particle diameter is decreased by a factor of 5. (Recall that alpha is a function of the particle diameter and P_0.) What did you learn from your plot? What should be your next settings of α and P_0 to learn more? Assume turbulent flow.

(e) Consider adding an inert to the reaction in Example 4-9, keeping the total molar flow rate at a constant. Plot the exit conversion and the equilibrium conversion as a function of the mole fraction of an inert. What are the advantages and disadvantages of adding an inert?

(f) Rework Example 4-10. Plot the molar flow rates of A, B, and C as a function of reactor length (i.e., volume) for different values of k_c between $k_c = 0.0$ (a conventional PFR) and $k_c = 7.0$ min^{-1}. What parameters would you expect to affect your results the most? Vary the parameters k, k_c, K_c, F_{A0} to study how the reaction might be optimized. Ask such questions as: What is the effect of the ratio of k to k_c, or of $k_c \, \tau \, C_{A0}$ to K_e? What generalizations can you make? How would your answer change if the reactor temperature were raised significantly? What if someone claimed that membrane reactors were not as safe as semibatch reactors? What would you tell them?

(g) In Example 4-11, plot the time at which maximum rate of reaction occurs (e.g., in the example the maximum rate is 0.00017 mol/dm$^3 \cdot$s at $t \simeq 25$ s) as a function of the entering molar feed rate of B. What if you were asked to obtain the maximum concentration of C and D in Example 4-11? What would you do?

(h) Vary some of the operating costs, conversions, and separations in Figure 4-11 to learn how the profit changes. Ethylene oxide, used to make ethylene glycol, sells for $0.56/lb. while ethylene glycol sells for $0.38/lb. Is this a money-losing proposition? Explain.

(i) What if you assumed that the reaction in Example 4-11 was first-order in methyl amine (B) and zero-order in bromine cyanide. Since it is in excess at the start of the reaction, show that the concentration of methyl amine at any time t is

$$C_B = \left[C_{B0} \Big/ \left(1 + \frac{V_0}{v_0} \right) \right] \left[1 - \left(\frac{V_0 + v_0 t}{V_0} \right)^{-[1 + k \, V_0/v_0]} \right]$$

Up to what time and under what conditions is the assumption valid? [See W. Ernst, *AIChE J.*, *43*, p. 1114 (1997).]

(j) What should you do if some of the ethylene glycol splashed out of the reactor onto your face and clothing? (*Hint*: Recall http://www.siri.org/.)

(k) What safety precautions should you take with the ethylene oxide formation discussed in Example 4-6? With the bromine cyanide discussed in Example 4-11?

P4-3$_A$ If it takes 11 minutes to cook spaghetti in Ann Arbor, Michigan, and 14 minutes in Boulder, Colorado, how long would it take in Cuzco, Peru? Discuss ways to make the spaghetti more tasty. If you prefer to make a creative spaghetti dinner for family or friends rather than answering this question, that's OK, too; you'll get full credit. (*Ans. t* = 21 min)

P4-4$_A$ Nutrition is an important part of ready-to-eat cereal. To make cereal healthier, many nutrients are added. Unfortunately, nutrients degrade over time, making it necessary to add more than the declared amount to assure enough for the life of the cereal. Vitamin *X* is declared at a level of 20% of the Recommended Daily Allowance per serving size (serving size = 30 g). The Recommended Daily Allowance is 6500 IU (1.7×10^6 IU = 1 g). It has been found that the degradation of this nutrient is first-order in the amount of nutrients. Accelerated storage tests have been conducted on this cereal, with the following results

Temperature (°C)	45	55	65
k (week^{-1})	0.0061	0.0097	0.0185

(a) Given the information above and that the cereal needs to have a vitamin level above the declared value of 6500 IU for 1 year at 25°C, what IU should be present in the cereal at the time it is manufactured? Your answer may also be reported in percent overuse: (*Ans.* 12%)

$$\%OU = \frac{C\,(t=0) - C\,(t=1\ \text{yr})}{C\,(t=1\ \text{yr})} \times 100$$

(b) At what percent of declared value of 6500 IU must you apply the vitamin? If 10,000,000 lb/yr of the cereal is made and the nutrient cost is $5 per pound, how much will this overuse cost?

(c) If this were your factory, what percent overuse would you actually apply and why?

(d) How would your answers change if you stored the material in a Bangkok warehouse for 6 months, where the daily temperature is 40°C, before moving it to the supermarket? (Table of results of accelerated storage tests on cereal; and Problem of vitamin level of cereal after storage courtesy of General Mills, Minneapolis, MN.)

<table>
<tr><td>Application
Pending
for Problem
Hall of
Fame</td></tr>
</table>

P4-5$_A$ The liquid-phase reaction

$$A + B \longrightarrow C$$

follows an elementary rate law and is carried out isothermally in a flow system. The concentrations of the A and B feed streams are 2 *M* before mixing. The volumetric flow rate of each stream is 5 dm³/min and the entering temperature is 300 K. The streams are mixed immediately before entering. Two reactors are available. One is a gray 200.0-dm³ CSTR that can be heated to 77°C or cooled to 0°C, and the other is a white 800.0-dm³ PFR operated at 300 K that cannot be heated or cooled but can be painted red or black. Note k = 0.07 dm³/mol·min at 300 K and *E* = 20 kcal/mol)

(a) Which reactor and what conditions do you recommend? Explain the reason for your choice (e.g., color, cost, space available, weather conditions). Back up your reasoning with the appropriate calculations.

(b) How long would it take to achieve 90% conversion in a 200-dm^3 batch reactor with $C_{A0} = C_{B0} = 1\ M$ after mixing at a temperature of 77°C?

(c) How would your answer to part (b) change if the reactor were cooled to 0°C? (*Ans.* 2.5 days)

(d) What concerns would you have to carry out the reaction at higher temperatures?

(e) Keeping Table 4-1 in mind, what batch reactor volume would be necessary to process the same amount of species A per day as the flow reactors while achieving not less than 90% conversion? Referring to Table 1-1, estimate the cost of the batch reactor.

(f) Write a couple of sentences describing what you learned from the problem and what you believe to be the point of the problem.

P4-6$_B$ Dibutyl phthalate (DBP), a plasticizer, has a potential market of 12 million lb/year (*AIChE 1984 Student Contest Problem*) and is to be produced by reaction of *n*-butanol with monobutyl phthalate (MBP). The reaction follows an elementary rate law and is catalyzed by H_2SO_4 (Figure P4-6). A stream containing MBP and butanol is to be mixed with the H_2SO_4 catalyst immediately before the stream enters the reactor. The concentration of MBP in the stream entering the reactor is 0.2 lb mol/ft^3 and the molar feed rate of butanol is five times that of MBP. The specific reaction rate at 100°F is 1.2 ft^3/lb mol·hr. There is a 1000-gallon CSTR and associated peripheral equipment available for use on this project for 30 days a year (operating 24 h/day).

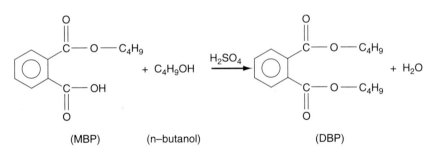

(MBP)   (n–butanol)   (DBP)

Figure P4-6

(a) Determine the exit conversion in the available 1000-gallon reactor if you were to produce 33% of the share (i.e., 4 million lb/yr) of the predicted market. (Ans.: $X = 0.33$).

(b) How might you increase the conversion and decrease the time of operation? For example, what conversion would be achieved if a second 1000-gal CSTR were placed either in series or in parallel with the CSTR?

(c) For the same temperature and feed conditions as part (a), what CSTR volume would be necessary to achieve a conversion of 85% for a molar feed rate of A of 1 lb mol/min?

(d) Compare your results for part (c) with those of the PFR necessary to achieve 85% conversion.

(e) Keeping in mind the times given in Table 4-1 for filling, and other operations, how many 1000-gallon reactors operated in the batch mode would be necessary to meet the required production of 4 million pounds in a 30-day period? Estimate the cost of the reactors in the system. [*Note*: Present in the feed stream may be some trace impurities, which you may lump as a hexanol. The activation energy is believed to be somewhere around 25 kcal/mol. The PFR in part (d) is more oblong than cylindrical, with a major-to-minor axis ratio of 1.3:1.] (*An Ans.:* 5 reactors)

P4-7$_A$ The elementary gas-phase reaction

$$(CH_3)_3COOC(CH_3)_3 \rightarrow C_2H_6 + 2CH_3COCH_3$$

is carried out isothermally in a flow reactor with no pressure drop. The specific reaction rate at 50°C is 10^{-4} min^{-1} (from pericosity data) and the activation energy is 85 kJ/mol. Pure di-*tert*-butyl peroxide enters the reactor at 10 atm and 127°C and a molar flow rate of 2.5 mol/min. Calculate the reactor volume and space time to achieve 90% conversion in:

(a) a CSTR (*Ans.:* 4700 dm^3)

(b) a PFR (*Ans.:* 967dm^3)

(c) If this reaction is to be carried out at 10 atm and 127°C in a batch mode with 90% conversion, what reactor size and cost would be required to process (2.5 mol/min × 60 min/h × 24 h/day) 3600 mol of di-*tert*-butyl peroxide per day? (*Hint:* Recall Table 4-1.)

(d) Assume that the reaction is reversible with $K_C = 0.025$ mol^2/dm^6 and calculate the equilibrium conversion and then redo (a) through (c) to achieve a conversion that is 90% of the equilibrium conversion.

(e) What CSTR temperature would you recommend for a 500-dm^3 CSTR to obtain the maximum conversion if $\Delta H_{Rx} = -100,000$ cal/mol ?
(*Hint:* Remember that the equilibrium conversion will be different for a flow reactor and a constant-volume batch reactor for a gas-phase reaction that has a change in the total number of moles.)

P4-8$_B$ A liquid-phase isomerization A $\longrightarrow$ B is carried out in a 1000-gal CSTR that has a single impeller located halfway down the reactor. The liquid enters at the top of the reactor and exits at the bottom. The reaction is second-order. Experimental data taken in a batch reactor predicted the CSTR conversion should be 50%. However, the conversion measured in the actual CSTR was 57%.

(a) Suggest reasons for the discrepancy and suggest something that would give closer agreement between the predicted and measured conversions. Back your suggestions with calculations.

(b) Consider the case where the reaction is reversible with $K_C = 15$ at 300 K, and $\Delta H_{rx} = -25,000$ cal/mol. Assuming that the batch data taken at 300 K are accurate and that $E = 15,000$ cal/mol, what CSTR temperature do you recommend for maximum conversion?

P4-9   *Sargent Nigel Ambercromby.* Scoundrels Incorporated, a small R&D company has developed a laboratory scale process for the elementary, solid-catalyzed-gas-phase reaction A + B → C + D (names coded for proprietary reasons). The feed is equal molar in A and B with the entering molar flow rate of A is 25 mol/min and the volumetric feed is 50 dm^3/min. Engineers at Scoundrels calculated that an industrial scale packed bed reactor with 500 kg of a *very* rare and expensive metal catalyst will yield a 66% conversion when run at 32°C and a feed pressure of 25 atm. At these conditions the specific reaction rate is 0.4 dm^3/mol·min·kg catalyst. Scoundrels sells this process and catalyst to Clueless Chemicals who then manufactured the packed bed. When Clueless put the process onstream at the specifications provided by Scoundrels, they could only achieve 60% conversion with 500 kg catalyst. Unfortunately the reaction was carried out at 31.5°C rather than 32°C. The corresponding 2160 mol/day of lost product made the process uneconomical. Scoundrels, Inc. say that they can guarantee 66% conversion if Clueless will purchase 500 more kg of catalyst at one and a half times the original cost. As a result, Clueless contacted Sgt. Ambercromby from Scotland Yard (on loan to the L.A.P.D.) about possible industrial fraud. What are the first three questions Sargent Ambercromby asks? What are potential causes for this lost conversion (support with calculations)? What do you think Sgt. Ambercromby suggests to rectify the situation (with Dan Dixon, Reaction Engineering Alumni W'97).

P4-10$_B$ The formation of diphenyl discussed in Section 3.1.4 is to be carried out at 760°C. The feed is to be pure benzene in the gas phase at a total pressure of 5 atm and 760°C. The specific reaction rate is 1800 ft^3/lb mol·s and the concentration equilibrium constant is 0.3 as estimated by the Davenport trispeed oscimeter. The batch reactor volume [part (d)] is 1500 dm^3. Also, the inside of the batch reactor is corroding badly, as evidenced by the particulate material that is falling off the sides onto the bottom of the reactor, and you may not need to address this effect.

(a) What is the equilibrium conversion?

Calculate the reactor volume necessary to achieve 98% of the equilibrium conversion of benzene in a

(b) **PFR** (with a benzene feed of 10 lb mol/min)

(c) **CSTR** (with a benzene feed of 10 lb mol/min)

(d) Calculate the volume of a constant-volume *batch reactor* that processes the same amount of benzene each day as the CSTR. What is the corresponding reactor cost? (*Hint*: Recall Table 4-1.)

(e) If the activation energy is 30,202 Btu/lb mol, what is the ratio of the initial rate of reaction (i.e., $X = 0$) at 1400°F to that at 800°F?

P4-11$_B$ The gaseous reaction A $\longrightarrow$ B has a unimolecular reaction rate constant of 0.0015 min^{-1} at 80°F. This reaction is to be carried out in *parallel tubes* 10 ft long and 1 in. inside diameter under a pressure of 132 psig at 260°F. A production rate of 1000 lb/h of B is required. Assuming an activation energy of 25,000 cal/g mol, how many tubes are needed if the conversion of A is to be 90%? Assume perfect gas laws. A and B each have molecular weights of 58. (From California Professional Engineers Exam.)

P4-12$_B$ The irreversible elementary reaction 2A $\longrightarrow$ B takes place in the gas phase in an *isothermal tubular (plug-flow) reactor*. Reactant A and a diluent C are fed in equimolar ratio, and conversion of A is 80%. If the molar feed rate of A is cut in half, what is the conversion of A assuming that the feed rate of C is left unchanged? Assume ideal behavior and that the reactor temperature remains unchanged. (From California Professional Engineers Exam.)

P4-13$_B$ Compound A undergoes a reversible isomerization reaction, A $\rightleftharpoons$ B, over a supported metal catalyst. Under pertinent conditions, A and B are liquid, miscible, and of nearly identical density; the equilibrium constant for the reaction (in concentration units) is 5.8. In a *fixed-bed isothermal flow reactor* in which backmixing is negligible (i.e., plug flow), a feed of pure A undergoes a net conversion to B of 55%. The reaction is elementary. If a second, identical flow reactor at the same temperature is placed downstream from the first, what overall conversion of A would you expect if:

(a) The reactors are directly connected in series? (*Ans.: X* = 0.74.)

(b) The products from the first reactor are separated by appropriate processing and only the unconverted A is fed to the second reactor?

(From California Professional Engineers Exam.)

P4-14$_C$ A total of 2500 gal/h of metaxylene is being isomerized to a mixture of orthoxylene, metaxylene, and paraxylene in a reactor containing 1000 ft^3 of catalyst. The reaction is being carried out at 750°F and 300 psig. Under these conditions, 37% of the metaxylene fed to the reactor is isomerized. At a flow rate of 1667 gal/h, 50% of the metaxylene is isomerized at the same temperature and pressure. Energy changes are negligible.

It is now proposed that a second plant be built to process 5500 gal/h of metaxylene at the same temperature and pressure as described above. What size reactor (i.e., what volume of catalyst) is required if conversion in the new plant is to be 46% instead of 37%? Justify any assumptions made for the scale-up calculation. (*Ans.:* 2931 ft^3 of catalyst.) (From California Professional Engineers Exam.)

P4-15$_A$ It is desired to carry out the gaseous reaction A $\longrightarrow$ B in an existing *tubular reactor* consisting of 50 parallel tubes 40 ft long with a 0.75-in. inside diameter. Bench-scale experiments have given the reaction rate constant for this first-order reaction as 0.00152 s^{-1} at 200°F and 0.0740 s^{-1} at 300°F. At what temperature should the reactor be operated to give a conversion of A of 80% with a feed rate of 500 lb/h of pure A and an operating pressure of 100 psig? A has a molecular weight of 73. Departures from perfect gas behavior may be neglected, and the reverse reaction is insignificant at these conditions. (*Ans.:* $T = 275$°F.) (From California Professional Engineers Exam.)

P4-16$_B$ An isothermal, constant-pressure *plug-flow reactor* is designed to give a conversion of 63.2% of A to B for the first-order gas-phase decomposition

$$A \xrightarrow{k} B$$

for a feed of pure A at a rate of 5 ft^3/h. At the chosen operating temperature, the first-order rate constant $k = 5.0$ h^{-1}. However, after the reactor is installed and in operation, it is found that conversion is 92.7% of the desired conversion. This discrepancy is thought to be due to a flow disturbance in the reactor that gives rise to a zone of intense backmixing. Assuming that this zone behaves like a perfectly mixed stirred-tank reactor in series and *in between* two plug-flow reactors, what fraction of the total reactor volume is occupied by this zone? (*Ans.:* 57%.)

P4-17$_B$ Currently, the herbicide atrozine found in the Des Plaines River is being treated by passing part of the river through a marsh, where it is degraded (Figure P4-17). The rate of degradation of atrazine, A, is assumed irreversible and to follow first-order homogeneous kinetics.

$$A \xrightarrow{k_1} \text{products}$$

As the wastewater flows and reacts, it also evaporates at a constant rate ($Q =$ kmol water/h · m^2) from the surface. None of the toxic species are lost to the air by evaporation. You may assume that the reactor (marsh) is rectangular and that the gentle downhill flow of the water can be modeled as plug flow.

(a) Derive an equation for C_A as a function of X and z.
(*Ans.:* $C_A = C_{A0}(1 - X)/(1 - az)$; $a = QW/\rho_0 v_0$.)

(b) Derive an equation for X as a function of distance, z, down the wetlands.
(*Ans.:* $X = 1 - (1 - az)^n$; $n = kD\rho_0/Q$.)

(c) Plot the conversion and rate of reaction as a function of distance for:
(1) no evaporation or condensation
(2) evaporation but no condensation

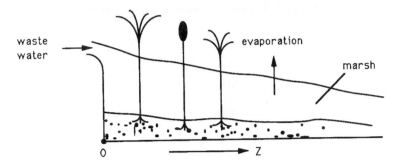

Figure P4-17   [See R. H. Kadlec and R. L. Knight, *Treatment of Wetlands*, CRC Press, Lewis Publishers, Boca Raton, Fla., 1996.]

   (3) condensation at a rate of 0.5 kmol/m²·h
   Compare your results for each of the three cases.

(d) Use POLYMATH to carry out a sensitivity analysis of the various param-
     eter values and their ratios (e.g., what is the effect of this ratio Q/kC_{A0}
     on the conversion)?

(e) A heated political battle is currently raging between those who want to
     preserve wetlands (which are currently protected by law) in their natural
     state and those who want to make them available for commercial devel-
     opment. List arguments on both sides of this issue, including one or more
     arising from a consideration of this problem. Then take a position one
     way or the other and justify it briefly. (Your grade doesn't depend on
     what position you take, only on how well you defend it.) [Part (e) sug-
     gested by Prof. Richard Felder, North Carolina State University.]

(f) Without solving any equations, suggest a better model for describing the
     wetlands. (*Hint:* An extended version of this problem can be found on
     the CD-ROM.)

Additional information:
W = width = 100 m, v_0 = entering volumetric flow rate = 2 m³/h

L = length = 1000 m, C_{A0} = entering concentration of toxic = 10^{-5} mol/dm³

D = average depth = 0.25 m, ρ_m = molar density of water = 55.5 kmol H_2O/m³

Q = evaporation rate = 1.11×10^{-3} kmol/h·m²,

k_1 = specific reaction rate = 16×10^{-5} h⁻¹

P4-18$_B$ Ethyl acetate is an extensively used solvent and can be formed by the vapor-phase
     esterification of acetic acid and ethanol. The reaction was studied using a
     microporous resin as a catalyst in a *packed-bed reactor* [*Ind. Eng. Chem. Res.*,
     26(2), 198(1987)]. The reaction is first-order in ethanol and pseudo-zero-order
     in acetic acid. For an equal molar feed rate of acetic acid and ethanol the specific
     reaction rate is 1.2 dm³*/g cat·min. The total molar feed rate is 10 mol/min,
     the initial pressure is 10 atm, the temperature is 118°C, and the pressure drop
     parameter, α, equals 0.01 g⁻¹.

(a) Calculate the maximum weight of catalyst that one could use and main-
     tain an exit pressure above 1 atm. (*Ans.:* W = 99g)

(b) Determine the catalyst weight necessary to achieve 90% conversion.
     What is the ratio of catalyst needed to achieve the last 5% (85 to 90%)

conversion to the weight necessary to achieve the first 5% conversion (0 to 5%) in the reactor? Vary α and write a few sentences describing and explaining your findings. (*Ans.*: $W = 82g$)

P4-19$_B$ The gas-phase dimerization

$$2A \longrightarrow B$$

follows an elementary rate law and takes place isothermally in a PBR charged with 1.0 kg of catalyst. The feed, consisting of pure A, enters the PBR at a pressure of 20 atm. The conversion exiting the PBR is 0.3, and the pressure at the exit of the PBR is 5 atm.

(a) If the PBR were replaced by a CSTR, what will be the conversion at the exit of the CSTR? You may assume that there is no pressure drop in the CSTR. (*Ans.*: $X = 0.4$.)

(b) What would be the conversion in the PBR if the mass flow rate were decreased by a factor of 4 and particle size were doubled? Assume turbulent flow. (*Final exam, Winter 1993*)

(c) Discuss the strengths and weaknesses of using this as a final exam problem.

P4-20$_D$ The decomposition of cumene,

$$C_6H_5CH(CH_3)_2 \longrightarrow C_6H_6 + C_3H_6$$

is to be carried out at a high temperature in a packed-bed reactor. At this temperature the reaction is internal diffusion limited and apparent first-order in cumene. Currently, 1000 kg of catalyst is packed in a 4-cm diameter pipe. The catalyst particles are 0.5 cm in diameter and the bulk density of the packed catalyst is 1000 kg/m^3. Currently, 6.4% conversion is realized when pure cumene enters the reactor. The entering pressure is 20 atm and the pressure at the exit of the reactor is 2.46 atm.

(a) What conversion would be achieved if the PBR were replaced by a fluidized CSTR containing 8000 kg catalyst with negligible pressure drop?

(b) We know from Chapter 12 [e.g., Equation (12-35)] that for internal diffusion limitations, the rate of reaction varies inversely with the catalyst particle size. Consequently, one of the engineers suggests that the catalyst be ground up into a smaller size. She also notes there are three other pipe sizes available into which the catalyst could be packed. These non-corrosive heat-resistant pipes, which can be cut to any length, are 2 cm, 6 cm, and 8 cm in diameter. Should you change the catalyst size and pipe diameter in which the catalyst is packed? If so, what are the appropriate catalyst particle size, the appropriate pipe diameter, and the exiting conversion? If nothing should be changed, explain your reasoning. Assume that the flow is highly turbulent and that the bulk catalyst density is the same for all catalyst and pipe sizes. Explain how your answers would change if the flow were "laminar."

(c) Discuss what you learned from this problem and what you believe to be the point of the problem.

P4-21$_C$ The first order irreversible gas phase reaction

$$\text{Normal pentane} \longrightarrow \text{Iso-pentane}$$

is to be carried out in a packed bed reactor. Currently 1000 kg of reform-ing catalyst are packed in a 4 cm diameter pipe. The catalyst particles are 0.5 cm in diameter and the bulk density of the packed catalyst is 1,000 kg/m^3. Currently 14.1% conversion is realized. The entering pressure is 20 atm and the pressure at the exit of the reactor is 9.0 atmospheres. It is believed that this reaction is internal diffusion limited. We know from Chapter 12 of Elements of CRE (e.g. P4-23$_C$ or Eqn. 12-35, page 751) that for internal diffusion limitations the rate of reaction varies inversely with the catalyst particle size. Consequently one of the engineers suggests that the catalyst be ground up into a smaller size. She also notes the smallest size to which the catalyst may be ground is 0.01 cm. and that there are 3 other pipe sizes available into which the catalyst could be packed. These non-corrosive heat-resistant pipes, which can be cut to any length, are 2 cm, 3 cm, and 6 cm in diameter.

(a) What conversion could be achieved in a CSTR with the same catalyst weight and no ΔP? (*Ans.*: $X = 0.18$.)

(b) Calculate the maximum value of the pressure drop parameter, α, that you can have and still maintain an exit pressure of 1 atm. (*Ans.*: $\alpha = 9.975 \times 10^{-4} \text{ kg}^{-1}$.)

(c) Should you change the catalyst size and pipe diameter in which 1000 kg of the catalyst is packed while maintaining the catalyst weight?

(d) Next consider how α would change if you changed both pipe size and particle size. Can you change pipe size and particle size at the same time such that α remains constant at the value calculated in part (b)?

(e) For the conditions of part (a) [i.e., maintain α constant at the value in part (a)], pick a pipe size and calculate a new particle size. (*Ans.*: $D_P = 0.044$ cm.) Assume turbulent flow.

(f) Calculate a new specific reaction rate ratio assuming (i.e., recall the effectiveness factor from Chapter 12) that

$$k \sim \frac{1}{D_P} \quad \text{then} \quad k_2 = k_1 \left(\frac{D_{P1}}{D_{P2}} \right)$$

(g) Using the new values of k and α, calculate the conversion for a PBR for the new particle size for an exit pressure of 1 atm. (*Ans.*: $X = 0.78$.)

P4-22$_B$ Alkylated cyclohexanols are important intermediates in the fragrance and perfume industry [*Ind. Eng. Chem. Res.*, **28**, 693 (1989)]. Recent work has focused on gas-phase catalyzed hydrogenation of o-cresol to 2-methylcyclohexanone, which is then hydrogenated to 2-methylcyclohexanol. In this problem we focus on only the first step in the reaction (Figure P4-22). The reaction on a nickel–silica catalyst was found to be zero-order in o-cresol and first-order in hydrogen with a specific reaction rate at 170°C of 1.74 mol of o-cresol/(kg cat · min · atm). The reaction mixture enters the packed-bed reactor at a total pressure of 5 atm. The molar feed consists of 67% H$_2$ and 33% o-cresol at a total molar rate of 40 mol/min.

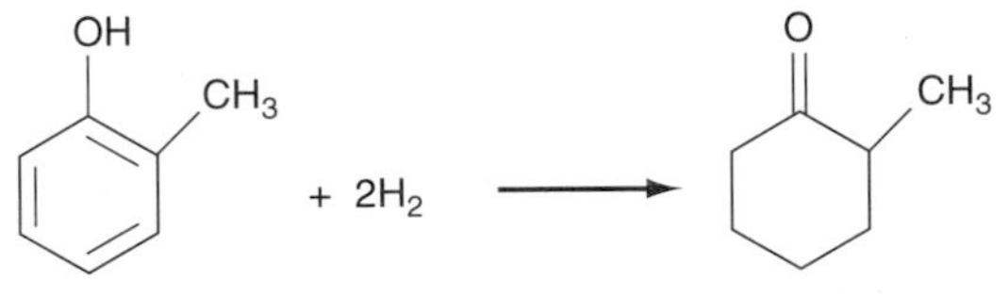

Figure P4-22

(a) Neglecting pressure drop, plot the rate of reaction of o-cresol and the concentrations of each species as a function of catalyst weight. What is the ratio of catalyst weight needed to achieve the last 5% conversion to the weight necessary to achieve the first 5% conversion (0 to 5%) in the plug-flow reactor?

(b) Accounting for the pressure drop in the packed bed using a value of $\alpha = 0.34$ kg^{-1}, redo part (a) along with a plot of pressure versus catalyst weight.

(c) Another engineer suggests that instead of changing catalyst size it would be better to pack the catalyst in a shorter reactor with twice the pipe diameter. If all other conditions remain the same, is this suggestion better?

P4-23$_C$ The elementary gas-phase reaction

$$A + B \longrightarrow C + D$$

is carried out in a packed-bed reactor. Currently, catalyst particles 1 mm in diameter are packed into 4-in. schedule 40 pipe ($A_C = 0.82126$ dm^2). The value of β_0 in the pressure drop equation is 0.001 atm/dm. A stoichiometric mixture of A and B enters the reactor at a total molar flowrate of 10 gmol/min, a temperature of 590 K, and a pressure of 20 atm. Flow is turbulent throughout the bed. Currently, only 12% conversion is achieved with 100 kg of catalyst.

It is suggested that conversion could be increased by changing the catalyst particle diameter. Use the data below to correlate the specific reaction rate as a function of particle diameter. Then use this correlation to determine the catalyst size that gives the highest conversion. As you will see in Chapter 12, k' for first-order reaction is expected to vary according to the following relationship

MEMBER
PROBLEM
HALL OF
FAME

$$k' = \eta k = \frac{3}{\Phi^2}(\Phi \coth\Phi - 1)\, k \qquad\qquad (P4\text{-}23.1)$$

where Φ varies directly with particle diameter, $\Phi = cD_p$. Although the reaction is not first-order, one notes from Figure 12-5 the functionality for a second-order reaction is similar to Equation (P4-23.1). (*Ans.: c = 75*)

(a) Make a plot of conversion as a function of catalyst size.

(b) Discuss how your answer would change if you had used the effectiveness factor for a second-order reaction rather than a first-order reaction.

(c) Discuss what you learned from this problem and what you believe to be the point of the problem.

Additional information:
Void fraction = 0.35   Bulk catalyst density = 2.35 kg/dm^3

Catalyst Diameter, d_p (mm)	2	1	0.4	0.1	0.02	0.002
k' (dm^6/mol·min·kg cat)	0.06	0.12	0.30	1.2	2.64	3.00

[*Hint*: You could use Equation (P4.23-1), which would include d_P and an unknown proportionality constant which you could evaluate from the data. For very small values of the Thiele modulus we know $\eta = 1$ and for very large values of the Thiele modulus we know that $\eta = 3/\Phi = 3/Cd_p$.]

P4-24$_C$ (*Spherical reactor*) Because it is readily available from coal, methanol has been investigated as an alternative raw material for producing valuable olefins such as ethene and propene. One of the first steps in the reforming process involves the dehydration of methanol.

$$2CH_3OH \rightarrow CH_2 = CH_2 + 2H_2O$$

This gas-phase reaction is carried out over a zeolite catalyst and follows an elementary rate law. The catalyst is packed in a tubular PBR that is 2 m in diameter and 22 m in length. Pure methanol is fed at a molar flow rate of 950 mol/s, a pressure of 1500 kPa, a concentration of 0.4 mol/ dm³, and 490°C. The conversion and pressure at the exit are 0.5 and 375 kPa, respectively. Flow throughout the bed is known to be such that the turbulent contribution in the Ergun equation (i.e., G^2) can be neglected. The tubular PBR is to be replaced with a spherical PBR containing an equal amount of identical catalyst. The spherical PBR measures 5.2 m in diameter and has screens placed 2 dm from each end (i.e., $L = L' = 24$ dm).

(a) What conversion and exit pressure can we expect from the spherical reactor? [*Hint*: What are the parameter values (e.g., α, k) for the PBR $(0.1 < k < 1.0)$? (*Ans.*: $X = 0.63$.)

(b) By how much can the feed rate to the spherical PBR be increased and still achieve the same conversion that was attained in the tubular PBR? Assume that the flow is completely laminar up to $F_{A0} = 2000$ mol/s.

(c) It is desired to minimize the pumping requirement for the feed to the spherical PBR. How low can the entry pressure, P_0, be and still achieve a conversion of 0.5?

(d) Put two spherical PBRs in series. What is the conversion and pressure at the exit? Experiment with putting more than two reactors in series. What is the maximum attainable conversion?

[The data given in this problem are based on kinetics data given in H. Schoenfelder, J. Hinderer, J. Werther, F.J. Keil, Methanol to olefins-—prediction of the performance of a circulating fluidized-bed reactor on the basis of kinetic experiments in a fixed-bed reactor. *Chem. Eng. Sci. 49*, 5377 (1994).]

4-25$_A$ A very proprietary industrial waste reaction which we'll code as $A \rightarrow B + S$ to be carried out in a 10-dm³ CSTR followed by 10-dm³ PFR. The reaction is elementary, but A, which enters at a concentration of 0.001 mol/dm³ and a molar flow rate of 20 mol/min, has trouble decomposing. The specific reaction rate at 42°C (i.e., room temperature in the Mojave desert) is 0.0001 s⁻¹. However, we don't know the activation energy; therefore, we cannot carry out this reaction in the winter in Michigan. Consequently this reaction, while important, is not worth your time to study. Therefore, perhaps you want to take a break and go watch a movie such as *Dances with Wolves* (a favorite of the author) or *Evita* or the *Sixth Sense*.

P4-26$_B$ Pure butanol is to be fed into a *semibatch reactor* containing pure ethyl acetate to produce butyl acetate and ethanol. The reaction

$$CH_3COOC_2H_5 + C_4H_9OH \; \rightleftharpoons \; CH_3COOC_4H_9 + C_2H_5OH$$

is elementary and reversible. The reaction is carried out isothermally at 300 K. At this temperature the equilibrium constant is 1.08 and the specific reaction rate is 9×10^{-5} dm³/mol·s. Initially, there is 200 dm³ of ethyl acetate in the vat and butanol is fed at a rate of 0.05 dm³/s. The feed and initial concentrations of butanol and ethyl acetate are 10.93 mol/dm³ and 7.72 mol/dm³, respectively.

(a) Plot the equilibrium conversion of ethyl acetate as a function of time.

(b) Plot the conversion of ethyl acetate, the rate of reaction, and the concentration of butanol as a function of time.

(c) Repeat parts (a) and (b) for different values of the butanol feed rate and of the amount of ethyl acetate in the vat.

(d) Rework part (b) assuming that ethanol evaporates (reactive distillation) as soon as it forms.

(e) Use POLYMATH or some other ODE solver to learn the sensitivity of conversion to various combinations of parameters.

(f) Discuss what you learned from this problem and what you believe to be the point of this problem.

P4-27$_B$ A catalyst is a material that affects the rate of a chemical reaction, yet emerges unchanged from the reaction. For example, consider the reaction

$$A + B \longrightarrow C + B$$

in which

$$r_A = -kC_A C_B$$

where B is a catalyst. The reaction is taking place in a *semibatch reactor*, in which 100 ft^3 of a solution containing 2 lb mol/ft^3 of A is initially present. No B is present initially. Starting at time $t = 0$, 5 ft^3/min of a solution containing 0.5 lb mol/ft^3 of B is fed into the reactor. The reactor is isothermal, and $k = 0.2$ ft^3/lb mol · min.

(a) How many moles of C are present in the reactor after half an hour?

(b) Plot the conversion as a function of time.

(c) Repeat parts (a) and (b) for the case when the reactor initially contains 100 ft^3 of B and A is fed to the reactor at a concentration of 0.02 lb mol/ft^3 and at a rate of 5 ft^3/min. The feed concentration of B is 0.5 lb mol/ft^3.

(d) Discuss what you learned from this problem and what you believe to be the point of this problem.

P4-28$_C$ An isothermal reversible reaction A $\rightleftharpoons$ B is carried out in an aqueous solution. The reaction is first-order in both directions. The forward rate constant is 0.4 h^{-1} and the equilibrium constant is 4.0. The feed to the plant contains 100 kg/m^3 of A and enters at the rate of 12 m^3/h. Reactor effluents pass to a separator, where B is completely recovered. The reactor is a *stirred tank* of volume 60 m^3. A fraction of the unreacted effluent is recycled as a solution containing 100 kg/m^3 of A and the remainder is discarded. Product B is worth \$2 per kilogram and operating costs are \$50 per cubic meter of solution entering the separator. What value of f maximizes the operational profit of the plant? What fraction A fed to the plant is converted at the optimum? (*H. S. Shankar, IIT Bombay*)

P4-29$_B$ (*CSTR train*) The elementary liquid-phase reaction

$$A + B \longrightarrow C$$

is to be carried out in a *CSTR* with three impellers (Figure P4-29). The mixing patterns in the CSTR are such that it is modeled as three equal-sized CSTRs in series. Species A and B are fed in separate lines to the CSTR, which is initially filled with inert material. Each CSTR is 200 dm^3 and the volumetric flow to the first reactor is 10 dm^3/min of A and 10 dm^3/min of B.

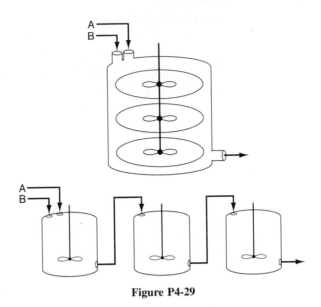

Figure P4-29

(a) What is the steady-state conversion of A? (*Ans.*: $X \cong 0.55$)
(b) Determine the time necessary to reach steady state (i.e., when C_A exiting the third reactor is 99% of the steady-state value).
(c) Plot the concentration of A exiting each tank as a function of time.
(d) Suppose that the feed for species B is split so that half is fed to the first tank and half to the second tank. Repeat parts (a), (b), and (c).
(e) Vary the system parameters, v_0, V, k, and so on, to determine their effects on startup. Write a paragraph describing the trends you found which includes a discussion of the parameter that most effects the results.

Additional information:

$$C_{A0} = C_{B0} = 2.0 \text{ mol/dm}^3 \qquad k = 0.025 \text{ dm}^3/\text{mol} \cdot \text{min}$$

P4-30$_B$ The reversible isomerization

$$A \rightleftharpoons B$$

is to be carried out in a *membrane reactor* (IMRCF). Owing to the configuration of species B, it is able to diffuse out the walls of the membrane, while A cannot.
(a) What is the equilibrium conversion assuming that B does not diffuse out of the reactor walls?
(b) Plot the conversion profiles to compare a 100-dm³ conventional PFR with a 100-dm³ membrane reactor. What statements or generalizations can you make? What parameters have the greatest effect on the exit conversion shape of the plots in part (a)?
(c) Plot the conversion and the species concentrations and the molar flow rates down the length of the reactor.
(d) Vary some of the parameters (e.g., $k = 0.8 \text{ s}^{-1}$, $k_c = 0.03 \text{ s}^{-1}$, $K_C = 1.0$, etc.) and write a paragraph describing your results.
(e) Discuss how your curves would change if the temperature were increased significantly or decreased significantly for an exothermic reaction and for an endothermic reaction.

Additional information:
    Specific reaction rate $= 0.05$ s^{-1}
    Transport coefficient $k_c = 0.3$ s^{-1}
    Equilibrium constant $= K_e = 0.5$
    Entering volumetric flow rate $v_0 = 10$ dm^3/s
    $C_{A0} = 0.2$ mol/dm^3

P4-31$_C$ (*Membrane reactor*) The first-order, reversible reaction

$$A \;\rightleftharpoons\; B + 2C$$

is taking place in a membrane reactor. Pure A enters the reactor, and B diffuses through the membrane. Unfortunately, some of the reactant A also diffuses through the membrane.

(a) Plot the flow rates of A, B, and C down the reactor, as well as the flow rates of A and B through the membrane.

(b) Compare the conversion profiles of a conventional PFR with those of an IMRCF. What generalizations can you make?

(c) Discuss how your curves would change if the temperature were increased significantly or decreased significantly for an exothermic reaction and for an endothermic reaction.

Additional information:

$k = 10$ min^{-1}	$F_{A0} = 100$ mol/min
$K_C = 0.01$ mol^2/dm^6	$v_0 = 100$ dm^3/min
$k_{CA} = 1$ min^{-1}	$V_{reactor} = 20$ dm^3
$k_{CB} = 40$ min^{-1}	

NOTE TO INSTRUCTORS: Additional problems (cf. those from the preceding edition) can be found in the solutions manual and on the CD-ROM. These problems could be photocopied and used to help reinforce the fundamental principles discussed in this chapter.

JOURNAL CRITIQUE PROBLEMS

P4C-1   In the article describing the liquid reaction of isoprene and maleic anhydride under pressure [*AIChE J.*, *16*(5), 766 (1970)], the authors show the reaction rate to be greatly accelerated by the application of pressure. For an equimolar feed they write the second-order reaction rate expression in terms of the mole fraction y:

$$\frac{dy}{dt} = -k_y y^2$$

and then show the effect of pressure on k_y (s^{-1}). Derive this expression from first principles and suggest a possible logical explanation for the increase in the true specific reaction rate constant k (dm^3/mol · s) with pressure that is different from the authors'. Make a quick check to verify your challenge.

P4C-2   The reduction of NO by char was carried out in a fixed bed between 500 and 845°C [*Int. Chem. Eng.*, *20*(2), 239, (1980)]. It was concluded that the reaction is first-order with respect to the concentration of NO feed (300 to 1000 ppm) over the temperature range studied. It was also found that activation energy begins to increase at about 680°C. Is first-order the true reaction order? If there were discrepancies in this article, what might be the reasons for them?

P4C-3 In the article describing vapor phase esterification of acetic acid with ethanol to form ethyl acetate and water [*Ind. Eng. Chem. Res., 26*(2), 198 (1987)], the pressure drop in the reactor was accounted for in a most unusual manner [i.e., $P = P_0(1 - fX)$, where f is a constant].

    (a) Using the Ergun equation along with estimating some of the parameter values (e.g., $\phi = 0.4$), calculate the value of α in the packed-bed reactor (2 cm ID by 67 cm long).

    (b) Using the value of α, redo part (a) accounting for pressure drop along the lines described in this chapter.

    (c) Finally, if possible, estimate the value of f used in these equations.

SOME THOUGHTS ON CRITIQUING WHAT YOU READ

Your textbooks after your graduation will be, in part, the professional journals that you read. As you read the journals, it is important that you study them with a critical eye. You need to learn if the author's conclusion is supported by the data, if the article is new or novel, if it advances our understanding, and to learn if the analysis is current. To develop this technique, one of the major assignments used in the graduate course in chemical reaction engineering at the University of Michigan for the past 20 years has been an in-depth analysis and critique of a journal article related to the course material. Significant effort is made to ensure that a cursory or superficial review is not carried out. Students are asked to analyze and critique ideas rather than ask questions such as: Was the pressure measured accurately? They have been told that they are not required to find an error or inconsistency in the article to receive a good grade, but if they do find such things, it just makes the assignment that much more enjoyable. Beginning with Chapter 4, a number of the problems at the end of each chapter in this book are based on students' analyses and critiques of journal articles and are designated with a C (e.g., P4C-1). These problems involve the analysis of journal articles that may have minor or major inconsistencies. A discussion on critiquing journal articles can be found on the CD-ROM.

CD-ROM MATERIAL

- **Learning Resources**
  1. *Summary Notes for Lectures 4, 5, 6, 7, 8 and 9*
  2. *Web Modules*
     - A. Wetlands
     - B. Membrane Reactors
     - C. Reactive Distillation
  3. *Interactive Computer Modules*
     - A. Mystery Theater
     - B. Tic-Tac
  4. *Solved Problems*
     - A. CDP4-A_B A Sinister Gentleman Messing with a Batch Reactor

    B. Solution to California Registration Exam Problem
    C. Ten Types of Home Problems: 20 Solved Problems
  5. *Analogy of CRE Algorithms to a Menu in a Fine French Restaurant*
  6. *Algorithm for Gas Phase Reaction*
- **Living Example Problems**
  1. *Example 4–7 Pressure Drop with Reaction-Numerical Solution*
  2. *Example 4–8 Dehydrogenation in a Spherical Reactor*
  3. *Example 4–9 Working in Terms of Molar Flow Rate in a PFR*
  4. *Example 4–10 Membrane Reactor*
  5. *Example 4–11 Isothermal Semibatch Reactor with a Second-order Reaction*
- **Professional Reference Shelf**
  1. *Time to Reach Steady State for a 1st Order Reaction in a CSTR*
  2. *Recycle Reactors*
  3. *Critiquing Journal Articles*
- **FAQ [Frequently Asked Questions]**– In Updates/FAQ icon section
- **Additional Homework Problems**

CDP4-A$_B$	A sinister gentlemen is interested in producing methyl perchlorate in a *batch reactor*. The reactor has a strange and unsettling rate law. [2nd Ed. P4-28]
CDP4-B$_C$	(*Ecological Engineering*) A much more complicated version of Problem 4-17 uses actual pond (*CSTR*) sizes and flow rates in modeling the site with CSTRs for the Des Plaines River experimental wetlands site (EW3) in order to degrade atrazine.
CDP4-C$_B$	The rate of binding ligands to receptors is studied in this application of reaction kinetics to *bioengineering*. The time to bind 50% of the ligands to the receptors is required. [2nd Ed. P4-34]
CDP4-D$_B$	A *batch reactor* is used for the bromination of p-chlorophenyl isopropyl ether. calculate the batch reaction time. [2nd Ed. P4-29]
CDP4-E$_B$	California Professional Engineers Exam Problem in which the reaction

$$B + H_2 \longrightarrow A$$

is carried out in a batch reactor. [2nd Ed. P4-15]

CDP4-F$_A$	The gas-phase reaction

$$A + 2B \longrightarrow 2D$$

has the rate law $-r_A = 2.5 C_A^{1/2} C_B$. Reactor volumes of *PFRs* and *CSTRs* are required in this multipart problem. [2nd Ed. P4-21]

CDP4-G$_B$	What type and arrangement of *flow reactors* should you use for a decomposition reaction with the rate law $-r_A = k_1 C_A^{1/2}/(1 + k_2 C_A)$? [2nd Ed. P4-14]
CDP4-H$_A$	Verify that the liquid-phase reaction of 5, 6-benzoquinoline with hydrogen is pseudo-first-order. [2nd Ed. P4-7]
CDP4-I$_B$	The liquid-phase reaction

$$2A + B \rightleftharpoons C + D$$

is carried out in a *semibatch reactor*. Plot the conversion, volume, and species concentration as a function of time. Reactive distillation is also considered in part (e). [2nd Ed. P4-27]

CDP4-J$_B$	Designed to reinforce the basic CRE principles through very straight forward calculations of CSTR and PFR volumes and a batch reactor

time. This problem was one of the most often assigned problems from the 2nd Edition. [2nd Ed. P4-4]

CDP4-K$_B$ Calculate the overall conversion for *PFR with recycle*. [2nd Ed. P4-28]

CDP4-L$_B$ The overall conversion is required in a *packed-bed reactor with recycle*. [2nd Ed. P4-22]

CDP4-M$_B$ A recycle reactor is used for the reaction

$$A + B \longrightarrow C$$

in which species C is partially condensed. The PFR reactor volume is required for 50% conversion. [2nd Ed. P4-32]

CDP4-N$_B$ Radial flow reactors can be used to good advantage for exothermic reactions with large heats of reaction. The radical velocity varies as

$$U = \frac{U_0 R_0}{r}\,(1 + \varepsilon X)\,\frac{P_0}{P}\,\frac{T}{T_0}$$

Vary the parameters and plot X as a function of r. [2nd Ed. P4-31]

CDP4-O$_B$ The growth of a bacterium is to be carried out in excess nutrient.

$$\text{nutrient} + \text{cells} \longrightarrow \text{more cells} + \text{product}$$

The growth rate law is

$$r_B = \mu_m C_B \left(1 - \frac{C_B}{C_{Bmax}} \right)$$

CDP4-P$_B$ A not very good semibatch problem, but it does require assessing what equation to use.

SUPPLEMENTARY READING

HILL, C. G., *An Introduction to Chemical Engineering Kinetics and Reactor Design*. New York: Wiley, 1977, Chap. 8.

LEVENSPIEL, O., *Chemical Reaction Engineering*, 2nd ed. New York: Wiley, 1972, Chaps. 4 and 5.

SMITH, J. M., *Chemical Engineering Kinetics*, 3rd ed. New York: McGraw-Hill, 1981.

STEPHENS, B., *Chemical Kinetics*, 2nd ed. London: Chapman & Hall, 1970, Chaps. 2 and 3.

ULRICH, G. D., *A Guide to Chemical Engineering Reactor Design and Kinetics*. Printed and bound by Braun-Brumfield, Inc., Ann Arbor, Mich., 1993.

WALAS, S. M., *Reaction Kinetics for Chemical Engineers*. New York: McGraw-Hill, 1970.

Recent information on reactor design can usually be found in the following journals: *Chemical Engineering Science, Chemical Engineering Communications, Industrial and Engineering Chemistry Research, Canadian Journal of Chemical Engineering, AIChE Journal, Chemical Engineering Progress.*

Collection and Analysis   5
of Rate Data

You can observe a lot just by watching.
                              Yogi Berra, New York Yankees

In Chapter 4 we showed that once the rate law is known, it can be substituted into the appropriate design equation, and through the use of the appropriate stoichiometric relationships we can size any isothermal reaction system. In this chapter we focus on ways of obtaining and analyzing reaction rate data to obtain the rate law for a specific reaction. In particular, we discuss two common types of reactors for obtaining rate data: the batch reactor, which is used primarily for homogeneous reactions, and the differential reactor, which is used for solid–fluid reactions. In batch reactor experiments, concentration, pressure, and/or volume are usually measured and recorded at different times during the course of the reaction. Data are collected from the batch reactor during unsteady-state operation, whereas measurements on the differential reactor are made during steady-state operation. In experiments with a differential reactor, the product concentration is usually monitored for different feed conditions.

Two techniques of data acquisition are presented: concentration-time measurements in a batch reactor and concentration measurements in a differential reactor. Six different methods of analyzing the data collected are used: the differential method, the integral method, the method of half-lives, method of initial rates, and linear and nonlinear regression (least-squares analysis). The differential and integral methods are used primarily in analyzing batch reactor data. Because a number of software packages (e.g., POLYMATH, MATLAB) are now available to analyze data, a rather extensive discussion of linear and nonlinear regression is included. We close the chapter with a discussion of experimental planning and of laboratory reactors (CD-ROM).

5.1 Batch Reactor Data

Batch reactors are used primarily to determine rate law parameters for homo-geneous reactions. This determination is usually achieved by measuring con-centration as a function of time and then using either the differential, integral, or least squares method of data analysis to determine the reaction order, α, and specific reaction rate, k. If some reaction parameter other than concentration is monitored, such as pressure, the mole balance must be rewritten in terms of the measured variable (e.g., pressure).

Process data in terms of the measured variable

5.1.1 Differential Method of Rate Analysis

When a reaction is *irreversible*, it is possible in many cases to determine the reaction order α and the specific rate constant by numerically differentiat-ing *concentration versus time data*. This method is applicable when reaction conditions are such that the rate is essentially a function of the concentration of only one reactant: for example, if, for the decomposition reaction

Assume that the rate law is of the form $-r_A = kC_A^\alpha$

$$A \rightarrow products$$

$$-r_A = kC_A^\alpha \tag{5-1}$$

then the differential method may be used.

However, by utilizing the method of excess, it is also possible to deter-mine the relationship between $-r_A$ and the concentration of other reactants. That is, for the irreversible reaction

$$A + B \rightarrow products$$

with the rate law

$$-r_A = k_A C_A^\alpha C_B^\beta \tag{5-2}$$

where α and β are both unknown, the reaction could first be run in an excess of B so that C_B remains essentially unchanged during the course of the reac-tion and

$$-r_A = k' C_A^\alpha \tag{5-3}$$

where

Method of excess

$$k' = kC_B^\beta \approx kC_{B0}^\beta \tag{5-4}$$

After determining α, the reaction is carried out in an excess of A, for which the rate law is approximated as

$$-r_A = k'' C_B^\beta \tag{5-5}$$

Once α and β are determined, k_A can be calculated from the measure-ment of $-r_A$ at known concentrations of A and B:

$$k_A = \frac{-r_A}{C_A^\alpha C_B^\beta} = (dm^3/mol)^{\alpha+\beta-1}/s$$

Both α and β can be determined by using the method of excess, coupled with a differential analysis of data for batch systems.

To outline the procedure used in the differential method of analysis, we consider a reaction carried out isothermally in a constant-volume batch reactor and the concentration recorded as a function of time. *By combining the mole balance with the rate law given by Equation* (5-1), we obtain

Constant-volume batch reactor

$$-\frac{dC_A}{dt} = k_A C_A^\alpha \qquad (5\text{-}6)$$

After taking the natural logarithm of both sides of Equation (5-6),

$$\ln\left(-\frac{dC_A}{dt}\right) = \ln k_A + \alpha \ln C_A \qquad (5\text{-}7)$$

observe that the slope of a plot of $\ln(-dC_A/dt)$ as a function of $(\ln C_A)$ is the reaction order (Figure 5-1).

Plot
$$\ln\left(-\frac{dC_A}{dt}\right)$$
versus $\ln C_A$
to find
α **and** k_A

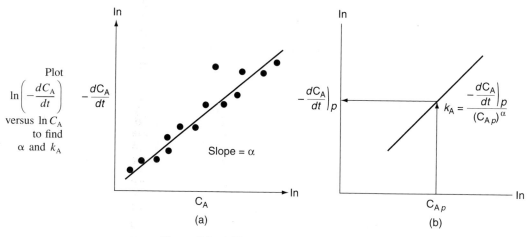

Figure 5-1   Differential method to determine reaction order.

Figure 5-1(a) shows a plot of $-(dC_A/dt)$ versus C_A on log-log paper where the slope is equal to the reaction order α. The specific reaction rate, k_A, can be found by first choosing a concentration in the plot, say C_{Ap}, and then finding the corresponding value of $-(dC_A/dt)$ as shown in Figure 5-1(b). After raising C_{Ap} to the α power, we divide it into $-(dC_A/dt)_p$ to determine k_A:

$$k_A = \frac{-(dC_A/dt)_p}{(C_{Ap})^\alpha}$$

To obtain the derivative $-dC_A/dt$ used in this plot, we must differentiate the concentration–time data either numerically or graphically. We describe three methods to determine the derivative from data giving the concentration as a function of time. These methods are:

- Graphical differentiation
- Numerical differentiation formulas
- Differentiation of a polynomial fit to the data

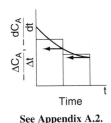

Graphical Method.   With this method disparities in the data are easily seen. As explained in Appendix A.2, the graphical method involves plotting $\Delta C_A/\Delta t$ as a function of t and then using equal-area differentiation to obtain dC_A/dt. An illustrative example is also given in Appendix A.2.

See Appendix A.2.

In addition to the graphical technique used to differentiate the data, two other methods are commonly used: differentiation formulas and polynomial fitting.

Numerical Method.   Numerical differentiation formulas can be used when the data points in the independent variable are *equally spaced*, such as $t_1 - t_0 = t_2 - t_1 = \Delta t$:

Time (min)	t_0	t_1	t_2	t_3	t_4	t_5
Concentration (mol/dm³)	C_{A0}	C_{A1}	C_{A2}	C_{A3}	C_{A4}	C_{A5}

The three-point differentiation formulas[1]

Initial point:
$$\left(\frac{dC_A}{dt}\right)_{t_0} = \frac{-3C_{A0} + 4C_{A1} - C_{A2}}{2\Delta t} \tag{5-8}$$

Interior points:
$$\left(\frac{dC_A}{dt}\right)_{t_i} = \frac{1}{2\Delta t}[(C_{A(i+1)} - C_{A(i-1)})] \tag{5-9}$$

$$\left[\text{e.g.,} \left(\frac{dC_A}{dt}\right)_{t_3} = \frac{1}{2\Delta t}[C_{A4} - C_{A2}]\right]$$

Last point:
$$\left(\frac{dC_A}{dt}\right)_{t_5} = \frac{1}{2\Delta t}[C_{A3} - 4C_{A4} + 3C_{A5}] \tag{5-10}$$

Methods for finding $-\dfrac{dC_A}{dt}$ from concentration–time data

can be used to calculate dC_A/dt. Equations (5-8) and (5-10) are used for the first and last data points, respectively, while Equation (5-9) is used for all intermediate data points.

---

[1] B. Carnahan, H. A. Luther, and J. O. Wilkes, *Applied Numerical Methods* (New York: Wiley, 1969), p. 129.

Polynomial Fit.   Another technique to differentiate the data is to first fit the concentration–time data to an nth-order polynomial:

$$C_A = a_0 + a_1 t + a_2 t^2 + \cdots + a_n t^n \qquad (5\text{-}11)$$

Many personal computer software packages contain programs that will calculate the best values for the constants a_i. One has only to enter the concentration–time data and choose the order of the polynomial. After determining the constants, a_i, one has only to differentiate Equation (5-11) with respect to time:

$$\frac{dC_A}{dt} = a_1 + 2a_2 t + 3a_3 t^2 + \cdots + na_n t^{n-1} \qquad (5\text{-}12)$$

Thus concentration and the time rate of change of concentration are both known at any time t.

Care must be taken in choosing the order of the polynomial. If the order is too low, the polynomial fit will not capture the trends in the data and not go through many of the points. If too large an order is chosen, the fitted curve can have peaks and valleys as it goes through most all of the data points, thereby producing significant errors when the derivatives, dC_A/dt, are generated at the various points. An example of this is shown in Figure 5-2, where the same pressure–time data fit to a third-order polynomial (a) and to a fifth-order polynomial (b). Observe how the derivative for the fifth order changes from a positive value at 15 minutes to a negative value at 20 minutes.

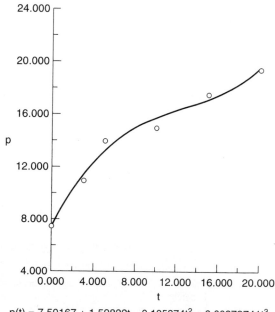

$$p(t) = 7.50167 + 1.59822t - 0.105874t^2 + 0.00279741t^3$$

(a)   variance = 0.738814

Figure 5-2   Polynomial fit of concentration–time data.

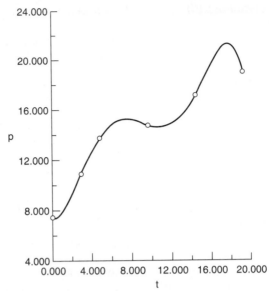

(b) $p(t) = 7.5 - 0.305644t + 0.86538t^2 - 0.151976t^3 + 0.00965104t^4 - 0.000205154t^5$

Figure 5-2   (*continued*)

Finding α and k.   Now, using either the graphical method, differentiation formulas or the polynomial derivative, the following table can be set up:

Time	t_0	t_1	t_2	t_3
Concentration	C_{A0}	C_{A1}	C_{A2}	C_{A3}
Derivative	$\left(-\dfrac{dC_A}{dt}\right)_0$	$\left(-\dfrac{dC_A}{dt}\right)_1$	$\left(-\dfrac{dC_A}{dt}\right)_2$	$\left(-\dfrac{dC_A}{dt}\right)_3$

The reaction order can now be found from a plot of $\ln(-dC_A/dt)$ as a function of $\ln C_A$, as shown in Figure 5-1(a), since

$$\ln\left(-\frac{dC_A}{dt}\right) = \ln k_A + \alpha \ln C_A \qquad (5\text{-}7)$$

Before solving an example problem let's list the steps to determine the reaction rate law from a set of data points (Table 5-1).

TABLE 5-1.   STEPS IN ANALYZING RATE DATA

1. Postulate a rate law.
2. Process your data in terms of the measured variable (rewrite the mole balance in terms of the measured variable).
3. Look for simplifications.
4. Calculate $-r_A$ as a function of reactant concentration to determine the reaction order.
    a.   If batch:
         Determine $-dC_A/dt$.
    b.   If differential packed-bed reactor (Section 5.3):
         Calculate $-r'_A = F_{A0}X/W = C_P v_0/W$.
5. Determine specific reaction rate, k.

Example 5–1 Differential Method of Analysis of Pressure–Time Data

Determine the reaction order for the gas-phase decomposition of di-*tert*-butyl peroxide,

$$(CH_3)_3COOC(CH_3)_3 \rightarrow C_2H_6 + 2CH_3 \overset{\overset{\textstyle O}{\textstyle \|}}{C} CH_3$$

This reaction was carried out in the laboratory in an isothermal batch system in which the total pressure was recorded at various times during the reaction.[2] The data given in Table E5-1.1 apply to this reaction. Only pure di-*tert*-butyl peroxide was initially present in the reaction vessel.

TABLE E5-1.1   PRESSURE–TIME DATA

Time (min)	*Total Pressure* (mmHg)
0.0	7.5
2.5	10.5
5.0	12.5
10.0	15.8
15.0	17.9
20.0	19.4

While the proper SI units of pressure are pascal (Pa) or kilopascal (kPa), a significant amount of kinetic data in past literature are reported in units of mmHg (torr), atmospheres, or psi. Consequently, we must be able to analyze pressure–time rate data in any one of these units.

$$1 \text{ atm} \equiv 14.7 \text{ psi} \equiv 1.103 \text{ bar} \equiv 101.3 \text{ kPa} \equiv 760 \text{ mmHg}$$

Solution

Let A represent di-*tert*-butyl peroxide.

1. Postulate a rate law.

$$-r_A = kC_A^\alpha$$

A combination of the *mole balance* on a constant-volume batch reactor and the *rate law* gives

$$\frac{-dC_A}{dt} = -r_A = kC_A^\alpha \qquad (E5-1.1)$$

Need
$C_A = f(P)$

where α and k are to be determined from the data listed in Table E5-1.1.

---

[2] A. F. Trotman-Dickenson, *J. Chem. Educ., 46*, 396 (1969).

2. Rewrite the design equation in terms of the measured variable. When there is a net increase or decrease in the total number of moles in a gas phase reaction, the reaction order may be determined from experiments performed with a constant-volume batch reactor by monitoring the total pressure as a function of time. The total pressure data **should not be converted** to conversion and then analyzed as conversion-time data just because the design equations are written in terms of the variable conversions. Rather, **transform the design equation to the measured variable**, which in this case is pressure. Consequently, we need to *express the concentration in terms of total pressure* and then substitute for the concentration of A in Equation (E5-1.1).

Processing data in terms of the measured variable, P

For the case of a constant-volume batch reactor, we recall Equations (3-26) and (3-38):

$$C_A = C_{A0}(1 - X) \tag{3-26}$$

$$V = V_0 \frac{P_0}{P}(1 + \varepsilon X)\frac{T}{T_0} \tag{3-38}$$

For isothermal operation and constant volume, Equation (3-38) solves to

$$X = \frac{1}{\varepsilon P_0}(P - P_0) = \frac{1}{y_{A0}\delta P_0}(P - P_0)$$

$$X = \frac{1}{\delta P_{A0}}(P - P_0) \tag{E5-1.2}$$

where

$$P_{A0} = y_{A0}P_0$$

Combining Equations (3-26) and (E5-1.2) gives

$$C_A = \frac{P_{A0} - [(P - P_0)/\delta]}{RT} \tag{E5-1.3}$$

For pure di-*tert*-butyl peroxide, initially $y_{A0} = 1.0$ and therefore $P_{A0} = P_0$. Stoichiometry gives $\delta = 1 + 2 - 1 = 2$.

$$C_A = \frac{P_0 - [(P - P_0)/2]}{RT} = \frac{3P_0 - P}{2RT} \tag{E5-1.4}$$

Substitute Equation (E5-1.4) into (E5-1.1) to get

$$\frac{1}{2RT}\frac{dP}{dt} = k\left(\frac{3P_0 - P}{2RT}\right)^\alpha$$

Let $k' = k(2RT)^{1-\alpha}$; then

$$\boxed{\frac{dP}{dt} = k'(3P_0 - P)^\alpha} \tag{E5-1.5}$$

Mole balance
in terms of the
measured
variable, P

Taking the natural logarithm of both sides gives us

$$\boxed{\ln \frac{dP}{dt} = \alpha \, \ln(3P_0 - P) + \ln k'}$$ (E5-1.6)

Observe that the reaction order, α, can be determined from the slope of a plot of $\ln(dP/dt)$ versus $\ln(3P_0 - P)$. Once α is known, the constant k' may be calculated from the ratio

$$k' = \frac{dP/dt}{(3P_0 - P)^\alpha}$$ (E5-1.7)

at any point.

3. Look for simplifications. We have assumed the reaction is reversible. Check to see if any terms in the equation can be neglected [e.g., the term εX in $(1 + \varepsilon X)$ when $(\varepsilon X \ll 1)$].

First find $\dfrac{dP}{dt}$

4. Determine dP/dt **from the pressure–time data and then the reaction order** α. The data are reported in terms of total pressure as a function of time; *consequently*, we must differentiate the data either numerically or graphically before we can use Equations (E5-1.6) and (E5-1.7) to evaluate the reaction order and specific reaction rate. First we shall evaluate dP/dt by graphical differentiation. Many prefer the graphical analysis because they use it to visualize the discrepancies in their data.

Graphical method is used to visualize discrepancies in data

Graphical Method. The derivative dP/dt is determined by calculating and plotting $\Delta P / \Delta t$ as a function of time, t, and then using the equal-area differentiation technique (Appendix A.2) to determine (dP/dt) as a function of P and t. First we calculate the ratio $\Delta P / \Delta t$ from the first two columns of Table E5-1.2; the result is written in the third column. Next we use Table E5-1.2 to plot the third column as a function of the first column in Figure E5-1.1. Using equal-area differentiation, the value of dP/dt is read off the figure (represented by the arrows) and then it is used to complete the fourth column of Table E5-1.2.

TABLE E5-1.2.   PROCESSED DATA

t (min)	P (mmHg)	$\dfrac{\Delta P}{\Delta t}$ (mmHg/min)	$\dfrac{dP}{dt}$ (mmHg/min)
0.0	7.5		1.44
		1.20	
2.5	10.5		0.95
		0.80	
5.0	12.5		0.74
		0.66	
10.0	15.8		0.53
		0.42	
15.0	17.9		0.34
		0.30	
20.0	19.4		0.25

Temperature = 170°C

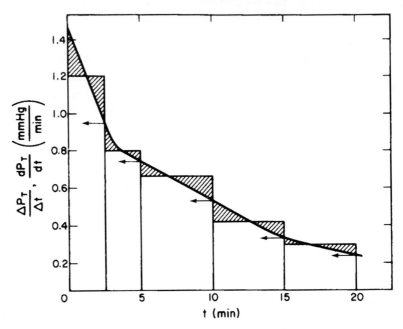

Figure E5-1.1 Graphical differentiation.

Finite Difference. Next we calculate dP/dt from finite difference formulas (5-8) through (5-10):

$$t = 0: \quad \left(\frac{dP}{dt}\right)_0 = \frac{-3P_0 + 4P_1 - P_2}{2\Delta t} = \frac{-3(7.5) + 4(10.5) - 12.5}{2(2.5)} = 1.40$$

Calculating

$$t = 2.5: \quad \left(\frac{dP}{dt}\right)_1 = \frac{P_2 - P_0}{2\Delta t} = \frac{12.5 - 7.5}{2(2.5)} = 1.0$$

$t = 5$: Here we have a change in time increments Δt, between P_1 and P_2 and between P_2 and P_3. Consequently, we have two choices for evaluating $(dP/dt)_2$:

(a)
$$\left(\frac{dP}{dt}\right)_2 = \frac{P_3 - P_0}{2(\Delta t)} = \frac{15.8 - 7.5}{2(5)} = 0.83$$

(b)
$$\left(\frac{dP}{dt}\right)_2 = \frac{-3P_2 + 4P_3 - P_4}{2(\Delta t)} = \frac{-3(12.5) + 4(15.8) - 17.9}{2(5)} = 0.78$$

$$t = 10: \quad \left(\frac{dP}{dt}\right)_3 = \frac{17.9 - 12.5}{2(5)} = 0.54$$

$$t = 15: \quad \left(\frac{dP}{dt}\right)_4 = \frac{19.4 - 15.8}{2(5)} = 0.36$$

$$t = 20: \quad \left(\frac{dP}{dt}\right)_5 = \frac{P_3 + 4P_4 - 3P_5}{2\Delta t} = \frac{15.8 - 4(17.9) + 3(19.4)}{2(5)} = 0.24$$

Polynomial (POLYMATH). Another method to determine dP/dt is to fit the total pressure to a polynomial in time and then to differentiate the resulting polynomial. Choosing a fourth-order polynomial

$$P = a_0 + a_1 t + a_2 t^2 + a_3 t^3 + a_4 t^4 \qquad \text{(E5-1.8)}$$

we use the POLYMATH software package to express pressure as a function of time. Here we first choose the polynomial order (in this case fourth order) and then type in the values of P at various times t to obtain

$$P(t) = 7.53 + 1.31t - 0.0718t^2 + 0.00276t^3 - 4.83 \times 10^{-5}t^4 \qquad \text{(E5-1.9)}$$

A plot of P versus t and the corresponding fourth-order polynomial fit is shown in Figure E5-1.2. Differentiating Equation (E5-1.9) yields

$$\frac{dP}{dt} = 1.31 - 0.144t + 0.00828t^2 - 0.000193t^3 \qquad \text{(E5-1.10)}$$

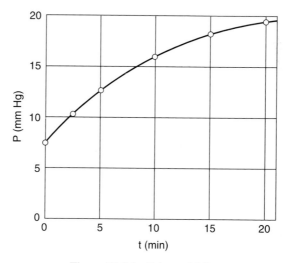

Figure E5-1.2   Polynomial fit.

To find the derivative at various times we substitute the appropriate time into Equation (E5-1.10) to arrive at the fourth column in Table E5-1.3. We can see that there is quite a close agreement between the graphical technique, finite difference, and the

TABLE E5-1.3.   SUMMARY OF PROCESSED DATA

t (min)	dP/dt (mmHg/min) Graphical	dP/dt (mmHg/min) Finite Difference	dP/dt (mmHg/min) POLYMATH	$3P_0 - P$ (mmHg)
0.0	1.44	1.40	1.31	15.0
2.5	0.95	1.00	1.0	12.0
5.0	0.74	0.78	0.78	10.0
10.0	0.53	0.54	0.51	6.7
15.0	0.34	0.36	0.37	4.6
20.0	0.25	0.24	0.21	3.1

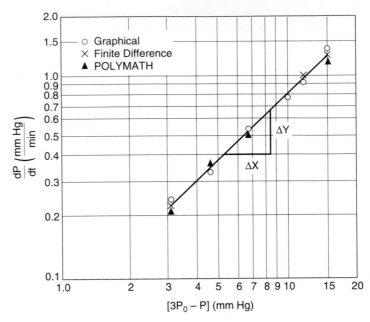

Figure E5-1.3.   Plot to determine k and α.

POLYMATH polynomial analysis. The second, third, fourth, and fifth columns of the processed data in Table E5-1.3 are plotted in Figure E5-1.3 to determine the reaction order and specific reaction rate.

Next, plot $\dfrac{dP}{dt}$ versus the appropriate function of total pressure

We shall determine the reaction order, α, from the slope of a log-log plot of dP/dt as a function of the appropriate function of pressure $f(P)$, which for the initial conditions and stoichiometry of this reaction turns out to be $(3P_0 - P)$ (see Table E5-1.3).[3] Recall that

$$\ln \frac{dP}{dt} = \alpha \ln (3P_0 - P) + \ln k' \qquad (E5\text{-}1.6)$$

Using the line through the data points in Figure E5-1.3 yields

$$\alpha = \text{slope} = \frac{\Delta y}{\Delta x} = \frac{1.2 \text{ cm}}{1.2 \text{ cm}} = 1.0$$

The reaction order is

$$\alpha = 1.0$$

$$k' = k(2RT)^{1-\alpha} = k \qquad (E5\text{-}1.11)$$

$$-r_A = kC_A \qquad (E5\text{-}1.12)$$

---

[3] If you are unfamiliar with any method of obtaining slopes from plots on log-log or semilog graphs, read Appendix D before proceeding.

5. Determine the specific reaction rate. The specific reaction rate can be determined using Equation (E5-1.7) with $\alpha = 1$ and then evaluating the numerator and denominator at any point, p.

$$k' = k = \frac{(dP/dt)_p}{(3P_0 - P)_p}$$

At $(3P_0 - P) = 5.0$ mmHg,

$$\frac{dP}{dt} = 0.4 \text{ mmHg/min}$$

From Equation (E5-1.5),

$$k = \left. \frac{\left(\dfrac{dP}{dt} \right)_p}{(3P_0 - P)_p} \right| = \frac{0.4 \text{ mmHg/min}}{5.0 \text{ mmHg}} = 0.08 \text{ min}^{-1}$$

The rate law is

$$\boxed{-r_A = \frac{0.08}{\text{min}} C_A} \tag{E5-1.13}$$

5.1.2 Integral Method

To determine the reaction order by the integral method, we guess the reaction order and integrate the differential equation used to model the batch system. If the order we assume is correct, the appropriate plot (determined from this integration) of the concentration–time data should be linear. The integral method is used most often when the reaction order is known and it is desired to evaluate the specific reaction rate constants at different temperatures to determine the activation energy.

The integral method uses a trial-and-error procedure to find reaction order

In the integral method of analysis of rate data we are looking for the appropriate function of concentration corresponding to a particular rate law that is linear with time. You should be thoroughly familiar with the methods of obtaining these linear plots for reactions of zero, first, and second order.

For the reaction

$$A \rightarrow \text{products}$$

It is important to know how to generate *linear* plots of C_A versus t for zero-, first-, and second-order reactions

carried out in a constant-volume batch reactor, the mole balance is

$$\frac{dC_A}{dt} = r_A$$

For a zero-order reaction, $r_A = -k$, and the combined rate law and mole balance is

$$\frac{dC_A}{dt} = -k \tag{5-13}$$

Integrating with $C_A = C_{A0}$ at $t = 0$, we have

Zero-order

$$\boxed{C_A = C_{A0} - kt}$$

(5-14)

A plot of the concentration of A as a function of time will be linear (Figure 5-3) with slope $(-k)$ for a zero-order reaction carried out in a constant-volume batch reactor.

$\alpha = 0$

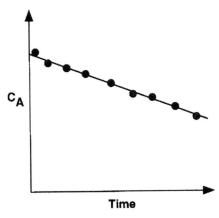

Figure 5-3 Zero-order reaction.

If the reaction is first order (Figure 5-4), integration of the combined *mole balance and the rate law*

$$-\frac{dC_A}{dt} = kC_A$$

with the limit $C_A = C_{A0}$ at $t = 0$ gives

First-order

$$\boxed{\ln \frac{C_{A0}}{C_A} = kt}$$

(5-15)

$\alpha = 1$

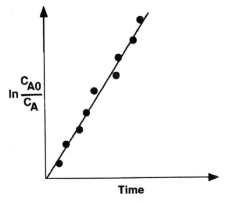

Figure 5-4 First-order reaction.

Consequently, we see that the slope of a plot of $[\ln(C_{A0}/C_A)]$ as a function of time is linear with slope k.

If the reaction is second order (Figure 5-5), then

$$-\frac{dC_A}{dt} = kC_A^2$$

Integrating with $C_A = C_{A0}$ initially yields

Second-order

$$\boxed{\frac{1}{C_A} - \frac{1}{C_{A0}} = kt}$$

(5-16)

$\alpha = 2$

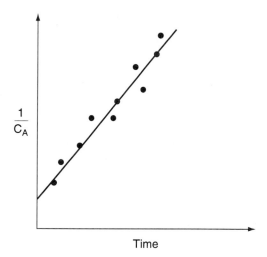

Figure 5-5 Second-order reaction.

We see that for a second-order reaction a plot of $1/C_A$ as a function of time should be linear with slope k.

In the three figures just discussed, we saw that when we plotted the appropriate function of concentration (i.e., C_A, $\ln C_A$ or $1/C_A$) versus time, the plots were linear and we concluded that the reactions were zero, first, or second order, respectively. However, if the plots of concentration data versus time had turned out not to be linear such as shown in Figure 5-6, we would say that the proposed reaction order did not fit the data. In the case of Figure 5-6, we would conclude the reaction is not second order.

The idea is to arrange the data so that a linear relationship is obtained

It is important to restate that given a reaction rate law, you should be able to choose quickly the appropriate function of concentration or conversion that yields a straight line when plotted against time or space time.

$\alpha \ne 2$

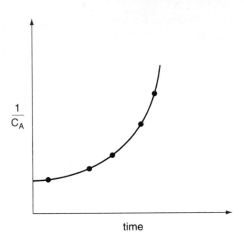

Figure 5-6   Plot of reciprocal concentration as a function of time.

Example 5–2 Integral Method of Analysis of Pressure–Time Data

Use the integral method to confirm that the reaction order for the di-*tert*-butyl peroxide decomposition described in Example 5-1 is first order.

Solution

Recalling Example 5-1, the combined mole balance and rate law for a constant-volume batch reactor can be expressed in the form

$$\frac{dP}{dt} = k'(3P_0 - P)^\alpha \tag{E5-1.7}$$

For $\alpha = 1$,

$$\frac{dP}{dt} = k'(3P_0 - P) \tag{E5-2.1}$$

Integrating with limits $P = P_0$ when $t = 0$ yields

Assuming a first-order reaction

$$\ln\frac{2P_0}{3P_0 - P} = k't \tag{E5-2.2}$$

If the reaction is first order, a plot of $\ln[2P_0/(3P_0 - P)]$ versus t should be linear.

TABLE E5-2.1.   PROCESSED DATA

t (min)	P (mmHg)	$2P_0/(3P_0 - P)$ (−)
0.0	7.5	1.00
2.5	10.5	1.25
5.0	12.5	1.50
10.0	15.8	2.24
15.0	17.9	3.26
20.0	19.4	4.84

After completing Table E5-2.1, using the raw data, a plot of $2P_0/(3P_0 - P)$ as a function of time was made using semilog paper as shown in Figure E5-2.1. From

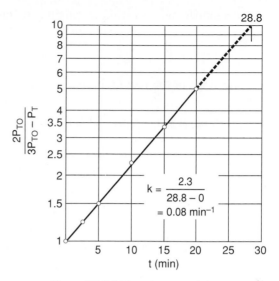

Figure E5-2.1 Plot of processed data.

the plot we see that $\ln[2P_0/(3P_0 - P)]$ is indeed linear with time, and we therefore conclude that the decomposition of di-*tert*-butyl peroxide follows first-order kinetics. From the slope of the plot in Figure E5-2.1, we can determine the specific reaction rate, $k = 0.08$ min^{-1}. [Recall $k' = k$ because $\alpha = 1$ (E5-1.5 p. 230)]

We found the plot of $\ln[2P_{t0}/(3P_{t0} - P_t)]$ versus t was linear, indicating that the reaction is first order (i.e., $\alpha = 1$). If we try zero, first, or second order as shown on the CD-ROM, and they do not seem to describe the reaction rate equation, it is usually best to try some other method of determining the reaction order, such as the differential method.

Integral method normally used to find k when order is known

By comparing the methods of analysis of the rate data presented above, we note that the differential method tends to accentuate the uncertainties in the data, while the integral method tends to smooth the data, thereby disguising the uncertainties in it. In most analyses it is imperative that the engineer know the limits and uncertainties in the data. This prior knowledge is necessary to provide for a safety factor when scaling up a process from laboratory experiments to design either a pilot plant or full-scale industrial plant.

5.2 Method of Initial Rates

The use of the differential method of data analysis to determine reaction orders and specific reaction rates is clearly one of the easiest, since it requires only one experiment. However, other effects, such as the presence of a significant

Used when reactions are reversible

reverse reaction, could render the differential method ineffective. In these cases, the method of initial rates could be used to determine the reaction order and the specific rate constant. Here, a series of experiments is carried out at different initial concentrations, C_{A0}, and the initial rate of reaction, $-r_{A0}$, is determined for each run. The initial rate, $-r_{A0}$, can be found by differentiating the data and extrapolating to zero time. For example, in the di-*tert*-butyl peroxide decomposition shown in Example 5-1, the initial rate was found to be

1.4 mmHg/min. By various plotting or numerical analysis techniques relating $-r_{A0}$ to C_{A0}, we can obtain the appropriate rate law. If the rate law is in the form

$$-r_{A0} = kC_{A0}^{\alpha}$$

the slope of the plot of $\ln(-r_{A0})$ versus $\ln C_{A0}$ will give the reaction order α.

Example 5–3 Method of Initial Rates in Solid–Liquid Dissolution Kinetics

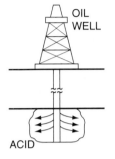

OIL
WELL

ACID

An important reaction for enhancement of oil flow in carbonate reservoirs

The dissolution of dolomite, calcium magnesium carbonate, in hydrochloric acid is a reaction of particular importance in the acid stimulation of dolomite oil reservoirs.[4] The oil is contained in pore space of the carbonate material and must flow through the small pores to reach the well bore. In matrix stimulation, HCl is injected into a well bore to dissolve the porous carbonate matrix. By dissolving the solid carbonate the pores will increase in size, and the oil and gas will be able to flow out at faster rates, thereby increasing the productivity of the well.[5] The dissolution reaction is

$$4HCl + CaMg(CO_3)_2 \rightarrow Mg^{2+} + Ca^{2+} + 4Cl^- + 2CO_2 + 2H_2O$$

The concentration of HCl at various times was determined from atomic absorption spectrophotometer measurements of the calcium and magnesium ions.

Determine the reaction order with respect to HCl from the data presented in Figure E5-3.1 for this batch reaction. Assume that the rate law is in the form given by Equation (5-1) and that the combined rate law and mole balance for HCl can be given by Equation (5-6).

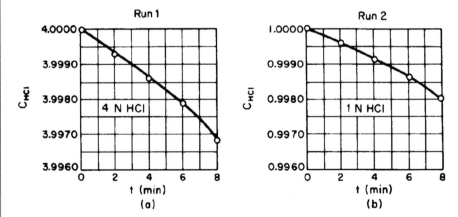

Figure E5-3.1   Concentration–time data.

Solution

Evaluating the mole balance on a constant-volume batch reactor at time $t = 0$ gives

$$\left(-\frac{dC_{HCl}}{dt}\right)_0 = -(r_{HCl})_0 = kC_{HCl,0}^{\alpha} \qquad (E5\text{-}3.1)$$

---

[4] K. Lund, H. S. Fogler, and C. C. McCune, *Chem. Eng. Sci., 28*, 691 (1973).

[5] M. Hoefner and H. S. Fogler, *AIChE Journal, 34*(1), 45 (1988).

Taking the log of both sides of Equation (E5-3.1), we have

$$\ln\left(-\frac{dC_{HCl}}{dt}\right)_0 = \ln k + \alpha \ln C_{HCl,0} \qquad (E5\text{-}3.2)$$

The derivative at time $t = 0$ can be found from the slope of the plot of concentration versus time evaluated at $t = 0$. Figure E5-3.1(a) and (b) give

4 N HCl solution	1 N HCl solution
$-r_{HCl,0} = -\dfrac{3.9982 - 4.0000}{5 - 0}$	$-r_{HCl,0} = -\dfrac{0.9987 - 1.0000}{6 - 0}$
$-r_{HCl,0} = 3.6 \times 10^{-4}$ g mol/L·min	$-r_{HCl,0} = 2.2 \times 10^{-4}$ g mol/L·min

Converting to a rate per unit area, $-r_A''$, and to seconds (30 cm² of solid per liter of solution), the rates at 1 N and 4 N become 1.2×10^{-7} mol/cm²·s and 2.0×10^{-7} mol/cm²·s, respectively. We also could have used either POLYMATH or the differentiation formulas to find the derivative at $t = 0$.

If we were to continue in this manner, we would generate the following data set.[6]

TABLE 5-3.

$C_{HCl,0}$ (mol/cm³)	1.0	4.0	2.0	0.1	0.5
$-r_{HCl,0}''$ (mol/cm²·s) $\times 10^7$	1.2	2.0	1.36	0.36	0.74

These data are plotted on Figure E5-3.2. The slope of this ln-ln plot of $-r_{HCl,0}''$ versus $C_{HCl,0}$ shown in Figure E5-3.2 gives a reaction order of 0.44. The rate law is

$$-r_{HCl,0}'' = kC_{HCl}^{0.44} \qquad (E5\text{-}3.3)$$

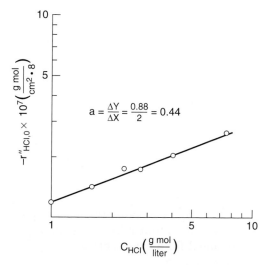

$$a = \frac{\Delta Y}{\Delta X} = \frac{0.88}{2} = 0.44$$

Figure E5-3.2   Initial rate as a function of initial HCl concentration.

[6] K. Lund, H. S. Fogler, and C. C. McCune, *Chem. Eng. Sci., 28*, 691 (1973).

For this dissolution of dolomite in HCl, the reaction order was also found to vary with temperature.

5.3 Method of Half-Lives

The method of half-lives requires many experiments

The half-life of a reaction, $t_{1/2}$, is defined as the time it takes for the concentration of the reactant to fall to half of its initial value. By determining the half-life of a reaction as a function of the initial concentration, the reaction order and specific reaction rate can be determined. If two reactants are involved in the chemical reaction, the experimenter will use the method of excess in conjunction with the method of half-lives to arrange the rate law in the form

$$-r_A = kC_A^\alpha \tag{5-1}$$

For the irreversible reaction

$$A \longrightarrow \text{products}$$

a mole balance on species A in a constant-volume batch reaction system results in the following expression:

$$-\frac{dC_A}{dt} = -r_A = kC_A^\alpha \tag{E5-1.1}$$

Integrating with the initial condition $C_A = C_{A0}$ when $t = 0$, we find that

$$t = \frac{1}{k(\alpha - 1)} \left(\frac{1}{C_A^{\alpha - 1}} - \frac{1}{C_{A0}^{\alpha - 1}} \right)$$

$$= \frac{1}{kC_{A0}^{\alpha - 1}(\alpha - 1)} \left[\left(\frac{C_{A0}}{C_A} \right)^{\alpha - 1} - 1 \right] \tag{5-17}$$

The half-life is defined as the time required for concentration to drop to half of its initial value; that is,

$$t = t_{1/2} \quad \text{when} \quad C_A = \tfrac{1}{2} C_{A0}$$

Substituting for C_A in Equation (5-17) gives us

$$t_{1/2} = \frac{2^{\alpha - 1} - 1}{k(\alpha - 1)} \left(\frac{1}{C_{A0}^{\alpha - 1}} \right) \tag{5-18}$$

There is nothing special about using the time required for the concentration to drop to one-half of its initial value. We could just as well use the time required for the concentration to fall to $1/n$ of the initial value, in which case

$$t_{1/n} = \frac{n^{\alpha - 1} - 1}{k(\alpha - 1)} \left(\frac{1}{C_{A0}^{\alpha - 1}} \right) \tag{5-19}$$

For the method of half-lives, taking the natural log of both sides of Equation (5-18),

Plot $t_{1/2}$ as a function of C_{A0} or use regressional software

$$\ln t_{1/2} = \ln \frac{2^{\alpha-1} - 1}{(\alpha - 1)k} + (1 - \alpha) \ln C_{A0}$$

we see that the slope of the plot of $\ln t_{1/2}$ as a function of $\ln C_{A0}$ is equal to 1 minus the reaction order:

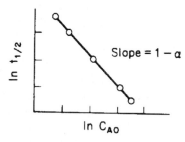

Figure 5-7   Method of half-lives.

Rearranging:

$$\alpha = 1 - \text{slope}$$

For the plot shown in Figure 5-7 the slope is -1:

$$\alpha = 1 - (-1) = 2$$

The corresponding rate law is

$$-r_A = kC_A^2$$

5.4 Differential Reactors

Data acquisition using the method of initial rates and a differential reactor are similar in that the rate of reaction is determined for a specified number of predetermined initial or entering reactant concentrations. A differential reactor is normally used to determine the rate of reaction as a function of either concentration or partial pressure. It consists of a tube containing a very small amount of catalyst usually arranged in the form of a thin wafer or disk. A typical arrangement is shown schematically in Figure 5-8. The criterion for a reactor being differential is that the conversion of the reactants in the bed is extremely small, as is the change in reactant concentration through the bed. As a result, the reactant concentration through the reactor is essentially constant and approximately equal to the inlet concentration. That is, the reactor is considered to be gradientless,[7] and the reaction rate is considered spatially uniform within the bed.

---

[7] B. Anderson, ed., *Experimental Methods in Catalytic Research* (San Diego, Calif.: Academic Press, 1968).

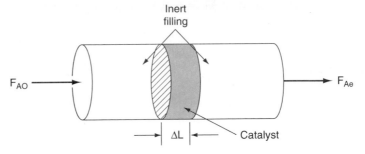

Most commonly
used catalytic
reactor to obtain
experimental data

Figure 5-8   Differential reactor.

      The differential reactor is relatively easy to construct at a low cost. Owing to the low conversion achieved in this reactor, the heat release per unit volume will be small (or can be made small by diluting the bed with inert solids) so that the reactor operates essentially in an isothermal manner. When operating this reactor, precautions must be taken so that the reactant gas or liquid does not bypass or channel through the packed catalyst, but instead flows uniformly across the catalyst. If the catalyst under investigation decays rapidly, the differential reactor is not a good choice because the reaction rate parameters at the start of a run will be different from those at the end of the run. In some cases sampling and analysis of the product stream may be difficult for small conversions in multicomponent systems.

Limitations of the
differential reactor

      The volumetric flow rate through the catalyst bed is monitored, as are the entering and exiting concentrations (Figure 5-9). Therefore, if the weight of

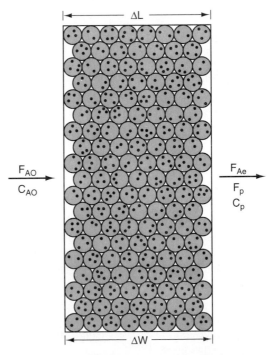

Figure 5-9   Differential catalyst bed.

catalyst, W, is known, the rate of reaction per unit mass of catalyst, r'_A, can be calculated. Since the differential reactor is assumed to be *gradientless*, the design equation will be similar to the CSTR design equation. A steady-state mole balance on reactant A gives

$$\begin{bmatrix} \text{flow} \\ \text{rate} \\ \text{in} \end{bmatrix} - \begin{bmatrix} \text{flow} \\ \text{rate} \\ \text{out} \end{bmatrix} + \begin{bmatrix} \text{rate of} \\ \text{generation} \end{bmatrix} = \begin{bmatrix} \text{rate of} \\ \text{accumulation} \end{bmatrix}$$

$$[F_{A0}] - [F_{Ae}] + \left[\left(\frac{\text{rate of reaction}}{\text{mass of cat.}} \right)(\text{mass of cat.}) \right] = 0$$

$$F_{A0} - F_{Ae} + (r'_A)(W) = 0$$

The subscript e refers to the exit of the reactor. Solving for $-r'_A$, we have

$$-r'_A = \frac{F_{A0} - F_{Ae}}{W} \tag{5-20}$$

The mole balance equation can also be written in terms of the concentration

Differential reactor design equation

$$\boxed{-r'_A = \frac{v_0 C_{A0} - v C_{Ae}}{W}} \tag{5-21}$$

or in terms of the conversion or product flow rate F_P:

$$\boxed{-r'_A = \frac{F_{A0} X}{W} = \frac{F_P}{W}} \tag{5-22}$$

The term $F_{A0} X$ gives the rate of formation of the product, F_P, when the stoichiometric coefficients of A and of P are identical.

For constant volumetric flow, Equation (5-21) reduces to

$$-r'_A = \frac{v_0 (C_{A0} - C_{Ae})}{W} = \frac{v_0 C_p}{W} \tag{5-23}$$

Consequently, we see that the reaction rate, $-r'_A$, can be determined by measuring the product concentration, C_p.

By using very little catalyst and large volumetric flow rates, the concentration difference, $(C_{A0} - C_{Ae})$, can be made quite small. The rate of reaction determined from Equation (5-23) can be obtained as a function of the reactant concentration in the catalyst bed, C_{Ab}:

$$-r'_A = -r'_A(C_{Ab}) \tag{5-24}$$

by varying the inlet concentration. One approximation of the concentration of A within the bed, C_{Ab}, would be the arithmetic mean of the inlet and outlet concentrations:

$$C_{Ab} = \frac{C_{A0} + C_{Ae}}{2} \qquad (5\text{-}25)$$

However, since very little reaction takes place within the bed, the bed concentration is essentially equal to the inlet concentration,

$$C_{Ab} \approx C_{A0}$$

so $-r_A'$ is a function of C_{A0}:

$$-r_A' = -r_A'(C_{A0}) \qquad (5\text{-}26)$$

As with the method of initial rates, various numerical and graphical techniques can be used to determine the appropriate algebraic equation for the rate law.

Example 5–4 Differential Reactor

The formation of methane from carbon monoxide and hydrogen using a nickel catalyst was studied by Pursley.[8] The reaction

$$3H_2 + CO \rightarrow CH_4 + 2H_2O$$

was carried out at 500°F in a differential reactor where the effluent concentration of methane was measured.

(a) Relate the rate of reaction to the exit methane concentration.

(b) The reaction rate law is assumed to be the product of a function of the partial pressure of CO, $f(CO)$, and a function of the partial pressure of H_2, $g(H_2)$:

$$r_{CH_4}' = f(CO) \cdot g(H_2) \qquad (E5\text{-}4.1)$$

Determine the reaction order with respect to carbon monoxide, using the data in Table E5-4.1. Assume that the functional dependence of r_{CH_4}' on P_{CO} is of the form

$$r_{CH_4}' \sim P_{CO}^\alpha \qquad (E5\text{-}4.2)$$

TABLE E5-4.1   RAW DATA

Run	P_{CO} (atm)	P_{H_2} (atm)	C_{CH_4} (g mol/dm^3)
1	1	1.0	2.44×10^{-4}
2	1.8	1.0	4.40×10^{-4}
3	4.08	1.0	10.0×10^{-4}
4	1.0	0.1	1.65×10^{-4}
5	1.0	0.5	2.47×10^{-4}
6	1.0	4.0	1.75×10^{-4}

P_{H_2} is constant in Runs 1, 2, 3
P_{CO} is constant in Runs 4, 5, 6

---

[8] J. A. Pursley, An Investigation of the Reaction between Carbon Monoxide and Hydrogen on a Nickel Catalyst above One Atmosphere, Ph.D. thesis, University of Michigan.

The exit volumetric flow rate from a differential packed bed containing 10 g of catalyst was maintained at 300 dm³/min for each run. The partial pressures of H_2 and CO were determined at the entrance to the reactor, and the methane concentration was measured at the reactor exit.

Solution

(a) In this example the product composition, rather than the reactant concentration, is being monitored. $-r'_A$ can be written in terms of the flow rate of methane from the reaction,

$$-r'_A = r'_{CH_4} = \frac{F_{CH_4}}{\Delta W}$$

Substituting for F_{CH_4} in terms of the volumetric flow rate and the concentration of methane gives

$$-r'_A = \frac{v_0 C_{CH_4}}{\Delta W} \tag{E5-4.3}$$

Since v_0, C_{CH_4}, and ΔW are known for each run, we can calculate the rate of reaction. For run 1:

$$-r'_A = \left(\frac{300 \text{ dm}^3}{\text{min}} \right) \frac{2.44 \times 10^{-4}}{10 \text{ g cat.}} \text{ g mol/dm}^3 = 7.33 \times 10^{-3} \frac{\text{g mol CH}_4}{\text{g cat.} \times \text{min}}$$

The rate for runs 2 through 6 can be calculated in a similar manner (Table E5-4.2).

TABLE E5-4.2.   RAW AND CALCULATED DATA

Run	P_{CO} (atm)	P_{H_2} (atm)	C_{CH_4}(g mol/dm³)	$r'_{CH_4} \left(\dfrac{\text{g mol CH}_4}{\text{g cat.} \times \text{min}} \right)$
1	1.0	1.0	2.44×10^{-4}	7.33×10^{-3}
2	1.8	1.0	4.40×10^{-4}	13.2×10^{-3}
3	4.08	1.0	10.0×10^{-4}	30.0×10^{-3}
4	1.0	0.1	1.65×10^{-4}	4.95×10^{-3}
5	1.0	0.5	2.47×10^{-4}	7.42×10^{-3}
6	1.0	4.0	1.75×10^{-4}	5.25×10^{-3}

For constant hydrogen concentration, the rate law

$$r'_{CH_4} = k P_{CO}^\alpha \cdot g(P_{H_2})$$

can be written as

$$r'_{CH_4} = k' P_{CO}^\alpha \tag{E5-4.4}$$

Taking the log of Equation (E5-4.4) gives us

$$\ln(r'_{CH_4}) = \ln k' + \alpha \ln P_{CO}$$

We now plot $\ln(r'_{CH_4})$ versus $\ln P_{CO}$ for runs 1, 2, and 3.

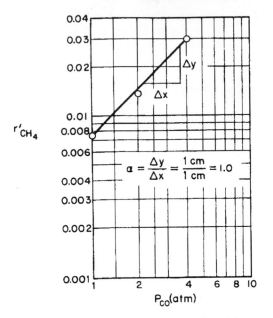

Figure E5-4.1   Reaction rate as a function of partial pressure.

(b) Runs 1, 2, and 3, for which the H_2 concentration is constant, are plotted in Figure E5-4.1. From the slope of the plot in Figure E5-5.1, we find that $\alpha = 1$.

$$-r'_A = k'P_{CO} \qquad (\text{E5-4.5})$$

Determining the Rate Law Dependence on H_2

From Table E5-4.2 it appears that the dependence of $-r'_{CH_4}$ on P_{H_2} cannot be represented by a power law. Comparing run 4 with run 5 and run 1 with run 6, we see that the reaction rate first increases with increasing partial pressure of hydrogen, and subsequently decreases with increasing P_{H_2}. That is, there appears to be a concentration of hydrogen at which the rate is maximum. One set of rate laws that is consistent with these observations is:

1. At low H_2 concentrations where r'_{CH_4} increases as P_{H_2} increases, the rate law may be of the form

$$r'_{CH_4} \sim P_{H_2}^{\beta_1} \qquad (\text{E5-4.6})$$

2. At high H_2 concentrations where r'_{CH_4} decreases as P_{H_2} increases,

$$r'_{CH_4} \sim \frac{1}{P_{H_2}^{\beta_2}} \qquad (\text{E5-4.7})$$

We would like to find one rate law that is consistent with reaction rate data at both high and low hydrogen concentrations. Experience suggests Equations (E5-4.6) and (E5-4.7) can be combined into the form

Typical form of the rate law for heterogeneous catalysis

$$r'_{CH_4} \sim \frac{P_{H_2}^{\beta_1}}{1 + bP_{H_2}^{\beta_2}} \qquad (\text{E5-4.8})$$

We will see in Chapter 10 that this combination and similar rate laws which have reactant concentrations (or partial pressures) in the numerator and denominator are common in *heterogeneous catalysis*.

Let's see if the resulting rate law (E5-4.8) is qualitatively consistent with the rate observed.

1. *For condition 1*: At low P_{H_2}, $b((P_{H_2})^{\beta_2} \ll 1)$ and Equation (E5-4.8) reduces to

$$r'_{CH_4} \sim P_{H_2}^{\beta_1} \qquad (E5\text{-}4.9)$$

Equation (E5-4.9) is consistent with the trend in comparing runs 4 and 5.

2. *For condition 2*: At high P_{H_2}, $b((P_{H_2})^{\beta_2} \gg 1)$ and Equation (E5-4.8) reduces to

$$r'_{CH_4} \sim \frac{(P_{H_2})^{\beta_1}}{(P_{H_2})^{\beta_2}} \sim \frac{1}{(P_{H_2})^{\beta_2 - \beta_1}} \qquad (E5\text{-}4.10)$$

where $\beta_2 > \beta_1$. Equation (E5-4.10) is consistent with the trends in comparing runs 5 and 6.

Theoretical considerations of the type to be discussed in Chapter 10 predict that if the rate-limiting step in the overall reaction is the reaction between atomic hydrogen absorbed on the nickel surface and CO in the gas phase, then the rate law will be in the form

$$r'_{CH_4} = \frac{aP_{CO}P_{H_2}^{1/2}}{1 + bP_{H_2}} \qquad (E5\text{-}4.11)$$

This rate law is qualitatively consistent with experimental observations. To obtain the parameter a and b, we rearrange Equation (E5-4.11) in the form

$$\frac{P_{CO}P_{H_2}^{1/2}}{r'_{CH_4}} = \frac{1}{a} + \frac{b}{a}P_{H_2} \qquad (E5\text{-}4.12)$$

Linearizing the rate law to determine the rate law parameters

A plot of $P_{CO}P_{H_2}^{1/2}/r'_{CH_4}$ as a function of P_{H_2} should be a straight line with an intercept of $1/a$ and a slope of b/a. From the plot in Figure E5-4.2 we see that the rate law is indeed consistent with the rate law data.

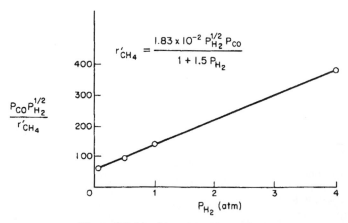

Figure E5-4.2 Linearizing plot of data.

As an exercise, use the analysis in section 5.5 to find the rate law shown in Figure E5-4.2.

5.5 Least-Squares Analysis

5.5.1 Linearization of the Rate Law

If a rate law depends on the concentration of more than one species and it is not possible to use the method of excess, we may choose to use a linearized least-squares method. This method of data analysis is also useful to determine the best values of the rate law parameters from a series of measurements when three or more parameters are involved (e.g., reaction order, α; frequency factor, A; and activation energy, E).

A mole balance on a constant-volume batch reactor gives

$$-\frac{dC_A}{dt} = -r_A = kC_A^{\alpha}C_B^{\beta} \tag{5-27}$$

If we now use the method of initial rates, then

$$\left(-\frac{dC_A}{dt}\right)_0 = -r_{A0} = kC_{A0}^{\alpha}C_{B0}^{\beta}$$

Taking the log of both sides, we have

Used when C_{A0} and C_{B0} are varied simultaneously

$$\boxed{\ln\left(-\frac{dC_A}{dt}\right)_0 = \ln k + \alpha \ln C_{A0} + \beta \ln C_{B0}} \tag{5-28}$$

Let $Y = \ln(-dC_A/dt)_0$, $X_1 = \ln C_{A0}$, $X_2 = \ln C_{B0}$, $a_0 = \ln k$, $a_1 = \alpha$, and $a_2 = \beta$. Then

$$Y = a_0 + a_1X_1 + a_2X_2 \tag{5-29}$$

If we now carry out N experimental runs, for the jth run, Equation (5-29) takes the form

$$\boxed{Y_j = a_0 + a_1X_{1j} + a_2X_{2j}} \tag{5-30}$$

where $X_{1j} = \ln C_{A0j}$, with C_{A0j} being the initial concentration of A for the jth run. The best values of the parameters a_0, a_1, and a_2 are found by solving Equations (5-31) through (5-33) simultaneously.

For N runs, 1, 2, ..., N,

$$\sum_{j=1}^{N} Y_j = Na_0 + a_1\sum_{j=1}^{N} X_{1j} + a_2\sum_{j=1}^{N} X_{2j} \tag{5-31}$$

Three equations, three unknowns (a_0, a_1, a_2)

$$\sum_{j=1}^{N} X_{1j}Y_j = a_0\sum_{j=1}^{N} X_{1j} + a_1\sum_{j=1}^{N} X_{1j}^2 + a_2\sum_{j=1}^{N} X_{1j}X_{2j} \tag{5-32}$$

$$\sum_{j=1}^{N} X_{2j}Y_j = a_0\sum_{j=1}^{N} X_{2j} + a_1\sum_{j=1}^{N} X_{1j}X_{2j} + a_2\sum_{j=1}^{N} X_{2j}^2 \tag{5-33}$$

We have three linear equations and three unknowns which we can solve for: a_0, a_1, and a_2. A detailed example delineating the kinetics of the reaction

$$H_2(g) + C_8H_{16}(g) \rightarrow C_8H_{18}(g)$$

using linear least-squares analysis can be found in Example 10-2. If we set $a_2 = 0$ and consider only two variables, Y and X, Equations (5-31) and (5-32) reduce to the familiar least-squares equations for two unknowns.

Example 5–5   Using Least-Squares Analysis to Determine Rate Law Parameters

The etching of semiconductors in the manufacture of computer chips is another important solid–liquid dissolution reaction (see Problem P5-12 and Section 12.10). The dissolution of the semiconductor MnO_2 was studied using a number of different acids and salts. The rate of dissolution was found to be a function of the reacting liquid solution redox potential relative to the energy-level conduction band of the semiconductor. It was found that the reaction rate could be increased by a factor of 10^5 simply by changing the anion of the acid[9]!! From the data below, determine the reaction order and specific reaction rate for the dissolution of MnO_2 in HBr.

A 10^5 fold increase in reaction rate!!!

C_{A0} (mol HBr/dm^3)	0.1	0.5	1.0	2.0	4.0
$-r''_{A0}$ (mol HBr/m$^2 \cdot$h) $\times 10^2$	0.073	0.70	1.84	4.86	12.84

Solution

We assume a rate law of the form

$$-r''_{HB\,r} = kC^{\alpha}_{HBr} \tag{E5-5.1}$$

Letting A = HBr, taking the ln of both sides of (E5-5.1), and using the initial rate and concentration gives

$$\ln(-r''_{A0}) = \ln k + \alpha \ln C_{A0} \tag{E5-5.2}$$

Let $Y = \ln(-r''_{A0})$, $a = \ln k$, $b = \alpha$, and $X = \ln C_{A0}$. Then

$$Y = a + bX \tag{E5-5.3}$$

The least-squares equations to be solved for the best values of a and b are for N runs

$$\sum_{i=1}^{N} Y_i = Na + b \sum_{i=1}^{N} X_i \tag{E5-5.4}$$

$$\sum_{i=1}^{N} X_i Y_i = a \sum_{i=1}^{N} X_i + b \sum_{i=1}^{N} X_i^2 \tag{E5-5.5}$$

where i = run number. Substituting the appropriate values from Table E5-5.1 into Equations (E5-5.4) and (E5-5.5) gives

$$-21.26 = 5a + -0.92b \tag{E5-5.6}$$

$$15.10 = -0.92a + 8.15b \tag{E5-5.7}$$

---

[9] S. E. Le Blanc and H. S. Fogler, *AIChE J., 32*, 1702 (1986).

TABLE E5-5.1

Run	C_{A0}	X_i	$-r''_{A0}$	Y_i	X_iY_i	X_i^2
1	0.1	−2.302	0.00073	−7.22	16.61	5.29
2	0.5	−0.693	0.007	−4.96	3.42	0.48
3	1.0	0.0	0.0184	−4.0	0.0	0.0
4	2.0	0.693	0.0486	−3.02	−2.09	0.48
5	4.0	1.38	0.128	−2.06	−2.84	1.90

$$\sum_{i=1}^{5} X_i = -0.92 \qquad \sum_{i=1}^{5} Y_i = -21.26 \quad \sum_{i=1}^{5} X_iY_i = 15.1 \quad \sum_{i=1}^{5} X_i^2 = 8.15$$

Solving for a and b yields

$$b = 1.4 \quad \text{therefore} \quad \alpha = 1.4$$

and

$$a = -3.99 \qquad k = 1.84 \times 10^{-2}\,(\text{dm}^3/\text{mol})^{0.4}/\text{m}^2 \cdot \text{h}$$

$$\boxed{r''_{\text{HBr}} = 0.0184 C_{\text{HBr}}^{1.4}} \tag{E5-5.8}$$

5.5.2 Nonlinear Least-Squares[10] Analysis

In nonlinear least squares analysis we search for those parameter values that minimize the sum of squares of the differences between the measured values and the calculated values for all the data points. Many software programs are available to find these parameter values and all one has to do is to enter the data. The POLYMATH software will be used to illustrate this technique. In order to carry out the search efficiently, in some cases one has to enter initial estimates of the parameter values close to the actual values. These estimates can be obtained using the linear-least-squares technique just discussed.

We will now apply nonlinear least-squares analysis to reaction rate data to determine the rate law parameters. Here we make estimates of the parameter values (e.g., reaction order, specific rate constants) in order to calculate the rate of reaction, r_c. We then search for those values that will minimize the sum of the squared differences of the measured reaction rates, r_m, and the calculated reaction rates, r_c. That is, we want the sum of $(r_m - r_c)^2$ for all data points to be minimum. If we carried out N experiments, we would want to find the parameter values (e.g., E, activation energy, reaction orders) that would minimize the quantity

$$\sigma^2 = \frac{s^2}{N-K} = \sum_{i=1}^{N} \frac{(r_{im} - r_{ic})^2}{N-K} \tag{5-34}$$

---

[10]See also R. Mezakiki and J. R. Kittrell, *AIChE J.*, *14*, 513 (1968), and J. R. Kittrell, *Ind. Eng. Chem.*, *61*, (5), 76–78 (1969).

where

$$s^2 = \sum (r_{im} - r_{ie})^2$$

N = number of runs

K = number of parameters to be determined

r_{im} = measured reaction rate for run i (i.e., $-r_{Aim}$)

r_{ic} = calculated reaction rate for run i (i.e., $-r_{Aic}$)

To illustrate this technique, let's consider the first-order reaction

$$A \longrightarrow Product$$

for which we want to learn the reaction order, α, and the specific reaction rate, k,

$$r = kC_A^{\alpha}$$

The reaction rate will be measured at a number of different concentrations and these measurements are shown on the left of Table 5-2.

We now choose values of k and α and calculate the rate of reaction (r_{ic}) at each concentration at which an experimental point was taken. We then subtract the calculated value from the measured value (r_{im}), square the result, and sum the squares for all the runs for the values of k and α we have choosen. For example, consider the data set given for runs 1 through 4 in the second and third columns. In trial 1 we first guess $k = 1$ and $\alpha = 1$ and then calculate the rate based on these values. For run 1 the calculated value of the rate is $r_i = (1)(0.6)^1 = 0.6$. The difference between the measured rate and the calculated rate is $r_{im} - r_{ic} = 1.9 - 0.6 = 1.3$. The squared difference $(r_{im} - r_{im})^2$ is 1.69. We make similar calculations for runs 2 through 4 and they are shown in the sixth column. Next we sum up all the squared differences $[s^2 = \sum_1^4 (r_{im} - r_{ic})^2]$ for all the runs and obtain $s^2 = 114.04$ for the values chosen: $\alpha = 1$, $k = 1$. Next choose new values of α and k. In the seventh and eighth columns the calculated rate and the differences $(r_{im} - r_{ic})^2$ are given for $\alpha = 1$, $k = 4$. Next, new values for k and α are chosen and the procedure is repeated. Initial estimates of k and α can be obtained by a linearized least-squares analysis. Table 5-2 shows an example of how the sum of the squares (σ_1^2 and σ_2^2) is calculated for N.

Finding values of α and k to minimize σ^2

TABLE 5-2.   MINIMIZING THE SUM OF THE SQUARE DIFFERENCES

Data			Trial 1 $k = 1$, $\alpha = 1$			Trial 2 $k = 4$, $\alpha = 1$		Trial 3 $k = 4$, $\alpha = 1.5$		Trial 4 $k = 5$, $\alpha = 1.5$		Trial 5 $k = 5$, $\alpha = 2$	
Run	C_A	r_m	r_c	$r_m - r_c$	$(r_m - r_c)^2$	r_c	$(r_m - r_c)^2$	r_c	$(r_m - r_c)^2$	r_c	$(r_m - r_c)^2$	r_c	$(r_m - r_c)^2$
1	0.6	1.9	0.6	1.3	1.69	2.4	0.25	1.86	0.0016	2.32	0.18	1.80	0.01
2	0.8	3.1	0.8	2.3	5.29	3.2	0.01	2.86	0.06	3.58	0.23	3.2	0.01
3	1.0	5.1	1.0	4.1	16.81	4.0	1.21	4.0	1.21	5.0	0.01	5.0	0.01
4	1.5	11.0	1.5	9.5	90.25	6.0	25.0	7.35	13.32	9.19	3.28	11.25	0.06
					$s_1^2 = 114$		$s_2^2 = 26.5$		$s_3^2 = 14.6$		$s_4^2 = 3.7$		$s_5^2 = 0.09$
					$\sigma_1^2 = 57$		$\sigma_1^2 = 13.25$		$\sigma_3^2 = 7.3$		$\sigma_4^2 = 1.85$		$\sigma_5^2 = 0.045$

This procedure is continued by further varying α and k until we find their best values, that is, those values that minimize the sum of the squares. Many well-known searching techniques are available to obtain the minimum value σ_m^2.[11] Figure 5-10 shows a hypothetical plot of the sum of the squares as a function of the parameters α and k:

$$\sigma^2 = f(k, \alpha) \tag{5-35}$$

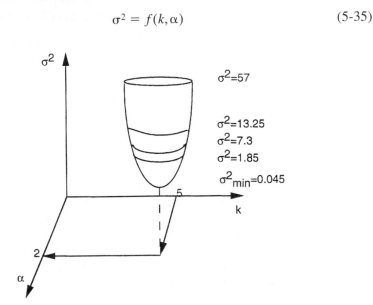

Figure 5-10   Minimum sum of squares.

In searching to find the parameter values that give the minimum of the sum of squares σ^2, one can use a number of optimization techniques or software packages. The procedure begins by guessing parameter values [e.g., Table 5-2 ($\alpha = 1$, $k = 1\,\text{s}^{-1}$)] and then calculating r_c and then σ^2 for these values (see, e.g., the sixth column in Table 5-2). Next a few sets of parameters are chosen around the initial guess, and σ^2 is calculated for these sets as well. The search technique looks for the smallest value of σ^2 in the vicinity of the initial guess and then proceeds along a trajectory in the direction of decreasing σ^2 to choose different parameter values and determine the corresponding σ^2. The trajectory is continually adjusted so as always to proceed in the direction of decreasing σ^2 until the minimum value of σ^2 is reached. A schematic of this procedure is shown in Figure 5-11, where the parameter values at the minimum are $\alpha = 2$ and $k = 5\,\text{s}^{-1}$. If the equations are highly nonlinear, the initial guess is extremely important. In some cases it is useful to try different initial guesses of the parameter to make sure that the software program converges on

Vary the initial guesses of parameters to make sure you find the true minimum

[11](a) B. Carnahan and J. O. Wilkes, *Digital Computing and Numerical Methods* (New York: Wiley, 1973), p. 405. (b) D. J. Wilde and C. S. Beightler, *Foundations of Optimization* (Upper Saddle River, N.J.: Prentice Hall, 1967). (c) D. Miller and M. Frenklach, *Int. J. Chem. Kinet.*, *15*, 677 (1983).

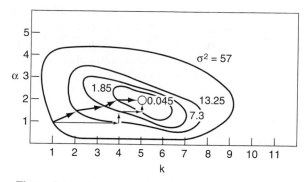

Figure 5-11   Trajectory to find the best values of k and α.

the same minimum for the different initial guesses. The dark lines and heavy arrows represent a computer trajector, and the light lines and arrows represent the hand calculations shown in Table 5-2.

A number of software packages are available to carry out the procedure to determine the best estimates of the parameter values and the corresponding confidence limits. All one has to do is to type the experimental values in the computer, specify the model, enter the initial guesses of the parameters, and then push the computer button, and the best estimates of the parameter values along with 95% confidence limits appear. If the confidence limits for a given parameter are larger than the parameter itself, the parameter is probably not significant and should be dropped from the model. After the appropriate model parameters are eliminated, the software is run again to determine the best fit with the new model equation.

Model Discrimination.   One can also determine which model or equation best fits the experimental data by comparing the sums of the squares for each model and then choosing the equation with a smaller sum of squares and/or carrying out an F-test. Alternatively, we can compare the residual plots for each model. These plots show the error associated with each data point and one looks to see if the error is randomly distributed or if there is a trend in the error. When the error is randomly distributed, this is an additional indication that the correct rate law has been chosen. To illustrate these principles, let's look at the following example.

Example 5–6   Hydrogenation of Ethylene to Ethane

The hydrogenation (H) of ethylene (E) to form ethane (EA),

$$H_2 + C_2H_4 \rightarrow C_2H_6$$

is carried out over a cobalt molybdenum catalyst [*Collect. Czech. Chem. Commun.*, *51*, 2760 (1988)]. Carry out a nonlinear least-squares analysis on the data given in Table E5-6.1, and determine which rate law best describes the data.

TABLE E5-6.1.   DIFFERENTIAL REACTOR DATA

Run Number	Reaction Rate (mol/kg cat.·s)	P_E (atm)	P_{EA} (atm)	P_H (atm)
1	1.04	1	1	1
2	3.13	1	1	3
3	5.21	1	1	5
4	3.82	3	1	3
5	4.19	5	1	3
6	2.391	0.5	1	3
7	3.867	0.5	0.5	5
8	2.199	0.5	3	3
9	0.75	0.5	5	1

Determine which of the following rate laws best describes the data.

(a) $$-r_A = \frac{kP_EP_H}{1 + K_AP_{EA} + K_EP_E}$$ (c) $$-r_A = \frac{kP_EP_H}{(1 + K_EP_E)^2}$$

(b) $$-r_A = \frac{kP_EP_H}{1 + K_EP_E}$$ (d) $$-r_A = kP_E^aP_H^b$$

Solution

POLYMATH was chosen as the software package to solve this problem. The data in Table E5-6.1 were typed into the system. The following copies of the computer screens illustrate the procedure.

Procedure
- Enter data
- Enter model
- Make initial estimates of parameters
- Run regression
- Examine parameters and variance
- Observe error distribution
- Choose model

1. First we enter the data in Table E5-6.1 into the POLYMATH table shown in Table E5-6.2. After entering the data and pressing shift F7, the following solution options appear on the screen:

```
                           SOLUTION OPTIONS
    1. Do linear regression.   L. Linear regression without free parameter.
    p. Fit a polynomial.       P. Polynomial passing through origin.
    s. Fit a cubic spline.     R. Do nonlinear regression.
                               ⇧ ⏎ or F8 for problem options
```

Next we type R and choose a model to fit the data.

2. Specify the model and determine the model parameters and σ^2. The first model we choose is model (a).

$$-r_A = \frac{kP_EP_H}{1 + K_AP_{EA} + K_EP_E} \qquad (E5\text{-}6.1)$$

Next we enter the model Equation (E5-6.1) and initial estimates of the parameter values.

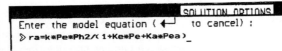

```
                           SOLUTION OPTIONS
 Enter the model equation ( ⏎    to cancel) :
 ▷ ra=k*Pe*Ph2/( 1+Ke*Pe+Ka*Pea )_
```

TABLE E5-6.2.   ANALYZING RATE DATA

Name	Pe	Ph2	Pea	ra			
1	1	1	1	1.04			
2	1	3	1	3.13			
3	1	5	1	5.21			
4	3	3	1	3.82			
5	5	3	1	4.19			
6	0.5	3	1	2.391			
7	0.5	5	0.5	3.867			
8	0.5	3	3	2.199			
9	0.5	1	5	0.75			
10							
11							
12							
13							
14							
15							
16							
17							
18							

PROBLEM OPTIONS

```
 ←┘  to edit current box.     ↑,↓,→,←, PgUp, PgDn, Home, End to move pointer.
⇧ ←┘ for row/column options.                           T.  to change title.
⇧ F7 to fit a curve or do regression.            F6 for helpful information.
⇧ F8 for file and library options.                      F7 to print problem.
```

Details for using
regression software

After entering the estimates the following screen appears:

Nonlinear Regression of Rate Data

Nonlinear regression model equation:

ra=k*Pe*Ph2/(1+Ke*Pe+Ka*Pea)

Initial estimates:

k = 1
Ke = 1
Ka = 0.01

REGRESSION OPTIONS

```
     e.  Change the model equation.
     i.  Change the initial guesses.
            ⇧ F7 to solve.
        F8 for the problem options.
```

after which we press shift ⇧F7 to solve the problem and the following screen
appears:

Param.	Converged Value	0.95 conf. interval	lower limit	upper limit
k	3.34788	0.391055	2.95682	3.73893
Ke	2.21108	0.318885	1.89219	2.52996
Ka	0.0428414	0.0713596	-0.0285181	0.114201

Model: ra=k*Pe*Ph2/(1+Ke*Pe+Ka*Pea)

k = 3.34788   Ka = 0.0428414
Ke = 2.21108

6 positive residuals, 3 negative residuals. Sum of squares = 0.0296167

$$-r_A = \frac{3.35 P_E P_H}{1 + 0.043 P_{EA} + 2.2 P_E} \tag{E5-6.2}$$

3. Next we examine the estimated parameters. We see from this last output that the mean value of K_A is 0.043 atm^{-1}, with the 95% confidence limits being ± 0.0712. The 95% confidence limit on K_A means essentially that if the experiment were performed 100 times, the calculated value of K_A would fall between -0.028 and 0.114 ninety-five out of the hundred times, that is,

$$K_A = 0.043 \pm 0.071 \tag{E5-6.3}$$

For this model, the value of the 95% confidence interval is greater than the value of the parameter itself! Consequently, we are going to set the parameter value K_A equal to zero. When we set K_A equal to zero this yields the second model, model (b).

$$-r_A = \frac{k P_E P_H}{1 + K_E P_E} \tag{E5-6.4}$$

4. Determine the model parameters and σ^2 for the second model. When this model is entered, the following results are obtained:

$$-r_A = \frac{3.19 P_E P_H}{1 + 2.1 P_E} \tag{E5-6.5}$$

The value of the minimum sum of squares is $\sigma_B^2 = 0.042$.

Param.	Converged Value	0.95 conf. interval	lower limit	upper limit
k	3.18678	0.288026	2.89876	3.47481
Ke	2.10133	0.263925	1.83741	2.36526

Model: ra=k*Pe*Ph2/(1+Ke*Pe)
k = 3.18678
Ke = 2.10133
5 positive residuals, 4 negative residuals. Sum of squares = 0.0423735

5. Determine the parameters and σ^2 for a third model. We now proceed to model (c),

$$-r_A = \frac{k P_E P_H}{(1 + K_E P_E)^2} \tag{E5-6.6}$$

for which the following results are obtained:

```
                 Converged    0.95 conf.       lower        upper
       Param.      Value       interval        limit        limit
        k         2.00878      0.266198       1.74259      2.27498
        Ke        0.361667     0.0623113      0.299356     0.423979
```

```
    Model:  ra=k*Pe*Ph2/(1+Ke*Pe)^2
    k = 2.00878
    Ke = 0.361667
    5 positive residuals, 4 negative residuals.  Sum of squares = 0.436122
```

$$-r_A = \frac{2P_E P_H}{(1 + 0.36 P_E)^2} \tag{E5-6.7}$$

Comparing the sums of squares for models (b) and (c), we see that model (b) gives the smaller sum ($\sigma_B^2 = 0.042$ versus $\sigma_C^2 = 0.436$) by an order of magnitude, that is,

$$\frac{\sigma_C^2}{\sigma_B^2} = \frac{0.436}{0.042} = 10.38$$

Therefore, we eliminate model (c) from consideration.[12]

6. Determine the parameters and σ^2 for a power law model. Finally, we enter in model (d).

$$-r_A = k P_E^a P_H^b \tag{E5-6.8}$$

The following results were obtained.

```
                 Converged    0.95 conf.       lower        upper
       Param.      Value       interval        limit        limit
        k         0.894025     0.256901       0.637124     1.15093
        a         0.258441     0.0708914      0.18755      0.329332
        b         1.06155      0.209307       0.852246     1.27086
```

```
    Model:  ra=k*Pe^a*Ph2^b
    k = 0.894025              b = 1.06155
    a = 0.258441
    5 positive residuals, 4 negative residuals.  Sum of squares = 0.297223
```

One observes the error ($r_{im} - r_{ic}$) is indeed randomly distributed, indicating the model chosen is most likely the correct one.

$$-r_A = 0.89 P_E^{0.26} P_H^{1.06} \tag{E5-6.9}$$

7. Choose the best model. Again, the sum of squares for model (d) is significantly higher than in model (b) ($\sigma_B = 0.042$ versus $\sigma_D = 0.297$). Hence, we choose model (b) as our choice to fit the data. For cases when the sums of squares are relatively close together, we can use the F-test to discriminate between models to learn if one model is statistically better than another.[13]

---

[12]See G. F. Froment and K. B. Bishoff, *Chemical Reaction Analysis and Design*, 2nd ed. (New York: Wiley, (1990), p. 96.

[13]Ibid.

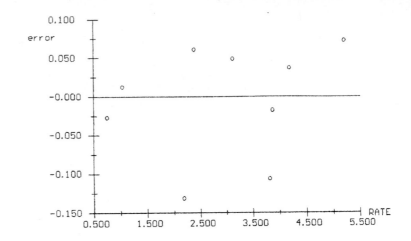

Note that the error is
not a function of
rate and appears to
be randomly
distributed

```
error = RATE - (k*Pe*PH2/(1+Ke*Pe))
k = 3.18678
Ke = 2.10133
```

Figure E5-6.1   Error as a function of calculated rate.

However, there is a caution! One cannot simply carry out a regression and then pick the model with the lowest value of the sums of squares. If this were the case, we would have chosen model (a) with $\sigma^2 = 0.03$. One must consider the physical realism of the parameters. In model (a) the 95% confidence interval was greater than the parameter itself, thereby yielding negative values of the parameter, K_A, which is physically impossible.

We can also use nonlinear regression to determine the rate law parameters from concentration–time data obtained in batch experiments. We recall that the combined rate law-stoichiometry-mole balance for a constant-volume batch reactor is

$$\frac{dC_A}{dt} = -kC_A^{\alpha} \tag{5-6}$$

We now integrate Equation (5-6) to give

$$C_{A0}^{1-\alpha} - C_A^{1-\alpha} = (1-\alpha)kt$$

Rearranging to obtain the concentration as a function of time, we obtain

$$C_A = [C_{A0}^{1-\alpha} - (1-\alpha)kt]^{1/1-\alpha} \tag{5-36}$$

Now we could use POLYMATH or MATLAB to find the values of α and k that would minimize the sum of squares of the differences between the measured and calculated concentrations. That is, for N data points,

$$s^2 = \sum_{i=1}^{N} (C_{Ami} - C_{Aci})^2 = \sum_{i=1}^{N} (C_{Ami} - [C_{A0}^{1-\alpha} - (1-\alpha)kt_i]^{1/1-\alpha})^2 \quad (5\text{-}37)$$

we want the values of α and k that will make s^2 a minimum.

If POLYMATH is used, one should use the absolute value for C_{Ac}, which is the term in brackets in Equation (5-33), that is,

$$s^2 = \sum_{i=1}^{n} [C_{Ami} - (\mathrm{abs}[C_{A0}^{1-\alpha} - (1-\alpha)kt_i])^{1/(1-\alpha)}]^2 \quad (5\text{-}38)$$

Another way to solve for the parameter values is to use time rather than concentrations:

$$t_c = \frac{C_{A0}^{1-\alpha} - C_A^{1-\alpha}}{k(1-\alpha)} \quad (5\text{-}39)$$

That is, we find the values of k and α that minimize

$$s^2 = \sum_{i=1}^{N} (t_{mi} - t_{ci})^2 = \sum_{i=1}^{N} \left[t_{mi} - \frac{C_{A0}^{1-\alpha} - C_{Ai}^{1-\alpha}}{k(1-\alpha)} \right]^2 \quad (5\text{-}40)$$

Discussion of weighted least squares as applied to a first-order reaction is provided on the CD-ROM.

5.5.3 Weighted Least-Squares Analysis

Both the linear and nonlinear least-squares analyses presented above assume that the variance is constant throughout the range of the measured variables. If this is not the case, a weighted least-squares analysis must be used to obtain better estimates of the rate law parameters. If the error in measurement is at a fixed level, the relative error in the dependent variable will increase as the independent variable increases (decreases). For example, in a first-order decay reaction ($C_A = C_{A0}e^{-kt}$), if the error in concentration measurement is $0.01C_{A0}$, the relative error in the concentration measurement $[0.01C_{A0}/C_A(t)]$ will increase with time. When this error condition occurs, the sum to be minimized for N measurements is

$$\sigma^2 = \sum_{i=1}^{N} W_i [y_i(\mathrm{exptl}) - y_i(\mathrm{calc})]^2$$

where W_i is a weighting factor.

For parameter estimation involving exponents, it has been shown that a weighted least-squares analysis is usually necessary.[14] Further discussion on weighted least squares as applied to a first-order reaction is given on the CD-ROM.

[14] A. C. Norris, *Computational Chemistry: An Introduction to Numerical Solution* (New York: Wiley, 1981), and D. M. Himmelblau, *Process Analysis by Statistical Methods* (New York: Wiley, 1970), p. 195.

5.6 Experimental Planning

> Four to six weeks in the lab can save you an hour in
> the library.
>
> G. C. Quarderer, Dow Chemical Co.

So far, this chapter has presented various methods of analyzing rate data. It is
just as important to know in which circumstances to use each method as it is
to know the mechanics of these methods. In this section we discuss a heuristic
to plan experiments to generate the data necessary for reactor design. However,
only a thumbnail sketch is presented; for a more thorough discussion the
reader is referred to the books and articles by Box and Hunter.[15]

Figure 5-12 provides a road map to help plan an experimental program.
A discussion of each of the items in Figure 5-12 appears on the CD-ROM
along with an example of an experimental design to study the kinetics of an
enzymatic reaction that depends on pH, temperature (T), and concentration
(C). Figure 5-13 shows the placement of high and low settings of each of these
variables.

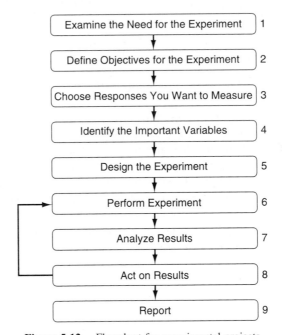

Figure 5-12   Flowchart for experimental projects.

Enzyme degradation is believed to occur at temperatures above 50°C and
pH values above 9.5 and below 3.0. The rate of reaction is negligible at temperatures below 6°C. For an urea concentration below 0.001 M, the reaction

---

[15]G. E. P. Box, W. G. Hunter, and J. S. Hunter, *Statistics for Experimenters: An Introduction to Design, Data Analysis, and Model Building* (New York: Wiley, 1978).

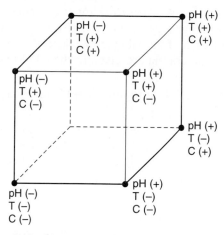

Figure 5-13   Placement (high/low) of controlled variables.

will not proceed at a measurable rate and the rate appears to be independent of concentration above 0.1 M. Consequently, the following high/low values of the parameters were chosen:

$$
\begin{array}{lllll}
A & (-) & \text{pH } 4 & (+) & \text{pH } 8 \\
B & (-) & 10°C & (+) & 40°C \\
C & (-) & 0.005 \ M & (+) & 0.1 \ M
\end{array}
$$

This example is discussed further on the CD-ROM.

5.7 Evaluation of Laboratory Reactors

The successful design of industrial reactors lies primarily with the *reliability of the experimentally determined parameters used in the scale-up*. Consequently, it is imperative to design equipment and experiments that will generate accurate and meaningful data. Unfortunately, there is usually no single comprehensive laboratory reactor that could be used for all types of reactions and catalysts. In this section we discuss the various types of reactors that can be chosen to obtain the kinetic parameters for a specific reaction system. We closely follow the excellent strategy presented in the article by V. W. Weekman of Mobil Oil.[16] The criteria used to evaluate various types of laboratory reactors are listed in Table 5-3.

TABLE 5-3.   CRITERIA USED TO EVALUATE LABORATORY REACTORS

1. Ease of sampling and product analysis
2. Degree of isothermality
3. Effectiveness of contact between catalyst and reactant
4. Handling of catalyst decay
5. Reactor cost and ease of construction

[16] V. W. Weekman, *AIChE J.*, *20*, 833 (1974).

Each type of reactor is examined with respect to these criteria and given a rating of good (G), fair (F), or poor (P). What follows is a brief description of each of the laboratory reactors. The reasons for rating each reactor for each of the criteria are given in the CD-ROM.

5.7.1 Integral (Fixed-Bed) Reactor

One advantage of the integral reactor is its ease of construction (see Figure 5-14). On the other hand, while channeling or bypassing of some of the catalyst by the reactant stream may not be as fatal to data interpretation in the case of this reactor as in that of the differential reactor, it may still be a problem.

Easy to construct

Figure 5-14   Integral reactor.

5.7.2 Stirred Batch Reactor

In the stirred batch reactor the catalyst is dispersed as a slurry, as shown in Figure 5-15. Although there will be better contacting between the catalyst and the fluid in this reactor than either the differential or integral reactors, there is a sampling problem in this reactor.

Good–fluid solid
contact

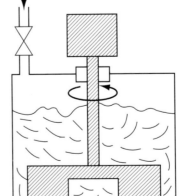

Figure 5-15   Stirred batch reactor. [From V. Weekman, *AIChE J. 20*, 833, (1974) with permission of the AIChE. Copyright © 1974 AIChE. All rights reserved.]

5.7.3 Stirred Contained Solids Reactor (SCSR)

Although there are a number of designs for contained solids reactors, all are essentially equivalent in terms of performance. A typical design is shown in Figure 5-16

Minimizes external
mass transfer
resistance

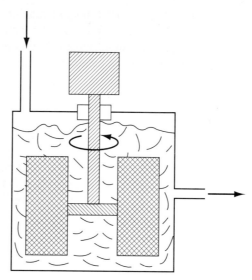

Figure 5-16   Stirred contained solids reactor. [From V. Weekman, *AIChE J. 20*, 833, (1974) with permission of the AIChE. Copyright © 1974 AIChE. All rights reserved.]

5.7.4 Continuous-Stirred Tank Reactor (CSTR)

The CSTR reactor (Figure 5-17) is used when there is significant catalyst decay. Fresh catalyst is fed to the reactor along with the fluid feed, and the cat-

One of the best
reactors for
isothermal operation

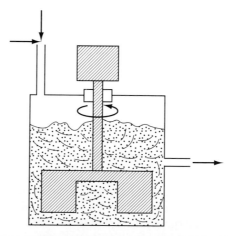

Figure 5-17   [From V. Weekman, *AIChE J. 20*, 833, (1974) with permission of the AIChE. Copyright © 1974 AIChE. All rights reserved.]

alyst leaves the reactor in the product stream at the same rate at which it is fed, to offset catalyst decay with time.

5.7.5 Straight-Through Transport Reactor

Commercially, the transport reactor (Figure 5-18) is used widely in the production of gasoline from heavier petroleum fractions. In addition, it has found use in the drying of grains. In this reactor, either an inert gas or the reactant itself transports the catalyst through the reactor.

Best for catalyst
decay

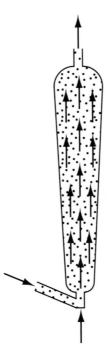

Figure 5-18   Straight-through transport reactor. [From V. Weekman, *AIChE J. 20*, 833, (1974) with permission of the AIChE. Copyright © 1974 AIChE. All rights reserved.]

5.7.6 Recirculating Transport Reactor

By recirculating the gas and catalyst through the transport reactor (Figure 5-19), one can achieve a well-mixed condition provided that the recirculation rate is large with respect to the feed rate.

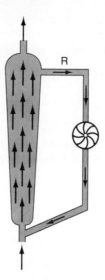

Figure 5-19   Recirculating transport reactor. [From V. Weekman, *AIChE J. 20*, 833, (1974) with permission of the AIChE. Copyright © 1974 AIChE. All rights reserved.]

5.7.7 Summary of Reactor Ratings

The ratings of the various reactors are summarized in Table 5-4. From this table one notes that the CSTR and recirculating transport reactor appear to be the best choices because they are satisfactory in every category except for construction. However, if the catalyst under study does not decay, the stirred batch and contained solids reactors appear to be the best choices. If the system is not limited by internal diffusion in the catalyst pellet, larger pellets could be used and the stirred-contained solids is the best choice. If the catalyst is nondecaying and heat effects are negligible, the fixed-bed (integral) reactor would be the top choice, owing to its ease of construction and operation. However, in practice, usually *more than one* reactor type is used in determining the reaction rate law parameters.

TABLE 5-4.   SUMMARY OF REACTOR RATINGS: GAS–LIQUID, POWDERED CATALYST, DECAYING CATALYST SYSTEM[a]

Reactor Type	Sampling and Analysis	Isothermality	Fluid–Solid Contact	Decaying Catalyst	Ease of Construction
Differential	P–F	F–G	F	P	G
Fixed bed	G	P–F	F	P	G
Stirred batch	F	G	G	P	G
Stirred-contained solids	G	G	F–G	P	F–G
Continuous-stirred tank	F	G	F–G	F–G	P-F
Straight-through transport	F–G	P–F	F–G	G	F–G
Recirculating transport	F–G	G	G	F–G	P–F
Pulse	G	F–G	P	F–G	G

[a]G, good; F, fair; P, poor.

SUMMARY

1. *Differential method for constant-volume systems*

$$-\frac{dC_A}{dt} = kC_A^\alpha \qquad \text{(S5-1)}$$

a. Plot $\Delta C_A/\Delta t$ as a function of t.
b. Determine dC_A/dt from this plot.
c. Taking the ln of both sides of (S5-1) gives

$$\ln\left(-\frac{dC_A}{dt}\right) = \ln k + \alpha \ln C_A \qquad \text{(S5-2)}$$

Plot $\ln(-dC_A/dt)$ versus $\ln C_A$. The slope will be the reaction order α. We could also use finite-difference formulas or software packages to evaluate $-dC_A/dt$ as a function of time and concentration.

2. *Integral method*
a. Guess the reaction order and integrate the mole balance equation.
b. Calculate the resulting function of concentration for the data and plot it as a function of time. If the resulting plot is linear, you have probably guessed the correct reaction order.
c. If the plot is not linear, guess another order and repeat the procedure.

3. *Method of initial rates*: In this method of analysis of rate data, the slope of a plot of $\ln(-r_{A0})$ versus $\ln C_{A0}$ will be the reaction order.

4. *Modeling the differential reactor*: The rate of reaction is calculated from the equation

$$-r_A' = \frac{F_{A0}X}{W} = \frac{F_P}{W} = \frac{v_0(C_{A0} - C_{Ae})}{W} = \frac{C_P v_0}{W} \qquad \text{(S5-3)}$$

In calculating the reaction order, α,

$$-r_A' = kC_A^\alpha$$

the concentration of A is evaluated either at the entrance conditions or at a mean value between C_{A0} and C_{Ae}.

5. *Least-squares analysis*
a. *Linear least-squares*: Linearize the rate law and solve the resulting algebraic equations [i.e., Equations (5-26) through (5-28)] for the reaction rate law parameters.

$$Y = a_0 + a_1 X_1 + a_2 X_2 \qquad \text{(S5-4)}$$

b. *Nonlinear least-squares*: Search for the parameters of the rate law that will minimize the sum of the squares of the difference between the measured rate of reaction and the rate of reaction calculated from the parameter values chosen. For N experimental runs and K parameters to be determined,

$$\sigma^2_{min} = \sum_{i=1}^{N} \frac{[r_i(\text{measured}) - r_i(\text{calculated})]^2}{N - K} \qquad (S5\text{-}5)$$

6. *Experimental planning* (from CD-ROM)
   a. Why?
   b. Within what limits?
   c. Are you choosing the correct parameters?
   d. Standards!
   e. Can you do it again?
   f. Be careful: you can lie with statistics.
   g. We don't believe an experiment until it's proved by theory.
   h. Milk your data for all its information.
   i. Tell the world!

Review Figure 5-13.

QUESTIONS AND PROBLEMS

The subscript to each of the problem numbers indicates the level of difficulty: A, least difficult; D, most difficult.

$$A = \bullet \qquad B = \blacksquare \qquad C = \blacklozenge \qquad D = \blacklozenge\blacklozenge$$

In each of the questions and problems below, rather than just drawing a box around your answer, write a sentence or two describing how you solved the problem, the assumptions you made, the reasonableness of your answer, what you learned, and any other facts that you want to include. You may wish to refer to W. Strunk and E. B. White, *The Elements of Style* (New York: Macmillian, 1979), and Joseph M. Williams, *Style: Ten Lessons in Clarity & Grace* (Glenview, Ill.: Scott, Foresman, 1989) to enhance the quality of your sentences.

Creative Problem

P5-1$_A$ (a) Compare Table 5-4 on laboratory reactors with a similar table on page 269 of Bisio and Kabel (see Supplementary Reading, listing 4). What are the similarities and differences?

(b) Develop an experimental plan that would determine the mechanism and rate law of an unknown reaction.

(c) Create an original problem based on Chapter 5 material.

(d) Design an experiment for the undergraduate laboratory that demonstrates the principles of chemical reaction engineering and will cost less than $500 in purchased parts to build. (*1998 AIChE National Student Chapter Competition*) Rules are provided on the CD-ROM.

(e) Plant a number of seeds in different pots (corn works well). The plant and soil of each pot will be subjected to different conditions. Measure the height of the plant as a function of time and fertilizer concentration. Other variables might include lighting, pH, and room temperature.

P5-2$_A$ **What if...**

(a) the reaction you were asked to apply the differential method of analysis to determine the rate law for a reaction that is reversible at the temperature of interest?

(b) the gas-phase reaction you are studying in your laboratory were extremely toxic. What precautions would you take?

P5-3$_A$ The irreversible isomerization

$$A \rightarrow B$$

was carried out in a *batch reactor* and the following concentration–time data were obtained:

t (min)	0	3	5	8	10	12	15	17.5
C_A (mol/dm^3)	4.0	2.89	2.25	1.45	1.0	0.65	0.25	0.07

(a) Determine the reaction order, α, and the specific reaction rate, k_A.

(b) If you were to repeat this experiment to determine the kinetics, what would you do differently? Would you run at a higher, lower, or the same temperature? Take different data points? Explain.

(c) It is believed that the technician made a dilution error in the concentration measured at 17.5 min. What do you think? How do your answers compare using regression (POLYMATH or other software) with those obtained by graphical methods? (*Ans.*: (a) $k = 0.2$ (mol/dm^3)$^{1/2}$/min)

P5-4$_A$ The liquid-phase irreversible reaction

$$A \rightarrow B + C$$

is carried out in a CSTR. To learn the rate law the volumetric flow rate, v_0, (hence $\tau = V/v_0$) is varied and the effluent concentrations of species A recorded as a function of the space time τ. Pure A enters the reactor at a concentration of 2 mol/dm^3. Steady-state conditions exist when the measurements are recorded.

Run	1	2	3	4	5
τ (min)	15	38	100	300	1200
C_A (mol/dm^3)	1.5	1.25	1.0	0.75	0.5

(a) Determine the reaction order and specific reaction rate.

(b) If you were to repeat this experiment to determine the kinetics, what would you do differently? Would you run at a higher, lower, or the same temperature? If you were to take more data, where would you place the measurements (e.g., τ)?

(c) It is believed that the technician may have made a dilution factor-of-10 error in one of the concentration measurements. What do you think? How do your answers compare using regression (POLYMATH or other software) compare with those obtained by graphical methods?

Note: All measurements were taken at steady-state conditions.

P5-5$_B$ The reaction

$$A \rightarrow B + C$$

was carried out in a constant-volume batch reactor where the following concentration measurements were recorded as a function of time.

t (min)	0	5	9	15	22	30	40	60
C_A (mol/dm³)	2	1.6	1.35	1.1	0.87	0.70	0.53	0.35

(a) Use nonlinear least squares (i.e., regression) and one other method to determine the reaction order α and the specific reaction rate.

(b) If you were to take more data, where would you place the points? Why?

P5-6$_B$ GaInAs films are important materials in fiber optic communication and in high-speed microelectronic devices. A preliminary reaction between triethylindium and arisine is carried out to form an intermediate, which is then used in the deposition to form GaInAs. The reaction is

$$Et_3In(g) + AsH_3(g) \rightleftharpoons adduct(g)$$

The data in Table P5-6 were obtained in a *plug-flow reactor* [*JECS, 135*(6), 1530 (1988)]. Total pressure (including inerts) = 152.0 torr.

TABLE P5-6

Distance Down Reactor (cm)	Et$_3$In (torr) for AsH$_3$ (torr inlet) of:		
	1.5	0.25	3.0
0	0.129	0.129	0.129
1.5	0.075	0.095	0.045
2.5	0.05	0.085	0.022
4.0	0.03	0.08	0.01
6.5	0.018	0.042	0.01
9.0	0.016	0.04	0.01

Find a rate law consistent with the experimental data.

P5-7$_B$ The following data were reported [C. N. Hinshelwood and P. J. Ackey, *Proc. R. Soc. (Lond)., A115*, 215 (1927)] for a gas-phase constant-volume decomposition of dimethyl ether at 504°C in a *batch reactor*. Initially, only $(CH_3)_2O$ was present.

Time (s)	390	777	1195	3155	∞
Total Pressure (mmHg)	408	488	562	799	931

(a) Why do you think the total pressure measurment at $t = 0$ is missing? Can you estimate it?

(b) Assuming that the reaction

$$(CH_3)_2O \rightarrow CH_4 + H_2 + CO$$

is irreversible and goes to completion, determine the reaction order and specific reaction rate k.

(c) What experimental conditions would you suggest if you were to obtain more data?

(d) How would the data and your answers change if the reaction were run at a higher or lower temperature?

P5-8$_B$ In order to study the photochemical decay of aqueous bromine in bright sunlight, a small quantity of liquid bromine was dissolved in water contained in a glass battery jar and placed in direct sunlight. The following data were obtained:

Time (min)	10	20	30	40	50	60
ppm Br$_2$	2.45	1.74	1.23	0.88	0.62	0.44

temperature = 25°C

(a) Determine whether the reaction rate is zero, first, or second order in bromine, and calculate the reaction rate constant in units of your choice.

(b) Assuming identical exposure conditions, calculate the required hourly rate of injection of bromine (in pounds) into a sunlit body of water, 25,000 gal in volume, in order to maintain a sterilizing level of bromine of 1.0 ppm. (*Ans.*: 0.43 lb/h)

(c) What experimental conditions would you suggest if you were to obtain more data?

(*Note*: ppm = parts of bromine per million parts of brominated water by weight. In dilute aqueous solutions, 1 ppm ≡ 1 milligram per liter.) (**California Professional Engineers Exam**)

P5-9$_C$ The gas-phase decomposition

$$A \longrightarrow B + 2C$$

is carried out in a *constant-volume batch reactor*. Runs 1 through 5 were carried out at 100°C while run 6 was carried out at 110°C.

(a) From the data in Table P5-9, determine the reaction order and specific reaction rate.

(b) What is the activation energy for this reaction?

TABLE P5-9   RAW DATA

Run	Initial Concentration, C_{A0} (g mol/L)	Half-Life, $t_{1/2}$ (min)
1	0.0250	4.1
2	0.0133	7.7
3	0.010	9.8
4	0.05	1.96
5	0.075	1.3
6	0.025	2.0

P5-10$_C$ The reactions of ozone were studied in the presence of alkenes [R. Atkinson et al., *Int. J. Chem. Kinet.*, 15(8), 721 (1983)]. The data in Table P5-10 are for one of the alkenes studied, *cis*-2-butene. The reaction was carried out isothermally at 297 K. Determine the rate law and the values of the rate law parameters.

TABLE P5-10   RAW DATA

Run	Ozone Rate (mol/s·dm³ × 10⁷)	Ozone Concentration (mol/dm³)	Butene Concentration (mol/dm³)
1	1.5	0.01	10^{-12}
2	3.2	0.02	10^{-11}
3	3.5	0.015	10^{-10}
4	5.0	0.005	10^{-9}
5	8.8	0.001	10^{-8}
6	4.7	0.018	10^{-9}

(Hint: Ozone also decomposes by collision with the wall)

P5-11$_A$ Tests were run on a small experimental reactor used for decomposing nitrogen oxides in an automobile exhaust stream. In one series of tests, a nitrogen stream containing various concentrations of NO_2 was fed to a reactor and the kinetic data obtained are shown in Figure P5-11. Each point represents one complete run. The reactor operates essentially as an *isothermal backmix reactor (CSTR)*. What can you deduce about the apparent order of the reaction over the temperature range studied?

   The plot gives the fractional decomposition of NO_2 fed versus the ratio of reactor volume V (in cm³) to the NO_2 feed rate, $F_{NO_{2,0}}$ (g mol/h), at different feed concentrations of NO_2 (in parts per million by weight).

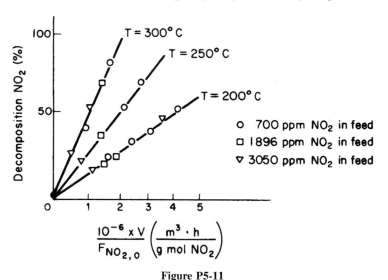

Figure P5-11

P5-12$_B$ *Microelectronic devices* are formed by first forming SiO_2 on a silicon wafer by chemical vapor deposition (Figure P5-12; cf. Problem P3-25). This procedure is followed by coating the SiO_2 with a polymer called a photoresist. The pattern of the electronic circuit is then placed on the polymer and the sample is irradiated with ultraviolet light. If the polymer is a positive photoresist, the sections that were irradiated will dissolve in the appropriate solvent and those sections not irradiated will protect the SiO_2 from further treatment. The wafer is then exposed to strong acids, such as HF, which etch (i.e., dissolve) the

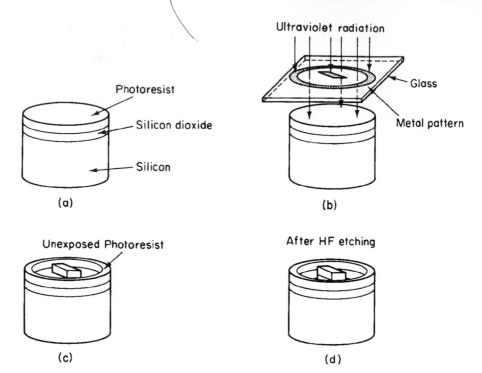

Figure P5-12

exposed SiO_2. It is extremely important to know the kinetics of the reaction so that the proper depth of the channel can be achieved. The dissolution reaction is

$$SiO_2 + 6HF \rightarrow H_2SiF_6 + 2H_2O$$

From the following initial rate data, determine the rate law.

Etching Rate (nm/min)	60	200	600	1000	1400
HF (wt %)	8	20	33	40	48

A total of 1000 thin wafer chips are to be placed in 0.5 dm³ of 20% HF. If a spiral channel 10 μm wide and 10 m in length were to be etched to a depth of 50 μm on both sides of each wafer, how long should the chips be left in the solution? Assume that the solution is well mixed. (*Ans.*: 330 min)

P5-13$_B$ The oxidation of propene (P) to acrolein (A) was carried out over a Mo–Pr–Bi catalyst [*Ind. Eng. Chem. Res.*, *26*, 1419 (1987)].

$$CH_3CH = CH_2 + O_2 \rightarrow CH_2 = CHCHO + H_2O$$

It has been proposed to correlate the data using the power law model for the rate law [cf. Equation (5-2)].

$$r_{\text{acrolein}} = kP_P^\alpha P_{O_2}^\beta$$

The reaction was carried out in a differential reactor with 0.5 g of catalyst at 623 K. From the data below, determine the reaction orders with respect to propene (α) and oxygen (β) and the specific reaction rate, k.

F_A (mmol/h)	0.21	0.48	0.09	0.39	0.6	0.14	1.44
P_P (atm)	0.1	0.2	0.05	0.3	0.4	0.05	0.5
P_{O_2} (atm)	0.1	0.2	0.05	0.01	0.02	0.4	0.5

where F_A = exiting molar flow rate of acrolein, mmol/h

$\qquad P_P$ = entering partial pressure of propene, atm

$\qquad P_{O_2}$ = entering partial pressure of oxygen, atm

P5-14$_B$ The ethane hydrogenolysis over a commercial nickel catalyst was studied in a *stirred contained solids reactor.*

$$H_2 + C_2H_6 \rightarrow 2CH_4$$

(a) Determine the rate law parameters from the data in Table P5-14. There are four *spinning baskets*, each with 10 g of catalyst. Only hydrogen and ethane are fed to the reactor at 300°C. (*Ans.:* $k = 0.48$ (mol·atm/kg·h))

TABLE P5-14   RAW DATA

Total Molar Feed Rate to Reactor (g mol/h)	Partial Pressure (atm) in Feed		Mole Fraction CH$_4$ in Exit Stream
	Ethane, P_{A0}	Hydrogen, P_{B0}	
1.7	0.5	0.5	0.05
1.2	0.5	0.5	0.07
0.6	0.5	0.5	0.16
0.3	0.4	0.6	0.16
0.75	0.6	0.6	0.10
2.75	0.6	0.4	0.06

(b) What experimental conditions would you suggest if you were to obtain more data?

P5-15$_C$ The thermal decomposition of isopropyl isocyanate was studied in a *differential packed-bed reactor.* From the data in Table P5-15, determine the reaction rate law parameters.

TABLE P5-15   RAW DATA

Run	Rate (mol/s·dm^3)	Concentration (mol/dm^3)	Temperature (K)
1	4.9×10^{-4}	0.2	700
2	1.1×10^{-4}	0.02	750
3	2.4×10^{-3}	0.05	800
4	2.2×10^{-2}	0.08	850
5	1.18×10^{-1}	0.1	900
6	1.82×10^{-2}	0.06	950

P5-16$_B$ Mixtures of hydrocarbons (e.g., petroleum feedstocks) that undergo cracking reactions or hydrodemethylation can sometimes be *lumped* as just one reactant or as two or more reactants. In many cases it is difficult to distinguish lumping as a single reactant with second-order kinetics from lumping as two reactants each with first-order kinetics [*Ind. Eng. Chem. Process Des. Dev., 19*, 197 (1980)]. To distinguish between these two cases, the initial reactant concentration must be varied in more than one run and conversions greater than 92% should be sought in taking the data.

From the *batch reactor* data below, determine whether first-order kinetics for lumping as two reactants A and B or second-order kinetics for lumping one reactant D best describes the system. Experimental conditions are such that one can neglect volume change.

Mechanism I	Mechanism II
Two species	One species
A $\xrightarrow{\ k_A\ }$ products	D $\xrightarrow{\ k_D\ }$ products
B $\xrightarrow{\ k_B\ }$ products	
$C(t) = C_A(t) + C_B(t)$	$C(t) = C_D(t)$

Only the total concentration of the lumped reactant, $C(t)$, can be monitored as a function of time. For two-parameter lumping, estimates of the initial concentrations are $C_{A0} = 0.008$ kmol/m^3 and $C_{B0} = 0.006$ kmol/m^3.

t (s)	0	10	20	30	40	60	80
Concentration (kmol/m^3)	0.014	0.0115	0.0097	0.0084	0.0074	0.0060	0.0051

Where would you place additional data points?

P5-17$_D$ Prepare a detailed experimental plan to learn the rate law for
  (a) The hydrogenation of cyclopentane on a Pt/Al$_2$O$_3$ catalyst.
  (b) The liquid-phase production of methyl bromide from an aqueous solution of methyl amine and bromine cyanide.
  (c) The acid-catalyzed production of ethylene glycol from an aqueous solution of ethylene oxide.

P5-18$_B$ The irreversible liquid-phase reaction

$$A \rightarrow B + C$$

is carried out in a batch reactor. The following data were collected during the course of the reaction:

t (min)	0.0	2.0	4.0	6.0
C_A (mol/dm^3)	2.00	1.31	0.95	0.73

Determine the order of reaction and the specific reaction rate using methods to differentiate your data.
  (a) Numerical technique–differentiation formulas. (Use Δ's to represent these points on any graphs you make.)

Graphical technique–equal area differentiation. (Use O's to represent these points on any graphs you make.)

Differentiating a polynomial. (use x for these points)

(b) Determine the reaction order.

(c) Assume a rate law of the form

$$-r_A = kC_A^\alpha$$

Integrate the equation for the combined mole balance and rate law and then use nonlinear least-squares analysis to determine α and k.

(d) Where would you place additional data points?

P5-19$_C$ The dehydrogenation of methylcyclohexane (M) to produce toluene (T) was carried out over a 0.3% PT/Al_2O_3 catalyst in a differential catalytic reactor. The reaction is carried out in the presence of hydrogen (H_2) to avoid coking [*J. Phys. Chem., 64,* 1559 (1960)].

(a) Determine the model parameters for each of the following rate laws.

$$(1)\ -r'_M = kP_M^\alpha P_{H_2}^\beta \qquad (3)\ -r'_M = \frac{kP_M P_{H_2}}{(1 + K_M P_M)^2}$$

$$(2)\ -r'_M = \frac{kP_M}{1 + K_M P_M} \qquad (4)\ -r'_M = \frac{kP_M P_{H_2}}{1 + K_M P_M + K_{H_2} P_{H_2}}$$

Use the data in Table P5-19.

TABLE P5-19   DEHYDROGENATION OF METHYLCYCLOHEXANE

P_{H_2} (atm)	P_M (atm)	$r'_T \left(\dfrac{\text{mol toluene}}{\text{s} \cdot \text{kg cat.}} \right)$
1	1	1.2
1.5	1	1.25
0.5	1	1.30
0.5	0.5	1.1
1	0.25	0.92
0.5	0.1	0.64
3	3	1.27
1	4	1.28
3	2	1.25
4	1	1.30
0.5	0.25	0.94
2	0.05	0.41

(b) Which rate law best describes the data? (*Hint:* We will learn in Chapter 10 that neither K_{H_2} or K_M can take on negative values.)

(c) Where would you place additional data points?

P5-20$_C$ In the production of ammonia

$$NO + \tfrac{5}{2}H_2 \rightleftarrows H_2O + NH_3 \qquad\qquad (1)$$

the following side reaction occurs:

$$NO + H_2 \rightleftarrows H_2O + \tfrac{1}{2}N_2 \qquad\qquad (2)$$

Ayen and Peters [*Ind. Eng. Chem. Process Des. Dev., 1*, 204 (1962)] studied catalytic reaction of nitric oxide with Girdler G–50 catalyst in a differential reactor at atmospheric pressure. Table P5-20 shows the reaction rate of the side reaction as a function of P_{H_2} and P_{NO} at a temperature of 375°C.

TABLE P5-20   FORMATION OF AMMONIA

P_{H_2} (atm)	P_{NO} (atm)	Reaction Rate $r_{H_2O} \times 10^5$ (g mol/min · g cat.) $T = 375°C,\ W = 2.39$ g
0.00922	0.0500	1.60
0.0136	0.0500	2.56
0.0197	0.0500	3.27
0.0280	0.0500	3.64
0.0291	0.0500	3.48
0.0389	0.0500	4.46
0.0485	0.0500	4.75
0.0500	0.00918	1.47
0.0500	0.0184	2.48
0.0500	0.0298	3.45
0.0500	0.0378	4.06
0.0500	0.0491	4.75

The following rate laws for side reaction (2), based on various catalytic mechanisms, were suggested:

$$r_{H_2O} = \frac{k K_{NO} P_{NO} P_{H_2}}{1 + K_{NO} P_{NO} + K_{H_2} P_{H_2}} \tag{3}$$

$$r_{H_2O} = \frac{k K_{H_2} K_{NO} P_{NO}}{1 + K_{NO} P_{NO} + K_{H_2} P_{H_2}} \tag{4}$$

$$r_{H_2O} = \frac{k_1 K_{H_2} K_{NO} P_{NO} P_{H_2}}{(1 + K_{NO} P_{NO} + K_{H_2} P_{H_2})^2} \tag{5}$$

Find the parameter values of the different rate laws and determine which rate law best represents the experimental data.

P5-21$_B$ For the reaction

$$2A + B \rightarrow 2C$$

the experimental rate data listed in Table P5-21 have been obtained.
(a) Is this reaction elementary? Defend your conclusion with supporting reasoning and/or analysis.
(b) Why is the sequence specified the way it is?
(c) Where would you place additional experiments?

TABLE P5-21   RAW DATA

Run	r_C (mol/dm$^3 \cdot$s)	P_A (atm)	P_B (atm)
1	0.6	1	0.5
4	1.2	1	1
9	4.2	1	4
8	8.1	1	10
7	9.6	2	2
3	11.0	3	1
5	53	2	30
6	59.3	5	2
10	200	5	10
2	250	5	15

JOURNAL CRITIQUE PROBLEMS

P5C-1   A *packed-bed reactor* was used to study the reduction of nitric oxide with ethylene on a copper–silica catalyst [*Ind. Eng. Chem. Process Des. Dev., 9*, 455 (1970)]. Develop the integral design equation in terms of the conversion at various initial pressures and temperatures. Is there a significant discrepancy between the experimental results shown in Figures 2 and 3 in the article and the calculated results based on the proposed rate law? If so, what is the possible source of this deviation?

P5C-2   Equation (3) in the article [*J. Chem. Technol. Biotechnol., 31*, 273 (1981)] is the rate of reaction and is incorporated into design equation (2). Rederive the design equation in terms of conversion. Determine the rate dependence on H_2 based on this new equation. How does the order obtained compare with that found by the authors?

P5C-3   In "The kinetics of the oxidation of hydrogen chloride over molten salt catalysts," *Chem. Eng. Sci., 23*, 981 (1968), use Figure 2 in the article to determine the initial rate of HCl oxidation for the various oxygen concentrations. Include these data in Figure 3 together with the other data. Is it possible to explain the curvature in the line at small partial pressures of chlorine (the square root of the partial pressure of Cl_2)?

P5C-4   See "Kinetics of catalytic esterification of terephthalic acid with methanol vapour," *Chem. Eng. Sci., 28*, 337 (1973). When one observes the data points in Figure 2 of this paper for large times, it is noted that the last data point always falls significantly off the straight-line interpretation. Is it possible to reanalyze these data to determine if the chosen reaction order is indeed correct? Substituting your new rate law into equation (3), derive a new form of equation (10) in the paper relating time and particle radius.

P5C-5   The kinetics of vapor-phase ammoxidation of 3-methylpyridine over a promoted $V_2O_5 - Al_2O_3$ catalyst was reported in *Chem. Eng. Sci., 35*, 1425 (1980). Suggest a mechanism and rate-limiting step consistent with each of the two mechanisms proposed. In each case how would you plot the data to extract the rate law parameters? Which mechanism is supported by the data?

P5C-6   The selective oxidation of toluene and methanol over vanadium pentoxide-supported alkali metal sulfate catalysts was studied recently [*AIChE J., 27*(1), 41

(1981)]. Examine the experimental technique used (equipment, variables, etc.) in light of the mechanism proposed. Comment on the shortcomings of the analysis and compare with another study of this system presented in *AIChE J.*, *28*(5), 855 (1982).

CD-ROM MATERIAL

- **Learning Resources**
  1. Summary Notes for Lectures 9 and 10
  3. Interactive Computer Modules
     A. Ecology
  4. Solved Problems
     A. CDP5-B_B Oxygenating Blood
     B. Example CD5–1 Integral Method of Analysis of Pressure-Time Data
- **Living Example Problems**
  1. Example 5–6 Hydrogenation of Ethylene to Ethane
- **Professional Reference Shelf**
  1. Weighted Least Squares Analysis
  2. Experimental Planning
  3. Evaluation of Laboratory Reactors
- **FAQ [Frequently Asked Questions]–** In Updates/FAQ icon section
- **Additional Homework Problems**

CDP5-A_B	The reaction of penicillin G with NH_2OH is carried out in a batch reactor. A colorimeter was used to measure the absorbency as a function of time. [1st Ed. P5-10]
CDP5-B_B	The kinetics of the deoxygenating of hemoglobin in the blood was studied with the aid of a tubular reactor. [1st Ed. P5-3]
CDP5-C_C	The kinetics of the formation of an important propellant ingredient, triaminoguandine, were studied in a batch reactor where the ammonia concentration was measured as a function of time. [2nd Ed. P5-6]
CDP5-D_B	The half-life of one of the pollutants, NO, in autoexhaust is required. [1st Ed. P5-11]
CDP5-E_B	The kinetics of a gas-phase reaction $A_2 \rightarrow 2A$ were studied in a constant-pressure batch reactor in which the volume was measured as a function of time. [1st Ed. P5-6]

SUPPLEMENTARY READING

1. A wide variety of techniques for measuring the concentrations of the reacting species may be found in

    ROBINSON, J. W., *Undergraduate Instrumental Analysis*, 5th ed. New York: Marcel Dekker, 1995.

    SKOOG, D. A., *Principles of Instrumental Analysis*, 3rd ed. Philadelphia: Holt, Rinehart and Winston, 1985.

2. A discussion on the methods of interpretation of batch reaction data can be found
   in

   CRYNES, B. L., and H. S. FOGLER, eds., *AIChE Modular Instruction Series E:
   Kinetics*, Vol. 2. New York: American Institute of Chemical Engineers,
   1981, pp. 51–74.

3. The interpretation of data obtained from flow reactors is also discussed in

   CHURCHILL, S. W., *The Interpretation and Use of Rate Data*. New York:
   McGraw-Hill, 1974.

   SMITH, J. M., *Chemical Engineering Kinetics*, 3rd ed. New York: McGraw-Hill,
   1981, Chap. 4.

4. The design of laboratory catalytic reactors for obtaining rate data is presented in

   RASE, H. F., *Chemical Reactor Design for Process Plants*, Vol. 1. New York:
   Wiley, 1983, Chap. 5.

   Most of these types of reactors are also discussed in

   ANDERSON, R. B., ed. *Experimental Methods in Catalytic Research*. New York:
   Academic Press, 1968.

5. Model building and current statistical methods applied to interpretation of rate data
   are presented in

   FROMENT, G. F., and K. B. BISCHOFF, *Chemical Reactor Analysis and Design*.
   New York: Wiley, 1979.

   KITTRELL, J. R., in *Advances in Chemical Engineering*, Vol. 8, T. B. Draw et al.,
   eds. New York: Academic Press, 1970, pp. 97–183.

   JOHANSEN, S., *Functional Relations, Random Coefficients, and Nonlinear
   Regression with Application to Kinetic Data*, New York: Springer-Verlag,
   1984.

   MARKERT, B. A., *Instrumental Element and Multi-Element Analysis of Plant
   Samples: Methods and Applications*, New York: Wiley, 1996.

6. The sequential design of experiments and parameter estimation is covered in

   BOX, G. E. P., W. G. HUNTER, and J. S. HUNTER, *Statistics for Experimenters:
   An Introduction to Design, Data Analysis, and Model Building*. New York:
   Wiley, 1978.

   GRAHAM, R. J., and F. D. STEVENSON, *Ind. Eng. Chem. Process Des. Dev., 11*,
   160 (1972).

Multiple Reactions   **6**

The breakfast of champions is not cereal, it's your opposition.

Nick Seitz

Seldom is the reaction of interest the *only one* that occurs in a chemical reactor. Typically, multiple reactions will occur, some desired and some undesired. One of the key factors in the economic success of a chemical plant is the minimization of undesired side reactions that occur along with the desired reaction.

In this chapter we discuss reactor selection and general mole balances for multiple reactions. There are three basic types of multiple reactions: series, parallel, and independent. In *parallel reactions* (also called *competing reactions*) the reactant is consumed by two different reaction pathways to form different products:

Parallel reactions

$$A \xrightarrow{\quad k_1 \quad} B$$
$$A \xrightarrow{\quad k_2 \quad} C$$

An example of an industrially significant parallel reaction is the oxidation of ethylene to ethylene oxide while avoiding complete combustion to carbon dioxide and water.

Serious chemistry

$$CH_2{=}CH_2 + O_2 \longrightarrow 2CO_2 + 2H_2O$$
$$CH_2{=}CH_2 + O_2 \longrightarrow \underset{CH_2-CH_2}{O}$$

In *series reactions*, also called *consecutive reactions*, the reactant forms an intermediate product, which reacts further to form another product:

Series reactions

$$A \xrightarrow{k_1} B \xrightarrow{k_2} C$$

An example of a series reaction involving ethylene oxide (E.O.) is its reaction with ammonia to form mono-, di-, and triethanolamine:

$$CH_2\!-\!CH_2 + NH_3 \longrightarrow HOCH_2CH_2NH_2$$

$$\xrightarrow{\text{E.O.}} (HOCH_2CH_2)_2\,NH \xrightarrow{\text{E.O.}} (HOCH_2CH_2)_3\,N$$

In recent years the shift has been toward the production of diethanolamine as the *desired* product rather than triethanolamine.

Multiple reactions involve a combination of both series and parallel reactions, such as

$$A + B \longrightarrow C + D$$
$$A + C \longrightarrow E$$

An example of a combination of parallel and series reactions is the formation of butadiene from ethanol:

Simultaneous series and parallel reactions

$$C_2H_5OH \longrightarrow C_2H_4 + H_2O$$
$$C_2H_5OH \longrightarrow CH_3CHO + H_2$$
$$C_2H_4 + CH_3CHO \longrightarrow C_4H_6 + H_2O$$

Independent reactions are of the type

Independent reactions

$$A \longrightarrow B$$
$$C \longrightarrow D + E$$

and occur in feed stocks containing many reactants. The cracking of crude oil to form gasoline is an example where independent reactions take place.

The first part of this chapter will be concerned primarily with parallel reactions. Of particular interest are reactants that are consumed in the formation of a *desired product*, D, and the formation of an *undesired product*, U, in a competing or side reaction. In the reaction sequence

$$A \xrightarrow{k_D} D$$
$$A \xrightarrow{k_U} U$$

we want to minimize the formation of U and maximize the formation of D, because the greater the amount of undesired product formed, the greater the cost of separating the undesired product U from the desired product D (Figure 6-1).

In a highly efficient and costly reactor scheme in which very little of undesired product U is formed in the reactor, the cost of the separation process could be quite low. On the other hand, even if a reactor scheme is inexpensive and inefficient, resulting in the formation of substantial amounts of U, the cost

The economic incentive

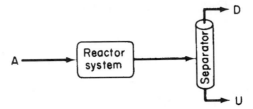

Figure 6-1 Reaction-separation system producing both desired and undesired products.

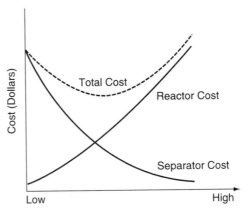

Figure 6-2 Efficiency of a reactor system.

of the separation system could be quite high. Normally, as the cost of a reactor system increases in an attempt to minimize U, the cost of separating species U from D decreases (Figure 6-2).

6.1 Maximizing the Desired Product in Parallel Reactions

In this section we discuss various means of minimizing the undesired product, U, through the selection of reactor type and conditions. We also discuss the development of efficient reactor schemes.

For the competing reactions

$$A \xrightarrow{k_D} D \qquad \text{(desired)}$$

$$A \xrightarrow{k_U} U \qquad \text{(undesired)}$$

the rate laws are

Rate laws for
formation of
desired and
undesired products

$$r_D = k_D C_A^{\alpha_1} \tag{6-1}$$

$$r_U = k_U C_A^{\alpha_2} \tag{6-2}$$

The rate of disappearance of A for this reaction sequence is the sum of the rates of formation of U and D:

$$-r_A = r_D + r_U \tag{6-3}$$

$$-r_A = k_D C_A^{\alpha_1} + k_U C_A^{\alpha_2} \tag{6-4}$$

where α_1 and α_2 are positive constants. We want the rate of formation of D, r_D, to be high with respect to the rate of formation of U, r_U. Taking the ratio of these rates [i.e., Equation (6-1) to Equation (6-2)], we obtain a *rate selectivity parameter*, S, which is to be maximized:

instantaneous selectivity

$$S_{DU} = \frac{r_D}{r_U} = \frac{k_D}{k_U} C_A^{\alpha_1 - \alpha_2} \tag{6-5}$$

6.1.1 Maximizing the Rate Selectivity Parameter S for One Reactant

Maximize the rate selectivity parameter

In this section we examine ways to maximize S, which is sometimes referred to as the instantaneous selectivity, for different reaction orders of the desired and undesired products.

α_1 is the order of the desired reaction; α_2, of the undesired reaction

Case 1: $\alpha_1 > \alpha_2$ For the case where the reaction order of the desired product is greater than the reaction order of the undesired product, let *a* be a positive number that is the difference between these reaction orders:

$$\alpha_1 - \alpha_2 = a$$

Then

$$S_{DU} = \frac{r_D}{r_U} = \frac{k_D}{k_U} C_A^a \tag{6-6}$$

For $\alpha_1 > \alpha_2$, make C_A as large as possible

To make this ratio as large as possible, we want to carry out the reaction in a manner that will keep the concentration of reactant A as high as possible during the reaction. If the reaction is carried out in the gas phase, we should run it without inerts and at high pressures to keep C_A high. If the reaction is in the liquid phase, the use of diluents should be kept to a minimum.[1]

A batch or plug-flow reactor should be used in this case, because in these two reactors, the concentration of A starts at a high value and drops progressively during the course of the reaction. In a *perfectly mixed* CSTR, the concentration of reactant within the reactor is always at its lowest value (i.e., that of the outlet concentration) and therefore not be chosen under these circumstances.

---

[1] For a number of liquid-phase reactions, the proper choice of a solvent can enhance selectivity. See, for example, *Ind. Eng. Chem.*, *62*(9), 16 (1970). In gas-phase heterogeneous catalytic reactions, selectivity is an important parameter of any particular catalyst.

Case 2: $\alpha_2 > \alpha_1$ When the reaction order of the undesired product is greater than that of the desired product,

$$A \xrightarrow{\;k_D\;} D$$

$$A \xrightarrow{\;k_U\;} U$$

Let $a = \alpha_2 - \alpha_1$, where a is a positive number; then

$$S_{DU} = \frac{r_D}{r_U} = \frac{k_D C_A^{\alpha_1}}{k_U C_A^{\alpha_2}} = \frac{k_D}{k_U C_A^{\alpha_2 - \alpha_1}} = \frac{k_D}{k_U C_A^{a}} \tag{6-7}$$

For the ratio r_D/r_U to be high, the concentration of A should be as low as possible.

For $\alpha_2 > \alpha_1$ use a CSTR and dilute the feed stream

This low concentration may be accomplished by diluting the feed with inerts and running the reactor at low concentrations of A. A CSTR should be used, because the concentrations of reactant are maintained at a low level. A recycle reactor in which the product stream acts as a diluent could be used to maintain the entering concentrations of A at a low value.

Because the activation energies of the two reactions in cases 1 and 2 are not known, it cannot be determined whether the reaction should be run at high or low temperatures. The sensitivity of the rate selectivity parameter to temperature can be determined from the ratio of the specific reaction rates,

$$\frac{k_D}{k_U} = \frac{A_D}{A_U} e^{-[(E_D - E_U)/RT]} \tag{6-8}$$

where A is the frequency factor, E the activation energy, and the subscripts D and U refer to desired and undesired product, respectively.

Case 3: $E_D > E_U$ In this case the specific reaction rate of the desired reaction k_D (and therefore the overall rate r_D) increases more rapidly with increasing temperature than does the specific rate of the undesired reaction k_U. Consequently, the reaction system should be operated at the highest possible temperature to maximize S_{DU}.

Case 4: $E_U > E_D$ In this case the reaction should be carried out at a low temperature to maximize S_{DU}, but not so low that the desired reaction does not proceed to any significant extent.

Example 6–1 Minimizing Unwanted Products for a Single Reactant

Reactant A decomposes by three simultaneous reactions to form three products: one that is desired, D, and two that are undesired, Q and U. These gas-phase reactions, together with their corresponding rate laws, are:

Desired product:

$$A \longrightarrow D$$

$$r_D = \left\{ 0.0012 \exp\left[26,000 \left(\frac{1}{300} - \frac{1}{T} \right) \right] \right\} C_A \qquad \text{(E6-1.1)}$$

Unwanted product U:

$$A \longrightarrow U$$

$$r_U = \left\{ 0.0018 \exp\left[25,000 \left(\frac{1}{300} - \frac{1}{T} \right) \right] \right\} C_A^{1.5} \qquad \text{(E6-1.2)}$$

Unwanted product Q:

$$A \longrightarrow Q$$

$$r_Q = \left\{ 0.00452 \exp\left[5000 \left(\frac{1}{300} - \frac{1}{T} \right) \right] \right\} C_A^{0.5} \qquad \text{(E6-1.3)}$$

How and under what conditions (e.g., reactor type, pressure, temperature, etc.) should the reactions above be carried out to minimize the concentrations of the unwanted products U and Q?

Solution

Because pre exponential factors are comparable, but the activation energies of reactions (E6-1.1) and (E6-1.2) are much greater than the activation energy of reaction (E6-1.3), the rate of formation of Q will be negligible with respect to the rates of formation of D and U at high temperatures:

$$S_{DQ} = \frac{r_D}{r_Q} \cong \text{very large} \qquad \text{(E6-1.4)}$$

Now we need only to consider the relative rates of formation of D and U at high temperatures:

$$S_{DU} = \frac{r_D}{r_U} = \frac{0.66 e^{1000\,[(1/300) - 1/T]}}{C_A^{0.5}} \qquad \text{(E6-1.5)}$$

From Equation (E6-1.5) we observe that the amount of undesired product, U, can be minimized by carrying out the reaction at low concentrations. Therefore, to maximize the conversion of A to D we would want to operate our reactor at high temperatures (to minimize the formation of Q) and at low concentrations of A (to minimize the formation of U). That is, carry out the reaction at

1. High temperatures.
2. Low concentrations of A, which may be accomplished by:
   a. Adding inerts.
   b. Using low pressures (if gas phase).
   c. Using a CSTR or a recycle reactor.

(margin, left)

$$A \longrightarrow D$$
$$A \longrightarrow X$$
$$A \longrightarrow Y$$

$$S_{D/XY} = \frac{r_D}{r_X + r_Y}$$

6.1.2 Maximizing the Rate Selectivity Parameter S for Two Reactants

Next consider two simultaneous reactions in which two reactants, A and B, are being consumed to produce a desired product, D, and an unwanted product, U, resulting from a side reaction. The rate laws for the reactions

$$A + B \xrightarrow{\;k_1\;} D$$

$$A + B \xrightarrow{\;k_2\;} U$$

are

$$r_D = k_1 C_A^{\alpha_1} C_B^{\beta_1} \tag{6-9}$$

$$r_U = k_2 C_A^{\alpha_2} C_B^{\beta_2} \tag{6-10}$$

The rate selectivity parameter

Instantaneous
selectivity

$$S_{DU} = \frac{r_D}{r_U} = \frac{k_1}{k_2} C_A^{\alpha_1 - \alpha_2} C_B^{\beta_1 - \beta_2} \tag{6-11}$$

is to be maximized. Shown in Figure 6-3 are various reactor schemes and conditions that might be used to maximize S_{DU}.

Choose from these
or similar schemes
to obtain the
greatest amount of
desired product and
least amount of
undesired product

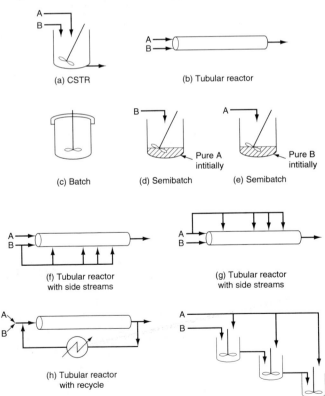

Figure 6-3   Different reactors and schemes for minimizing the unwanted product.

Example 6–2 Minimizing Unwanted Products for Two Reactants

For the parallel reactions

$$A + B \longrightarrow D$$
$$A + B \longrightarrow U$$

consider all possible combinations of reaction orders and select the reaction scheme that will maximize S_{DU}.

Solution

Case I: $\alpha_1 > \alpha_2$, $\beta_1 > \beta_2$. Let $a = \alpha_1 - \alpha_2$ and $b = \beta_1 - \beta_2$, where a and b are positive constants. Using these definitions we can write Equation (6-12) in the form

$$\boxed{S_{DU} = \frac{r_D}{r_U} = \frac{k_1}{k_2} C_A^a C_B^b} \tag{E6-2.1}$$

To maximize the ratio r_D/r_U, maintain the concentrations of both A and B as high as possible. To do this, use:

- A tubular reactor.
- A batch reactor.
- High pressures (if gas phase).

Case II: $\alpha_1 > \alpha_2$, $\beta_1 < \beta_2$. Let $a = \alpha_1 - \alpha_2$ and $b = \beta_2 - \beta_1$, where a and b are positive constants. Using these definitions we can write Equation (E6-2.1) in the form

$$S_{DU} = \frac{r_D}{r_U} = \frac{k_1 C_A^a}{k_2 C_B^b} \tag{E6-2.2}$$

To make S_{DU} as large as possible we want to make the concentration of A high and the concentration of B low. To achieve this result, use:

- A semibatch reactor in which B is fed slowly into a large amount of A (Figure 6-3d).
- A tubular reactor with side streams of B continually fed to the reactor (Figure 6-3f).
- A series of small CSTRs with A fed only to the first reactor and small amounts of B fed to each reactor. In this way B is mostly consumed before the CSTR exit stream flows into the next reactor (Reverse of Figure 6-3i).

Case III: $\alpha_1 < \alpha_2$, $\beta_1 < \beta_2$. Let $a = \alpha_2 - \alpha_1$ and $b = \beta_2 - \beta_1$, where a and b are positive constants. Using these definitions we can write Equation (E6-2.1) in the form

$$S_{DU} = \frac{r_D}{r_U} = \frac{k_1}{k_2 C_A^a C_B^b} \tag{E6-2.3}$$

To make S_{DU} as large as possible, the reaction should be carried out at low concentrations of A and of B. Use:

- A CSTR.
- A tubular reactor in which there is a large recycle ratio.
- A feed diluted with inerts.
- Low pressure (if gas-phase).

Case IV: $\alpha_1 < \alpha_2$, $\beta_1 > \beta_2$. Let $a = \alpha_2 - \alpha_1$ and $b = \beta_1 - \beta_2$, where a and b are positive constants. Using these definitions we can write Equation (E6-2.1) in the form

$$S_{DU} = \frac{r_D}{r_U} = \frac{k_1 C_B^b}{k_2 C_A^a} \tag{E6-2.4}$$

To maximize S_{DU}, run the reaction at high concentrations of B and low concentrations of A. Use:

- A semibatch reactor in which A is slowly fed to a large amount of B.
- A tubular reactor with side streams of A.
- A series of small CSTRs with fresh A fed to each reactor.

Another definition of selectivity used in the current literature, $\tilde{S}_{DU}$, is given in terms of the flow rates leaving the reactor. $\tilde{S}_{DU}$ is the overall selectivity.

Overall selectivity

$$\boxed{\tilde{S}_{DU} = \text{selectivity} = \frac{F_D}{F_U} = \frac{\text{exit molar flow rate of desired product}}{\text{exit molar flow rate of undesired product}}} \tag{6-12}$$

For a batch reactor, the overall selectivity is given in terms of the number of moles of D and U at the end of the reaction time:

$$\tilde{S}_{DU} = \frac{N_D}{N_U}$$

Two definitions for selectivity and yield are found in the literature

One also finds that the reaction yield, like the selectivity, has two definitions: one based on the ratio of reaction rates and one based on the ratio of molar flow rates. In the first case, the yield at a point can be defined as the

Instantaneous yield

ratio of the reaction rate of a given product to the reaction rate of the key reactant A. This is sometimes referred to as the instantaneous yield.[2]

Yield based on rates

$$Y_D = \frac{r_D}{-r_A} \tag{6-13}$$

In the case of reaction yield based on molar flow rates, the overall yield, $\tilde{Y}_D$, is defined as the ratio of moles of product formed at the end of the reaction to the number of moles of the key reactant, A, that have been consumed.

For a batch system:

Overall yield

$$\tilde{Y}_D = \frac{N_D}{N_{A0} - N_A} \tag{6-14}$$

For a flow system:

Overall yield

$$\tilde{Y}_D = \frac{F_D}{F_{A0} - F_A} \tag{6-15}$$

$

As a consequence of the different definitions for selectivity and yield, when reading literature dealing with multiple reactions, check carefully to ascertain the definition intended by the author. From an economic standpoint it is the *overall* selectivities, $\tilde{S}$, and yields, $\tilde{Y}$, that are important in determining profits. However, the rate-based selectivities give insights in choosing reactors and reaction schemes that will help maximize the profit. However, many times there is a conflict between selectivity and conversion (yield) because you want to make a lot of your desired product (D) and at the same time minimize the undesired product (U). However, in many instances the greater conversion you achieve, not only do you make more D, you also form more U.

6.2 Maximizing the Desired Product in Series Reactions

In Section 6.1 we saw that the undesired product could be minimized by adjusting the reaction conditions (e.g., concentration) and by choosing the proper reactor. For series of consecutive reactions, the most important variable is time: space-time for a flow reactor and real-time for a batch reactor. To illustrate the importance of the time factor, we consider the sequence

---

[2] J. J. Carberry, in *Applied Kinetics and Chemical Reaction Engineering*, R. L. Gorring and V. W. Weekman, eds. (Washington, D.C.: American Chemical Society, 1967), p. 89.

$$A \xrightarrow{k_1} B \xrightarrow{k_2} C$$

in which species B is the desired product.

If the first reaction is slow and the second reaction is fast, it will be extremely difficult to produce species B. If the first reaction (formation of B) is fast and the reaction to form C is slow, a large yield of B can be achieved. However, if the reaction is allowed to proceed for a long time in a batch reactor, or if the tubular flow reactor is too long, the desired product B will be converted to C. In no other type of reaction is exactness in the calculation of the time needed to carry out the reaction more important than in consecutive reactions.

Example 6–3 Maximizing the Yield of the Intermediate Product

The oxidation of ethanol to form acetaldehyde is carried out on a catalyst of 4 wt % Cu–2 wt % Cr on Al_2O_3.[3] Unfortunately, acetaldehyde is also oxidized on this catalyst to form carbon dioxide. The reaction is carried out in a threefold excess of oxygen and in dilute concentrations (ca. 0.1% ethanol, 1% O_2, and 98.9% N_2). Consequently, the volume change with the reaction can be neglected. Determine the concentration of acetaldehyde as a function of space-time,

$$CH_3CH_2OH(g) \xrightarrow[-H_2O]{+\frac{1}{2}O_2} CH_3CHO \xrightarrow[-2H_2O]{+\frac{5}{2}O_2} 2CO_2$$

The reactions are irreversible and first-order in ethanol and acetaldehyde, respectively.

Solution

Because O_2 is in excess, we can write the equation above as

$$A \xrightarrow{k_1} B \xrightarrow{k_2} C$$

1. **Mole balance on A:**

$$\frac{dF_A}{dW} = r'_A \qquad \text{(E6-3.1)}$$

a. **Rate law:**

$$-r'_A = k_1 C_A$$

b. **Stoichiometry** $(\varepsilon \ll 1)$:

$$F_A = C_A v_0$$

c. **Combining**, we have

$$v_0 \frac{dC_A}{dW} = -k_1 C_A \qquad \text{(E6-3.2)}$$

Let $\tau' = W/v_0 = \rho_b V/v_0 = \rho_b \tau$, where ρ_b is the bulk density of the catalyst.

[3] R. W. McCabe and P. J. Mitchell, *Ind. Eng. Chem. Process Res. Dev.*, 22, 212 (1983).

d. Integrating with $C_A = C_{A0}$ at $W = 0$ gives us

$$\boxed{C_A = C_{A0}e^{-k_1\tau'}} \tag{E6-3.3}$$

2. **Mole balance on B:**

$$\frac{dF_B}{dW} = r'_{B_{net}} \tag{E6-3.4}$$

a. **Rate law** (net):

$$r'_{B_{net}} = r'_{B_{rxn1}} + r'_{B_{rxn2}}$$
$$r'_{B_{net}} = k_1 C_A - k_2 C_B \tag{E6-3.5}$$

b. **Stoichiometry:**

$$F_B = v_0 C_B$$

c. **Combining** yields

$$v_0 \frac{dC_B}{dW} = k_1 C_A - k_2 C_B \tag{E6-3.6}$$

Substituting for C_A and rearranging, we have

$$\frac{dC_B}{d\tau'} + k_2 C_B = k_1 C_{A0} e^{-k_1\tau'}$$

d. **Using** the integrating factor gives us

$$\frac{d(C_B e^{+k_2\tau'})}{d\tau'} = k_1 C_{A0} e^{(k_2 - k_1)\tau'}$$

At the entrance to the reactor, $W = 0$, $\tau' = W/v_0 = 0$, and $C_B = 0$. Integrating, we get

$$C_B = k_1 C_{A0}\left(\frac{e^{-k_1\tau'} - e^{-k_2\tau'}}{k_2 - k_1}\right) \tag{E6-3.7}$$

The concentrations of A, B, and C are shown in Figure E6-3.1.

There is a space
time at which B is a
maximum

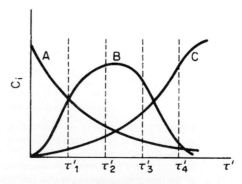

Figure E6-3.1

3. **Optimum yield.** The concentration of B goes through a maximum at a point along the reactor. Consequently, to find the optimum reactor length, we need to differentiate Equation (E6-3.7):

$$\frac{dC_B}{d\tau'} = 0 = \frac{k_1 C_{A0}}{k_2 - k_1} \left(-k_1 e^{-k_1 \tau'} + k_2 e^{-k_2 \tau'} \right) \tag{E6-3.8}$$

Solving for τ'_{opt} gives

$$\tau'_{opt} = \frac{1}{k_1 - k_2} \ln \frac{k_1}{k_2} \tag{E6-3.9}$$

$$\boxed{W_{opt} = \frac{v_0}{k_1 - k_2} \ln \frac{k_1}{k_2}} \tag{E6-3.10}$$

The corresponding conversion of A is

$$X_{opt} = \frac{C_{A0} - C_A}{C_{A0}} = 1 - e^{-k_1 \tau'}$$

$$= 1 - \exp \left[-\ln \left(\frac{k_1}{k_2} \right)^{k_1/(k_1 - k_2)} \right] \tag{E6-3.11}$$

$$= 1 - \left(\frac{k_1}{k_2} \right)^{k_1/(k_2 - k_1)}$$

The yield has been defined as

$$\tilde{Y}_A = \frac{\text{moles of acetaldehyde in exit}}{\text{moles of ethanol fed}}$$

and is shown as a function of conversion in Figure E6-3.2.

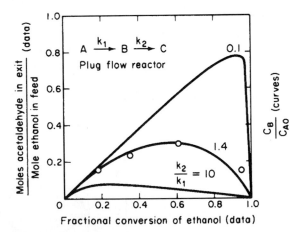

Figure E6-3.2   Yield of acetaldehyde as a function of ethanol conversion. Data were obtained at 518 K. Data points (in order of increasing ethanol conversion) were obtained at space velocities of 26,000, 52,000, 104,000, and 208,000 h^{-1}. The curves were calculated for a first-order series reaction in a plug-flow reactor and show yield of the intermediate species B as a function of the conversion of reactant for various ratios of rate constants k_2 and k_1. [Reprinted with permission from *Ind. Eng. Chem. Prod. Res. Dev.*, 22, 212 (1983). Copyright © 1983 American Chemical Society.]

Another technique is often used to follow the progress for two reactions in series. The concentrations of A, B and C are plotted as a singular point at different space times (e.g., τ_1', τ_2') on a triangular diagram (see Figure 6-4). The vertices correspond to pure A, B, and C.

For $(k_1/k_2) \gg 1$ a large quantity of B can be obtained

For $(k_1/k_2) \ll 1$ very little B can be obtained

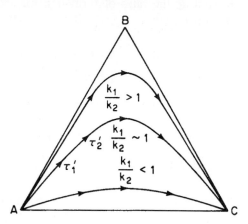

Figure 6-4   Reaction paths for different values of the specific rates.

6.3 Algorithm for Solution to Complex Reactions

6.3.1 Mole Balances

In complex reaction systems consisting of combinations of parallel and series reactions the availability of software packages (ODE solvers) makes it much easier to solve problems using moles N_j or molar flow rates F_j rather than conversion. For liquid systems, concentration may be the preferred variable used in the mole balance equations. The resulting coupled differential equations can be easily solved using an ODE solver. In fact, this section has been developed to take advantage of the vast number of computational techniques now available on mainframe (e.g., Simulsolv) and personal computers (POLYMATH).

Table 6-1 gives the forms of the mole balances we shall use for complex reactions where r_A and r_B are the *net* rates of reaction.

TABLE 6-1.   MOLE BALANCES FOR MULTIPLE REACTIONS

These are the forms of the mole balances we will use for multiple reactions

Reactor	Gas	Liquid
Batch	$\dfrac{dN_A}{dt} = r_A V$	$\dfrac{dC_A}{dt} = r_A$
CSTR	$V = \dfrac{F_{A0} - F_A}{-r_A}$	$V = v_0 \dfrac{C_{A0} - C_A}{-r_A}$
PFR/PBR	$\dfrac{dF_A}{dV} = r_A$	$v_0 \dfrac{dC_A}{dV} = r_A$
Semibatch	$\dfrac{dN_A}{dt} = r_A V$	$\dfrac{dC_A}{dt} = r_A - \dfrac{v_0 C_A}{V}$
	$\dfrac{dN_B}{dt} = r_B V + F_{B0}$	$\dfrac{dC_B}{dt} = r_B + \dfrac{v_0 [C_{B0} - C_B]}{V}$

6.3.2 Net Rates of Reaction

Having written the mole balances, the key point for multiple reactions is to write the **net** rate of formation of each species (e.g., A, B). That is, we have to sum up the rates of formation for each reaction in order to obtain the net rate of formation, e.g. r_A. If q reactions are taking place

Reaction 1: $A + B \xrightarrow{k_{1A}} 3C + D$

Reaction 2: $A + 2C \xrightarrow{k_{2A}} 3E$

Reaction 3: $2B + 3E \xrightarrow{k_{3B}} 4F$

$$\vdots$$

Reaction q: $A + \dfrac{1}{2} B \xrightarrow{k_{qA}} G$

Then the net rates of reaction of A and B are

Net rates of reaction

$$r_A = r_{1A} + r_{2A} + r_{3A} + \cdots + r_{qA} = \sum_{i=1}^{q} r_{iA}$$

$$r_B = r_{1B} + r_{2B} + r_{3B} + \cdots + r_{qB} = \sum_{i=1}^{q} r_{iB}$$

When we sum the rates of the individual reaction for a species, we note that for those reactions in which a species (e.g., A, B) does not appear, the rate is zero. For the first three reactions above, $r_{3A} = 0$, $r_{2B} = 0$, and $r_{2D} = 0$.

To write the reactions above in more compact notation we could let $A_1 = A$, $A_2 = B$, and so on, to arrive at the generic sequence of q reactions shown in Table 6-2. The letter A_j represents a chemical species (e.g., $A_1 = HCl$, $A_2 = NaOH$). The first subscript, i, in the stoichiometric coefficient ν_{ij} and in the reaction rate r_{ij} refers to the reaction number while the second subscript, j, refers to the particular species in the reaction. We are now in a position to evaluate the total rate of formation of each species from all reac-

r_{ij}
└ species
└ reaction number

TABLE 6-2

Reaction Number	Reaction Stoichiometry
1	$\nu_{11}A_1 + \nu_{12}A_2 \longrightarrow \nu_{1j}A_j$
2	$\nu_{21}A_1 + \nu_{2j}A_j \longrightarrow \nu_{23}A_3 + \nu_{24}A_4$
.	.
.	.
.	.
i	$\nu_{ij}A_j + \nu_{ik}A_k \longrightarrow \nu_{ip}A_p$
.	.
.	.
q	$\nu_{qj}A_j + \nu_{q3}A_3 \longrightarrow \nu_{q5}A_5$

tions. That is, the net rate of reaction for species A_j is the sum of all rates of reaction in which species A_j appears. For q reactions taking place,

$$r_j = \sum_{i=1}^{q} r_{ij} \qquad (6\text{-}16)$$

6.3.3 Rate Laws

The rate laws for each of the individual reactions are expressed in terms of concentrations, C_j, of the reacting species. For example, if reaction 2 above (i.e., $A + 2C \rightarrow 3E$) followed an elementary rate law, then the rate of disappearance of A could be

$$-r_{2A} = k_{2A} C_A C_C^2$$

or in terms of the rate of formation of A in reaction 2,

$$r_{2A} = -k_{2A} C_A C_C^2$$

For the general reaction set given in Table 6-2, the rate law for the rate of formation of reactant species A_j in reaction i might depend on the concentration of species A_k and species A_j, for example,

$$r_{ij} = -k_{ij} C_k^2 C_j^3$$

We need to determine the rate law for at least one species in each reaction.

6.3.4 Stoichiometry: Relative Rates of Reaction

The next step is to relate the rate law for a particular reaction and species to other species participating in that reaction. To achieve this relationship we simply recall the generic reaction from Chapters 2 and 3,

$$aA + bB \longrightarrow cC + dD \qquad (2\text{-}1)$$

and use Equation (2-20) to relate the rates of disappearance of A and B to the rates of formation of C and D:

$$\frac{-r_A}{a} = \frac{-r_B}{b} = \frac{r_C}{c} = \frac{r_D}{d} \qquad (2\text{-}3)$$

In working with multiple reactions it is usually more advantageous to relate the *rates of formation* of each species to one another. This can be achieved by rewriting (2-20) in the form for reaction i

Relative rates of reaction

$$\frac{r_{iA}}{-a_i} = \frac{r_{iB}}{-b_i} = \frac{r_{iC}}{c_i} = \frac{r_{iD}}{d_i} \qquad (6\text{-}17)$$

$$\left[\text{e.g. for reaction 2: } \quad r_{2C} = \frac{c_2}{-a_2} r_{2A} = \frac{c_2}{a_2} (-r_{2A}) \right]$$

Applying Equation (6-17) to reaction 2 above [i.e., $(A + 2C \xrightarrow{k_{2A}} 3E)$ where $r_{2A} = -k_{2A}C_AC_C^2$], the rate of formation of species E, r_{2E}, is

$$r_{2E} = \frac{3}{-1}(r_{2A}) = -3(-k_{2A}C_AC_C^2) = 3k_{2A}C_AC_C^2$$

and the rate of formation of C is

r_{2E}
— species
— reaction number

$$r_{2C} = \frac{-2}{-1}r_{2A} = -2k_{2A}C_AC_C^2$$

To relate the relative rates of formation in more compact notation suppose the rate law for the rate of formation of species A_k is given in reaction i as

$$r_{ik} = k_{ik}C_k^2C_2$$

To find the rate of formation of species A_j in reaction i, r_{ij}, we multiply the rate law for species A_k in reaction i by the ratio of stoichiometric coefficients of species A_j, ν_{ij}, and species A_k, ν_{ik}, in reaction i:

Relative rates of reaction in reaction i in compact notation

$$\boxed{\frac{r_{ij}}{\nu_{ij}} = \frac{r_{ik}}{\nu_{ik}}} \qquad (6\text{-}18)$$

This relationship only holds for relative rates in the same reaction (i.e., reaction i). When relating relative *rates of formation* the stoichiometric coefficients, ν_{ij}, of reactants are taken as negative and the coefficients of products as positive.

In analyzing the multiple reactions in Table 6-2, we carry out the procedure shown in Table 6-3 (not necessarily in exact order) when the rate law is known for *at least one species in each* of the individual reactions.

TABLE 6-3.   STEPS IN ANALYZING MULTIPLE REACTIONS

---

1. Number each reaction.
2. Write the mole balances for each species.
3. Determine the rate laws for *each* species in *each* reaction.
4. Relate the rate of reaction of each species to the species for which the rate law is given for each reaction.
5. Determine the net rate of formation of each species.
6. Express rate laws as a function of concentration, C_j, for the case of *no* volume change.
7. Express the rate laws as a function of moles, N_j (batch), or molar flow rates, F_j (flow) when there is volume change with reaction.
8. Combine all of the above and solve the resulting set of coupled differential (PFR, PBR, batch) or algebraic (CSTR) equations.

The multiple-reaction algorithm for isothermal reactions

Example 6–4 Stoichiometry and Rate Laws for Multiple Reactions

Consider the following set of reactions:

Rate Laws[4]

Reaction 1: $4NH_3 + 6NO \longrightarrow 5N_2 + 6H_2O$ $-r_{1NO} = k_{1NO}C_{NH_3}C_{NO}^{1.5}$

Reaction 2: $2NO \longrightarrow N_2 + O_2$ $r_{2N_2} = k_{2N_2}C_{NO}^2$

Reaction 3: $N_2 + 2O_2 \longrightarrow 2NO_2$ $-r_{3O_2} = k_{3O_2}C_{N_2}C_{O_2}^2$

Write the rate law for each species in each reaction and then write the net rates of formation of NO, O_2, and N_2.

Solution

The rate laws for reactions 1, 2, and 3 are given in terms of species NO, N_2, and O_2, respectively. Consequently, to relate each reacting species in each reaction to its rate law more clearly, we divide each reaction through by the stoichiometric coefficient of the species for which the rate law is given.

1. $NO + \dfrac{2}{3}NH_3 \longrightarrow \dfrac{5}{6}N_2 + H_2O$ $-r_{1NO} = k_{1NO}C_{NH_3}C_{NO}^{1.5}$ (E6-4.1)

2. $2NO \longrightarrow N_2 + O_2$ $r_{2N_2} = k_{2N_2}C_{NO}^2$ (E6-4.2)

3. $O_2 + \dfrac{1}{2}N_2 \longrightarrow NO_2$ $-r_{3O_2} = k_{3O_2}C_{N_2}C_{O_2}^2$ (E6-4.3)

The corresponding rate laws are related by:

Recalling Eqn. (6-17)

$$\frac{r_{iA}}{-a_i} = \frac{r_{iB}}{-b_i} = \frac{r_{iC}}{c_i} = \frac{r_{iD}}{d_i}$$

Reaction 1: The rate law w.r.t. NO is

$$-r_{1NO} = k_{1NO}C_{NH_3}C_{NO}^{1.5}$$

The relative rates are

$$\frac{r_{1NO}}{(-1)} = \frac{r_{1NH_3}}{(-2/3)} = \frac{r_{1N_2}}{(5/6)} = \frac{r_{1H_2O}}{(1)}$$

Then the rate of disappearance of NH_3 is

$$-r_{1NH_3} = \frac{2}{3}(-r_{1NO}) = \frac{2}{3}k_{1NO}C_{NH_3}C_{NO}^{1.5}$$ (E6-4.4)

---

[4] From tortusimetry data (11/2/19).

$$r_{1N_2} = \frac{5}{6}(-r_{1NO}) = \frac{5}{6}k_{1NO}C_{NH_3}C_{NO}^{1.5} \qquad \text{(E6-4.5)}$$

$$r_{1H_2O} = -r_{1NO} \qquad \text{(E6-4.6)}$$

Reaction 2:

$$-r_{2NO} = 2r_{2N_2} = 2k_{2N_2}C_{NO}^2 \qquad \text{(E6-4.7)}$$

$$r_{2O_2} = r_{2N_2} \qquad \text{(E6-4.8)}$$

Reaction 3:

$$-r_{3N_2} = \frac{1}{2}(-r_{3O_2}) = \frac{1}{2}k_{3O_2}C_{N_2}C_{O_2}^2 \qquad \text{(E6-4.9)}$$

$$r_{3NO_2} = -r_{3O_2} \qquad \text{(E6-4.10)}$$

Next, let us examine the *net* rate of formations. The net rates of formation of NO is:

$$r_{NO} = \sum_{i=1}^{3} r_{iNO} = r_{1NO} + r_{2NO} + 0 \qquad \text{(E6-4.11)}$$

$$\boxed{r_{NO} = -k_{1NO}C_{NH_3}C_{NO}^{1.5} - 2k_{2N_2}C_{NO}^2} \qquad \text{(E6-4.12)}$$

Next consider N_2

$$r_{N_2} = \sum_{i=1}^{3} r_{iN_2} = r_{1N_2} + r_{2N_2} + r_{3N_2} \qquad \text{(E6-4.13)}$$

$$\boxed{r_{N_2} = \frac{5}{6}k_{1NO}C_{NH_3}C_{NO}^{1.5} + k_{2N_2}C_{NO}^2 - \frac{1}{2}k_{3O_2}C_{N_2}C_{O_2}^2} \qquad \text{(E6-4.14)}$$

Finally O_2

$$r_{O_2} = r_{2O_2} + r_{3O_2} = r_{2N_2} + r_{3O_2} \qquad \text{(E6-4.15)}$$

$$\boxed{r_{O_2} = k_{2N_2}C_{NO}^2 - k_{3O_2}C_{N_2}C_{O_2}^2} \qquad \text{(E6-4.16)}$$

6.3.5 Stoichiometry: Concentrations

Now to express the concentrations in terms of molar flow rates we recall that for liquids

Liquid phase

$$C_j = \frac{F_j}{v_0} \qquad \text{(6-19)}$$

For ideal gases recall Equation (3-45):

Gas phase

$$C_j = \frac{F_{T0}}{v_0}\left(\frac{F_j}{F_T}\right)\frac{P}{P_0}\frac{T_0}{T} = C_{T0}\left(\frac{F_j}{F_T}\right)\frac{P}{P_0}\frac{T_0}{T} \qquad \text{(3-45)}$$

where

$$F_T = \sum_{j=1}^{n} F_j \qquad (6\text{-}20)$$

and

$$C_{T0} = \frac{P_0}{RT_0} \qquad (6\text{-}21)$$

For isothermal systems with no pressure drop

Gas phase

$$C_j = C_{T0} \left(\frac{F_j}{F_T} \right) \qquad (6\text{-}22)$$

and we can express the rates of disappearance of each species as a function of the molar flow rates $(F_1, \ldots, F_j)$:

$$r_1 = fn_1 \left(C_{T0} \frac{F_1}{F_T}, \ C_{T0} \frac{F_2}{F_T}, \ldots, \ C_{T0} \frac{F_j}{F_T} \right) \qquad (6\text{-}23)$$

$$r_2 = fn_2 \left(C_{T0} \frac{F_1}{F_T}, \ldots, \ C_{T0} \frac{F_j}{F_T} \right) \qquad (6\text{-}24)$$

where *fn* represents the functional dependence on concentration of the net rate of formation such as that given in Equation (E6-4.12) for N_2.

6.3.6 Combining Step

We now insert rate laws written in terms of molar flow rates [e.g., Equation (3-45)] into the mole balances (Table 6-1). After performing this operation for each species we arrive at a coupled set of first-order ordinary differential equations to be solved for the molar flow rates as a function of reactor volume (i.e., distance along the length of the reactor). In liquid-phase reactions, incorporating and solving for total molar flow rate is not necessary at each step along the solution pathway because there is no volume change with reaction.

Combining mole balance, rate laws, and stoichiometry for species 1 through species j in the gas phase and for isothermal operation with no pressure drop gives us

Coupled ODEs

$$\frac{dF_1}{dV} = r_1 = \sum_{i=1}^{m} r_{i1} = fn_1 \left(C_{T0} \frac{F_1}{F_T}, \ldots, \ C_{T0} \frac{F_j}{F_T} \right) \qquad (6\text{-}25)$$

$$\vdots$$

$$\frac{dF_j}{dV} = r_j = \sum_{i=1}^{q} r_{ij} = fn_j \left(C_{T0} \frac{F_1}{F_T}, \ldots, \ C_{T0} \frac{F_j}{F_T} \right) \qquad (6\text{-}26)$$

For constant-pressure batch systems we would simply substitute N_i for F_i in the equations above. For constant-volume batch systems we would use concentrations:

$$C_j = N_j / V_0 \tag{6-27}$$

We see that we have j coupled ordinary differential equations that must be solved simultaneously with either a numerical package or by writing an ODE solver. In fact, this procedure has been developed to take advantage of the vast number of computation techniques now available on mainframe (e.g., Simusolv) and personal computers (POLYMATH, Mathematica, MATLAB).

Example 6–5   Combining Mole Balances, Rate Laws, and Stoichiometry for Multiple Reactions

Consider again the reaction in Example 6-4. Write the mole balances on a PFR in terms of molar flow rates for each species.

Reaction 1: $NO + \dfrac{2}{3}NH_3 \longrightarrow \dfrac{5}{6}N_2 + H_2O$

Reaction 2: $2NO \longrightarrow N_2 + O_2$

Reaction 3: $O_2 + \dfrac{1}{2}N_2 \longrightarrow NO_2$

Solution

For gas-phase reactions, the concentration of species j is

$$C_j = C_{T0} \frac{F_j}{F_T} \frac{P}{P_0} \frac{T_0}{T}$$

For no pressure drop and isothermal operation,

$$C_j = C_{T0} \frac{F_j}{F_T} \tag{E6-5.1}$$

In combining the mole balance, rate laws, and stoichiometry, we will use our results from Example 6-4. The total molar flow rate of all the gases is

$$\boxed{F_T = F_{NO} + F_{NH_3} + F_{N_2} + F_{H_2O} + F_{O_2} + F_{NO_2}} \tag{E6-5.2}$$

We now rewrite mole balances on each species in the total molar flow rate.

Using the results of Example 6-4

(1) Mole balance on NO:

$$\frac{dF_{NO}}{dV} = r_{NO} = -k_{1NO}C_{NH_3}C_{NO}^{1.5} - 2k_{2N_2}C_{NO}^2$$

$$\tag{E6-5.3}$$

$$\boxed{\frac{dF_{NO}}{dV} = -k_{1NO}C_{TO}^{2.5}\left(\frac{F_{NH_3}}{F_T}\right)\left(\frac{F_{NO}}{F_T}\right)^{1.5} - 2k_{2N_2}C_{TO}^2\left(\frac{F_{NO}}{F_T}\right)^2}$$

(2) Mole balance on NH_3:

$$\frac{dF_{NH_3}}{dV} = r_{NH_3} = r_{1NH_3} = \frac{2}{3} r_{1NO} = -\frac{2}{3} k_{1NO} C_{NH_3} C_{NO}^{1.5}$$

(E6-5.4)

$$\boxed{\frac{dF_{NH_3}}{dV} = -\frac{2}{3} k_{1NO} C_{T0}^{2.5} \left(\frac{F_{NH_3}}{F_T}\right) \left(\frac{F_{NO}}{F_T}\right)^{1.5}}$$

(3) Mole balance on H_2O:

$$\frac{dF_{H_2O}}{dV} = r_{H_2O} = r_{1H_2O} = -r_{1NO} = k_{1NO} C_{NH_3} C_{NO}^{1.5}$$

(E6-5.5)

$$\boxed{\frac{dF_{H_2O}}{dV} = k_{1NO} C_{T0}^{2.5} \left(\frac{F_{NH_3}}{F_T}\right) \left(\frac{F_{NO}}{F_T}\right)^{1.5}}$$

(4) Mole balance on N_2:

$$\frac{dF_{N_2}}{dV} = r_{N_2} = \frac{5}{6} k_{1NO} C_{NH_3} C_{NO}^{1.5} + k_{2N_2} C_{NO}^2 - \frac{1}{2} k_{3O_2} C_{N_2} C_{O_2}^2$$

$$\boxed{\frac{dF_{N_2}}{dV} = \frac{5}{6} k_{1NO} C_{T0}^{2.5} \left(\frac{F_{NH_3}}{F_T}\right) \left(\frac{F_{NO}}{F_T}\right)^{1.5} + k_{2N_2} C_{T0}^2 \left(\frac{F_{NO}}{F_T}\right)^2 - \frac{1}{2} k_{3O_2} C_{T0}^3 \left(\frac{F_{N_2}}{F_T}\right) \left(\frac{F_{O_2}}{F_T}\right)^2}$$

(E6-5.6)

(5) Mole balance on O_2:

$$\frac{dF_{O_2}}{dV} = r_{O_2} = r_{2O_2} + r_{3O_2} = r_{2N_2} + r_{3O_2}$$

(E6-5.7)

$$\boxed{\frac{dF_{O_2}}{dV} = k_{2N_2} C_{T0}^2 \left(\frac{F_{NO}}{F_T}\right)^2 - k_{3O_2} C_{T0}^3 \left(\frac{F_{N_2}}{F_T}\right) \left(\frac{F_{O_2}}{F_T}\right)^2}$$

(6) Mole balance on NO_2:

$$\boxed{\frac{dF_{NO_2}}{dV} = r_{NO_2} = r_{3NO_2} = -r_{3O_2} = k_{3O_2} C_{T0}^3 \left(\frac{F_{N_2}}{F_T}\right) \left(\frac{F_{O_2}}{F_T}\right)^2}$$ (E6-5.8)

The entering molar flow rates, F_{j0}, along with the entering temperature, T_0, and pressure, $P_0 (C_{T0} = P_0/RT_0)$, are specified as are the specific reaction rates k_{ij} [e.g., $k_{1NO} = 0.43$ $(dm^3/mol)^{1.5}/s$, $k_{2N_2} = 2.7$ $dm^3/mol \cdot s$, etc.]. Consequently, Equations (E6-5.1) through (E6-5.8) can be solved simultaneously with an ODE solver (e.g., POLYMATH, MATLAB).

Summarizing to this point, we show in Table 6-4 the equations for species j and reaction i that are to be combined when we have q reactions and n species.

TABLE 6-4.   SUMMARY OF RELATIONSHIP
FOR MULTIPLE REACTIONS OCCURING IN A PFR

The basic equations

Mole balance:	$\dfrac{dF_j}{dV} = r_j$	(6-26)
Rate laws:	$r_{ij} = k_{ij} f_i(C_1, C_j, C_n)$	
Stoichiometry:	$r_j = \displaystyle\sum_{i=1}^{q} r_{ij}$	(6-16)
	$\dfrac{r_{iA}}{-a_i} = \dfrac{r_{iB}}{-b_i} = \dfrac{r_{iC}}{c_i} = \dfrac{r_{iD}}{d_i}$	(6-17)
Stoichiometry:	$F_T = \displaystyle\sum_{j=1}^{n} F_j$	(6-20)
Stoichiometry:		
(gas-phase)	$C_j = C_{T0} \dfrac{F_j}{F_T} \dfrac{P}{P_0} \dfrac{T_0}{T}$	(3-45)
(liquid-phase)	$C_j = \dfrac{F_j}{v_0}$	(6-19)

Example 6–6   Hydrodealkylation of Mesitylene in a PFR

The production of m-xylene by the hydrodealkylation of mesitylene over a Houdry Detrol catalyst[5] involves the following reactions:

$$\text{(mesitylene)} + H_2 \longrightarrow \text{(}m\text{-xylene)} + CH_4 \qquad \text{(E6-6.1)}$$

m-Xylene can also undergo hydrodealkylation to form toluene:

A significant
economic incentive

$$\text{(}m\text{-xylene)} + H_2 \longrightarrow \text{(toluene)} + CH_4 \qquad \text{(E6-6.2)}$$

The second reaction is undesirable, because m-xylene sells for a higher price than toluene (65 cents/lb vs. 11.4 cents/lb).[6] Thus we see that there is a significant incentive to maximize the production of m-xylene.

---

[5] *Ind. Eng. Chem. Process Des. Dev.*, 4, 92 (1965); 5, 146 (1966).

[6] September 1996 prices, from *Chemical Market Reporter* (Schnell Publishing Co.), *252, 29* (July 7, 1997). Also see http://www.chemweek.com/

The hydrodealkylation of mesitylene is to be carried out isothermally at 1500°R and 35 atm in a packed-bed reactor in which the feed is 66.7 mol% hydrogen and 33.3 mol% mesitylene. The volumetric feed rate is 476 ft³/h and the reactor volume (i.e., $V = W/\rho_b$) is 238 ft³.

The rate laws for reactions 1 and 2 are, respectively,

$$-r_{1M} = k_1 C_M C_H^{0.5} \tag{E6-6.3}$$

$$r_{2T} = k_2 C_X C_H^{0.5} \tag{E6-6.4}$$

where the subscripts are: M = mesitylene, X = *m*-xylene, T = toluene, Me = methane, and H = hydrogen (H_2).

At 1500°R the specific reaction rates are:

Reaction 1: k_1 = 55.20 (ft³/lb mol)$^{0.5}$/h

Reaction 2: k_2 = 30.20 (ft³/lb mol)$^{0.5}$/h

The bulk density of the catalyst has been included in the specific reaction rate (i.e., $k_1 = k_1' \rho_b$).

Plot the concentrations of hydrogen, mesitylene, and xylene as a function of space-time. Calculate the space-time where the production of xylene is a maximum (i.e., τ_{opt}).

Solution

$$\text{Reaction 1:} \quad M + H \longrightarrow X + Me \tag{E6-6.1}$$

$$\text{Reaction 2:} \quad X + H \longrightarrow T + Me \tag{E6-6.2}$$

1. **Mole balances:**

Hydrogen: $\qquad \dfrac{dF_H}{dV} = r_{1H} + r_{2H} \tag{E6-6.5}$

Mesitylene: $\qquad \dfrac{dF_M}{dV} = r_{1M} \tag{E6-6.6}$

Xylene: $\qquad \dfrac{dF_X}{dV} = r_{1X} + r_{2X} \tag{E6-6.7}$

Toluene: $\qquad \dfrac{dF_T}{dV} = r_{2T} \tag{E6-6.8}$

Methane: $\qquad \dfrac{dF_{Me}}{dV} = r_{1Me} + r_{2Me} \tag{E6-6.9}$

2. **Rate laws:**

Reaction 1: $\quad -r_{1H} = k_1 C_H^{1/2} C_M \tag{E6-6.3}$

Reaction 2: $\quad r_{2T} = k_2 C_H^{1/2} C_X \tag{E6-6.4}$

3. **Stoichiometry** (no volume change with reaction, $v = v_0$)

   a. Reaction rates:

$$\text{Reaction 1:} \quad -r_{1H} = -r_{1M} = r_{1X} = r_{1Me} \tag{E6-6.10}$$

$$\text{Reaction 2:} \quad -r_{2H} = -r_{2X} = r_{2T} = r_{2Me} \tag{E6-6.11}$$

   b. Flow rates:

$$F_H = v_0 C_H \tag{E6-6.12}$$

$$F_M = v_0 C_M \tag{E6-6.13}$$

$$F_X = v_0 C_X \tag{E6-6.14}$$

$$F_{Me} = v_0 C_{Me} = F_{H0} - F_H = v_0(C_{H0} - C_H) \tag{E6-6.15}$$

$$F_T = F_{M0} - F_M - F_X = v_0(C_{M0} - C_M - C_X) \tag{E6-6.16}$$

4. **Combining** and substituting in terms of the space-time yields

$$\tau = \frac{V}{v_0}$$

If we know C_M, C_H, and C_X, then C_{Me} and C_T can be calculated from the reaction stoichiometry. Consequently, we need only to solve the following three equations:

The emergence of user-friendly *ODE solvers* favors this approach over fractional conversion

$$\frac{dC_H}{d\tau} = -k_1 C_H^{1/2} C_M - k_2 C_X C_H^{1/2} \tag{E6-6.17}$$

$$\frac{dC_M}{d\tau} = -k_1 C_M C_H^{1/2} \tag{E6-6.18}$$

$$\frac{dC_X}{d\tau} = k_1 C_M C_H^{1/2} - k_2 C_X C_H^{1/2} \tag{E6-6.19}$$

5. **Parameter evaluation:**

$$C_{H0} = \frac{y_{H0} P_0}{RT} = \frac{(0.667)(35)}{(0.73)(1500)} = 0.021 \text{ lb mol/ft}^3$$

$$C_{M0} = \frac{1}{2} C_{H0} = 0.0105 \text{ lb mol/ft}^3$$

$$C_{X0} = 0$$

$$\tau = \frac{V}{v_0} = \frac{238 \text{ ft}^3}{476 \text{ ft}^3/\text{hr}} = 0.5 \text{ h}$$

We now solve these three equations simultaneously using POLYMATH. The program and output in graphical form are shown in Table E6-6.1 and Figure E6-6.1, respectively. However, I hasten to point out that these equations can be solved analytically and the solution was given in the first edition of this text.

TABLE E6-6.1.   POLYMATH PROGRAM

Equations	Initial Values
d(ch)/d(t)=r1+r2	0.021
d(cm)/d(t)=r1	0.0105
d(cx)/d(t)=-r1+r2	0
k1=55.2	
k2=30.2	
r1=-k1*cm*(ch**.5)	
r2=-k2*cx*(ch**.5)	
$t_0 = 0$, $t_f = 0.5$	

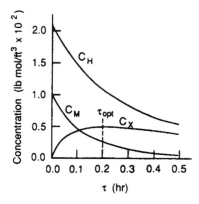

Figure E6-6.1   Concentration profiles in a PFR.

6.3.7 Multiple Reactions in a CSTR

For a CSTR, a coupled set of algebraic equations analogous to PFR differential equations must be solved.

$$V = \frac{F_{j0} - F_j}{-r_j} \qquad (6\text{-}28)$$

Rearranging yields

$$F_{j0} - F_j = -r_j V \qquad (6\text{-}29)$$

Recall that r_j in Equation (6-16) is a function (f_j) of the species concentrations

$$r_j = \sum_{i=1}^{q} r_{ij} = f_j(C_1, C_2, \ldots, C_N) \qquad (6\text{-}16)$$

After writing a mole balance on each species in the reaction set, we substitute for concentrations in the respective rate laws. If there is no volume change with reaction, we use concentrations, C_j, as variables. If the reactions are gas-phase and there is volume change, we use molar flow rates, F_j, as variables. The total molar flow rate for n species is

$$F_T = \sum_{j=1}^{n} F_j \qquad (6\text{-}30)$$

For q reactions occurring in the gas phase, where N different species are present, we have the following set of algebraic equations:

$$F_{10} - F_1 = -r_1 V = V \sum_{i=1}^{q} -r_{i1} = V \cdot f_1 \left(\frac{F_1}{F_T} C_{T0}, \dots, \frac{F_N}{F_T} C_{T0} \right) \quad (6\text{-}31)$$

$$F_{j0} - F_j = -r_j V = V \cdot f_j \left(\frac{F_1}{F_T} C_{T0}, \dots, \frac{F_N}{F_T} C_{T0} \right) \qquad (6\text{-}32)$$

$$F_{N0} - F_N = -r_N V = V \cdot f_N \left(\frac{F_1}{F_T} C_{T0}, \dots, \frac{F_N}{F_T} C_{T0} \right) \qquad (6\text{-}33)$$

We can use an equation solver in POLYMATH or similar to solve Equations (6-31) through (6-33).

Example 6–7 Hydrodealkylation of Mesitylene in a CSTR

For the multiple reactions and conditions described in Example 6-6, calculate the conversion of hydrogen and mesitylene along with the exiting concentrations of mesitylene, hydrogen, and xylene in a CSTR.

Solution

1. **Mole balances:**

   Hydrogen: $F_{H0} - F_H = (-r_{1H} + -r_{2H}) V$ \qquad (E6-7.1)

   Mesitylene: $F_{M0} - F_M = -r_{1M} V$ \qquad (E6-7.2)

   Xylene: $F_X = (r_{1X} + r_{2X}) V$ \qquad (E6-7.3)

   Toluene: $F_T = r_{2T} V$ \qquad (E6-7.4)

   Methane: $F_{Me} = (r_{1Me} + r_{2Me}) V$ \qquad (E6-7.5)

2. **Rate laws:**

   Reaction 1: $-r_{1H} = -r_{1M} = r_{1X} = r_{1Me} = k_1 C_H^{1/2} C_M$ \qquad (E6-7.6)

   Reaction 2: $-r_{2H} = -r_{2X} = r_{2T} = r_{2Me} = k_2 C_H^{1/2} C_X$ \qquad (E6-7.7)

3. **Stoichiometry** $(v = v_0)$:   (E6-7.8)

$$F_H = v_0 C_H$$ (E6-7.9)

$$F_M = v_0 C_M$$ (E6-7.10)

$$F_X = v_0 C_X$$ (E6-7.11)

$$F_T = v_0 C_T = (F_{M0} - F_M) - F_X$$ (E6-7.12)

$$F_{Me} = v_0 C_{Me} = v_0 (C_{H0} - C_H)$$ (E6-7.13)

4. **Combining** and letting $\tau = V/v_0$ (space-time) yields:

$$C_{H0} - C_H = (k_1 C_H^{1/2} C_M + k_2 C_H^{1/2} C_X)\tau$$ (E6-7.14)

$$C_{M0} - C_M = (k_1 C_H^{1/2} C_M)\tau$$ (E6-7.15)

$$C_X = (k_1 C_H^{1/2} C_M - k_2 C_H^{1/2} C_X)\tau$$ (E6-7.16)

Next, we put these equations in a form such that they can be readily solved using POLYMATH.

$$f(C_H) = 0 = C_H - C_{H0} + (k_1 C_H^{1/2} C_M + k_2 C_H^{1/2} C_X)\tau$$ (E6-7.17)

$$f(C_M) = 0 = C_M - C_{M0} + k_1 C_H^{1/2} C_M \tau$$ (E6-7.18)

$$f(C_X) = 0 = (k_1 C_H^{1/2} C_M - k_2 C_H^{1/2} C_X)\tau - C_X$$ (E6-7.19)

The POLYMATH program and solution are shown in Table E6-7.1. The problem was solved for different values of τ and the results are plotted in Figure E6-7.1.

For a space-time of $\tau = 0.5$, the exiting concentrations are $C_H = 0.0089$, $C_M = 0.0029$, and $C_X - 0.0033$. The overall conversion is

Hydrogen: $X_M = \dfrac{F_{H0} - F_H}{F_{H0}} = \dfrac{C_{H0} - C_H}{C_{H0}} = \dfrac{0.021 - 0.0089}{0.021} = 0.58$

Mesitylene: $X_H = \dfrac{F_{M0} - F_M}{F_{M0}} = \dfrac{C_{M0} - C_M}{C_{M0}} = \dfrac{0.0105 - 0.0029}{0.0105} = 0.72$

TABLE E6-7.1.   POLYMATH PROGRAM AND SOLUTION

Equations	Initial Values
f(ch)=ch-.021+(55.2*cm*ch**.5+30.2*cx*ch**.5)*tau	0.006
f(cm)=cm-.0105+(55.2*cm*ch**.5)*tau	0.0033
f(cx)=(55.2*cm*ch**.5-30.2*cx*ch**.5)*tau-cx	0.005
tau=0.5	

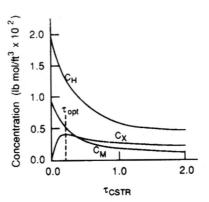

Figure E6-7.1   Concentrations as a function of space time.

We resolve Equations (E6-7.17) through (E6-7.19) for different values of τ to arrive at Figure E6-7.1.

The moles of hydrogen consumed in reaction 1 are equal to the moles of mesitylene consumed. Therefore, the conversion of hydrogen in reaction 1 is

$$X_{1H} = \frac{C_{M0} - C_M}{C_{H0}} = \frac{0.0105 - 0.0029}{.021} = X_{1H} = 0.36$$

The conversion of hydrogen in reaction 2 is

$$X_{2H} = X_H - X_{1H} = 0.58 - 0.36 = 0.22$$

Next, we determine the selectivity and yield. First, consider the rate selectivity parameter, S_{XT}, at the optimum space-time. At τ_{opt} (see Figure E6-7.1), the concentration of xylene is a maximum. Therefore,

$$\frac{dC_X}{d\tau} = 0 = r_X$$

Thus, the rate (i.e., *instantaneous*) selectivity parameter of xylene relative to toluene is

$$S_{XT} = \frac{r_X}{r_T} = \frac{0}{r_T} = 0$$

Similarly, the xylene yield based on reaction rates is also zero. Consequently, we see that under these conditions (τ_{opt}) the instantaneous selectivity and instantaneous yield, which are based on reaction rates, are not very meaningful parameters and we must use the overall selectivity $\tilde{S}_{XT}$ and the overall yield $\tilde{Y}_{XT}$, which are based on molar flow rates. The yield of xylene from mesitylene based on molar flow rates exiting the CSTR for $\tau = 0.5$ is

Overall selectivity, $\tilde{S}$, and yield, $\tilde{Y}$.

$$\tilde{Y}_{MX} = \frac{F_X}{F_{M0} - F_M} = \frac{C_X}{C_{M0} - C_M} = \frac{0.00313}{0.0105 - 0.0029}$$

$$\tilde{Y}_{MX} = \frac{0.41 \text{ mol xylene produced}}{\text{mole mesitylene reacted}}$$

To vary τ_{CSTR} one can vary either v_0 for a fixed V or vary V for a fixed v_0

The overall selectivity of xylene relative to toluene is

$$\tilde{S}_{XT} = \frac{F_X}{F_T} = \frac{C_X}{C_T} = \frac{C_X}{C_{M0} - C_M - C_X} = \frac{0.00313}{0.0105 - 0.0029 - 0.00313}$$

$$\tilde{S}_{XT} = \frac{0.7 \text{ mol xylene produced}}{\text{mole toluene produced}}$$

In the two preceding examples there was no volume change with reaction; consequently, we could use concentration as our dependent variable. We now consider a gas-phase reaction with volume change taking place in a PFR. Under these conditions, we must use the molar flow rates as our dependent variables.

***Example 6–8 Calculating Concentrations as Functions
of Position for NH₃ Oxidation in a PFR***

The following gas-phase reactions take place simultaneously on a metal oxide–supported catalyst:

1. $4NH_3 + 5O_2 \longrightarrow 4NO + 6H_2O$

2. $2NH_3 + 1.5O_2 \longrightarrow N_2 + 3H_2O$

3. $2NO + O_2 \longrightarrow 2NO_2$

4. $4NH_3 + 6NO \longrightarrow 5N_2 + 6H_2O$

Writing these equations in terms of symbols yields

Reaction 1: $4A + 5B \longrightarrow 4C + 6D$ $-r_{1A} = k_{1A}C_A C_B^2$ (E6-8.1)

Reaction 2: $2A + 1.5B \longrightarrow E + 3D$ $-r_{2A} = k_{2A}C_A C_B$ (E6-8.2)

Reaction 3: $2C + B \longrightarrow 2F$ $-r_{3B} = k_{3B}C_C^2 C_B$ (E6-8.3)

Reaction 4: $4A + 6C \longrightarrow 5E + 6D$ $-r_{4C} = k_{4C}C_C C_A^{2/3}$ (E6-8.4)

with[7] $k_{1A} = 5.0 \text{ (m}^3/\text{kmol)}^2/\text{min}$ $k_{2C} = 2.0 \text{ m}^3/\text{kmol} \cdot \text{min}$

$k_{3B} = 10.0 \text{ (m}^3/\text{kmol)}^2/\text{min}$ $k_{4C} = 5.0 \text{ (m}^3/\text{kmol)}^{2/3}/\text{min}$

Note: We have converted the specific reaction rates to a per unit volume basis by multiplying the k' on a per mass of catalyst basis by the bulk density of the packed bed.

Determine the concentrations as a function of position (i.e., volume) in a PFR.

Additional information: Feed rate = 10 dm³/min; volume of reactor = 10 dm³; and

$$C_{A0} = C_{B0} = 1.0 \text{ mol/dm}^3$$

---

[7] Reaction orders and rate constants were estimated from periscosity measurements for a bulk catalyst density of 1.2 kg/m³.

Solution

First, we divide each equation through by the stoichiometric coefficient of the species for which the rate law is given:

$$1: \quad A + 1.25B \longrightarrow C + 1.5D \qquad -r_{1A} = k_{1A}C_A C_B^2 \qquad (E6\text{-}8.5)$$

$$2: \quad A + 0.75B \longrightarrow 0.5E + 1.5D \qquad -r_{2A} = k_{2A}C_A C_B \qquad (E6\text{-}8.6)$$

$$3: \quad B + 2C \longrightarrow 2F \qquad -r_{3B} = k_{3B}C_C^2 C_B \qquad (E6\text{-}8.7)$$

$$4: \quad C + \tfrac{2}{3}A \longrightarrow \tfrac{5}{6}E + D \qquad -r_{4C} = k_{4C}C_C C_A^{2/3} \qquad (E6\text{-}8.8)$$

Stoichiometry. We will express the concentrations in terms of the molar flow rates:

$$C_j = \frac{F_j}{F_T} C_{T0}$$

and then substitute for the concentration of each reaction species in the rate laws. Writing the rate law for species A in reaction 1 in terms of the rate of formation, r_{1A}, and molar flow rates, F_A and F_B we obtain

$$r_{1A} = -k_1 C_A C_B^2 = -k_{1A}\left(C_{T0}\frac{F_A}{F_T}\right)\left(C_{T0}\frac{F_B}{F_T}\right)^2$$

Thus

$$r_{1A} = -k_{1A}C_{T0}^3 \frac{F_A F_B^2}{F_T^3} \qquad (E6\text{-}8.9)$$

Similarly for the other reactions,

$$r_{2A} = -k_{2A}C_{T0}^2 \frac{F_A F_B}{F_T^2} \qquad (E6\text{-}8.10)$$

$$r_{3B} = -k_{3B}C_{T0}^3 \frac{F_C^2 F_B}{F_T^3} \qquad (E6\text{-}8.11)$$

$$r_{4C} = -k_{4C}C_{T0}^{5/3} \frac{F_C F_A^{2/3}}{F_T^{5/3}} \qquad (E6\text{-}8.12)$$

Next, we determine the *net* rate of reaction for each species by using the appropriate stoichiometric coefficients and then summing the rates of the individual reactions.

Net rates of formation:

$$\text{Species A:} \qquad r_A = r_{1A} + r_{2A} + \frac{2}{3}r_{4C} \qquad (E6\text{-}8.13)$$

$$\text{Species B:} \qquad r_B = 1.25\,r_{1A} + 0.75\,r_{2A} + r_{3B} \qquad \text{(E6-8.14)}$$

$$\text{Species C:} \qquad r_C = -r_{1A} + 2\,r_{3B} + r_{4C} \qquad \text{(E6-8.15)}$$

$$\text{Species D:} \qquad r_D = -1.5\,r_{1A} - 1.5\,r_{2A} - r_{4C} \qquad \text{(E6-8.16)}$$

$$\text{Species E:} \qquad r_E = -\frac{r_{2A}}{2} - \frac{5}{6}r_{4C} \qquad \text{(E6-8.17)}$$

$$\text{Species F:} \qquad r_F = -2r_{3B} = 2k_{3B}C_C^2 C_B \qquad \text{(E6-8.18)}$$

Finally, we write mole balances on each species.

Mole balances:

$$\text{Species A:} \qquad \frac{dF_A}{dV} = r_A = r_{1A} + r_{2A} + \frac{2}{3}r_{4C} \qquad \text{(E6-8.19)}$$

$$\text{Species B:} \qquad \frac{dF_B}{dV} = r_B = 1.25\,r_{1A} + 0.75\,r_{2A} + r_{3B} \qquad \text{(E6-8.20)}$$

$$\text{Species C:} \qquad \frac{dF_C}{dV} = r_C = -r_{1A} + 2r_{3B} + r_{4C} \qquad \text{(E6-8.21)}$$

Solutions to these equations are most easily obtained with an ODE solver

$$\text{Species D:} \qquad \frac{dF_D}{dV} = r_D = -1.5\,r_{1A} - 1.5\,r_{2A} - r_{4C} \qquad \text{(E6-8.22)}$$

$$\text{Species E:} \qquad \frac{dF_E}{dV} = r_E = -\frac{r_{2A}}{2} - \frac{5}{6}r_{4C} \qquad \text{(E6-8.23)}$$

$$\text{Species F:} \qquad \frac{dF_F}{dV} = r_F = -2r_{3B} \qquad \text{(E6-8.24)}$$

$$\text{Total:} \qquad F_T = F_A + F_B + F_C + F_D + F_E + F_F \qquad \text{(E6-8.25)}$$

Combining

Rather than combining the concentrations, rate laws, and mole balances to write everything in terms of the molar flow rate as we did in the past, it is more convenient here to write our computer solution (either POLYMATH or our own program) using equations for r_{1A}, F_A, and so on. Consequently, we shall write Equations (E6-8.9) through (E6-8.12) and (E6-8.19) through (E6-8.25) as individual lines and let the computer combine them to obtain a solution.

The corresponding POLYMATH program written for this problem is shown in Table E6-8.1 and a plot of the output is shown in Figure E6-8.1. One notes that there is a maximum in the concentration of NO (i.e., C) at approximately 1.5 dm³.

TABLE E6-8.1.   POLYMATH PROGRAM

Equations	Initial Values
d(fb)/d(v)=1.25*r1a+.75*r2a+r3b	10
d(fa)/d(v)=r1a+r2a+2*r4c/3	10
d(fc)/d(v)=-r1a+2*r3b+r4c	0
d(fd)/d(v)=-1.5*r1a-1.5*r2a-r4c	0
d(fe)/d(v)=.5*r2a-5*r4c/6	0
d(ff)/d(v)=-2*r3b	0
ft=fa+fb+fc+fd+fe+ff	
r1a=-5*8*(fa/ft)*(fb/ft)**2	
r2a=-2*4*(fa/ft)*(fb/ft)	
r4c=-5*3.175*(fc/ft)*(fa/ft)**(2/3)	
r3b=-10*8*(fc/ft)**2*(fb/ft)	
ca=2*fa/ft	
$v_0 = 0,$ $v_f = 10$	

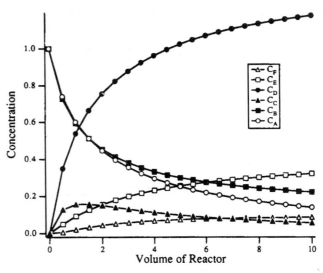

Figure E6-8.1 Concentration profiles.

However, there is one fly in the ointment here: It may not be possible to determine the rate laws for each of the reactions. In this case it may be necessary to work with the minimum number of reactions and hope that a rate law can be found for each reaction. That is, you need to find the number of linearly independent reactions in your reaction set. In Example 6-8 just discussed, there are four reactions given [(E6-8.5) through (E6-8.8)]. However, only three of these reactions are independent, as the fourth can be formed from a linear combination of the other three. Techniques for determining the number of independent reactions are given by Aris.[8]

---

[8] R. Aris, *Elementary Chemical Reactor Analysis* (Upper Saddle River, N.J.: Prentice Hall, 1969).

6.4 Sorting It All Out

In Example 6-8 we were given the rate laws and asked to calculate the product distribution. The inverse of the problem described in Example 6-8 must frequently be solved. Specifically, the rate laws often must be determined from the variation in the product distribution generated by changing the feed concentrations. In some instances this determination may not be possible without carrying out independent experiments on some of the reactions in the sequence. The best strategy to use to *sort out* all of the rate law parameters will vary from reaction sequence to reaction sequence. Consequently, the strategy developed for one system may not be the best approach for other multiple-reaction systems. One general rule is to start an analysis by looking for species produced in only one reaction; next, study the species involved in only two reactions, then three, and so on.

Nonlinear least-squares

When the intermediate products (e.g., species C) are free radicals, it may not be possible to perform independent experiments to determine the rate law parameters. Consequently, we must deduce the rate law parameters from changes in the distribution of reaction products with feed conditions. Under these circumstances, the analysis turns into an optimization problem to estimate the best values of the parameters that will minimize the sums of the squares between the calculated variables and measured variables. This process is basically the same as that described in Section 5.4.2, but more complex, owing to the larger number of parameters to be determined. We begin by estimating the 12 parameter values using some of the methods just discussed. Next, we use our estimates to use nonlinear regression techniques to determine the best estimates of our parameter values from the data for all of the experiments.[9] Software packages such as SimuSolv[10] are becoming available for an analysis such as this one.

6.5 The Fun Part

I'm not talking about fun you can have at an amusement park, but CRE fun. Now that we have an understanding on how to solve for the exit concentrations of multiple reactions in a CSTR and how to plot the species concentration down the length of a PFR or PBR, we can address one of the most important and fun areas of chemical reaction engineering. This area, discussed in Section 6.1, is learning how to maximize the desired product and minimize the undesired product. It is this area that can make or break a chemical process financially. It is also an area that requires creativity in designing the reactor schemes and feed conditions that will maximize profits. Here you can mix and match reactors, feed steams, and side streams as well as vary the ratios of feed concentration in order to maximize or minimize the selectivity of a particular species. Problems of this type are what I call *digital-age problems*[11] because

---

[9] See, for example, Y. Bard, *Nonlinear Parameter Estimation*, (Academic Press, San Diego, Calif.: 1974).

[10] The SimuSolv Computer Program is a proprietary product of The Dow Chemical Company that is leased with restricted rights according to license terms and conditions. SimuSolv is a trademark of The Dow Chemical Company.

[11] H. Scott Fogler, *Teaching Critical Thinking, Creative Thinking, and Problem Solving in the Digital Age*, Phillips Lecture (Stillwater, Okla.: OSU Press, 1997).

we normally need to use ODE solvers along with critical and creative thinking skills to find the best answer. A number of problems at the end of this chapter will allow you to practice these critical and creative thinking skills. These problems offer opportunity to explore many different solution alternatives to enhance selectivity and have fun doing it.

However, to carry CRE to the next level and to have a lot more fun solving multiple reaction problems, we will have to be patient a little longer. The reason is that in this chapter we consider only isothermal multiple reactions, and it is nonisothermal multiple reactions where things really get interesting. Consequently, we will have to wait to carry out schemes to maximize the desired product in nonisothermal multiple reactions until we study heat effects in Chapters 8 and 9. After studying these chapters we will add a new dimension to multiple reactions, as we now have another variable, temperature, that we may or may not be able to use to affect selectivity and yield. One particularly interesting **problem (P8-30)** we will study is the production of styrene from ethylbenzene in which two side reactions, one endothermic, and one exothermic, must be taken into account. Here we may vary a whole slew of variables, such as entering temperature, diluent rate, and observe optima, in the production of styrene. However, we will have to delay gratification of the styrene study until we have mastered Chapter 8.

6.6 The Attainable Region

A technique developed by Professors Glasser and Hildebrandt[12] allows one to find the optimum reaction system for certain types of rate laws. The WWW[12] uses modified van de Vusse kinetics, that is,

$$A \underset{k_2}{\overset{k_1}{\rightleftharpoons}} B \xrightarrow{k_3} C$$

$$2A \xrightarrow{k_4} D$$

to illustrate what combination of reactors PFR/CSTR should be used to obtain the maximum amount of B. The combined mole balance and rate laws for these liquid phase reactions can be written in terms of space-time as

van de Vusse kinetics

$$\frac{dC_A}{d\tau} = -k_1 C_A + k_2 C_B - k_4 C_A^2$$

$$\frac{dC_B}{d\tau} = k_1 C_A - k_2 C_B - k_3 C_B$$

$$\frac{dC_C}{d\tau} = k_3 C_B$$

PFR

$$\frac{dC_D}{d\tau} = \frac{k_4}{2} C_A^2$$

---

[12]Department of Chemical Engineering, Witswatersrand University, Johannesburg, South Africa. See also D. Glasser, D. Hildebrandt, and C. Crowe, *IEC Res., 26,* 1803 (1987). http://www.engin.umich.edu/~cre/Chapters/ARpages/Intro/intro.htm **and** http://sunsite.wits.ac.za/wits/fac/engineering/procmat/ARHomepage/frame.htm

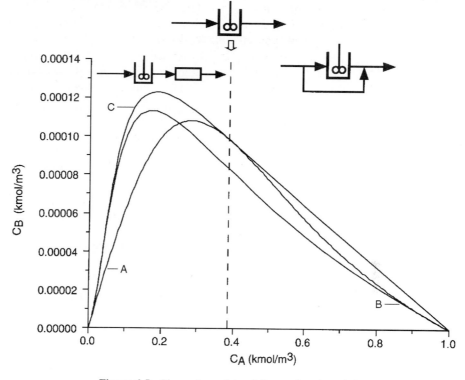

A: CSTR
B: PFR
C: CSTR & PFR

Figure 6-5   Phase plane plots of C_B as a function of C_A.

One can solve this set of ODEs to obtain the plot of C_B as a function of C_A shown in Figure 6-5.

In a similar fashion one can solve the combined CSTR mole balances and rate laws, that is,

CSTR

$$C_{A0} - C_A = \tau[k_1 C_A - k_2 C_B + k_4 C_A^2]$$
$$C_B = \tau[k_1 C_A - k_2 C_B - k_3 C_B]$$

These equations can be solved to give C_A and C_B as a function of space time and also C_B as a function of C_A. The latter is shown as the dashed line in Figure 6-5. The values of the specific reaction rates or $k_1 = 0.01\ \mathrm{s}^{-1}$, $k_2 = 5\ \mathrm{s}^{-1}$,

$k_3 = 10\ \mathrm{s}^{-1}$, $k_4 = 100\ \dfrac{\mathrm{m}^3}{\mathrm{kmol} \cdot \mathrm{s}}$

The WWW[12] shows how to use these plots along with the attainable region technique to maximize the amount of B produced.

From Figure 6-5, we see that if the space-time is such that the effluent concentration of A is between 0.38 and 1.0 kmol/m², a CSTR with by-pass will give us the maximum concentration of B. If the effluent concentration A is exactly 0.38, then a single CSTR is the best choice. Finally, if the total space-time ($\tau = \tau_{\mathrm{PFR}} + \tau_{\mathrm{CSTR}}$) is such that the effluent concentration is below 0.38 kmol/m³, then a CSTR followed by a PFR will give the maximum amount of B.

SUMMARY

For the competing reactions

$$\text{Reaction 1:} \qquad A + B \xrightarrow{\ k_D\ } D \qquad\qquad \text{(S6-1)}$$

$$\text{Reaction 2:} \qquad A + B \xrightarrow{\ k_U\ } U \qquad\qquad \text{(S6-2)}$$

1. Typical rate expressions are

$$r_D = A_D e^{-E_D/RT} C_A^{\alpha 1} C_B^{\beta 1} \qquad\qquad \text{(S6-3)}$$

$$r_U = A_U e^{-E_U/RT} C_A^{\alpha 2} C_B^{\beta 2} \qquad\qquad \text{(S6-4)}$$

and the instantaneous selectivity parameter is defined as

$$S_{DU} = \frac{r_D}{r_U} = \frac{A_D}{A_U} \exp\left(-\frac{(E_D - E_U)}{RT}\right) C_A^{\alpha 1 - \alpha 2} C_B^{\beta 1 - \beta 2} \qquad \text{(S6-5)}$$

a. If $E_D > E_U$, the selectivity parameter S_{DU} will increase with increasing temperature.

b. If $\alpha_1 > \alpha_2$ and $\beta_2 > \beta_1$, the reaction should be carried out at high concentrations of A and low concentrations of B to maintain the selectivity parameter S_{DU} at a high value. Use a semibatch reactor with pure A initially or a tubular reactor in which B is fed at different locations down the reactor. Other cases discussed in the text are $\alpha_1 > \alpha_2$, $\beta_2 > \beta_1$ and $\alpha_1 > \alpha_2$, $\beta_1 > \beta_2$.

2. The *instantaneous yield* at a point is defined as the ratio of the rate of formation of a specified product D to the rate of depletion of the key reactant A:

$$Y_D = \frac{r_D}{-r_A} = \frac{r_D}{-r_{A1} - r_{A2}} \qquad\qquad \text{(S6-6)}$$

The overall yield is the ratio of the number of moles of a product at the end of a reaction to the number of moles of the key reactant that have been consumed:

$$\tilde{Y}_D = \frac{N_D}{N_{A0} - N_A} \qquad\qquad \text{(S6-7)}$$

For a flow system, this yield is

$$\tilde{Y}_D = \frac{F_D}{F_{A0} - F_A} \qquad\qquad \text{(S6-8)}$$

The overall selectivity, based on molar flow rates leaving the reactor, for the reactions given by Equations (S6-1) and (S6-2) is

$$\tilde{S}_{DU} = \frac{F_D}{F_U} \tag{S6-9}$$

3. The algorithm:

Mole balances:

$$\frac{dF_j}{dV} = r_j \qquad \text{PFR} \tag{S6-10}$$

$$F_{j0} - F_j = -r_j V \qquad \text{CSTR} \tag{S6-11}$$

$$\frac{dN_j}{dt} = r_j V \qquad \text{Batch} \tag{S6-12}$$

$$\frac{dC_j}{dt} = r_j - \frac{v_0(C_{j0} - C_j)}{V} \qquad \text{Liquid-semibatch} \tag{S6-13}$$

Rate laws:

$$r_{ij} = k_{ij} f_i(C_1, C_j, C_n) \tag{S6-14}$$

Stoichiometry:

$$F_T = \sum_{j=1}^{n} F_j \tag{S6-15}$$

$$r_j = \sum_{i=1}^{q} r_{ij} \tag{S6-16}$$

$$\frac{r_{iA}}{-a_i} = \frac{r_{iB}}{-b_i} = \frac{r_{iC}}{c_i} = \frac{r_{iD}}{d_i} \tag{S6-17}$$

Gas phase:

$$C_j = C_{T0} \frac{F_j}{F_T} \frac{P}{P_0} \frac{T_0}{T} \tag{S6-18}$$

$$\frac{dP}{dW} = -\frac{\alpha}{2} \left(\frac{T}{T_0}\right) \frac{P_0}{P/P_0} \left(\frac{F_T}{F_{T0}}\right) \tag{4-28}$$

Let $y = P/P_0$:

$$\frac{dy}{dW} = -\frac{\alpha}{2y} \left(\frac{F_T}{F_{T0}}\right) \frac{T}{T_0} \tag{S6-19}$$

O.D.E. SOLVER ALGORITHM

MULTIPLE ELEMENTARY REACTIONS IN A PFR

$$HCHO + \frac{1}{2}O_2 \xrightarrow{k_1} HCOOH \xrightarrow{k_3} CO + H_2O$$

$$2HCHO \xrightarrow{k_2} HCOOCH_3$$

$$HCOOCH_3 + H_2O \xrightarrow{k_4} CH_3OH + HCOOH$$

Let A = HCHO, B = O_2, C = HCOOH, D = $HCOOCH_3$, E = CO, W = H_2O, G = CH_3OH

$$\frac{dF_A}{dV} = -k_1 C_{T0}^{3/2}\left(\frac{F_A}{F_T}\right)\left(\frac{F_B}{F_T}\right)^{1/2} - k_2 C_{T0}^2\left(\frac{F_A}{F_T}\right)^2$$

$$\frac{dF_B}{dV} = -\frac{k_1}{2} C_{T0}^{3/2}\left(\frac{F_A}{F_T}\right)\left(\frac{F_B}{F_T}\right)^{1/2}$$

$$\frac{dF_C}{dV} = k_1 C_{T0}^{3/2}\left(\frac{F_A}{F_T}\right)\left(\frac{F_B}{F_T}\right)^{1/2} - k_3 C_{T0}\left(\frac{F_C}{F_T}\right) + k_4 C_{T0}^2\left(\frac{F_W}{F_T}\right)\left(\frac{F_D}{F_T}\right)$$

$$\frac{dF_D}{dV} = \frac{k_2}{2} C_{T0}^2\left(\frac{F_A}{F_T}\right)^2 - k_4 C_{T0}^2\left(\frac{F_D}{F_T}\right)\left(\frac{F_W}{F_T}\right)$$

$$\frac{dF_E}{dV} = k_3 C_{T0}\left(\frac{F_C}{F_T}\right)$$

$$\frac{dF_W}{dV} = k_3 C_{T0}\left(\frac{F_C}{F_T}\right) - k_4 C_{T0}^2\left(\frac{F_W}{F_T}\right)\left(\frac{F_D}{F_T}\right)$$

$$\frac{dF_G}{dV} = k_4 C_{T0}^2\left(\frac{F_W}{F_T}\right)\left(\frac{F_D}{F_T}\right)$$

$$F_T = F_A + F_B + F_C + F_D + F_E + F_W + F_G$$

$F_{A0} = 10$, $F_{B0} = 5$, $V_F = 1000$, $k_1 C_{T0}^{3/2} = 0.04$, $k_2 C_{T0}^2 = 0.007$, $k_3 C_{T0} = 0.014$, $k_4 C_{T0}^2 = 0.45$

QUESTIONS AND PROBLEMS

The subscript to each of the problem numbers indicates the level of difficulty: A, least difficult; D, most difficult.

$$A = \bullet \quad B = \blacksquare \quad C = \blacklozenge \quad D = \blacklozenge\blacklozenge$$

In each of the questions and problems below, rather than just drawing a box around your answer, write a sentence or two describing how you solved the problem, the assumptions you made, the reasonableness of your answer, what you learned, and any other facts that you want to include. You may wish to refer to W. Strunk and E. B. White, *The Elements of Style* (New York: Macmillan, 1979), and Joseph M. Williams, *Style: Ten Lessons in Clarity & Grace* (Glenview, Ill.: Scott, Foresman, 1989) to enhance the quality of your sentences.

P6-1   Make up and solve an original problem to illustrate the principles of this chapter. See Problem P4-1 for guidelines.

P6-2   **(a)** **What if** you could vary the temperature in Example 6-1? What temperature and CSTR space-time would you use to maximize Y_D for an entering concentration of A of 1 mol/dm^3? What temperature would you choose?

    (b) **What if** the reactions in Example 6-3 were carried out in a CSTR instead of a PBR? What would be the corresponding optimum conversion and space-time, τ'_{opt}?

    (c) **What if** you varied the catalyst size in Example 6-6? Would it increase or decrease $\tilde{S}_{XT}$? [*Hint:* Use a form of Equation (S6-18) and make a plot of $\tilde{S}_{XT}$ versus α.]

    (d) **What** feed conditions and reactors or combination of reactors shown in Figure 6-3 would you use to

        (1) maximize the overall selectivity $\tilde{S}_{CF}$ in Example 6-8? Start by plotting $\tilde{S}_{CF}$ as a function of Θ_B in a 10-dm^3 PFR;

        (2) maximize the yield of Y_{FA} with and the overall selectivity of $\tilde{S}_{FC}$?

    (e) **What if** you could vary the ratio of hydrogen to mesitylene in the feed $(0.2 < \Theta_H < 5)$ in Example 6-6. What is the effect of Θ_H on τ_{opt}? Plot the optimum yield of xylene as a function of Θ_H. Plot the selectivity $\tilde{S}_{XT}$ as a function of Θ_H. Suppose that the reactions could be run at different temperatures. What would be the effect of the ratio of k_1 to k_2 on τ_{opt} and on the selectivity $\tilde{S}_{XT}$ and the yield?

    (f) Repeat part (d) for Example 6-7.

    (g) How would pressure drop affect the results shown in Figure E6-8.1?

P6-3$_B$ The hydrogenation of *o*-cresol 2-methylphenol (MP) is carried out over a Ni-catalyst [*Ind. Eng. Chem. Res., 28*, 693 (1989)] to form 2-methylcyclohexanone (ON), which then reacts to form two stereoisomers, *cis-* (cs-OL) and *trans* (tr-OL)-2-methylcyclohexanol. The equilibrium compositions (on a hydrogen-free basis) are shown in Figure P6-3.

    (a) Plot (sketch) the selectivities of ON to cs-OL and of tr-OL to cs-OL as a function of temperature in the range 100 to 300°C.

    (b) Estimate the heat of reaction for the isomerization

$$\text{tr-OL} \rightleftharpoons \text{cs-OL}$$

    (c) Plot (sketch) the yields of MP to tr-OL and of MP to ON as functions of temperature assuming a stoichiometric feed.

P6-4$_A$ **(a)** What reaction schemes and conditions would you use to maximize the selectivity parameters S for the following parallel reactions:

$$A + C \longrightarrow D \quad r_D = 800e^{(-2000/T)}C_A^{0.5}C_C$$

$$A + C \longrightarrow U_1 \quad r_U = 10e^{(-300/T)}C_AC_C$$

where D is the desired product and U$_1$ is the undesired product?

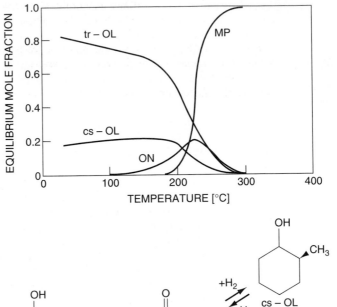

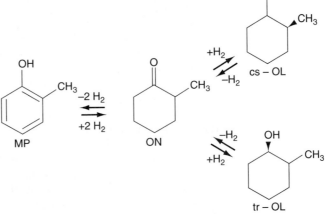

Figure P6-3 [Reprinted with permission from W. K. Schumann, O. K. Kut, and A. Baiker, [*Ind. Eng. Chem. Res., 28,* 693 (1989). Copyright © (1989) American Chemical Society.]

(b) State how your answer to part (a) would change if C were to react with D to form another undesired product,

$$D + C \longrightarrow U_2 \qquad r_{U_2} = 10^6 e^{(-8000/T)} C_C C_D$$

At what temperature should the reactor be operated **if** the concentrations of A and D in the reactor were 1 mol/dm³?

(c) For a 2-dm³ laboratory CSTR with $C_{C0} = C_{A0} = 1$ mol/dm³ and $v_0 = 1$ dm³/min, what temperature would you recommend to maximize $\tilde{Y}_D$?

(d) Two gas-phase reactions are occurring in a plug-flow tubular reactor, which is operated isothermally at a temperature of 440°F and a pressure of 5 atm. The first reaction is first-order:

$$A \longrightarrow B \qquad -r_A = k_1 C_A \qquad k_1 = 10 \text{ s}^{-1}$$

and the second reaction is zero-order:

$$C \longrightarrow D + E \qquad -r_C = k_2 = 0.03 \text{ lb mol/ft}^3 \cdot \text{s}$$

The feed, which is equimolar in A and C, enters at a flow rate of 10 lb mol/s. What reactor volume is required for a 50% conversion of A to B? (*Ans.: V* = 105 ft^3.)

P6-5$_B$ A mixture of 50% A, 50% B is charged to a constant-volume batch reactor in which equilibrium is rapidly achieved. The initial total concentration is 3.0 mol/dm^3.

(a) Calculate the equilibrium concentrations and conversion of A at 330 K for the reaction sequence

Reaction 1: A + B $\rightleftharpoons$ C + D K_{c1} (330 K) = 4.0, K_{c1} (350 K) = 2.63

Reaction 2: C + B $\rightleftharpoons$ X + Y K_{c2} (330 K) = 1.0, K_{c2} (350 K) = 1.51

(b) Suppose now the temperature is increased to 350 K. As a result, a third reaction must now be considered in addition to the reactions above:

Reaction 3: A + X $\rightleftharpoons$ Z K_{c3} (350 K) = 5.0 dm^3/mol

Calculate the equilibrium concentrations, conversion of A and overall selectivities $\tilde{S}_{CX}$, $\tilde{S}_{DZ}$, and $\tilde{S}_{YZ}$.

(c) Vary the temperature over the range 300 to 500 K to learn the effect of selectivities $\tilde{S}_{CX}$, $\tilde{S}_{DZ}$, and $\tilde{S}_{YZ}$ on temperature.

Additional information:

$$\Delta H_{R1} = -20{,}000 \text{ J/mol A}   \quad \Delta H_{R2} = +20{,}000 \text{ J/mol B}$$

$$\Delta H_{R3} = -40{,}000 \text{ J/mol A}$$

P6-6$_A$ Consider the following system of gas-phase reactions:

$$A \longrightarrow X \quad r_X = k_1 \quad k_1 = 0.002 \text{ mol/dm}^3 \cdot \text{min}$$

$$A \longrightarrow B \quad r_B = k_2 C_A \quad k_2 = 0.06 \text{ min}^{-1}$$

$$A \longrightarrow Y \quad r_Y = k_3 C_A^2 \quad k_3 = 0.3 \text{ dm}^3/\text{mol} \cdot \text{min}$$

B is the desired product, and X and Y are foul pollutants that are expensive to get rid of. The specific reaction rates are at 27°C. The reaction system is to be operated at 27°C and 4 atm. Pure A enters the system at a volumetric flow rate of 10 dm^3/min.

(a) Sketch the instantaneous selectivities (S_{BX}, S_{BY}, and $S_{B/XY} = r_B/(r_X + r_Y)$) as a function of the concentration of C_A.

(b) Consider a series of reactors. What should be the volume of the first reactor?

(c) What are the effluent concentrations of A, B, X, and Y from the first reactor.

(d) What is the conversion of A in the first reactor?

(e) If 90% conversion of A is desired, what reaction scheme and reactor sizes should you use?

(f) Suppose that E_1 = 10,000 cal/mol, E_2 = 20,000 cal/mol, and E_3 = 30,000 cal/mol. What temperature would you recommend for a single CSTR with a space-time of 10 min and an entering concentration of A of 0.1 mol/dm^3?

P6-7$_B$ Pharmacokinetics concerns the ingestion, distribution, reaction, and elimination reaction of drugs in the body. Consider the application of pharmacokinetics to one of the major problems we have in the United States, drinking and driving. Here we shall model how long one must wait to drive after having a tall martini. In most states the legal intoxication limit is 1.0 g of ethanol per liter of body fluid. (In Sweden it is 0.5 g/L, and in Eastern Europe and Russia it is any value above 0.0 g/L.) The ingestion of ethanol into the bloodstream and subsequent elimination can be modeled as a series reaction. The rate of absorption from the gastrointestinal tract into the bloodstream and body is a first-order reaction with a specific reaction rate constant of 10 h^{-1}. The rate at which ethanol is broken down in the bloodstream is limited by regeneration of a coenzyme. Consequently, the process may be modeled as a zero-order reaction with a specific reaction rate of 0.192 g/h·L of body fluid. How long would a person have to wait (a) in the United States; (b) in Sweden; and (c) in Russia if they drank two tall martinis immediately after arriving at a party? How would your answer change if (d) the drinks were taken $\frac{1}{2}$ h apart; (e) the two drinks were consumed at a uniform rate during the first hour? (f) Suppose that one went to a party, had one and a half tall martinis right away, and then received a phone call saying an emergency had come up and they needed to drive home immediately. How many minutes would they have to reach home before he/she became legally intoxicated, assuming that the person had nothing further to drink? (g) How would your answers be different for a thin person? A heavy person? For each case make a plot of concentration as a function of time. (*Hint:* Base all ethanol concentrations on the volume of body fluid. Plot the concentration of ethanol in the blood as a function of time.)

Additional information:

Ethanol in a tall martini: 40 g

Volume of body fluid: 40 L   (**SADD-MADD problem**)

P6-8$_B$ (*Pharmacokinetics*) Tarzlon is a liquid antibiotic that is taken orally to treat infections of the spleen. It is effective only if it can maintain a concentration in the blood-stream (based on volume of body fluid) above 0.4 mg per dm^3 of body fluid. Ideally, a concentration of 1.0 mg/dm^3 in the blood would like to be realized. However, if the concentration in the blood exceeds 1.5 mg/dm^3, harmful side effects can occur. Once the Tarzlon reaches the stomach it can proceed in two pathways, both of which are first order: (1) It can be absorbed into the bloodstream through the stomach walls; (2) it can pass out through the gastrointestinal tract and not be adsorbed into the blood. Both these processes are first order in Tarzlon concentration in the stomach. Once in the bloodstream, Tarzlon attacks bacterial cells and is subsequently degraded by a zero-order process. Tarzlon can also be removed from the blood and excreted in urine through a first-order process within the kidneys. In the stomach:

Adsorption into blood $k_1 = 0.15$ h^{-1}

Elimination through gastrointestine $k_2 = 0.6$ h^{-1}

In the bloodstream:

Degradation of Tarzlon $k_3 = 0.1$ mg/dm^3·h

Elimination through urine $k_4 = 0.2$ h^{-1}

(a) Plot the concentration of Tarzlon in the blood as a function of time when 1 dose (i.e. one liquid capsule) of Tarzlon is taken.

(b) How should the Tarzlon be administered (dosage and frequency) over a 48-h period to be most effective?

(c) Comment on the dose concentrations and potential hazards.

(d) How would your answers change if the drug were taken on a full or empty stomach?

One dose of Tarzlon is 250 mg. in liquid form: Volume of body fluid = 40 dm^3

P6-9$_B$ The elementary liquid-phase-series reaction

$$A \xrightarrow{k_1} B \xrightarrow{k_2} C$$

is carried out in a 500-dm³ batch reactor. The initial concentration of A is 1.6 mol/dm³. The desired product is B and separation of the undesired product C is very difficult and costly. Because the reaction is carried out at a relatively high temperature, the reaction is easily quenched.

Additional information:

Cost of pure reactant A = \$10/mol A
Selling price of pure B = \$50/mol B
Separation cost of A from B = \$50/mol A
Separation cost of C from B = \$30 $(e^{0.5C_C} - 1)$
      $k_1 = 0.4$ h^{-1}
      $k_2 = 0.01$ h^{-1} at 100°C

(a) Assuming that each reaction is irreversible, plot the concentrations of A, B, and C as a function of time.
(b) Calculate the time the reaction should be quenched to achieve the maximum profit.
(c) For a CSTR space-time of 0.5 h, what temperature would you recommend to maximize B? ($E_1 = 10,000$ cal/mol, $E_2 = 20,000$ cal/mol)
(d) Assume that the first reaction is reversible with $k_{-1} = 0.3$ h^{-1}. Plot the concentrations of A, B, and C as a function of time.
(e) Plot the concentrations of A, B, and C as a function of time for the case where both reactions are reversible with $k_{-2} = 0.005$ h^{-1}.
(f) Vary k_1, k_2, k_{-1}, and k_{-2}. Explain the consequence of $k_1 > 100$ and $k_2 < 0.1$ with $k_{-1} = k_{-2} = 0$ and with $k_{-2} = 1$, $k_{-1} = 0$, and $k_{-2} = 0.25$.
(g) Reconsider part (a) for reactions are carried out in a packed-bed reactor with 100 kg of catalyst for which $k_1 = 0.25$ dm³/kg cat.·min and $k_2 = 0.15$ dm³/kg cat.·min. The flow is turbulent with $v_0 = 10$ dm³ and $C_{A0} = 1$ mol/dm³. It has been suggested to vary the particle size, keeping $W = 100$ kg, in order to increase F_B, $\tilde{Y}_B$, and $\tilde{S}_{BC}$. If the particle size could be varied between 2 and 0.1 cm, what particle size would you choose? The pressure drop parameter for particles 1 cm in diameter is $\alpha = 0.00098$ kg^{-1}. Is there a better way to improve the selectivity?

P6-10$_B$ You are designing a plug-flow reactor for the following gas-phase reaction:

$$A \longrightarrow B \qquad r_B = k_1 C_A^2 \quad k_1 = 15 \text{ ft}^3/\text{lb mol}\cdot\text{s}$$

Unfortunately, there is also a side reaction:

$$A \longrightarrow C \qquad r_C = k_2 C_A \quad k_2 = 0.015 \text{ s}^{-1}$$

C is a pollutant and costs money to dispose of; B is the desired product.
(a) What size of reactor will provide an effluent stream at the maximum dollar value? B has a value of \$60/lb mol; it costs \$15 per lb mol to dispose of C. A has a value of \$10 per lb mol. (*Ans.: V = 896* ft³.)
(b) Suppose that $E_1 = 10,000$ Btu/lb mol and $E_2 = 20,000$ Btu/lb mol. What temperature (400 to 700°F) would you recommend for a 400-ft³ CSTR?

Additional information:

        Feed: 22.5 SCF/s of pure A
        Reaction conditions: 460°F, 3 atm pressure
        Volumetric flow rate: 15 ft³/s at reaction conditions
        Concentration of A in feed: 4.47×10^{-3} lb mol/ft³

P6-11$_B$ The following liquid-phase reactions were carried out in a CSTR at 325 K.

$$3A \longrightarrow B + C \qquad -r_{1A} = k_{1A}C_A \qquad k_{1A} = 0.7 \text{min}^{-1}$$

$$2C + A \longrightarrow 3D \qquad r_{2D} = k_{2D}C_C^2 C_A \qquad k_{2D} = 0.3 \, \frac{\text{dm}^6}{\text{mol}^2 \cdot \text{min}}$$

$$4D + 3C \longrightarrow 3E \qquad r_{3E} = k_{3E}C_D C_C \qquad k_{3E} = 0.2 \, \frac{\text{dm}^3}{\text{mol} \cdot \text{min}}$$

The concentrations measured *inside* the reactor were $C_A = 0.10$, $C_B = 0.93$, $C_C = 0.51$, and $C_D = 0.049$ all in mol/dm³.

(a) What are r_{1A}, r_{2A}, and r_{3A}? ($r_{1A} = -0.07$ mol/dm³·min)
(b) What are r_{1B}, r_{2B}, and r_{3B}?
(c) What are r_{1C}, r_{2C}, and r_{3C}? ($r_{1C} = 0.023$ mol/dm³·min)
(d) What are r_{1D}, r_{2D}, and r_{3D}?
(e) What are r_{1E}, r_{2E}, and r_{3E}?
(f) What are the net rates of formation of A, B, C, D, and E?
(g) The entering volumetric flow rate is 100 dm³/min and the entering concentration of A is 3 *M*. What is the CSTR reactor volume? (*Ans.:* 4000 dm³.)

P6-12 Calculating the space-time for parallel reactions. *m*-Xylene is reacted over a ZMS-5 zeolite catalyst. The following *parallel* elementary reactions were found to occur [*Ind. End. Chem. Res., 27,* 942 (1988)]:

$$m\text{-xylene} \xrightarrow{k_1} \text{benzene} + \text{methane}$$

$$m\text{-xylene} \xrightarrow{k_2} p\text{-xylene}$$

(a) Calculate the space-time to achieve 90% conversion of *m*-xylene in a packed-bed reactor. Plot the overall selectivity and yields as a function of τ. The specific reaction rates are $k_1 = 0.22$ s^{-1} and $k_2 = 0.71$ s^{-1} at 673°C. A mixture of 75% *m*-xylene and 25% inerts is fed to a tubular reactor at volumetric flow rate of 2000 dm³/min and a total concentration of 0.05 mol/dm³. As a first approximation, neglect any other reactions such as the reverse reactions and isomerization to *o*-xylene.
(b) Suppose that $E_1 = 20{,}000$ cal/mol and $E_2 = 10{,}000$ cal/mol, what temperature would you recommend to maximize the formation of *p*-xylene in a CSTR with a space-time of 0.5 s?

P6-13$_B$ The following liquid phase reactions are carried out isothermally in a 50 dm³ PFR:

$$A + 2B \longrightarrow C + D \qquad r_{D1} = k_{D1}C_A C_B^2$$

$$2D + 3A \longrightarrow C + E \qquad r_{E2} = k_{E2}C_A C_D$$

$$B + 2C \longrightarrow D + F \qquad r_{F3} = k_{F3}C_B C_C^2$$

Additional information:

$$k_{D1} = 0.25 \ \text{dm}^6/\text{mol}^2 \cdot \text{min} \qquad v_0 = 10 \ \text{dm}^3/\text{min}$$

$$k_{E2} = 0.1 \ \text{dm}^3/\text{mol} \cdot \text{min} \qquad C_{A0} = 1.5 \ \text{mol/dm}^3$$

$$k_{F3} = 5.0 \ \text{dm}^6/\text{mol}^2 \cdot \text{min} \qquad C_{B0} = 2.0 \ \text{mol/dm}^3$$

(a) Plot the species concentrations and the conversion of A as a function of the distance (i.e., volume) down a 50-dm³ PFR. Note any maxima.

(b) Determine the effluent concentrations and conversion from a 50-dm³ CSTR. (*Ans.:* $C_A = 0.61$, $C_B = 0.79$, $C_F = 0.25$, and $C_D = 0.45 \ \text{mol/dm}^3$)

(c) Plot the species concentrations and the conversion of A as a function of time when the reaction is carried out in a semibatch reactor initially containing 40 dm³ of liquid. Consider two cases: (1) A is fed to B, and (2) B is fed to A. What differences do you observe for these two cases?

(d) Vary the ratio of B to A ($1 < \Theta_B < 10$) in the feed to the PFR and describe what you find.

P6-14$_B$ The production of maleic anhydride by the air oxidation of benzene was recently studied using a vanadium pentoxide catalyst [*Chem. Eng. Sci., 43,* 1051 (1988)]. The reactions that occur are:

Reaction 1: $\qquad C_6H_6 + \dfrac{9}{2}O_2 \longrightarrow C_4H_2O_3 + 2CO_2 + 2H_2O$

Reaction 2: $\quad C_4H_2O_3 + 3O_2 \longrightarrow 4CO_2 + H_2O$

Reaction 3: $\qquad C_6H_6 + \dfrac{15}{2}O_2 \longrightarrow 6CO_2 + 3H_2O$

Because these reactions were carried out in excess air, volume change with reaction can be neglected, and the reactions can be written symbolically as a pseudo-first-order reaction sequence

$$A \xrightarrow{\ k_1\ } B \xrightarrow{\ k_2\ } C$$
$$\searrow{\scriptstyle k_3}$$
$$D$$

where A = benzene, B = maleic anhydride, C = products (H_2O, CO_2), D = products (CO_2, H_2O). The corresponding pseudo specific reaction rates, k_i, are (in all $\text{m}^3/\text{kg cat.} \cdot \text{s}$):

$$k_1 = 4280 \ \exp[-12{,}660/T(\text{K})] \qquad k_2 = 70{,}100 \ \exp[-15{,}000/T(\text{K})]$$

$$k_3 = 26 \ \exp[-10{,}800/T(\text{K})]$$

At 848 K, $k_1 = 1.4 \times 10^{-3}$, $k_2 = 1.46 \times 10^{-3}$, $k_3 = 7.65 \times 10^{-5}$. These reactions are carried out isothermally in both a CSTR and a PBR. Benzene enters the reactor at a concentration of 0.01 mol/dm³. The total volumetric flow rate is 0.0025 m³/s.

(a) Which reactions will dominate at low temperatures and which will dominate at high temperatures? For the sake of comparison, assume that 848 K is a moderate temperature.

(b) For a catalytic weight of 50 kg, determine the exit concentrations from a "fluidized" CSTR at 848 K. (*Ans.:* $C_B = 0.3$ mol/dm^3)

(c) What is the selectivity of B to C and of B to D in the CSTR?

(d) Plot the concentrations of all species as a function of PBR catalyst weight (up to 10 kg) assuming isothermal operation at 848 K.

(e) What feed conditions and reactor or combinations of reactors shown in Figure 6-3 would you use to maximize the production of maleic anhydride?

(f) How would your results in part (d) change if pressure drop were taken into account with $\alpha = 0.099$ kg cat.$^{-1}$ in PBR? Make a plot similar to that in part (d) and describe any differences.

P6-15$_B$ **(a)** Rework Examples 6-6 and 6-7 for the case where the toluene formed in reaction (E6-6.2) can also undergo hydrodealkylation to yield benzene, B:

$$H + T \longrightarrow B + Me \qquad (P6\text{-}15.1)$$

The rate law is

$$r_B = k_3 C_T C_H^{1/2} \qquad (P6\text{-}15.2)$$

with $k_3 = 11.2\,(\text{ft}^3/\text{lb mol})^{0.5}/\text{h}$.

Include the concentrations of toluene and benzene in your results. The feed conditions given for Examples 6-6 and 6-7 apply to the case where all three reactions are taking place.

(b) Vary the ratio of hydrogen to mesitylene ($1 < \Theta_H < 20$) and describe what you find.

(c) What reactor schemes and feed conditions do you suggest to maximize $\tilde{S}_{XT}$? $\tilde{S}_{TB}$?

(d) Suppose that the activation energies were $E_1 = 20,000$ cal/mol, $E_2 = 10,000$ cal/mol, and $E_3 = 30,000$ cal/mol. What temperature would you recommend for the CSTR size in Example 6-7?

P6-16$_C$ The following hydrodealkylation reactions occur over a Houdry Detol catalyst near 800 K and 3500 kPa:

(1) $\quad H_2 + C_6H(CH_3)_5 \longrightarrow C_6H_2(CH_3)_4 + CH_4 \qquad r_{C11} = k_1 C_{H_2}^{1/2} C_{11}$

(2) $\quad H_2 + C_6H_2(CH_3)_4 \longrightarrow C_6H_3(CH_3)_3 + CH_4 \qquad r_{C10} = k_2 C_{H_2}^{1/2} C_{10}$

(3) $\quad H_2 + C_6H_3(CH_3)_3 \longrightarrow C_6H_4(CH_3)_2 + CH_4 \qquad r_{C9} = k_3 C_{H_2}^{1/2} C_9$

(4) $\quad H_2 + C_6H_4(CH_3)_2 \longrightarrow C_6H_5(CH_3) + CH_4 \qquad r_{C8} = k_4 C_{H_2}^{1/2} C_8$

(5) $\quad H_2 + C_6H_5CH_3 \longrightarrow C_6H_6 + CH_4 \qquad r_{C7} = k_5 C_{H_2}^{1/2} C_7$

$$k_5 = 2.1\,(\text{kmol/m}^3)^{-1/2}/\text{s}$$

$$\frac{k_1}{k_5} = 17.6 \qquad \frac{k_2}{k_5} = 10 \qquad \frac{k_3}{k_5} = 4.4 \qquad \frac{k_4}{k_5} = 2.7$$

The feed is equimolar in hydrogen and pentamethylbenzene.

(a) For an entering volumetric flow rate of 1 m^3/s, what ratio of hydrogen to pentamethylbenzene and what PFR reactor volume would you recommend to maximize the formation of $C_6H_4(CH_3)_2$ (i.e. C_8)? [Hint: Plot the overall selectivity as a function of reactor volume.]

(b) How would your answer change if you were to maximize the overall selectivity to C_8 to C_9, i.e. $\tilde{S}_{89}$? To C_8 to C_7?

(c) What do you think the point of this problem to be?

Make a plot of the mole fraction of each component as a function of conversion of pentamethylbenzene. Make a plot of the mole fraction of each component as a function of plug-flow reactor volume. Discuss any optimization that could be done.

P6-17$_B$ Review the oxidation of formaldehyde to formic acid reactions over a vandium titanium oxide catalyst [*Ind. Eng. Chem. Res., 28*, 387 (1989)] shown in the ODE solver algorithm in the Summary.

(a) Plot the species concentrations as a function of distance down the PFR for an entering flow rate of 100 dm^3/min at 5 atm and 140°C. The feed is 66.7% HCHO and 33.3% O_2. Note any maximum in species concentrations.

(b) Plot the yield of overall HCOOH yield and overall selectivity of HCOH to CO, of HCOOCH$_3$ to CH$_3$OH and of HCOOH to HCOOCH$_3$ as a function of the Θ_{O_2}. Suggest some conditions to best produce formic acid. Write a paragraph describing what you find.

(c) Compare your plot in part (a) with a similar plot when pressure drop is taken into account with $\alpha = 0.002$ dm^{-3}.

(d) Suppose that $E_1 = 10{,}000$ cal/mol, $E_2 = 30{,}000$ cal/mol, $E_3 = 20{,}000$ cal/mol, and $E_4 = 10{,}000$ cal/mol, what temperature would you recommend for a 1000-dm^3 PFR?

P6-18$_B$ The liquefaction of Kentucky Coal No. 9 was carried out in a slurry reactor [D. D. Gertenbach, R. M. Baldwin, and R. L. Bain, *Ind. Eng. Chem. Process Des. Dev., 21*, 490 (1982)]. The coal particles, which were less than 200 mesh, were dissolved in a ~250°C vacuum cut of recycle oil saturated with hydrogen at 400°C. Consider the reaction sequence

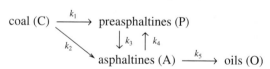

which is a modified version of the one given by Gertenbach et al. All reactions are first order. Calculate the molar flow rate of each species as a function of space-time in

(a) A plug-flow reactor.

(b) A 3-m^3 CSTR.

Additional information:

Entering concentration of coal: 2 kmol/m^3
Entering flow rate: 10 dm^3/min

At 400°F, $k_1 = 0.12$ min^{-1}, $k_2 = 0.046$ min^{-1}, $k_3 = 0.020$ min^{-1}, $k_4 = 0.034$ min^{-1}, $k_5 = 0.04$ min^{-1}.

P6-19$_B$ A liquid feed to a well-mixed reactor consists of 0.4 g mol/dm^3 of A and the same molar concentration of F. The product C is formed from A by two different reaction mechanisms: either by direct transformation or through intermediate B. The intermediate is also formed from F. Together with C, which remains in solution, an insoluble gas D is formed, which separates in the reactor.

All reaction steps are irreversible and first-order, except for the formation of B from F, which is second-order in F. The liquid carrier for reactants and products is an inert solvent, and no volume change results from the reaction:

$$k_1 = 0.01 \text{ min}^{-1} \qquad k_2 = 0.02 \text{ min}^{-1}$$
$$k_3 = 0.07 \text{ min}^{-1} \qquad k_4 = 0.50 \text{ dm}^3/\text{g mol} \cdot \text{min}$$

$$\text{reactor volume} = 120 \text{ L}$$

(a) What is the maximum possible molar concentration of C in the product?
(b) If the feed rate is 2.0 dm³/min, what is the yield of C (expressed as a percentage of the maximum), and what is the mole fraction of C in the product on a solvent-free basis? (*Ans.:* 0.583.)

(California Professional Engineers Exam)

P6-20$_C$ The reaction sequence discussed in Example 6-8 is to be carried out in a microscale "fluidized" CSTR. The CSTR catalyst weight is 3 g. All of the conditions in the problem statement are the same except that the reaction is to be carried out at a slightly higher temperature and the rate constants are:

$$k_1 = 10 \text{ m}^9/(\text{kmol}^2 \cdot \text{kg cat.} \cdot \text{min}) \qquad k_2 = 4 \text{ m}^6/(\text{kmol} \cdot \text{kg cat.} \cdot \text{min})$$
$$k_3 = 15 \text{ m}^9/(\text{kmol}^2 \cdot \text{kg cat.} \cdot \text{min}) \qquad k_4 = 10 \text{ m}^3(\text{m}^3/\text{kmol})^{2/3}/\text{kg cat.} \cdot \text{min}$$

(a) What are the exiting molar flow rates of each species.
(b) What is the overall yield of
    (1) A to E?
    (2) B to F?
    (3) A to C?

P6-21$_D$ *(Flame retardants)* We now reconsider a more comprehensive version of the combustion of CO discussed in P7-3. The reactions and their corresponding rate law parameters are given in Table P6-23. All reactions are assumed to be elementary [*Combustion and Flame*, **69**, 113 (1987)]. The precombustion compositions (mol %) are:

TABLE P6-21   RATE LAW PARAMETERS

Reaction	Rate Parameters, $k = AT^n \exp(-E/RT)$ (cm³, s, cal, mol units)		
	A	n	E
A. Uninhibited			
1. $OH + CO \rightarrow CO_2 + H$	4.40E06	1.5	-740
2. $CO + O + M \rightarrow CO_2 + M$	5.30E13	0	-4,538
3. $H + O_2 \rightarrow O + OH$	1.20E17	-0.91	16,504
4. $O + H_2 \rightarrow H + OH$	1.50E07	2.0	7,547
B. Inhibited			
5. $Cl_2 + M \rightarrow 2Cl + M$	1.00E14	0	48,180
6. $HCl + H \rightarrow H_2 + Cl$	7.94E12	0	3,400
7. $H + Cl_2 \rightarrow HCl + Cl$	8.51E13	0	1,170

A. Without HCl:

$$CO: 20\%, \quad O_2: 10\% \quad H_2: 1\%, \quad N_2: 69\%$$

B. With HCl:

$$CO: 20\%, \quad O_2: 10\%, \quad H_2: 1\%, \quad N_2: 67\%, \quad HCl: 2\%$$

The combustion takes place at 1000 K and 1 atm and the initial hydrogen radical mole fraction is 10^{-7}. Treat the flame as a batch reactor.

(a) Plot the mole fraction of each species and free radicals as a function of time for both uninhibited and inhibited combustions.

(b) Repeat part (a) for different temperatures and pressures.

(S. Senkan, UCLA)

P6-22$_B$ The production of acetylene is described by R. W. Wansbaugh [*Chem. Eng.*, *92*(16), 95 (1985)]. Using the reaction and data in this article, develop a problem and solution.

P6-23$_B$ The hydrogenation of benzene (B) is carried out in a CSTR slurry reactor where the desired product is cyclohexene (C) [*Chem. Eng. Sci.*, *51*, 2873 (1996)].

The rate laws for reactions (1) and (2) at 403 K are

$$-r_{1B} = \frac{k_1 K_1 P_{C_6H_6} P_{H_2}^{1/2}}{1 + K_1 P_{C_6H_6}}$$

$$K_1 = 0.0264 \text{ kPa}^{-1}$$

$$k_1 = 2.7 \frac{\text{mol}}{\text{h} \cdot \text{kg} \cdot \text{cat.} \cdot \text{kPa}^{0.5}}$$

$$-r_{2C} = \frac{k_2 K_2 P_{C_6H_{10}} P_{H_2}}{1 + K_2 P_{C_6H_{10}}}$$

$$K_2 = 0.04 \text{ kPa}^{-1}$$

$$k_2 = 0.07 \frac{\text{mol}}{\text{h} \cdot \text{kg} \cdot \text{cat.} \cdot \text{kPa}}$$

with

$$130 \text{ kPa} < P_{H_2} < 2 \text{ MPa}$$

$$30 \text{ kPa} < P_{C_6H_6} < 450 \text{ MPa}$$

The entering molar flow rate of benzene is 2 mol/s and it is necessary to achieve at least 40% conversion of benzene at 403 K.

At what conditions should the reaction be carried out to maximize the yield of cyclohexene?

P6-24$_C$ *(Methanol synthesis)* A new catalyst has been proposed for the synthesis of methanol from carbon monoxide and hydrogen gas. This catalyst is reasonably active between temperatures of 330 K to about 430 K. The isothermal reactions involved in the synthesis include

$$CO + 2H_2 \rightleftharpoons CH_3OH$$

$$CO + H_2O \rightleftharpoons CO_2 + H_2$$

$$CH_3OH \longrightarrow CH_2O + H_2$$

The reactions are elementary and take place in the gas phase. The reaction is to be carried out isothermally and as a first approximating pressure drop will be neglected. The feed consists of $\frac{7}{15}$ hydrogen gas, $\frac{1}{5}$ carbon monoxide, $\frac{1}{5}$ carbon dioxide, and $\frac{2}{15}$ steam. The total molar flow rate is 300 mol/s. The entering pressure may be varied between 1 atm and 160 atm and the entering temperature between 300 K and 400 K. Tubular (PFR) reactor volumes between 0.1 m³ and 2 m³ are available for use.

(a) Determine the entering conditions of temperature and pressure and reactor volume that will optimize the production of methanol. (Hint: First try $T_o = 330$ at $P_o = 40$ atm, then try $T_o = 380$ $P_o = 1$ atm.)

(b) Vary the ratios of the entering reactants to CO (i.e., Θ_{H_2} and Θ_{H_2O}) to maximize methanol production. How do your results compare with those in part (a)? Describe what you find.

Additional information:

$$V = 40 \text{ dm}^3, \quad T \text{ is in kelvin.} \qquad R = 1.987 \text{ cal/mol}$$

$$K_1 = \frac{131{,}667}{\exp\left[\dfrac{-30{,}620}{R} \cdot \left(\dfrac{1}{T} - \dfrac{1}{298}\right)\right]} (0.001987T)^2 \left(\frac{\text{dm}^3}{\text{mol}}\right)^2$$

$$K_2 = \frac{103{,}943}{\exp\left[\dfrac{-9834}{R} \cdot \left(\dfrac{1}{T} - \dfrac{1}{298}\right)\right]}$$

$$k_1 = 0.933 \exp\left[2.5 \cdot \left(\frac{31{,}400}{R} \cdot \left(\frac{1}{330} - \frac{1}{T}\right)\right)\right] \left(\frac{\text{dm}^3}{\text{mol}}\right)^2 \text{s}^{-1}$$

$$k_2 = 0.636 \exp\left[\frac{18{,}000}{R} \cdot \left(\frac{1}{300} - \frac{1}{T}\right)\right] \frac{\text{dm}^3}{\text{mol} \cdot \text{s}}$$

$$k_3 = 0.244 \exp\left[1.5 \cdot \left(\frac{28{,}956}{R} \cdot \left(\frac{1}{325} - \frac{1}{T}\right)\right)\right] \text{s}^{-1}$$

P6-25$_C$ Olealic acid epoxide (E) is produced by the catalytic epoxidation of olealic acid (OA) [J. Fotopoulos, C. Georgakis, and H. Stenger, *Chem. Eng. Sci., 51*, 1899 (1996)]. The raw materials are pure benzaldehyde (B) and oleic acid (OA). Unfortunately, undesired products are also formed, including benzoic (BA) and perbenzoic acids (PBA). The reaction sequence is

$$(1) \quad O_2 + B \xrightarrow{k_1} PBA$$

$$(2) \quad PBA + OA \xrightarrow{\ k_2\ } E + BA$$

$$(3) \quad B + PBA \xrightarrow{\ k_3\ } 2BA$$

Additional information:

Concentration of pure B = 9.8 kmol/m³
Concentration of pure OA = 3.15 kmol/m³
Catalyst concentration = 4×10^{-5} kmol/m³

$$r_1 = \frac{k_1[\text{cat}]^{0.5}[B]^{1.7}}{(1 + K_1[OA])^{0.7}}$$

$k_1 = 1.745 \left(\dfrac{\text{dm}^3}{\text{mol}}\right)^{1.2}$ min⁻¹ at 47°C with $E_1 = 18{,}500$ cal/mol and

$K_1 = 64.03$ m³/kmol

$$r_2 = k_2[PBA]^{0.5}[OA]^{0.6}$$ $k_2 = 0.178 \left(\dfrac{\text{dm}^3}{\text{mol}}\right)^{0.1}$ min⁻¹ at 47°C with $E_2 = 9177$ cal/mol

$$r_3 = \frac{k_3[PBA]^{1.0}}{(1 + K_3[OA])^{1.0}}$$

$k_3 = 0.08$ min⁻¹ at 47°C with $E_3 = 17{,}132$ cal/mol and

$K_3 = 8.064$ m³/kmol

How should the reaction be carried out (e.g. type of reactor(s), volume, temperature, feed rate) to produce 100 kmol of oleic acid expoxide (E) per day? There will be a prize for the solution that meets this criteria and minimizes the undesirable products. If equipment costs are available they should be included.

Additional information:

Concentration of pure oleic acid = 3.16 mol/dm³ = 3.16 kmol/m³
Concentration of pure benzaldehide = 9.8 mol/dm³ = 9.8 kmol/m³
Typical catalyze concentration = 4×10^{-5} kmol/m³

P6-26$_C$ A new catalytic pathway for an important intermediate in the production of the nonsteroidal anti-inflammatory drug ibuprofen (e.g., Advil) has been developed [*Chem. Eng. Sci.*, *51*, 10, 1663 (1996)]. The pathway involves the hydrogenation of p-isobutyl acetophenone (B) in a solution of methanol containing a HY zeolite catalyst and saturated with hydrogen. The intermediate products are *p*-isobutylphenyl ithanol (C), *p*-isobutylphenylethylmethyl ether (E) and *p*-isobutylethyl benzene (F). The reaction scheme is shown below. The following rate laws apply to the above equation.

$$(1) \quad A + B \longrightarrow C$$

$$(2) \quad C + A \longrightarrow F + H_2O$$

$$(3) \quad C + M \longrightarrow E + H_2O$$

$$(4) \quad E + A \longrightarrow F + M$$

$$-r_B = r_B = r_1 = \frac{wk_1 A^* B}{1 + k_A A^*} = \frac{wk_{1B} C_A C_B}{1 + K_A C_A} \tag{1}$$

$$-r_{2C} = r_2 = \frac{wk_2 A^* C}{1 + K_A A^*} = \frac{wk_{2C} C_A C_C}{1 + K_A C_A} \tag{2}$$

$$-r_{2C} = r_3 = wk_3C = wk_{2C}C_C \tag{3}$$

$$-r_{4E} = -r_4 = \frac{wk_4A*E}{1 + k_A A*} = \frac{wk_{4E}C_A C_E}{1 + K_A C_A} \tag{4}$$

Because methanol, M, is in excess, it does not appear in the rate law. The rate and equilibrium constants are given in Table P6-26. The catalyst charge, w, is 10 kg/m3 and the initial concentration of p-isobutyl acetephenone, (B) is 0.54 kmol/dm3, the partial pressure of hydrogen is 5.9 MPa, and the temperature is 393 K. Plot the concentrations of A, B, C, D, E, and F as a function of time.

TABLE P6-26.   RATE AND EQUILIBRIUM CONSTANTS

Temperature (K)	Rate parameters, $[(m^3/kg)(m^3/kmol \cdot s)] = \dfrac{m^6}{kg \cdot kmol \cdot s}$					Henry's Law Constant $\dfrac{kmol}{m^3 \cdot MPa}$
	$k_1 \times 10^4$	$k_2 \times 10^4$	$k_3 \times 10^4$	$k_4 \times 10^4$	k_A (m³/kmol)	
373	3.2	1.5	9	7	26	0.055
393	4.68	2.27	28.2	14.7	22.76	0.058
413	8.5	4	95.2	30.4	18.03	0.061

*k_3 is in (m³/kg·s)

P6-27$_C$ For the van de Vusse elementary reactions

$$A \underset{k_2}{\overset{k_1}{\rightleftharpoons}} B \overset{k_3}{\longrightarrow} C$$

$$2A \overset{k_4}{\longrightarrow} D$$

determine the reactor or combination of reactors that maximize the amount of B formed.

Additional information:

$$k_1 = 0.01 \text{ s}^{-1} \qquad k_2 = 0.05 \text{ s}^{-1} \qquad k_3 = 10 \text{ s}^{-1} \qquad k_4 = 100 \text{ m}^3/\text{kmol}$$

$$C_{A0} = 2 \text{ kmol/m}^3 \text{ and } v_0 = 0.2 \text{ m}^3/\text{s}$$

Repeat for $k_2 = 0.002 \text{ s}^{-1}$.

P6-28$_B$ The gas phase reactions take place isothermally in a membrane reactor packed with catalyst. Pure A enters the reactor a 24.6 atm and 500K and a flow rate of A of 10 mol/min

$$A \rightleftharpoons B + C \qquad r'_{1C} = k_{1C}\left[C_A - \frac{C_B C_C}{K_{1C}}\right]$$

$$A \longrightarrow D \qquad r'_{2D} = k_{2D}C_A$$

$$2C + D \longrightarrow 2E \qquad r'_{3E} = k_{3E}C^2_C C_D$$

Only species B diffuses out of the reactor through the membrane.
(a) Plot the concentrations down the length of the reactor.
(b) Explain why your curves look the way they do.
(c) Vary some of the parameters (e.g., k_B, k_{1C}, K_{1C}) and write a paragraph describing what you find.

Additional Information
Overall mass transfer coefficient $k_B = 1.0 \text{ dm}^3 / \text{kg cat} \cdot \text{min}$
    $k_{1C} = 2 \text{ dm}^3 / \text{kg cat} \cdot \text{min}$
    $K_{1C} = 0.2 \text{ mol} / \text{dm}^3$
    $k_{2D} = 0.4 \text{ dm}^3 / \text{kg cat} \cdot \text{min}$
    $k_{3E} = 5 \text{ dm}^3 / \text{mol}^2 \cdot \text{kg cat} \cdot \text{min}$
    $W_f = 100 \text{ kg}$

JOURNAL CRITIQUE PROBLEMS

P6C-1 Is it possible to extrapolate the curves on Figure 2 [*AIChE J.*, *17*, 856 (1971)] to obtain the initial rate of reaction? Use the Wiesz–Prater criterion to determine if there are any diffusion limitations in this reaction. Determine the partial pressure of the products on the surface based on a selectivity for ethylene oxide ranging between 51 and 65% with conversions between 2.3 and 3.5%.

P6C-2 Equation 5 [*Chem. Eng. Sci.*, *35*, 619 (1980)] is written to express the formation rate of C (olefins). As described in equation 2, there is no change in the concentration of C in the third reaction of the series:

$$\left.\begin{array}{l} A \xrightarrow{k_1} B \\ A + B \xrightarrow{k_2} C \\ B + C \xrightarrow{k_3} C \\ C \xrightarrow{k_4} D \end{array}\right\} \text{ equation 2}$$

(a) Determine if the rate law given in equation 5 is correct.
(b) Can equations 8, 9, and 12 be derived from equation 5?
(c) Is equation 14 correct?
(d) Are the adsorption coefficients b_i and b_j calculated correctly?

CD-ROM MATERIAL

- **Learning Resources**
  1. *Summary Notes for Lectures 11, 12, and 13*
  2. *Web Modules*
     A. Cobra Bites
  4. *Solved Problems*
     A. CDP6-B$_B$ All You Wanted to Know About Making Malic
        Anhydride and More.
  5. *Clarification: PFR with feed streams along the length of the reactor.*
- **Living Example Problems**
  1. *Example 6–6 Hydrodealkylation of Mesitylene in a PFR*
  2. *Example 6–7 Hydrodealkylation of Mesitylene in a CSTR*
  3. *Example 6–8 Calculating Concentrations as a Function of Position*
     for NH$_3$ Oxidation in a PFR
- **FAQ [Frequently Asked Questions]**– In Updates/FAQ icon section
- **Additional Homework Problems**

CDP6-A$_B$ Suggest a reaction system and conditions to minimize X and Y for the
parallel reactions A $\longrightarrow$ X, A $\longrightarrow$ B, and A $\longrightarrow$ Y.
[2nd Ed. P9-5].

CDP6-B$_B$ Rework maleic anhydride problem, P6-14, for the case when reaction
1 is second order. [2nd Ed. P9-8]

CDP6-C$_B$ The reaction sequence A $\longrightarrow$ B, B $\longrightarrow$ C, B $\longrightarrow$ D is
carried out in a batch reactor and in a CSTR. [2nd Ed. P9-12]

CDP6-D$_B$ Isobutylene is oxidized to methacrolum, CO, and CO$_2$. [1st Ed. P9-16]

CDP6-E$_B$ Given a batch reactor with A $\rightleftharpoons$ B $\rightleftharpoons$ D, calculate the
composition after 6.5 h. [1st Ed. P9-11]

CDP6-F$_B$ Chlorination of benzene to monochlorobenzene and dichlorobenzene
in a CSTR. [1st Ed. P9-14]

CDP6-G$_C$ Determine the number of independent reactions in the oxidation of
ammonia. [1st Ed. P9-17]

CDP6-H$_B$ Oxidation of formaldehyde:

$$HCHO + \frac{1}{2} O_2 \xrightarrow{k_1} HCOOH$$

$$2HCHO \xrightarrow{k_2} HCOOCH_3 \quad [\text{2nd Ed. P9-13}_B]$$

CDP6-I$_B$ Continuation of CDP6-H:

$$HCOOH \xrightarrow{k_3} CO_2 + H_2$$

$$HCOOH \xrightarrow{k_4} CO + H_2O \quad [\text{2nd Ed. P9-14}_B]$$

CDP6-J$_B$ Continuation of CDP6-H and I:

$$HCOOCH_3 \xrightarrow{k_4} CH_3OH + HCOOH \quad [\text{2nd Ed. P9-15}_C]$$

CDP6-K$_C$ Design a reactor for the alkylation of benzene with propylene to maximize the selectivity of isopropylbenzene. [*Proc. 2nd Joint China/USA
Chem. Eng. Conf. III, 51, (1997)*].

CDP6-L$_D$ Reactions between paraffins and olefins to form highly branched paraffins are carried out in a slurry reactor to increase the octane number in gasoline. [*Chem. Eng. Sci. 51*, 10, 2053 (1996)].

CDP6-M$_A$ Design a reaction system to maximize the production of alkyl chloride. [1st Ed. P9-19]

CDP6-N$_C$ Design a reaction system to maximize the selectivity of *p*-xylene from methanol and toluene over a HZSM-8 zeolite catalyst. [2nd Ed. P9-17]

CDP6-O$_B$ Rework maleic anhydride problem, P6-14, for the case when reaction 1 is second order. [2nd Ed. P9-8]

CDP6-P$_C$ Oxidation of propylene to acrolein (*Chem. Eng. Sci. 51*, 2189 (1996)).

SUPPLEMENTARY READING

1. Selectivity, reactor schemes, and staging for multiple reactions, together with evaluation of the corresponding design equations, are presented in

> DENBIGH, K. G., and J. C. R. TURNER, *Chemical Reactor Theory*, 2nd ed. Cambridge: Cambridge University Press, 1971, Chap. 6.

> LEVENSPIEL, O., *Chemical Reaction Engineering*, 2nd ed. New York: Wiley, 1972, Chap. 7.

Some example problems on reactor design for multiple reactions are presented in

> HOUGEN, O. A., and K. M. WATSON, *Chemical Process Principles*, Part 3: *Kinetics and Catalysis*. New York: Wiley, 1947, Chap. XVIII.

> SMITH, J. M., *Chemical Engineering Kinetics*, 3rd ed. New York: McGraw-Hill, 1980, Chap. 4.

2. Books that have many analytical solutions for parallel, series, and combination reactions are

> CAPELLOS, C., and B. H. J. BIELSKI, *Kinetic Systems*. New York: Wiley, 1972.

> WALAS, S. M., *Chemical Reaction Engineering Handbook of Solved Problems*. Newark, N.J.: Gordon and Breach, 1995.

3. A brief discussion of a number of pertinent references on parallel and series reactions is given in

> ARIS, R., *Elementary Chemical Reactor Analysis*. Upper Saddle River, N.J.: Prentice Hall, 1969, Chap. 5.

4. An excellent example of the determination of the specific reaction rates, k_i, in multiple reactions is given in

> BOUDART, M., and G. DJEGA-MARIADASSOU, *Kinetics of Heterogeneous Catalytic Reactions*. Princeton, N.J.: Princeton University Press, 1984.

Nonelementary Reaction Kinetics

7

The next best thing to knowing something is knowing where to find it.

Samuel Johnson (1709–1784)

Until now, we have been discussing homogeneous reaction rate laws in which the concentration is raised to some power n, which is an integer. That is, the rate law (i.e., kinetic rate expression) is

$$-r_A = kC_A^n \qquad (7\text{-}1)$$

We said that if $n = 1$, the reaction was first-order with respect to A; if $n = 2$, the reaction was second-order with respect to A; and so on. However, a large number of homogeneous reactions involve the formation and subsequent reaction of an intermediate species. When this is the case it is not uncommon to find a reaction order that is not an integer. For example, the rate law for the decomposition of acetaldehyde,

$$CH_3CHO \longrightarrow CH_4 + CO$$

at approximately 500°C is

$$-r_{CH_3CHO} = kC_{CH_3CHO}^{3/2} \qquad (7\text{-}2)$$

Another common form of the rate law resulting from reactions involving active intermediates is one in which the rate is directly proportional to the reactant concentration and inversely proportional to the sum of a constant and the reactant concentration. An example of this type of kinetic expression is observed for the formation of hydrogen iodide,

$$H_2 + I_2 \longrightarrow 2HI$$

339

The rate law for this reaction is

$$r_{HI} = \frac{k_1 k_3 C_{I_2} C_{H_2}}{k_2 + k_3 C_{H_2}} \tag{7-3}$$

For rate expressions similar or equivalent to those given by Equation (7-3), reaction orders cannot be defined. That is, for rate laws where the denominator is a polynomial function of the species concentrations, reaction orders are described only for limiting values of the reactant and/or product concentrations. Reactions of this type are nonelementary in that there is no direct correspondence between reaction order and stoichiometry.

PSSH, Polymers,
Enzymes, Bacteria   In this chapter we discuss four topics: the pseudo-steady-state hypothesis, polymerization, enzymes, and bioreactors. The pseudo-steady-state hypothesis (PSSH) plays an important role in developing nonelementary rate laws. Consequently, we will first discuss the fundamentals of the PSSH, followed by its use of polymerization reactions and enzymatic reactions. Because enzymes are involved in all living organisms, we close the chapter with a discussion on bioreactions and reactors.

7.1 Fundamentals

Nonelementary rate laws similar to Equations (7-2) and (7-3) come about as a result of the overall reaction taking place by a mechanism consisting of a series of reaction steps. In our analysis, we assume each reaction step in the reaction mechanism to be *elementary*; the reaction orders and stoichiometric coefficients are identical.

To illustrate how rate laws of this type are formed, we shall first consider the gas-phase decomposition of azomethane, AZO, to give ethane and nitrogen:

$$(CH_3)_2N_2 \longrightarrow C_2H_6 + N_2 \tag{7-4}$$

Experimental observations show that the rate law for N_2 is first-order with respect to AZO at pressures greater than 1 atm (relatively high concentrations)

$$r_{N_2} \propto C_{AZO}$$

and second-order at pressures below 50 mmHg (low concentrations):[1]

$$r_{N_2} \propto C_{AZO}^2$$

7.1.1 Active Intermediates

This apparent change in reaction order can be explained by the theory developed by Lindemann.[2] An activated molecule, $[(CH_3)_2N_2]^*$, results from collision or interaction between molecules:

---

[1] H. C. Ramsperger, *J. Am. Chem. Soc.*, **49**, 912 (1927).

[2] F. A. Lindemann, *Trans. Faraday Soc.*, **17**, 598 (1922).

$$(CH_3)_2N_2 + (CH_3)_2N_2 \xrightarrow{k_1} (CH_3)_2N_2 + [(CH_3)_2N_2]^* \qquad (7\text{-}5)$$

This activation can occur when translational kinetic energy is transferred into energy stored in internal degrees of freedom, particularly vibrational degrees of freedom.[3] An unstable molecule (i.e., active intermediate) is not formed solely as a consequence of the molecule moving at a high velocity (high translational kinetic energy). The energy must be absorbed into the chemical bonds where high-amplitude oscillations will lead to bond ruptures, molecular rearrangement, and decomposition. In the absence of photochemical effects or similar phenomena, the transfer of translational energy to vibrational energy to produce an active intermediate can occur only as a consequence of molecular collision or interaction. Other types of active intermediates that can be formed are *free radicals* (one or more unpaired electrons, e.g., H·), ionic intermediates (e.g., carbonium ion), and enzyme-substrate complexes, to mention a few.

Properties of an active intermediate A*

In Lindemann's theory of active intermediates, decomposition of the intermediate does not occur instantaneously after internal activation of the molecule; rather, there is a time lag, although infinitesimally small, during which the species remains activated. For the azomethane reaction, the active intermediate is formed by the reaction

$$(CH_3)_2N_2 + (CH_3)_2N_2 \xrightarrow{k_1} (CH_3)_2N_2 + [(CH_3)_2N_2]^* \qquad (7\text{-}5)$$

Because the reaction is elementary, the rate of formation of the active intermediate in Equation (7-5) is

Nonelementary reaction is seen as a sequence of elementary reactions

$$r_{AZO*(7-5)} = k_1 C_{AZO}^2 \qquad (7\text{-}6)$$

where

$$AZO \equiv [(CH_3)_2N_2]$$

There are two reaction paths that the active intermediate (activated complex) may follow after being formed. In one path the activated molecule may become deactivated through collision with another molecule,

$$[(CH_3)_2N_2]^* + (CH_3)_2N_2 \xrightarrow{k_2} (CH_3)_2N_2 + (CH_3)_2N_2 \qquad (7\text{-}7)$$

with

$$r_{AZO*(7-7)} = -k_2 C_{AZO} C_{AZO*} \qquad (7\text{-}8)$$

This reaction is, of course, just the reverse reaction of that given by Equation (7-5). In the alternative path the active intermediate decomposes spontaneously to form ethane and nitrogen:

$$[(CH_3)_2N_2]^* \xrightarrow{k_3} C_2H_6 + N_2 \qquad (7\text{-}9)$$

$$r_{AZO*(7-9)} = -k_3 C_{AZO*} \qquad (7\text{-}10)$$

---

[3] W. J. Moore, *Physical Chemistry*, 5th ed., Prentice Hall, Upper Saddle River, N.J., 1972.

The overall reaction [Equation (7-4)], for which the rate expression is nonelementary, consists of the sequence of elementary reactions, Equations (7-5), (7-7), and (7-9).

Nitrogen and ethane are only formed in the reaction given by Equation (7-9). Consequently, the net rate of formation of nitrogen is

$$r_{N_2} = k_3 C_{AZO^*} \tag{7-11}$$

Concentration of A*
is difficult to
measure and needs
to be replaced in
the rate law

The concentration of the active intermediate, AZO*, is very difficult to measure, because it is highly reactive and very short-lived ($\sim 10^{-9}$ s). Consequently, evaluation of the reaction rate laws, (7-8), (7-10), and (7-11), in their present forms becomes quite difficult, if not impossible. To overcome this difficulty, we need to express the concentration of the active intermediate, C_{AZO^*}, in terms of the concentration of azomethane, C_{AZO}. As mentioned in Chapter 3, the total or net rate of formation of a particular species involved in many simultaneous reactions is the sum of the rates of formation of each reaction for that species.

We can generalize the rate of formation of species j occurring in n different reactions as

The total rate
of formation
of species j from all
reactions

$$r_j = \sum_{i=1}^{n} r_{ji} \tag{7-12}$$

Because the active intermediate, AZO*, is present in all three reactions in the decomposition mechanism, the net rate of formation of AZO* is the sum of the rates of each of the reaction equations, (7-5), (7-7), and (7-9):

$$
\begin{bmatrix} \text{net rate} \\ \text{of} \\ \text{formation} \\ \text{of AZO*} \end{bmatrix} = \begin{bmatrix} \text{rate of} \\ \text{formation} \\ \text{of AZO* in} \\ \text{Equation (7-5)} \end{bmatrix} + \begin{bmatrix} \text{rate of} \\ \text{formation} \\ \text{of AZO* in} \\ \text{Equation (7-7)} \end{bmatrix} + \begin{bmatrix} \text{rate of} \\ \text{formation} \\ \text{of AZO* in} \\ \text{Equation (7-9)} \end{bmatrix}
$$

$$r_{AZO^*} = r_{AZO^*(7-5)} + r_{AZO^*(7-7)} + r_{AZO^*(7-9)}$$

$$\tag{7-13}$$

By substituting Equations (7-6), (7-8), and (7-10) into Equation (7-13), we obtain

Rate of formation
of active
intermediate

$$r_{AZO^*} = k_1 C_{AZO}^2 - k_2 C_{AZO} C_{AZO^*} - k_3 C_{AZO^*} \tag{7-14}$$

To express C_{AZO^*} in terms of measurable concentrations, we use the pseudo-steady-state hypothesis (PSSH).

7.1.2 Pseudo-Steady-State Hypothesis (PSSH)

In most instances it is not possible to eliminate the concentration of the active intermediate in the differential forms of the mole balance equations to obtain closed-form solutions. However, an approximate solution may be obtained. The active intermediate molecule has a very short lifetime because of its high reactivity (i.e., large specific reaction rates). We shall also consider it to be present only in low concentrations. These two conditions lead to the pseudo-steady-state approximation, in which the rate of formation of the active intermediate is assumed to be equal to its rate of disappearance.[4] As a result, the net rate of formation of the active intermediate, r^*, is zero:

The PSSH assumes that the net rate of formation of A* is zero

$$r^* = 0$$

We found that the rate of formation of the product, nitrogen, was

$$r_{N_2} = k_3 C_{AZO^*} \tag{7-11}$$

and that the rate of formation of AZO* was

$$r_{AZO^*} = k_1 C_{AZO}^2 - k_2 C_{AZO} C_{AZO^*} - k_3 C_{AZO} \tag{7-14}$$

Using the pseudo-steady-state hypothesis (PSSH), Equations (7-11) and (7-14) can be combined to obtain a rate law for N_2 solely in terms of the concentration of azomethane. First we solve for the concentration of the active intermediate AZO* in terms of the concentration of azomethane, AZO. From the PSSH,

$$r_{AZO^*} = 0 \tag{7-15}$$

$$r_{AZO^*} = k_1 C_{AZO}^2 - k_2 C_{AZO} C_{AZO^*} - k_3 C_{AZO^*} = 0 \tag{7-16}$$

we can solve Equation (7-16) for C_{AZO^*} in terms of C_{AZO}:

$$C_{AZO^*} = \frac{k_1 C_{AZO}^2}{k_3 + k_2 C_{AZO}} \tag{7-17}$$

Substituting Equation (7-17) into Equation (7-11) gives

The final form of the rate law

$$r_{N_2} = \frac{k_1 k_3 C_{AZO}^2}{k_3 + k_2 C_{AZO}} \tag{7-18}$$

At low concentrations

$$k_2 C_{AZO} \ll k_3$$

for which case we obtain the following second-order rate law:

$$r_{N_2} = k_1 C_{AZO}^2 \tag{7-19}$$

At high concentrations

$$k_2 C_{AZO} \gg k_3$$

---

[4] For further elaboration on this section, see R. Aris, *Am. Sci., 58*, 419 (1970).

in which case the rate expression follows first-order kinetics,

$$r_{N_2} = \frac{k_1 k_3}{k_2} C_{AZO} = k C_{AZO}$$ (7-20)

In describing reaction orders for this equation one would say the reaction is *apparent first-order* at high azomethane concentrations and *apparent second-order* at low azomethane concentrations.

7.2 Searching for a Mechanism

In many instances the rate data are correlated before a mechanism is found. It is a normal procedure to reduce the additive constant in the denominator to 1. We therefore divide the numerator and denominator of Equation (7-18) by k_3 to obtain

$$r_{N_2} = \frac{k_1 C_{AZO}^2}{1 + k' C_{AZO}}$$ (7-21)

7.2.1 General Considerations

The rules of thumb listed in Table 7-1 may be of some help in the development of a mechanism that is consistent with the experimental rate law. Upon application of Table 7-1 to the azomethane example just discussed, we see from rate equation (7-18) that:

1. The active intermediate, AZO*, collides with azomethane, AZO [Equation (7-7)], resulting in the appearance of the concentration of AZO in the denominator.
2. AZO* decomposes spontaneously [Equation (7-9)], resulting in a constant in the denominator of the rate expression.
3. The appearance of AZO in the numerator suggests that the active intermediate AZO* is formed from AZO. Referring to Equation (7-5), we see that this case is indeed true.

TABLE 7-1.   RULES OF THUMB FOR DEVELOPMENT OF A MECHANISM

---

1. Species having the concentration(s) appearing in the denominator of the rate law probably collide with the active intermediate, e.g.,

$$A + A^* \longrightarrow [\text{collision products}]$$

2. If a constant appears in the denominator, one of the reaction steps is probably the spontaneous decomposition of the active intermediate, e.g.,

$$A^* \longrightarrow [\text{decomposition products}]$$

3. Species having the concentration(s) appearing in the numerator of the rate law probably produce the active intermediate in one of the reaction steps, e.g.,

$$[\text{reactant}] \longrightarrow A^* + [\text{other products}]$$

Example 7–1   The Stern–Volmer Equation

Light is given off when a high-intensity ultrasonic wave is applied to water.[5] This light results from microsize bubbles being formed by the wave and then being compressed by it. During the compression stage of the wave, the contents of the bubble (e.g., water and whatever is dissolved in the water) are compressed adiabatically. This compression gives rise to high temperatures, which generate active intermediates and cause chemical reactions to occur in the bubble. The intensity of the light given off, I, is proportional to the rate of reaction of an activated water molecule that has been formed in the microbubble.

$$H_2O^* \xrightarrow{\ k_3\ } H_2O + h\nu$$

$$\text{intensity} \propto (-r_{H_2O^*}) = k_3 C_{H_2O^*}$$

An order-of-magnitude increase in the intensity of sonoluminescence is observed when either carbon disulfide or carbon tetrachloride is added to the water. The intensity of luminescence, I, for the reaction

$$CS_2^* \xrightarrow{\ k_4\ } CS_2 + h\nu$$

is

$$I \propto (-r_{CS_2^*}) = k_4 C_{CS_2^*}$$

A similar result exists for CCl_4.

However, when an aliphatic alcohol, X, is added to the solution, the intensity decreases with increasing concentration of alcohol. The data are usually reported in terms of a Stern–Volmer plot in which relative intensity is given as a function of alcohol concentration, C_X. (See Figure E7-1.1, where I_0 is the sonoluminescence intensity in the absence of alcohol and I is the sonoluminescence intensity in the presence of alcohol.) Suggest a mechanism consistent with experimental observation.

Stern–Volmer plot

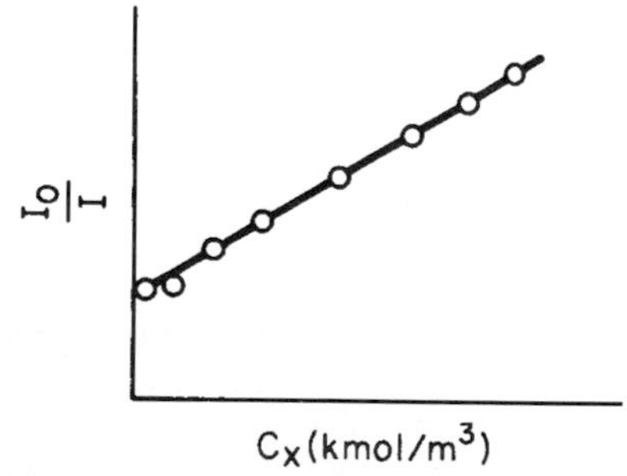

Figure E7-1.1

[5] P. K. Chendke and H. S. Fogler, *J. Phys. Chem.*, 87, 1362 (1983).

Solution

From the linear plot we know that

$$\frac{I_0}{I} = A + BC_X \equiv A + B(X) \tag{E7-1.1}$$

where $C_X \equiv (X)$. Inverting yields

$$\frac{I}{I_0} = \frac{1}{A + B(X)} \tag{E7-1.2}$$

From rule 1 of Table 7-1, the denominator suggests that alcohol collides with the active intermediate:

$$X + \text{intermediate} \longrightarrow \text{deactivation products} \tag{E7-1.3}$$

The alcohol acts as what is called a scavenger to deactivate the active intermediate. The fact that the addition of CCl_4 or CS_2 increases the intensity of the luminescence,

$$I \propto (CS_2) \tag{E7-1.4}$$

leads us to postulate (rule 3 of Table 7-1) that the active intermediate was probably formed from CS_2:

$$M + CS_2 \longrightarrow CS_2^* + M \tag{E7-1.5}$$

where M is a third body (CS_2, H_2O, etc.).

We also know that deactivation can occur by the reverse of Reaction (E7-1.5). Combining this information, we have as our mechanism:

Activation: $M + CS_2 \xrightarrow{k_1} CS_2^* + M$ (E7-1.5)

Deactivation: $M + CS_2^* \xrightarrow{k_2} CS_2 + M$ (E7-1.6)

The mechanism

Deactivation: $X + CS_2^* \xrightarrow{k_3} CS_2 + X$ (E7-1.3)

Luminescence: $CS_2^* \xrightarrow{k_4} CS_2 + h\nu$ (E7-1.7)

$$I = k_4(CS_2^*) \tag{E7-1.8}$$

Using the PSSH on CS_2^* yields

$$r_{CS_2^*} = 0 = k_1(CS_2)(M) - k_2(CS_2^*)(M) - k_3(X)(CS_2^*) - k_4(CS_2^*)$$

Solving for CS_2^* and substituting into Equation (E7-1.8) gives us

$$I = \frac{k_4 k_1 (CS_2)(M)}{k_2 (M) + k_3 (X) + k_4} \tag{E7-1.9}$$

In the absence of alcohol,

$$I_0 = \frac{k_4 k_1 (CS_2)(M)}{k_2 (M) + k_4} \tag{E7-1.10}$$

For constant concentrations of CS_2 and the third body, M, we take a ratio of Equation (E7-1.10) to (E7-1.9):

$$\frac{I_0}{I} = 1 + \frac{k_3}{k_2 (M) + k_4} (X) = 1 + k'(X) \tag{E7-1.11}$$

which is of the same form as that suggested by Figure E7-1.1. Equation (E7-1.11) and similar equations involving scavengers are called *Stern–Volmer equations*.

Now, let us proceed to some slightly more complex examples involving chain reactions. A chain reaction consists of the following sequence:

Steps in a chain
reaction

1. *Initiation:* formation of an active intermediate.
2. *Propagation or chain transfer:* interaction of an active intermediate with the reactant or product to produce another active intermediate.
3. *Termination:* deactivation of the active intermediate.

Example 7–2 PSSH Applied to Thermal Cracking of Ethane

The thermal decomposition of ethane to ethylene, methane, butane, and hydrogen is believed to proceed in the following sequence:

Initiation:

(1) $\quad\quad C_2H_6 \xrightarrow{k_{C_2H_6}} 2CH_3\bullet$ $\qquad\qquad -r_{1C_2H_6} = k_{1C_2H_6} [C_2H_6]$

$\qquad\qquad\qquad\qquad\qquad\qquad\qquad$ Let $k_1 = k_{1C_2H_6}$

Propagation:

(2) $CH_3\bullet + C_2H_6 \xrightarrow{k_2} CH_4 + C_2H_5\bullet$ $\qquad -r_{2C_2H_6} = k_2 [CH_3\bullet][C_2H_6]$

(3) $\quad\quad C_2H_5\bullet \xrightarrow{k_3} C_2H_4 + H\bullet$ $\qquad\qquad r_{3C_2H_4} = k_3 [C_2H_5\bullet]$

(4) $\quad H\bullet + C_2H_6 \xrightarrow{k_4} C_2H_5\bullet + H_2$ $\qquad -r_{4C_2H_6} = k_4 [H\bullet][C_2H_6]$

Termination:

(5) $\quad\quad 2C_2H_5\bullet \xrightarrow{k_5} C_4H_{10}$ $\qquad\qquad -r_{5C_2H_5\bullet} = k_{5C_2H_5} [C_2H_5\bullet]^2$

$\qquad\qquad\qquad\qquad\qquad\qquad\qquad$ Let $k_5 \equiv k_{5C_2H_5\bullet}$

(a) Use the PSSH to derive a rate law for the rate of formation of ethylene.
(b) Compare the PSSH solution in Part (a) to that obtained by solving the complete set of ODE mole balances.

Solution

Part (a) *Developing the Rate Law*

The rate of formation of ethylene is

$$\boxed{r_{3C_2H_4} = k_3 [C_2H_5\bullet]} \tag{E7-2.1}$$

Given the following reaction sequence:

For the active intermediates: $CH_3 \bullet$, $C_2H_5 \bullet$, $H \bullet$ the net rates of reaction are

$$r_{C_2H_5 \bullet} = r_{2C_2H_5 \bullet} + r_{3C_2H_5 \bullet} + r_{4C_2H_5 \bullet} + r_{5C_2H_5 \bullet} = 0$$

$$= -r_{2C_2H_6} - r_{3C_2H_4} - r_{4C_2H_6} + r_{5C_2H_5 \bullet} = 0 \tag{E7-2.2}$$

$$r_{H \bullet} = r_{3C_2H_4} + r_{4C_2H_6} = 0 \tag{E7-2.3}$$

$$r_{CH_3 \bullet} = -2r_{1C_2H_6} + r_{2C_2H_6} = 0 \tag{E7-2.4}$$

Substituting the rate laws into Equation (E7-2.4) gives

$$2k_1[C_2H_6] - k_2[CH_3 \bullet][C_2H_6] = 0 \tag{E7-2.5}$$

$$[CH_3 \bullet] = \frac{2k_1}{k_2} \tag{E7-2.6}$$

Adding Equations (E7-2.2) and (E7-2.3) yields

$$-r_{2C_2H_6} + r_{5C_2H_5 \bullet} = 0$$

$$k_2[CH_3 \bullet][C_2H_6] - k_5[C_2H_5 \bullet]^2 = 0 \tag{E7-2.7}$$

Solving for $[C_2H_5 \bullet]$ gives us

$$[C_2H_5 \bullet] = \left\{ \frac{k_2}{k_5}[CH_3 \bullet][C_2H_6] \right\}^{1/2} = \left\{ \frac{2k_1k_2}{k_2k_5}[C_2H_6] \right\}^{1/2}$$

$$\tag{E7-2.8}$$

$$= \left\{ \frac{2k_1}{k_5}[C_2H_6] \right\}^{1/2}$$

Substituting for $C_2H_5 \bullet$ in Equation (E7-2.1) yields

$$\boxed{r_{C_2H_4} = k_3[C_2H_5 \bullet] = k_3 \left(\frac{2k_1}{k_5} \right)^{1/2} [C_2H_6]^{1/2}} \tag{E7-2.9}$$

$$r_{C_2H_6} = -k_1[C_2H_6] - k_2[CH_3 \bullet][C_2H_6] - k_4[H \bullet][C_2H_6] \tag{E7-2.10}$$

Substituting the rate laws in Equation (E7-2.3), we find that

$$k_3[C_2H_5 \bullet] - k_4[H \bullet][C_2H_6] = 0$$

Using Equation (E7-2.8) to substitute for $C_2H_5 \bullet$ gives

$$[H \bullet] = \frac{k_3}{k_4} \left(\frac{2k_1}{k_5} \right)^{1/2} [C_2H_6]^{-1/2} \tag{E7-2.11}$$

The rate of disappearance of ethane now becomes

$$\boxed{-r_{C_2H_6} = (k_1 + 2k_1)(C_2H_6) + k_3 \left(\frac{2k_1}{k_5} \right)^{1/2} C_2H_6^{1/2}} \tag{E7-2.12}$$

For a constant-volume batch reactor, the combined mole balances and rate laws for disappearance of ethane (P1) and the formation of ethylene (P5) are

$$\frac{dC_{P1}}{dt} = -\left[(3k_1 C_{P1}) + k_3 \left(\frac{2k_1}{k_5} \right)^{1/2} C_{P1}^{1/2} \right] \tag{E7-2.13}$$

$$\frac{dC_{P5}}{dt} = k_3 \left(\frac{2k_1}{k_5} \right)^{1/2} C_{P1}^{1/2} \tag{E7-2.14}$$

The P in $P1$ (i.e., C_{P1}) and $P5$ (i.e., C_{P5}) is to remind us that we have used the PSSH in arriving at these balances.

At 1000 K the specific reaction rates are $k_1 = 1.5 \times 10^{-3}$ s^{-1}, $k_2 = 2.3 \times 10^6$ dm^3/mol·s, $k_3 = 5.71 \times 10^4$ s^{-1}, $k_4 = 9.53 \times 10^8$ dm^3/mol·s, and $k_5 = 3.98 \times 10^9$ dm^3/mol·s.

For an entering ethane concentration of 0.1 mol/dm^3 and a temperature of 1000 K, Equations (E7-2.13) and (E7-2.14) were solved and the concentrations of ethane, C_{P1}, and ethylene, C_{P5}, are shown as a function of time in Figures E7-2.1 and E7-2.2.

In developing the above concentration–time relationship, we used PSSH. However, we can now utilize the techniques described in Chapter 6 to solve the full set of equations for ethane cracking and then compare these results with the much simpler PSSH solutions.

Part (b) *Testing the PSSH for Ethane Cracking*

The thermal cracking of ethane is believed to occur by the reaction sequence given in **Part (a)**. The specific reaction rates are given as a function of temperature:

$$k_1 = 10e^{(87,000/R)(1/1250 - 1/T)} \text{s}^{-1} \qquad k_2 = 8.45 \times 10^6 e^{(13,000/R)(1/1250 - 1/T)} \text{dm}^3/\text{mol·s}$$

$$k_3 = 3.2 \times 10^6 e^{(40,000/R)(1/1250 - 1/T)} \text{s}^{-1} \quad k_4 = 2.53 \times 10^9 e^{(9700/R)(1/1250 - 1/T)} \text{dm}^3/\text{mol·s}$$

$$k_5 = 3.98 \times 10^9 \text{ dm}^3/\text{mol·s} \qquad E = 0$$

Part (b): Plot the concentrations of ethane and ethylene as a function of time and compare with the PSSH concentration–time measurements. The initial concentration of ethane is 0.1 mol/dm^3 and the temperature is 1000 K.

Solution Part (b)

Let $1 = C_2H_6$, $2 = CH_3\bullet$, $3 = CH_4$, $4 = C_2H_5\bullet$, $5 = C_2H_4$, $6 = H\bullet$, $7 = H_2$, and $8 = C_4H_{10}$. The combined mole balances and rate laws become

$$(C_2H_6): \quad \frac{dC_1}{dt} = -k_1 C_1 - k_2 C_1 C_2 - k_4 C_1 C_6 \tag{E7-2.13}$$

$$(CH_3\bullet): \quad \frac{dC_2}{dt} = 2k_1 C_1 - k_2 C_2 C_1 \tag{E7-2.14}$$

$$(CH_4): \quad \frac{dC_3}{dt} = k_2 C_1 C_2 \tag{E7-2.15}$$

$$(C_2H_5\bullet): \quad \frac{dC_4}{dt} = k_2C_1C_2 - k_3C_4 + k_4C_1C_6 - k_5C_4^2 \qquad \text{(E7-2.16)}$$

$$(C_2H_4): \quad \frac{dC_5}{dt} = k_3C_4 \qquad \text{(E7-2.17)}$$

$$(H\bullet): \quad \frac{dC_6}{dt} = k_3C_4 - k_4C_1C_6 \qquad \text{(E7-2.18)}$$

$$(H_2): \quad \frac{dC_7}{dt} = k_4C_1C_6 \qquad \text{(E7-2.19)}$$

$$(C_4H_{10}): \quad \frac{dC_8}{dt} = \frac{1}{2}k_5C_4^2 \qquad \text{(E7-2.20)}$$

The POLYMATH program is given in Table E7-2.1.

TABLE E7-2.1.   POLYMATH PROGRAM

Equations:	*Initial Values:*
`d(C1)/d(t)=-k1*C1-k2*C1*C2-k4*C1*C6`	0.1
`d(C2)/d(t)=2*k1*C1-k2*C1*C2`	0
`d(C6)/d(t)=k3*C4-k4*C6*C1`	0
`d(C7)/d(t)=k4*C1*C6`	0
`d(C3)/d(t)=k2*C1*C2`	0
`d(C4)/d(t)=k2*C1*C2-k3*C4+k4*C6*C1-k5*C4^2`	0
`d(C5)/d(t)=k3*C4`	0
`d(C8)/d(t)=0.5*k5*C4^2`	0
`d(CP5)/d(t)=k3*(2*k1/k5)^0.5*CP1^0.5`	0
`d(CP1)/d(t)=-k1*CP1-2*k1*CP1-(k3*(2*k1/k5)^0.5)*(CP1^0.5)`	0.1
`k5=3980000000`	
`T=1000`	
`k1=10*exp((87500/1.987)*(1/1250-1/T))`	
`k2=8450000*exp((13000/1.987)*(1/1250-1/T))`	
`k4=2530000000*exp((9700/1.987)*(1/1250-1/T))`	
`k3=3200000*exp((40000/1.987)*(1/1250-1/T))`	
$t_0 = 0, \quad t_f = 12$	

Figure E7-2.1 shows a comparison of the concentration-time trajectory for ethane calculated from the PSSH (CP1) with the ethane trajectory (C1) calculated from solving the mole balance Equations (E7-2.14) through (E7-2.20). Figure E7-2.2 shows a similar comparison for ethylene (CP5) and (C5). One notes that the curves are identical, indicating the validity of the PSSH under these conditions. Figure E7-2.3 shows a comparison the concentration-time trajectories for methane (C3) and butane (C8). Problem P7-2(a) explores the temperature for which the PSSH is valid for the cracking of ethane.

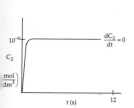

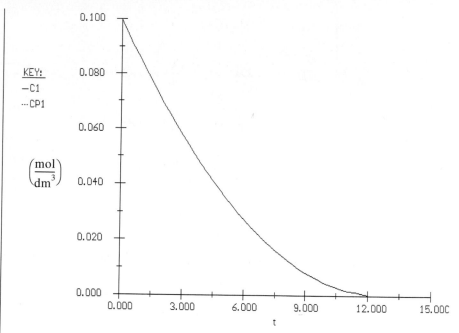

Figure E7-2.1   Comparison of concentration-time trajectories for ethane.

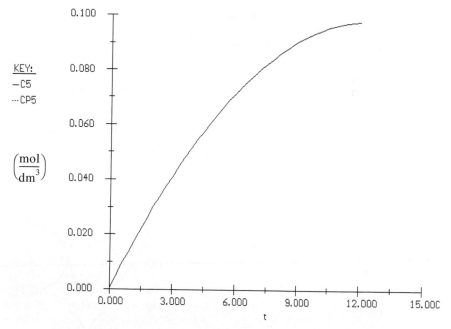

Figure E7-2.2   Comparison for temperature-time trajectory for ethylene.

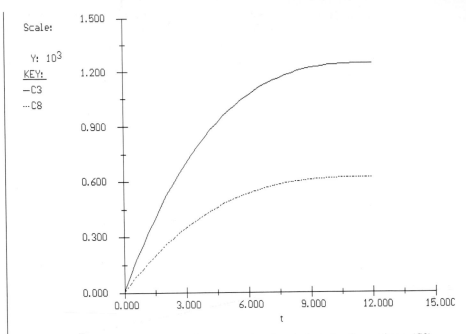

Figure E7-2.3   Comparison of concentration-time trajectories for methane (C3) and butane (C8).

7.2.2 Reaction Pathways

Ethane Cracking.   With the increase in computing power, more and more analyses involving free-radical reactions as intermediates are carried out using the coupled sets of differential equations (cf. Example 7-2). The key in any such analyses is to identify which intermediate reactions are important in the overall sequence in predicting the end products. Once the key reactions are identified, one can sketch the pathways in a manner similar to that shown for the ethane cracking in Example 7-2 (see Figure 7-1).

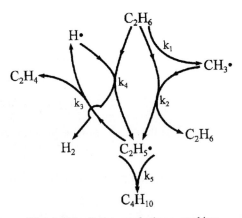

Figure 7-1   Pathway of ethane cracking.

Smog Formation.   Nitrogen and oxygen react to form nitric oxide in the cylinder of automobile engines. The NO from automobile exhaust is oxidized to NO_2 in the presence of peroxide radicals.

$$R\dot{O}O + NO \longrightarrow \dot{R}O + NO_2 \tag{R1}$$

Nitrogen dioxide is then decomposed photochemically to give nascent oxygen,

$$NO_2 + h\nu \longrightarrow NO + O \tag{R2}$$

which reacts to form ozone:

$$O + O_2 \longrightarrow O_3 \tag{R3}$$

The ozone then becomes involved in a whole series of reactions with hydrocarbons in the atmosphere to form aldehydes, various free radicals, and other intermediates, which react further to produce undesirable products in air pollution:

$$\text{ozone} + \text{olefin} \longrightarrow \text{aldehydes} + \text{free radicals}$$

$$O_3 + RCH{=}CHR \overset{\longrightarrow}{\underset{h\nu \longrightarrow \dot{R} + H\dot{C}O}{\quad}} RCHO + \dot{R}O + H\dot{C}O \tag{R4}$$

$$\quad \overset{}{\underset{}{\qquad}} h\nu \longrightarrow \dot{R} + H\dot{C}O \tag{R5}$$

One specific example is the reaction of ozone with 1,3–butadiene to form acrolein and formaldehyde, which are *severe eye irritants*.

Eye irritants

$$\tfrac{2}{3}O_3 + CH_2{=}CHCH{=}CH_2 \overset{h\nu}{\longrightarrow} CH_2{=}CHCHO + HCHO \tag{R6}$$

By regenerating NO_2, more ozone can be formed and the cycle continued. One means by which this regeneration may be accomplished is through the reaction of NO with the free radicals in the atmosphere (RI). For example, the free radical formed in Reaction (R4) can react with O_2 to give the peroxy free radical,

$$\dot{R} + O_2 \longrightarrow R\dot{O}O \tag{R7}$$

The coupling of all the reactions above is shown schematically in Figure 7-2.
    We see that the cycle has been completed and that with a relatively small amount of nitrogen oxides, a large amount of pollutants can be produced. Of course, many other reactions are taking place, so do not be misled by the brevity of the preceding discussion; it does, however, serve to present, in rough outline, the role of nitrogen oxides in air pollution.

Finding the Reaction Mechanism.   Now that a rate law has been synthesized from the experimental data, we shall try to propose a mechanism that is consistent with this equation. The method of attack will be as given in Table 7-2.

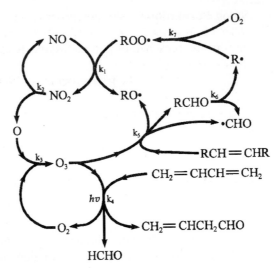

Figure 7-2   Reaction pathways in smog formation.

TABLE 7-2.   STEPS TO DEDUCE A RATE LAW

1. Assume an activated intermediate(s).
2. Postulate a mechanism, utilizing the rate law obtained from experimental data, if possible.
3. Model each reaction in the mechanism sequence as an elementary reaction.
4. After writing rate laws for the rate of formation of desired product, write the rate laws for each of the active intermediates.
5. Use the PSSH.
6. Eliminate the concentration of the intermediate species in the rate laws by solving the simultaneous equations developed in steps 4 and 5.
7. If the derived rate law does not agree with experimental observation, assume a new mechanism and/or intermediates and go to step 3. A strong background in organic and inorganic chemistry is helpful in predicting the activated intermediates for the reaction under consideration.

Once the rate law is found, the search for the mechanism begins

7.3 Polymerization

Polymers are finding increasing use throughout our society. Well over 100 billion pounds of polymer are produced each year and it is expected that this figure will double in the coming years as higher-strength plastics and composite materials replace metals in automobiles and other products. Consequently, the field of polymerization reaction engineering will have an even more prominent place in the chemical engineering profession. Since there are entire books on this field (see Supplementary Reading) it is the intention here to give only the most rudimentary thumbnail sketch of some of the principles of polymerization.

10^{11} lb/yr

A polymer is a molecule made up of repeating structural (monomer) units. For example, polyethylene, which is used for such things as tubing, and electrical insulation is made up of repeating units of ethylene

$$nCH_2{=}CH_2 \longrightarrow -[CH_2-CH_2-]_n$$

where n may be 25,000 or higher.

Polymerization is the process in which monomer units are linked together by chemical reaction to form long chains. It is these long chains that set polymers apart from other chemical species and gives them their unique characteristic properties.

The polymer chains can be linear, branched, or cross-linked (Figure 7-3).

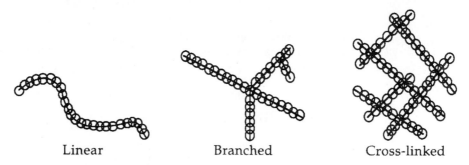

| Linear | Branched | Cross-linked |

Figure 7-3   Types of polymer chains.

Homopolymers are polymers consisting of a single repeating unit, such as $[—CH_2—CH_2—]$. Homopolymers can also be made from two different monomers whose structural units form the repeating unit such as the formation of a polyamide (e.g., Nylon) from a diamine and a diacid.

Polymerization reactions are divided into two groups known as **step reactions** (also called condensation reactions) and **chain reactions,** also known as addition reactions. Step reactions require bifunctional or polyfunctional monomers, while chain reactions require the presence of an initiator.

$$n \overbrace{NH_2RNH_2}^{Monomer} + n \overbrace{HOOCR'COOH}^{Monomer} \longrightarrow H\underbrace{[\; \overbrace{NHRNH}^{\substack{Structural \\ Unit\ 1}} \; \overbrace{OCR'CO}^{\substack{Structural \\ Unit\ 2}} \;]_n}_{Repeating\ Unit} OH + (2n-1)H_2O$$

Copolymers are polymers made up of two or more repeating units. There are five basic categories of copolymers that have two different repeating units Q and S. They are

1. *Alternating*:   –Q–S–Q–S–Q–S–Q–S–Q–S–
2. *Block*:   –Q–Q–Q–Q–Q–S–S–S–S–S–
3. *Random*:   –Q–Q–S–Q–S–S–Q–S–S–S–
4. *Graft*:   –Q–Q–Q–Q–Q–Q–Q–Q–Q–Q–
              └ S–S–S–S–S–S–
5. *Statistical* (follow certain addition laws)

Examples of each can be found in Young and Lovell.[6]

---

[6] R. J. Young and P. A. Lovell, *Introduction to Polymers*, 2nd ed., Chapman & Hall, New York, 1991.

7.3.1 Step Polymerization

Step polymerization requires that there is at least a reactive functional group on each end of the monomer that will react with functional groups with other monomers. For example, amino-caproic acid

$$NH_2\text{—}(CH_2)_5\text{—}COOH$$

has an amine group at one end and a carboxyl group at the other. Some common functional groups are —OH, —COOH, —COCl, —NH$_2$.

In step polymerization the molecular weight usually builds up slowly

$$\text{Dimer} \quad 2H\ \overbrace{(NH\text{—}R\text{—}CO)}^{\text{Structural Unit}}\ OH \longrightarrow H\ \overbrace{(NH\text{—}R\text{—}CO)_2}^{\text{Repeating Unit}}\ OH + H_2O$$

For the case shown above the structural unit and the repeating unit are the same.

Letting $A \equiv H$, $R \equiv NH\text{—}R_1\text{—}CO$, and $B \equiv OH$, $AB \equiv H_2O$. We can write the above reaction as

Dimer	$ARB + ARB$	$\longrightarrow A\text{—}R_2\text{—}B + AB$
Trimer	$ARB + A\text{—}R_2\text{—}B$	$\longrightarrow A\text{—}R_3\text{—}B + AB$
Tetramer	$ARB + A\text{—}R_3\text{—}B$	$\longrightarrow A\text{—}R_4\text{—}B + AB$
	$A\text{—}R_2\text{—}B + A\text{—}R_2\text{—}B$	$\longrightarrow A\text{—}R_4\text{—}B + AB$
Pentamer	$ARB + A\text{—}R_4\text{—}B$	$\longrightarrow A\text{—}R_5\text{—}B + AB$
	$A\text{—}R_2\text{—}B + A\text{—}R_3\text{—}B$	$\longrightarrow A\text{—}R_5\text{—}B + AB$
Hexamer	$ARB + A\text{—}R_5\text{—}B$	$\longrightarrow A\text{—}R_6\text{—}B + AB$
	$A\text{—}R_2\text{—}B + A\text{—}R_4\text{—}B$	$\longrightarrow A\text{—}R_6\text{—}B + AB$
	$A\text{—}R_3\text{—}B + A\text{—}R_6\text{—}B$	$\longrightarrow A\text{—}R_6\text{—}B + AB$

etc.

overall: $n\ NH_2RCOOH \longrightarrow H(NHRCO)_n\ OH + (n-1)H_2O$

$$n\ ARB \longrightarrow A\text{—}R_n\text{—}B + (n-1)AB$$

We see that from tetramers on, the –mer can be formed by a number of different pathways.

The A and B functional groups can also be on different monomers such as the reaction for the formation of polyester (shirts) from diols and dibasic acids.

$$n\ HO\overbrace{R_1}^{\text{Unit 1}}OH + n\ H\overbrace{OOCR_2COO}^{\text{Unit 2}}H \longrightarrow HO\underbrace{(\overbrace{R_1}^{\text{Unit 1}}\overbrace{OOCR_2COO}^{\text{Unit 2}})_n}_{\text{Repeating Unit}}H + (2n-1)H_2O$$

$$n\ AR_1A + n\ BR_2B \longrightarrow A(R_1 - R_2)_n\ B + (2n-1)AB$$

By using diols and diacids we can form polymers with two different structural units which together become the repeating unit. An example of an AR_1A plus BR_2B reaction is that used to make Coca-cola® bottles, i.e. terephthalic acid plus ethylene glycol to form poly (ethylene glycol terephthalate).

When discussing the progress of step polymerization it is not meaningful to use conversion of monomer as a measure because the reaction will still proceed even though all the monomer has been consumed. For example, if the monomer A—R—B has been consumed. The polymerization is still continuing with

$$A\text{---}R_2\text{---}B + A\text{---}R_3\text{---}B \longrightarrow A\text{---}R_5\text{---}B + AB$$

$$A\text{---}R_5\text{---}B + A\text{---}R_5\text{---}B \longrightarrow A\text{---}R_{10}\text{---}B + AB$$

because there are both A and B functional groups that can react. Consequently, we measure the progress by the parameter p which is the fraction of functional groups, A, B, that have reacted. We shall only consider reaction with equal molar feed of functional groups. In this case

$$\boxed{p = \frac{M_o - M}{M_o} = \begin{array}{l}\text{fraction of functional groups of either A or B that} \\ \text{have reacted}\end{array}}$$

M = concentration of either A or B functional groups (mol/dm³)

As an example of step polymerization, consider the polyester reaction in which sulfuric acid is used as a catalyst in a batch reactor. Assuming the rate of disappearance is first order in A, B, and catalyst concentration (which is constant for an externally added catalyst). The balance on A is

$$\frac{-d[A]}{dt} = k[A][B] \tag{7-22}$$

For equal molar feed we have

$$[A] = [B] = M$$

$$\frac{dM}{dt} = -kM^2$$

$$M = \frac{M_o}{1 + M_o kt} \tag{7-23}$$

In terms of the fractional conversion of functional groups, p,

$$\frac{1}{1-p} = M_o kt + 1 \tag{7-24}$$

The number average degree of polymerization, $\overline{X}_n$, is the average number of structural units per chain

Degree
of Polymerization

$$\boxed{\overline{X}_n = \frac{1}{1-p}} \tag{7-25}$$

The number average molecular weight, $\overline{M}_n$ is just the average molecular weight of a structural unit $\overline{M}_s$, times the average number of structural unit per chain, $\overline{X}_n$ plus the molecular weight of the end groups, M_{eg}

$$\overline{M}_n = \overline{X}_n \overline{M}_s + M_{eg}$$

Since M_{eg} is usually small (18 for the polyester reaction), it is neglected and

$$\boxed{\overline{M}_n = \overline{X}_n \overline{M}_s} \tag{7-26}$$

In addition to the conversion of the functional groups, the degree of polymerization, and the number average molecular weight we are interested in the distribution of chain lengths, n, (i.e. molecular weights M_n).

Example 7–3 Determining the concentrations of polymers
*   for step polymerization*

Determine the concentration and mole fraction of polymers of chain length j in terms of initial concentration of ARB, M_o, the concentration of unreacted functional groups M, the propagation constant k and time t.

Solution

Letting $P_1 = A$—R—B, $P_2 = A$—R_2—B, ... $P_j = A$—R_j—B and omitting the water condensation products AB for each reaction we have

	Reaction	Rate Laws
(1)	$2P_1 \rightarrow P_2$	$-r_{1P_1} = 2kP_1^2, \quad r_{1P_2} = -\dfrac{r_{1P_1}}{2} = kP_1^2$
(2)	$P_1 + P_2 \rightarrow P_3$	$-r_{2P_1} = -r_{2P_2} = r_{2P_3} = 2kP_1P_2$
(3)	$P_1 + P_3 \rightarrow P_4$	$-r_{3P_1} = -r_{3P_3} = r_{3P_4} = 2kP_1P_3$
(4)	$P_2 + P_2 \rightarrow P_4$	$-r_{4P_2} = 2kP_2^2, \quad r_{4P_4} = -\dfrac{r_{4P_2}}{2} = kP_2^2$

The factor of 2 in the disappearance term (e.g. $-r_{3P_3} = 2kP_1P_3$) comes about because there are two ways A and B can react.

$$A - R_n - B$$
$$\times$$
$$A - R_m - B$$

The net rate of reaction of P_1, P_2 and P_3 for reactions (1) through (4) are

$$r_1 \equiv r_{P_1} = -2kP_1^2 - 2kP_1P_2 - 2kP_1P_3 \tag{E7-3.1}$$

$$r_2 \equiv r_{P_2} = kP_1^2 - 2kP_1P_2 - 2kP_2^2 \tag{E7-3.2}$$

$$r_3 \equiv r_{P_3} = 2kP_1P_2 - 2kP_1P_3 - 2kP_2P_3 \tag{E7-3.3}$$

If we continue in this way we would find that the net rate of formation of the P_1 is

$$r_{P_1} = -2kP_1 \sum_{j=1}^{\infty} P_j \tag{E7-3.4}$$

However, we note that $\sum_{j=1}^{\infty} P_j$ is just the total concentration of functional groups of either A or B, which is M $\left(M = \sum_{j=1}^{\infty} P_j \right)$.

$$r_{P_1} = -2kP_1 M \tag{E7-3.5}$$

Similarly we can generalize reactions (1) through (4) to obtain the net rate of formation of the j-mer, for $j \geq 2$.

$$\boxed{r_j = k \sum_{i=1}^{j-1} P_i P_{j-1} - 2kP_j M} \tag{E7-3.6}$$

For a batch reactor the mole balance on P_1 and using Equation (7-23) to eliminate M gives

$$\frac{dP_1}{dt} = -2kP_1 M = -2kP_1 \frac{M_o}{1 + M_o kt} \tag{E7-3.7}$$

which solves to

$$P_1 = M_o \left(\frac{1}{1 + M_o kt} \right)^2 \tag{E7-3.8}$$

Having solved for P_1 we can now use r_j to solve successively for P_j

$$\frac{dP_2}{dt} = r_2 = kP_1^2 - 2kP_2 M \tag{E7-3.9}$$

$$= kM_o^2 \left(\frac{1}{1 + M_o kt} \right)^4 - 2M_o k \left(\frac{1}{1 + M_o kt} \right) P_2 \tag{E7-3.10}$$

with $P_2 = 0$ at $t = 0$

$$P_2 = M_o \left(\frac{1}{1 + M_o kt} \right)^2 \left(\frac{M_o kt}{1 + M_o kt} \right) \tag{E7-3.11}$$

continuing we find, that in general[7]

$$\boxed{P_j = M_o \left(\frac{1}{1 + M_o kt} \right)^2 \left(\frac{M_o kt}{1 + M_o kt} \right)^{j-1}} \tag{E7-3.12}$$

---

[7] N. A. Dotson, R. Galván, R. L. Lawrence, and M. Tirrell, *Polymerization Process Modeling*, VCH Publishers, New York, NY (1996).

Recalling $p = \dfrac{M_o - M}{M_o}$

$$P_j = M_o(1-p)^2 p^{j-1} \qquad \text{(E7-3.13)}$$

The mole fraction of polymer with a chain length j is just

$$y_j = \frac{P_j}{M}$$

Recalling $M = M_o(1 - p)$, we obtain

$$y_j = (1-p)\,p^{j-1} \qquad \text{(7-27)}$$

This is the Flory–Schulz distribution. We discuss this distribution further after we discuss chain reactions.

7.3.2 Chain Polymerizations Reactions

Chains (i.e., addition) polymerization requires an initiator (I) and proceeds by adding one repeating unit at a time.

$$I + M \longrightarrow R_1$$
$$M + R_1 \longrightarrow R_2$$
$$M + R_2 \longrightarrow R_3$$
$$M + R_3 \longrightarrow R_4$$
$$M + R_4 \longrightarrow R_5, \text{ etc.}$$

Here the molecular weight in a chain usually builds up rapidly once a chain is initiated. The formation of polystyrene,

$$n\ C_6H_5CH{=}CH_2 \longrightarrow [-\underset{\underset{C_6H_5}{|}}{C}HCH_2-]_n$$

is an example of chain polymerization. A batch process to produce polystyrene for use in a number of molded objects is shown in Figure 7-4.

We can easily extend the concepts described in the preceding section to polymerization reactions. In this section we show how the rate laws are formulated so that one can use the techniques developed in Chapter 6 for multiple reactions to determine the molecular weight distribution and other properties. In the material that follows we focus on *free-radical polymerization*.

7.3.2.1 Steps in Free-Radical Polymerization

The basic steps in free-radical polymerization are initiation, propagation, chain transfer, and termination.

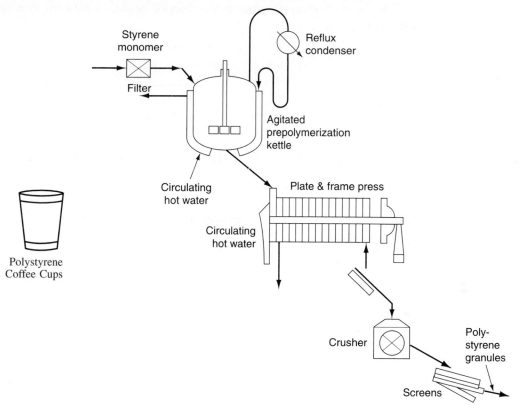

Figure 7-4 Batch bulk polystyrene process. (From *Chemical Reactor Theory*, p. 543, Copyright © 1977, Prentice Hall. Reprinted by permission of Prentice Hall, Upper Saddle River, NJ)

Initiation.   Chain polymerization reactions are different because an initiation step is needed to start the polymer chain growth. Initiation can be achieved by adding a small amount of a chemical that decomposes easily to form free radicals. Initiators can be monofunctional and form the same free radical:

Initiation

$$I_2 \xrightarrow{\ k_0\ } 2I$$

for example, 2,2-azobisisobutyronitrile:

$$(CH_3)_2 \underset{CN}{C} = N \underset{CN}{C}(CH_3)_2 \longrightarrow 2(CH_3)_2 \underset{CN}{C} \cdot + N_2$$

or they can be multifunctional and form different radicals. Multifunctional initiators contain more than one labile group[8] [e.g., 2,5 dimethyl-2,5-bis(benzoylperoxy)hexane].

---

[8] J. J. Kiu and K. Y. Choi, *Chem. Eng. Sci., 43*, 65 (1988); K. Y. Choi and G. D. Lei, *AIChE J., 33*, 2067 (1987).

For monofunctional initiators the reaction sequence between monomer M and initiator I is

$$I + M \xrightarrow{\ k_i\ } R_1$$

for example,

$$(CH_3)_2C\cdot + CH_2\!\!=\!\!CHCl \longrightarrow (CH_3)_2C\ \overset{H}{\underset{Cl}{\overset{|}{C}}}\cdot$$
$$\underset{CN}{|} \qquad\qquad\qquad\qquad\qquad \underset{CN}{|}$$

Propagation. The propagation sequence between a free radical R_1 with a monomer unit is

Propagation

$$R_1 + M \xrightarrow{\ k_p\ } R_2$$
$$R_2 + M \xrightarrow{\ k_p\ } R_3$$

In general,

Assumption of equal reactivity

$$R_j + M \xrightarrow{\ k_p\ } R_{j+1}$$

for example,

$$(CH_3)_2C(CH_2CHCl)_j\, CH_2\overset{H}{\underset{Cl}{\overset{|}{C}}}\cdot + CH_2\!\!=\!\!CHCl \longrightarrow (CH_3)_2C(CH_2CHCl)_{j+1}CH_2\overset{H}{\underset{Cl}{\overset{|}{C}}}\cdot$$
$$\underset{CN}{|}$$

The specific reaction rates k_p are assumed to be identical for the addition of each monomer to the growing chain. This is usually an excellent assumption once two or more monomers have been added to R_1 and for low conversions of monomer. The specific reaction rate k_i is often taken to be equal to k_p.

Chain Transfer. The transfer of a radical from a growing polymer chain can occur in the following ways:

1. Transfer to a monomer:

$$R_j + M \xrightarrow{\ k_m\ } P_j + R_1$$

Here a *live* polymer chain of j monomer units tranfers its free radical to the monomer to form the radical R_1 and a *dead* polymer chain of j monomer units.

2. Transfer to another species:

Chain
transfer

$$R_j + C \xrightarrow{\ k_c\ } P_j + R_1$$

3. Transfer of the radical to the solvent:

$$R_j + S \xrightarrow{\ k_s\ } P_j + R_1$$

The species involved in the various chain transfer reactions such as $CCl_3\cdot$ and $C_6H_5CH_2\cdot$ are all assumed to have the same reactivity as R_1. In other words, all the R_1's produced in chain transfer reactions are taken to be the same. However, in some cases the chain transfer agent may be too large or unreactive to propagate the chain. The choice of solvent in which to carry out the polymerization is important. For example, the solvent transfer specific reaction rate k_s is 10,000 times greater in CCl_4 than in benzene.

The specific reaction rates in chain transfer are all assumed to be independent of the chain length. We also note that while the radicals R_1 produced in each of the chain transfer steps are different, they function in essentially the same manner as the radical R_1 in the propagation step to form radical R_2.

Termination.   Termination to form dead polymer occurs primarily by two mechanisms:

1. Addition (coupling) of two growing polymers:

$$R_j + R_k \xrightarrow{\ k_a\ } P_{j+k}$$

2. Termination by disproportionation:

$$R_j + R_k \xrightarrow{\ k_d\ } P_j + P_k$$

for example,

$$(CH_3)_2C(CH_2CHCl)_j\,CH_2\overset{\overset{\displaystyle H}{|}}{\underset{\underset{\displaystyle Cl}{|}}{C}}\cdot + \cdot\overset{\overset{\displaystyle H}{|}}{\underset{\underset{\displaystyle Cl}{|}}{C}}-CH_2(CH_2CHCl)_k(CH_3)_2C\overset{}{\underset{\underset{\displaystyle CN}{|}}{}}$$

Termination

$$(CH_3)_2C(CH_2CHCl)_j\,\overset{\overset{\displaystyle H}{|}}{\underset{\underset{\displaystyle Cl}{|}}{CH}} + \overset{\overset{\displaystyle H}{|}}{\underset{\underset{\displaystyle Cl}{|}}{C}}{=}CH(CH_2CHCl)_k(CH_3)_2C\overset{}{\underset{\underset{\displaystyle CN}{|}}{}}$$

The steps in free-radical polymerization reaction and the corresponding rate laws are summarized in Table 7-3. For the polymerization of styrene at 80°C initiated by 2,2-azobisisobutyronitrile the rate constants[9] are

Initiation
Propagation
Transfer
Termination

$$k_0 = 1.4 \times 10^{-3} \text{ s}^{-1} \qquad\qquad k_m = 3.2 \times 10^{-2} \text{ dm}^3/\text{mol} \cdot \text{s}$$

$$k_p = 4.4 \times 10^2 \text{ dm}^3/\text{mol} \cdot \text{s} \qquad k_a = 1.2 \times 10^8 \text{ dm}^3/\text{mol} \cdot \text{s}$$

$$k_s = 2.9 \times 10^{-3} \text{ dm}^3/\text{mol} \cdot \text{s} \qquad k_d = 0$$

TABLE 7-3

	Rate Law
Initiation: $I_2 \xrightarrow{k_0} 2I$ $I + M \xrightarrow{k_i} R_1$	$\begin{cases} -r_{I_2} = k_0 I_2 \\ r_{If} = 2fk_0 I_2 \\ -r_i = k_i M I \end{cases}$
Propagation: $R_j + M \xrightarrow{k_p} R_{j+1}$	$-r_j = k_p M R_j$
Chain transfer to: Monomer: $R_j + M \xrightarrow{k_m} P_j + R_1$	$-r_{mj} = k_m M R_j$
Another species: $R_j + C \xrightarrow{k_c} P_j + R_1$	$-r_{cj} = k_c C R_j$
Solvent: $R_j + S \xrightarrow{k_s} P_j + R_1$	$-r_{sj} = k_s S R_j$
Termination: Addition: $R_j + R_k \xrightarrow{k_a} P_{j+k}$	$-r_{aj} = k_a R_j R_k$
Disproportionation: $R_j + R_k \xrightarrow{k_d} P_j + P_k$	$-r_{dj} = k_d R_j R_k$

Typical initial concentrations for the solution polymerization of styrene are 0.01 M for the initiator, 3 M for the monomer, and 7 M for the solvent.

7.3.2.2 Developing the Rate Laws for the Net Rate of Reaction

We begin by considering the rate of formation of the initiator radical I. Because there will always be scavenging or recombining of the primary radicals, only a certain fraction f will be successful in initiating polymer chains.

[9] D. C. Timm and J. W. Rachow, *ACS Symposium Series 133*, H. M. Hulburt, ed., 1974, p. 122.

Since each reaction step is assumed to be elementary, the rate law for the formation of the initiator free radicals, r_{If}, is

$$r_{If} = 2fk_0(I_2)$$

where f is the fraction of initiator free radicals successful in initiating chaining and has a typical value in the range 0.2 to 0.7. The rate law for the formation of R_1 in the initiation step is

$$r_{R1} = -r_i = k_i(M)(I) \tag{7-28}$$

Using the PSSH for the initiator free radical, I, we have

$$r_I = 2fk_0(I_2) - k_i(M)(I) = 0$$

$$(I) = \frac{2fk_0(I_2)}{(M)k_i} \tag{7-29}$$

Then

Rate of initiation
$$-r_i = 2fk_0(I_2) \tag{7-30}$$

Before writing the rate of disappearance of R_1, we need to make a couple of points. First, the radical R_1 can undergo the following termination sequence by addition.

$$R_1 + R_1 \xrightarrow{k_a} P_2$$

$$R_1 + R_2 \xrightarrow{k_a} P_3$$

Termination of R_1 In general,

$$R_1 + R_j \xrightarrow{k_a} P_{j+1}$$

Consequently, the total loss of R_1 radicals in the above reactions is found by adding the loss of R_1 radicals in each reaction so that the rate of disappearance by termination addition is given by

$$-r_{1t} = k_a R_1^2 + k_a R_1 R_2 + k_a R_1 R_3 + \cdots + k_a R_1 R_j + \cdots$$

$$-r_{1t} = k_a R_1 \sum_{j=1}^{\infty} R_j$$

Net rate of disappearance of radicals of chain length one   Free radicals usually have concentrations in the range 10^{-6} to 10^{-8} mol/dm³. We can now proceed to write the net rate of disappearance of the free radical, R_1. $[R_1 \equiv (R_1) \equiv C_{R_1}.]$

$$-r_1 = -r_i + k_p R_1 M + k_a R_1 \sum_{j=1}^{\infty} R_j + k_d R_1 \sum_{j=1}^{\infty} R_j$$

$$- k_m M \sum_{j=2}^{\infty} R_j - k_c C \sum_{j=2}^{\infty} R_j - k_s S \sum_{j=2}^{\infty} R_j \tag{7-31}$$

Net rate of
disappearance of
radicals of chain
length j

In general, the net rate of disappearance of live polymer chains with j monomer units (i.e., species j) for $(j \geq 2)$ is

$$-r_j = k_p M (R_j - R_{j-1}) + (k_a + k_d) R_j \sum_{i=1}^{\infty} R_i$$
$$+ k_m M R_j + k_c C R_j + k_s S R_j$$

(7-32)

At this point one could use the techniques developed in Chapter 6 on multiple reactions to follow polymerization process. However, by using the PSSH, we can manipulate the rate law into a form that allows closed-form solutions for a number of polymerization reactions.

First, we let R^* be the total concentration of the radicals R_j:

$$R^* = \sum_{j=1}^{\infty} R_j$$

(7-33)

and k_t be the termination constant $k_t = (k_a + k_d)$. Next we sum Equation (7-32) over all free-radical chain lengths from $j = 2$ to $j = \infty$, and then add the result to Equation (7-31) to get

$$\sum_{j=1}^{\infty} -r_j = -r_i + k_t (R^*)^2$$

The total rate of termination is just

$$r_t = k_t (R^*)^2$$

(7-34)

Using the PSSH for all free radicals, that is, $\sum_{j=1}^{\infty} -r_j = 0$, the total free-radical concentration solves to

$$R^* = \sqrt{\frac{-r_i}{k_t}} = \sqrt{\frac{2k_0 (I_2) f}{k_t}}$$

(7-35)

We now use this result in writing the net rate of monomer consumption. As a first approximation we will neglect the monomer consumed by monomer chain transfer. The net rate of monomer consumption, $-r_M$, is the rate of consumption by the initiator plus the rate of consumption by all the radicals R_j in each of the propagation steps (r_p).

$$-r_M = -r_i + -r_p = -r_i + k_p M \sum_{j=1}^{\infty} R_j$$

We now use the long-chain approximation (LCA). The LCA is that the rate of propagation is much greater than the rate of initiation:

Long-chain
approximation
(LCA)

$$\frac{r_p}{r_i} \gg 1$$

Substituting for r_p and r_i, we obtain

$$\frac{r_p}{r_i} = \frac{-k_p M R^*}{-k_i M I} = \frac{k_p (2k_0 f (I_2)/k_t)^{1/2}}{k_i (2k_0 f (I_2)/M k_i)}$$

$$= \frac{M}{I_2^{1/2}} \sqrt{\frac{k_p^2}{2k_0 f k_t}}$$

Consequently, we see that the LCA is valid when both the ratio of monomer concentration to initiator concentration and the ratio of k_p^2 to $(k_0 f k_t)$ are high. Assuming the LCA gives

Rate of
disappearance
of monomer

$$-r_M = k_p M \sum_{j=1}^{\infty} R_j = k_p M R^* \qquad (7\text{-}36)$$

Using Equation (7-35) to substitute for R^*, the rate of disappearance of monomer is

$$-r_M = k_p M \sqrt{\frac{2k_0 (I_2) f}{k_t}} \qquad (7\text{-}37)$$

The rate of disappearance of monomer, $-r_M$, is also equal to the rate of propagation, r_p:

$$r_p = -r_M$$

Finally, the net rate of formation of dead polymer P_j by addition is

$$r_{P_j} = 0.5 k_a \sum_{k=1}^{k=j-1} R_k R_{j-k} \qquad (7\text{-}38)$$

The rate of formation of all dead polymers

$$r_P = \sum_{j=1}^{\infty} r_{P_j}$$

Rate of formation of
dead polymers

$$r_P = 0.5 k_a (R^*)^2$$

7.3.3 Modeling a Batch Polymerization Reactor

To conclude this section we determine the concentration of monomer as a function of time in a batch reactor. A balance on the monomer combined with the LCA gives

Monomer balance

$$-\frac{dM}{dt} = k_p M \sum R_j = k_p M R^* = k_p M \sqrt{\frac{2k_0(I_2)f}{k_t}} \qquad (7\text{-}39)$$

A balance on the initiator I_2 gives

Initiator balance

$$-\frac{dI_2}{dt} = k_0 I_2$$

Integrating and using the initial condition $I_2 = I_{20}$ at $t = 0$, we obtain the equation of the initiator concentration profile:

$$I_2 = I_{20} \exp(-k_0 t) \qquad (7\text{-}40)$$

Substituting for the initiator concentration in Equation (7-39), we get

$$\frac{dM}{dt} = -k_p M \left(\frac{2k_0 I_{20} f}{k_t}\right)^{1/2} \exp\left(-\frac{k_0}{2} t\right) \qquad (7\text{-}41)$$

Integration of Equation (7-41) gives

$$\ln \frac{M}{M_0} = \left(\frac{8k_p^2 f I_{20}}{k_0 k_t}\right)^{1/2} \left[\exp\left(-\frac{k_0 t}{2}\right) - 1\right] \qquad (7\text{-}42)$$

One notes that as $t \longrightarrow \infty$, there will still be some monomer left unreacted. Why?

A plot of monomer concentration is shown as a function of time in Figure 7-5 for different initiator concentrations.

The fractional conversion of a monofunctional monomer is

$$X = \frac{M_0 - M}{M_0}$$

We see from Figure 7-5 that for an initiator concentration 0.001 M, the monomer concentration starts at 3 M and levels off at a concentration of 0.6 M, corresponding to a maximum conversion of 80%.

Now that we can determine the monomer concentration as a function of time, we will focus on determining the distribution of dead polymer, P_j. The concentrations of dead polymer and the molecular weight distribution can be derived in the following manner.[10] The probability of propagation is

---

[10] E. J. Schork, P. B. Deshpande, K. W. Leffew, *Control of Polymerization Reactor*, New York: Marcel Dekker (1993).

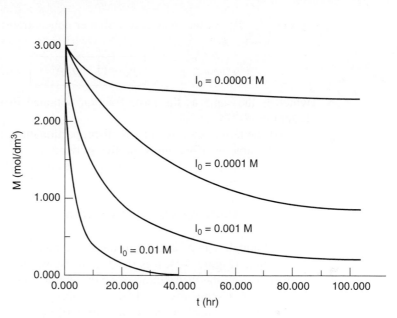

Figure 7-5   Monomer concentration as a functional time.

$$\beta = \frac{\text{rate of propagation}}{\text{rate of propagation} + \text{rate of termination}} = \frac{r_p}{r_p + r_t}$$

$$\beta = \frac{k_p M R^*}{k_p M R^* + k_s S R^* + k_m M R^* + k_c C R^* + k_t (R^*)^2}$$

Simplifying

$$\beta = \frac{k_p M}{k_p M + k_m M + k_c C + k_s S + \sqrt{2 k_t k_o f(I_2)}} \qquad (7\text{-}43)$$

In the absence of chain transfer, the monomer concentration, M, can be determined from Equation (7-43) and concentration of initiator, I_2, from Equation (7-40). Consequently we have β as a function of time.

It can be shown that in the absence of termination by combination, the mole fractions y_j and weight fraction w_j are exactly the same as those for step polymerization. That is if we set

$$\beta = p$$

we can determine the dead polymer concentrations and molecular weight distribution of dead polymer in free radial polymerization for the Flory distributions. For example, the concentration of dead polymer of chain length n is

$$P_n = y_n \left(\sum_{j=2}^{\infty} P_j \right) = \left(\sum_{j=2}^{\infty} P_j \right) (1-p) p^{n-1}$$

where $\left(\displaystyle\sum_{n=2}^{\infty} P_n\right)$ is the total dead polymer concentration and

Dead

Live

$$\boxed{y_n = (1 - p)\, p^{n-1}} \tag{7-27}$$

which is the same as the mole fraction obtained in step polymerization, i.e. Equation (7-27).

If the termination is only by disproportionation, the dead polymer P_j will have the same distribution as the live polymer R_j.

We will discuss the use of the Flory Equation after we discuss molecular weight distributions.

7.3.4 Molecular Weight Distribution

Although it is of interest to know the monomer concentration as a function of time (Figure 7-5), it is the polymer concentration, the average molecular weight, and the distribution of chain lengths that give a polymer its unique properties. Consequently, to obtain such things as the average chain length of the polymer, we need to determine the molecular weight distribution of radicals, (live polymer) R_j, and then dead polymers P_j as well as the molecular weight distribution. Consequently, we need to quantify these parameters. A typical distribution of chain lengths for all the P_j ($j = 1$ to $j = n$) is shown in Figure 7-6. Gel permeation chromatography is commonly used to determine the molecular weight distribution. We will now explore some properties of these distributions. If one divides the y-axis by the total concentration of polymer (i.e., ΣP_j), that axis simply becomes the mole fraction of polymer with j repeating units embedded in it (i.e., y_j).

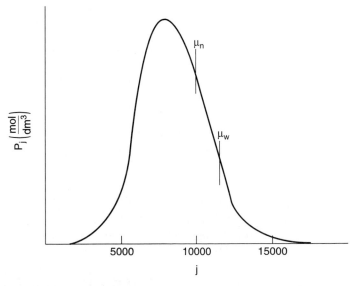

Figure 7-6 Distribution of concentration of dead polymers of length j.

Properties of the Distribution.   From the distribution of molecular weights of polymers, some of the parameters one can use to quantify the distribution shown in Figure 7-6 and their relationships are given below.

1. The moments of the distribution

$$\lambda_n = \sum_{n=1}^{\infty} j^n P_j \tag{7-44}$$

2. The zeroth moment is just the total polymer concentration:

$$\lambda_0 = \sum_{j=1}^{\infty} P_j = P \tag{7-45}$$

3. The first moment is related to total number of monomer units (i.e., mass):

$$\lambda_1 = \sum_{n=1}^{\infty} j P_j \tag{7-46}$$

4. The first moment divided by the zeroth moment gives the *number-average chain length* (NACL), μ_n:

$$\text{NACL} = \mu_n = \frac{\lambda_1}{\lambda_0} = \frac{\Sigma j P_j}{\Sigma P_j} \tag{7-47}$$

   For step-reaction polymerization, the NACL is also sometimes referred to as the *degree of polymerization*. It is the average number of structural units per chain and can also be calculated from

$$\mu_n \equiv \overline{X}_n = \frac{1}{1-p}$$

5. The number-average molecular weight,

$$\overline{M}_n = \mu_n \overline{M}_s \tag{7-48}$$

   where $\overline{M}_s$ is the average molecular weight of the structural units. In chain polymerization, the average molecular weight of the structural unit is just the molecular weight of the monomer, M_M.

6. The second moment gives emphasis to the larger chains:

$$\lambda_2 = \sum_{j=1}^{\infty} j^2 P_j \tag{7-49}$$

7. The mass per unit volume of each polymer species is just $\overline{M}_s j P_j$. The mass average chain length is just the ratio of moment 2 to moment 1:

$$\text{WACL} = \frac{\lambda_2}{\lambda_1} = \mu_w = \frac{\Sigma j^2 P_j}{\Sigma j P_j} \tag{7-50}$$

8. The weight-average molecular weight is

$$\overline{M}_w = \overline{M}_s \, \mu_w \tag{7-51}$$

9. The number-average variance is

$$\sigma_n^2 = \frac{\lambda_2}{\lambda_0} - \left(\frac{\lambda_1}{\lambda_0}\right)^2 \tag{7-52}$$

10. The polydispersity index (D) is

$$D = \frac{\mu_w}{\mu_n} = \frac{\lambda_0 \lambda_2}{\lambda_1^2} \tag{7-53}$$

A polydispersity of 1 means that the polymers are all the same length and a polydispersity of 3 means that there is a wide distribution of polymer sizes. The polydispersity of typical polymers ranges form 2 to 10.

Example 7–4   Parameters Distributions of Polymers

A polymer was fractionated into the following six fractions:

Fraction	Molecular Weight	Mole Fraction
1	10,000	0.1
2	15,000	0.2
3	20,000	0.4
4	25,000	0.15
5	30,000	0.1
6	35,000	0.05

The molecular weight of the monomer was 25 Daltons.
Calculate NACL, WACL, the number variance, and the polydispersity.

Solution

MW	j	y	jy	j^2y
10,000	400	0.1	40	16,000
15,000	600	0.2	120	72,000
20,000	800	0.4	320	256,000
25,000	1000	0.15	150	150,000
30,000	1200	0.1	120	144,000
35,000	1400	0.05	70	98,000
			750	736,000

The number-average chain length, Equation (7-47), can be rearranged as

$$\text{NACL} = \frac{\Sigma j P_j}{\Sigma P_j} = \Sigma j \frac{P_j}{\Sigma P_j} = \Sigma j y_j$$

$$= \mu_n = 750 \text{ structural (monomer) units}$$

(E7-4.1)

The number-average molecular weight is

$$\overline{M}_n = \mu_n M_M = 750 \times 25 = 18,750$$

Recalling Equation (7-50) and rearranging, we have

$$\text{WACL} = \mu_w = \frac{\Sigma j^2 P_j}{\Sigma j P_j} = \frac{\Sigma j^2 (P_j / \Sigma P_j)}{\Sigma j (P_j / \Sigma P_j)}$$

$$= \frac{\Sigma j^2 y}{\Sigma j y} = \frac{736,000}{750} = 981.3 \text{ monomer units.}$$

(E7-4.2)

The mass average molecular weight is

$$\overline{M}_w = M_M \mu_w = 25 \times 981.33 = 24,533$$

The variance is

$$\sigma_n^2 = \frac{\lambda_2}{\lambda_0} - \left(\frac{\lambda_1}{\lambda_0} \right)^2 = 736,000 - (750)^2$$

$$= 173,500$$

$$\sigma_n = 416$$

(E7-4.3)

The polydispersity index D is

$$D = \frac{\overline{M}_w}{\overline{M}_n} = \frac{24,533}{18,750} = 1.31$$

(E7-4.4)

Flory Statistics of the Molecular Weight Distribution.   The solution to the complete set ($j = 1$ to $j = 100,000$) of coupled-nonlinear ordinary differential equations needed to calculate the distribution is an enormous undertaking even with the fastest computers. However, we can use probability theory to estimate the distribution. This theory was developed by Nobel laureate Paul Flory. We have shown that for step polymerization and for free radical polymerization in which termination is by disproportionation the mole fraction of polymer with chain length j is

Flory mole fraction
distribution

$$y_j = (1 - p) p^{j-1}$$

(7-27)

In terms of the polymer concentration

$$P_j = y_j M = M_o (1 - p)^2 p^{j-1}$$

(7-54)

The number average molecular weight

$$\overline{M}_n = \sum_{j=1}^{\infty} y_i M_j = \sum_{j=1}^{\infty} y_j j \overline{M}_s$$

$$= \overline{M}_s (1-p) \sum_{j=1}^{\infty} j p^{j-1} = \overline{M}_s (1-p) \frac{1}{(1-p)^2}$$

and we see that the number average molecular weight is identical to that given by Equation (7-26)

$$\overline{M}_n = \overline{X}_n \overline{M}_s = \frac{\overline{M}_s}{1-p} \qquad (7\text{-}26)$$

The weight fraction of polymer of chain length j is

$$w_j = \frac{P_j M_j}{\sum\limits_{j=1}^{\infty} P_j M_j} = \frac{P_j j \overline{M}_s}{\overline{M}_s \sum\limits_{j=1}^{\infty} j P_j} = \frac{j P_j}{\sum\limits_{j=1}^{\infty} j P_j}$$

$$w_j = \frac{j(1-p)^2 p^{j-1}}{(1-p)^2 \underbrace{\sum\limits_{j=1}^{\infty} j p^{j-1}}_{\frac{1}{(1-p)^2}}}$$

$$\boxed{w_j = j(1-p)^2 p^{j-1}} \qquad (7\text{-}56)$$

The weight fraction is shown in Figure 7-7 as a function of chain length.
The weight average molecular weight is

$$\overline{M}_w = \sum_{j=1}^{\infty} w_j M_j = \overline{M}_s \sum_{j=1}^{\infty} j w_j$$

$$\boxed{\overline{M}_w = \overline{M}_s \frac{(1+p)}{(1-p)}}$$

These equations will also apply for AR_1A and BR_2B polymers if the monomers are fed in stoichiometric portions. Equations (7-54) through (7-56) also can be used to obtain the distribution of concentration and molecular weights for radical reactions where termination is by chain transfer or by disproportionation if by p is given by Equation (7-43). However, they cannot be used for termination by combination.

Figure 7-7 compares the molecular weight distribution for poly(hexamethylene adipamide) calculated from Flory's most probable distribution[11] [Equa-

---

[11] P. J. Flory, *Principles of Polymer Chemistry*, Cornell University Press, Ithaca, N.Y., 1953.

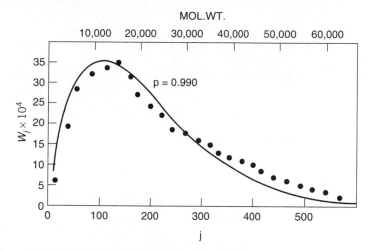

Figure 7-7 Molecular distribution. [Adapted from G. Tayler, *Journal of the American Chemical Society, 69*, p. 638, 1947. Reprinted by permission.]

tion (7-56)] for a conversion of 99% with the experimental values obtained by fractionation. One observes that the comparison is reasonably favorable.

For termination by combination, the mole fraction of polymers with j repeating units is

Termination by combination

$$y_j = (j-1)(1-p)^2 p^{j-2} \qquad (7\text{-}57)$$

while the corresponding weight fraction is

$$w_j = \tfrac{1}{2} j(1-p)^3 (j-1) p^{j-2} \qquad (7\text{-}58)$$

where p is given by Equation (7-43).

7.3.5 Anionic Polymerization

To illustrate the development of the growth of live polymer chains with time, we will use anionic polymerization. In anionic polymerization, initiation takes place by the addition of an anion, which is formed by dissociation of strong bases such as hydroxides, alkyllithium or alkoxides to which reacts with the monomer to form a active center, R_1^-. The dissociation of the initiator is very rapid and essentially at equilibrium. The propagation proceeds by the addition of monomer units to the end of the chain with the negative charge. Because the live ends of the polymer are negatively charged, termination can occur only by charge transfer to either the monomer or the solvent or by the addition of a neutralizing agent to the solution. Let $R_j^- \equiv R_j$ and the sequence of reactions for anionic polymerization becomes

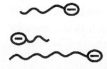

Initiation:
$$\begin{cases} AB \underset{k_{-i}}{\overset{k_i}{\rightleftharpoons}} A^- + B^+ \\ A^- + M \xrightarrow{k_p} R_1 \end{cases}$$

Propagation:
$$\begin{cases} R_1 + M \xrightarrow{\;k_p\;} R_2 \\ R_j + M \xrightarrow{\;k_p\;} R_{j+1} \end{cases}$$

Chain transfer to solvent:

$$R_j + S \xrightarrow{\;k_{tS}\;} P_j + S^-$$

Transfer to monomer:

Batch reactor
calculations

$$R_j + M \xrightarrow{\;k_{tm}\;} P_j + R_1$$

The corresponding combined batch reactor mole balances and rate laws are:
For the initiator:

$$\frac{dA^-}{dt} = k_i AB - k_{-i} A^- B^+ - k_p A^- M$$

Live

For the **live** polymer:

$$\frac{dR_1}{dt} = k_p A^- M - k_p R_1 M + k_{tm} M \sum_{j=1}^{n} R_j$$

$$\frac{dR_j}{dt} = k_p (R_{j-1} - R_j) M - k_{tS} S R_j - k_{tm} M R_j$$

Dead

For the **dead** polymer:

$$\frac{dP_j}{dt} = k_{tS} S R_j + k_{tm} M R_j$$

In theory one could solve this coupled set of differential equations. However, this process is very tedious and almost insurmountable if one were to carry it through for molecular weights of tens of thousands of Daltons, even with the fastest of computers. Fortunately, for some polymerization reactions there is a way out of this dilemma.

"Houston, we have a
problem!"
—Apollo 13

Some Approximations.   To solve this set of coupled ODEs we need to make some approximations. There are a number of approximations that could be made, but we are going to make ones that allow us to obtain solutions that provide insight on how the live polymerization chains grow and dead polymer chains form. First we neglect the termination terms ($k_{tS} S R_j$ and $k_{tm} R_j M$) with respect to propagation terms in the mole balances. This assumption is an excellent one as long as the monomer concentration remains greater than the live polymer concentration.

For this point there are several assumptions that we can make. We could assume that the initiator ($I = A^-$) reacts slowly to form R_1 (such is the case in Problem P7-22).

$$I + M \xrightarrow{\ k_0\ } R_1$$

Initiation
$$\frac{dR_1}{dt} = k_0 M I - k_p R_1 M$$

Another assumption is that the rate of formation of R_1 from the initiator is instantaneous and that at time $t = 0$ the initial concentration of live polymer is $R_{10} = I_0$. This assumption is very reasonable for this initiation mechanism. Under the latter assumption the mole balances become

Propagation
$$\frac{dR_1}{dt} = -k_p M R_1 \tag{7-59}$$

$$\frac{dR_2}{dt} = k_p M (R_1 - R_2) \tag{7-60}$$

$$\vdots$$

$$\frac{dR_j}{dt} = k_p (R_{j-1} - R_j) M \tag{7-61}$$

For the live polymer with the largest chain length that will exist, the mole balance is

$$\frac{dR_n}{dt} = k_p M R_{n-1} \tag{7-62}$$

If we sum Equations (7-59) through (7-62), we find that

$$\sum_{j=1}^{n} \frac{dR_j}{dt} = \frac{dR^*}{dt} = 0$$

Consequently, we see the total free live polymer concentration is a constant at $R^* = R_{10} = I_0$.

There are a number of different techniques that can be used to solve this set of equations, such as use of Laplace transforms, generating functions, statistical methods, and numerical and analytical techniques. We can obtain an analytical solution by using the following transformation. Let

$$d\Theta = k_p M \, dt \tag{7-63}$$

Then Equation (7-59) becomes

$$\frac{dR_1}{d\Theta} = -R_1 \tag{7-64}$$

Using the initial conditions that when $t = 0$, then $\Theta = 0$ and $R_1 = R_{10} = I_0$. Equation (7-64) solves to

$$R_1 = I_0 e^{-\Theta} \tag{7-65}$$

Next we transform Equation (7-60) to

$$\frac{dR_2}{d\Theta} = R_1 - R_2$$

and then substitute for R_1:

$$\frac{dR_2}{d\Theta} + R_2 = I_0 e^{-\Theta}$$

With the aid of the integrating factor, e^{Θ}, along with the initial condition that at $t = 0$, $\Theta = 0$, $R_2 = 0$, we obtain

$$R_2 = I_0(\Theta e^{-\Theta})$$

In a similar fashion,

$$R_3 = I_0\left(\frac{\Theta^2}{2\cdot 1} e^{-\Theta}\right)$$

$$R_4 = I_0\left(\frac{\Theta^3}{3\cdot 2\cdot 1} e^{-\Theta}\right)$$

In general,

Concentration of
live polymer of
chain length j

$$R_j = I_0 \frac{\Theta^{j-1}}{(j-1)!} e^{-\Theta} \qquad (7\text{-}66)$$

The live polymer concentrations are shown as a function of time and of chain length and time in Figures 7-8 and 7-9, respectively.

Anionic
polymerization

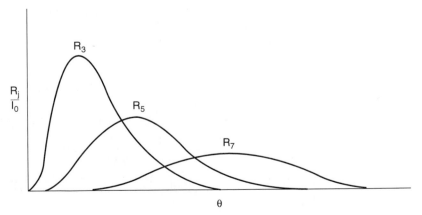

Figure 7-8 Live polymer concentration as a function of scaled time.

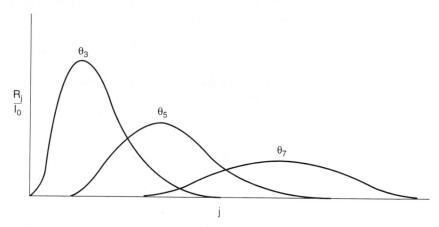

Figure 7-9 Live polymer concentration as a function of chain length at different scaled times.

Neglecting the rate of chain transfer to the monomer with respect to the rate of propagation, a mole balance on the monomer gives

$$\frac{dM}{dt} = -k_p M \sum_{j=1}^{n} R_j = -k_p M R_{10} = -k_p M I_0 \qquad (7\text{-}67)$$

Knowing the initial monomer concentration, M_0, we can solve for the monomer concentration at any time:

$$M = M_0 e^{-I_0 k_p t} \qquad (7\text{-}68)$$

We can also evaluate the scaled time Θ:

$$\Theta = \int_0^t k_p M \, dt = M_0 k_p \int_0^t e^{-I_0 k_p t} \, dt$$

$$= \left. + \frac{M_0 k_p}{k_p I_0} (e^{-I_0 k_p t}) \right|_t^0$$

Relationship between the scaled time, Θ, and real time t

$$\boxed{\Theta = \frac{M_0}{I_0}(1 - e^{-I_0 k_p t})} \qquad (7\text{-}69)$$

One can now substitute Equation (7-69) into Equation (7-66) to determine the live polymer concentrations at any time t.

For anionic polymerization, termination can occur by neutralizing the live polymer R_j to P_j.

Example 7–5 Calculating the Distribution Parameters from Analytical Expressions for Anionic Polymerization

Calculate μ_n, μ_m, and D for the live polymer chains R_j.

Solution

$$R_j = I_0 \frac{\Theta^{j-1}}{(j-1)!} e^{-\Theta} \tag{7-66}$$

We recall that the zero moment is just the total radical concentrations:

$$\lambda_0 = \sum_{j=1}^{\infty} R_j = I_0 \tag{E7-5.1}$$

The first moment is

$$\lambda_1 = \sum_{j=1}^{\infty} jR_j = I_0 \sum_{j=1}^{\infty} \frac{j\Theta^{j-1}e^{-\Theta}}{(j-1)!} \tag{E7-5.2}$$

Let $k = j - 1$:

$$\lambda_1 = I_0 \sum_{k=0}^{\infty} (k+1)\frac{\Theta^k e^{-\Theta}}{k!} \tag{E7-5.3}$$

Expanding the $(k + 1)$ term gives

$$\lambda_1 = I_0 \left(\sum_{k=0}^{\infty} \frac{\Theta^k e^{-\Theta}}{k!} + \sum_{k=0}^{\infty} \frac{k\Theta^k e^{-\Theta}}{k!} \right) \tag{E7-5.4}$$

Recall that

$$\sum_{k=0}^{\infty} \frac{\Theta^k}{k!} = e^{+\Theta} \tag{E7-5.5}$$

Therefore,

$$\lambda_1 = I_0 \left(1 + e^{-\Theta} \sum_{k=0}^{\infty} \frac{k\Theta^k}{k!} \right) \tag{E7-5.6}$$

Let $l = k - 1$:

$$\lambda_1 = I_0 \left(1 + e^{-\Theta}\Theta \sum_{l=0}^{\infty} \frac{\Theta^l}{l!} \right) = I_0(1 + e^{-\Theta}\Theta e^{\Theta}) \tag{E7-5.7}$$

The first moment is

$$\boxed{\lambda_1 = I_0(1 + \Theta)} \tag{E7-5.8}$$

The number-average length of growing polymer radical (i.e., live polymer) is

$$\boxed{\mu_n = \frac{\lambda_1}{\lambda_0} = 1 + \Theta} \tag{E7-5.9}$$

$$\lambda_2 = I_0 \sum_{j=0}^{\infty} j^2 R_j = I_0 \sum_{j=0}^{\infty} j^2 \frac{\Theta^{j-1}}{(j-1)!} \tag{E7-5.10}$$

Realizing that the $j = 0$ term in the summation is zero and after changing the index of the summation and some manipulation, we obtain

$$\lambda_2 = I_0(1 + 3\Theta + \Theta^2) \qquad \text{(E7-5.11)}$$

$$\mu_w = \frac{\lambda_2}{\lambda_1} = \frac{1 + 3\Theta + \Theta^2}{1 + \Theta} \qquad \text{(E7-5.12)}$$

$$D = \frac{\mu_w}{\mu_n} = \frac{1 + 3\Theta + \Theta^2}{(1 + \Theta)^2} \qquad \text{(E7-5.13)}$$

Plots of μ_n and μ_w along with the polydispersity, D, are shown in Figure E7-5.1.

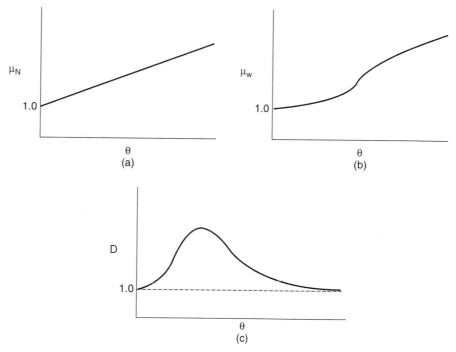

Figure E7-5.1   Moments of live polymer chain lengths: (a) number-average chain length; (b) weight-average chain length; (c) polydispersity.

We note from Equation (7-69) that after a long time the maximum value of Θ, Θ_M, will be reached:

$$\Theta_M = \frac{M_0}{I_0}$$

The distributions of live polymer species for an anionic polymerization carried out in a CSTR are developed in Problem P7-19.

Example 7–6 Determination of Dead Polymer Distribution When Transfer
to Monomer Is the Main Termination Mechanism

Determine an equation for the concentration of polymer as a function of scaled time. Once we have the live polymer concentration as a function of time, we can determine the dead polymer concentration as a function of time. If transfer to monomer is the main mechanism for termination,

$$R_j + M \xrightarrow{\ k_{tm}\ } P_j + R_1$$

A balance of dead polymer of chain length j is

Anionic
polymerization

$$\frac{dP_j}{dt} = k_{tm} R_j M \tag{E7-6.1}$$

As a *very* first approximation, we neglect the rate of transfer to dead polymer from the live polymer with respect to the rate of propagation:

$$(k_{tm} MR_j \ll k_p MR_j)$$

so that the analytical solution obtained in Equation (7-65) can be used. Then

$$\frac{dP_j}{d\Theta} = \frac{k_{tm}}{k_p} R_j = \frac{k_{tm}}{k_p} \frac{I_0 \Theta^{j-1} e^{-\Theta}}{(j-1)!} \tag{E7-6.2}$$

Integrating, we obtain the dead polymer concentrations as a function of scaled time from CRC Mathematical Tables integral number 521:

$$P_j = \frac{k_{tm}}{k_p} I_0 \left[1 - e^{-\theta} \left[\sum_{\ell=0}^{j-1} \frac{\theta^{(j-1-\ell)}}{(j-1-\ell)!} \right] \right] \tag{E7-6.3}$$

We recall that the scaled time Θ can be calculated from

$$\Theta = \frac{M_0}{I_0} (1 - e^{-I_0 k_p t})$$

In many instances termination of anionic polymerization is brought about by adding a base to neutralize the propagating end of the polymer chain.

Other Useful Definitions.   The number-average kinetic chain length, V_N, is the ratio of the rate of the propagation rate to the rate of termination:

$$V_N = \frac{r_p}{r_t} \tag{7-70}$$

Most often, the PSSH is used so that $r_t = r_i$:

$$V_N = \frac{r_p}{r_i}$$

The long-chain approximation holds when V_N is large.

For the free-radical polymerization in which termination is by transfer to the monomer or a chain transfer agent and by addition, the kinetic chain length is

$$V_N = \frac{r_p}{r_t} = \frac{k_p M R^*}{k_{tm} M R^* + k_t (R^*)^2 + k_{Ct} R^* C} = \frac{k_p M}{k_{tm} M + k_t R^* + k_{ct} C}$$

$$\boxed{V_N = \frac{k_p M}{k_{tm} M + (2 k_t k_o I_2 f)^{1/2} + k_{ct} C}} \qquad (7\text{-}71)$$

For termination by combination

$$\overline{M}_n = 2 V_N M_M$$

and for termination by disproportionation

$$\overline{M}_n = V_N M_M$$

Excellent examples that will reinforce and expand the principles discussed in this section can be found in Holland and Anthony,[12] and the reader is encouraged to consult this text as the next step in studying polymer reaction engineering.

7.4 Enzymatic Reaction Fundamentals

7.4.1 Definitions and Mechanisms

Another class of reactions in which the PSSH is used is the enzymatically catalyzed reaction, which is characteristic of most biological reactions. An enzyme, E, is a protein or proteinlike substance with catalytic properties. A substrate, S, is the substance that is chemically transformed at an accelerated rate because of the action of the enzyme on it. An important property of enzymes is that they are specific in that *one* enzyme can catalyze only *one* reaction. For example, a protease hydrolyzes *only* bonds specific between specific amino acids in proteins, an amylase works on bonds between glucose molecules in starch, and liptase attacks fats, degrading them to fatty acids and glycerol. Consequently, unwanted products are easily controlled. Enzymes are produced only by living organisms, and commercial enzymes are generally produced by bacteria. Enzymes usually work (i.e., catalyze reactions) under mild conditions: pH 4 to 9 and temperatures 75 to 160°F.

Figure 7-10 shows the schematic of the enzyme chymotrypsin. In many cases the enzyme's active catalytic sites are found where the various loops interact. For chymotrypsin the catalytic sites are noted by the numbers 57, 102, and 195 in Figure 7-10. A number of structures of enzymes or pertinent information can be found on the following WWW sites:

---

[12]C. D. Holland and R. G. Anthony, *Fundamentals of Chemical Reaction Engineering*, 2nd ed., Prentice Hall, Upper Saddle River, N.J., 1977, p. 457.

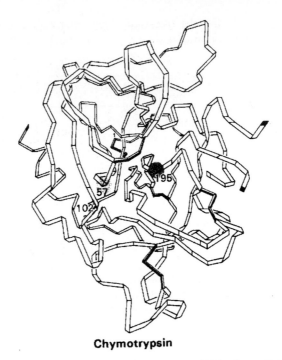

Chymotrypsin

Figure 7-10   Enzyme unease. [From *Biochemistry*, 3/E by Stryer © 1988 by Lubert Stryer. Used with permission of W. H. Freeman and Company.]

http://expasy.hcuge.ch/sprot/enzyme.html
http://www.wcslc.edu/pers_pages/w-pool/chem350/chap6/

These sites also give information about enzymatic reactions in general.

Most enzymes are named in terms of the reactions they catalyze. It is a customary practice to add the suffix *-ase* to a major part of the name of the substrate on which the enzyme acts. For example, the enzyme that catalyzes the decomposition of urea is urease and the enzyme that attacks tyrosine is tyrosinase.

There are three major types of enzyme reactions:

I. Soluble enzyme–insoluble substrate
II. Insoluble enzyme–soluble substrate
III. Soluble enzyme–soluble substrate

Types of enzyme reactions

An example of a type I reaction is the use of enzymes such as proteases or amylases in laundry detergents; however, this enzyme reaction has caused some controversy in relation to water pollution. Once in solution, the soluble enzyme may digest (i.e., break down) an insoluble substrate such as a blood stain.

A major research effort is currently being directed at type II reactions. By attaching active enzyme groups to solid surfaces, continuous processing units similar to the packed catalytic bed reactor discussed in Chapter 10 can be developed.

Clearly, the greatest activity in the study of enzymes has been in relation to biological reactions, because virtually every synthetic and degradation reaction in all living cells has been shown to be controlled and catalyzed by specific enzymes.[13] Many of these reactions are homogeneous in the liquid phase; that is, they are type III reactions (soluble enzyme–soluble substrate). In the following brief presentation we limit our discussion to type III reactions, although the resulting equations have been found to be applicable to type I and type II reactions in certain instances.

In developing some of the elementary principles of the kinetics of enzyme reactions, we shall discuss an enzymatic reaction that has been suggested by Levine and LaCourse as part of a system that would reduce the size of an artificial kidney.[14] The desired result is the production of an artificial kidney that could be worn by the patient and would incorporate a replaceable unit for the elimination of the nitrogenous waste products such as uric acid and creatinine. In the microencapsulation scheme proposed by Levine and LaCourse, the enzyme urease would be used in the removal of urea from the bloodstream. Here, the catalytic action of urease would cause urea to decompose into ammonia and carbon dioxide. The mechanism of the reaction is believed to proceed by the following sequence of elementary reactions:

1. The enzyme urease reacts with the substrate urea to form an enzyme–substrate complex, $E \cdot S$:

The reaction mechanism

$$NH_2CONH_2 + urease \xrightarrow{k_1} [NH_2CONH_2 \cdot urease]^* \qquad (7\text{-}72)$$

2. This complex can decompose back to urea and urease:

$$[NH_2CONH_2 \cdot urease]^* \xrightarrow{k_2} urease + NH_2CONH_2 \qquad (7\text{-}73)$$

3. Or it can react with water to give ammonia, carbon dioxide, and urease:

$$[NH_2CONH_2 \cdot urease]^* + H_2O \xrightarrow{k_3} 2NH_3 + CO_2 + urease \quad (7\text{-}74)$$

We see that some of the enzyme added to the solution binds to the urea, and some remains unbound. Although we can easily measure the total concentration of enzyme, (E_t), it is difficult to measure the concentration of free enzyme, (E).

Letting E, S, W, $E \cdot S$, and P represent the enzyme, substrate, water, the enzyme–substrate complex, and the reaction products, respectively, we can write Reactions (7-72), (7-73), and (7-74) symbolically in the forms

$$E + S \xrightarrow{k_1} E \cdot S \qquad (7\text{-}75)$$

$$E \cdot S \xrightarrow{k_2} E + S \qquad (7\text{-}76)$$

$$E \cdot S + W \xrightarrow{k_3} P + E \qquad (7\text{-}77)$$

Here $P = 2NH_3 + CO_2$.

---

[13]R. G. Denkewalter and R. Hirschmann, *Am. Sci.*, *57*(4), 389 (1969).
[14]N. Levine and W. C. LaCourse, *J. Biomed. Mater. Res.*, *1*, 275 (1967).

We need to replace
unbound enzyme
concentration (E) in
the rate law

The rate of disappearance of the substrate, $-r_s$, is

$$-r_s = k_1(E)(S) - k_2(E \cdot S) \tag{7-78}$$

The net rate of formation of the enzyme–substrate complex is

$$r_{E \cdot S} = k_1(E)(S) - k_2(E \cdot S) - k_3(W)(E \cdot S) \tag{7-79}$$

We note from the reaction sequence that the enzyme is not consumed by the reaction. The total concentration of the enzyme in the system, (E_t), is constant and equal to the sum of the concentrations of the free or unbonded enzyme E and the enzyme–substrate complex E·S:

Total enzyme
concentration,
bound + free

$$(E_t) = (E) + (E \cdot S) \tag{7-80}$$

Rearranging Equation (7-80), the enzyme concentration becomes

$$(E) = (E_t) - (E \cdot S) \tag{7-81}$$

Substituting Equation (7-81) into Equation (7-79) and using the PSSH for the enzyme complex gives

$$r_{E \cdot S} = 0 = k_1[(E_t) - (E \cdot S)](S) - k_2(E \cdot S) - k_3(E \cdot S)(W) \tag{7-82}$$

Solving for (E·S) yields

$$(E \cdot S) = \frac{k_1(E_t)(S)}{k_1(S) + k_2 + k_3(W)} \tag{7-83}$$

Next, substituting Equation (7-81) into Equation (7-78) yields

$$-r_s = k_1[(E_t) - (E \cdot S)](S) - k_2(E \cdot S) \tag{7-84}$$

Subtracting Equation (7-82) from Equation (7-84), we get

$$-r_s = k_3(W)(E \cdot S) \tag{7-85}$$

Substituting for (E·S) gives us

The final form of
the rate law

$$-r_s = \frac{k_1 k_3(W)(E_t)(S)}{k_1(S) + k_2 + k_3(W)} \tag{7-86}$$

Note: Throughout, $E_t \equiv (E_t)$ = total concentration of enzyme with typical units $(kmol/m^3)$.

7.4.2 Michaelis–Menten Equation

Because the reaction of urea and urease is carried out in aqueous solution, water is, of course, in excess, and the concentration of water is therefore considered constant. Let

$$k_3' = k_3(W) \qquad \text{and} \qquad K_m = \frac{k_3' + k_2}{k_1}$$

Dividing the numerator and denominator of Equation (7-86) by k_1, we obtain a form of the *Michaelis–Menten equation*:

$$-r_s = \frac{k_3'(S)(E_t)}{(S) + K_m} \tag{7-87}$$

where K_m is called the Michaelis constant. If, in addition, we let V_{max} represent the maximum rate of reaction for a given total enzyme concentration,

$$V_{max} = k_3'(E_t) \tag{7-88}$$

the Michaelis–Menten equation takes the familiar form

Michaelis–Menten
equation

$$\boxed{-r_s = \frac{V_{max}(S)}{K_m + (S)}} \tag{7-89}$$

For a given enzyme concentration, a sketch of the rate of disappearance of the substrate is shown as a function of the substrate concentration in Figure 7-11. At low substrate concentration,

$$-r_s \cong \frac{V_{max}(S)}{K_m}$$

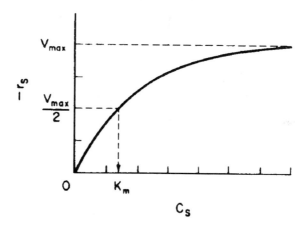

Figure 7-11   Identifying the Michaelis–Menten parameters.

At high substrate concentration,

$$(S) \gg K_m$$

and

$$-r_s \cong V_{max}$$

Consider the case when the substrate concentration is such that the reaction rate is equal to one-half the maximum rate,

$$-r_s = \frac{V_{max}}{2}$$

then

$$\frac{V_{\max}}{2} = \frac{V_{\max}(S_{1/2})}{K_m + (S_{1/2})} \tag{7-90}$$

Solving Equation (7-90) for the Michaelis constant yields

Interpretation of
Michaelis constant

$$K_m = (S_{1/2}) \tag{7-91}$$

The Michaelis constant is equal to the substrate concentration at which the rate of reaction is equal to one-half the maximum rate.

The parameters $V_{\max}$ and K_m characterize the enzymatic reactions that are described by Michaelis–Menten kinetics. $V_{\max}$ is dependent on total enzyme concentration, whereas K_m is not.

Example 7–7   Evaluation of Michaelis–Menten Parameters $V_{\max}$ and K_m

Determine the Michaelis–Menten parameters $V_{\max}$ and K_m for the reaction

$$\text{urea} + \text{urease} \underset{k_2}{\overset{k_1}{\rightleftharpoons}} [\text{urea} \cdot \text{urease}]^* \overset{k_3}{\underset{-H_2O}{\longrightarrow}} 2NH_3 + CO_2 + \text{urease}$$

The rate of reaction is given as a function of urea concentration in the following table:

C_{urea} (kmol/m³)	0.2	0.02	0.01	0.005	0.002
$-r_{urea}$ (kmol/m³·s)	1.08	0.55	0.38	0.2	0.09

Solution

Inverting Equation (7-89) gives us

$$\frac{1}{-r_s} = \frac{(S) + K_m}{V_{\max}(S)} = \frac{1}{V_{\max}} + \frac{K_m}{V_{\max}} \frac{1}{(S)} \tag{E7-7.1}$$

or

$$\frac{1}{-r_{urea}} = \frac{1}{V_{\max}} + \frac{K_m}{V_{\max}}\left(\frac{1}{C_{urea}}\right) \tag{E7-7.2}$$

A plot of the reciprocal reaction rate versus the reciprocal urea concentration should be a straight line with an intercept $1/V_{\max}$ and slope $K_m/V_{\max}$. This type of plot is called a *Lineweaver–Burk plot*. The data in Table E7-7.1 are presented in Figure E7-7.1 in the form of a Lineweaver–Burk plot. The intercept is 0.75, so

$$\frac{1}{V_{\max}} = 0.75 \text{ m}^3 \cdot \text{s/kmol}$$

Therefore, the maximum rate of reaction is

$$V_{\max} = 1.33 \text{ kmol/m}^3 \cdot \text{s} = 1.33 \text{ mol/dm}^3 \cdot \text{s}$$

TABLE E7-7.1   RAW AND PROCESSED DATA

C_{urea} (kmol/m^3)	$-r_{\text{urea}}$ (kmol/m$^3\cdot$s)	$1/C_{\text{urea}}$ (m^3/kmol)	$1/-r_{\text{urea}}$ (m$^3\cdot$s/kmol)
0.20	1.08	5.0	0.93
0.02	0.55	50.0	1.82
0.01	0.38	100.0	2.63
0.005	0.20	200.0	5.00
0.002	0.09	500.0	11.11

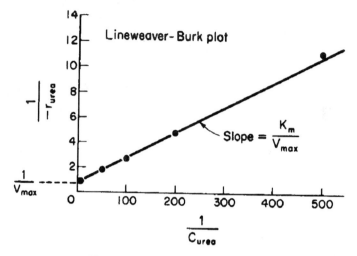

Figure E7-7.1   Lineweaver-Burk Plot.

From the slope, which is 0.02 s, we can calculate the Michaelis constant, K_m:

For enzymatic reactions the two key rate-law parameters are V_{max} and K_m

$$\frac{K_m}{V_{\text{max}}} = \text{slope} = 0.02 \text{ s}$$

$$K_m = 0.0266 \text{ kmol/m}^3$$

Substituting K_m and V_{max} into Equation (7-89) gives us

$$-r_s = \frac{1.33 C_{\text{urea}}}{0.0266 + C_{\text{urea}}} \qquad \text{(E7-7.3)}$$

where C_{urea} has units of kmol/m^3 and $-r_s$ has units of kmol/m$^3\cdot$s. Levine and LaCourse suggest that the total concentration of urease, (E_t), corresponding to the value of V_{max} above is approximately 5 g/dm^3.

7.4.3 Batch Reactor Calculations

A mole balance on urea in the batch reactor gives

Mole balance

$$-\frac{dN_{\text{urea}}}{dt} = -r_{\text{urea}}V$$

Because this reaction is liquid phase, the mole balance can be put in the following form:

$$-\frac{dC_{urea}}{dt} = -r_{urea} \tag{7-92}$$

The rate law for urea decomposition is

Rate law
$$-r_{urea} = \frac{V_{max}C_{urea}}{K_m + C_{urea}} \tag{7-93}$$

Substituting Equation (7-93) into Equation (7-92) and then rearranging and integrating, we get

Combine
$$t = \int_{C_{urea}}^{C_{urea0}} \frac{dC_{urea}}{-r_{urea}} = \int_{C_{urea}}^{C_{urea0}} \frac{K_m + C_{urea}}{V_{max}C_{urea}} dC_{urea}$$

Integrate
$$t = \frac{K_m}{V_{max}} \ln \frac{C_{urea0}}{C_{urea}} + \frac{C_{urea0} - C_{urea}}{V_{max}} \tag{7-94}$$

We can write Equation (7-94) in terms of conversion as

$$C_{urea} = C_{urea0}(1 - X)$$

Time to achieve a conversion X in a batch enzymatic reaction

$$\boxed{t = \frac{K_m}{V_{max}} \ln \frac{1}{1 - X} + \frac{C_{urea0}X}{V_{max}}} \tag{7-95}$$

The parameters K_m and V_{max} can readily be determined from batch reactor data by using the integral method of analysis. Dividing both sides of Equation (7-95) by tK_m/V_{max} and rearranging yields

$$\frac{1}{t} \ln \frac{1}{1 - X} = \frac{V_{max}}{K_m} - \frac{C_{urea0}X}{K_m t} \tag{7-96}$$

We see that K_m and V_{max} can be determined from the slope and intercept of a plot of $1/t \ln[1/(1 - X)]$ versus X/t. We could also express the Michaelis–Menten equation in terms of the substrate concentration S:

$$\frac{1}{t} \ln \frac{S_0}{S} = \frac{V_{max}}{K_m} - \frac{S_0 - S}{K_m t} \tag{7-97}$$

where S_0 is the initial concentration of substrate. In cases similar to Equation (7-97) where there is no possibility of confusion, we shall not bother to enclose the substrate or other species in parentheses to represent concentration [i.e., $C_S \equiv (S) \equiv S$]. The corresponding plot in terms of substrate concentration is shown in Figure 7-12.

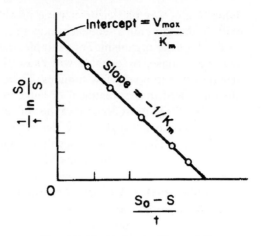

Figure 7-12   Evaluating $V_{\max}$ and K_m.

Example 7–8   Batch Enzymatic Reactors

Calculate the time needed to convert 80% of the urea to ammonia and carbon dioxide in a 0.5-dm³ batch reactor. The initial concentration of urea is 0.1 mol/dm³, and the urease concentration is 0.001 g/dm³. The reaction is to be carried out isothermally at the same temperature at which the data in Table E7-7.1 were obtained.

Solution

We can use Equation (7-95),

$$t = \frac{K_m}{V_{\max}} \ln \frac{1}{1-X} + \frac{C_{\text{urea}0} X}{V_{\max}} \tag{7-95}$$

where $K_m = 0.0266$ g mol/dm³, $X = 0.8$, and $C_{\text{urea}0} = 0.1$ g mol/dm³, $V_{\max}$ was 1.33 g mol/dm³·s. However, for the conditions in the batch reactor, the enzyme concentration is only 0.001 g/dm³. Because $V_{\max} = E_t \cdot k_3$, $V_{\max}$ for the second enzyme concentration is

$$V_{\max2} = \frac{E_{t2}}{E_{t1}} V_{\max1} = \frac{0.001}{5} \times 1.33 = 2.66 \times 10^{-4}\ \text{mol/s} \cdot \text{dm}^3$$

$$X = 0.8$$

$$t = \frac{0.0266}{0.000266} \ln \frac{1}{0.2} + \frac{(0.8)(0.1)}{0.000266}$$

$$= 160.9 + 300.8$$

$$= 461.7\ \text{s}$$

7.4.4 Inhibition of Enzyme Reactions

Another factor that greatly influences the rates of enzyme-catalyzed reactions in addition to pH is the presence of an inhibitor. The most dramatic consequences of enzyme inhibition are found in living organisms, where the inhibition of any particular enzyme involved in a primary metabolic sequence will render the entire sequence inoperative, resulting in either serious damage or death of the organism. For example, the inhibition of a single enzyme, cytochrome oxidase, by cyanide will cause the aerobic oxidation process to stop; death occurs in a very few minutes. There are also beneficial inhibitors such as the ones used in the treatment of leukemia and other neoplastic diseases.

The three most common types of reversible inhibition occurring in enzymatic reactions are *competitive*, *uncompetitive*, and *noncompetitive*. (See Problem P7-12$_B$) The enzyme molecule is analogous to the heterogeneous catalytic surface in that it contains active sites. When competitive inhibition occurs, the substrate and inhibitor are usually similar molecules that compete for the same site on the enzyme. Uncompetitive inhibition occurs when the inhibitor deactivates the enzyme–substrate complex, usually by attaching itself to both the substrate and enzyme molecules of the complex. Noncompetitive inhibition occurs with enzymes containing at least two different types of sites. The inhibitor attaches to only one type of site and the substrate only to the other. Derivation of the rate laws for these three types of inhibition is shown on the CD-ROM.

7.4.5 Multiple Enzyme and Substrate Systems

In the preceding section we discussed how the addition of a second substrate, I, to enzyme-catalyzed reactions could deactivate the enzyme and greatly inhibit the reaction. In the present section we look not only at systems in which the addition of a second substrate is necessary to activate the enzyme, but also other multiple-enzyme and multiple-substrate systems in which cyclic regeneration of the activated enzyme occurs.

Enzyme Regeneration.   The first example considered is the oxidation of glucose (S_r) with the aid of the enzyme glucose oxidase [represented as either G.O. or (E_o)] to give δ-gluconolactone (P):

$$\text{glucose} + \text{G.O.} \rightleftharpoons (\text{glucose} \cdot \text{G.O.}) \rightleftharpoons (\delta\text{-lactone} \cdot \text{G.O.H}_2)$$

$$\rightleftharpoons \delta\text{-lactone} + \text{G.O.H}_2$$

In this reaction, the reduced form of glucose oxidase (G.O.H_2), which will be represented by E_r, cannot catalyze further reactions until it is oxidized back to E_o. This oxidation is usually carried out by adding molecular oxygen to the system so that glucose oxidase, E_o, is regenerated. Hydrogen peroxide is also produced in this oxidation regeneration step:

$$\text{G.O.H}_2 + \text{O}_2 \longrightarrow \text{G.O.} + \text{H}_2\text{O}_2$$

Overall, the reaction is written

$$\text{glucose} + \text{O}_2 \xrightarrow[\text{oxidase}]{\text{glucose}} \text{H}_2\text{O}_2 + \delta\text{-gluconolactone}$$

In biochemistry texts, reactions of this type involving regeneration are usually written in the form

$$\text{glucose}(S_r) \quad \begin{matrix} \\ \end{matrix} \begin{matrix} \text{G.O.}(E_o) \\ \text{G.O.H}_2(E_r) \end{matrix} \quad \begin{matrix} \text{H}_2\text{O}_2(P_2) \\ \text{O}_2(S_2) \end{matrix}$$
$$\delta\text{-lactone}(P_1)$$

Derivation of the rate laws for this reaction sequence is given on the CD-ROM.

Enzyme Cofactors. In many enzymatic reactions, and in particular biological reactions, a second substrate (i.e., species) must be introduced to activate the enzyme. This substrate, which is referred to as a *cofactor* or *coenzyme* even though it is not an enzyme as such, attaches to the enzyme and is most often either reduced or oxidized during the course of the reaction. The enzyme–cofactor complex is referred to as a *holoenzyme*. The inactive form of the enzyme–cofactor complex for a specific reaction and reaction direction is called an *apoenzyme*. An example of the type of system in which a cofactor is used is the formation of ethanol from acetaldehyde in the presence of the enzyme alcohol dehydrogenase (ADH) and the cofactor nicotinamide adenine dinucleotide (NAD):

$$
\begin{matrix}
\text{alcohol dehydrogenase} \\
\text{acetaldehyde }(S_1) \qquad\qquad \text{NADH }(S_2) \\
\text{H}^+ \\
\text{ethanol }(P_1) \qquad\qquad \text{NAD}^+ (S_2^*)
\end{matrix}
$$

Derivation of the rate laws for this reaction sequence is given on the CD-ROM.

7.5 Bioreactors

Because enzymatic reactions are involved in the growth of microorganisms, we now proceed to study microbial growth and bioreactors. Not surprisingly, the Monod equation, which describes the growth law for a number of bacteria, is similar to the Michaelis–Menton equation. Consequently, even though bioreactors are not truly homogeneous because of the presence of living cells, we include them in this chapter as a logical progression from enzymatic reactions.

The use of living cells to produce marketable chemical products is becoming increasingly important. By the year 2000, chemicals, agricultural products, and food products produced by biosynthesis will have risen from the 1990 market of $275 million to around $17 billion.[15] Both microorganisms and mammalian cells are being used to produce a variety of products, such as insulin, most antibiotics, and polymers. It is expected that in the future a number of organic chemicals currently derived from petroleum will be produced by living cells. The advantages of bioconversions are mild reaction conditions, high yields (e.g., 100% conversion of glucose to gluconic acid with *Aspergillus niger*), that organisms contain several enzymes that can catalyze successive steps in a reaction, and most important, that organisms act as stereospecific

The growth of biotechnology

---

[15] *Frontiers in Chemical Engineering*, National Academy Press, Washington, D.C., 1988.

catalysts. A common example of specificity in bioconversion production of a *single* desired isomer that when produced chemically yields a mixture of isomers is the conversion of *cis*-propenylphonic acid to the antibiotic (−) *cis*-1,2-epoxypropyl-phosphonic acid.

In biosynthesis, the cells, also referred to as the *biomass*, consume nutrients to grow and produce more cells and important products. Internally, a cell uses its nutrients to produce energy and more cells. This transformation of nutrients to energy and bioproducts is accomplished through a cell's use of a number of different enzymes (catalysts) in a series of reactions to produce metabolic products. These products can either remain in the cell (intracellular) or be secreted from the cells (extracellular). In the former case the cells must be lysed (ruptured) and the product purified from the whole broth (reaction mixture).

In general, the growth of an aerobic organism follows the equation

Cell multiplication

$$[\text{cells}] + \begin{bmatrix} \text{carbon} \\ \text{source} \end{bmatrix} + \begin{bmatrix} \text{nitrogen} \\ \text{source} \end{bmatrix} + \begin{bmatrix} \text{oxygen} \\ \text{source} \end{bmatrix} + \begin{bmatrix} \text{phosphate} \\ \text{source} \end{bmatrix} + \cdots$$

$$[CO_2] + [H_2O] + [\text{products}] + \begin{bmatrix} \text{more} \\ \text{cells} \end{bmatrix} \xleftarrow{\begin{array}{c} \text{culture media} \\ \text{conditions} \\ \text{(pH, temperature, etc.)} \end{array}} \qquad (7\text{-}98)$$

A more abbreviated form generally used is

$$\text{substrate} \xrightarrow{\text{Cells}} \text{more cells} + \text{product} \qquad (7\text{-}99)$$

The products in Equation (7-99) include CO_2, water, proteins, and other species specific to the particular reaction. An excellent discussion of the stoichiometry (atom and mole balances) of Equation (7-98) can be found in Wang[16] and in Bailey and Ollis.[17] The substrate culture medium contains all the nutrients (carbon, nitrogen, etc.) along with other chemicals necessary for growth. Because, as we will soon see, the rate of this reaction is proportional to the cell concentration, the reaction is autocatalytic. A rough schematic of a simple batch biochemical reactor and the growth of two types of microorganisms, cocci (i.e., spherical) bacteria and yeast, is shown in Figure 7-13.

7.5.1 Cell Growth

Stages of cell growth in a batch reactor are shown schematically in Figure 7-14. Here, the log of the number of living cells is shown as a function of time. Initially, a small number of cells is inoculated into (i.e., added to) the batch reactor containing the nutrients and the growth process begins.

Lag phase

In phase I, called the lag phase, there is little increase in cell concentration. During the lag phase the cells are adjusting to their new environment,

[16]D. C. Wang et al., *Fermentation and Enzyme Technology*, Wiley, New York, 1979.

[17]T. J. Bailey and D. Ollis, *Biochemical Engineering*, 2nd ed., McGraw-Hill, New York, 1987.

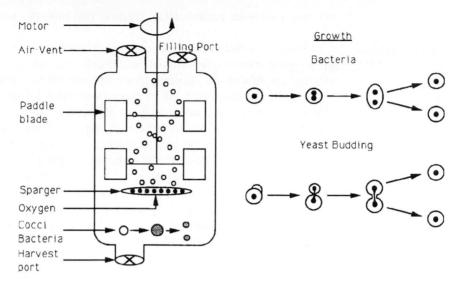

Batch Bioreactor

Figure 7-13   Batch bioreactor.

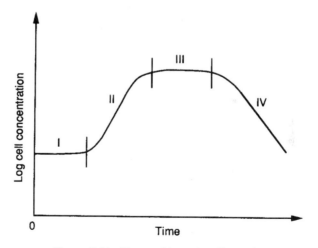

Figure 7-14   Phases of bacteria cell growth.

synthesizing enzymes, and getting ready to begin reproducing. It is during this time that the cells carry out such functions as synthesizing transport proteins for moving the substrate into the cell, synthesizing enzymes for utilizing the new substrate, and beginning the work for replicating the cells' genetic material. The duration of the lag phase depends upon the growth medium (i.e., reactor) from which the inoculum was taken relative to the reaction medium in which it is placed. If the inoculum is similar to the medium of the batch reactor, the lag phase will be almost nonexistent. If, however, the inoculum were placed in a medium with a different nutrient or other contents, or if the inoculum culture were in the stationary or death phase, the cells would have to readjust their metabolic path to allow them to consume the nutrients in their new environment.

Phase II is called the exponential growth phase owing to the fact that the cell's growth rate is proportional to the cell concentration. In this phase the cells are dividing at the maximum rate because all of the enzyme's pathways for metabolizing the media are in place (as a result of the lag phase) and the cells are able to use the nutrients most efficiently.

Phase III is the stationary phase, during which the cells reach a minimum biological space where the lack of one or more nutrients limits cell growth. During the stationary phase, the growth rate is zero as a result of the depletion of nutrients and essential metabolites. Many important fermentation products, including most antibiotics, are produced in the stationary phase. For example,

penicillin produced commercially using the fungus *Penicillium chrysogenum* is formed only after cell growth has ceased. Cell growth is also slowed by the buildup of organic acids and toxic materials generated during the growth phase.

The final phase, Phase IV, is the death phase where a decrease in live cell concentration occurs. This decline is a result of either the toxic by-products and/or the depletion of nutrient supply.

7.5.2 Rate Laws

While many laws exist for the cell growth rate of new cells, that is,

$$\text{cells} + \text{substrate} \longrightarrow \text{more cells} + \text{product}$$

the most commonly used expression is the *Monod* equation for exponential growth:

$$r_g = \mu C_c \tag{7-100}$$

where r_g = cell growth rate, g/dm^3·s
C_c = cell concentration, g/dm^3
μ = specific growth rate, s^{-1}

The specific cell growth rate can be expressed as

$$\mu = \mu_{max} \frac{C_s}{K_s + C_s} \quad \text{s}^{-1} \tag{7-101}$$

where μ_{max} = a maximum specific growth reaction rate, s^{-1}
K_s = the *Monod* constant, g/dm^3
C_s = substrate concentration, g/dm^3

For a number of different bacteria, the constant K_s is small, in which case the rate law reduces to

$$r_g = \mu_{max} C_c \tag{7-102}$$

The growth rate, r_g, often depends on more than one nutrient concentration; however, the nutrient that is limiting is usually the one used in Equation (7-101). Combining Equations (7-100) and (7-101) yields

$$r_g = \mu_{max} \frac{C_c C_s}{K_s + C_s} \qquad (7\text{-}103)$$

In many systems the product inhibits the rate of growth. A classic example of this inhibition is in wine-making, where the fermentation of glucose to produce ethanol is inhibited by the product ethanol. There are a number of different equations to account for inhibition; one such rate law takes the form

$$r_g = k_{obs} \frac{\mu_{max} C_s C_c}{K_s + C_s} \qquad (7\text{-}104)$$

where

Product inhibition

$$k_{obs} = \left(1 - \frac{C_p}{C_p^*}\right)^n \qquad (7\text{-}105)$$

with

C_p^* = product concentration at which all metabolism ceases, g/dm^3
n = empirical constant

For the glucose-to-ethanol fermentation, typical inhibition parameters are

$$n = 0.5 \quad \text{and} \quad C_p^* = 93 \text{ g/dm}^3$$

In addition to the Monod equation, two other equations are also commonly used to describe the cell growth rate; they are the *Tessier* equation,

$$r_g = \mu_{max}\left[1 - \exp\left(-\frac{C_s}{k}\right)\right] C_c \qquad (7\text{-}106)$$

and the *Moser* equation,

$$r_g = \frac{\mu_{max} C_c}{(1 + k C_s^{-\lambda})} \qquad (7\text{-}107)$$

where λ and k are empirical constants determined by a best fit of the data.

The Moser and Tessier growth laws are often used because they have been found to better fit experimental data at the beginning or end of fermentation. Other growth equations can be found in Dean.[18]

The cell death rate is given by

$$r_d = (k_d + k_t C_t) C_c \qquad (7\text{-}108)$$

where C_t is the concentration of a substance toxic to the cell. The specific death rate constants k_d and k_t refer to the natural death and death due to a toxic

---

[18]A. R. C. Dean, *Growth, Function, and Regulation in Bacterial Cells*, Oxford University Press, London, 1964.

substance, respectively. Representative values of k_d range from 0.1 h^{-1} to less than 0.0005 h^{-1}. The value of k_t depends on the nature of the toxin.

Doubling times   Microbial growth rates are measured in terms of *doubling times*. Doubling time is the time required for a mass of an organism to double. Typical doubling times for bacteria range from 45 minutes to 1 hour but can be as fast as 15 minutes. Doubling times for simple eukaryotes, such as yeast, range from 1.5 to 2 hours but may be as fast as 45 minutes.

7.5.3 Stoichiometry

The stoichiometry for cell growth is very complex and varies with microorganism/nutrient system and environmental conditions such as pH, temperature, and redox potential. This complexity is especially true when more than one nutrient contributes to cell growth, as is usually the case. We shall focus our discussion on a simplified version for cell growth, one that is limited by only one nutrient in the medium. In general, we have

$$\text{cells} + \text{substrate} \longrightarrow \text{more cells} + \text{product}$$
$$\text{S} \xrightarrow{\text{cells}} Y_{c/s}\text{C} + Y_{p/s}\text{P} \tag{7-109}$$

where the yield coefficients are

$$Y_{c/s} = \frac{\text{mass of new cells formed}}{\text{mass of substrate consumed to produce new cells}}$$

with

$$Y_{c/s} = \frac{1}{Y_{s/c}}$$

The stoichiometric yield coefficient that relates the amount of product formed per mass of substrate consumed is

$$Y_{p/s} = \frac{\text{mass of product formed}}{\text{mass of substrate consumed to form product}}$$

In addition to consuming substrate to produce new cells, part of the substrate must be used just to maintain a cell's daily activities. The corresponding maintenance utilization term is

Cell maintenance $$m = \frac{\text{mass of substrate consumed for maintenance}}{\text{mass of cells} \cdot \text{time}}$$

A typical value is

$$m = 0.05 \, \frac{\text{g substrate}}{\text{g dry weight}} \frac{1}{\text{h}} = 0.05 \text{ h}^{-1}$$

The rate of substrate consumption for maintenance whether or not the cells are growing is

$$\boxed{r_{sm} = mC_c}$$

(7-110)

The yield coefficient $Y'_{c/s}$ accounts for substrate consumption for maintenance:

$$Y'_{c/s} = \frac{\text{mass of new cells formed}}{\text{mass of substrate consumed}}$$

Product formation can take place during different phases of cell growth. When product is produced only during the growth phase, we can write

$$r_p = Y_{p/c} r_g$$

However, when product is produced during the stationary phase, we can relate product formation to substrate consumption by

$$r_p = Y_{p/s}(-r_s)$$

We now come to the difficult task of relating the rate of nutrient consumption, $-r_s$, to the rates of cell growth, product generation, and cell maintenance. In general, we can write

Substrate accounting

$$\begin{bmatrix} \text{net rate of} \\ \text{substrate} \\ \text{consumption} \end{bmatrix} = \begin{bmatrix} \text{rate} \\ \text{consumed} \\ \text{by cells} \end{bmatrix} + \begin{bmatrix} \text{rate} \\ \text{consumed to} \\ \text{form product} \end{bmatrix} + \begin{bmatrix} \text{rate} \\ \text{consumed for} \\ \text{maintenance} \end{bmatrix}$$

$$-r_s \quad = \quad Y_{s/c} r_g \quad + \quad Y_{s/p} r_p \quad + \quad mC_c$$

(7-111)

In a number of cases extra attention must be paid to the substrate balance. If product is produced during the growth phase, it may not be possible to separate out the amount of substrate consumed for growth from that consumed to produce the product. Under these circumstances all the substrate consumed is lumped into the stoichiometric coefficient, $Y_{s/c}$, and the rate of substrate disappearance is

Product formation in the growth phase

$$-r_s = Y_{s/c} r_g + mC_c$$

(7-112)

The corresponding rate of product formation is

The stationary phase

$$r_p = r_g Y_{p/c}$$

(7-113)

Because there is no growth during the stationary phase, it is clear that Equation (7-112) cannot be used to account for substrate consumption, nor can the rate of product formation be related to the growth rate [e.g., Equation (7-113)]. Many antibiotics, such as penicillin, are produced in the stationary phase. In this phase, the nutrient consumed for growth has become virtually exhausted and a different nutrient, called the secondary nutrients, is used for cell maintenance and to produce the desired product. Usually, the rate law for product formation during the stationary phase is similar in form to the Monod equation, that is,

Product formation in the stationary phase

$$r_p = \frac{k_p C_{sn} C_c}{K_{sn} + C_{sn}} \tag{7-114}$$

where C_{sn} = concentration of the secondary nutrient, g/dm^3
k_p = specific rate constant with product, s^{-1}
C_c = cell concentration, g/dm^3
K_{sn} = constant, g/dm^3
$r_p = Y_{p/sn}(-r_{sn})$

The net rate of substrate consumption during the stationary phase is

$$-r_{sn} = mC_c + Y_{sn/p}r_p$$

$$= mC_c + \frac{Y_{sn/p}k_p C_{sn} C_c}{K_{sn} + C_{sn}} \tag{7-115}$$

Because the desired product can be produced when there is no cell growth, it is always best to relate the product concentration to the change in substrate concentration. For a batch system the concentration of product, C_p, formed after a time t can be related to the substrate concentration, C_s, at that time.

$$C_p = Y_{p/s}(C_{s0} - C_s) \tag{7-116}$$

We have considered two limiting situations for relating substrate consumption to cell growth and product formation; product formation only during the growth phase and product formation only during the stationary phase. An example where neither of these situations apply is fermentation using lactobacillus, where lactic acid is produced during both the logarithmic growth and stationary phase.

7.5.4 Mass Balances

There are two ways that we could account for the growth of microorganisms. One is to account for the number of living cells and the other is to account for the mass of the living cells. We shall use the latter. A mass balance on the microorganism in a CSTR (chemostat) (shown in Figure 7-15) of constant volume is

$$\begin{bmatrix} \text{rate of} \\ \text{accumulation} \\ \text{of cells,} \\ \text{g/s} \end{bmatrix} = \begin{bmatrix} \text{rate of} \\ \text{cells} \\ \text{entering,} \\ \text{g/s} \end{bmatrix} - \begin{bmatrix} \text{rate of} \\ \text{cells} \\ \text{leaving,} \\ \text{g/s} \end{bmatrix} + \begin{bmatrix} \text{net rate of} \\ \text{generation} \\ \text{of live cells,} \\ \text{g/s} \end{bmatrix} \tag{7-117}$$

$$V\frac{dC_c}{dt} = v_0 C_{c0} - v C_c + (r_g - r_d)V$$

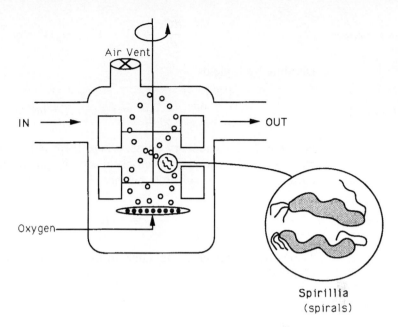

Figure 7-15   Bioreactor.

The corresponding substrate balance is

$$
\begin{bmatrix} \text{rate of} \\ \text{accumulation} \\ \text{of substrate,} \\ \text{g/s} \end{bmatrix} = \begin{bmatrix} \text{rate of} \\ \text{substrate} \\ \text{entering,} \\ \text{g/s} \end{bmatrix} - \begin{bmatrix} \text{rate of} \\ \text{substrate} \\ \text{leaving,} \\ \text{g/s} \end{bmatrix} + \begin{bmatrix} \text{rate of} \\ \text{substrate} \\ \text{generation,} \\ \text{g/s} \end{bmatrix} \qquad (7\text{-}118)
$$

$$
V\frac{dC_s}{dt} \quad = \quad v_0 C_{s0} \quad - \quad v C_s \quad + \quad r_s V
$$

In most systems the entering microorganism concentration C_{c0} is zero. For a batch system the mass balances develop are as follows:

Cell

Batch
$$
V\frac{dC_c}{dt} = r_g V - r_d V
$$

Dividing by the reactor volume V gives

$$
\frac{dC_c}{dt} = r_g - r_d \qquad (7\text{-}119)
$$

Substrate

The rate of disappearance of substrate, $-r_s$, results from substrate used for cell growth and substrate used for cell maintenance,

The mass balances

$$V\frac{dC_s}{dt} = r_s V = Y_{s/c}(-r_g)V - mC_c V \qquad (7\text{-}120)$$

Dividing by V yields

Growth phase

$$\frac{dC_s}{dt} = Y_{s/c}(-r_g) - mC_c$$

For cells in the stationary phase, where there is no growth, cell maintenance and product formation are the only reactions to consume the substrate. Under these conditions the substrate balance, Equation (7-118), reduces to

Stationary phase

$$V\frac{dC_s}{dt} = -mC_c V + Y_{s/p}(-r_p)V \qquad (7\text{-}121)$$

Typically, r_p will have the same form of the rate law as r_g [e.g., Equation (7-114)].

Batch stationary growth phase

Product

The rate of product formation, r_p, can be related to the rate of substrate consumption through the following balance:

$$V\frac{dC_p}{dt} = r_p V = Y_{p/s}(-r_s)V \qquad (7\text{-}122)$$

During the growth phase we could also relate the rate of formation of product, r_p, to the cell growth rate, r_g. The coupled first-order ordinary differential equations above can be solved by a variety of numerical techniques.

Example 7–9   Bacteria Growth in a Batch Reactor

Glucose-to-ethanol fermentation is to be carried out in a batch reactor using an organism such as *Saccharomyces cerevisiae*. Plot the concentrations of cells, substrate, and product and growth rates as functions of time. The initial cell concentration is 1.0 g/dm³ and the substrate (glucose) concentration is 250 g/dm³.

Additional data [partial source: R. Miller and M. Melick, *Chem. Eng.*, Feb. 16, p. 113 (1987)]:

$$C_p^* = 93 \text{ g/dm}^3 \qquad\qquad Y_{c/s} = 0.08 \text{ g/g}$$

$$n = 0.52 \qquad\qquad Y_{p/s} = 0.45 \text{ g/g (est.)}$$

$$\mu_{max} = 0.33 \text{ h}^{-1} \qquad\qquad Y_{p/c} = 5.6 \text{ g/g (est.)}$$

$$K_s = 1.7 \text{ g/dm}^3 \qquad\qquad k_d = 0.01 \text{ h}^{-1}$$

$$m = 0.03 \text{ (g substrate)}/\text{(g cells} \cdot \text{h)}$$

Solution

1. **Mass balances:**

The algorithm

Cells:
$$V\frac{dC_c}{dt} = (r_g - r_d)V \tag{E7-9.1}$$

Substrate:
$$V\frac{dC_s}{dt} = Y_{s/c}(-r_g)V - r_{sm}V \tag{E7-9.2}$$

Product:
$$V\frac{dC_p}{dt} = Y_{p/c}(r_gV) \tag{E7-9.3}$$

2. **Rate laws:**

$$r_g = \mu_{max}\left(1 - \frac{C_p}{C_p^*}\right)^{0.52}\frac{C_cC_s}{K_s + C_s} \tag{E7-9.4}$$

$$r_d = k_dC_c \tag{E7-9.5}$$

$$r_{sm} = mC_c \tag{7-110}$$

3. **Stoichiometry:**

$$r_p = Y_{p/c}r_g \tag{E7-9.6}$$

4. **Combining gives**

$$\frac{dC_c}{dt} = \mu_{max}\left(1 - \frac{C_p}{C_p^*}\right)^{0.52}\frac{C_cC_s}{K_s + C_s} - k_dC_c \tag{E7-9.7}$$

Cells
Substrate
Product

$$\frac{dC_s}{dt} = -Y_{s/c}\mu_{max}\left(1 - \frac{C_p}{C_p^*}\right)^{0.52}\frac{C_cC_s}{K_m + C_s} - mC_c \tag{E7-9.8}$$

$$\frac{dC_p}{dt} = Y_{p/c}r_g$$

These equations were solved on an ODE equation solver (see Table E7-9.1). The results are shown in Figure E7-9.1 for the parameter values given in the problem statement.

TABLE E7-9.1.   POLYMATH PROGRAM

Equations:	*Initial Values:*
d(cc)/d(t)=rg-rd	1
d(cs)/d(t)=ysc*(-rg)-rsm	250
d(cp)/d(t)=rg*ypc	0
rd=cc*.01	
ysc=1/.08	
ypc=5.6	
ks=1.7	
m=.03	
umax=.33	
rsm=m*cc	
kobs=(umax*(1-cp/93)^.52)	
rg=kobs*cc*cs/(ks+cs)	
$t_0 = 0$, $t_f = 12$	

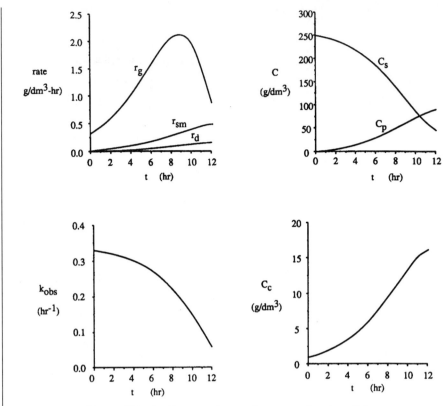

Figure E7-9.1   Concentrations and rates as a function of time.

7.5.5 Chemostats

Chemostats are essentially CSTRs that contain microorganisms. A typical chemostat is shown in Figure 7-16, along with the associated monitoring equipment and pH controller. One of the most important features of the chemostat is that it allows the operator to control the cell growth rate. This control of the growth rate is achieved by adjusting the volumetric feed rate (dilution rate).

7.5.6 Design Equations

CSTR

In this section we return to mass equations on the cells [Equation (7-117)] and substrate [Equation (7-118)] and consider the case where the volumetric flow rates in and out are the same and that no live (i.e., viable) cells enter the chemostat. We next define a parameter common to bioreactors called the dilution rate, D. The dilution rate is

$$D = \frac{v_0}{V}$$

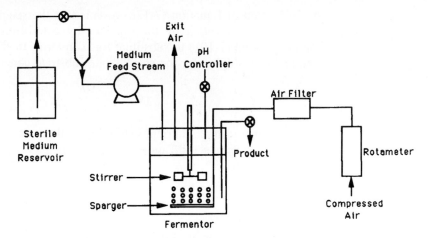

Figure 7-16 Chemostat system.

and is simply the reciprocal of the space time τ. Dividing Equations (7-117) and (7-118) by V and using the definition of the dilution rate, we have

$$\text{accumulation} = \text{in} \quad -\text{out} \quad +\text{generation}$$

CSTR mass balances

Cell: $$\frac{dC_c}{dt} = 0 \quad -DC_c + (r_g - r_d) \qquad (7\text{-}123)$$

Substrate: $$\frac{dC_s}{dt} = DC_{s0} - DC_s + r_s \qquad (7\text{-}124)$$

Using the Monod equation, the growth rate is determined to be

$$r_g = \mu C_c \qquad (7\text{-}100)$$

where

$$\mu = \frac{\mu_{max}C_s}{K_s + C_s} \qquad (7\text{-}101)$$

For steady-state operation we have

$$DC_c = r_g - r_d \qquad (7\text{-}125)$$

and

$$D(C_{s0} - C_s) = r_s \qquad (7\text{-}126)$$

We now neglect the death rate, r_d, and combine Equations (7-100) and (7-125) for steady-state operation to obtain the mass flow rate of cells out of the system, $\dot{m}_c$.

$$\dot{m}_c = C_c v_0 = r_g V = \mu C_c V \qquad (7\text{-}127)$$

After we divide by $C_c V$,

Dilution rate

$$\boxed{D = \mu} \qquad (7\text{-}128)$$

An inspection of Equation (7-128) reveals that the specific growth rate of the cells *can be controlled* by the operator by controlling the dilution rate D. Using Equation (7-101) to substitute for μ in terms of the substrate concentration and then solving for the steady-state substrate concentration yields

$$C_s = \frac{DK_s}{\mu_{max} - D} \tag{7-129}$$

Assuming that a single nutrient is limiting, cell growth is the only process contributing to substrate utilization, and that cell maintenance can be neglected, the stoichiometry is

$$-r_s = r_g Y_{s/c} \tag{7-130}$$

$$C_c = Y_{c/s}(C_{s0} - C_s) \tag{7-131}$$

Substituting for C_s using Equation (7-129), we obtain

$$C_c = \frac{Y_{c/s}(C_{s0} + K_s)}{\mu_{max} - D}\left(\frac{\mu_{max} C_{s0}}{K_s + C_{s0}} - D\right) \tag{7-132}$$

7.5.7 Wash-out

To learn the effect of increasing the dilution rate, we combine Equations (7-123) and (7-100) and set $r_d = 0$ to get

$$\frac{dC_c}{dt} = (\mu - D)C_c \tag{7-133}$$

We see that if $D > \mu$, then dC_c/dt will be negative and the cell concentration will continue to decrease until we reach a point where all cells will be washed out:

$$C_c = 0$$

The dilution rate at which wash-out will occur is obtained from Equation (7-132) by setting $C_c = 0$.

Flow rate at which
wash-out occurs

$$D_{max} = \frac{\mu_{max} C_{s0}}{K_s + C_{s0}} \tag{7-134}$$

We next want to determine the other extreme for the dilution rate, which is the rate of maximum cell production. The cell production rate per unit volume of reactor is the mass flow rate of cells out of the reactor (i.e., $\dot{m}_c = C_c v_0$) divided by the volume V, or

$$\frac{v_0 C_c}{V} = DC_c$$

Substituting for C_c yields

$$DC_c = DY_{c/s}\left(C_{s0} - \frac{DK_s}{\mu_{max} - D}\right) \tag{7-135}$$

Figure 7-17 shows production rate, cell concentration, and substrate concentration as functions of dilution rate. We observe a maximum in the production rate and this maximum can be found by differentiating production with respect to the dilution rate D:

$$\frac{d(DC_c)}{dD} = 0$$

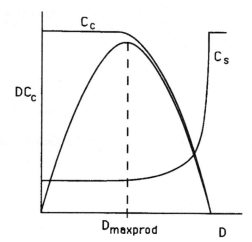

Maximum rate of
cell production
(DC$_c$)

Figure 7-17   Cell concentration and production rate as a function of dilution rate.

Then

Maximum rate of
cell production

$$D_{\text{maxprod}} = \mu_{\text{max}} \left(1 - \sqrt{\frac{K_s}{K_s + C_{s0}}} \right) \qquad (7\text{-}136)$$

7.5.8 Oxygen-Limited Fermentation

Oxygen is necessary for all aerobic fermentations (by definition). Maintaining the appropriate concentration of dissolved oxygen in fermentation is important for efficient operation of a fermentor. For oxygen-limited systems, it is necessary to design a fermentor to maximize the oxygen transfer between the injected air bubble and the cell. Typically, a fermentor contains a gas sparger, heat transfer surfaces, and an impeller, for a batch reactor. A chemostat has a similar configuration with the addition of inlet and outlet streams. The CD-ROM discusses the transport steps from the bulk liquid to and within the microorganism. A series of mass transfer correlations area also given.

7.5.9 Scale-up

Scale-up for the growth of microorganisms is usually based on maintaining a constant dissolved oxygen concentration in the liquid (broth), independent of reactor size. Guidelines for scaling from a pilot-plant bioreactor to a commercial plant reactor are given on the CD-ROM. One key to a scale-up is to have the speed of the end (tip) of the impeller equal to the velocity in both the laboratory pilot reactor and the full-scale plant reactor. If the impeller speed is too rapid, it can lyse the bacteria; if the speed is too slow, the reactor contents will not be well mixed. Typical tip speeds range from 5 to 7 m/s.

SUMMARY

1. The azomethane (AZO) decomposition mechanism is

$$2\text{AZO} \underset{k_2}{\overset{k_1}{\rightleftharpoons}} \text{AZO} + \text{AZO}^*$$

$$\text{AZO}^* \xrightarrow{k_3} \text{N}_2 + \text{ethane} \tag{S7-1}$$

The rate expression for the mechanism of this decomposition,

$$r_{\text{N}_2} = \frac{k(\text{AZO})^2}{1 + k'(\text{AZO})} \tag{S7-2}$$

exhibits first-order dependence with respect to AZO at high AZO concentrations and second-order dependence with respect to AZO at low AZO concentrations.

2. In the PSSH, we set the rate of formation of the active intermediates equal to zero. If the active intermediate A^* is involved in m different reactions, we set it to

$$r_{A^*,\text{net}} \equiv \sum_{i=1}^{m} r_{A^*i} = 0 \tag{S7-3}$$

This approximation is justified when the active intermediate is highly reactive and present in low concentrations.

3. Polymerization: (step reactions and chain reactions)

   (a) Steps: initiation, propagation, transfer, termination.
   (b) The fraction of functional groups that have reacted is

$$p = \frac{M_o - M}{M_o}$$

   (c) The degree of polymerization is

$$\bar{X}_n = \frac{1}{1 - p} \tag{S7-4}$$

   (d) The Flory distribution for the mole fraction is

$$y_j = (1 - p)\, p^{j-1} \tag{S7-5}$$

and for the weight fraction is

$$w_j = j(1-p)^2 p^{j-1} \qquad (S7\text{-}6)$$

4. The enzymatic reaction for the decomposition of urea, S, catalyzed by urease, E, is

$$
\begin{array}{c}
E + S \underset{k_2}{\overset{k_1}{\rightleftharpoons}} E \cdot S \\[6pt]
W + E \cdot S \xrightarrow{k_3} P + E
\end{array}
\qquad (S7\text{-}7)
$$

It follows Michaelis–Menten kinetics and the rate expression is

$$-r_s = \frac{V_{max}\, S}{S + K_m} \qquad (S7\text{-}8)$$

where V_{max} is the maximum rate of reaction for a given enzyme concentration and K_m is the Michaelis constant.

5. The total amount of a given enzyme in the system is the sum of the free enzyme, E, and the bound enzyme, $E \cdot S$:

$$E_t = E \cdot S + E \qquad (S7\text{-}9)$$

To arrive at Equation (S7-8) we treat each reaction as elementary, apply the PSSH to the complex, and use Equation (S7-9).

6. Bioreactors:

$$\text{cells + substrate} \longrightarrow \text{more cells + product}$$

(a) Phases of bacteria growth:

     I. Lag   II. Exponential   III. Stationary   IV. Death

(b) Monod growth rate law:

$$r_g = \mu_{max}\, \frac{C_c C_s}{K_s + C_s} \qquad (S7\text{-}10)$$

(c) Stoichiometry:

$$Y_{c/s} = \frac{\text{mass of new cells formed}}{\text{substrate consumed to produce new cells}}$$

(d) Unsteady-state mass balance on a chemostat:

$$\frac{dC_c}{dt} = \frac{C_{c0} - C_c}{\tau} + r_g - r_d \qquad (S7\text{-}11)$$

$$\frac{dC_s}{dt} = \frac{C_{s0} - C_s}{\tau} + r_s \qquad (S7\text{-}12)$$

$$-r_s = Y_{s/c} r_g + Y_{s/p} r_p + m C_c \qquad (S7\text{-}13)$$

QUESTIONS AND PROBLEMS

The subscript to each of the problem numbers indicates the level of difficulty: A, least difficult; D, most difficult.

$$A = \bullet \quad B = \blacksquare \quad C = \blacklozenge \quad D = \blacklozenge\blacklozenge$$

In each of the questions and problems below, rather than just drawing a box around your answer, write a sentence or two describing how you solved the problem, the assumptions you made, the reasonableness of your answer, what you learned, and any other facts that you want to include. You may wish to refer to W. Strunk and E. B. White, *The Elements of Style* (New York: Macmillan, 1979) and Joseph M. Williams, *Style: Ten Lessons in Clarity & Grace* (Glenview, Ill.: Scott, Foresman, 1989) to enhance the quality of your sentences. See the Preface for additional generic parts (x), (y), (z) to the home problems.

P7-1$_C$ Read over all of this chapter's problems. Make up an original problem that uses the concepts presented in this chapter. To obtain a solution:
   (a) Make up your data and reaction.
   (b) Use a real reaction and real data.
   See R. M. Felder, *Chem. Eng. Educ.*, *19*(4), 176 (1985).

P7-2$_A$ **What if...**
   (a) you carried out the ethane reaction in Example 7-2 at 1500 K or 2000 K? Would the PSSH still be valid? Can you find a temperature at which the PSSH is not a good approximation? Explain.
   [*Hint:* Calculate different ratios of radicals (e.g., $CH_3\bullet/H\bullet$, $CH_3\bullet/C_2H_5\bullet$, etc.).]

   (b) you could choose from a number of different initiators ($I_2 \xrightarrow{k_0} 2I$): fast to slow, $k_0 = 0.01$ s^{-1}, $k_0 = 0.0001$ s^{-1}, and $k_0 = 0.00001$ s^{-1}, and initiator concentrations, 10^{-1} to 10^{-5} *m*? What would guide your selection? (*Hint:* Plot *M* vs. *t*.)

   (c) the enzymatic reaction in Example 7-7 were exothermic? What would be the effect of raising or lowering the temperature of the rate law parameters on the overall rate?

   (d) the enzymatic reaction in Example 7-8 were carried out in a CSTR with a space time of 400 s? What conversion would be achieved? How would your answer change for two CSTRs in series, with $\tau = 200$ s for each? What if both the total enzyme concentration and the substrate concentration were increased by a factor of 4 in the CSTR with $\tau = 400$ s?

   (e) you were asked to carry out the bioreaction in Example 7-9 at higher and lower temperatures? What do you think the substrate, cell, and product concentrations would look like? Sketch each as a function of time for different temperatures. Discuss the reasonableness of assumptions you made to arrive at your curves. What is the relationship between C_p^* and $Y_{P/S}$ for which the final cell concentration is invariant? Can an equation for the washout dilution rate be derived for the Tessier Eqn.? If so what is it?

P7-3$_B$ (*Flame retardants*) Hydrogen radicals are important to sustaining combustion reactions. Consequently, if chemical compounds are introduced that can scavange the hydrogen radicals, the flames can be extinguished. While many reactions occur during the combustion process, we shall choose CO flames as a model system to illustrate the process [S. Senkan et al., *Combustion and Flame, 69*, p. 113 (1987)]. In the absence of inhibitors

$$O_2 \longrightarrow O\bullet + O\bullet \qquad\qquad (P7\text{-}3.1)$$

$$H_2O + O\cdot \longrightarrow 2OH\cdot \qquad (P7\text{-}3.2)$$

$$CO + OH\cdot \longrightarrow CO_2 + H\cdot \qquad (P7\text{-}3.3)$$

$$H\cdot + O_2 \longrightarrow OH\cdot + O\cdot \qquad (P7\text{-}3.4)$$

The last two reactions are rapid compared to the first two. When HCl is introduced to the flame, the following additional reactions occur:

$$H\cdot + HCl \longrightarrow H_2 + Cl\cdot$$

$$H\cdot + Cl\cdot \longrightarrow HCl$$

Assume that all reactions are elementary and that the PSSH holds for the $O\cdot$, $OH\cdot$, and $Cl\cdot$ radicals.
(a) Derive a rate law for the consumption of CO when no retardant is present.
(b) Derive an equation for the concentration of $H\cdot$ as a function of time assuming constant concentration of O_2, CO, and H_2O for both uninhibited combustion and combustion with HCl present. Sketch $H\cdot$ versus time for both cases.
(c) Sketch a reaction pathway diagram for this reaction.
More elaborate forms of this problem can be found in Chapter 6, where the PSSH is not invoked.

P7-4$_A$ The pyrolysis of acetaldehyde is believed to take place according to the following sequence:

$$CH_3CHO \xrightarrow{k_1} CH_3\cdot + CHO\cdot$$

$$CH_3\cdot + CH_3CHO \xrightarrow{k_2} CH_3\cdot + CO + CH_4$$

$$CHO\cdot + CH_3CHO \xrightarrow{k_3} CH_3\cdot + 2CO + H_2$$

$$2CH_3\cdot \xrightarrow{k_4} C_2H_6$$

(a) Derive the rate expression for the rate of disappearance of acetaldehyde, $-r_{Ac}$.
(b) Under what conditions does it reduce to Equation (7-2)?
(c) Sketch a reaction pathway diagram for this reaction.

P7-5$_B$ The gas-phase homogeneous oxidation of nitrogen monoxide (NO) to dioxide (NO_2),

$$2NO + O_2 \xrightarrow{k} 2NO_2$$

is known to have a form of third-order kinetics which suggests that the reaction is elementary as written, at least for low partial pressures of the nitrogen oxides. However, the rate constant k actually *decreases* with increasing absolute temperature, indicating an apparently *negative* activation energy. Because the activation energy of any elementary reaction must be positive, some explanation is in order.

Provide an explanation, starting from the fact that an active intermediate species, NO_3, is a participant in some other known reactions that involve oxides of nitrogen.

P7-6$_B$ For the decomposition of ozone in an inert gas M, the rate expression is

$$-r_{O_3} = \frac{k(O_3)^2(M)}{(O_2)(M) + k'(O_3)}$$

Suggest a mechanism.

P7-7$_C$ (*Tribiology*) One of the major reasons for engine oil degradation is the oxidation of the motor oil. To retard the degradation process, most oils contain an antioxidant [see *Ind. Eng. Chem.* **26**, 902 (1987)]. Without an inhibitor to oxidation present, the suggested mechanism at low temperatures is

$$I_2 \xrightarrow{k_0} 2I\cdot$$

$$I\cdot + RH \xrightarrow{k_i} R\cdot + HI$$

$$R\cdot + O_2 \xrightarrow{k_{p1}} RO_2^{\cdot}$$

$$RO_2^{\cdot} + RH \xrightarrow{k_{p2}} ROOH + R\cdot$$

$$2RO_2^{\cdot} \xrightarrow{k_t} \text{inactive}$$

where I_2 is an initiator and RH is the hydrocarbon in the oil. When the temperature is raised to 100°C, the following additional reaction occurs as a result of the decomposition of the unstable ROOH:

$$ROOH \xrightarrow{k_{i3}} RO\cdot + \cdot OH$$

$$RO\cdot + RH \xrightarrow{k_{p4}} ROH + R\cdot$$

$$\cdot OH + RH \xrightarrow{k_{p5}} H_2O + R\cdot$$

Goblu Motor Oil

Effective Lubricant Design

When an antioxidant is added to retard degradation at low temperatures, the following additional termination steps occur:

$$RO_2\cdot + AH \xrightarrow{k_{A1}} ROOH + A\cdot$$

$$A\cdot + RO_2^{\cdot} \xrightarrow{k_{A2}} \text{inactive}$$

$$\left(\text{for example, AH} = \begin{array}{c} OH \\ \bigcirc \\ CH_3 \end{array}, \quad \text{inactive} = \begin{array}{c} OH \\ \bigcirc \\ CH_2OOR \end{array} \right)$$

Derive a rate law for the degradation of the motor oil in the absence of an antioxidant at
(a) Low temperatures.
(b) High temperatures.
Derive a rate law for the rate of degradation of the motor oil in the presence of an antioxidant for
(c) Low temperatures.
(d) High temperatures. Here assume that the inactive products formed with antioxidant do not decompose (probably a bad assumption).
(e) How would your answer to part (a) change if the radicals $I\cdot$ were produced at a constant rate in the engine and then found their way into the oil?
(f) Sketch a reaction pathway diagram for both high and low temperatures, with and without antioxidant.

P7-8$_A$ Consider the application of the PSSH to epidemiology. We shall treat each of the following steps as elementary in that the rate will be proportional to the number of people in a particular state of health. A healthy person, H, can become ill, I, spontaneously,

$$H \xrightarrow{k_1} I \qquad \text{(P7-8.1)}$$

or he may become ill through contact with another ill person:

$$I + H \xrightarrow{k_2} 2I \qquad \text{(P7-8.2)}$$

The ill person may become healthy:

$$I \xrightarrow{k_3} H \qquad \text{(P7-8.3)}$$

or he may expire:

$$I \xrightarrow{k_4} D \qquad \text{(P7-8.4)}$$

The reaction given in Equation (P7-8.4) is normally considered completely irreversible, although the reverse reaction has been reported to occur.
(a) Derive an equation for the death rate.
(b) At what concentration of healthy people does the death rate become critical? [*Ans.:* When $[H] = (k_3 + k_4)/k_2$.]
(c) Comment on the validity of the PSSH under the conditions of part (b).
(d) If $k_1 = 10^{-5}$ h^{-1}, $k_2 = 10^{-7}$ (people·h)$^{-1}$, $k_3 = 5 \times 10^{-6}$ h, $k_4 = 10^{-7}$ h, and H$_o = 10^9$ people, plot H, I, and D versus time. Vary k_i and describe what you find. Check with your local *disease control center* or search the WWW to modify the model and/or substitute appropriate values of k_i.

P7-9$_C$ (*Postacidification in yogurt*) Yogurt is produced by adding two strains of bacteria (*Lactobacillus bulgaricus* and *Streptococcus thermophilus*) to pasteurized milk. At temperatures of 110°F, the bacteria grow and produce lactic acid. The acid contributes flavor and causes the proteins to coagulate, giving the characteristic properties of yogurt. When sufficient acid has been produced (about 0.90%), the yogurt is cooled and stored until eaten by consumers. A lactic acid level of 1.10% is the limit of acceptability. One limit on the shelf life of yogurt is "postacidification," or continued production of acid by the yogurt cultures during storage. The table that follows shows acid production (% lactic acid) in yogurt versus time at four different temperatures.

Time (days)	35°F	40°F	45°F	50°F
1	1.02	1.02	1.02	1.02
14	1.03	1.05	1.14	1.19
28	1.05	1.06	1.15	1.24
35	1.09	1.10	1.22	1.26
42	1.09	1.12	1.22	1.31
49	1.10	1.12	1.22	1.32
56	1.09	1.13	1.24	1.32
63	1.10	1.14	1.25	1.32
70	1.10	1.16	1.26	1.34

cre™
yogurt

Chemical Engineering
in the Food Industry

Acid production by yogurt cultures is a complex biochemical process. For the purpose of this problem, assume that acid production follows first-order kinetics with respect to the consumption of lactose in the yogurt to produce lactic acid. At the start of acid production the lactose concentration is about 1.5%, the bacteria concentration is 10^{11} cells/dm^3, and the acid concentration at which all metabolic activity ceases is 1.4% lactic acid.

(a) Determine the activation energy for the reaction.

(b) How long would it take to reach 1.10% acid at 38°F?

(c) If you left yogurt out at room temperature, 77°F, how long would it take to reach 1.10% lactic acid?

(d) Assuming that the lactic acid is produced in the stationary state, do the data fit any of the modules developed in this chapter?

[Problem developed by General Mills, Minneapolis, Minnesota]

P7-10$_C$ The enzymatic hydrolization of fish oil extracted from crude eel oil has been carried out using lipase L (*Proc. 2nd Joint China/USA Chemical Engineering Conference*, Vol. III, p. 1082, 1997). One of the desired products is docosahexaenic acid, which is used as a medicine in China. For 40 mg of enzyme the Michaelis constant is 6.2×10^{-2} (mL/mL) and V_{max} is 5.6 μmol/mL· min. Calculate the time necessary to reduce the concentration of fish oil from 1.4% to 0.2 vol %. Note: There may be an inconsistency in the article. The half life for an initial volume of 25% fish oil is stated to be 4.5 days. However, this yields a different initial fish oil concentration one finds from looking in the literature. Search the web for *fish oil*. Suggest a way to resolve this controversy.

Creative Problem

P7-11$_B$ Beef catalase has been used to accelerate the decomposition of hydrogen peroxide to yield water and oxygen [*Chem. Eng. Educ.*, 5, 141 (1971)]. The concentration of hydrogen peroxide is given as a function of time for a reaction mixture with a pH of 6.76 maintained at 30°C.

t (min)	0	10	20	50	100
$C_{H_2O_2}$ (mol/L)	0.02	0.01775	0.0158	0.0106	0.005

(a) Determine the Michaelis–Menten parameters V_{max} and K_m.

(b) If the total enzyme concentration is tripled, what will the substrate concentration be after 20 min?

(c) How could you make this problem more difficult?

P7-12$_B$ In this problem three different types of reaction inhibition are explored:

(a) In competitive inhibition, an inhibitor adsorbs on the same type of site as the substrate. The resulting inhibitor–enzyme complex is inactive. Show that the rate law for competitive inhibition

$$E + S \rightleftharpoons E{\cdot}S$$
$$E + I \rightleftharpoons E{\cdot}I$$
$$E{\cdot}S \longrightarrow E + P$$

is

$$r_p = \frac{V_{max}\, S}{S + K_m(1 + I/K_i)}$$

(b) In uncompetitive inhibition the inhibitor attaches itself to enzyme–substrate complex, rendering it inactive. Show that for uncompetitive inhibition,

$$E + S \rightleftharpoons E{\cdot}S$$
$$I + E{\cdot}S \rightleftharpoons IES$$
$$E{\cdot}S \longrightarrow P + E$$

the rate law is

$$r_p = \frac{V_{max} S}{K_m + S(1 + I/K_i)}$$

(c) In noncompetitive inhibition, the inhibitor adsorbs itself to a different type of site than the substrate to render the enzyme–substrate complex inactive. Assuming that all concentrations of a species can be expressed by their equilibrium concentrations, show that for noncompetitive inhibition

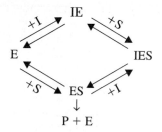

the rate law is

$$r_p = \frac{V_{max} S}{(S + K_m)(1 + I/K_i)}$$

(d) Sketch three different inhibitions on a Lineweaver–Burk plot.

P7-13$_B$ It has been observed that substrate inhibition occurs in the following enzymatic reaction:

$$E + S \rightleftharpoons E{\cdot}S$$

$$E{\cdot}S \longrightarrow P + E$$

$$E{\cdot}S + S \rightleftharpoons S{\cdot}E{\cdot}S$$

(a) Show that the rate law for the sequence above is consistent with the plot in Figure P7-13 of $-r_s$ (mmol/L·min) versus the substrate concentration S (mmol/L). (Refer to Problem P7-12.)

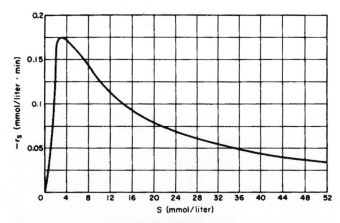

Figure P7-13

    (b) If this reaction is carried out in a CSTR that has a volume of 1000 L, to which the volumetric flow rate is 3.2 L/min, determine the three possible steady states, noting, if possible, which are stable. The entrance concentration of the substrate is 50 mmol/L.

    (c) What is the highest conversion possible for this CSTR when operated at the conditions specified above?

P7-14$_B$ The following data on bakers' yeast in a particular medium at 23.4°C and various oxygen partial pressures were obtained:

P_{O_2}	Q_{O_2} (no sulfanilamide)	Q_{O_2} (20 mg sulfanilamide/mL added to medium)
0.0	0.0	0.0
0.5	23.5	17.4
1.0	33.0	25.6
1.5	37.5	30.8
2.5	42.0	36.4
3.5	43.0	39.6
5.0	43.0	40.0

P_{O_2} = oxygen partial pressure, mmHg; Q_{O_2} = oxygen uptake rate, μL of O_2 per hour per mg of cells.

    (a) Calculate the Q_{O_2} maximum (V_{max}), and the Michaelis–Menten constant K_m. (*Ans.:* $V_{max} \doteq 52.63 \ \mu L \ O_2/h \cdot mg$ cells.)

    (b) By use of the Lineweaver–Burk plot, determine whether sulfanilamide is a competitive or noncompetitive inhibitor to O_2 uptake. (Refer to Problem P7-12).

                                              (University of Pennsylvania)

P7-15   **(a)** Use the supplementary reading to give three examples each of polymers formed by step polymerization and by chain polymerization. Describe the monomers, structural unit(s) and repeating unit in each case.

    (b) Plot y_j, w_j as a function of j for degrees of polymerization of 5, 10, and 20. Plot P_{10} as a function of conversion of function groups, P.

    (c) Plot y_j and w_j as a function of j for free radical termination by combination for $P = 0.8$, $p = 0.9$ and $p = 0.95$.

    (d) Determine the molecular weight distribution for the formation of polystyrene for an initiator concentration of 10^{-3} molar and a monomer concentration of 3 molar. What are $\overline{M}_n$, $\overline{M}_w$, and the polydispersity, D, after 40 hrs? How would the $\overline{M}_n$ change if chain transfer were neglected?

    (e) The polymerization of styrene is carried out in a batch reactor. Plot the mole fraction of polystyrene of chain length 10 as a function of time for an initial concentration of 3M and 0.01M of monomer an initiator respectively. The solvent concentration is 10 molar.

P7-16$_B$ The instantaneous number-average degree of polymerization, X_N, can be expressed as

$$\overline{X}_N = \frac{\text{moles monomer consumed/dm}^3 \cdot s}{\text{dead polymer molecules produced/dm}^3 \cdot s} = \frac{-r_M}{\Sigma \ r_j} \qquad \text{(P7-16.1)}$$

    (a) Consider an additional termination step resulting from the combination of the initiator molecule and a free radical R_j:

$$R_j + I \xrightarrow{\;k_{ti}\;} P_j \qquad\qquad (P7\text{-}16.2)$$

(b) Show that

$$\overline{X}_N = \left[\left(\frac{fk_0 k_t}{k_p^2} \right)^{1/2} \frac{(I_2)^{1/2}}{(M)} + \frac{k_m}{k_p} + \frac{k_s(S)}{k_p(M)} + \frac{k_{ti}}{k_p} \frac{(I)}{(M)} \right]^{-1} \qquad (P7\text{-}16.3)$$

where

$$k_t = k_a + k_d + k_{ti}$$

(c) Neglecting solvent transfer, show that

$$\frac{1}{\overline{X}_N} = \frac{k_t(-r_M)}{k_p^2 (M)^2} + \frac{k_m}{k_p} + \frac{k_{ti}}{k_p} \frac{(I)}{(M)}$$

(d) Explain how you can determine rate law parameters from experimental data obtained in a CSTR. Use sketches to elucidate your explanation.

(e) Typical activation energies for the initiation, propagation, and termination steps are: 20 kJ/mol, 20 kJ/mol, and 9 kJ/mol. Discuss the effect of temperature on free-radical polymerization.

P7-17$_B$ The free-radical polymerization reaction discussed in Section 7.3.3 for a batch reactor is to be carried out in a CSTR and a PFR.

(a) Plot the effluent initiator and monomer concentrations as a function of the space-time τ.

(b) Compare your results for two equal-sized CSTRs in series with one CSTR. The total volume is the same in both cases.

(c) Vary the parameter values (i.e., k_0, k_p) and discuss your results.

Additional information:

$$M_0 = 3 \text{ mol/dm}^3 \qquad I_{20} = 0.01 \text{ mol/dm}^3$$

$$k_0 = 10^{-3} \text{ s}^{-1} \qquad k_p = 10 \text{ dm}^3/\text{mol} \cdot \text{s} \qquad k_t = 5 \times 10^7 \text{ dm}^3/\text{mol} \cdot \text{s}$$

P7-18$_A$ *(Anionic polymerization)*

(a) Determine the final number-average and weight-average chain lengths and molecular weights along with the extent of polymerization and polydispersity in an anionic polymerization for an initial monomer concentration of 2 M and for initiator concentrations of 1, 0.01, and 0.0001 M.

(b) Calculate the radical concentration and polymer concentration as a function of time for $j = 3, 7, 10$ for the same monomer and initiator concentrations. $k_i = \infty$, $k_p = 100 \text{ dm}^3/\text{mol} \cdot \text{s}$

(c) Repeat (b) for the case

$$r_i = k_i [A^-] M$$

and

$$K_i = \frac{[A^-][B^+]}{[AB]}$$

with $k_i = 10 \text{ dm}^3/\text{mol} \cdot \text{s}$ and $K_i = 10^{-8} \text{ mol/dm}^3$.

P7-19$_A$ An anionic polymerization is to be carried out in a CSTR. The reaction steps are

$$I + M \xrightarrow{\;k_i\;} R_1$$

$$R_j + M \xrightarrow{\;k_p\;} R_{j+1}$$

The entering concentration of monomer and initiator are M_0 and I_0, respectively.

(a) Derive an equation involving the monomer concentration and only the variables k_p, k_1, I_0, τ (i.e., $\tau = V/v$), and M_0.

(b) Derive an equation of the radical concentration as a function of time.

(Ans. $R_j = \dfrac{I_0}{(1 + \tau k_i M)} \dfrac{k_i}{k_p} \left(\dfrac{k_p M \tau}{1 + k_p M \tau} \right)^j$).

(c) Choose representative values of k_i (0.015 dm³/mol·s) and k_p (10^3 dm³/mol·s) to plot I and M as a function of τ. What if $k_i \gg k_p$ and $R_1 = I_0$

(d) Derive equations for the first and second moments and μ_n, μ_w, and D.

P7-20 Polyesters (shirts) can be formed by the reaction of diacids and diols

$$n\,\text{HOR}_1\text{OH} + n\,\text{HOOCR}_2\text{COOH} \xrightarrow{\text{cat}} \text{HO}(\text{R}_1\text{OOCR}_2\text{COO})_n\,\text{H} + 2(n-1)\text{H}_2\text{O}$$

Let [COOH] represent the concentration of carboxyl functional groups and [—OH] the concentration of hydroxyl groups. The feed is equimolar in [OH] and [COOH].

The combined mole balance and rate law is

$$-\frac{d[\text{COOH}]}{dt} = k[\text{COOH}][\text{OH}][\text{cat}]$$

(a) Assume that the polymerization is self catalyzed [cat] = [COOH], plot the appropriate function of the fraction of functional groups reacted, p, as a function of time in order to obtain a linear plot. Plot the experimental data on the same plot. In what regions do the theory and experiment agree, and in which region do they disagree?

(b) Assume that the reaction is catalyzed by the (H^+) ion and that the (H^+) ion is supplied by the dissociation of the weak acid [COOH]. Show that the overall rate of reaction is 5/2 order. Plot the appropriate function of p versus time so that the plot is linear. In which regions do the theory and experiment agree, and which regions do they disagree?

(c) It is proposed that the mechanism for the polymerization is

$$(1) \quad --[\text{COOH}] + \text{HA} \;\rightleftharpoons\; --\overset{\text{H}}{\underset{+}{\text{COH}}} + \text{A}^-$$

$$(2) \quad --\overset{\text{OH}}{\underset{+}{\text{COH}}} + {\sim}{\sim}\text{OH} \;\longrightarrow\; --\overset{\text{OH}}{\underset{{\sim}{\sim}\text{OH}\;+}{\text{COH}}} \;\longrightarrow$$

$$(3) \quad \text{A}^- + --\overset{\text{OH}}{\underset{{\sim}{\sim}\text{OH}\;+}{\text{COH}}} \;\longrightarrow\; --\overset{\text{O}}{\text{CO}}{\sim}{\sim} + \text{HA}$$

Can this mechanism be made to be consistent with both rate laws?

Additional information:

Experimental Data

t (min)	0	50	100	200	400	700	1200	1600
p	0	.049	0.68	0.8	0.88	0.917	0.936	0.944

P7-21$_B$ Sketch the polymer concentration, P_j, mole fraction of polymer with j monomer units, y_j, and the corresponding weight fraction, w_j, for $j = 2, 10, 20$ as a function of monomer conversion in Styrene polymerization for
 (a) Termination by means other than combination.
 (b) Termination by combination. The molecular weight of the monomer is 25 and its initial concentration is 3 M and the initiator concentration is 0.01 M.
 (c) How would your answers to parts (a) and (b) change if the initiator concentration were 0.0001 M?
 (d) What are the corresponding average molecular weights at $X = 0.2, 0.8$, and 0.999?

P7-22$_B$ Rework Example 7-5 for the case when the initiator does not react immediately with monomer to form the radical R_1^- (i.e., R_1), but instead reacts at a finite rate with a specific reaction rate k_0:

$$I + M \xrightarrow{k_0} R_1$$

The rate law is

$$-r_i = k_0 IM$$

The initiator concentration at time at $t = 0$ is I_0.
 (a) Derive an equation for R_j as a function of Θ.
 (b) For $M_o = 3$ mol/dm^3, $I_o = 10^{-3}$ mol/dm^3, $k_o = 0.1$ dm^3/mol·s, $k_p = 10$ dm^3/mol·s. Plot R_8 and P_8 as a function of real time t.

P7-23$_B$ Rework Problem P7-22 for the case in which the reaction is carried out in a CSTR. Derive an equation for R_j as a function of the space-time, τ.

P7-24$_B$ The growth of a bacteria *Stepinpoopi* can be described by the logistic growth law

$$r_g = \mu_m \left(1 - \frac{C_c}{C_m}\right) C_c$$

with $\mu_m = 0.5$ h^{-1} and $C_m = 20$ g/dm^3. The substrate is in excess.
 (a) The cell growth is to be carried out in a 2-dm^3 batch reactor. Plot the growth rate and cell concentration (g/dm^3) as functions of time after inoculation of 0.4 g of cells into the reactor (ignore the lag period).
 (b) The batch vessel in part (a) is to be turned into a CSTR. Derive an equation for the wash-out rate. Choose values for the volumetric flow rate of the entering substrate and plot the cell concentration as a function of time after inoculation.

P7-25$_B$ The following data were obtained for *Pyrodictium occultum* at 98°C. Run 1 was carried out in the absence of yeast extract and run 2 with yeast extract. Both runs initially contained Na$_2$S. The vol % of the growth product H$_2$S collected above the broth was reported as a function of time. [*Ann. N. Y. Acad. Sci.*, *506*, 51 (1987)].

Run 1:

Time (h)	0	10	15	20	30	40	50	60	70
Cell Density (cells/mL) $\times 10^{-4}$	2.7	2.8	15	70	400	600	775	600	525
% H_2S	0.5	0.8	1.0	1.2	6.8	4.7	7.5	8.0	8.2

Run 2:

Time (h)	0	5	10	15	20	30	40	50	60
Cell Density (cells/mL) $\times 10^{-4}$	2.7	7	11	80	250	350	350	250	—
% H_2S	0.1	0.7	0.7	0.8	1.2	4.3	7.5	11.0	12.3

(a) What is the lag time with and without the yeast extract?
(b) What is the difference in the specific growth rates, μ_{max}, of the bacteria with and without the yeast extract?
(c) How long is the stationary phase?
(d) During which phase does the majority production of H_2S occur?
(e) The liquid reactor volume in which these batch experiments were carried out was 0.2 dm³. If this reactor were converted to a continuous-flow reactor, what would be the corresponding wash-out rate?

P7-26$_C$ Cell growth with uncompetitive substrate inhibition is taking place in a CSTR. The cell growth rate law for this system is

$$r_g = \frac{\mu_{max} C_s C_c}{K_s + C_s \left(1 + C_s / K_I\right)}$$

with $\mu_{max} = 1.5$ h^{-1}, $K_s = 1$ g/dm³, $K_I = 50$ g/dm³, $C_{s0} = 30$ g/dm³, $Y_{c/s} = 0.08$, $C_{c0} = 0.5$ g/dm³, and $D = 0.75$ h^{-1}.

(a) Make a plot of the steady-state cell concentration C_c as a function of D.
(b) Make a plot of the substrate concentration C_s as a function of D on the same graph as that used for part (a).
(c) Initially, 0.5 g/dm³ of bacteria was placed in the tank containing the substrate and the flow to the tank started. Plot the concentrations of bacteria and substrate as functions of time.

P7-27$_B$ A solution containing bacteria at a concentration of 0.001 g/dm³ was fed to a semibatch reactor. The nutrient was in excess and the growth rate law is first order in the cell concentration. The reactor was empty at the start of the experiment. If the concentration of bacteria in the reactor at the end of 2 h is 0.025 g/dm³, what is the specific growth rate k in min^{-1}?

P7-28$_A$ An understanding of bacteria transport in porous media is vital to the efficient operation of the water flooding of petroleum reservoirs. Bacteria can have both beneficial and harmful effects on the reservoir. In enhanced microbial oil recovery, EMOR, bacteria are injected to secrete surfactants to reduce the interfacial tension at the oil–water interface so that the oil will flow out more easily. However, under some circumstances the bacteria can be harmful, by plugging the pore space and thereby block the flow of water and oil. One bacteria that has been studied, *Leuconostoc mesentroides*, has the unusual behavior that when it is injected into a porous medium and fed sucrose, it greatly reduces the flow (i.e., damages the formation and reduces permeability). When the bacteria are fed fructose or glucose, there is no damage to the porous medium. [R. Lappan and H. S. Fogler, *SPE Prod. Eng.*, 7(2), 167–171 (1992)]. The cell concentration, C_c, is given below as a function of time for different initial sucrose concentrations.

(a) From the data below, determine the lag time, the time to reach the stationary phase, the Michaelis constant, K_s, and the reaction velocity, μ, as a function of sucrose concentration.

(b) Will an inhibition model of the form

$$r_g = \mu C_C \left(1 - \frac{C_c - C_{c0}}{C_p^*} \right)^n$$

where n and C_p^* are parameters, fit your data?

Cell Concentration Data

Sucrose Conc. Time (h)	1 g/cm³ $C_c \times 10^{-7}$ (no./cm³)	5 g/cm³ $C_c \times 10^{-7}$ (no./cm³)	10 g/cm³ $C_c \times 10^{-7}$ (no./cm³)	15 g/cm³ $C_c \times 10^{-7}$ (no./cm³)
0.00	3.00	2.00	2.00	1.33
1.00	4.16	3.78	6.71	5.27
2.00	5.34	5.79	1.11	0.30
3.00	7.35	—	5.72	3.78
4.00	6.01	9.36	3.71	7.65
5.00	8.61	6.68	8.32	10.3
6.00	10.1	17.6	21.1	17.0
7.00	18.8	35.5	37.6	38.4
8.00	28.9	66.1	74.2	70.8
9.00	36.2	143	180	194
10.0	42.4	160	269	283
11.0	44.4	170	237	279
12.0	46.9	165	256	306
13.0	46.9	163	149	289

P7-29$_A$ A CSTR is being operated at steady state. The cell growth follows the Monod growth law without inhibition. The exiting substrate and cell concentrations are measured as a function of the volumetric flow rate (represented as the dilution rate), and the results are shown below. Of course, measurements are not taken until steady state is achieved after each change in the flow rate. Neglect substrate consumption for maintenance and the death rate, and assume that $Y_{p/c}$ is zero. For run 4, the entering substrate concentration was 50 g/dm³ and the volumetric flow rate of the substrate was 2 dm³/s.

Run	C_S (g/dm³)	D (s⁻¹)	C_C (g/dm³)
1	1	1	0.9
2	3	1.5	0.7
3	4	1.6	0.6
4	10	1.8	4

(a) Determine the Monod growth parameters μ_{max} and K_S.

(b) Estimate the stoichiometric coefficients, $Y_{c/s}$ and $Y_{s/c}$.

P7-30$_B$ The production of glycerol from corn amlylum/asylum is to be carried out by fermentation using yeast cells (*Proc. 2nd Joint China/USA Chemical Engineering Conference*, Beijing, China, Vol. III, p. 1094, 1997). The growth law is

$$r_g = \mu\, C_C$$

with

$$\mu = \mu_0 \frac{C_S}{K_S + C_S + C_S^2/K_{SI}} \left(1 - \frac{C_P}{C_P^*}\right) \exp(-K_{PI}C_P)$$

$\mu_0 = 0.25$ L mg/mL·s

$K_S = 0.018$ mg/mL

$K_{SI} = 11.8$ mg/mL

$C_P^* = 32.4$ mg/mL

$K_{PI} = 0.06$ mL/mg

$$\frac{dP}{dt} = (\alpha\mu + \beta)X \qquad \alpha = 34.5 \quad \beta = -0.147$$

$$\frac{dS}{dt} = \frac{1}{Y_{P/S}}\frac{dP}{dt} \qquad Y_{P/S} = 1.33$$

Plot the concentration of cells, substrate, and product as a function of time for initial concentrations of cells of 10^{-8} g/dm^3 and a substrate concentration of 50 g/dm^3.

P7-31$_B$ Wastewater containing terephthalic acid (TA) is treated using two aerobic sludge tanks in series (*Proc. 2nd Joint China/USA Chemical Engineering Conference*, Beijing, China, Vol. III, p. 970, 1997). The first tank was 12 dm^3 and the second was 24 dm^3. It is determined to reduce the TA concentration (reported in chemical oxygen demand, COD, mg/dm^3) from 5000 mg/dm^3 to below 100 mg/dm^3). The results of the experiments were reported in the following manner:

Tank 1:
$$\frac{C_C V}{v_0(C_{S0} - C_S)} = 0.16\frac{1}{C_S} + 0.62$$

with $\mu_0 = 1.61$ day^{-1} (g substrate/g cells) and $K_S = 0.25$ g/dm^3 for the Monod equation.

Tank 2:
$$\frac{C_C V}{v_0(C_{S0} - C_S)} = 0.3\frac{1}{C_S} + 5.4$$

where C_C is the concentration of biomass in the tank and C_{S0} and C_S represent the entering and exiting concentration of TA in the waste stream.

Design a CSTR sludge system to handle a wastewater flow of 1000 m^3/day:

$$\text{loading} = 5\text{–}6 \frac{\text{kg COD}}{\text{m}^3 \cdot \text{h}}$$

P7-32$_B$ The production of L-malic acid (used in medicines and food additives) was produced over immobilized cells of *Bacillus flavum* MA-3 (*Proc. 2nd Joint China/USA Chemical Engineering Conference*, Beijing, China, Vol. III, p. 1033, 1997].

$$HOOCCH = CHCOOH + H_2O \xrightarrow{\text{fumarase}} HOOCH_3\ CHCOOH$$
$$\underset{\displaystyle OH}{\big|}$$

The following rate law was obtained for the rate of formation of product:

$$r_p = \frac{V_{max}C_S}{K_m + C_S}\left(1 - \frac{C_P}{C_P^*}\right)C_C$$

where $V_{max} = 76$, $K_m = 0.048$ mol/dm^3, and $C_P^* = 1.69$ mol/dm^3. Design a reactor to process 10 m^3/day of 2 mol/dm^3 of fumaric acid.

CD-ROM MATERIAL

CD-ROM

LINKS

- **Learning Resources**
  1. *Summary Notes for Lectures 25, 26, 36, 37, 38, and 39*
  4. *Solved Problems*
     A. Hydrogen Bromide
        Example CD7–1 Deducing the Rate Law
        Example CD7–2 Find a Mechanism
- **Living Example Problems**
  1. *Example 7–2 PSSH Applied to Thermal Cracking of Ethane*
- **Professional Reference Shelf**
  1. *Enzyme Inhibition*
     A. Competitive
        Example CD7–3 Derive a Rate Law For Competitive Inhibition
     B. Uncompetitive
     C. Non-Competitive
        Example CD7–4 Derive a Rate Law For Non-Competitive Inhibition
        Example CD7–5 Match Eadie Plots to the Different Types of Inhibition
  2. *Multiple Enzymes and Substrate Systems*
     A. Enzyme Regeneration
        Example CD7–6 Construct a Lineweaver-Burk Plot for Different Oxygen
                       Concentrations
     B. Enzyme Co-fractors
        Example CD7–7 Derive an Initial Rate Law for Alcohol Dehydrogenates
     C. Multiple-Substrate Systems
        Example CD7–8 Derive a Rate Law for a Multiple Substrate System
        Example CD7–9 Calculate the Initial Rate of Formation of Ethanol
                       in the Presence of Porpanediol
     D. Multiple Enzyme Systems
  3. *Oxidation Limited Fermentation*
  4. *Fermentation Scale up*
- **Additonal Homework Problems**

CDP7-A$_A$ Determine the rate law and mechanism for the reaction

$$2GCH_3 \longrightarrow 2G = CH_2 + H_2$$

[2nd Ed. P7-6$_A$]

CDP7-B$_A$ Suggest a mechanism for the reaction

$$I^- + OCl^- \longrightarrow OI^- + Cl^-$$

[2nd Ed. P7-8$_B$]

CDP7-C$_A$ Develop a rate law for substrate inhibition of an enzymatic reaction. [2nd Ed. P7-16$_A$]

CDP7-D$_B$ Use POLYMATH to analyze an enzymatic reaction. [2nd Ed. P7-19$_B$].

CDP7-E$_B$ Redo Problem P7-17 to include chain transfer. [2nd Ed. P7-23$_B$]

CDP7-F$_B$ Determine the rate of diffusion of oxygen to cells. [2nd Ed. P12-12$_B$]

CDP7-G$_B$ Determine the growth rate of ameoba predatory on bacteria. [2nd Ed. P12-15$_C$]

CDP7-H$_C$ Plan the scale-up of an oxygen fermentor. [2nd Ed. P12-16$_B$]

CDP7-I$_B$ Assess the effectiveness of bacteria used for denitrification in a batch reactor. [2nd Ed. P12-18$_B$]

CDP7-J$_A$ Determine rate law parameters for the Monod equation. [2nd Ed. P12-19$_A$]

JOURNAL CRITIQUE PROBLEMS

P7C-1   Compare the theoretical curve with actual data points in Figure 5b [*Biotechnol. Bioeng.*, *24*, 329 (1982)], a normalized residence-time curve. Note that the two curves do not coincide at higher conversions. First, rederive the rate equation and the normalized residence-time equations used by the authors, and then, using the values for kinetic constants and lactase concentration cited by the authors, see if the theoretical curve can be duplicated. Linearize the normalized residence-time equation and replot the data, the theoretical curve in Figure 5b, and a theoretical curve that is obtained by using the constants given in the paper. What is the simplest explanation for the results observed?

P7C-2   In Figure 3 [*Biotechnol. Bioeng.*, *23*, 361 (1981)], $1/V$ was plotted against $(1/S)(1/PGM)$ at three constant 7-ADCA concentrations, with an attempt to extract V_{max} for the reaction. Does the V_{max} obtained in this way conform to the true value? How is the experimental V_{max} affected by the level of PGM in the medium?

P7C-3   In *J. Catal.*, *79*, 132 (1983), a mechanism was proposed for the catalyzed hydrogenation of pyridine in slurry reactors. Reexamine the data and model using an Eley–Rideal adsorption mechanism and comment on the appropriateness of this new analysis.

SUPPLEMENTARY READING

1. A discussion of complex reactions involving active intermediates is given in

FROST, A. A., and R. G. PEARSON, *Kinetics and Mechanism*, 2nd ed. New York: Wiley, 1961, Chap. 10.

LAIDLER, K. J., *Chemical Kinetics*, 3rd ed. New York: HarperCollins, 1987.

PILLING, M. J., *Reaction Kinetics*, New York: Oxford University Press, 1995.

SEINFELD, J. H., and S. N. PANDIS, *Atmospheric Chemistry and Physics*, New York: Wiley, 1998.

TEMKIN, O. N., *Chemical Reaction Networks: A Graph-Theoretical Approach*, Boca Raton, Fla: CRC Press, 1996.

2. Further discussion of enzymatic reactions is presented in

CORNISH-BOWDEN, A., *Analysis of Enzyme Kinetic Data*, New York: Oxford University Press, 1995.

VAN SANTEN, R. A. and J. W. NIEMANTSVERDRIET, *Chemical Kinetics and Catalysis*, New York: Plenum Press, 1995.

WARSHEL, A., *Computer Modeling of Chemical Reactions in Enzymes and Solutions*, New York: Wiley, 1991.

WINGARD, L. B., *Enzyme Engineering*. New York: Wiley-Interscience, 1972.

3. The following references concern polymerization reaction engineering:

BILLMEYER, F. W., *Textbook of Polymer Science*, 3rd ed. New York: Wiley, 1984.

DOTSON, N. A., R. GALVÁN, R. L. LAWRENCE, and M. TIRRELL, *Polymerization Process Modeling*, New York: VCH Publishers, 1996.

HOLLAND, C. D., and R. G. ANTHONY, *Fundamentals of Chemical Reaction Engineering*, 2nd ed. Upper Saddle River, N.J.: Prentice Hall, 1989, Chap. 10.

ODIAN, G., *Principles of Polymerization*, 3rd ed. New York: Wiley, 1983.

SCHORK, F. J., P. B. DESHPANDE, and K. W. LEFFEW, *Control of Polymerization Reactors*, New York: Marcel Dekker, 1993.

YOUNG, R. J. and P. A. LOVELL, *Introduction to Polymers*, 2nd ed., New York: Chapman & Hall, 1991.

4. Material on bioreactors can be found in

AIBA, S., A. E. HUMPHREY, N. F. MILLIS, *Biochemical Engineering*, 2nd ed. San Diego, Calif.: Academic Press, 1973.

BAILEY, T. J., and D. OLLIS, *Biochemical Engineering*, 2nd ed. New York: McGraw-Hill, 1987.

CRUEGER, W., and A. CRUEGER, *Biotechnology: A Textbook of Industrial Microbiology*. Madison, Wisc.: Science Tech., 1982.

SCRAGG, A. H., ed., *Biotechnology for Engineers*. New York: Wiley, 1988.

Steady-State Nonisothermal Reactor Design   8

> If you can't stand the heat, get out of the kitchen.
>
> Harry S Truman

Heat effects

We now focus our attention on heat effects in chemical reactors. The basic design equations, rate laws, and stoichiometric relationships derived and used in Chapter 4 for isothermal reactor design are still valid for the design of nonisothermal reactors. The major difference lies in the method of evaluating the design equation when temperature varies along the length of a PFR or when heat is removed from a CSTR. In Section 8.1 we show why we need the energy balance and how it will be used to solve reactor design problems. Section 8.2 concerns the derivation and manipulation of the energy balance for its application to various reactor types. In Sections 8.3 and 8.4, the energy balance is coupled with the mole balance, rate laws, and stoichiometry to design nonisothermal reactors. In Section 8.5 a typical nonisothermal industrial reactor and reaction, the SO_2 oxidation, is discussed in detail. We address the multiplicity of steady states in Section 8.6 and close the chapter with Section 8.7, nonisothermal multiple reactions.

8.1 Rationale

To identify the additional information necessary to design nonisothermal reactors, we consider the following example, in which a highly exothermic reaction is carried out adiabatically in a plug-flow reactor.

Example 8–1 What Additional Information Is Required?

Calculate the reactor volume necessary for 70% conversion.

$$A \longrightarrow B$$

426

The reaction is exothermic and the reactor is operated adiabatically. As a result, the temperature will increase with conversion down the length of the reactor.

Solution

1. **Design equation**:

$$\frac{dX}{dV} = \frac{-r_A}{F_{A0}} \tag{E8-1.1}$$

2. **Rate law**: $-r_A = kC_A \tag{E8-1.2}$

3. **Stoichiometry** (liquid phase):

$$v = v_0$$

$$F_A = C_A v_0$$

$$F_{A0} = C_{A0} v_0$$

$$C_A = C_{A0}(1 - X) \tag{E8-1.3}$$

4. **Combining** and canceling the entering concentration, C_{A0}, yields

$$\frac{dX}{dV} = \frac{k(1 - X)}{v_0} \tag{E8-1.4}$$

Recalling the Arrhenius equation,

$$k = k_1 \exp\left[\frac{E}{R}\left(\frac{1}{T_1} - \frac{1}{T}\right)\right] \tag{E8-1.5}$$

we know that k is a function of temperature, T. Consequently, because T varies along the length of the reactor, k will also vary, which was not the case for isothermal plug-flow reactors. Combining Equations (E8-1.4) and (E8-1.5) gives us

Why we need the
energy balance

$$\frac{dX}{dV} = k_1 \exp\left[\frac{E}{R}\left(\frac{1}{T_1} - \frac{1}{T}\right)\right]\frac{1 - X}{v_0} \tag{E8-1.6}$$

We see that we need another relationship relating X and T or T and V to solve this equation. *The energy balance will provide us with this relationship.*

8.2 The Energy Balance

8.2.1 First Law of Thermodynamics

We begin with the application of the first law of thermodynamics first to a closed system and then to an open system. A system is any bounded portion of the universe, moving or stationary, which is chosen for the application of the various thermodynamic equations. For a closed system, in which no mass crosses the system boundaries, the change in total energy of the system, $d\hat{E}$, is equal to the heat flow to the system, δQ, minus the work done *by* the system on the surroundings, δW. For a *closed system*, the energy balance is

$$d\hat{E} = \delta Q - \delta W \tag{8-1}$$

The δ's signify that δQ and δW are not exact differentials of a state function.

　　The continuous-flow reactors we have been discussing are considered to be *open systems* in that mass crosses the system boundary. We shall carry out an energy balance on the open system as shown in Figure 8-1. For an open system in which some of the energy exchange is brought about by the flow of mass across the system boundaries, the energy balance for the case of *only one* species entering and leaving becomes

rate of accumulation of energy within the system		rate of flow of heat to the system from the surroundings		rate of work done by the system on the surroundings		rate of energy added to the system by mass flow *into* the system		rate of energy leaving system by mass flow *out* of the system
$\dfrac{d\hat{E}_{sys}}{dt}$	$=$	$\dot{Q}$	$-$	$\dot{W}$	$+$	$F_{in}E_{in}$	$-$	$F_{out}E_{out}$

$$\tag{8-2}$$

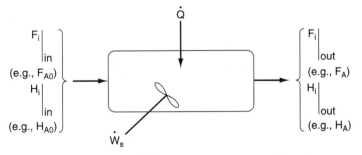

Figure 8-1　Energy balance on an open system: schematic.

　　The unsteady-state energy balance for an open system that has n species, each entering and leaving the system at its respective molar flow rates F_i (moles of i per time) and with its respective energy E_i (joules per mole of i), is

The starting point

$$\frac{d\hat{E}_{sys}}{dt} = \dot{Q} - \dot{W} + \sum_{i=1}^{n} E_i F_i \bigg|_{in} - \sum_{i=1}^{n} E_i F_i \bigg|_{out} \tag{8-3}$$

We will now discuss each of the terms in Equation (8-3).

8.2.2 Evaluating the Work Term

It is customary to separate the work term, $\dot{W}$, into *flow work* and *other work*, $\dot{W}_s$. Flow work is work that is necessary to get the mass *into* and *out of* the system. For example, when shear stresses are absent, we write

Flow work and shaft work

$$\dot{W} = \overbrace{-\sum_{i=1}^{n} F_i P \mathbf{V}_i \bigg|_{\text{in}} + \sum_{i=1}^{n} F_i P \mathbf{V}_i \bigg|_{\text{out}}}^{[\text{rate of flow work}]} + \dot{W}_s \qquad (8\text{-}4)$$

where P is the pressure (Pa) and $\mathbf{V}_i$ is the specific volume (m³/mol of i).

The term $\dot{W}_s$, often referred to as the *shaft work*, could be produced from such things as a stirrer in a CSTR or a turbine in a PFR. In most instances, the flow work term is combined with those terms in the energy balance that represent the energy exchange by mass flow across the system boundaries. Substituting Equation (8-4) into (8-3) and grouping terms, we have

$$\frac{d\hat{E}_{\text{sys}}}{dt} = \dot{Q} - \dot{W}_s + \sum_{i=1}^{n} F_i(E_i + P\mathbf{V}_i)\bigg|_{\text{in}} - \sum_{i=1}^{n} F_i(E_i + P\mathbf{V}_i)\bigg|_{\text{out}} \qquad (8\text{-}5)$$

The energy E_i is the sum of the internal energy (U_i), the kinetic energy ($u_i^2/2$), the potential energy (gz_i), and any other energies, such as electric or magnetic energy or light:

$$E_i = U_i + \frac{u_i^2}{2} + gz_i + \text{other} \qquad (8\text{-}6)$$

In almost all chemical reactor situations, the kinetic, potential, and "other" energy terms are negligible in comparison with the enthalpy, heat transfer, and work terms, and hence will be omitted; that is,

$$E_i = U_i \qquad (8\text{-}7)$$

We recall that the enthalpy, H_i (J/mol), is defined in terms of the internal energy U_i (J/mol), and the product $P\mathbf{V}_i$ (1 Pa·m³/mol = 1 J/mol·):

Enthalpy

$$H_i = U_i + P\mathbf{V}_i \qquad (8\text{-}8)$$

Typical units of H_i are

$$(H_i) = \frac{\text{J}}{\text{mol } i} \text{ or } \frac{\text{Btu}}{\text{lb mol } i} \text{ or } \frac{\text{cal}}{\text{mol } i}$$

Enthalpy carried into (or out of) the system can be expressed as the sum of the net internal energy carried into (or out of) the system by mass flow plus the flow work:

$$F_i H_i = F_i(U_i + P\mathbf{V}_i)$$

Combining Equations (8-5), (8-7), and (8-8), we can now write the energy balance in the form

$$\frac{d\hat{E}_{sys}}{dt} = \dot{Q} - \dot{W}_s + \sum_{i=1}^{n} F_i H_i \bigg|_{in} - \sum_{i=1}^{n} F_i H_i \bigg|_{out}$$

The energy of the system at any instant in time, $\hat{E}_{sys}$, is the sum of the products of the number of moles of each species in the system multiplied by their respective energies. This term will be discussed in more detail when unsteady-state reactor operation is considered in Chapter 9.

We shall let the subscript "0" represent the inlet conditions. Unsubscripted variables represent the conditions at the outlet of the chosen system volume.

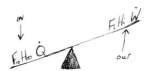

$$\dot{Q} - \dot{W}_s + \sum_{i=1}^{n} F_{i0} H_{i0} - \sum_{i=1}^{n} F_i H_i = \frac{d\hat{E}_{sys}}{dt} \qquad (8\text{-}9)$$

To put this equation in a more applicable form, *there are two items to dissect*:

1. The molar flow rates, F_i and F_{i0}
2. The molar enthalpies, H_i, $H_{i0} [H_i \equiv H_i(T)$, and $H_{i0} \equiv H_i(T_0)]$

CD-ROM animation

An animated version of what follows for the derivation of the energy balance can be found in the reaction engineering modules "Heat Effects 1" and "Heat Effects 2" on the CD-ROM.

8.2.3 Dissecting the Steady-State Molar Flow Rates to Obtain the Heat of Reaction

We will now consider flow systems that are operated at steady state. The steady-state energy balance is obtained by setting $(d\hat{E}_{sys}/dt)$ equal to zero in Equation (8-9) in order to yield

Steady-state
energy balance

$$\dot{Q} - \dot{W}_s + \sum_{i=1}^{n} F_{i0} H_{i0} - \sum_{i=1}^{n} F_i H_i = 0 \qquad (8\text{-}10)$$

To carry out the manipulations to write Equation (8-10) in terms of the heat of reaction we shall use the generalized reaction

$$A + \frac{b}{a}B \longrightarrow \frac{c}{a}C + \frac{d}{a}D \qquad (2\text{-}2)$$

The inlet and outlet terms in Equation (8-10) are expanded, respectively, to:

In: $\sum H_{i0} F_{i0} = H_{A0} F_{A0} + H_{B0} F_{B0} + H_{C0} F_{C0} + H_{D0} F_{D0} + H_{I0} F_{I0}$ (8-11)

and

Out: $\sum H_i F_i = H_A F_A + H_B F_B + H_C F_C + H_D F_D + H_I F_I$ (8-12)

We first express the molar flow rates in terms of conversion.

In general, the molar flow rate of species i for the case of no accumulation and a stoichiometric coefficient v_i is

$$F_i = F_{A0}(\Theta_i + v_i X)$$

Specifically, for Reaction (2-2) we have

$$F_A = F_{A0}(1 - X)$$

$$F_B = F_{A0}\left(\Theta_B - \frac{b}{a}X\right)$$

Steady-state operation

$$F_C = F_{A0}\left(\Theta_C + \frac{c}{a}X\right)$$

$$F_D = F_{A0}\left(\Theta_D + \frac{d}{a}X\right)$$

$$F_I = \Theta_I F_{A0}$$

We can substitute these values into Equations (8-11) and (8-12), then subtract Equation (8-12) from (8-11) to give

$$\sum_{i=1}^{n} H_{i0} F_{i0} - \sum_{i=1}^{n} F_i H_i = F_{A0}[(H_{A0} - H_A) + (H_{B0} - H_B)\Theta_B$$

$$+ [(H_{C0} - H_C)\Theta_C + (H_{D0} - H_D)\Theta_D + (H_{I0} - H_I)\Theta_I]]$$

$$- \underbrace{\left(\frac{d}{a}H_D + \frac{c}{a}H_C - \frac{b}{a}H_B - H_A\right)}_{\Delta H_{Rx}} F_{A0} X \qquad (8\text{-}13)$$

The term in parentheses that is multiplied by $F_{A0}X$ is called the **heat of reaction** at temperature T and is designated ΔH_{Rx}.

Heat of reaction at temperature T

$$\Delta H_{Rx}(T) = \frac{d}{a}H_D(T) + \frac{c}{a}H_C(T) - \frac{b}{a}H_B(T) - H_A(T) \qquad (8\text{-}14)$$

All of the enthalpies (e.g., H_A, H_B) are evaluated at the temperature at the outlet of the system volume, and consequently, $[\Delta H_{Rx}(T)]$ is the heat of reaction at the specific temperature T. The heat of reaction is always given per mole of the species that is the basis of calculation [i.e., species A (joules per mole of A reacted)].

Substituting Equation (8-14) into (8-13) and reverting to summation notation for the species, Equation (8-13) becomes

$$\sum_{i=1}^{n} F_{i0} H_{i0} - \sum_{i=1}^{n} F_i H_i = F_{A0} \sum_{i=1}^{n} \Theta_i (H_{i0} - H_i) - \Delta H_{Rx}(T) F_{A0} X \quad (8\text{-}15)$$

Combining Equations (8-10) and (8-15), we can now write the *steady-state* [i.e., $(d\hat{E}_{sys}/dt = 0)$] energy balance in a more usable form:

One can use this form of the steady-state energy balance if the enthalpies are available

$$\boxed{\dot{Q} - \dot{W}_s + F_{A0} \sum_{i=1}^{n} \Theta_i (H_{i0} - H_i) - \Delta H_{Rx}(T) F_{A0} X = 0} \quad (8\text{-}16)$$

If a *phase change* takes place during the course of a reaction, this form of the energy balance [i.e., Equation (8-16)] should be used, e.g., Problem 8-3.

8.2.4 Dissecting the Enthalpies

We are neglecting any enthalpy changes on mixing so that the partial molal enthalpies are equal to the molal enthalpies of the pure components. The molal enthalpy of species i at a particular temperature and pressure, H_i, is usually expressed in terms of an *enthalpy of formation* of species i at some reference temperature T_R, $H_i^\circ(T_R)$, plus the change in enthalpy that results when the temperature is raised from the reference temperature to some temperature T, ΔH_{Qi}:

$$H_i = H_i^\circ(T_R) + \Delta H_{Qi} \quad (8\text{-}17)$$

The reference temperature at which H_i° is given is usually 25°C. For any substance i that is being heated from T_1 to T_2 in the *absence* of phase change,

$$\Delta H_{Qi} = \int_{T_1}^{T_2} C_{pi}\, dT \quad (8\text{-}18)$$

Typical units of the heat capacity, C_{pi}, are

$$(C_{pi}) = \frac{J}{(\text{mol of } i)(K)} \quad \text{or} \quad \frac{Btu}{(\text{lb mol of } i)(R)} \quad \text{or} \quad \frac{cal}{(\text{mol of } i)(K)}$$

Example 8–2   Relating $H_A(T)$ to $H_A^\circ(T_R)$

Species A is a solid at 25°C. Its enthalpy of formation is H_A° (298 K). Write an expression for the enthalpy of substance A in the gaseous state at temperature T.

Solution

Calculating the enthalpy when phase changes are involved

Here, in addition to the increase in enthalpy of the solid, liquid, and gas from the temperature increase, one must include the heat of melting at the melting point, $\Delta H_{mA}(T_m)$, and the heat of vaporization at the boiling point, $\Delta H_{vA}(T_b)$.

$$
\begin{bmatrix} \text{enthalpy of} \\ \text{species} \\ \text{A at } T \end{bmatrix} = \begin{bmatrix} \text{enthalpy of} \\ \text{formation} \\ \text{of species} \\ \text{A at } T_R \end{bmatrix} + \begin{bmatrix} \Delta H_Q \text{ in heating} \\ \text{solid from} \\ T_R \text{ to } T_m \end{bmatrix} + \begin{bmatrix} \text{heat of} \\ \text{melting} \\ \text{at } T_m \end{bmatrix} \quad \text{(E8-2.1)}
$$

$$
+ \begin{bmatrix} \Delta H_Q \text{ in heating} \\ \text{liquid from} \\ T_m \text{ to } T_b \end{bmatrix} + \begin{bmatrix} \text{heat of} \\ \text{vaporization} \\ \text{at } T_b \end{bmatrix} + \begin{bmatrix} \Delta H_Q \text{ in heating} \\ \text{gas from} \\ T_b \text{ to } T \end{bmatrix}
$$

$$
H_{\text{A}}(T) = H_{\text{A}}^{\circ}(T_R) + \int_{T_R}^{T_m} C_{ps\text{A}}\, dT + \Delta H_{m\text{A}}(T_m)
$$

$$
+ \int_{T_m}^{T_b} C_{pl\text{A}}\, dT + \Delta H_{v\text{A}}(T_b) + \int_{T_b}^{T} C_{pv\text{A}}\, dT \quad \text{(E8-2.2)}
$$

(See Problems P8-3 and P9-4.)

A large number of chemical reactions carried out in industry do not involve phase change. Consequently, we shall further refine our energy balance to apply to *single-phase* chemical reactions. Under these conditions the enthalpy of species i at temperature T is related to the enthalpy of formation at the reference temperature T_R by

$$
H_i = H_i^{\circ}(T_R) + \int_{T_R}^{T} C_{pi}\, dT \quad \text{(8-19)}
$$

If phase changes do take place in going from the temperature for which the enthalpy of formation is given and the reaction temperature T, Equation (E8-2.2) must be used instead of Equation (8-19).

The heat capacity at temperature T is frequently expressed as a quadratic function of temperature, that is,

$$
C_{pi} = \alpha_i + \beta_i T + \gamma_i T^2 \quad \text{(8-20)}
$$

To calculate the change in enthalpy $(H_i - H_{i0})$ when the reacting fluid is heated without phase change from its entrance temperature T_{i0} to a temperature T, we use Equation (8-19) to write

$$
H_i - H_{i0} = \left[H_i^{\circ}(T_R) + \int_{T_R}^{T} C_{pi}\, dT \right] - \left[H_i^{\circ}(T_R) + \int_{T_R}^{T_{i0}} C_{pi}\, dT \right]
$$

$$
= \int_{T_{i0}}^{T} C_{pi}\, dT \quad \text{(8-21)}
$$

Substituting for H_i and H_{i0} in Equation (8-16) yields

Result of dissecting the enthalpies

$$
\boxed{\dot{Q} - \dot{W}_s - F_{A0} \sum_{i=1}^{n} \int_{T_{i0}}^{T} \Theta_i C_{pi}\, dT - \Delta H_{\text{Rx}}(T) F_{A0} X = 0} \quad \text{(8-22)}
$$

8.2.5 Relating $\Delta H_{Rx}(T)$, $\Delta H^\circ_{Rx}(T_R)$, and ΔC_p

The heat of reaction at temperature T is given in terms of the enthalpy of each species at temperature T, that is,

$$\Delta H_{Rx}(T) = \frac{d}{a}H_D(T) + \frac{c}{a}H_C(T) - \frac{b}{a}H_B(T) - H_A(T) \qquad (8\text{-}14)$$

where the enthalpy of each species is given by

$$H_i = H^\circ_i(T_R) + \int_{T_R}^{T} C_{pi}\, dT \qquad (8\text{-}19)$$

If we now substitute for the enthalpy of each species, we have

$$\Delta H_{Rx}(T) = \left[\frac{d}{a}H^\circ_D(T_R) + \frac{c}{a}H^\circ_C(T_R) - \frac{b}{a}H^\circ_B(T_R) - H^\circ_A(T_R)\right]$$
$$+ \int_{T_R}^{T}\left[\frac{d}{a}C_{p_D} + \frac{c}{a}C_{p_C} - \frac{b}{a}C_{p_B} - C_{p_A}\right]dT \qquad (8\text{-}23)$$

The first set of terms on the right-hand side of Equation (8-23) is the heat of reaction at the reference temperature T_R,

$$\Delta H^\circ_{Rx}(T_R) = \frac{d}{a}H^\circ_D(T_R) + \frac{c}{a}H^\circ_C(T_R) - \frac{b}{a}H^\circ_B(T_R) - H^\circ_A(T_R) \qquad (8\text{-}24)$$

One can look up the heats of formation at T_R, then calculate the heat of reaction at this reference temperature

The enthalpies of formation of many compounds, $H^\circ_i(T_R)$, are usually tabulated at 25°C and can readily be found in the *Handbook of Chemistry and Physics*[1] and similar handbooks. For other substances, the heat of combustion (also available in these handbooks) can be used to determine the enthalpy of formation. The method of calculation is described in these handbooks. From these values of the standard heat of formation, $H^\circ_i(T_R)$, we can calculate the heat of reaction at the reference temperature T_R from Equation (8-24).

The second term in brackets on the right-hand side of Equation (8-23) is the overall change in the heat capacity per mole of A reacted, ΔC_p,

$$\Delta C_p = \frac{d}{a}C_{p_D} + \frac{c}{a}C_{p_C} - \frac{b}{a}C_{p_B} - C_{p_A} \qquad (8\text{-}25)$$

Combining Equations (8-25), (8-24), and (8-23) gives us

Heat of reaction at temperature T

$$\Delta H_{Rx}(T) = \Delta H^\circ_{Rx}(T_R) + \int_{T_R}^{T}\Delta C_p\, dT \qquad (8\text{-}26)$$

---

[1] *CRC Handbook of Chemistry and Physics* (Boca Raton, Fla.: CRC Press, 1996). http://webbook.nist.gov

Equation (8-26) gives the heat of reaction at any temperature T in terms of the heat of reaction at a reference temperature (usually 298 K) and an integral involving the ΔC_p term. Techniques for determining the heat of reaction at pressures above atmospheric can be found in Chen.[2] For the reaction of hydrogen and nitrogen at 400°C, it was shown that the heat of reaction increased by only 6% as the pressure was raised from 1 atm to 200 atm.

8.2.6 Constant or Mean Heat Capacities

For the case of constant or mean heat capacities, Equation (8-26) becomes

$$\Delta H_{\text{Rx}}(T) = \Delta H_{\text{Rx}}^{\circ}(T_R) + \Delta \hat{C}_p(T - T_R) \tag{8-27}$$

The circumflex denotes that the heat capacities are evaluated at some mean temperature value between T_R and T.

$T_R \xrightarrow{\Delta \hat{C}_p} T$

$$\Delta \hat{C}_p = \frac{\displaystyle\int_{T_R}^{T} \Delta C_p \, dT}{T - T_R} \tag{8-28}$$

In a similar fashion we can write the integral involving Θ_i and C_{pi} in Equation (8-22) as

$$\Sigma \, \Theta_i \int_{T_{i0}}^{T} C_{pi} \, dT = \Sigma \, \Theta_i \tilde{C}_{pi}(T - T_{i0})$$

$\tilde{C}_{pi}$ is the mean heat capacity of species i between T_{i0} and T:

$T_{i0} \xrightarrow{\tilde{C}_{pi}} T$

$$\tilde{C}_{pi} = \frac{\displaystyle\int_{T_{i0}}^{T} C_{pi} \, dT}{T - T_{i0}} \tag{8-29}$$

Substituting the mean heat capacities into Equation (8-22), the steady-state energy balance becomes

Energy balance in terms of mean or constant heat capacities

$$\boxed{\dot{Q} - \dot{W}_s - F_{\text{A0}} \, \Sigma \, \Theta_i \tilde{C}_{pi}(T - T_{i0}) - F_{\text{A0}} X [\Delta H_{\text{Rx}}^{\circ}(T_R) + \Delta \hat{C}_p(T - T_R)] = 0}$$

$$\tag{8-30}$$

In almost all of the systems we will study, the reactants will be entering the system at the same temperature; therefore, $T_{i0} = T_0$.

---

[2] N. H. Chen, *Process Reactor Design* (Needham Heights, Mass.: Allyn and Bacon, 1983), p. 26.

8.2.7 Variable Heat Capacities

We next want to arrive at a form of the energy balance for the case where heat capacities are strong functions of temperature over a wide temperature range. Under these conditions the mean values used in Equation (8-30) may not be adequate for the relationship between conversion and temperature. Combining Equation (8-23) with the quadratic form of the heat capacity, Equation (8-20),

$$C_{pi} = \alpha_i + \beta_i T + \gamma_i T^2 \tag{8-20}$$

we find that

$$\Delta H_{\text{Rx}}(T) = \Delta H^\circ_{\text{Rx}}(T_R) + \int_{T_R}^{T} (\Delta\alpha + \Delta\beta T + \Delta\gamma T^2)\, dT$$

Integrating gives us

Heat capacity as a function of temperature

$$\boxed{\Delta H_{\text{Rx}}(T) = \Delta H^\circ_{\text{Rx}}(T_R) + \Delta\alpha\,(T - T_R) + \frac{\Delta\beta}{2}\,(T^2 - T_R^2) + \frac{\Delta\gamma}{3}\,(T^3 - T_R^3)}$$

$$\tag{8-31}$$

where

$$\Delta\alpha = \frac{d}{a}\alpha_{\text{D}} + \frac{c}{a}\alpha_{\text{C}} - \frac{b}{a}\alpha_{\text{B}} - \alpha_{\text{A}}$$

$$\Delta\beta = \frac{d}{a}\beta_{\text{D}} + \frac{c}{a}\beta_{\text{C}} - \frac{b}{a}\beta_{\text{B}} - \beta_{\text{A}}$$

$$\Delta\gamma = \frac{d}{a}\gamma_{\text{D}} + \frac{c}{a}\gamma_{\text{C}} - \frac{b}{a}\gamma_{\text{B}} - \gamma_{\text{A}}$$

In a similar fashion, we can evaluate the heat capacity term in Equation (8-22):

$$\sum_{i=1}^{n} \Theta_i \int_{T_0}^{T} C_{pi}\, dT = \int_{T_0}^{T} \left(\sum \alpha_i \Theta_i + \sum \beta_i \Theta_i T + \sum \gamma_i \Theta_i T^2 \right) dT$$

$$= \sum \alpha_i \Theta_i (T - T_0) + \frac{\sum \beta_i \Theta_i}{2}\,(T^2 - T_0^2) + \frac{\sum \gamma_i \Theta_i}{3}\,(T^3 - T_0^3) \tag{8-32}$$

Substituting Equations (8-31) and (8-32) into Equation (8-22), the form of the energy balance is

Energy balance for the case of highly temperature-sensitive heat capacities

$$\dot{Q} - \dot{W}_s - F_{\text{A0}} \left[\sum \alpha_i \Theta_i (T - T_0) + \frac{\sum \beta_i \Theta_i}{2}\,(T^2 - T_0^2) + \frac{\sum \gamma_i \Theta_i}{3}\,(T^3 - T_0^3) \right]$$

$$- F_{\text{A0}} X \left[\Delta H^\circ_{\text{Rx}}(T_R) + \Delta\alpha\,(T - T_R) + \frac{\Delta\beta}{2}\,(T^2 - T_R^2) + \frac{\Delta\gamma}{3}\,(T^3 - T_R^3) \right] = 0$$

$$\tag{8-33}$$

Example 8–3 Heat of Reaction

Calculate the heat of reaction for the synthesis of ammonia from hydrogen and nitrogen at 150°C in kcal/mol of N_2 reacted *and* in kJ/mol of H_2 reacted.

Solution

$$N_2 + 3H_2 \longrightarrow 2NH_3$$

Calculate the heat of reaction at the reference temperature using the heats of formation of the reacting species obtained from *Perry's Handbook*[3] or the *Handbook of Chemistry and Physics*.

$$\Delta H_{Rx}^{\circ}(T_R) = 2H_{NH_3}^{\circ}(T_R) - 3H_{H_2}^{\circ}(T_R) - H_{N_2}^{\circ}(T_R) \tag{E8-3.1}$$

The heats of formation of the elements (H_2, N_2) are zero at 25°C.

$$\Delta H_{Rx}^{\circ}(T_R) = 2H_{NH_3}^{\circ}(T_R) - 3(0) - 0 = 2H_{NH_3}^{\circ}$$

$$= 2(-11,020) \frac{cal}{mol\ N_2}$$

$$= -22,040 \text{ cal/mol } N_2 \text{ reacted}$$

or

$$\Delta H_{Rx}^{\circ}(298 \text{ K}) = -22.04 \text{ kcal/mol } N_2 \text{ reacted}$$

$$= -92.22 \text{ kJ/mol } N_2 \text{ reacted}$$

Exothermic reaction

The minus sign indicates the reaction is *exothermic*. If the heat capacities are constant or if the mean heat capacities over the range 25 to 150°C are readily available, the determination of ΔH_{Rx} at 150°C is quite simple.

$$\hat{C}_{P_{H_2}} = 6.992 \text{ cal/mol } H_2 \cdot K$$

$$\hat{C}_{P_{N_2}} = 6.984 \text{ cal/mol } N_2 \cdot K$$

$$\hat{C}_{P_{NH_3}} = 8.92 \text{ cal/mol } NH_3 \cdot K$$

$$\Delta\hat{C}_p = 2\hat{C}_{P_{NH_3}} - 3\hat{C}_{P_{H_2}} - \hat{C}_{P_{N_2}} \tag{E8-3.2}$$

$$= 2(8.92) - 3(6.992) - 6.984$$

$$= -10.12 \text{ cal/g mol } N_2 \text{ reacted} \cdot K$$

$$\Delta H_{Rx}(T) = \Delta H_{Rx}^{\circ}(T_R) + \Delta\hat{C}_p(T - T_R) \tag{8-27}$$

$$\Delta H_{Rx}(423 \text{ K}) = -22,040 + (-10.12)(423 - 298)$$

$$= -23,310 \text{ cal/mol } N_2 = -23.31 \text{ kcal/mol } N_2$$

$$= -97,530 \text{ J/mol } N_2$$

(Recall: 1 cal = 4.184 J)

---

[3] R. H. Perry, D. W. Green, and J. O. Maloney, eds., *Perry's Chemical Engineers' Handbook*, 6th ed. (New York: McGraw-Hill, 1984), pp. 3–147.

The heat of reaction based on the moles of H_2 reacted is

$$\Delta H_{Rx}(423 \text{ K}) = \frac{1 \text{ g mol N}_2}{3 \text{ g mol H}_2}\left(-97.53 \ \frac{\text{kJ}}{\text{mol N}_2}\right)$$

$$= -32.51 \ \frac{\text{kJ}}{\text{mol H}_2} \text{ at 423 K}$$

8.2.8 Heat Added to the Reactor, $\dot{Q}$

The heat flow *to* the reactor, $\dot{Q}$, is given in many instances in terms of the overall heat-transfer coefficient, U, the heat-exchange area, A, and the difference between the ambient temperature, T_a, and the reaction temperature, T.

CSTRs. Figure 8-2 shows schematics of a CSTR with a heat exchanger. The coolant enters the exchanger at a mass flow rate $\dot{m}_C$ at a temperature T_{a1} and leaves at a temperature T_{a2}. The rate of heat transfer *from* the exchanger *to* the reactor is[4]

$$\dot{Q} = \frac{UA(T_{a1}-T_{a2})}{\ln\left[(T-T_{a1})/(T-T_{a2})\right]} \tag{8-34}$$

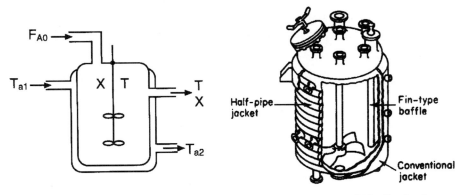

Figure 8-2   CSTR tank reactor with heat exchanger. [(b) Courtesy of Pfaudler, Inc.]

As a first approximation, we assume a quasi-steady state for the coolant flow and neglect the accumulation term (i.e., $dT_a/dt = 0$). An energy balance on the coolant fluid entering and leaving the exchanger is

---

[4] Information on the overall heat-transfer coefficient may be found in C. O. Bennett and J. E. Myers, *Momentum, Heat, and Mass Transfer*, 2nd ed. (New York: McGraw-Hill, 1974), p. 316.

Energy balance on
heat exchanger

$$\begin{bmatrix} \text{rate of} \\ \text{energy} \\ \text{in} \\ \text{by flow} \end{bmatrix} - \begin{bmatrix} \text{rate of} \\ \text{energy} \\ \text{out} \\ \text{by flow} \end{bmatrix} - \begin{bmatrix} \text{rate of} \\ \text{heat transfer} \\ \text{\textit{from} exchanger} \\ \text{\textit{to} reactor} \end{bmatrix} = 0 \quad (8\text{-}35)$$

$$\dot{m}_C C_{P_C}(T_{a1} - T_R) - \dot{m}_C C_{P_C}(T_{a2} - T_R) - \frac{UA(T_{a1} - T_{a2})}{\ln(T - T_{a1})/(T - T_{a2})} = 0 \quad (8\text{-}36)$$

where C_{P_C} is the heat capacity of the coolant fluid and T_R is the reference temperature. Simplifying gives us

$$Q = \dot{m}_C C_{P_C}(T_{a1} - T_{a2}) = \frac{UA(T_{a1} - T_{a2})}{\ln(T - T_{a1})/(T - T_{a2})} \quad (8\text{-}37)$$

Solving Equation (8-37) for the exit temperature of the coolant fluid yields

$$T_{a2} = T - (T - T_{a1}) \exp\left(\frac{-UA}{\dot{m}_C C_{P_C}}\right) \quad (8\text{-}38)$$

From Equation (8-37)

$$Q = \dot{m}_C C_{P_C}(T_{a1} - T_{a2}) \quad (8\text{-}39)$$

Substituting for T_{a2} in Equation (8-39), we obtain

$$\dot{Q} = \dot{m}_C C_{P_C}\left\{(T_{a1} - T)\left[1 - \exp\left(\frac{-UA}{\dot{m}_C C_{P_C}}\right)\right]\right\} \quad (8\text{-}40)$$

For large values of the coolant flow rate, the exponent can be expanded in a Taylor series where second-order terms are neglected in order to give

Heat transfer to a
CSTR

$$\dot{Q} = \dot{m}_C C_{P_C}(T_{a1} - T)\left[1 - \left(1 - \frac{UA}{\dot{m}_C C_{P_C}}\right)\right] \quad (8\text{-}41)$$

Then

$$\dot{Q} = UA(T_a - T) \quad (8\text{-}42)$$

where $T_{a1} \approx T_{a2} = T_a$.

Tubular Reactors (PFR/PBR).   When the heat flow varies along the length of the reactor, such as would be the case in a tubular flow reactor, we must integrate the heat flux equation along the length of the reactor to obtain the total heat added to the reactor,

$$\dot{Q} = \int^A U(T_a - T)\, dA = \int^V Ua(T_a - T)\, dV \quad (8\text{-}43)$$

where a is the heat-exchange area per unit volume of reactor. The variation in heat added along the reactor length (i.e., volume) is found by differentiating $\dot{Q}$ with respect to V.

Heat transfer to a
PFR

$$\frac{d\dot{Q}}{dV} = Ua(T_a - T) \tag{8-44}$$

For a tubular reactor of diameter D,

$$a = \frac{4}{D}$$

For a packed-bed reactor, we can write Equation (8-44) in terms of catalyst weight by simply dividing by the bulk catalyst density

$$\frac{1}{\rho_b}\frac{d\dot{Q}}{dV} = \frac{Ua}{\rho_b}(T_a - T) \tag{8-45}$$

Recalling $dW = \rho_b\,dV$, then

Heat transfer to a
PBR

$$\frac{d\dot{Q}}{dW} = \frac{Ua}{\rho_b}(T_a - T) \tag{8-46}$$

8.3 Nonisothermal Continuous-Flow Reactors

In this section we apply the general energy balance [Equation (8-22)] to the CSTR and to the tubular reactor operated at steady state. We then present example problems showing how the mole and energy balances are combined to size reactors operating adiabatically.

Substituting Equation (8-26) into Equation (8-22), the steady-state energy balance becomes

$$\dot{Q} - \dot{W}_s - F_{A0}\sum_{i=1}^{n}\int_{T_{i0}}^{T}\Theta_i C_{pi}\,dT - \left[\Delta H_{Rx}^\circ(T_R) + \int_{T_R}^{T}\Delta C_p\,dT\right]F_{A0}X = 0$$

$$\tag{8-47}$$

These are the forms
of the steady-state
balance we will use

[**Note:** In many calculations the CSTR mole balance ($F_{A0}X = -r_A V$) will be used to replace the term following the brackets in Equation (8-47); that is, ($F_{A0}X$) will be replaced by ($-r_A V$).] Rearranging yields the steady-state balance for the case of *constant or mean heat capacities* in the form

$$\dot{Q} - \dot{W}_s - [\Delta H_{Rx}^\circ(T_R) + \Delta\hat{C}_p(T - T_R)]F_{A0}X = F_{A0}\sum_{i=1}^{n}\Theta_i\tilde{C}_{pi}(T - T_{i0}) \tag{8-48}$$

How can we use this information?   Let's stop a minute and consider a system with the special set of conditions of no work, $\dot{W}_s = 0$, adiabatic operation $\dot{Q} = 0$, and then rearrange (8-48) into the form

$$X = \frac{\Sigma\,\Theta_i\,\tilde{C}_{pi}(T - T_{io})}{-[\Delta H^{\circ}_{Rx}(T_R) + \Delta\hat{C}_p(T - T_R)]} \tag{8-49}$$

In many instances the $\Delta\hat{C}_p(T - T_R)$ term in the denominator of Equation (8-49) is negligible with respect to the ΔH°_{Rx} term, so that a plot of X vs. T will usually be linear, as shown in Figure 8-3. To remind us that the conversion in this plot was obtained from the energy balance rather than the mole balance it is given the subscript EB (i.e., X_{EB}) in Figure 8-3. Equation (8-49) applies to a CSTR, PFR, PBR, and also to a batch (as will be shown in Chapter 9). For $\dot{Q} = 0$ and $\dot{W}_s = 0$, Equation (8-49) gives us the explicit relationship between X and T needed to be used in conjunction with the mole balance to solve reaction engineering problems as discussed in Section 8.1.

Relationship
between X and T
for *adiabatic*
exothermic
reactions

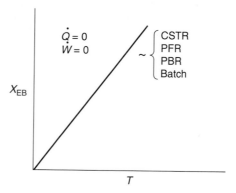

Figure 8-3   Adiabatic temperature–conversion relationship.

8.3.1 Application to the CSTR

Although the CSTR is well mixed and the temperature is uniform throughout the reaction vessel, these conditions do not mean that the reaction is carried out isothermally. Isothermal operation occurs when the feed temperature is identical to the temperature of the fluid inside the CSTR. The design equation for a CSTR in which there is no spatial variation in the rate of reaction is

$$V = \frac{F_{A0}X}{-r_A} \tag{2-13}$$

Equation (2-13) is coupled with a slight rearrangement of Equation (8-48):

$$\frac{\dot{Q} - \dot{W}_s}{F_{A0}} - X[\Delta H^{\circ}_{Rx}(T_R) + \Delta\hat{C}_p(T - T_R)] = \Sigma\,\Theta_i\,\tilde{C}_{pi}(T - T_{i0}) \tag{8-50}$$

TABLE 8-1.   CSTR ALGORITHM

The first-order irreversible liquid-phase reaction

$$A \longrightarrow B$$

is carried out adiabatically.

1. **CSTR design equation:**

$$V = \frac{F_{A0}X}{-r_A} \qquad (T8\text{-}1.1)$$

2. **Rate law:** $\qquad -r_A = kC_A \qquad (T8\text{-}1.2)$

with $\qquad k = Ae^{-E/RT} \qquad (T8\text{-}1.3)$

An algorithm

3. **Stoichiometry** (liquid-phase, $v = v_0$):

$$C_A = C_{A0}(1 - X)$$

4. **Combining** yields

$$V = \frac{v_0}{Ae^{-E/RT}}\left(\frac{X}{1-X}\right) \qquad (T8\text{-}1.4)$$

Divide and conquer

Case A. The variables X, v_0, C_{A0}, and F_{i0} are specified and the reactor volume, V, must be determined. The procedure is:

5A. Solve for the temperature, T, for pure A entering, and $\tilde{C}_{P_A} = \tilde{C}_{P_B}(\Delta\tilde{C}_p = 0)$. For the adiabatic case, solve Equation (8-52) for T:

$$T = T_0 + \frac{X(-\Delta H^{\circ}_{Rx})}{\tilde{C}_{P_A}} \qquad (T8\text{-}1.5)$$

For the nonadiabatic case with $Q = UA(T_a - T)$, solve Equation (8-51) for T:

$$T = \frac{F_{A0}X(-\Delta H^{\circ}_{Rx}) + F_{A0}\tilde{C}_{P_A}T_0 + UAT_a}{F_{A0}\tilde{C}_{P_A} + UA} \qquad (T8\text{-}1.6)$$

6A. **Calculate** k from the Arrhenius equation.

7A. **Calculate the reactor volume,** V, from Equation (T8-1.4).

Case B. The variables v_0, C_{A0}, V, and F_{i0} are specified and the exit temperature, T, and conversion, X, are unknown quantities. The procedure is:

5B. **Solve the energy balance (adiabatic) for** X **as a function of** T.

Energy balance

$$X_{EB} = \frac{\tilde{C}_{P_A}(T - T_0)}{-[\Delta H^{\circ}_{Rx}(T_R)]} \qquad (T8\text{-}1.7)$$

For the nonadiabatic with $Q = UA(T_a - T)$ case, solve Equation (8-51) for X_{EB}:

$$X_{EB} = \frac{UA(T - T_a)/F_{A0} + \tilde{C}_{P_A}(T - T_0)}{-\Delta H^{\circ}_{Rx}} \qquad (T8\text{-}1.8)$$

6B. **Solve Equation (T8-1.4) for** X **as a function of** T.

Mole balance

$$X_{MB} = \frac{\tau Ae^{-E/RT}}{1 + \tau Ae^{-E/RT}} \qquad \text{where } \tau = V/v_0 \qquad (T8\text{-}1.9)$$

7B. **Find the values of** X **and** T **that satisfy both the energy balance [Equation (T8-1.7)] and the mole balance [Equation (T8-1.9)].** This result can be achieved either numerically or graphically [plotting X vs. T using Equations (T8-1.7) and (T8-1.9) on the same graph].

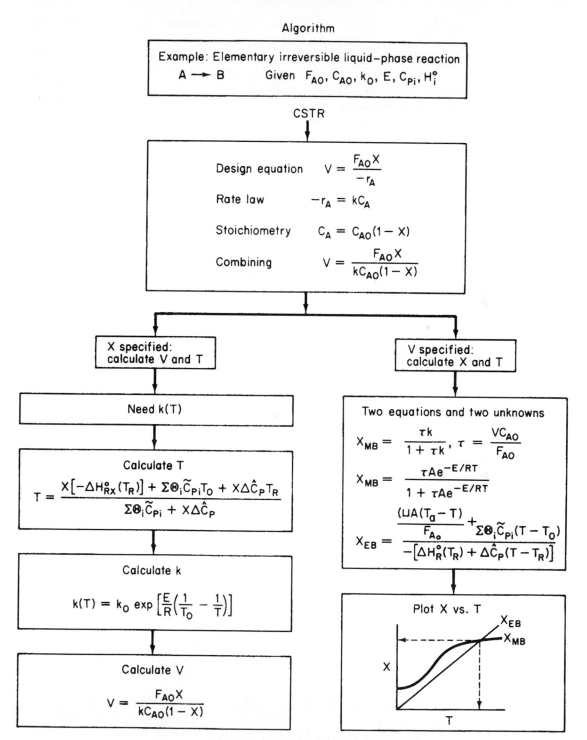

Figure 8-4 Algorithm for adiabatic CSTR design.

and used to design CSTRs (i.e., to obtain the reactor volume or operating temperature). If necessary, the CSTR is either heated or cooled by a heating or cooling jacket as shown in Figure 8-2, or by a coil placed inside the reactor.

Reactions are frequently carried out adiabatically, often with heating or cooling provided upstream or downstream of the reaction vessel. With the exception of processes involving highly viscous materials such as in Problem P8-4, the work done by the stirrer can usually be neglected. After substituting Equation (8-42) for $\dot{Q}$, the energy balance can be written as

CSTR with heat exchange
$$\boxed{\frac{UA(T_a - T)}{F_{A0}} - X[\Delta H^\circ_{Rx}(T_R) + \Delta \hat{C}_p(T - T_R)] = \sum_{i=1}^{n} \Theta_i \tilde{C}_{p_i}(T - T_{i0})} \quad (8\text{-}51)$$

Under conditions of adiabatic operation and negligible stirring work, both $\dot{Q}$ and $\dot{W}_s$ are zero, and the energy balance becomes

Adiabatic operation of a CSTR
$$\boxed{-X[\Delta H^\circ_{Rx}(T_R) + \Delta \hat{C}_p(T - T_R)] = \sum_{i=1}^{n} \Theta_i \tilde{C}_{pi}(T - T_{i0})} \quad (8\text{-}52)$$

The procedure for nonisothermal reactor design can be illustrated by considering the first-order irreversible liquid-phase reaction shown in Table 8-1. The algorithm for working through either case A or B is summarized in Figure 8-4. Its application is illustrated in the following example.

From here on, for the sake of brevity we will let

$$\Sigma = \sum_{i=1}^{n}$$

unless otherwise specified.

Example 8–4 Production of Propylene Glycol in an Adiabatic CSTR

Propylene glycol is produced by the hydrolysis of propylene oxide:

$$CH_2\text{—}CH\text{—}CH_3 + H_2O \xrightarrow{H_2SO_4} CH_2\text{—}CH\text{—}CH_3$$
$$\underset{O}{\diagdown\diagup} \qquad\qquad \underset{OH \quad OH}{|\quad\ |}$$

Production, uses, and economics

Over 800 million pounds of propylene glycol were produced in 1997 and the selling price was approximately $0.67 per pound. Propylene glycol makes up about 25% of the major derivatives of propylene oxide. The reaction takes place readily at room temperature when catalyzed by sulfuric acid.

You are the engineer in charge of an adiabatic CSTR producing propylene glycol by this method. Unfortunately, the reactor is beginning to leak, and you must replace it. (You told your boss several times that sulfuric acid was corrosive and that mild steel was a poor material for construction.) There is a nice overflow CSTR of 300-gal capacity standing idle; it is glass-lined and you would like to use it.

You are feeding 2500 lb/h (43.04 lb mol/h) of propylene oxide (P.O.) to the reactor. The feed stream consists of (1) an equivolumetric mixture of propylene oxide (46.62 ft³/h) and methanol (46.62 ft³/h), and (2) water containing 0.1 wt % H_2SO_4. The volumetric flow rate of water is 233.1 ft³/h, which is 2.5 times the methanol–P.O. flow rate. The corresponding molar feed rates of methanol and water are 71.87 and 802.8 lb mol/h, respectively. The water–propylene oxide–methanol mixture undergoes a slight decrease in volume upon mixing (approximately 3%), but you neglect this decrease in your calculations. The temperature of both feed streams is 58°F prior to mixing, but there is an immediate 17°F temperature rise upon mixing of the two feed streams caused by the heat of mixing. The entering temperature of all feed streams is thus taken to be 75°F (Figure E8-4.1).

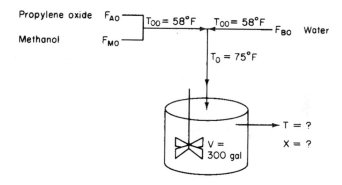

Figure E8-4.1

Furosawa et al.[5] state that under conditions similar to those at which you are operating, the reaction is first-order in propylene oxide concentration and apparent zero-order in excess of water with the specific reaction rate

$$k = Ae^{-E/RT} = 16.96 \times 10^{12} (e^{-32,400/RT})\ h^{-1}$$

The units of E are Btu/lb mol.

There is an important constraint on your operation. Propylene oxide is a rather low-boiling substance (b.p. at 1 atm, 93.7°F). With the mixture you are using, you feel that you cannot exceed an operating temperature of 125°F, or you will lose too much oxide by vaporization through the vent system.

Can you use the idle CSTR as a replacement for the leaking one if it will be operated adiabatically? If so, what will be the conversion of oxide to glycol?

Solution

(All data used in this problem were obtained from the *Handbook of Chemistry and Physics* unless otherwise noted.) Let the reaction be represented by

$$A + B \longrightarrow C$$

where

---

[5] T. Furusawa, H. Nishimura, and T. Miyauchi, *J. Chem. Eng. Jpn.*, 2, 95 (1969).

A is propylene oxide ($C_{P_A} = 35$ Btu/lb mol $\cdot$ °F) [6]

B is water ($C_{P_B} = 18$ Btu/lb mol $\cdot$ °F)

C is propylene glycol ($C_{P_C} = 46$ Btu/lb mol $\cdot$ °F)

M is methanol ($C_{P_M} = 19.5$ Btu/lb mol $\cdot$ °F)

In this problem neither the exit conversion nor the temperature of the adiabatic reactor is given. By application of the material and energy balances we can solve two equations with two unknowns (X and T). Solving these coupled equations, we determine the exit conversion and temperature for the glass-lined reactor to see if it can be used to replace the present reactor.

1. **Mole balance and design equation**:

$$F_{A0} - F_A + r_A V = 0$$

The design equation in terms of X is

$$V = \frac{F_{A0}X}{-r_A} \tag{E8-4.1}$$

2. **Rate law**:

$$-r_A = kC_A \tag{E8-4.2}$$

3. **Stoichiometry** (liquid-phase, $v = v_0$):

$$C_A = C_{A0}(1 - X) \tag{E8-4.3}$$

4. **Combining** yields

$$V = \frac{F_{A0}X}{kC_{A0}(1-X)} = \frac{v_0 X}{k(1-X)} \tag{E8-4.4}$$

5. **Solving** for X as a function of T and recalling that $\tau = V/v_0$ gives

$$X_{MB} = \frac{\tau k}{1 + \tau k} = \frac{\tau A e^{-E/RT}}{1 + \tau A e^{-E/RT}} \tag{E8-4.5}$$

This equation relates temperature and conversion through the **mole balance**.

6. The **energy balance** for this adiabatic reaction in which there is negligible energy input provided by the stirrer is

$$-X[\Delta H_{Rx}^{\circ}(T_R) + \Delta \hat{C}_p(T - T_R)] = \Sigma \Theta_i \tilde{C}_{pi}(T - T_{i0}) \tag{8-52}$$

Solving for X, we obtain

$$X_{EB} = \frac{\Sigma \Theta_i \tilde{C}_{pi}(T - T_{i0})}{-[\Delta H_{Rx}^{\circ}(T_R) + \Delta \hat{C}_p(T - T_R)]} \tag{E8-4.6}$$

---

[6] C_{P_A} and C_{P_C} are estimated from the observation that the great majority of low-molecular-weight oxygen-containing organic liquids have a mass heat capacity of 0.6 cal/g $\cdot$ °C $\pm 15\%$.

This equation relates X and T through the energy balance. We see that there are two equations [Equations (E8-4.5) and (E8-4.6)] that must be solved for the two unknowns, X and T.

7. **Calculations**:

a. *Heat of reaction at temperature* T:[7]

$$\Delta H_{Rx}(T) = \Delta H^{\circ}_{Rx}(T_R) + \Delta \hat{C}_p(T - T_R) \tag{8-27}$$

$$H^{\circ}_A(68°F) : -66{,}600 \text{ Btu/lb mol}$$

$$H^{\circ}_B(68°F) : -123{,}000 \text{ Btu/lb mol}$$

$$H^{\circ}_C(68°F) : -226{,}000 \text{ Btu/lb mol}$$

$$H^{\circ}_{Rx}(68°F) = -226{,}000 - (-123{,}000) - (-66{,}600)$$

$$= -36{,}400 \text{ Btu/lb mol propylene oxide}$$

Calculating the parameter values

ΔH_{Rx}
$\Delta \hat{C}_p$
v_o
τ
C_{A0}
θ_M
θ_B
$\sum\limits_{i=1}^{n} C_{p_i} \theta_i$

$$\Delta \hat{C}_p = \hat{C}_{P_C} - \hat{C}_{P_B} - \hat{C}_{P_A} \tag{E8-4.7}$$

$$= 46 - 18 - 35 = -7 \text{ Btu/lb mol} \cdot °F$$

$$\Delta H^{\circ}_{Rx}(T) = -36{,}400 - (7)(T - 528)$$

b. *Stoichiometry* (C_{A0}, Θ_I, τ): The total liquid volumetric flow rate entering the reactor is

$$v_0 = v_{A0} + v_{M0} + v_{B0}$$

$$= 46.62 + 46.62 + 233.1 = 326.3 \text{ ft}^3/\text{h} \tag{E8-4.8}$$

$$V = 300 \text{ gal} = 40.1 \text{ ft}^3$$

$$\tau = \frac{V}{v_0} = \frac{40.1 \text{ ft}^3}{326.3 \text{ ft}^3/\text{h}} = 0.1229 \text{ h}$$

$$C_{A0} = \frac{F_{A0}}{v_0} = \frac{43.04 \text{ lb mol/h}}{326.3 \text{ ft}^3/\text{h}}$$

$$= 0.132 \text{ lb mol/ft}^3 \tag{E8-4.9}$$

For methanol: $\Theta_M = \dfrac{F_{M0}}{F_{A0}} = \dfrac{71.87 \text{ lb mol/h}}{43.04 \text{ lb mol/h}} = 1.67$

For water: $\Theta_B = \dfrac{F_{B0}}{F_{A0}} = \dfrac{802.8 \text{ lb mol/h}}{43.04 \text{ lb mol/h}} = 18.65$

c. *Energy balance terms:*

$$\sum \Theta_i \tilde{C}_{pi} = \tilde{C}_{P_A} + \Theta_B \tilde{C}_{P_B} + \Theta_M \tilde{C}_{P_M}$$

$$= 35 + (18.65)(18) + (1.67)(19.5)$$

$$= 403.3 \text{ Btu/lb mol} \cdot °F \tag{E8-4.10}$$

---

[7] H°_A and H°_C are calculated from heat-of-combustion data.

$$T_0 = T_{00} + \Delta T_{mix} = 58°F + 17°F = 75°F$$

$$= 535°R \qquad \text{(E8-4.11)}$$

$$T_R = 68°F = 528°R$$

The conversion calculated from the energy balance, X_{EB}, for an adiabatic reaction is found by rearranging Equation (8-52):

$$X_{EB} = -\frac{\Sigma \Theta_i \tilde{C}_{pi}(T - T_{i0})}{\Delta H^\circ_{Rx}(T_R) + \Delta \hat{C}_p(T - T_R)} \qquad \text{(E8-4.6)}$$

Substituting all the known quantities into the mole and energy balances gives us

Plot X_{EB} as a function of temperature

$$X_{EB} = \frac{(403.3 \text{ Btu/lb mol} \cdot °F)(T - 535)°F}{-[-36,400 - 7(T - 528)] \text{ Btu/lb mol}}$$

$$= \frac{403.3(T - 535)}{36,400 + 7(T - 528)} \qquad \text{(E8-4.12)}$$

The conversion calculated from the mole balance, X_{MB}, is found from Equation (E8-4.5).

Plot X_{MB} as a function of temperature

$$X_{MB} = \frac{(16.96 \times 10^{12} \text{ h}^{-1})(0.1229 \text{ h}) \exp(-32,400/1.987T)}{1 + (16.96 \times 10^{12} \text{ h}^{-1})(0.1229 \text{ h}) \exp(-32,400/1.987T)}$$

$$= \frac{(2.084 \times 10^{12}) \exp(-16,306/T)}{1 + (2.084 \times 10^{12}) \exp(-16,306/T)} \qquad \text{(E8-4.13)}$$

8. **Solving**. There are a number of different ways to solve these two simultaneous equations [e.g., substituting Equation (E8-4.12) into (E8-4.13)]. To give insight into the functional relationship between X and T for the mole and energy balances, we shall obtain a graphical solution. Here X is plotted as a function of T for the mole and energy balances, and the intersection of the two curves gives the solution where both the mole and energy balance solutions are satisfied. In addition, by plotting these two curves we can learn if there is more than one intersection (i.e., multiple steady states) for which both the energy balance and mole balance are satisfied. If numerical root-finding techniques were used to solve for X and T, it would be quite possible to obtain only one root when there is actually more than one. We shall discuss multiple steady states further in Section 8.6. We choose T and then calculate X (Table E8-4.1). The calculations are plotted in Figure E8-4.2. The

TABLE E8-4.1

T (°R)	X_{MB} [Eq. (E8-4.13)]	X_{EB} [Eq. (E8-4.12)]
535	0.108	0.000
550	0.217	0.166
565	0.379	0.330
575	0.500	0.440
585	0.620	0.550
595	0.723	0.656
605	0.800	0.764
615	0.860	0.872
625	0.900	0.980

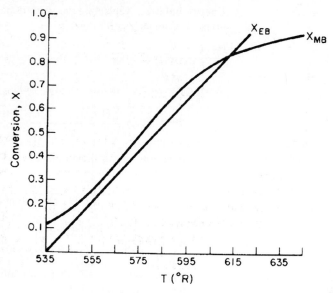

Figure E8-4.2

The reactor cannot be used because it will exceed the specified maximum temperature of 585 R

virtually straight line corresponds to the energy balance [Equation (E8-4.12)] and the curved line corresponds to the mole balance [Equation (E8-4.13)]. We observe from this plot that the only intersection point is at 85% conversion and 613°R. At this point both the energy balance and mole balance are satisfied. Because the temperature must remain below 125°F (585 R), we cannot use the 300-gal reactor as it is now.

Example 8–5 CSTR with a Cooling Coil

A cooling coil has been located for use in the hydration of propylene oxide discussed in Example 8-4. The cooling coil has 40 ft^2 of cooling surface and the cooling water flow rate inside the coil is sufficiently large that a constant coolant temperature of 85°F can be maintained. A typical overall heat-transfer coefficient for such a coil is 100 Btu/h·ft^2·°F. Will the reactor satisfy the previous constraint of 125°F maximum temperature if the cooling coil is used?

Solution

If we assume that the cooling coil takes up negligible reactor volume, the conversion calculated as a function of temperature from the mole balance is the same as that in Example 8-4 [Equation (E8-4.13)].

1. **Combining the mole balance, stoichiometry,** and **rate law**, we have

$$X_{MB} = \frac{(2.084 \times 10^{12}) \exp(-16{,}306/T)}{1 + (2.084 \times 10^{12}) \exp(-16{,}306/T)} \qquad \text{(E8-4.13)}$$

2. **Energy balance**. Neglecting the work done by the stirrer, we combine Equations (8-42) and (8-50) to obtain

$$\frac{UA(T_a - T)}{F_{A0}} - X[\Delta H^\circ_{Rx}(T_R) + \Delta \hat{C}_p(T - T_R)] = \sum \Theta_i \tilde{C}_{pi}(T - T_0) \qquad \text{(E8-5.1)}$$

Solving the energy balance for X_{EB} yields

$$X_{EB} = \frac{\sum \Theta_i \tilde{C}_{pi}(T - T_0) + [UA(T - T_a)/F_{A0}]}{-[\Delta H^\circ_{Rx}(T_R) + \Delta \hat{C}_p(T - T_R)]} \qquad \text{(E8-5.2)}$$

The cooling coil term in Equation (E8-5.2) is

$$\frac{UA}{F_{A0}} = 100 \; \frac{\text{Btu}}{\text{h} \cdot \text{ft}^2 \cdot \text{°F}} \left(\frac{40 \; \text{ft}^2}{43.04 \; \text{lb mol/h}} \right) = \frac{92.9 \; \text{Btu}}{\text{lb mol} \cdot \text{°F}} \qquad \text{(E8-5.3)}$$

Recall that the cooling temperature is

$$T_a = 85\text{°F} = 545 \; \text{R}$$

The numerical values of all other terms of Equation (E8-5.2) are identical to those given in Equation (E8-4.12):

$$\boxed{X_{EB} = \frac{403.3(T - 535) + 92.9(T - 545)}{36,400 + 7(T - 528)}} \qquad \text{(E8-5.4)}$$

We now have two equations [(E8-4.13) and (E8-5.4)] and two unknowns, X and T.

TABLE E8-5.1.   POLYMATH: CSTR WITH HEAT EXCHANGE

Equations:	Initial Values:
f(X)=X-(403.3*(T-535)+92.9*(T-545))/(36400+7*(T-528))	0.367
f(T)=X-tau*k/(1+tau*k)	564
tau=0.1229	
A=16.96*10**12	
E=32400	
R=1.987	
k=A*exp(-E/(R*T))	

We can now use the glass lined reactor

TABLE E8-5.2.   EXAMPLE 8-4 CSTR WITH HEAT EXCHANGE

	Solution	
Variable	Value	f()
X	0.363609	-6.779e-16
T	563.729	-6.855e-16
tau	0.1229	
A	1.696e+13	
E	32400	
R	1.987	
k	4.64898	

The POLYMATH program and solution to Equations (E8-4.13), X_{MB}, and (E8-5.4), X_{EB}, are given in Table E8-5.1. The exiting temperature and conversion are 103.7°F (563.7 R) and 36.4%, respectively.

8.3.2 Adiabatic Tubular Reactor

The energy balance given by Equation (8-48) relates the conversion at any point in the reactor to the temperature of the reaction mixture at the same point (i.e., it gives X as a function of T). Usually, there is a negligible amount of work done on or by the reacting mixture, so normally, the work term can be neglected in tubular reactor design. However, unless the reaction is carried out adiabatically, Equation (8-48) is still difficult to evaluate, because in nonadiabatic reactors, the heat added to or removed from the system varies along the length of the reactor. This problem does not occur in adiabatic reactors, which are frequently found in industry. Therefore, the adiabatic tubular reactor will be analyzed first.

Because $\dot{Q}$ and $\dot{W}_s$ are equal to zero for the reasons stated above, Equation (8-47) reduces to

Energy balance for adiabatic operation of PFR

$$X[-\Delta H_{Rx}(T)] = \int_{T_{i0}}^{T} \Sigma \Theta_i C_{pi} \, dT \qquad (8\text{-}53)$$

This equation can be combined with the differential mole balance

$$F_{A0} \frac{dX}{dV} = -r_A(X, T)$$

to obtain the temperature, conversion, and concentration profiles along the length of the reactor. One way of accomplishing this combination is to use Equation (8-53) to construct a table of T as a function of X. Once we have T as a function of X, we can obtain $k(T)$ as a function of X and hence $-r_A$ as a function of X alone. We then use the procedures detailed in Chapter 2 to size the different types of reactors.

The algorithm for solving PFRs and PBRs operated adiabatically is shown in Table 8-2.

TABLE 8-2A.   ADIABATIC PFR/PBR ALGORITHM

The elementary reversible gas-phase reaction

$$A \underset{\longleftarrow}{\longrightarrow} B$$

is carried out in a PFR in which pressure drop is neglected and pure A enters the reactor.

Mole balance:
$$\frac{dX}{dV} = \frac{-r_A}{F_{A0}} \tag{T8-2.1}$$

Rate law:
$$-r_A = k\left(C_A - \frac{C_B}{K_C}\right) \tag{T8-2.2}$$

with
$$k = k_1 \exp\left[\frac{E}{R}\left(\frac{1}{T_1} - \frac{1}{T}\right)\right] \tag{T8-2.3}$$

$$K_C = K(T_2) \exp\left[\frac{\Delta H_{Rx}^\circ}{R}\left(\frac{1}{T_2} - \frac{1}{T}\right)\right] \tag{T8-2.4}$$

Stoichiometry: Gas, $\varepsilon = 0$, $P = P_0$

$$C_A = C_{A0}(1 - X)\frac{T_0}{T} \tag{T8-2.5}$$

$$C_B = C_{A0}X\frac{T_0}{T} \tag{T8-2.6}$$

Combine:
$$-r_A = kC_{A0}\left[(1 - X) - \frac{X}{K_C}\right]\frac{T_0}{T} \tag{T8-2.7}$$

Energy balance:
     To relate temperature and conversion we apply the energy balance to an adiabatic PFR. If all species enter at the same temperature, $T_{i0} = T_0$.
Solving Equation (8-50) to obtain the function of conversion yields

$$T = \frac{X[-\Delta H_{Rx}^\circ(T_R)] + \sum \Theta_i \tilde{C}_{pi} T_0 + X \Delta \hat{C}_p T_R}{\sum \Theta_i \tilde{C}_{pi} + X \Delta \hat{C}_p} \tag{T8-2.8}$$

If pure A enters and $\Delta \hat{C}_p = 0$, then

$$T = T_0 + \frac{X[-\Delta H_{Rx}^\circ(T_R)]}{\tilde{C}_{p_A}} \tag{T8-2.9}$$

Equations (T8-2.1) through (T8-2.9) can easily be solved using either Simpson's rule or an ODE solver.

TABLE 8-2B.   SOLUTION PROCEDURES FOR ADIABATIC PFR/PBR REACTOR

A. Numerical Technique

Integrating the PFR mole balance,

$$V = F_{A0} \int_0^{x_3} \frac{dX}{-r_A} = F_{A0} \int_0^{x_3} \left(\frac{1}{-r_A} \right) dX \qquad \text{(T8-2.10)}$$

1. Set $X = 0$.
2. Calculate T using Equation (T8-2.9).
3. Calculate k using Equation (T8-2.3).
4. Calculate K_C using Equation (T8-2.4).
5. Calculate T_0/T (gas phase).
6. Calculate $-r_A$ using Equation (T8-2.7).
7. Calculate $(1/-r_A)$.
8. If X is less than the X_3 specified, increment X (i.e., $X_{i+1} = X_i + \Delta X$) and go to step 2.
9. Prepare table of X vs. $(1/-r_A)$.
10. Use numerical integration formulas, for example,

$$V = F_{A0} \int_0^{x_3} \left(\frac{1}{-r_A} \right) dX = \frac{3}{8} h \left[\frac{1}{-r_A(x=0)} + \frac{3}{-r_A(X_1)} + \frac{3}{-r_A(X_2)} + \frac{1}{-r_A(X_3)} \right]$$

(T8-2.11)

with

$$h = \frac{X_3}{3}$$

B. Ordinary Differential Equation (ODE) Solver

1. $$\frac{dX}{dV} = kC_{A0} \left[(1 - X) - \frac{X}{K_C} \right] \frac{T_0}{T} \qquad \text{(T8-2.12)}$$

2. $$k = k_1(T_1) \exp \left[\frac{E}{R} \left(\frac{1}{T_1} - \frac{1}{T} \right) \right] \qquad \text{(T8-2.13)}$$

3. $$K_C = K_{C2}(T_2) \exp \left[\frac{\Delta H_{Rx}}{R} \left(\frac{1}{T_2} - \frac{1}{T} \right) \right] \qquad \text{(T8-2.14)}$$

4. $$T = T_0 + \frac{X[-\Delta H_{Rx}(T_R)]}{\tilde{C}_{P_A}} \qquad \text{(T8-2.15)}$$

5. Enter parameter values k_1, E, R, K_{C2}, $\Delta H_{Rx}(T_R)$, $\tilde{C}_{P_A}$, C_{A0}, T_0, T_1, T_2.

6. Enter in intial values $X = 0$, $V = 0$ and final values $X = X_f$ and $V = V_f$.

Example 8–6 Liquid-Phase Isomerization of Normal Butane

Normal butane, C_4H_{10}, is to be isomerized to isobutane in a plug-flow reactor. Isobutane is a valuable product that is used in the manufacture of gasoline additives. For example, isobutane can be further reacted to form isooctane. The 1996 selling price of n-butane was 37.2 cents per gallon, while the price of isobutane was 48.5 cents per gallon.

    The reaction is to be carried out adiabatically in the liquid phase under high pressure using essentially trace amounts of a liquid catalyst which gives a specific reaction rate of 31.1 h^{-1} at 360 K. Calculate the PFR volume necessary to process 100,000 gal/day (163 kg mol/h) of a mixture 90 mol % n-butane and 10 mol % i-pentane, which is considered an inert. The feed enters at 330 K.

Additional information:

The economic
incentive
$ = 48.5¢/lb
vs. 37.2¢/lb

$$\Delta H_{Rx} = -6900 \text{ J/mol} \cdot \text{butane}$$

Butane	i-Pentane
$C_{p_{n\text{-B}}} = 141 \text{ J/mol} \cdot \text{K}$	$C_{p_{i\text{-P}}} = 161 \text{ J/mol} \cdot \text{K}$
$C_{p_{i\text{-B}}} = 141 \text{ J/mol} \cdot \text{K}$	Activation energy $= 65.7$ kJ/mol

$$K_C = 3.03 \text{ at } 60°C$$

$$C_{A0} = 9.3 \text{ g mol/dm}^3 = 9.3 \text{ kg mol/m}^3$$

Solution

$$n\text{-}C_4H_{10} \rightleftharpoons i\text{-}C_4H_{10}$$

$$A \rightleftharpoons B$$

Mole balance: $\quad F_{A0}\dfrac{dX}{dV} = -r_A$ $\hfill$ (E8-6.1)

The algorithm

Rate law: $\quad -r_A = k\left(C_A - \dfrac{C_B}{K_C}\right)$ $\hfill$ (E8-6.2)

$$k = k(T_1)e^{\left[\frac{E}{R}\left(\frac{1}{T_1}-\frac{1}{T}\right)\right]}$$ $\hfill$ (E8-6.3)

$$K_C = K_C(T_2)\,e^{\left[\frac{\Delta H}{R}\left(\frac{1}{T_2}-\frac{1}{T}\right)\right]}$$ $\hfill$ (E8-6.4)

Stoichiometry (liquid phase, $v = v_0$):

$$C_A = C_{A0}(1-X)$$ $\hfill$ (E8-6.5)

$$C_B = C_{A0}X$$ $\hfill$ (E8-6.6)

Combine:

$$-r_A = kC_{A0}\left[1-\left(1+\frac{1}{K_C}\right)X\right]$$ $\hfill$ (E8-6.7)

Integrating Equation (E8-6.1) yields

$$V = \int_0^X \frac{F_{A0}}{-r_A}\,dX$$ $\hfill$ (E8-6.8)

Energy balance. Recalling Equation (8-30), we have

$$\dot{Q} - \dot{W}_s - F_{A0} \sum (\Theta_i \tilde{C}_{pi}(T - T_0)) - F_{A0}X[\Delta H^\circ_{Rx}(T_R) + \Delta \hat{C}_p(T - T_R)] = 0 \quad (8\text{-}30)$$

From the problem statement

$$\text{Adiabatic:} \quad \dot{Q} = 0$$

$$\text{No work:} \quad \dot{W} = 0$$

$$\Delta \hat{C}_p = \tilde{C}_{P_B} - \tilde{C}_{P_A} = 141 - 141 = 0$$

Applying the conditions above to Equation (8-30) and rearranging gives

$$T = T_0 + \frac{(-\Delta H_{Rx})X}{\sum \Theta_i C_{pi}} \quad (E8\text{-}6.9)$$

$$\sum \Theta_i C_{pi} = \tilde{C}_{P_A} + \Theta_I \tilde{C}_{P_I} = \left(141 + \frac{0.1}{0.9}161\right) J/mol \cdot K$$

$$= 159.5 \ J/mol \cdot K$$

$$T = 330 + \frac{-(-6900)}{159.5}X$$

$$\boxed{T = 330 + 43.3X} \quad (E8\text{-}6.10)$$

Substituting for the activation energy, T_1, and k_1 in Equation (E8-6.3), we obtain

$$k = 31.1 \ \exp\left[\frac{65{,}700}{8.31}\left(\frac{1}{360} - \frac{1}{T}\right)\right]$$

$$\boxed{k = 31.1 \ \exp\left[7906\left(\frac{T-360}{360T}\right)\right]} \quad (E8\text{-}6.11)$$

Substituting for ΔH_{Rx}, T_2, and $K_C(T_2)$ in Equation (E8-6.4) yields

$$K_C = 3.03 \ \exp\left[\frac{-6900}{8.31}\left(\frac{1}{333} - \frac{1}{T}\right)\right]$$

$$\boxed{K_C = 3.03 \ \exp\left[-830.3\left(\frac{T-333}{333T}\right)\right]} \quad (E8\text{-}6.12)$$

Recalling the rate law gives us

$$-r_A = kC_{A0}\left[1 - \left(1 + \frac{1}{K_C}\right)X\right] \quad (E8\text{-}6.7)$$

At equilibrium

$$-r_A \equiv 0$$

and therefore we can solve Equation (E8-6.7) for the equilibrium conversion

$$X_e = \frac{K_C}{1 + K_C} \tag{E8-6.13}$$

Solution by Hand Calculation (you probably won't see this in the 4th edition)

We will now integrate Equation (E8-6.8) using Simpson's rule after forming a table (E8-6.1) to calculate $(F_{A0}/-r_A)$ as a function of X. This procedure is similar to that described in Chapter 2. We now carry out a sample calculation to show how Table E8-6.1 was constructed, for example, at $X = 0.2$.

(a) $T = 330 + 43.33(0.2) = 338.6$ K

Sample calculation for Table E8-6.1

(b) $k = 31.1 \exp\left[7906\left(\dfrac{338.6 - 360}{(360)(338.6)}\right)\right] = 31.1 \exp(-1.388) = 7.76 \text{ h}^{-1}$

(c) $K_C = 3.03 \exp\left[-830.3\left(\dfrac{338.6 - 333}{(333)(338.6)}\right)\right] = 3.03 e^{-0.0412} = 2.9$

(d) $X_e = \dfrac{2.9}{1 + 2.9} = 0.74$

(e) $-r_A = \left(\dfrac{7.76}{\text{h}}\right)(9.3)\dfrac{\text{mol}}{\text{dm}^3}\left[1 - \left(1 + \dfrac{1}{2.9}\right)(0.2)\right] = 52.8\,\dfrac{\text{mol}}{\text{dm}^3 \cdot \text{h}} = 52.8\,\dfrac{\text{kmol}}{\text{m}^3 \cdot \text{h}}$

(f) $\dfrac{F_{A0}}{-r_A} = \dfrac{(0.9 \text{ mol butane/mol total})(163. \text{ kmol total/h})}{52.8\,\dfrac{\text{kmol}}{\text{m}^3 \cdot \text{h}}} = 2.778 \text{ m}^3$

TABLE E8-6.1.   HAND CALCULATION

X	T (K)	k (h^{-1})	K_C	X_e	$-r_A$(kmol/m$^3 \cdot$h)	$\dfrac{F_{A0}}{-r_A}$ (m^3)
0	330	4.22	3.1	0.76	39.2	3.74
0.2	338.7	7.76	2.9	0.74	52.8	2.78
0.4	347.3	13.93	2.73	0.73	58.6	2.50
0.6	356.0	24.27	2.57	0.72	37.7	3.89
0.65	358.1	27.74	2.54	0.715	24.5	5.99
0.7	360.3	31.67	2.5	0.71	4.1	35.8

$$V = \int_0^{0.6} \frac{F_{A0}}{-r_A}\,dX + \int_{0.6}^{0.7} \frac{F_{A0}}{-r_A}\,dX \tag{E8-6.14}$$

Using Equations (A-24) and (A-22), we obtain

$$V = \frac{3}{8}(0.15)[3.74 + 3(2.78) + 3(2.5) + 3.89]\text{ m}^3 + \frac{0.1}{3}[3.89 + 4(5.99) + 35.8]\text{ m}^3$$

$$= 1.32 \text{ m}^3 + 2.12 \text{ m}^3$$

$$\boxed{V = 3.4 \text{ m}^3}$$

Computer Solution

We could have also solved this problem using POLYMATH or some other ODE solver. The POLYMATH program using Equations (E8-6.1), (E8-6.10), (E8-6.7), (E8-6.11), (E8-6.12), and (E8-6.13) is shown in Table E8-6.2. The graphical output is

TABLE E8-6.2.   POLYMATH PROGRAM

Equations:	*Initial Values:*
d(X)/d(V)=-ra/Fao	0
T=330+43.3*X	
Fao=.9*163	
Cao=9.3	
k=31.1*exp(7906*(T-360)/(T*360))	
Kc=3.03*exp(-830.3*((T-333)/(T*333)))	
Xe=Kc/(1+Kc)	
ra=-k*Cao*(1-(1+1/Kc)*X)	
rate=-ra	
$V_0 = 0,\qquad V_f = 3.3$	

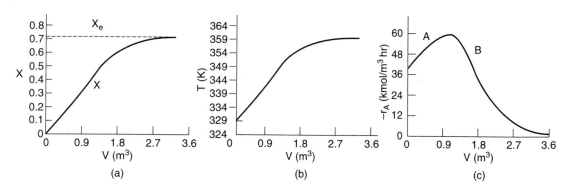

Figure E8-6.1   Conversion, temperature, and reaction rate profiles.

shown in Figure E8-6.1). We see from Figure E8-6.1a that 1.15 m³ is required to 40% conversion. The conversion, temperature, in Figure E8-6.1 and reaction rate profiles are shown. One observes that the rate of reaction

$$-r_A = \underbrace{kC_{A0}}_{A}\ \underbrace{\left[1-\left(1+\frac{1}{K_C}\right)X\right]}_{B} \qquad (E8\text{-}6.15)$$

goes through a maximum. Near the entrance to the reactor, term A increases more rapidly than term B decreases and thus the rate increases. Near the end of the reactor, term B is decreasing more rapidly than term A increases. Consequently, because of these two competing effects, we have a maximum in the rate of reaction.

Let's calculate the CSTR volume necessary to achieve 40% conversion. The mole balance is

$$V = \frac{F_{A0}X}{-r_A}$$

Using Equation (E8-6.2) in the mole balance, we obtain

$$V = \frac{F_{A0}X}{kC_{A0}\left[1 - \left(1 + \dfrac{1}{K_C}\right)\right]X} \qquad (E8\text{-}6.16)$$

From the energy balance we have Equation (E8-6.10):

$$T = 330 + 43.3X$$
$$= 330 + 43.3(0.4) = 347.3$$

Using Equations (E8-6.11) and (E8-6.12) or from Table E8-6.1,

$$k = 13.93 \ \text{h}^{-1}$$
$$K_C = 2.73$$

Then

$$-r_A = 58.6 \ \text{kmol/m}^3 \cdot \text{h}$$

The adiabatic CSTR volume is *less* than the PFR volume

$$V = \frac{(146.7 \ \text{kmol butane/h})(0.4)}{58.6 \ \text{kmol/m}^3 \cdot \text{h}}$$

$$V = 1.0 \ \text{m}^3$$

We see that the CSTR volume (1 m³) to achieve 40% conversion in this adiabatic reaction is less than the PFR volume (1.15 m³).

8.3.3 Steady-State Tubular Reactor with Heat Exchange

In this section we consider a tubular reactor in which heat is added or removed through the cylindrical walls of the reactor (Figure 8-5). In modeling the reactor we shall assume that there is no radial gradient in the reactor and that the heat flux through the wall per unit volume of reactor is as shown in Figure 8-5.

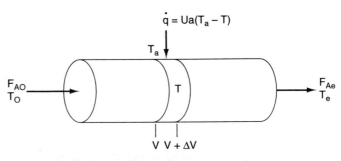

Figure 8-5 Tubular reactor with heat gain or loss.

Recalling Equation (8-47) and ignoring any work done on the reacting fluid, we obtain

$$\dot{Q} - F_{A0} \int_{T_0}^{T} \Sigma \Theta_i C_{pi} \, dT - \left[\Delta H_{Rx}^{\circ}(T_R) + \int_{T_R}^{T} \Delta C_p \, dT \right] F_{A0} X = 0 \quad (8\text{-}54)$$

Differentiating with respect to the volume V and collecting terms gives us

$$\frac{d\dot{Q}}{dV} - [F_{A0}(\Sigma \Theta_i C_{pi} + X \, \Delta C_p)] \frac{dT}{dV}$$

$$- \left[\Delta H_{Rx}^{\circ}(T_R) + \int_{T_R}^{T} \Delta C_p \, dT \right] F_{A0} \frac{dX}{dV} = 0 \quad (8\text{-}55)$$

Recalling that $-r_A = F_{A0}(dX/dV)$, and substituting Equation (8-44) for $(d\dot{Q}/dV)$, we can rearrange Equation (8-55) to obtain

Energy balance on PFR with heat transfer

$$\boxed{\frac{dT}{dV} = \frac{Ua(T_a - T) + (-r_A)[-\Delta H_{Rx}(T)]}{F_{A0}(\Sigma \Theta_i C_{pi} + X \, \Delta C_p)}} \quad (8\text{-}56)$$

Energy balance   The differential equation describing the change of temperature with volume (i.e., distance) down the reactor,

$$\boxed{\frac{dT}{dV} = g(X, T)}$$

Numerical integration of two coupled differential equations is required

must be coupled with the mole balance,

Mole balance

$$\boxed{\frac{dX}{dV} = \frac{-r_A}{F_{A0}} = f(X, T)}$$

and solved simultaneously. A variety of numerical integration schemes can be used to solve these two equations simultaneously.

Energy Balance in Measures Other Than Conversion.   If we work in terms other than conversion, we derive an equation similar to Equation (8-56) by first differentiating Equation (8-10),

$$\dot{Q} - \dot{W}_s + \sum_{i=1}^{n} F_{i0} H_{i0} - \sum_{i=1}^{n} F_i H_i = 0 \quad (8\text{-}10)$$

with respect to the volume, V:

$$\frac{d\dot{Q}}{dV} - \frac{d\dot{W}_s}{dV} + 0 - \Sigma \frac{dF_i}{dV} H_i - \Sigma F_i \frac{dH_i}{dV} = 0 \quad (8\text{-}57)$$

From a mole balance on species i, we have

$$\frac{dF_i}{dV} = r_i = v_i(-r_A) \tag{8-58}$$

and from (8-44),

$$\frac{d\dot{Q}}{dV} = Ua(T_a - T)$$

Differentiating Equation (8-19) with respect to volume, V:

$$\frac{dH_i}{dV} = C_{pi}\frac{dT}{dV} \tag{8-59}$$

Neglecting the work term and substituting Equation (8-58), (8-59) and (8-44) into Equation (8-57), we obtain

$$Ua(T_a - T) - \sum v_i H_i(-r_A) - \sum F_i C_{pi}\frac{dT}{dV} = 0$$

Rearranging, we have

PFR/PBR

$$\boxed{\frac{dT}{dV} = \frac{Ua(T_a - T) + (-\Delta H_{Rx})(-r_A)}{\sum F_i C_{pi}}} \tag{8-60}$$

This equation is coupled with the mole balances on each species [Equation (8-58)]. Next we express r_A as a function of either the concentrations for liquid systems or molar flow rates for gas systems as described in Section 3.4.

For a CSTR the energy balance can be written as

> These forms of the energy balance will be applied to multiple reactions

$$UA(T_a - T) - \sum F_{i0}C_{pi}(T - T_0) + (-\Delta H_{Rx})(-r_A V) = 0 \tag{8-61}$$

Rearranging gives

CSTR

$$\boxed{T = \frac{UAT_a + \sum F_{i0}C_{pi}T_0 + (-\Delta H_{Rx})(-r_A V)}{UA + \sum F_{i0}C_{pi}}} \tag{8-62}$$

Table 8-3 gives the algorithm for the design of PFRs and PBRs with heat exchange for case A: conversion as the reaction variable and case B: molar flow rates as the reaction variable. The procedure in case B must be used when multiple reactions are present.

Having considered the exothermic liquid-phase reaction with constant heat capacities, we now consider an endothermic gas-phase reaction with variable heat capacities.

TABLE 8-3.   PFR/PBR ALGORITHM FOR HEAT EFFECTS

A. Conversion as the reaction variable

$$A + B \rightleftarrows 2C$$

1. **Mole balance:**

$$\frac{dX}{dV} = \frac{-r_A}{F_{A0}} \tag{T8-3.1}$$

2. **Rate law:**

$$-r_A = k_A \left(C_A C_B - \frac{C_C^2}{K_C} \right) \tag{T8-3.2}$$

$$k = k_1(T_1) \exp\left[\frac{E}{R}\left(\frac{1}{T_1} - \frac{1}{T} \right) \right] \tag{T8-3.3}$$

for $\Delta \hat{C}_p \cong 0$.

$$K_C = K_{C2}(T_2) \exp\left[\frac{\Delta H_{Rx}^\circ}{R}\left(\frac{1}{T_2} - \frac{1}{T} \right) \right] \tag{T8-3.4}$$

3. **Stoichiometry** (gas phase, no ΔP):

$$C_A = C_{A0}(1 - X)\frac{T_0}{T} \tag{T8-3.5}$$

$$C_B = C_{A0}(\Theta_B - X)\frac{T_0}{T} \tag{T8-3.6}$$

$$C_C = 2C_{A0}X\frac{T_0}{T} \tag{T8-3.7}$$

4. **Energy balance:**

$$\frac{dT}{dV} = \frac{Ua(T_a - T) + (-r_A)(-\Delta H_{Rx})}{F_{A0}[C_{p_A} + \Theta_B C_{p_B} + X\,\Delta C_p]} \tag{T8-3.8}$$

B. Molar flow rates as the reaction variable

1. **Mole balances:**

$$\frac{dF_A}{dV} = r_A \tag{T8-3.9}$$

$$\frac{dF_B}{dV} = r_B \tag{T8-3.10}$$

$$\frac{dF_C}{dV} = r_C \tag{T8-3.11}$$

2. **Rate law:**

$$-r_A = k_A \left(C_A C_B - \frac{C_C^2}{K_C} \right) \tag{T8-3.2}$$

$$k = k_1(T_1) \exp\left[\frac{E}{R}\left(\frac{1}{T_1} - \frac{1}{T} \right) \right] \tag{T8-3.6}$$

for $\Delta \hat{C}_p \cong 0$.

$$K_C = K_{C2}(T_2) \exp\left[\frac{\Delta H_{Rx}^\circ}{R}\left(\frac{1}{T_2} - \frac{1}{T} \right) \right] \tag{T8-3.7}$$

TABLE 8-3.   (CONTINUED)   PFR/PBR ALGORITHM FOR HEAT EFFECTS

3. Stoichiometry (gas phase, no ΔP):

$$r_B = r_A \tag{T8-3.12}$$

$$r_C = 2r_A \tag{T8-3.13}$$

$$C_A = C_{T0} \frac{F_A}{F_T} \frac{T_0}{T} \tag{T8-3.14}$$

$$C_B = C_{T0} \frac{F_B}{F_T} \frac{T_0}{T} \tag{T8-3.15}$$

$$C_C = C_{T0} \frac{F_C}{F_T} \frac{T_0}{T} \tag{T8-3.16}$$

4. Energy balance:

$$\frac{dT}{dV} = \frac{Ua(T_a - T) + (-r_A)(-\Delta H_{Rx})}{F_A C_{P_A} + F_B C_{P_B} + F_C C_{P_C}} \tag{T8-3.17}$$

Enter parameter values:

$$k_1, E, R, C_{T0}, T_a, T_0, T_1, T_2, K_{C2}, \Theta_B, \Delta H^{\circ}_{Rx}, C_{P_A}, C_{P_B}, C_{P_C}, Ua$$

Enter intial values: $F_{A0}, F_{B0}, F_{C0}, F_{D0}, T_0$ and final values : $V_f =$ _____

Use ODE solver.

Example 8–7 Production of Acetic Anhydride

Jeffreys,[8] in a treatment of the design of an acetic anhydride manufacturing facility, states that one of the key steps is the vapor-phase cracking of acetone to ketene and methane:

$$CH_3COCH_3 \rightarrow CH_2CO + CH_4$$

He states further that this reaction is first-order with respect to acetone and that the specific reaction rate can be expressed by

$$\ln k = 34.34 - \frac{34,222}{T} \tag{E8-7.1}$$

where k is in reciprocal seconds and T is in kelvin. In this design it is desired to feed 8000 kg of acetone per hour to a tubular reactor. The reactor consists of a bank of 1000 1-inch schedule 40 tubes. We will consider two cases:

1. The reactor is operated *adiabatically.*
2. The reactor is surrounded by a *heat exchanger* where the heat-transfer coefficient is 110 $J/m^2 \cdot s \cdot K$, and the ambient temperature is 1150 K.

The inlet temperature and pressure are the same for both cases at 1035 K and 162 kPa (1.6 atm), respectively. Plot the conversion and temperature along the length of the reactor.

---

[8] G. V. Jeffreys, *A Problem in Chemical Engineering Design: The Manufacture of Acetic Anhyride,* 2nd ed. (London: Institution of Chemical Engineers, 1964).

Solution

Let $A = CH_3COCH_3$, $B = CH_2CO$, and $C = CH_4$. Rewriting the reaction symbolically gives us

$$A \rightarrow B + C$$

1. **Mole balance**:

$$\frac{dX}{dV} = -r_A / F_{A0} \qquad \text{(E8-7.2)}$$

2. **Rate law**:

$$-r_A = kC_A \qquad \text{(E8-7.3)}$$

3. **Stoichiometry** (gas-phase reaction with no pressure drop):

$$C_A = \frac{C_{A0}(1-X)T_0}{(1+\varepsilon X)T} \qquad \text{(E8-7.4)}$$

$$\varepsilon = y_{A0}\delta = 1(1 + 1 - 1) = 1$$

4. **Combining** yields

$$-r_A = \frac{kC_{A0}(1-X)}{1+X}\frac{T_0}{T} \qquad \text{(E8-7.5)}$$

$$\frac{dX}{dV} = \frac{-r_A}{F_{A0}} = \frac{k}{v_0}\left(\frac{1-X}{1+X}\right)\frac{T_0}{T} \qquad \text{(E8-7.6)}$$

To solve the differential equation above, it is first necessary to use the energy balance to determine T as a function of X.

5. **Energy balance:**

CASE I. ADIABATIC OPERATION

For no work done on the system, $\dot{W}_s = 0$, and adiabatic operation, $\dot{Q} = 0$ (i.e., $U \equiv 0$), Equation (8-56) becomes

$$\frac{dT}{dV} = \frac{(-r_A)\left\{-\left[\Delta H^{\circ}_{Rx}(T_R) + \int_{T_R}^{T}(\Delta\alpha + \Delta\beta T + \Delta\gamma T^2)\,dT\right]\right\}}{F_{A0}(\sum \Theta_i C_{pi} + X\,\Delta C_p)} \qquad \text{(E8-7.7)}$$

where in general

$$\Delta\alpha = \frac{d}{a}\alpha_D + \frac{c}{a}\alpha_C - \frac{b}{a}\alpha_B - \alpha_A$$

For acetone decomposition

$$\Delta\alpha = \alpha_B + \alpha_C - \alpha_A$$

Equivalent expressions exist for $\Delta\beta$ and $\Delta\gamma$.

Because only A enters,

$$\sum \Theta_i C_{pi} = C_{p_A}$$

and Equation (E8-7.7) becomes

$$\frac{dT}{dV} = \frac{(-r_A)\left\{-\left[\Delta H^\circ_{Rx}(T_R) + \int_{T_R}^{T} (\Delta\alpha + \Delta\beta T + \Delta\gamma T^2)\,dT\right]\right\}}{F_{A0}(C_{p_A} + X\,\Delta C_p)}$$

Integrating gives

$$\frac{dT}{dV} = \frac{(-r_A)\left\{-\left[\Delta H^\circ_{Rx} + \Delta\alpha(T - T_R) + \frac{\Delta\beta}{2}(T^2 - T_R^2) + \frac{\Delta\gamma}{3}(T^3 - T_R^3)\right]\right\}}{F_{A0}(\alpha_A + \beta_A T + \gamma_A T^2 + X\,\Delta C_p)} \qquad \text{(E8-7.8)}$$

Adiabatic PFR
with variable heat
capacities

6. **Calculation of mole balance parameters:**

$$F_{A0} = \frac{8000\ \text{kg/h}}{58\ \text{g/mol}} = 137.9\ \text{kmol/h} = 38.3\ \text{mol/s}$$

$$C_{A0} = \frac{P_{A0}}{RT} = \frac{162\ \text{kPa}}{8.31\ \dfrac{\text{kPa}\cdot\text{m}^3}{\text{kmol}\cdot\text{K}}\ (1035\ \text{K})} = 0.0188\ \frac{\text{kmol}}{\text{m}^3} = 18.8\ \text{mol/m}^3$$

$$v_0 = \frac{F_{A0}}{C_{A0}} = 2.037\ \text{m}^3/\text{s}$$

7. **Calculation of energy balance parameters**:
   a. $\Delta H^\circ_{Rx}(T_R)$: At 298 K, the standard heats of formation are

$$H^\circ_{Rx}(T_R)_{\text{acetone}} = -216.67\ \text{kJ/mol}$$

$$H^\circ_{Rx}(T_R)_{\text{ketene}} = -61.09\ \text{kJ/mol}$$

$$H^\circ_{Rx}(T_R)_{\text{methane}} = -74.81\ \text{kJ/mol}$$

$$\Delta H^\circ_{Rx}(T_R) = (-61.09) + (-74.81) - (-216.67)$$

$$= 80.77\ \text{kJ/mol}$$

   b. ΔC_p: The heat capacities are:

$$CH_3COCH_3:\quad C_{pA} = 26.63 + 0.183T - 45.86 \times 10^{-6} T^2\ \text{J/mol}\cdot\text{K}$$

$$CH_2CO:\quad C_{pB} = 20.04 + 0.0945T - 30.95 \times 10^{-6} T^2\ \text{J/mol}\cdot\text{K}$$

$$CH_4:\quad C_{pC} = 13.39 + 0.077T - 18.71 \times 10^{-6} T^2\ \text{J/mol}\cdot\text{K}$$

$$\Delta\alpha = \alpha_C + \alpha_B - \alpha_A = 13.39 + 20.04 - 26.63$$

$$= 6.8\ \text{J/mol}\cdot\text{K}$$

$$\Delta\beta = \beta_C + \beta_B - \beta_A = 0.077 + 0.0945 - 0.183$$

$$= -0.0115\ \text{J/mol}\cdot\text{K}^2$$

$$\Delta\gamma = \gamma_C + \gamma_B - \gamma_A = (-18.71 \times 10^{-6}) + (-30.95 \times 10^{-6}) - (-45.86 \times 10^{-6})$$

$$= -3.8 \times 10^{-6}\ \text{J/mol}\cdot\text{K}^3$$

$$\frac{\Delta\beta}{2} = -5.75 \times 10^{-3} \qquad \frac{\Delta\gamma}{3} = -1.27 \times 10^{-6}$$

See Table E8-7.1 for a summary of the calculations and Table E8-7.2 and Figure E8-7.1 for the POLYMATH program and its graphical output.

TABLE E8-7.1.   SUMMARY

Adiabatic PFR with variable heat capacities

$$\frac{dX}{dV} = \frac{-r_A}{F_{A0}} \tag{E8-7.2}$$

$$-r_A = -\frac{k C_{A0}(1-X)}{(1+X)} \frac{T_0}{T} \tag{E8-7.5}$$

$$\frac{dT}{dV} = \frac{(-r_A)\left\{-\left[\Delta H_{Rx}^\circ(T_R) + \Delta\alpha(T - T_R) + \frac{\Delta\beta}{2}(T^2 - T_R^2) + \frac{\Delta\gamma}{3}(T^3 - T_R^3)\right]\right\}}{F_{A0}(\alpha_A + \beta_A T + \gamma_A T^2 + X\,\Delta C_p)} \tag{E8-7.8}$$

$$\Delta C_p = 6.8 - 11.5 \times 10^{-3} T - 3.81 \times 10^{-6} T^2$$

$$\Delta H_{Rx}^\circ(T_R) = 80{,}770$$

$$\alpha_A = 26.63 \qquad\qquad\qquad \Delta\alpha = 6.8$$

$$\beta_A = 0.183 \qquad\qquad\qquad \frac{\Delta\beta}{2} = -5.75 \times 10^{-3}$$

$$\gamma_A = -45.86 \times 10^{-6} \qquad\qquad \frac{\Delta\gamma}{3} = -1.27 \times 10^{-6}$$

$$C_{A0} = 18.8$$

$$F_{A0} = 38.3$$

$$k = 8.2 \times 10^{14}\, \exp\left(\frac{-34{,}222}{T}\right) \tag{E8-7.12}$$

TABLE E8-7.2.   POLYMATH PROGRAM

Equations:	*Initial Values:*
`d(T)/d(V)=-ra*(-deltaH)/(Fao*(Cpa+X*delCp))`	1035
`d(X)/d(V)=-ra/Fao`	0

```
k=8.2*10**14*exp(-34222/T)
Fao=38.3
Cpa=26.63+.183*T-45.86*10**(-6)*T**2
delCp=6.8-11.5*10**(-3)*T-3.81*10**(-6)*T**2
Cao=18.8
To=1035
Tr=298
deltaH=80770+6.8*(T-Tr)-5.75*10**(-3)*(T**2-Tr**2)-1.27*10**(-6)*(T**3-Tr**3)
ra=-k*Cao*(1-X)/(1+X)*To/T
```

$V_0 = 0, \quad V_f = 5$

Adiabatic
endothermic
reaction in a PFR

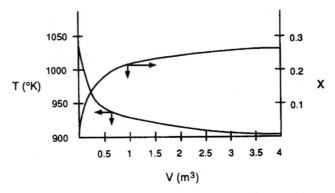

Figure E8-7.1

Death of a reaction

Note that for this adiabatic endothermic reaction, the reaction virtually *dies out* after 2.5 m³, owing to the large drop in temperature, and very little conversion is achieved beyond this point. One way to increase the conversion would be to add a diluent such as N_2, which could supply the sensible heat for this endothermic reaction. However, if too much diluent is added, the concentration and rate will be quite low. On the other hand, if too little diluent is added, the temperature will drop and virtually extinguish the reaction. How much diluent to add is left as an exercise (see Problem P8-2).

A bank of 1000 1-in. schedule 40 tubes 2.28 m in length corresponds to 1.27 m³ and gives 20% conversion. Ketene is unstable and tends to explode, which is a good reason to keep the conversion low. However, the pipe material and schedule size should be checked to learn if they are suitable for these temperatures and pressures.

CASE II. OPERATION OF A PFR WITH HEAT EXCHANGE

See Figure E8-7.2.

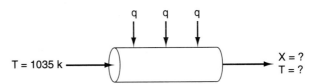

Figure E8-7.2　PFR with heat exchange.

1. **The mole balance**:

$$\text{(1)} \qquad \frac{dX}{dV} = \frac{-r_A}{F_{A0}} \qquad \text{(E8-7.2)}$$

Using (2) the **rate law** (E8-7.3) and (3) **stoichiometry** (E8-7.4) for the adiabatic case discussed previously, we (4) **combine** to obtain the reaction rate as

$$\text{(4)} \qquad -r_A = kC_{A0}\left(\frac{1-X}{1+X}\right)\frac{T_0}{T} \qquad \text{(E8-7.5)}$$

5. **Energy balance.** Equation (8-56):

$$\frac{dT}{dV} = \frac{Ua(T_a - T) + (-r_A)[-\Delta H_{Rx}(T)]}{F_{A0}(\sum \Theta_i C_{pi} + X \Delta C_p)} \tag{E8-7.9}$$

For the acetone reaction system,

$$\frac{dT}{dV} = \frac{Ua(T_a - T) + (r_A)[\Delta H_{Rx}(T)]}{F_{A0}(C_{p_A} + X \Delta C_p)} \tag{E8-7.10}$$

PFR with heat exchange

6. **Parameter evaluation**:
   a. *Mole balance.* On a per tube basis, $v_0 = 0.002$ m³/s. The concentration of acetone is 18.8 mol/m³, so the entering molar flow rate is

$$F_{A0} = C_{A0}v_0 = \left(18.8\ \frac{\text{mol}}{\text{m}^3}\right)\left(2 \times 10^{-3}\ \frac{\text{m}^3}{\text{s}}\right) = 0.0376\ \frac{\text{mol}}{\text{s}}$$

The value of k at 1035 K is 3.58 s⁻¹; consequently, we have

$$k(T) = 3.58\ \exp\left[34{,}222\left(\frac{1}{1035} - \frac{1}{T}\right)\right] \tag{E8-7.11}$$

   b. *Energy balance.* From the adiabatic case above we already have ΔC_p, C_{p_A}, α_A, β_A, γ_A, $\Delta\alpha$, $\Delta\beta$, and $\Delta\gamma$. The heat-transfer area per unit volume of pipe is

$$a = \frac{\pi DL}{(\pi D^2/4)L} = \frac{4}{D} = \frac{4}{0.0266\ \text{m}} = 150\ \text{m}^{-1}$$

$$U = 110\ \text{J/m}^2 \cdot \text{s} \cdot \text{K}$$

Combining the overall heat-transfer coefficient with the area yields

$$Ua = 16{,}500\ \text{J/m}^3 \cdot \text{s} \cdot \text{K}$$

We now use Equations (E8-7.1) through (E8-7.6), and Equations (E8-7.10) and (E8-7.11) along with the POLYMATH program (Table E8-7.4), to determine the conversion and temperature profiles shown in Figure E8-7.3.

The corresponding variables in the POLYMATH program are

$$t1 = T,\ dh = \Delta H_{Rx}(T),\ dcp = \Delta C_p,\ cpa = C_{p_A},\ ua = Ua$$

TABLE E8-7.4.   POLYMATH PROGRAM FOR PFR WITH HEAT EXCHANGE

Equations:	Initial Values:
`d(t)/d(v)=(ua×(ta-t)+ra×dh)/(fa0×(cpa+x×dcp))`	1035
`d(x)/d(v)=-ra/fa0`	0
`fa0=.0376`	
`ua=16500`	
`ta=1150`	
`cpa=26.6+.183×t-.0000459×t×t`	
`dcp=6.8-.0115×t-.00000381×t×t`	
`ca0=18.8`	
`t0=1035`	
`term=-.00000127×(t××3-298××3)`	
`dh=80770+6.8×(t-298)-.00575×(t××2-298××2)+term`	
`ra=-ca0×3.58×exp(34222×(1/t0-1/t))×(1-x)×(t0/t)/(1+x)`	
$v_0 = 0,\quad v_f = 0.001$	

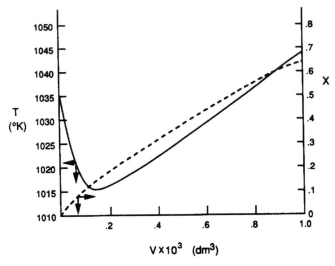

Figure E8-7.3   Temperature and conversion profiles in PFR.

One notes the reactor temperature goes through a minimum along the length of the reactor. At the front of the reactor the reaction takes place very rapidly, drawing energy from the sensible heat of the gas causing the gas temperature to drop because the heat exchanger cannot supply energy at the equal or greater rate. This drop in temperature, coupled with the consumption of reactants, slows the reaction rate as we move down the reactor. Because of this slower reaction rate, the heat exchanger supplies energy at a rate greater than reaction draws energy from the gases and as a result the temperature increases.

8.4 Equilibrium Conversion

The highest conversion that can be achieved in reversible reactions is the equilibrium conversion. For endothermic reactions, the equilibrium conversion increases with increasing temperature up to a maximum of 1.0. For exothermic reactions the equilibrium conversion decreases with increasing temperature.

8.4.1 Adiabatic Temperature and Equilibrium Conversion

Exothermic Reactions.   Figures 8-6 and 8-7 show typical plots of equilibrium conversion as a function of temperature for an exothermic reaction. To determine the maximum conversion that can be achieved in an exothermic reaction carried out adiabatically, we find the intersection of the equilibrium conversion as a function of temperature with temperature–conversion relationships from the energy balance (Figure 8-7). For $T_{i0} = T_0$,

$$X_{\text{EB}} = \frac{\sum \Theta_i \tilde{C}_{pi}(T - T_0)}{-\Delta H_{\text{Rx}}(T)} \qquad \text{(E8-4.6)}$$

PFR with heat exchange

For reversible reactions, the equilibrium conversion, X_e, is usually calculated first

For exothermic
reactions,
equilibrium
conversion
decreases with
increasing
temperature

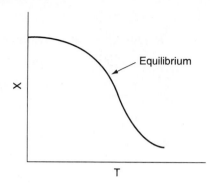

Figure 8-6   Variation of equilibrium constant and conversion with temperature for an exothermic reaction.

Adiabatic
equilibrium
conversion for
exothermic
reactions

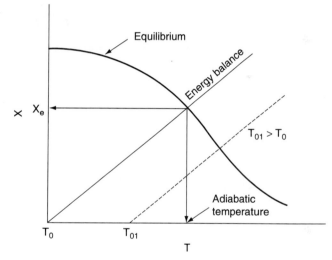

Figure 8-7   Graphical solution of equilibrium and energy balance equations to obtain adiabatic temperature and equilibrium conversion.

If the entering temperature is increased from T_0 to T_{01}, the energy balance line will be shifted to the right and will be parallel to the original line, as shown by the dashed line. Note that as the inlet temperature increases, the adiabatic equilibrium conversion decreases.

Example 8–8   Calculating the Adiabatic Equilibrium Temperature

For the elementary solid-cataobyzed liquid-phase reaction

$$A \; \rightleftharpoons \; B$$

make a plot of equilibrium conversion as a function of temperature. Determine the adiabatic equilibrium temperature and conversion when pure A is fed to the reactor at a temperature of 300 K.

Additional information:

$$H_A^\circ(298 \text{ K}) = -40,000 \text{ cal/mol} \qquad H_B^\circ(298 \text{ K}) = -60,000 \text{ cal/mol}$$

$$C_{p_A} = 50 \text{ cal/mol} \cdot \text{K} \qquad C_{p_B} = 50 \text{ cal/mol} \cdot \text{K}$$

$$K_e = 100,000 \text{ at } 298 \text{ K}$$

Solution

1. **Rate law**:

$$-r_A = k\left(C_A - \frac{C_B}{K_e} \right) \tag{E8-8.1}$$

2. **Equilibrium**, $-r_A = 0$; so

$$C_{Ae} = \frac{C_{Be}}{K_e}$$

3. **Stoichiometry** $(v = v_0)$ yields

$$C_{A0}(1 - X_e) = \frac{C_{A0}X_e}{K_e}$$

Solving for X_e gives

$$X_e = \frac{K_e(T)}{1 + K_e(T)} \tag{E8-8.2}$$

4. **Equilibrium constant**. Calculate ΔC_p, then $K_e(T)$:

$$\Delta C_p = C_{p_B} - C_{p_A} = 50 - 50 = 0 \text{ cal/mol} \cdot \text{K}$$

For $\Delta C_p = 0$, the equilibrium constant varies with temperature according to the relation

$$K_e(T) = K_e(T_1) \exp\left[\frac{\Delta H_{Rx}^\circ}{R} \left(\frac{1}{T_1} - \frac{1}{T} \right) \right] \tag{E8-8.3}$$

$$\Delta H_{Rx}^\circ = H_B^\circ - H_A^\circ = -20,000 \text{ cal/mol}$$

$$K_e(T) = 100,000 \exp\left[\frac{-20,000}{1.987} \left(\frac{1}{298} - \frac{1}{T} \right) \right]$$

$$K_e = 100,000 \exp\left[-33.78 \left(\frac{T - 298}{T} \right) \right] \tag{E8-8.4}$$

Substituting Equation (E8-8.4) into (E8-8.2) we can calculate equilibrium conversion as a function of temperature:

Conversion calculated from equilibrium relationship

$$X_e = \frac{100,000 \exp[-33.78(T - 298)/T]}{1 + 100,000 \exp[-33.78(T - 298)/T]} \tag{E8-8.5}$$

The calculations are shown in Table E8-8.1.

TABLE E8-8.1.   EQUILIBRIUM CONVERSION
AS A FUNCTION OF TEMPERATURE

T	K_e	X_e
298	100,000.00	1.00
350	661.60	1.00
400	18.17	0.95
425	4.14	0.80
450	1.11	0.53
475	0.34	0.25
500	0.12	0.11

For a reaction carried out adiabatically, the energy balance reduces to

$$X_{EB} = \frac{\sum \Theta_i \tilde{C}_{pi}(T - T_0)}{-\Delta H_{Rx}} = \frac{C_{P_A}(T - T_0)}{-\Delta H_{Rx}} \qquad \text{(E8-8.6)}$$

Conversion
calculated from
energy balance

$$\boxed{X_{EB} = \frac{50(T - 300)}{20,000} = 2.5 \times 10^{-3}(T - 300)} \qquad \text{(E8-8.7)}$$

Data from Table E8-8.1 and the following data are plotted in Figure E8-8.1.

T (K)	300	400	500	600
X_{EB}	0	0.25	0.50	0.75

$$X_e = 0.41 \quad T_e = 465 \text{ K}$$

For a feed temperature of 300 K, the adiabatic equilibrium temperature is 465 K and the corresponding adiabatic equilibrium conversion is 0.41.

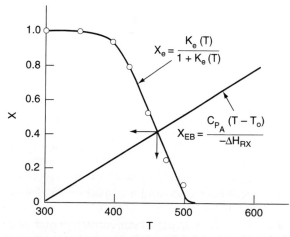

Figure E8-8.1   Finding the adiabatic equilibrium temperature (T_e) and conversion (X_e).

Higher conversions than those shown in Figure E8-8.1 can be achieved for adiabatic operations by connecting reactors in series with interstage cooling:

The conversion–temperature plot for this scheme is shown in Figure 8-8.

Interstage cooling used for exothermic reversible reactions

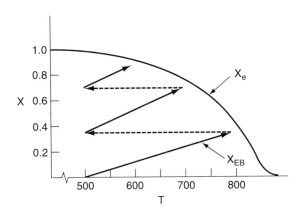

Figure 8-8   Increasing conversion by interstage cooling.

Endothermic Reactions.   Another example of the need for interstage heat transfer in a series of reactors can be found when upgrading the octane number of gasoline. The more compact the hydrocarbon molecule for a given number of carbon atoms, the higher the octane rating. Consequently, it is desirable to convert straight-chain hydrocarbons to branched isomers, naphthenes, and aromatics. The reaction sequence is

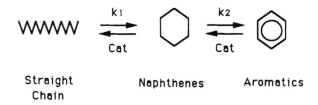

The first reaction step (k_1) is slow compared to the second step, and each step is highly endothermic. The allowable temperature range for which this reaction can be carried out is quite narrow: Above 530°C undesirable side reactions occur and below 430°C the reaction virtually does not take place. A typical feed stock might consist of 75% straight chains, 15% naphthas, and 10% aromatics.

One arrangement currently used to carry out these reactions is shown in Figure 8-9. Note that the reactors are not all the same size. Typical sizes are on the order of 10 to 20 m high and 2 to 5 m in diameter. A typical feed rate of

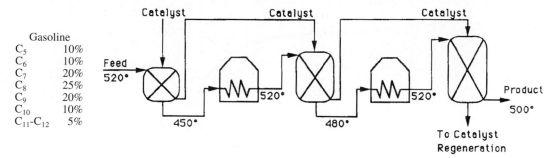

Gasoline

C_5	10%
C_6	10%
C_7	20%
C_8	25%
C_9	20%
C_{10}	10%
C_{11}-C_{12}	5%

Figure 8-9   Interstage heating for gasoline production in moving-bed reactors.

Typical values for gasoline production

gasoline is approximately 200 m³/h at 2 atm. Hydrogen is usually separated from the product stream and recycled.

Because the reaction is endothermic, equilibrium conversion increases with increasing temperature. A typical equilibrium curve and temperature conversion trajectory for the reactor sequence are shown in Figure 8-10.

Interstage heating

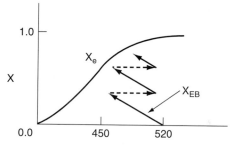

Figure 8-10   Temperature–conversion trajectory for interstage heating of an endothermic reaction corresponding to Figure 8-9.

Example 8–9 Interstage Cooling

What conversion could be achieved in Example 8-8 if two interstage coolers were available that had the capacity to cool the exit stream to 350°K? Also determine the heat duty of each exchanger for a molar feed rate of A of 40 mol/s. Assume that 95% of equilibrium conversion is achieved in each reactor. The feed temperature to the first reactor is 300 K.

We saw in Example 8-8 that for an entering temperature of 300 K the adiabatic equilibrium conversion was 0.41. For 95% of equilibrium conversion, the conversion exiting the first reactor is 0.4. The exit temperature is found from a rearrangement of Equation (E8-8.7):

$$T = 300 + 400X = 300 + (400)(0.4)$$

$$T_1 = 460 \text{ K}$$

We now cool the gas stream exiting the reactor at 460 K down to 350 K in a heat exchanger (Figure E8-9.1).

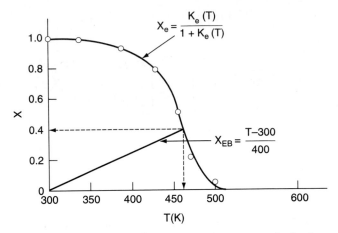

Figure E8-9.1   Determining exit conversion and temperature in the first stage.

There is no work done on the reaction gas mixture in the exchanger and the reaction does not take place in the exchanger. Under these conditions ($F_{i|in} = F_{i|out}$) the energy balance given by Equation (8-10),

$$\dot{Q} - \dot{W}_s + \sum F_{i0} H_i - \sum F_i H_i = 0 \tag{8-10}$$

becomes

Energy balance on
the reaction gas
mixture **in** the heat
exchanger

$$\dot{Q} = \sum F_i H_i - \sum F_{i0} H_{i0} = \sum F_{i0} (H_i - H_{i0}) \tag{E8-9.1}$$

$$= \sum F_i \hat{C}_{p_i}(T_2 - T_1) = (F_A \hat{C}_{P_A} + F_B \hat{C}_P)(T_2 - T_1) \tag{E8-9.2}$$

But $\hat{C}_{P_A} = \hat{C}_{p_B}$,

$$\dot{Q} = (F_A + F_B)(\hat{C}_{P_A})(T_2 - T_1) \tag{E8-9.3}$$

Also, $F_{A0} = F_A + F_B$,

$$\dot{Q} = F_{A0} \hat{C}_{P_A}(T_2 - T_1)$$

$$= \frac{40 \text{ mol}}{\text{s}} \cdot \frac{50 \text{ mol}}{\text{mol} \cdot \text{K}} (350 - 460)$$

$$= -220 \frac{\text{kcal}}{\text{s}} \tag{E8-9.4}$$

We see that 220 kcal/s is removed from the reaction system mixture. The rate at which energy must be absorbed by the coolant stream in the exchanger is

$$\dot{Q} = \dot{m}_C \hat{C}_{P_C}(T_{\text{out}} - T_{\text{in}}) \tag{E8-9.5}$$

We consider the case where the coolant is available at 270 K but cannot be heated above 400 K and calculate the coolant flow rate necessary to remove 220 kcal/s from the reaction mixture. Rearranging Equation (E8-9.5) and noting that the coolant heat capacity is 18 cal/mol·K gives

Sizing the interstage heat exchanger and coolant flow rate

$$\dot{m}_C = \frac{Q}{\hat{C}_{P_C}(T_{out} - T_{in})} = \frac{220{,}000 \text{ cal/s}}{18 \dfrac{\text{cal}}{\text{mol} \cdot \text{K}}(400 - 270)} \tag{E8-9.6}$$

$$= 94 \text{ mol/s} = 1692 \text{ g/s} = 1.69 \text{ kg/s}$$

The necessary coolant flow rate is 1.69 kg/s.

Let's next determine the countercurrent heat exchanger area. The exchanger inlet and outlet temperatures are shown in Figure E8-9.2. The rate of heat transfer in a countercurrent heat exchanger is given by the equation[9]

Bonding with Unit Operations

$$\dot{Q} = UA \frac{[(T_{h2} - T_{c2}) - (T_{h1} - T_{c1})]}{\ln\left(\dfrac{T_{h2} - T_{c2}}{T_{h1} - T_{c1}}\right)} \tag{E8-9.7}$$

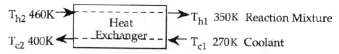

T_{h2} 460K → Heat Exchanger → T_{h1} 350K Reaction Mixture

T_{c2} 400K ← ← T_{c1} 270K Coolant

Figure E8-9.2   Countercurrent heat exchanger.

Rearranging Equation (E8-9.7) assuming a value of U of 1000 cal/s·m²·K, and then substituting the appropriate values gives

$$A = \frac{Q \ln\left(\dfrac{T_{h2} - T_{c2}}{T_{h1} - T_{c1}}\right)}{U[(T_{h2} - T_{c2}) - (T_{h1} - T_{c1})]} = \frac{220{,}000 \dfrac{\text{cal}}{\text{s}} \ln\left(\dfrac{460 - 400}{350 - 270}\right)}{1000 \dfrac{\text{cal}}{\text{s} \cdot \text{m}^2 \cdot \text{K}}[(460 - 400) - (350 - 270)]\,\text{K}}$$

$$= \frac{220 \ln(0.75)}{-20} \text{ m}^2$$

$$= 3.16 \text{ m}^2$$

Sizing the heat exchanger

The heat-exchanger surface area required to accomplish this rate of heat transfer is 3.16 m².

Now let's return to determine the conversion in the second reactor. The conditions entering the second reactor are $T = 350$ and $X = 0.4$. The energy balance starting from this point is shown in Figure E8-9.3. The corresponding adiabatic equilibrium conversion is 0.63. Ninety-five percent of the equilibrium conversion is 60% and the corresponding exit temperature is $T = 350 + (0.6 - 0.4)400 = 430$ K.

The heat-exchange duty to cool the reacting mixture from 430 K back to 350 K can be calculated from Equation (E8-9.4):

$$\dot{Q} = F_{A0}\hat{C}_{P_A}(350 - 430) = \left(\frac{40 \text{ mol}}{\text{s}}\right)\left(\frac{50 \text{ cal}}{\text{mol} \cdot \text{K}}\right)(-80)$$

$$= -160 \frac{\text{kcal}}{\text{s}}$$

[9] See page 271 of C. J. Geankoplis, *Transport Processes and Unit Operations* (Upper Saddle River, N.J.: Prentice Hall, 1993).

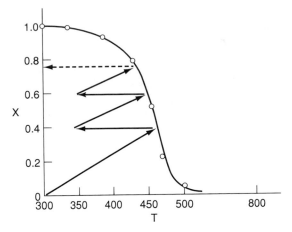

Figure E8-9.3   Three reactors in series with interstage cooling.

For the final reactor we begin at $T_0 = 350$ K and $X = 0.6$ and follow the line representing the equation for the energy balance along to the point of intersection with the equilibrium conversion, which is $X = 0.8$. Consequently, the final conversion achieved with three reactors and two interstage coolers is $(0.95)(0.8) = 0.76$.

8.4.2 Optimum Feed Temperature

We now consider an adiabatic reactor of fixed size or catalyst weight and investigate what happens as the feed temperature is varied. The reaction is reversible and exothermic. At one temperature extreme, using a very high feed temperature, the specific reaction rate will be large and the reaction will proceed rapidly, but the equilibrium conversion will be close to zero. Consequently, very little product will be formed. A plot of the equilibrium conversion and the conversion calculated from the adiabatic energy balance,

$$T = T_0 - \frac{\Delta H^\circ_{Rx}}{\hat{C}_{P_A} + \hat{C}_{P_B}} X = T_0 + 400X$$

is shown in Figure 8-11. We see that for an entering temperature of 600 K the adiabatic equilibrium conversion is 0.15. The corresponding conversion profile down the length of the reactor is shown in Figure 8-12. We see that because of the high entering temperature the rate is very rapid and equilibrium is achieved very near the reactor entrance.

We notice that the conversion and temperature increase very rapidly over a short distance (i.e., a small amount of catalyst). This sharp increase is sometimes referred to as the point/temperature at which the reaction ignites. If the inlet temperature were lowered to 500 K, the corresponding equilibrium conversion is increased to 0.33; however, the reaction rate is slower at this lower temperature, so that this conversion is not achieved until close to the end of the

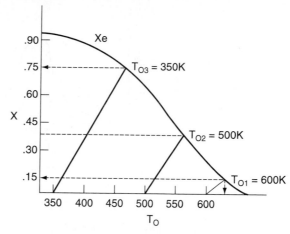

Figure 8-11   Equilibrium conversion for different feed temperatures.

Observe how the
temperature profile
changes as the
entering
temperature is
decreased from
600 K

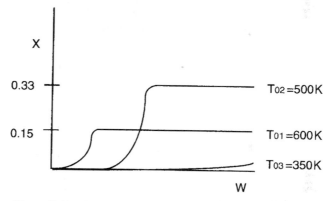

Figure 8-12   Conversion profiles for different feed temperatures.

reactor. If the entering temperature were lowered further to 350, the corresponding equilibrium conversion is 0.75, but the rate is so slow that a conversion of 0.05 is achieved for the catalyst weight in the reactor. At a very low
feed temperature, the specific reaction rate will be so small that virtually all of
the reactant will pass through the reactor without reacting. It is apparent that
with conversions close to zero for both high and low feed temperatures there
must be an optimum feed temperature that maximizes conversion. As the feed
temperature is increased from a very low value, the specific reaction rate will
increase, as will the conversion. The conversion will continue to increase with
increasing feed temperature until the equilibrium conversion is approached in
the reaction. Further increases in feed temperature will only decrease the conversion due to the decreasing equilibrium conversion. This optimum inlet temperature is shown in Figure 8-13.

Optimum inlet
temperature

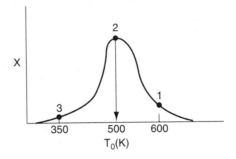

Figure 8-13   Finding the optimum feed temperature.

8.5 Nonadiabatic Reactor Operation: Oxidation of Sulfur Dioxide Example

8.5.1 Manufacture of Sulfuric Acid

In the manufacture of sulfuric acid from sulfur, the first step is the burning of sulfur in a furnace to form sulfur dioxide:

$$S + O_2 \rightarrow SO_2$$

Following this step, the sulfur dioxide is converted to sulfur trioxide, using a catalyst:

$$SO_2 + \tfrac{1}{2}O_2 \xrightarrow{\ \ V_2O_5\ \ } SO_3$$

A flowsheet of a typical sulfuric acid manufacturing plant is shown in Figure 8-14. It is the converter that we shall be treating in this section.

Although platinum catalysts once were used in the manufacture of sulfuric acid, the only catalysts presently in use employ supported vanadia.[10] For our problem we shall use a catalyst studied by Eklund, whose work was echoed extensively by Donovan[11] in his description of the kinetics of SO_2 oxidation. The catalyst studied by Eklund was a Reymersholm V_2O_5 catalyst deposited on a pumice carrier. The cylindrical pellets had a diameter of 8 mm and a length of 8 mm, with a bulk density of 33.8 lb/ft^3. Between 818 and 1029°F, the rate law for SO_2 oxidation over this particular catalyst was

$$-r'_{SO_2} = k \sqrt{\frac{P_{SO_2}}{P_{SO_3}}} \left[P_{O_2} - \left(\frac{P_{SO_3}}{K_p P_{SO_2}} \right)^2 \right] \tag{8-63}$$

[10]G. M. Cameron, *Chem. Eng. Prog.,* 78(2), 71 (1982).

[11]R. B. Eklund, Dissertation, Royal Institute of Technology, Stockholm, 1956, as quoted by J. R. Donovan, in *The Manufacture of Sulfuric Acid,* ACS Monograph Series 144, W. W. Duecker and J. R. West, eds. (New York: Reinhold, 1959), pp. 166–168.

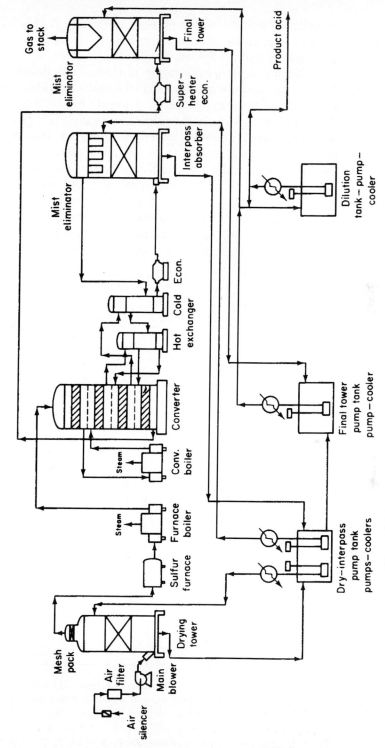

Figure 8-14 Flowsheet of a sulfuric acid manufacturing process. [Reprinted with permission of the AIChE and L. J. Friedman. Copyright © 1982 AIChE. All rights reserved.]

479

in which P_i was the partial pressure of species i. This equation can be used when the conversion is greater than 5%. At all conversions below 5%, the rate is essentially that for 5% conversion.

Sulfuric acid manufacturing processes use different types of reactors. Perhaps the most common type has the reactor divided into adiabatic sections with cooling between the sections (recall Figure 8-8). One such layout is shown in Figure 8-15. In the process in Figure 8-15, gas is brought out of the converter to cool it between stages, using the hot converter reaction mixture to preheat boiler feedwater, produce steam, superheat steam, and reheat the cold gas, all to increase the energy efficiency of the process. Another type has cooling tubes embedded in the reacting mixture. The one illustrated in Figure 8-16 uses incoming gas to cool the reacting mixture.

An SO_2 flow rate of 0.241 mol/s over 132.158 lb of catalyst can produce 1000 tons of acid per day

A typical sulfuric acid plant built in the 1970s produces 1000 to 2400 tons of acid/day.[12] Using the numbers of Kastens and Hutchinson,[13] a 1000-ton/day sulfuric acid plant might have a feed to the SO_2 converter of 7900 lb mol/h,

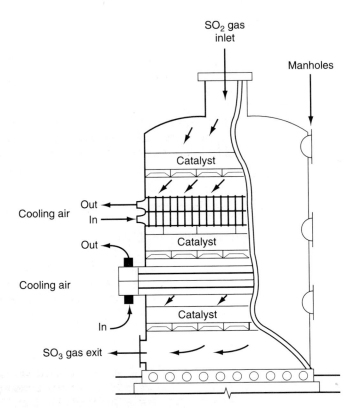

Figure 8-15 Sulfur dioxide converter with internal cooling between catalyst layers. [Reprinted with permission of Barnes & Noble Books.]

[12]L. F. Friedman, *Chem. Eng. Prog., 78*(2), 51 (1982).

[13]M. L. Kastens and J. C. Hutchinson, *Ind. Eng. Chem., 40,* 1340 (1948).

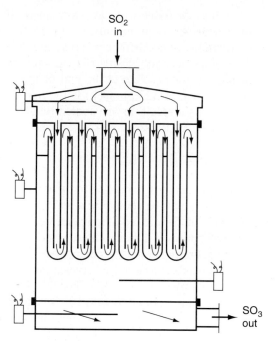

Figure 8-16 Sulfur dioxide converter with catalyst cooled by incoming reaction mixture. [Reprinted with permission of Barnes & Noble Books.]

consisting of 11% SO_2, 10% O_2, and 79% inerts (principally N_2). We shall use these values.

For preliminary design purposes, we shall calculate the conversions for two situations and compare the results. Only one of the situations will be presented in detail in this example. The other is given as a problem on the CD-ROM, but the answer will be used in the comparison.

1. The first situation concerns two stages of a typical commercial adiabatic reactor. The principles of calculating the conversion in an adiabatic reactor were covered earlier and illustrated in Section 8.3, so will not be presented here but as a problem at the end of the chapter.
2. The second case concerns a reactor with the catalyst in tubes, with the walls cooled by a constant-temperature boiling liquid. Calculations for this system are presented in detail below.

8.5.2 Catalyst Quantities

Harrer[14] states that the volumetric flow rate in an adiabatic SO_2 converter, measured at normal temperature and pressure, customarily is about 75 to 100 $ft^3/min \cdot ft^2$ of converter area. He also states that the catalyst beds in the converter may be from 20 to 50 in. deep.

---

[14]T. S. Harrer, in *Kirk-Othmer Encyclopedia of Chemical Technology,* 2nd ed., Vol. 19 (New York: Wiley-Interscience, 1969), p. 470.

It is desirable to have a low mass velocity through the bed to minimize blower energy requirements, so the 75 ft³/min · ft² value will be used. Normal conversions in adiabatic converters are 70% in the first stage and an additional 18% in the second.[15] Using Eklund's Reymersholm catalyst, solution of the adiabatic reactor problem at the end of the chapter shows that these conversions require 1550 ft³ (23 in. deep) in the first stage and 2360 ft³ (35 in. deep) in the second. As a result, in our cooled tubular reactor, we shall use a total catalyst volume of 3910 ft³.

8.5.3 Reactor Configuration

The catalyst is packed in tubes, and the tubes are put in heat exchangers where they will be cooled by a boiling liquid. The outside diameter of the tubes will be 3 in. Severe radial temperature gradients have been observed in SO_2 oxidation systems,[16] although these systems had platinum catalysts and greatly different operating conditions than those being considered here. The 3-in. diameter is chosen as a compromise between minimizing temperature gradients and keeping the number of tubes low. For this service, 12-gauge thickness is specified, which means a thickness of 0.109 in. and an inside diameter of 2.782 in. A 20-ft length will be used, as a compromise between decreasing blower energy requirements (shorter tube length) and lowering capital costs (fewer tubes from a longer tube length). For 3910 ft³ of catalyst, the number of tubes that will be required is

Optimizing capital and operating costs

$$N_t = \frac{\text{volume of catalyst}}{\text{volume per tube}} = \frac{3910}{(20)(\pi)(2.782/12)^2/4} = 4631 \text{ tubes}$$

The total cross-sectional area of the tubes is

$$A_c = \frac{3910 \text{ ft}^3}{20 \text{ ft}} = 195.5 \text{ ft}^2$$

The overall heat-transfer coefficient between the reacting gaseous mixture and the boiling coolant is assumed to be 10 Btu/ h · ft² · °F. This coefficient is toward the upper end of the range of heat-transfer coefficients for such situations as reported by Colburn and Bergelin.[17]

8.5.4 Operating Conditions

Sulfur dioxide converters operate at pressures only slightly higher than atmospheric. An absolute pressure of 2 atm will be used in our designs. The inlet

[15]J. R. Donovan and J. M. Salamone, in *Kirk-Othmer Encyclopedia of Chemical Technology,* 3rd ed., Vol. 22 (New York: Wiley-Interscience, 1978), p. 190.

[16]For example, R. W. Olson, R. W. Schuler, and J. M. Smith, *Chem Eng. Prog., 46,* 614 (1950); and R. W. Schuler, V. P. Stallings, and J. M. Smith, *Chem. Eng. Prog. Symp. Ser. 48*(4), 19 (1952).

[17]Colburn and Bergelin, in *Chemical Engineers' Handbook*; 3rd ed. (New York: McGraw-Hill, 1950).

temperature to the reactor will be adjusted so as to give the maximum conversion. Two constraints are present here. The reaction rate over V_2O_5 catalyst is negligible below ~750°F, and the reactor temperature should not exceed ~1125°F at any point.[18] A series of inlet temperatures should be tested, and the one above 760°F giving the maximum conversion, yet having no reactor temperature exceeding 1120°F, should be used.

The cooling substance should operate at a high temperature so as to improve thermal efficiency by reuse of heat. The most suitable substance appears to be Dowtherm A, with a normal operating limit of ~750°F but which on occasion has been used as the coolant in this preliminary design.[19]

Example 8–10   Oxidation of SO_2

The feed to an SO_2 converter is 7900 lb mol/h and consists of 11% SO_2, 10% O_2, and 79% inerts (principally N_2). The converter consists of 4631 tubes packed with catalyst, each 20 ft long. The tubes are 3 in. o.d. and 2.782 in. i.d. The tubes will be cooled by a boiling liquid at 805°F, so the coolant temperature is constant over this value. The entering pressure is 2 atm.

For inlet temperatures of 740 and 940°F, plot the conversion, temperature, equilibrium conversion, and reaction rate profile down the reactor.

Additional information:

$\phi = 0.45$	$U = 10 \text{ Btu/h} \cdot \text{ft}^2 \cdot \text{R}$
$\rho_0 = 0.054 \text{ lb/ft}^3$	$A_c = 0.0422 \text{ ft}^3$
$P_0 = 2 \text{ atm}$	$T_0 = 1400°\text{R (also } T_0 = 1200 \text{ R)}$
$D_p = 0.015 \text{ ft}$	$g_c = 4.17 \times 10^8 \text{ lb}_m \cdot \text{ft/lb}_f \cdot \text{h}^2$
$\mu = 0.090 \text{ lb/ft} \cdot \text{h at 1400 R}$	$\rho_b = 33.8 \text{ lb/ft}^3 \text{ (bulk density)}$

Using recent JANAF[20] values of K_p at 700 and 900 K, the equilibrium constant at any temperature T is

$$K_p = \exp\left(\frac{42{,}311}{RT} - 11.24\right) \quad (K_p \text{ in atm}^{-1/2}, T \text{ in R}) \qquad \text{(E8-10.1)}$$

at 1600°R,

$$K_p = 7.8 \text{ atm}^{-1/2}$$

---

[18]J. R. Donovan and J. M. Salamone, in *Kirk-Othmer Encyclopedia of Chemical Technology,* 3rd ed. (New York: Wiley, 1984).

[19]The vapor pressure of Dowtherm A at 805°F is very high, and this pressure would have to be maintained in the shell side of the reactor for boiling Dowtherm A to be used as a coolant at this temperature. This aspect will be included in the discussion of the problem results.

[20]D. R. Stull and H. Prophet, Project Directors, *JANAF Thermochemical Tables,* 2nd ed., NSRDS-NBS 37 (Washington, D.C.: U.S. Government Printing Office, 1971).

For rate constants, the data of Eklund[21] can be correlated very well by the equation

$$k = \exp\left[\frac{-176,008}{T} - (110.1 \ln T) + 912.8\right] \tag{E8-10.2}$$

where k is in lb mol SO_2/lb cat.$\cdot$s$\cdot$atm and T is in R.

Kinetic and thermodynamic properties

There are diffusional effects present in this catalyst at these temperatures, and Equation (E8-10.2) should be regarded as an empirical equation that predicts the effective reaction rate constant over the range of temperatures listed by Donovan (814 to 1138°F). The JANAF tables were used to give the following:

$$\Delta H_{Rx}(800°F) = -42,471 \text{ Btu/lb mol } SO_2 \tag{E8-10.3}$$

$$C_{P_{SO_2}} = 7.208 + 5.633 \times 10^{-3} T - 1.343 \times 10^{-6} T^2 \tag{E8-10.4}$$

$$C_{P_{O_2}} = 5.731 + 2.323 \times 10^{-3} T - 4.886 \times 10^{-7} T^2 \tag{E8-10.5}$$

$$C_{P_{SO_3}} = 8.511 + 9.517 \times 10^{-3} T - 2.325 \times 10^{-6} T^2 \tag{E8-10.6}$$

$$C_{P_{N_2}} = 6.248 + 8.778 \times 10^{-4} T - 2.13 \times 10^{-8} T^2 \tag{E8-10.7}$$

where C_p is in Btu/lb mol$\cdot$°R and T in °R.

Solution

1. **General procedure**:
   a. Apply the *plug-flow design equation* relating catalyst weight to the rate of reaction and conversion. Use stoichiometric relationships and feed specifications to express the rate law as a function of conversion.
   b. Apply the *energy balance* relating catalyst weight and temperature.
   c. Using the *Ergun equation* for pressure drop, determine the pressure as a function of catalyst weight.
   d. State *property values* [e.g., k, K_p, $\Delta H°_{Rx}(T_R)$, C_{pi}] and their respective temperature dependences necessary to carry out the calculations.
   e. *Numerically integrate* the design equation, energy balance, and Ergun equation simultaneously to determine the exit conversion and the temperature and concentration profiles.

2. **Design equations**. The general mole balance equations (design equations) based on the weight of catalyst were given in their differential and integral forms by

$$F_{A0}\frac{dX}{dW} = -r'_A$$

3. **Rate law**:

$$-r'_{SO_2} = k\sqrt{\frac{P_{SO_2}}{P_{SO_3}}}\left[P_{O_2} - \left(\frac{P_{SO_3}}{K_p P_{SO_2}}\right)^2\right]$$

---

[21]R. B. Eklund, as quoted by J. R. Donovan, in W. W. Duecker and J. R. West, *The Manufacture of Sulfuric Acid* (New York: Reinhold, 1959).

4. **Stoichiometric relationships and expressing $-r'_{SO_2}$ as a function of X:**

$$SO_2 + \tfrac{1}{2}O_2 \rightleftharpoons SO_3$$

$$A + \tfrac{1}{2}B \rightleftharpoons C$$

We let A represent SO_2 and v_i the stoichiometric coefficient for species i:

$$P_i = C_i(RT) = C_{A0}\frac{(\Theta_i + v_i X)(RT)P}{(1 + \varepsilon X)(T/T_0)P_0} = P_{A0}\frac{(\Theta_i + v_i X)P}{(1 + \varepsilon X)P_0} \quad \text{(E8-10.8)}$$

Substituting for partial pressures in the rate law and combining yields

$$\frac{dX}{dW} = \frac{-r'_A}{F_{A0}} = \frac{k}{F_{A0}}\sqrt{\frac{1-X}{\Theta_{SO_3}+X}}\left[\frac{P}{P_0}P_{A0}\frac{\Theta_{O_2}-\tfrac{1}{2}X}{1+\varepsilon X} - \left(\frac{\Theta_{SO_3}-X}{1+X}\right)^2\frac{1}{K_p^2}\right] \quad \text{(E8-10.9)}$$

where $\varepsilon = -0.055$, $P_{A0} = 0.22$ atm, $\Theta_{SO_2} = 1.0$, $\Theta_{O_2} = 0.91$, $\Theta_{SO_3} = 0.0$, and $\Theta_{N_2} = 7.17$; $F_{T0} = 7900$ lb mol/h, and $F_{A0} = 869$ lb mol/h.

Per tube:

Weight of catalyst in one tube $= W = \rho_b \pi D^2/4L = 28.54$ lb cat./tube

$$F_{A0} = \frac{869}{4631} = 0.188 \text{ lb mol/h} \cdot \text{tube}$$

Substituting the values above gives us

The combined mole balance, rate law, and stoichiometry

$$\boxed{\frac{dX}{dW} = \frac{-r'_A}{F_{A0}} = 5.32k\sqrt{\frac{1-X}{X}}\left\{\left(\frac{0.2-0.11X}{1-0.055X}\right)\frac{P}{P_0} - \left[\frac{X}{(1-X)K_p}\right]^2\right\}} \quad \text{(E8-10.10)}$$

that is,

$$\frac{dX}{dW} = f_1(X, T, P) \quad \text{(E8-10.11)}$$

The limits of integration are from zero to the weight of catalyst in one tube, 28.54 lb.

5. **Energy balance.** For steady-state operation and no shaft work, Equation (8-56) can be rewritten in terms of catalyst weight as the spatial variable, that is,

$$\frac{dT}{dW} = \frac{(4U/\rho_b D)(T_a - T) + (-r'_A)[-\Delta H_{Rx}(T)]}{F_{A0}(\Sigma\Theta_i C_{pi} + X\Delta C_p)} \quad \text{(E8-10.12)}$$

6. **Evaluating the energy balance parameters:**

Heat of reaction:

$$\Delta H_{Rx}(T) = \Delta H°_{Rx}(T_R) + \Delta\alpha(T - T_R) + \frac{\Delta\beta}{2}(T^2 - T_R^2) + \frac{\Delta\gamma}{3}(T^3 - T_R^3) \quad \text{(E8-10.13)}$$

For the SO_2 oxidation, $SO_2 + \frac{1}{2}O_2 \rightarrow SO_3$,

$$\Delta\alpha = \alpha_{SO_3} - \tfrac{1}{2}\alpha_{O_2} - \alpha_{SO_2} = 8.511 - (0.5)(5.731) - 7.208 = -1.563$$

Similarly,

$$\Delta\beta = 0.00262 \quad \text{and} \quad \Delta\gamma = -0.738 \times 10^{-6}$$

Substituting into Equation (E8-10.13) with $T_R = 1260°R$, we have

$$\Delta H_{Rx}(T) = -42{,}471 - (1.563)(T - 1260) + (1.36 \times 10^{-3})(T^2 - 1260^2)$$
$$- (2.459 \times 10^{-7})(T^3 - 1260^3) \qquad \text{(E8-10.14)}$$

$$\sum \Theta_i C_{pi} = 57.23 + 0.014T - 1.94 \times 10^{-6}T^2 \qquad \text{(E8-10.15)}$$

Heat-transfer coefficient term:

$$\frac{U\pi D}{\rho_b A_c} = \frac{4U}{\rho_b D} = \frac{4(10 \text{ Btu/h} \cdot \text{ft}^2 \cdot °R)}{(33.8 \text{ lb/ft}^3)[(2.78/12) \text{ ft}]}$$

$$= 5.11 \text{ Btu/h} \cdot \text{lb cat} \cdot R$$

Energy balance

$$\frac{dT}{dW} = \frac{5.11(T_a - T) + (-r'_A)[-\Delta H_{Rx}(T)]}{0.188(\sum \Theta_i C_{pi} + X \Delta C_p)} \qquad \text{(E8-10.16)}$$

that is,

$$\frac{dT}{dW} = f_2(T, P, X) \qquad \text{(E8-10.17)}$$

7. **Pressure drop**: After rearranging Equation (4-23), the pressure drop is given by

Momentum balance

$$\frac{dP}{dz} = -\frac{(1 - \phi)G(1 + \varepsilon X)}{\rho_0 (P/P_0)(T_0/T) g_c D_p \phi^3}\left[\frac{150\mu(1 - \phi)}{D_p} + 1.75G\right]$$

where

$$G = \frac{\sum F_{i0}M_i}{A_c} \quad (M_i = \text{molecular weight of } i)$$

$$= 1307.6 \text{ lb/ft}^2 \cdot \text{h}$$

$$A_c = \text{cross-sectional area } \pi D^2/4$$

Recalling that $W = \rho_b A_c z$, we obtain

$$\frac{dP}{dW} = -\frac{GTP_0(1 - \phi)(1 + \varepsilon X)}{\rho_b A_c \rho_0 T_0 P D_p g_c \phi^3}\left[\frac{150(1 - \phi)\mu}{D_p} + 1.75G\right] \qquad \text{(E8-10.18)}$$

8. **Evaluating the pressure-drop parameters**:

$$f_3(X, T, P) = -\frac{GTP_0(1 - \phi)(1 + \varepsilon X)}{\rho_b A_c \rho_0 T_0 \phi^3 D_p g_c P}\left[\frac{150(1 - \phi)\mu}{D_p} + 1.75G\right] \qquad \text{(E8-10.19)}$$

Substituting in Equation (E8-10.18), we get

$$\frac{dP}{dW} = \frac{-1.12 \times 10^{-8}(1 - 0.55X)T}{P}(5500\mu + 2288) \qquad \text{(E8-10.20)}$$

We wish to obtain an order-of-magnitude estimate of the pressure drop. To obtain this estimate, we consider the reaction to be carried out isothermally with $\varepsilon = 0$,

$$\boxed{\frac{dP}{dW} = \frac{-0.0432}{P}}$$

Back-of-the-enevelope calculation for ΔP

Integrating with limits $P_0 = 2$ atm at $W = 0$ and $P = P$ at $W = 28.54$ lb of catalyst yields

$$\frac{P^2 - 4}{2} = -0.0432(0 - 28.54)$$

$$P = 1.239 \text{ atm}$$

$$\Delta P = 2 - 1.24 = 0.76 \text{ atm}$$

Because the gas-phase viscosity is a weakly varying function of temperature (i.e., $\mu \sim \sqrt{T}$), we shall consider viscosity to be independent of temperature:

$$\frac{dP}{dW} = f_3(T, P, X) \qquad \text{(E8-10.21)}$$

9. **Solution procedure.** There are three coupled differential equations that must be solved simultaneously:

Mole balance: $\qquad \dfrac{dX}{dW} = f_1(T, P, X) \qquad$ (8-10.11)

The coupled differential equations to be solved with an ODE solver

Energy balance: $\qquad \dfrac{dT}{dW} = f_2(T, P, X) \qquad$ (8-10.17)

Momentum balance: $\qquad \dfrac{dP}{dW} = f_3(T, P, X) \qquad$ (8-10.21)

10. **Numerical procedure.** The rate equation is independent of conversion between $X = 0.0$ and $X = 0.05$ and the rate of disappearance of SO_2 over this range is equal to the rate of reaction at $X = 0.05$:

$$-r'_{SO_2} = k\left(0.848 - \frac{0.012}{K_p^2}\right) \qquad \text{(E8-10.22)}$$

a. Set $X = 0.00$, $T = T_0$, and $P = P_0$.
b. Calculate k from Equation (E8-10.2).
c. Calculate K_p from Equation (E8-10.1).
d. If $X < 0.05$, calculate $-r'_{SO_2}$ from Equation (E8-10.22). If $X > 0.05$, use Equation (E8-10.10).
e. Calculate X, T, and P from a numerical solution to Equations (E8-10.10), (E8-10.16) and (E8-10.20).

The POLYMATH program is given in Table E8-10.1.

TABLE E8-10.1.   SO$_3$ OXIDATION POLYMATH PROGRAM

Equations:	Initial Values:

```
d(P)/d(w)=(-1.12*10**(-8)*(1-.055*x)*T)*(5500*visc+2288)/P      2
d(x)/d(w)=-(ra)/fao                                             0
d(T)/d(w)=(5.11*(Ta-T)+(-ra)*(-deltah))/(fao*(sum+x*dcp))       1400
fao=.188
visc=.090
Ta=1264.67
deltah=-42471-1.563*(T-1260)+.00136*(T**2-1260**2)-2.459*10*
    *(-7)*(T**3-1260**3)
sum=57.23+.014*T-1.94*10**(-6)*T**2
dcp=-1.5625+2.72*10**(-3)*T-7.38*10**(-7)*T**2
k=3600*exp(-176008/T-(110.1*ln(T))+912.8)
thetaso=0
Po=2
Pao=.22
thetao=.91
eps=-.055
R=1.987
Kp=exp(42311/R/T-11.24)
ra=if(x<=.05)then(-k*(.848-.012/(Kp**2)))else(-k*((1-x)/(the
    taso+x))**.5*(P/Po*Pao*((thetao-.5*x)/
    (1+eps*x))-((thetaso+x)/(1-x))**2/(Kp**2)))
w₀ = 0,   wf = 28.54
```

11. **Discussion of results**. Figure E8-10.1 shows the profiles for inlet temperatures of 1200°R and 1400°R, respectively. Only 68.5% conversion is achieved for $T_0 = 1200°R$, even though $X_e = 0.99$. For an entering temperature of 1400°R, the major portion of the reaction takes place in the first 6 ft of the reactor. At this point, the conversion is 0.81, with only another 0.06 of the conversion occurring in the remaining 14 ft, as shown in Figure E8-10.1b. The cause of this low amount of conversion in the final 14 ft is the steadily dropping temperature in the reactor. Beyond the 6-ft point, the temperature is too low for much reaction to take place, which means that the reactor is cooled too much.

This detrimental situation indicates that the coolant temperature is too low for obtaining maximum conversion. Thus even boiling Dowtherm A at its highest possible operating temperature is not a suitable coolant. Perhaps a gas would give a better performance as a coolant in this reaction system. Two problems at the end of the chapter pursue this aspect. One of them seeks the optimum coolant temperature for a constant-coolant-temperature system, and the other uses inlet gas as a coolant.

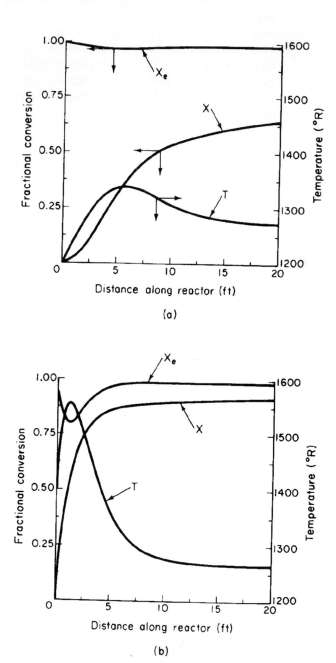

Figure E8-10.1 Conversion, temperature, and equilibrium conversion profiles within the reactor: (a) inlet temperature at 1200°R; (b) inlet temperature at 1400°R.

Another possible way to operate such a reactor is to use multiple-stage operation with progressively higher coolant temperatures. Because pressure drop over the reactor is small (~0.7 atm), neglecting the pressure drop does not affect the exit conversion significantly (Figure 8-17). The effect is more significant at lower reactor inlet temperatures because the rate of reaction is appreciable over a longer portion of the reactor bed. At higher inlet temperatures the conversion is limited by the approach to equilibrium, and hence the pressure drop has a negligible effect.

Analyzing the effects of pressure drop

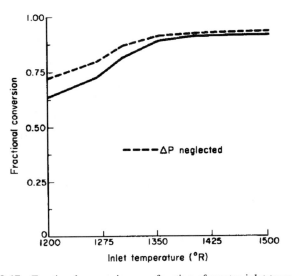

Figure 8-17   Fractional conversion as a function of reactor inlet temperature.

8.6 Multiple Steady States

In this section we consider the steady-state operation of a CSTR in which a first-order reaction is taking place. We begin by recalling the hydrolysis of propylene oxide, Example 8-4.

If one were to examine Figure E8-4.2, one would observe that if a parameter were changed slightly, there might be more than one intersection of the energy and mole balance curves. When more than one intersection occurs, there is more than one set of conditions that satisfy both the energy balance and mole balance, and consequently there will be multiple steady states at which the reactor may operate.

Neglecting shaft work in the CSTR, and setting $\Delta \hat{C}_p = 0$, [i.e., for $\Delta \hat{C}_p = 0, \Delta H^\circ_{\text{Rx}}(T) = \Delta H^\circ_{\text{Rx}}(T_R) \equiv \Delta H^\circ_{\text{Rx}}$] the energy balance [Equation (8-51)] can be rearranged in the form

$$-F_{A0}X\,\Delta H^\circ_{\text{Rx}}(T) = F_{A0}\left[\sum \Theta_i \tilde{C}_{pi}(T - T_0) + \frac{UA}{F_{A0}}(T - T_a)\right] \quad (8\text{-}64)$$

Substituting the mole balance on a CSTR,

$$F_{A0}X = -r_A V \quad (2\text{-}12)$$

into Equation (8-64) gives

$$(-r_A V)(-\Delta H_{Rx}) = F_{A0} C_{p0}(T - T_0) + UA(T - T_a) \tag{8-65}$$

where

$$C_{p0} = \Sigma \Theta_i \tilde{C}_{pi}$$

By letting

$$\kappa = \frac{UA}{C_{p0} F_{A0}}$$

and

$$\boxed{T_c = \frac{T_0 F_{A0} C_{p0} + UAT_a}{UA + C_{p0} F_{A0}} = \frac{\kappa T_a + T_0}{1 + \kappa}} \tag{8-66}$$

Equation (8-65) may be simplified to

$$(-r_A V / F_{A0})(-\Delta H_{Rx}^\circ) = C_{p0}(1 + \kappa)(T - T_c) \tag{8-67}$$

The left-hand side is referred to as the *heat-generated term*:

G(T) = heat-generated term

$$\boxed{G(T) = (-\Delta H_{Rx}^\circ)(-r_A V / F_{A0})} \tag{8-68}$$

The right-hand side of Equation (8-67) is referred to as the *heat-removed term* (by flow and heat exchange) R(T):

Heat-removed term

$$\boxed{R(T) = C_{p0}(1 + \kappa)(T - T_c)} \tag{8-69}$$

To study the multiplicity of steady states, we shall plot both $R(T)$ and $G(T)$ as a function of temperature on the same graph and analyze the circumstances under which we will obtain multiple intersections.

8.6.1 Heat-Removed Term, R(T)

From Equation (8-69) we see that $R(T)$ increases linearly with temperature, with slope $C_{p0}(1 + \kappa)$. As the entering temperature T_0 is increased, the line retains the same slope but shifts to the right as shown in Figure 8-18.

If one increases κ by either decreasing the molar flow rate F_{A0} or increasing the heat-exchange area, the slope increases and the ordinate intercept moves to the left as shown in Figure 8-19, for conditions of $T_a < T_0$:

$$\kappa = 0 \quad T_c = T_0$$

$$\kappa = \infty \quad T_c = T_a$$

If $T_a > T_0$, the intercept will move to the right as κ increases.

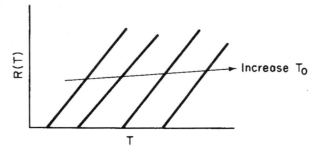

Heat-removed curve
R(T)

Figure 8-18   Variation of heat removal line with inlet temperature.

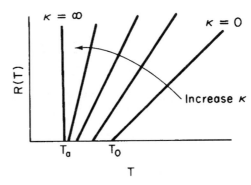

$$\kappa = \frac{UA}{C_{p0} F_{A0}}$$

$$T_c = \frac{T_0 + \kappa T_a}{1 + \kappa}$$

Figure 8-19   Variation of heat removal line with κ ($\kappa = UA/C_{p0} F_{A0}$).

8.6.2 Heat of Generation, $G(T)$

The heat-generated term, Equation (8-68), can be written in terms of conversion, (recall: $X = -r_A V/F_{A0}$)

$$G(T) = (-\Delta H^{\circ}_{Rx})X \qquad (8\text{-}70)$$

To obtain a plot of heat generated, $G(T)$, as a function of temperature, we must solve for X as a function of T using the CSTR mole balance, the rate law, and stoichiometry. For example, for a first-order liquid-phase reaction, the CSTR mole balance becomes

$$V = \frac{F_{A0}X}{kC_A} = \frac{v_0 C_{A0} X}{kC_{A0}(1 - X)}$$

Solving for X yields

1st order reaction

$$X = \frac{\tau k}{1 + \tau k}$$

Substituting for X in Equation (8-70), we obtain

$$G(T) = \frac{-\Delta H^{\circ}_{Rx}\tau k}{1 + \tau k} \qquad (8\text{-}71)$$

Finally, substituting for k in terms of the Arrhenius equation, we obtain

$$G(T) = \frac{-\Delta H_{Rx}^{\circ}\tau A e^{-E/RT}}{1 + \tau A e^{-E/RT}} \qquad (8\text{-}72)$$

Note that equations analogous to Equation (8-71) for $G(T)$ can be derived for other reaction orders and for reversible reactions simply by solving the CSTR mole balance for X. For example, for the second-order liquid-phase reaction

2nd order reaction

$$X = \frac{(2\tau k C_{A0} + 1) - \sqrt{4\tau k C_{A0} + 1}}{2\tau k C_{A0}}$$

and the corresponding heat generated is

$$G(T) = \frac{-\Delta H_{Rx}^{\circ}[(2\tau C_{A0} A e^{-E/RT} + 1) - \sqrt{4\tau C_{A0} A e^{-E/RT} + 1}]}{2\tau C_{A0} A e^{-E/RT}} \qquad (8\text{-}73)$$

At very low temperatures, the second term in the denominator of Equation (8-72) can be neglected, so that $G(T)$ varies as

Low T

$$G(T) = -\Delta H_{Rx}^{\circ}\tau A e^{-E/RT}$$

(Recall ΔH_{Rx}° means the heat of reaction is evaluated at T_R.)

At very high temperatures, the second term in the denominator dominates and $G(T)$ is reduced to

High T

$$G(T) = -\Delta H_{Rx}^{\circ}$$

$G(T)$ is shown as a function of T for two different activation energies, E_A, in Figure 8-20. If the flow rate is decreased or the reactor volume increased so as to increase τ, the heat of generation term, $G(T)$, changes as shown in Figure 8-21.

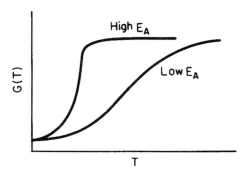

Figure 8-20   Heat generation curve.

8.6.3 Ignition-Extinction Curve

The points of intersection of $R(T)$ and $G(T)$ give us the temperature at which the reactor can operate at steady state. Suppose that we begin to feed our reactor at some relatively low temperature, T_{01}. If we construct our $G(T)$ and

Heat-generated
curves, $G(T)$

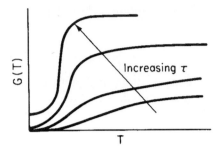

Figure 8-21   Variation of heat generation curve with space-time.

$R(T)$ curves, illustrated by curves y and a, respectively, in Figure 8-22, we see that there will be only one point of intersection, point 1. From this point of intersection, one can find the steady-state temperature in the reactor, T_{s1}, by following a vertical line to the T axis and reading off the temperature as shown in Figure 8-22.

   If one were now to increase the entering temperature to T_{02}, the $G(T)$ curve would remain unchanged but the $R(T)$ curve would move to the right, as shown by line b in Figure 8-22, and will now intersect the $G(T)$ at point 2 and be tangent at point 3. Consequently, we see from Figure 8-22 that there are two steady-state temperatures, T_{s2} and T_{s3}, that can be realized in the CSTR for an entering temperature T_{02}. If the entering temperature is increased to T_{03}, the $R(T)$ curve, line c (Figure 8-23), intersects the $G(T)$ three times and there are three steady-state temperatures. As we continue to increase T_0, we finally reach line e, in which there are only two steady-state temperatures. By further increasing T_0 we reach line f, corresponding to T_{06}, in which we have only one temperature that will satisfy both the mole and energy balances. For the six entering temperatures, we can form Table 8-4, relating the entering temperature to the

Both the mole and
energy balances are
satisfied at the
points of
intersection or
tangency

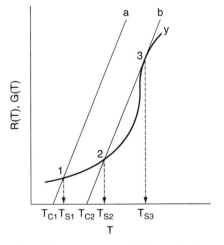

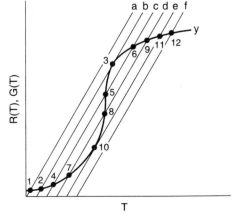

Figure 8-22   Finding multiple steady states
with T_0 varied.

Figure 8-23   Finding multiple steady states
with T_0 varied.

TABLE 8-4.   MULIPLE STEADY-STATE TEMPERATURES

Entering Temperature	Reactor Temperatures			
T_{01}			T_{s1}	
T_{02}		T_{s2}		T_{s3}
T_{03}	T_{s4}		T_{s5}		T_{s6}
T_{04}	T_{s7}		T_{s8}		T_{s9}
T_{05}		T_{s10}		T_{s11}
T_{06}			T_{s12}	

possible reactor operating temperatures. By plotting T_s as a function of T_0, we obtain the well-known *ignition-extinction curve* shown in Figure 8-24. From this figure we see that as the entering temperature is increased, the steady-state temperature increases along the bottom line until T_{05} is reached. Any fraction of a degree increase in temperature beyond T_{05} and the steady-state reactor temperature will jump up to T_{s11}, as shown in Figure 8-24. The temperature at which this jump occurs is called the *ignition temperature*. If a reactor were operating at T_{s12} and we began to cool the entering temperature down from T_{06}, the steady-state reactor temperature T_{s3} would eventually be reached, corresponding to an entering temperature T_{02}. Any slight decrease below T_{02} would drop the steady-state reactor temperature to T_{s2}. Consequently, T_{02} is called the *extinction temperature*.

> We must exceed a certain feed temperature to operate at the upper steady state where the temperature and conversion are higher

The middle points 5 and 8 in Figures 8-23 and 8-24 represent unstable steady-state temperatures. For example, if by some means one were operating at T_{s5} and a pulse increase in the temperature suddenly occurred, the heat gener-

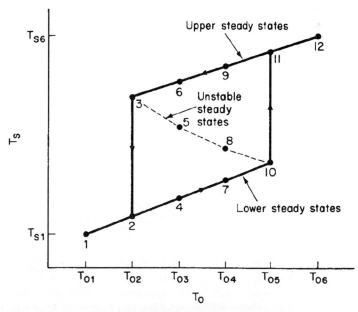

Figure 8-24 Temperature ignition-extinction curve.

Not all the multiple
steady states are
stable

ated would be greater than the heat removed and the steady-state reactor temperature would continue to increase until temperature T_{s6} was reached. On the other hand, if a sudden pulse decrease in temperature occurred at T_{s5}, the heat removed $R(T)$ would be greater than the heat generated $G(T)$ and the steady-state reactor temperature would continue to fall until T_{s4} was reached. Steady-state conditions that behave in this manner are said to be unstable.

In contrast to these unstable operating points, consider what would happen to the reactor temperature if a reactor operating at T_{s12} were subjected to very small temperature fluctuations. From Figure 8-24 we observe that a pulse increase in reactor temperature would make the heat of removal greater than the heat of generation [the $R(T)$ curve would be above the $G(T)$ curve] and the temperature would drop back down to T_{s12}. If a small pulse decrease in the reactor temperature occurred while the feed temperature remained constant at T_{06}, we would see that the curve $G(T)$ would be above the heat-removed curve $R(T)$ and the reactor temperature would continue to rise until T_{s12} was again reached. A similar analysis could be carried through for reactor temperatures T_{s1}, T_{s2}, T_{s4}, T_{s6}, T_{s7}, T_{s9}, T_{s11}, and T_{s12} and one would find that when these reactor temperatures are subjected to either a small positive or negative fluctuation, they will always return to their local steady-state values. Consequently, we say that these points are *locally stable steady states*. While these points are locally stable, they are not necessarily globally stable. That is, a perturbation in temperature or concentration, while small, may be sufficient to cause the reactor to fall from the

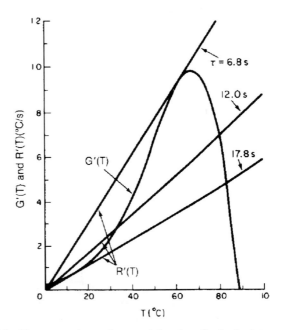

Figure 8-25 Heat generation and removal functions for feed mixture of 0.8 M $Na_2S_2O_3$ and 1.2 M H_2O_2 at 0°C. By S. A. Vejtasa and R. A. Schmitz, *AIChE J.*, *16*(3), 415, (1970). (Reproduced by permission of the American Institute of Chemical Engineers. Copyright © 1970 AIChE. All right reserved.) See Problem P8C-4.

upper steady state (corresponding to high conversion and temperature) to the lower steady state (corresponding to low temperature and conversion). We will examine this case in detail in Section 9.4 and in Problem P9-15.

An excellent experimental investigation that demonstrates the multiplicity of steady states was carried out by Vejtasa and Schmitz.(Figure 8-25)[22] They studied the reaction between sodium thiosulfate and hydrogen peroxide:

$$2Na_2S_2O_3 + 4H_2O_2 \rightarrow Na_2S_3O_6 + Na_2SO_4 + 4H_2O$$

in a CSTR operated adiabatically. The multiple steady-state temperatures were examined by varying the flow rate over a range of space times, τ, as shown in Figure 8-26. One observes from this figure that at a space-time of 12 s, steady-state reaction temperatures of 4, 33, and 80°C are possible. If one were operating on the higher steady-state temperature line and the volumetric flow rates were steadily increased (i.e., the space-time decreased), one notes that if the space velocity dropped below about 7 s, the reaction temperature would drop from 70°C to 2°C. The flow rate at which this drop occurs is referred to as the *blowout velocity.*

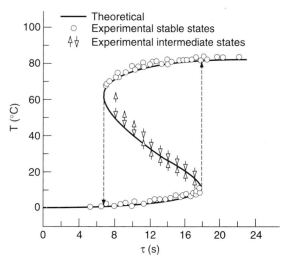

Figure 8-26   Multiple steady states.

8.6.4 Runaway Reactions

In many reacting systems the temperature of the upper steady state may be sufficiently high that it is undesirable or even dangerous to operate at this condition. From Figure 8-24 we saw that once we exceed the ignition temperature, we will proceed to the upper steady state. The ignition temperature occurs at point 10 on Figure 8-23, which is a point of tangency of the heat-removed curve with

---

[22]S. A. Vejtasa and R. A. Schmitz, *AIChE J., 16,* 410 (1970).

the heat-generated curve. At this point of tangency the slopes of the $R(T)$ and $G(T)$ curves are equal. That is, for the heat removal curve we have

$$\frac{dR(T)}{dT} = C_{p0}(1 + \kappa) \tag{8-74}$$

and for the heat-generated curve

$$\frac{dG(T)}{dT} = \frac{d\left[(-\Delta H_{Rx})\dfrac{-r_A V}{F_{A0}}\right]}{dT} = \frac{(-\Delta H_{Rx})V}{F_{A0}}\frac{d(-r_A)}{dT}$$

Assuming that the reaction is irreversible, follows a power law model, and that the concentrations of the reacting species are weak functions of temperature,

$$-r_A = (Ae^{-E/RT})\,\text{fn}(C_i)$$

then

$$\frac{d(-r_A)}{dT} = \frac{E}{RT^2}Ae^{-E/RT}\,\text{fn}(C_i) = \frac{E}{RT^2}(-r_A) \tag{8-75}$$

Equating Equations (8-74) and (8-75) yields

$$C_{p0}(1 + \kappa) = \frac{E}{RT^2}(-r_A)\frac{-\Delta H_{Rx}}{F_{A0}}V \tag{8-76}$$

Finally, we divide Equation (8-67) by Equation (8-76) to obtain the following ΔT value for a CSTR operating at $T = T_r$.

$$\boxed{\Delta T_{rc} = T_r - T_c = \frac{RT_r^2}{E}} \tag{8-77}$$

If this ΔT_{rc} is exceeded, transition to the upper steady state will occur. For many industrial reactions E/RT is typically between 16 and 24, and the reaction temperatures may be 300 to 500 K. Consequently, this critical temperature difference ΔT_{rc} will be somewhere around 15 to 30°C.

8.6.5 Steady-State Bifurcation Analysis

In reactor dynamics it is particularly important to find out if multiple stationary points exist or if sustained oscillations can arise. Bifurcation analysis is aimed at locating the set of parameter values for which multiple steady states will occur.[23] We apply bifurcation analysis to learn whether or not multiple steady states are possible. An outline of what is on the CD-ROM follows.

---

[23]V. Balakotaiah and D. Luss, in *Chemical Reaction Engineering*, Boston ACS Symposium Series 196 (Washington, D.C.: American Chemical Society, 1982), p. 65; M. Golubitsky and B. L. Keyfitz, *Siam. J. Math. Anal.*, *11*, 316 (1980); A. Uppal, W. H. Ray, and A. B. Poore, *Chem. Eng. Sci.*, *29*, 967 (1974).

Bifurcation analysis will be applied to the CSTR mole and energy balances. First, Equations (8-68) and (8-69) are combined and the energy balance is written as

$$F(T) = C_{p_0}(1 + \kappa)T - C_{p_0}(1 + \kappa)T_c - G(T) \qquad (8\text{-}78)$$

which is of the form

$$F(T) = \alpha T - \beta - G(T) \qquad (8\text{-}79)$$

Similarly, the CSTR mole balance can be written as

$$\frac{C_A}{\tau} - \frac{C_{A0}}{\tau} - r_A(C_A) = 0 \qquad (8\text{-}80)$$

which is of a similar nature for the energy balance,

$$F(C_A) = \alpha C_A - \beta - G(C_A) \qquad (8\text{-}81)$$

Both CSTR energy and mole balances are of the form

$$F(y) = \alpha y - \beta - G(y) \qquad (8\text{-}82)$$

The conditions for uniqueness are then shown to be those that satisfy the relationship

$$\max\left(\frac{\partial G}{\partial y}\right) < \alpha \qquad (8\text{-}83)$$

For example, if we use energy balance in the form given by Equation (8-78) and use Equations (8-75) and (8-76) we would find the criteria for uniqueness (i.e. No Multiple Steady States (MMS)) is

Criteria for no MSS

$$\boxed{\left(\frac{E}{RT^2}(r_A \Delta H_{Rx})\frac{V}{F_{A0}}\right)_{max} < C_{p_0}(1 + \kappa)}$$

However, if

$$\max\left(\frac{\partial G}{\partial y}\right) > \alpha \qquad (8\text{-}84)$$

we do not know if multiple steady solutions exist and we must carry the analysis further. Specifically, the conditions for which multiple steady states exist must satisfy the following set of equations:

(1)
$$\left.\frac{dF}{dy}\right|_{y*} = 0 = \alpha - \left.\frac{dG}{dy}\right|_{y*} \qquad (8\text{-}85)$$

(2)
$$F(y*) = 0 = \alpha y* - \beta - G(y*) \qquad (8\text{-}86)$$

An example is given on the CD-ROM which maps out the regions where multiple steady states are possible and not possible for the reaction

$$CO + \frac{1}{2} O_2 \xrightarrow{\text{Pt}} CO_2$$

with the rate law

$$-G(C_A) = -r_A = \frac{kC_A}{(1 + KC_A)^2}$$

A portion of the solution to the example problem is shown here (Figure CDE8-1.1), highlighting regions where multiple steady states are not possible.

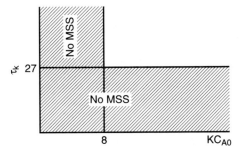

Figure CDE8-1.1   Mapping the regions of no multiple steady states.

8.7 Nonisothermal Multiple Chemical Reactions

8.7.1 Plug-Flow Reactors

In this section we give the energy balance for multiple reactions that are in parallel and/or in series. The energy balance for a single reaction taking place in a PFR was given by Equation (8-60)

$$\frac{dT}{dV} = \frac{Ua(T_a - T) + (-r_A)[-\Delta H_{Rx}(T)]}{\sum\limits_{i=1}^{m} F_i C_{pi}} \tag{8-60}$$

When q multiple reactions are taking place and there are m species, it is easily shown that Equation (8-60) can be generalized to

Energy balance for multiple reactions

$$\boxed{\frac{dT}{dV} = \frac{Ua(T_a - T) + \sum\limits_{i=1}^{q} (-r_{ij})[-\Delta H_{Rxij}(T)]}{\sum\limits_{j=1}^{m} F_j C_{pj}}} \tag{8-87}$$

The heat of reaction for reaction i must be referenced to the same species in the rate, r_{ij}, by which ΔH_{Rxij} is multiplied, that is,

$$[-r_{ij}][-\Delta H_{Rxij}] = \left[\frac{\text{moles of } j \text{ reacted in reaction } i}{\text{volume} \cdot \text{time}}\right] \times \left[\frac{\text{joules "released" in reaction } i}{\text{moles of } j \text{ reacted in reaction } i}\right]$$

$$= \left[\frac{\text{joules "released" in reaction } i}{\text{volume} \cdot \text{time}}\right]$$

where the subscript j refers to the species, the subscript i to the particular reaction, q is the number of **independent** reactions, and m is the number of species.

Consider the following reaction sequence carried out in a PFR:

$$\text{Reaction 1:} \qquad \text{A} \xrightarrow{k_1} \text{B} \tag{8-88}$$

$$\text{Reaction 2:} \qquad \text{B} \xrightarrow{k_2} \text{C} \tag{8-89}$$

The PFR energy balance becomes

$$\frac{dT}{dV} = \frac{Ua(T_a - T) + (-r_{1A})(-\Delta H_{Rx1A}) + (-r_{2B})(-\Delta H_{Rx2B})}{F_A C_{pA} + F_B C_{pB} + F_C C_{pC}} \tag{8-90}$$

where ΔH_{Rx1A} = [kJ/mol of A reacted in reaction 1] and [ΔH_{Rx2B} = kJ/mol of B reacted in reaction 2].

Example 8–11 Parallel Reactions in a PFR with Heat Effects

The following gas-phase reactions occur in a PFR:

$$\text{Reaction 1:} \quad \text{A} \xrightarrow{k_1} \text{B} \qquad -r_{1A} = k_{1A}C_A \tag{E8-11.1}$$

$$\text{Reaction 2:} \quad 2\text{A} \xrightarrow{k_2} \text{C} \qquad -r_{2A} = k_{2A}C_A^2 \tag{E8-11.2}$$

Pure A is fed at a rate of 100 mol/s, a temperature of 150°C and a concentration of 0.1 mol/dm^3. Determine the temperature and flow rate profiles down the reactor.

Additional information:

ΔH_{Rx1A} = J/mol of A reacted in reaction 1 = $-20{,}000$ J/mol

ΔH_{Rx2A} = J/mol of A reacted in reaction 2 = $-60{,}000$ J/mol

$$C_{PA} = 90 \text{ J/mol} \cdot \text{°C} \qquad k_{1A} = 10 \exp\left[\frac{E_1}{R}\left(\frac{1}{300} - \frac{1}{T}\right)\right] \text{ s}^{-1}$$

$$C_{PB} = 90 \text{ J/mol} \cdot \text{°C} \qquad E_1 = 8000 \text{ cal/mol}$$

$$C_{PC} = 180 \text{ J/mol} \cdot \text{°C} \qquad k_{2A} = 0.09 \exp\left[\frac{E_2}{R}\left(\frac{1}{300} - \frac{1}{T}\right)\right] \frac{\text{dm}^3}{\text{mol} \cdot \text{s}}$$

$$Ua = 4000 \text{ J/m}^3 \cdot \text{s} \cdot \text{°C} \qquad E_2 = 18{,}000 \text{ cal/mol}$$

$$T_a = 100\text{°C}$$

Solution

The PFR energy balance becomes (cf. Equation (8-87))

$$\frac{dT}{dV} = \frac{Ua(T_a - T) + (-r_{1A})(-\Delta H_{Rx1A}) + (-r_{2A})(-\Delta H_{Rx2A})}{F_A C_{pA} + F_B C_{pB} + F_C C_{pC}} \qquad \text{(E8-11.3)}$$

Mole balances:

$$\frac{dF_A}{dV} = r_{1A} + r_{2A} \qquad \text{(E8-11.4)}$$

$$\frac{dF_B}{dV} = -r_{1A} \qquad \text{(E8-11.5)}$$

$$\frac{dF_C}{dV} = -\frac{1}{2}r_{2A} \qquad \text{(E8-11.6)}$$

Rate laws:

$$C_A = C_{T0}\left(\frac{F_A}{F_T}\right)\left(\frac{T_0}{T}\right) \qquad \text{(E8-11.7)}$$

$$C_B = C_{T0}\left(\frac{F_B}{F_T}\right)\left(\frac{T_0}{T}\right) \qquad \text{(E8-11.8)}$$

$$C_C = C_{T0}\left(\frac{F_C}{F_T}\right)\left(\frac{T_0}{T}\right) \qquad \text{(E8-11.9)}$$

$$F_T = F_A + F_B + F_C \qquad \text{(E8-11.10)}$$

$$k_{1A} = 0.07 \exp\left[3000\left(\frac{1}{300} - \frac{1}{T}\right)\right]$$

$$k_{2A} = 5.0 \exp\left[9000\left(\frac{1}{300} - \frac{1}{T}\right)\right]$$

$$\frac{dT}{dV} = \frac{4000(375 - T) + (-r_{1A})(20,000) + (-r_{2A})(60,000)}{90F_A + 90F_B + 180F_C} \qquad \text{(E8-11.11)}$$

The algorithm for
multiple reactions
with heat effects

Stoichiometry (gas phase), $\Delta P = 0$:

$$r_{1A} = -k_{1A}C_A$$

$$r_{2A} = -k_{2A}C_A^2$$

The POLYMATH program and its graphical outputs are shown in Table E8-11.1 and Figures E8-11.1 and E8-11.2.

TABLE E8-11.1.   POLYMATH PROGRAM

Equations:	*Initial Values:*

```
d(Fb)/d(V)=-r1a                                                    0
d(Fa)/d(V)=r1a+r2a                                                 100
d(Fc)/d(V)=-r2a/2                                                  0
d(T)/d(V)=(4000*(373-T)+(-r1a)*20000+(-r2a)*60000)/(90*Fa+90       423
    *Fb+180*Fc)
k1a=10*exp(4000*(1/300-1/T))
k2a=0.09*exp(9000*(1/300-1/T))
Cto=0.1
Ft=Fa+Fb+Fc
To=423
Ca=Cto*(Fa/Ft)*(To/T)
Cb=Cto*(Fb/Ft)*(To/T)
Cc=Cto*(Fc/Ft)*(To/T)
r1a=-k1a*Ca
r2a=-k2a*Ca**2
```

$$V_0 = 0, \qquad V_f = 1$$

Why does the
temperature go
through a maximum
value?

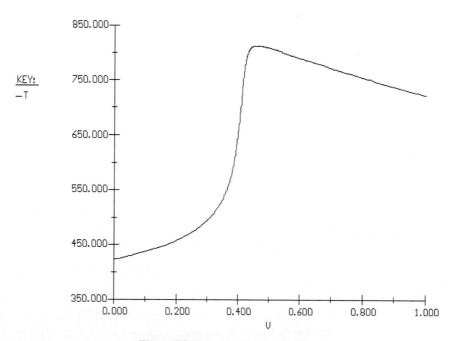

KEY:
—T

Figure E8-11.1   Temperature profile.

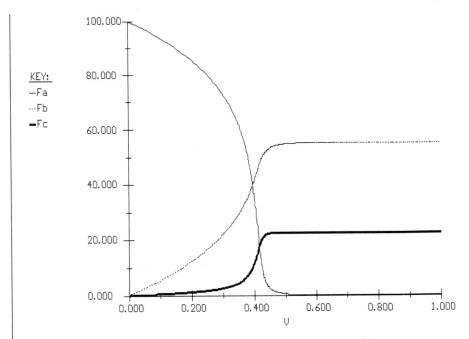

Figure E8-11.2 Profile of molar flow rates F_A, F_B, and F_C.

8.7.2 CSTR

For a CSTR, recall that $-F_{A0}X = r_A V$ and therefore the steady-state energy balance for a single reaction is

$$\dot{Q} - \dot{W}_s - F_{A0} \int_{T_0}^{T} \Sigma\,\Theta_i\,C_{pi}\,dT + \left[\Delta H_{Rx}^{\circ}(T_R) + \int_{T_R}^{T} \Delta C_p\,dT\right][r_A V] = 0 \quad (8\text{-}48)$$

For q multiple reactions and m species, the CSTR energy balance becomes

$$\boxed{\dot{Q} - \dot{W}_s - F_{A0} \int_{T_0}^{T} \Sigma\,\Theta_i\,C_{pi}\,dT + V \sum_{i=1}^{q} r_{ij}\,\Delta H_{Rxij}(T) = 0} \quad (8\text{-}91)$$

Energy Balance For Multiple Reactions in a CSTR

Substituting Equation (8-42) for $\dot{Q}$, neglecting the work term, and assuming constant heat capacities, Equation (8-91) becomes

$$\boxed{UA(T_a - T) - F_{A0} \Sigma\,C_{pi}\,\Theta_i(T - T_0) + V \sum_{i=1}^{q} r_{ij}\,\Delta H_{Rxij}(T) = 0} \quad (8\text{-}92)$$

For the two parallel reactions described in Example 8-11, the CSTR energy balance is

$$UA(T_a - T) - F_{A0} \sum_{i=1}^{m} \Theta_i C_{pi}(T - T_0) + Vr_{1A} \Delta H_{Rx1A}(T) + Vr_{2A} \Delta H_{Rx2A}(T) = 0$$

(8-92)

One of the major goals of this text is to have the reader solve problems involving multiple reactions with heat effects (**cf. P8-30**).

Example 8–12   Multiple Reactions in a CSTR

The elementary liquid-phase reactions

$$A \xrightarrow{k_1} B \xrightarrow{k_2} C$$

take place in a 10-dm³ CSTR. What are the effluent concentrations for a volumetric feed rate of 10 dm³/min at a concentration of A of 0.3 mol/dm³?

The inlet temperature is 283 K.

Additional information:

$$C_{P_A} = C_{P_B} = C_{P_C} = 200 \text{ J/mol} \cdot \text{K}$$

$k_1 = 3.03 \text{ min}^{-1}$ at 300 K, with $E_1 = 9900 \text{ cal/mol}$

$k_2 = 4.58 \text{ min}^{-1}$ at 500 K, with $E_2 = 27{,}000 \text{ cal/mol}$

$\Delta H_{Rx1A} = -55{,}000 \text{ J/mol A}$ $UA = 40{,}000 \text{ J/min} \cdot \text{K with } T_a = 57°\text{C}$

$\Delta H_{Rx2B} = -71{,}500 \text{ J/mol B}$

Solution

A: Combined mole balance and rate law for A:

$$V = \frac{v_0(C_{A0} - C_A)}{k_1 C_A}$$

(E8-12.1)

Solving for C_A gives us

$$C_A = \frac{C_{A0}}{1 + \tau k_1}$$

(E8-12.2)

B: Combined mole balance and rate law for B:

$$V = \frac{0 - C_B v_0}{-r_B} = \frac{C_B v_0}{(r_{1B} + r_{2B})} = \frac{C_B v_0}{k_1 C_A - k_2 C_B}$$

(E8-12.3)

Solving for C_B yields

$$C_B = \frac{\tau k_1 C_A}{1 + \tau k_2} = \frac{\tau k_1 C_{A0}}{(1 + \tau k_1)(1 + \tau k_2)}$$

(E8-12.4)

Rate laws:

$$-r_{1A} = k_1 C_A = \frac{k_1 C_{A0}}{1 + \tau k_1} \qquad \text{(E8-12.5)}$$

$$-r_{2B} = k_2 C_B = \frac{k_2 \tau k_1 C_{A0}}{(1 + \tau k_1)(1 + \tau k_2)} \qquad \text{(E8-12.6)}$$

Applying Equation (8-92) to this system gives

$$UA(T_a - T) - F_{A0} \tilde{C}_{P_A}(T - T_0) + V[r_{1A} \Delta H_{Rx1A} + r_{2B} \Delta H_{Rx2B}] = 0 \qquad \text{(E8-12.7)}$$

Substituting for r_{1A} and r_{2B} and rearranging, we have

$$\overbrace{\left[-\frac{\Delta H_{Rx1A} \tau k_1}{1 + \tau k_1} - \frac{\tau k_1 \tau k_2 \Delta H_{Rx2B}}{(1 + \tau k_1)(1 + \tau k_2)} \right]}^{G(T)} = \overbrace{C_p(1 + \kappa)[T - T_c]}^{R(T)} \qquad \text{(E8-12.8)}$$

$$\kappa = \frac{UA}{F_{A0}C_{P_A}} = \frac{40{,}000 \text{ J/min} \cdot \text{K}}{(0.3 \text{ mol/dm}^3)(1000 \text{ dm}^3/\text{min})\, 200 \text{ J/mol} \cdot \text{K}} = 0.667$$

$$T_c = \frac{T_0 + \kappa T_a}{1 + \kappa} = \frac{283 + (0.666)(330)}{1 + 0.667} = 301.8 \text{ K} \qquad \text{(E8-12.9)}$$

$$G(T) = \left[-\frac{\Delta H_{Rx1A} \tau k_1}{1 + \tau k_1} - \frac{\tau k_1 \tau k_2 \Delta H_{Rx2B}}{(1 + \tau k_1)(1 + \tau k_2)} \right] \qquad \text{(E8-12.10)}$$

$$R(T) = C_p(1 + \kappa)[T - T_c] \qquad \text{(E8-12.11)}$$

The POLYMATH program to plot $R(T)$ and $G(T)$ vs. T is shown in Table E8-12.1, and the resulting graph is shown in Figure E8-12.1.

TABLE E8-12.1.   POLYMATH

	Equations:	Initial Values:

```
d(T)/d(t)=2                                              273
Cp=200
Cao=0.3
To=283
tau=.01
DH1=-55000
DH2=-71500
vo=1000
E2=27000
E1=9900
UA=40000
Ta=330
k2=4.58*exp((E2/1.987)*(1/500-1/T))
k1=3.3*exp((E1/1.987)*(1/300-1/T))
Ca=Cao/(1+tau*k1)
kappa=UA/(vo*Cao)/Cp
G=-tau*k1/(1+k1*tau)*DH1-k1*tau*k2*tau*DH2/((1+tau*k1)*(1+ta
   u*k2))
Tc=(To+kappa*Ta)/(1+kappa)
Cb=tau*k1*Ca/(1+k2*tau)
R=Cp*(1+kappa)*(T-Tc)
Cc=Cao-Ca-Cb
F=G-R
t₀ = 0,    t_f = 225
```

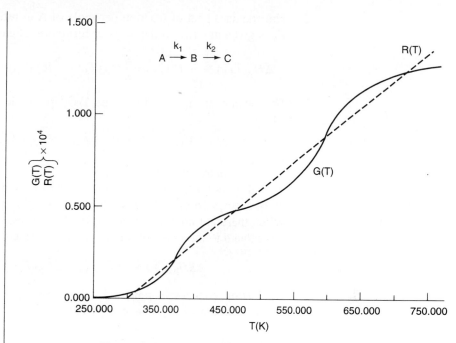

Figure E8-12.1   Heat-removed and heat-generated curves.

We see that five steady states (SS) exist. The exit concentrations and temperatures listed in Table E8-12.2 were interpreted from the tabular output of the POLY-MATH program.

TABLE E8-12.2.   EFFLUENT CONCENTRATIONS AND TEMPERATURES

SS	T	C_A	C_B	C_C
1	310	0.285	0.015	0
2	363	0.189	0.111	0.0
3	449	0.033	0.265	0.002
4	558	0.004	0.163	0.132
5	677	0.001	0.005	0.294

SUMMARY

For the reaction

$$A + \frac{b}{a}B \rightarrow \frac{c}{a}C + \frac{d}{a}D$$

1.   The heat of reaction at temperature T, per mole of A, is

$$\Delta H_{Rx}(T) = \frac{c}{a}H_C(T) + \frac{d}{a}H_D(T) - \frac{b}{a}H_B(T) - H_A(T) \quad \text{(S8-1)}$$

2. The standard heat of reaction per mole of A at reference temperature T_R is given in terms of the heats of formation of each species:

$$\Delta H^\circ_{\text{Rx}}(T_R) = \frac{c}{a} H^\circ_{\text{C}}(T_R) + \frac{d}{a} H^\circ_{\text{D}}(T_R) - \frac{b}{a} H^\circ_{\text{B}}(T_R) - H^\circ_{\text{A}}(T_R) \qquad \text{(S8-2)}$$

3. The mean heat capacity difference, $\Delta \hat{C}_p$, per mole of A is

$$\Delta \hat{C}_p = \frac{c}{a} \hat{C}_{p\text{C}} + \frac{d}{a} \hat{C}_{p\text{D}} - \frac{b}{a} \hat{C}_{p\text{B}} - \hat{C}_{p\text{A}} \qquad \text{(S8-3)}$$

where $\hat{C}_{pi}$ is the mean heat capacity of species i between temperatures T_R and T, not to be confused with $\tilde{C}_{pi}$, which is the mean heat capacity of species i between temperatures T_0 and T.

4. When there are no phase changes, the heat of reaction at temperature T is related to the standard reference heat of reaction by

$$\Delta H_{\text{Rx}}(T) = H^\circ_{\text{Rx}}(T_R) + \Delta \hat{C}_p (T - T_R) \qquad \text{(S8-4)}$$

5. Neglecting changes in potential energy, kinetic energy, and viscous dissipation, the steady-state energy balance is

$$\frac{\dot{Q}}{F_{\text{A0}}} - \frac{\dot{W}_s}{F_{\text{A0}}} - X[\Delta H^\circ_{\text{Rx}}(T) + \Delta \hat{C}_p (T - T_R)] = \sum_{i=1}^{n} \Theta_i \tilde{C}_{pi} (T - T_{i0}) \qquad \text{(S8-5)}$$

where n is the number of species entering the reactor.

If all species enter at the same temperature, $T_{i0} = T_0$, and no work is done on the system, the energy balance reduces to

$$\frac{\dot{Q}}{F_{\text{A0}}} - X[\Delta H^\circ_{\text{Rx}}(T) + \Delta \hat{C}_p (T - T_R)] = \left(\sum_{i=1}^{n} \Theta_i \tilde{C}_{pi} \right) (T - T_0) \qquad \text{(S8-6)}$$

For adiabatic operation of a PFR, PBR, CSTR, or batch reactor

$$T = \frac{X[- \Delta H^\circ_{\text{Rx}}(T_R) + \Sigma \Theta_i C_{pi} T_0 + X \Delta \hat{C}_p T_R]}{[\Sigma \Theta_i \tilde{C}_{pi} + X \Delta \hat{C}_p]}$$

6. The energy balance on a PFR/PBR

$$\frac{dT}{dV} = \frac{Ua(T_a - T) + (-r_{\text{A}})[-\Delta H_{\text{Rx}}(T)]}{\displaystyle\sum_{i=1}^{m} F_i C_{pi}} \qquad \text{(S8-7)}$$

In terms of conversion

$$\frac{dT}{dV} = \frac{Ua(T_a - T) + (-r_{\text{A}})[-\Delta H_{\text{Rx}}(T)]}{F_{\text{A0}}(\Sigma \Theta_i C_{pi} + X \Delta C_p)} \qquad \text{(S8-8)}$$

7. The CSTR energy balance is

$$\frac{UA}{F_{A0}}(T_a - T) - X[\Delta H^{\circ}_{Rx}(T_R) + \Delta \hat{C}_p(T - T_R)] = \Sigma \Theta_i \tilde{C}_{pi}(T - T_{i0}) \qquad \text{(S8-9)}$$

8. The temperature dependence of the specific reaction rate is given in the form

$$k(T) = k_1(T_1) \exp\left[\frac{E}{R}\left(\frac{T - T_1}{TT_1}\right)\right] \qquad \text{(S8-10)}$$

9. The temperature dependence of the equilibrium constant is given by van't Hoff's equation:

$$\frac{d \ln K_p}{dT} = \frac{\Delta H_{Rx}(T)}{RT^2}$$

If $\Delta C_p = 0$,

$$K_p(T) = K_p(T_1) \exp\left[\frac{\Delta H_{Rx}}{R}\left(\frac{1}{T_1} - \frac{1}{T}\right)\right] \qquad \text{(S8-11)}$$

10. Multiple steady states:

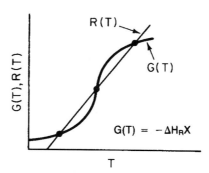

$$G(T) = (-\Delta H_{Rx})\left(\frac{-r_A V}{F_{A0}}\right) = (-\Delta H_{Rx})(X) \qquad \text{(S8-12)}$$

$$R(T) = C_{p0}(1 + \kappa)(T - T_c) \qquad \text{(S8-13)}$$

For an irreversible first-order reaction,

$$G(T) = -\Delta H_{Rx} \frac{\tau A \exp(-E/RT)}{1 + \tau A \exp(-E/RT)}$$

11. The criteria for *Runaway Reactions* is when $(T_r - T_c) > RT_r^2/E$, where T_r is the reactor temperature and $T_c = (T_o + \kappa T_a)/(1 + \kappa)$.

12. Bifurcation analysis (CD-ROM) is used to find multiple steady states. At the bifurcation point, y^*,

$$F(y^*) = 0 = \alpha y^* - \beta - G(y^*) \qquad \text{(S8-14)}$$

$$\left(\frac{dF}{dy}\right)_{y^*} = 0 = \alpha - \frac{dG}{dy}\bigg|_{y^*} \qquad \text{(S8-15)}$$

Multiple steady states will not exist if

$$\max\left(\frac{dG}{dy}\right) < \alpha \qquad \text{(S8-16)}$$

13. When q multiple reactions are taking place and there are m species,

$$\frac{dT}{dV} = \frac{Ua(T_a - T) + \sum\limits_{i=1}^{q} (-r_{ij})[-\Delta H_{\mathrm{R}xij}(T)]}{\sum\limits_{j=1}^{m} F_j C_{pj}} \qquad \text{(S8-17)}$$

ODE SOLVER ALGORITHM

Packed-Bed Reactor with Heat Exchange and Pressure Drop

$$2A \rightleftharpoons C$$

Pure A enters at 5 mol/min at 450 K.

$$\frac{dX}{dW} = \frac{-r_A'}{F_{A0}}$$

$$\frac{dT}{dW} = \frac{UA/\rho_c (T_a - T) + (r_A)(\Delta H_{\mathrm{Rx}}^{\circ})}{C_{p_A} F_{A0}}$$

$$\frac{dy}{dW} = -\frac{\alpha}{2y} (1 - 0.5X)(T/T_0)$$

$$r_A' = -k[C_A^2 - C_C/K_C]$$

$$C_A = C_{A0}[(1 - X)/(1 - 0.5X)](T_0/T)(y)$$

$$C_C = \tfrac{1}{2} C_{A0} X (T_0/T) y/(1 - 0.5X)$$

$$k = 0.5 \, \exp\left[5032((1/450) - (1/T))\right]$$

$$K_C = 25{,}000 \, \exp\left[-10{,}065\left[\left(\frac{1}{450}\right) - \left(\frac{1}{T}\right)\right]\right]$$

$\alpha = 0.019/\text{kg cat.}$

$C_{A0} = 0.25 \ \text{mol/dm}^3$

$UA/\rho_C = 0.8 \ \text{J/kg cat.} \cdot \text{s} \cdot \text{K}$

$T_a = 500 \ \text{K}$

$\Delta H_{\mathrm{Rx}}^{\circ} = -20{,}000 \ \text{J/mol}$

$C_{p_A} = 40 \ \text{J/mol} \cdot \text{K}$

$F_{A0} = 5.0 \ \text{mol/s}$

$T_0 = 450 \ \text{K}$

$W_{\mathrm{f}} = 90 \ \text{kg cat.}$

QUESTIONS AND PROBLEMS

The subscript to each of the problem numbers indicates the level of difficulty: A, least difficult; D, most difficult.

$$A = \bullet \quad B = \blacksquare \quad C = \blacklozenge \quad D = \blacklozenge\blacklozenge$$

In each of the questions and problems below, rather than just drawing a box around your answer, write a sentence or two describing how you solved the problem, the assumptions you made, the reasonableness of your answer, what you learned, and any other facts that you want to include. You may wish to refer to W. Strunk and E. B. White, *The Elements of Style* (New York: Macmillan, 1979) and Joseph M. Williams, *Style: Ten Lessons in Clarity & Grace* (Glenview, Ill.: Scott, Foresman, 1989) to enhance the quality of your sentences. See the Preface for additional generic parts **(x)**, **(y)**, **(z)** to the home problems.

P8-1$_A$ Read over the problems at the end of this chapter. Make up an original problem that uses the concepts presented in this chapter. To obtain a solution:
   (a) Make up your data and reaction.
   (b) Use a real reaction and real data.
   See Problem P4-1 for guidelines.
   (c) Prepare a list of safety considerations for designing and operating chemical reactors. (See http://www.siri.org/graphics)
   See R. M. Felder, *Chem. Eng. Educ., 19*(4), 176 (1985). The August 1985 issue of *Chemical Engineering Progress* may be useful for part (c).

Load the following POLYMATH/MATLAB programs from the CD-ROM:

P8-2$_A$ **What if...**
   (a) you were asked to prepare a list of safety considerations of redesigning and operating a chemical reactor, what would be the first four items on your list?
   (b) you were asked to give an everyday example that demonstrates the principles discussed in this chapter? (Would sipping a teaspoon of Tabasco or other hot sauce be one?)
   (c) the isomerization of butane reaction (Example 8-6) were carried out adiabatically in a 1.0-m^3 pressurized CSTR. What inlet temperature would you recommend between 300 and 600 K?
   (d) you could vary the entering temperature between 300 and 600 K for the butane isomerization in Example 8-6. What inlet temperature would you recommend? Plot the exit conversion as a function of the entering mole fraction of isopentane, keeping the total molar feed rate constant. Describe what you find in each case, noting any maximum in conversion.

   (e) you reconsider production of acetic anhydride in Example 8-7. For the adiabatic case, consider feeding nitrogen ($C_{p_N} = 6.25 + 8.78 \times 10^{-3}T - 2.1 \times 10^{-8}T^2$) along with acetone. Plot the conversion as a function of the mole fraction of nitrogen for the same total molar feed rate. Explain why your curve looks the way it does.
   (f) you reconsider Example 8-5 and then make a plot exit conversion as a function of F_{A0}. What would happen if the heat-exchange area were reduced to 4 ft^2 and the entering temperature decreased to 531°R?
   (g) for nonadiabatic operation of the acetone cracking in Example 8-7, for the same reactor volume, you made plots of the exit conversion and the temperature profiles as a function of reactor diameter, D (recall that

$a = 4/D$). Do you think there is a point at which some of the assumptions in your equations break down?

(h) you kept the total catalyst weight constant but varied the catalyst particle size in Example 8-10. At what entering temperature does pressure drop have a significant effect on the exit conversion?

(i) you were able to increase Ua in Example 8-11 by a factor of 5 or 10? What would be the effect on the selectivity S_{BC}?

(j) you were able to vary T_a between 0 and 200°C in Example 8-12. What would a plot of the reactor temperature versus T_a look like? (*Hint:* Compare with Figure 8-24.)

(k) you were to apply the runaway criteria [Equation (8-77)] to the CSTR in which the butane isomerization (Example 8-6) was taking place? At what inlet temperature would it run away? Under what condition would the SO_2 oxidation run away if it were not safe to exceed a reaction temperature of 1400°C?

P8-3$_C$  The following is an excerpt from *The Morning News*, Wilmington, Delaware (August 3, 1977): "Investigators sift through the debris from blast in quest for the cause [that destroyed the new nitrous oxide plant]. A company spokesman said it appears more likely that the [fatal] blast was caused by another gas—ammonium nitrate—used to produce nitrous oxide." An 83% (wt) ammonium nitrate and 17% water solution is fed at 200°F to the CSTR operated at a temperature of about 510°F. Molten ammonium nitrate decomposes directly to produce gaseous nitrous oxide and steam. It is believed that pressure fluctuations were observed in the system and as a result the molten ammonium nitrate feed to the reactor may have been shut off approximately 4 min prior to the explosion. **(a)** Can you explain the cause of the blast? [*Hint:* See P9-3 page 574 and Equation (8-77).] **(b)** If the feed rate to the reactor just before shutoff was 310 lb of solution per hour, what was the exact temperature in the reactor just prior to shutdown? **(c)** How would you start up or shut down and control such a reaction? **(d)** What do you learn when you apply the runaway reaction criteria?

Assume that at the time the feed to the CSTR stopped, there was 500 lb of ammonium nitrate in the reactor. The conversion in the reactor is virtually complete at about 99.99%.

Additional information (approximate but close to the real case):

$$\Delta H_{Rx}^{\circ} = -336 \text{ Btu/lb ammonium nitrate at } 500°F \text{ (constant)}$$

$$C_p = 0.38 \text{ Btu/lb ammonium nitrate} \cdot °F$$

$$C_p = 0.47 \text{ Btu/lb of steam} \cdot °F$$

$$-r_A V = k C_A V = k \frac{M}{V} V = kM(\text{lb/h})$$

where M is the mass of ammonium nitrate in the CSTR (lb) and k is given by the relationship below.

T (°F)	510	560
k (h^{-1})	0.307	2.912

The enthalpies of water and steam are

$$H_1(200°F) = 168 \text{ Btu/lb}$$

$$H_g(500°F) = 1202 \text{ Btu/lb}$$

(e) Explore this problem and describe what you find. (For example, can you plot a form of $R(T)$ versus $G(T)$?) **(f)** Discuss what you believe to be the point of the problem. The idea for this problem originated from an article by Ben Horowitz.

P8-4$_B$ The endothermic liquid-phase elementary reaction

$$A + B \rightarrow 2C$$

proceeds, substantially, to completion in a single steam-jacketed, continuous-stirred reactor (Table P8-4). From the following data, calculate the steady-state reactor temperature:

Reactor volume: 125 gal
Steam jacket area: 10 ft²
Jacket steam: 150 psig (365.9°F saturation temperature)
Overall heat-transfer coefficient of jacket, U: 150 Btu/h·ft²·°F
Agitator shaft horsepower: 25 hp
Heat of reaction, $\Delta H^{\circ}_{Rx} = +20{,}000$ Btu/lb mol of A (independent of temperature)

TABLE P8-4

	Component		
	A	B	C
Feed (mol/h)	10.0	10.0	0
Feed temperature (°F)	80	80	—
Specific heat (Btu/lb mol·°F*)	51.0	44.0	47.5
Molecular weight	128	94	—
Density (lb/ft³)	63.0	67.2	65.0

*Independent of temperature.
(Ans: $T = 199$°F)
(Courtesy of the California Board of Registration for Professional & Land Surveyors.)

P8-5$_A$ The elementary irreversible organic liquid-phase reaction

$$A + B \rightarrow C$$

is carried out adiabatically in a flow reactor. An equal molar feed in A and B enters at 27°C, and the volumetric flow rate is 2 dm³/s.
(a) Calculate the PFR and CSTR volumes necessary to achieve 85% conversion.
(b) What is the maximum inlet temperature one could have so that the boiling point of the liquid (550 K) would not be exceeded even for complete conversion?
(c) Plot the conversion and temperature as a function of PFR volume (i.e., distance down the reactor).
(d) Calculate the conversion that can be achieved in one 500-dm³ CSTR and in two 250-dm³ CSTRs in series.
(e) Vary the activation energy $1000 < E < 30{,}000$ (cal/mol) and heat of reaction $2000 < |\Delta H_{Rx}| < 25{,}000$ (cal/mol) to learn their effect on the PFR conversion profile.
(f) Discuss the application of the runaway reaction criteria. What value of T_c would you use to prevent runaway if $\kappa = 3$ at $T_a = 300$ K?

(g) What do you believe to be the point of this problem?
(h) Ask another question or suggest another calculation for this reaction.

Additional information:

$$H_A^\circ (273) = -20 \text{ kcal/mol}$$
$$H_B^\circ (273) = -15 \text{ kcal/mol}$$
$$H_C^\circ (273) = -41 \text{ kcal/mol}$$
$$C_{A0} = 0.1 \text{ kmol/m}^3$$
$$C_{P_A} = C_{P_B} = 15 \text{ cal/mol} \cdot \text{K} \qquad C_{P_C} = 30 \text{ cal/mol} \cdot \text{K}$$
$$k = 0.01 \frac{\text{dm}^3}{\text{mol} \cdot \text{s}} \text{ at } 300 \text{ K} \qquad E = 10{,}000 \text{ cal/mol}$$

P8-6$_A$ The elementary irreversible gas-phase reaction

$$A \rightarrow B + C$$

is carried out adiabatically in a PFR packed with a catalyst. Pure A enters the reactor at a volumetric flow rate of 20 dm³/s at a pressure of 10 atm and a temperature of 450 K.

(a) Plot the conversion and temperature down the plug-flow reactor until an 80% conversion (if possible) is reached. (The maximum catalyst weight that can be packed into the PFR is 50 kg.) Assume that $\Delta P = 0.0$.
(b) What catalyst weight is necessary to achieve 80% conversion in a CSTR?
(c) What do you believe to be the point of this problem?

Additional information:

$$C_{P_A} = 40 \text{ J/mol} \cdot \text{K} \qquad C_{P_B} = 25 \text{ J/mol} \cdot \text{K} \qquad C_{P_C} = 15 \text{ J/mol} \cdot \text{K}$$
$$H_A^\circ = -70 \text{ kJ/mol} \qquad H_B^\circ = -50 \text{ kJ/mol} \qquad H_C^\circ = -40 \text{ kJ/mol}$$

All heats of formation are referenced to 273 K.

$$k = 0.133 \exp\left[\frac{E}{R}\left(\frac{1}{450} - \frac{1}{T}\right)\right] \frac{\text{dm}^3}{\text{kg} \cdot \text{cat} \cdot \text{s}} \text{ with } E = 31.4 \text{ kJ/mol}$$

P8-7$_B$ Repeat Problem P8-6 by taking pressure drop into account in the PFR. The pressure-drop coefficient is

$$\frac{dP}{dW} = -\frac{\alpha}{2}\left(\frac{T}{T_0}\right)\frac{P_0}{(P/P_0)}(1 + \varepsilon X)$$

The reactor can be packed with one of two particle sizes. Choose one.

$$\alpha = 0.019/\text{kg cat. for particle diameter } D_1$$
$$\alpha = 0.0075/\text{kg cat. for particle diameter } D_2$$

(a) Plot the temperature, conversion, and pressure along the length of the reactor.
(b) Vary the parameters α and P_0 to learn the ranges of values in which they dramatically affect the conversion.

P8-8$_B$ Rework Problem P8-6 for the case when heat is removed by a heat exchanger jacketing the reactor. The flow rate of coolant through the jacket is sufficiently high that the ambient exchanger temperature is constant at 50°C.

(a) For the PBR:

$$\frac{Ua}{\rho_b} = 0.8 \, \frac{J}{s \cdot kg \, cat. \cdot K}$$

where

ρ_b = bulk density of the catalyst (kg/m³)

a = heat-exchange area per unit volume of reactor (m²/m³)

U = overall heat-transfer coefficient (J/s · m² · K)

What if UA/ρ_b were increased by a factor of 25?

(b) Make an estimate of the minimum entering temperature at which the reaction will "ignite" (i.e., achieve a reasonable conversion for a reasonable amount of catalyst).

(c) Find X and T for a "fluidized" CSTR with 80 kg of catalyst.

$$UA = 500 \, \frac{J}{s \cdot K}, \qquad \rho_b = 1 \, kg/m^3$$

(d) Repeat part (a) for $W = 80.0$ kg assuming a reversible reaction with a reverse specific reaction rate of

$$k_r = 0.2 \, \exp\left[\frac{E_r}{R}\left(\frac{1}{450} - \frac{1}{T}\right)\right]\left(\frac{dm^6}{kg \, cat. \cdot mol \cdot s}\right); \qquad E_r = 51.4 \, kJ/mol$$

(e) Carry out a series of computer experiments in which you vary (Ua/ρ_b) between values of 0.1 and 20 (J/s · kg cat. · K). In addition, study the effect of ambient temperatures (northern winter–summer operation) on your results. Write a paragraph describing your findings.

(f) What problems might you encounter if you were to include pressure drop (cf. Problem P8-7)?

(g) Use or modify the data in this problem to suggest another question or calculation.

P8-9$_B$ The formation of styrene from vinylacetylene is essentially irreversible and follows an elementary rate law:

$$2 \, vinylacetylene \rightarrow styrene$$

$$2A \rightarrow B$$

(a) Determine the conversion achieved in a 10-dm³ PFR for an entering temperature of 675 K. Plot the temperature and conversion down the length (volume) of the reactor.

(b) Vary the entering temperature and plot the conversion as a function of entering temperature.

(c) Vary the ambient temperature in the heat exchanger and find the maximum ambient temperature at which runaway will not occur in the reactor.

(d) Compare your answers with the case when the reaction is carried out adiabatically.

(e) Repeat part (b) assuming that the reaction is reversible, $K_C = 100,000$ at 675 K. In addition, a stream of inerts ($C_{p_I} = 100 \, J/mol \cdot °C$) with $F_I =$

$3F_{A0}$ enters the reactor. Plot the conversion as a function of the entering temperature. Is there a maximum? If so, why? If not, why not?

(f) Ask another question or suggest another calculation that can be made for this problem.

Additional information:

$$C_{A0} = 1 \text{ mol/dm}^3 \qquad\qquad\qquad Ua = 5.0 \text{ kJ/s} \cdot \text{dm}^3/\text{K}$$

$$F_{A0} = 5 \text{ mol/s} \qquad\qquad\qquad\qquad T_a = 700 \text{ K}$$

$$\Delta H_{Rx} = -231 - 0.012(T - 298) \text{ kJ/mol}$$

$$C_{P_A} = 0.1222 \text{ kJ/mol} \cdot \text{K}$$

$$k = 1.48 \times 10^{11} \exp(-19{,}124/T) \text{ dm}^3/\text{mol} \cdot \text{s}$$

$$T_0 = 675 \text{ K}$$

[Lundgard, *Int. J. Chem Kinet.*, *16*, 125 (1984).]

P8-10$_B$ The irreversible endothermic vapor-phase reaction follows an elementary rate law:

$$CH_3COCH_3 \rightarrow CH_2CO + CH_4$$

$$A \rightarrow B + C$$

and is carried out adiabatically in a 500-dm³ PFR. Species A is fed to the reactor at a rate of 10 mol/min and a pressure of 2 atm. An inert stream is also fed to the reactor at 2 atm. The entrance temperature of both streams is 1100 K.

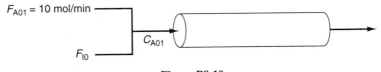

F_{A01} = 10 mol/min

C_{A01}

F_{I0}

Figure P8-10

(a) First derive an expression for C_{A01} as a function of C_{A0} and Θ_I.

(b) Plot the exit conversion as a function of Θ_I.

(c) Is there a ratio of inerts to the entering molar flow rate of A (i.e., $\Theta_I = F_{I0}/F_{A0}$) at which the conversion is at a maximum? Explain why there "is" or "is not" a maximum.

(d) How would your answer change if the entering temperature were increased or decreased by 200 K?

(e) What do you believe to be the point of this problem?

Additional information:

$$k = \exp(34.34 - 34{,}222/T) \text{ dm}^3/\text{mol} \cdot \text{min} \qquad C_{pI} = 200 \text{ J/mol} \cdot \text{K}$$

(T in degrees Kelvin)

$$C_{pA} = 170 \text{ J/mol} \cdot \text{K} \qquad\qquad\qquad C_{pB} = 90 \text{ J/mol} \cdot \text{K}$$

$$C_{pC} = 80 \text{ J/mol} \cdot \text{K} \qquad\qquad\qquad \Delta H_{Rx} = 80{,}000 \text{ J/mol}$$

P8-11$_C$ Derive the energy balance for a packed bed membrane reactor. Apply the balance to the reaction in Problem P8-6 for the case when it is reversible as described in Problem P8-8(d). Species C diffuses out of the membrane.
  (a) Plot the concentration profiles or different values of k_c when the reaction is carried out adiabatically.
  (b) Repeat part (a) when the heat transfer coefficient is the same as that given in P8-8(a). All other conditions are the same as those in Problem P8-6.

P8-12$_B$ The elementary irreversible *exothermic* gas phase reaction

$$A \rightarrow 2B$$

is carried out in a packed bed reactor. There is pressure drop in the reactor and the pressure-drop coefficient [(see Equation 4-32)] is 0.007 kg^{-1}. Pure A enters the reactor at a flow rate of 5 mol/s, at a concentration of 0.25 mol/dm^3, a temperature of 450 K, and a pressure of 9.22 atm. Heat is removed by a heat exchanger jacketing the reactor. The coolant flow rate in the jacket is sufficient to maintain the ambient temperature of the heat exchanger at 27°C. The term giving the product of the heat-transfer coefficient and area per unit volume divided by the bulk catalyst density is

$$\frac{Ua}{\rho_b} = \frac{5 \text{ J}}{\text{kg cat.} \cdot \text{s} \cdot \text{K}}$$

where

ρ_b = bulk density of catalyst (kg/m^3)

a = heat-exchange area per unit volume of reactor (m^2/m^3)

U = overall heat-transfer coefficient $\left(\dfrac{J}{\text{m}^2 \cdot \text{s} \cdot \text{K}} \right)$

The maximum weight of catalyst that can be packed in this reactor is 50 kg.
  (a) Plot the temperature, conversion X, and the pressure ratio $(y = P/P_0)$ as a function of catalyst weight.
  (b) At what catalyst weight down the reactor does the rate of reaction $(-r_A')$ reach its maximum value?
  (c) At what catalyst weight down the reactor does the temperature reach its maximum value?
  (d) What happens when the heat-transfer coefficient is doubled? What happens if the heat coefficient is halved? Discuss your observations on the effects on reactor performance (i.e., conversion, temperature, and pressure drop).

Additional information:

$$\Delta H_{Rx} = -20{,}000 \text{ J/mol A at 273 K}$$

$$C_{P_A} = 40 \text{ J/mol} \cdot \text{K} \quad C_{P_B} = 20 \text{ J/mol} \cdot \text{K}$$

$$k = \exp\left[\frac{E}{R}\left(\frac{1}{450} - \frac{1}{T} \right) \right] \frac{\text{dm}^3}{\text{kg} \cdot \text{cat.} \cdot \text{s}}$$

with

$$E = 31.4 \frac{\text{kJ}}{\text{mol}} \qquad R = 8.314 \frac{\text{J}}{\text{mol} \cdot \text{K}}$$

Hence $E/R = 3776.76$ K.

P8-13$_A$ The adiabatic equilibrium conversion for the reaction

$$A + B \; \rightleftharpoons \; C + D$$

is carried out in a series of staged packed-bed reactors with interstage cooling. The lowest temperature to which the reactant stream may be cooled is 27°C. The feed is equal molar in A and B and the catalyst weight in each reactor is sufficient to achieve 99.9% of the equilibrium conversion. The feed enters at 27°C and the reaction is carried out adiabatically. If four reactors and three coolers are available, what conversion may be achieved?

Additional information:

$$\Delta H_{\text{Rx}} = -30,000 \text{ cal/mol A} \qquad C_{P_A} = C_{P_B} = C_{P_C} = C_{P_D} = 25 \text{ cal/g mol} \cdot \text{K}$$

$$K_e(50°C) = 500,000 \qquad F_{A0} = 10 \text{ mol A/min}$$

First prepare a plot of equilibrium conversion as a function of temperature. [*Partial ans.:* $T = 360$ K, $X_e = 0.984$; $T = 520$ K, $X_e = 0.09$; $T = 540$ K, $X_e = 0.057$]

P8-14$_A$ Figure 8-8 shows the temperature–conversion trajectory for a train of reactors with interstage heating. Now consider replacing the interstage heating with injection of the feed stream in three equal portions as shown below:

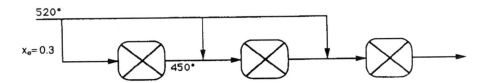

Sketch the temperature–conversion trajectories for **(a)** an endothermic reaction with entering temperatures as shown, and **(b)** an exothermic reaction with the temperatures to and from the first reactor reversed.

P8-15$_B$ The elementary reversible gas-phase reaction

$$A + B \; \rightleftharpoons \; 2C$$

is to be carried out in a PFR and a CSTR. The feed contains only A and B in stoichiometric proportions at 580.5 kPa and 77°C. The molar feed rate of A is 20 mol/s. The reaction is carried out adiabatically.

(a) Determine the plug-flow reactor volume necessary to achieve 85% of the adiabatic equilibrium conversion.

(b) Plot $-r_A$, X, and T as a function of reactor length if the cross-sectional area is 0.01 m^2. What do the plots suggest about changing the feed conditions?

(c) Determine the conversion that can be achieved in a 1.5-m^3 CSTR with a heat exchanger mounted inside.

(d) Repeat parts (a) and (c) for the case of an endothermic reaction with a $\Delta H^{\circ}_{\text{Rx}}$ of the same absolute magnitude but with an entering temperature of 277°C.

(e) Comment on the magnitudes of $\Delta H^{\circ}_{\text{Rx}}$, E, K_C, and U. When (or under what conditions) might "runaway" occur?

(f) Use a software package to study the effect of the ambient temperature T_a (winter–summer) on part (c).

Additional information:

Rate-law parameters:

$$k = 0.035 \text{ dm}^3/\text{mol} \cdot \text{min at 273 K}$$
$$E = 70,000 \text{ J/mol}$$

Thermodynamic parameters at 25°C:

$$H_A^\circ = -40 \text{ kJ/mol} \quad C_{P_A} = 25 \text{ J/mol} \cdot \text{K}$$
$$H_B^\circ = -30 \text{ kJ/mol} \quad C_{P_B} = 15 \text{ J/mol} \cdot \text{K}$$
$$H_C^\circ = -45 \text{ kJ/mol} \quad C_{P_C} = 20 \text{ J/mol} \cdot \text{K}$$
$$K_C = 25,000$$

Heat-exchanger data:

Overall heat-transfer coefficient = 10 W/m² · K

Exchanger area = 2 m²

Ambient temperature = 17°C

P8-16$_B$ The elementary liquid-phase reaction

$$A + B \rightleftarrows C$$

is to be carried out adiabatically in a set of tubular reactors packed with a resin catalyst and arranged in series. The feed consists of 25 mol % A, 25 mol % B, and 50 mol % inerts at 300 K. There are available to you a large number of interstage heat exchangers that can cool the reaction stream to temperatures as low as 300 K. The reactors have sufficient catalyst to reach 95% of the corresponding equilibrium conversion. What is the smallest number of reactors in series necessary to obtain an exiting conversion above 50%? What would you need to do to increase the conversion to 95%? What concerns would you have about reaching 95% conversion?

Additional information:

$$C_{P_A} = 2.0 \text{ cal/g mol} \cdot °C \quad C_{P_C} = 7.0 \text{ cal/g mol} \cdot °C$$
$$C_{P_B} = 5.0 \text{ cal/g mol} \cdot °C \quad C_{P_I} = 1.5 \text{ cal/g mol} \cdot °C$$
$$\Delta H_{Rx}^\circ (273 \text{ K}) = -10,000 \text{ cal/g mol A}$$
$$F_{A0} = 10 \text{ mol/min}$$

T (K)	350	400	450	500	550	600	650	700
X_e	1.00	0.97	0.74	0.4	0.2	0.1	0.06	0.04

P8-17$_B$ The first-order irreversible exothermic liquid-phase reaction

$$A \rightarrow B$$

is to be carried out in a jacketed CSTR. Species A and an inert I are fed to the reactor in equilmolar amounts. The molar feed rate of A is 80 mol/min.
(a) What is the reactor temperature for a feed temperature of 450 K?

(b) Plot the reactor temperature as a function of the feed temperature.

(c) To what inlet temperature must the fluid be preheated for the reactor to operate at a high conversion? What are the corresponding temperature and conversion of the fluid in the CSTR at this inlet temperature?

(d) Suppose that the fluid is now heated 5°C above the temperature in part (c) and then cooled 20°C, where it remains. What will be the conversion?

(e) What is the inlet extinction temperature for this reaction system? (*Ans.:* $T_0 = 87°C$.)

Additional information:

Heat capacity of the inert: 30 cal/g mol·°C

Heat capacity of A and B: 20 cal/g mol·°C

UA: 8000 cal/min·°C

Ambient temperature, T_a: 300 K

$\tau = 100$ min

$\Delta H_{Rx} = -7500$ cal/mol

$k = 6.6 \times 10^{-3}$ min^{-1} at 350 K

$E = 40{,}000$ cal/mol·K

P8-18$_A$ Radial flow reactors are used to help eliminate hot spots in highly exothermic reactions. The velocity is highest at the inlet and then decreases as $1/r$ as the fluid moves away from the inlet. The overall heat-transfer coefficient varies with the square root of the radial velocity:

$$U = U(r = r_0)\left(\frac{\text{velocity at } r_0}{\text{velocity at } r}\right)^{1/2}$$

and at the inlet $U = U(r = r_0) = 100$ Btu/h·ft^2·°F.

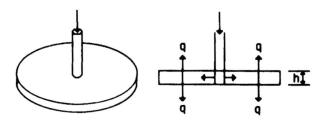

Figure P8-18   Radial flow reactor.

(a) Rework Problem P8-6 for a radial reactor.

(b) Rework Problem P8-12 for a radial reactor.

(c) Consider the flow conditions to one of the tubes for the SO$_2$ oxidation described in Example 8-10. Replace the tube with a radial flow reactor 1 cm in height with an inlet diameter of 0.5 cm. The reactor is immersed in the same boiling liquid as in Example 8-10. Plot the temperature and conversion as a function of radius and catalyst weight for three different inlet temperatures. Study the behavior of this reactor by varying a number of parameters, such as flow rate and gas composition.

(d) The reaction discussed in Problem P8-15 is to be carried out in a single plug-flow reactor immersed in a coolant and in a radial flow reactor immersed in the same coolant. The tubular flow reactor is 2 ft in diameter. The height of the radial flow reactor is $\frac{1}{2}$ in. and the diameter of the inlet is 1 in. In both cases the coolant temperature is 100°C and the overall heat-transfer coefficient is 2000 J/ m² · h · °C . Plot conversion and temperatures as a function of reactor radius.

P8-19$_C$ The zero-order exothermic liquid-phase reaction

$$A \rightarrow B$$

is carried out at 85°C in a jacketed 0.2-m³ CSTR. The coolant temperature in the reactor is 32°F. The heat transfer coefficient is 120 W/ m² · K . Determine the critical value of the heat-transfer area below which the reactor will run away and explode [*Chem. Eng.*, *91*(10), 54 (1984)].

Additional information:

Specific reaction rate:

$$k = 1.127 \text{ kmol/m}^3 \cdot \text{min at } 40°C$$

$$k = 1.421 \text{ kmol/m}^3 \cdot \text{min at } 50°C$$

The heat capacity of the solution is 4 J/g. The solution density is 0.90 kg/dm³. The heat of reaction is −500 J/g. The feed temperature is 40°C and the feed rate is 90 kg/min. MW of A = 90 g/mol.

P8-20$_B$ The elementary reversible liquid-phase reaction

$$A \overset{\longrightarrow}{\longleftarrow} B$$

takes place in a CSTR with a heat exchanger. Pure A enters the reactor.
(a) Derive an expression (or set of expressions) to calculate $G(T)$ as a function of heat of reaction, equilibrium constant, temperature, and so on. Show a sample calculation for $G(T)$ at $T = 400$ K.
(b) What are the steady-state temperatures? (*Ans.:* 310, 377, 418 K)
(c) Which steady states are locally stable?
(d) What is the conversion corresponding to the upper steady state?
(e) Vary the ambient temperature T_a and make a plot of the reactor temperature as a function of T_a, identifying the ignition and extinction temperatures.
(f) If the heat exchanger in the reactor suddenly fails (i.e., $UA = 0$), what would be the conversion and the reactor temperature when the new upper steady state is reached? (*Ans.:* 431 K)
(g) What is the adiabotic blow out flow rate, v_0.
(h) Suppose that you want to operate at the lower steady state. What parameter values would you suggest to prevent runaway?

Additional information:

$UA = 3600$ cal/min · K	$E/R = 20{,}000$ K
$C_{P_A} = C_{P_B} = 40$ cal/mol · K	$V = 10$ dm³
$\Delta H_{Rx} = -80{,}000$ cal/mol A	$v_0 = 1$ dm³/min
$K_{eq} = 100$ at 400 K	$F_{A0} = 10$ mol/min
$k = 1$ min^{-1} at 400 K	
Ambient temperature, $T_a = 37°C$	Feed temperature, $T_0 = 37°C$

P8-21$_C$ The first-order irreversible liquid-phase reaction

$$A \rightarrow B$$

is to be carried out in a jacketed CSTR. Pure A is fed to the reactor at a rate of 0.5 g mol/min. The heat-generation curve for this reaction and reactor system,

$$G(T) = \frac{-\Delta H^{\circ}_{Rx}}{1 + 1/\tau\kappa}$$

is shown in Figure P8-21.

(a) To what inlet temperature must the fluid be preheated for the reactor to operate at a high conversion? (*Ans.:* $T_0 \geq 214°C$.)

(b) What is the corresponding temperature of the fluid in the CSTR at this inlet temperature? (*Ans.:* $T_s = 164°C$, $184°C$.)

(c) Suppose that the fluid is now heated 5°C above the temperature in part (a) and then cooled 10°C, where it remains. What will be the conversion? (*Ans.:* $X = 0.9$.)

(d) What is the extinction temperature for this reaction system? (*Ans.:* $T_0 = 200°C$.)

Additional information:

     Heat of reaction (constant): -100 cal/g mol A

     Heat capacity of A and B: 2 cal/g mol · °C

     UA: 1 cal/min · °C

     Ambient temperature, T_a: 100°C

MSS

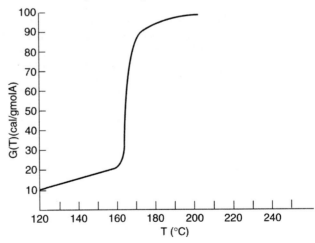

Figure P8-21 $G(T)$ curve.

P8-22$_A$ The reversible elementary reaction

$$A \rightleftharpoons B$$

is carried out in a CSTR. Plot the heat-generated and heat-removed curves on the same graph.

(a) How many multiple steady states are there?
(b) What is the effect of changing air temperatures, T_a (winter–summer) on multiple steady states?

Additional information:

$$\Delta H^\circ_{Rx} = -80 \text{ kJ/mol A} \quad F_{A0} = 10 \text{ mol/s} \quad C_{A0} = 2 \text{ mol/dm}^3$$

$$UA = 2000 \ \frac{\text{J}}{\text{K} \cdot \text{s}} \qquad\qquad V = 500 \text{ dm}^3$$

$$T_a = 40°\text{C} \qquad\qquad k = 0.001 \text{ s}^{-1} \text{ at 373 K}$$

$$C_{P_A} = C_{P_B} = 40 \text{ J/mol} \cdot \text{K} \qquad\qquad E = 150 \text{ kJ/mol}$$

$$K_e = 100 \text{ at 350 K} \qquad\qquad T_0 = 27°\text{C}$$

(c) Make a plot of the reactor temperature T as a function of inlet temperature T_0. What are the ignition and extinction temperatures?
(d) Repeat parts (a), (b), and (c) for the case when the reaction is irreversible with $K_e = \infty$.

P8-23$_B$ The vapor-phase cracking of acetone is to be carried out adiabatically in a bank of 1000 1-in. schedule 40 tubes 10 m in length. The molar feed rate of acetone is 6000 kg/h at a pressure of 500 kPa. The maximum feed temperature is 1050 K. Nitrogen is to be fed together with the acetone to provide the sensible heat of reaction. Determine the conversion as a function of nitrogen feed rate (in terms of Θ_{N_2}) for
(a) Fixed total molar flow rate.
(b) Molar flow rate increasing with increasing Θ_{N_2}.

P8-24$_C$ This chapter neglected radial variations in temperature and concentration.
(a) Use a shell balance on the segment shown in Figure P8-24a to arrive at the steady energy and mole balances that account for axial variations in concentration and both radial and axial variations in temperature.

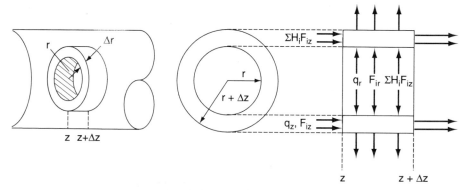

Figure P8-24a   Shell balance.

Show that

$$-\frac{\partial q_z}{\partial z} - \frac{1}{r}\frac{\partial(r q_r)}{\partial r} - \sum_{i=1}^{m} u C_i C_{pi} \frac{\partial T}{\partial z} + r_A \Delta H_{Rx} = 0 \qquad \text{(P8-24.1)}$$

where u is the superficial velocity (m/s) and q is the heat flux (J/m$^2 \cdot$ s).

(b) Let k_e be the effective thermal conductivity in both the axial and radial directions and use Fourier's law,

$$q_z = -k_e \frac{\partial T}{\partial z} \quad \text{and} \quad q_r = -k_e \frac{\partial T}{\partial r}$$

to show that

$$k_e \frac{\partial^2 T}{\partial z^2} + k_e \frac{\partial^2 T}{\partial r^2} + \frac{k_e}{r} \frac{\partial T}{\partial r} - u \sum_{i=1}^{m} C_i C_{pi} \frac{\partial T}{\partial z} + r_A(C_i, T)\, \Delta H_{Rx} = 0 \quad \text{(P8-24.2)}$$

(c) Explain the use of the following boundary conditions:

$$\text{At } r = R, \text{ then } U(T_R(z) - T_a) = -k_e \frac{\partial T}{\partial r} \qquad \text{(P8-24.3)}$$

$$\text{At } r = 0, \text{ then } \frac{\partial T}{\partial r} = 0 \qquad \text{(P8-24.4)}$$

(d) At the entrance and exit of the reactor, there are different boundary conditions that can be used depending on the degree of sophistication required. Explain why the simplest (see Chapter 14) set is

$$\text{At } z = 0, \quad T = T_0 \text{ and } C_i = C_{i0} \qquad \text{(P8-24.5)}$$

$$\text{At } z = L, \quad \frac{\partial T}{\partial z} = 0 \text{ and } \frac{\partial C_i}{\partial z} = 0 \qquad \text{(P8-24.6)}$$

One can usually neglect conduction in the axial directions with respect to convection.

(e) Sketch the radial temperature gradients down a PFR for
    (1) An exothermic reaction.
    (2) An endothermic reaction.
    (3) An exothermic reaction carried out adiabatically.

(f) Use the data in Problem P8-7 to account for radial temperature gradients in the reactor.

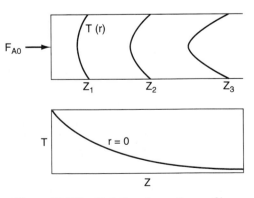

Figure P8-24b   Endothermic reaction profiles.

Additional information:

$$k_e = 0.4 \; \frac{\text{cal}}{\text{cm} \cdot \text{min} \cdot \text{K}}$$

Reactor diameter = 3 cm

Plot temperature reaction rate as a function of radius at specific locations down the reactor. Plot the centerline temperature as a function of conversion.

(g) What do you think about neglecting radial variations in concentration? Read Chapter 11, then show that

$$k_e \frac{\partial^2 T}{\partial z^2} + k_e \frac{\partial^2 T}{\partial r^2} + \frac{k_e}{r} \frac{\partial T}{\partial r} - u \sum_{j=1}^{m} C_j C_{pj} \frac{\partial T}{\partial z}$$

$$+ \left(\sum_{j=1}^{m} C_j D_e \frac{\partial C_j}{\partial r} \right) \frac{\partial T}{\partial r} + r_A(C_j, T) \, \Delta H_{\text{Rx}} = 0 \quad \text{(P8-24.7)}$$

where D_e is the effective diffusivity. *(Experts only)*

P8-25$_C$ A reaction is to be carried out in the packed-bed reactor shown in Figure P8-25.

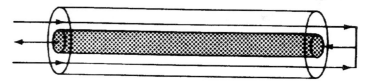

Figure P8-25

The reactants enter in the annular space between an outer insulated tube and an inner tube containing the catalyst. No reaction takes place in the annular region. Heat transfer between the gas in this packed-bed reactor and the gas flowing counter currently in the annular space occurs along the length of the reactor. The overall heat-transfer coefficient is 5 W/m² · K. Plot the conversion and temperature as a function of reactor length for the data given in

(a) Problem P8-5.

(b) Problem P8-6.

(c) Problem P8-15.

P8-26$_B$ *(Position of reaction front)* Using the computer program (or a slight variation thereof) given in Example 8-10, determine the effect on the temperature and conversion profiles of varying the inlet temperature when the reaction is carried out adiabatically. In particular, study the effect of the inlet temperature on the position of the reaction front (i.e., the position where the temperature rises very steeply).

P8-27$_C$ *(Optimum inlet temperature for sulfur dioxide oxidation)* Using the same data, feed flow rate, and composition as given in Example 8-10, determine the inlet temperature that will give the maximum conversion of SO_2 in an existing tubular reactor containing 2500 lb of V_2O_5 that is operated adiabatically. *(Hint: Only a slight modification of the computer program shown in Section 8.10 may be necessary to find this optimum T_0.)*

P8-28$_B$ You are an engineer who is to design a CSTR for the elementary consecutive gas-phase reactions

$$A \xrightarrow{k_1} B \xrightarrow{k_2} C$$

The feed conditions and desired product specifications are known, together with the temperature of the heating medium. It is your job to design the reactor, that is, to specify the reactor volume and the area of the heating coil inside the reactor.

(a) Calculate the desired operating temperature inside the reactor. (*Ans.:* 269°F.)

(b) Calculate the volume of the reactor. (*Ans.:* 6.58 ft³.)

(c) Calculate the area of the heating coil. (*Ans.:* 18.1 ft².)

(d) Find a set of conditions that will give multiple steady states if reaction 1 is exothermic rather than endothermic. How many steady states can you find? (*Ans.:* 5.) Make a plot of $G(T)$ vs. $R(T)$ and of T_s vs. T_0.

Additional information:

    Product:

        The ratio C_B/C_C is equal to 10.

        50% of A in the feed is converted.

    Feed:

        The feed is gas-phase and pure A.

        The molar flow rate is 0.05 lb mol/s.

        The volumetric flow rate is 7.85 ft³/s.

        The temperature is 400°F.

        The pressure in the reactor is 4 atm.

    Heat capacities:

        C_p's of A, B, and C are all 25 Btu/lb mol · °F.

Reaction 1:	Reaction 2:
$A_1 = 2 \times 10^9 \text{ s}^{-1}$	$A_2 = 1 \times 10^{11} \text{ s}^{-1}$
$E_1 = 31,000$ Btu/lb mol	$E_2 = 40,000$ Btu/lb mol
$\Delta H_{Rx1} = 15,000$ Btu/lb mol A reacted	$\Delta H_{Rx2} = -20,000$ Btu/lb mol C reacted

    Heat transfer:

        The heating medium is saturated high-pressure steam at 350°F.

        The overall heat-transfer coefficient between the heating medium and the reaction mixture is 400 Btu/h · ft² · °F.

P8-29$_C$ (*Multiple reactions with heat effects*) Xylene has three major isomers, *m*-xylene, *o*-xylene, and *p*-xylene. When *o*-xylene is passed over a Cryotite catalyst, the following elementary reactions are observed:

Application Pending for Problem Hall of Fame

The feed to the reactor is equal molar in in both *m*-xylene and *o*-xylene (species A and B). For a total feed rate of 2 mol/min and the reaction conditions below, plot the temperature and the molar flow rates of each species as a function of catalyst weight up to a weight of 100 kg.

(a) Find the lowest concentration of *o*-xylene achieved in the reactor.

(b) Find the highest concentration of *m*-xylene achieved in the reactor.

(c) Find the maximum concentration of *o*-xylene in the reactor.

(d) Repeat parts (a) to (d) for a pure feed of *o*-xylene.

(e) Vary some of the system parameters and describe what you learn.

(f) What do you believe to be the point of this problem?

Additional information:[24]

All heat capacities are virtually the same at 100 J/mol·K.

$$C_{T0} = 2 \text{ mol/dm}^3$$

$$\Delta H_{Rx1O} = -1800 \text{ J/mol } o\text{-xylene}[24]$$

$$\Delta H_{Rx3O} = -1100 \text{ J/mol } o\text{-xylene}[24]$$

$$k_1 = 0.5 \exp[2(1 - 320/T)] \text{ dm}^3/\text{kg cat.} \cdot \text{min}$$

$$k_2 = k_1/K_C$$

$$k_3 = 0.005 \exp\{[4.6(1 - (460/T))]\} \text{ dm}^3/\text{kg cat.} \cdot \text{min}$$

$$K_C = 10 \exp[4.8(430/T - 1.5)]$$

$$T_0 = 330 \text{ K}$$

$$T_a = 500 \text{ K}$$

$$Ua/\rho_b = 16 \text{ J/kg cat.} \cdot \text{min} \cdot °\text{C}$$

$$W = 100 \text{ kg}$$

P8-30$_C$ *(Multiple reactions with heat effects)* Styrene can be produced from ethylbenzene by the following reaction:

$$\text{ethylbenzene} \longleftrightarrow \text{styrene} + H_2 \qquad (1)$$

However, several irreversible side reactions also occur:

$$\text{ethylbenzene} \longrightarrow \text{benzene} + \text{ethylene} \qquad (2)$$

$$\text{ethylbenzene} + H_2 \longrightarrow \text{toluene} + \text{methane} \qquad (3)$$

[J. Snyder and B. Subramaniam, *Chem. Eng. Sci.*, *49*, 5585 (1994)]. Ethylbenzene is fed at a rate of 0.00344 kmol/s to a 10.0-m³ PFR reactor along with inert steam at a total pressure of 2.4 atm. The steam/ethylbenzene molar ratio is initially [i.e., parts (a) to (c)] 14.5:1 but can be varied. Given the following data, find the exiting molar flow rates of styrene, benzene, and toluene for the following inlet temperatures when the reactor is operated adiabatically.

(a) $T_0 = 800$ K

(b) $T_0 = 930$ K

(c) $T_0 = 1100$ K

---

[24] Obtained from inviscid pericosity measurements.

(d) Find the ideal inlet temperature for the production of styrene for a steam/ethylbenzene ratio of 58:1. (*Hint:* Plot the molar flow rate of styrene versus T_0. Explain why your curve looks the way it does.)

(e) Find the ideal steam/ethylbenzene ratio for the production of styrene at 900 K. [*Hint:* See part (d).]

(f) What do you believe to be the points of this problem?

(g) Ask another question or suggest another calculation that can be made for this problem.

Additional information:

Heat capacities

Methane	68 J/mol·K	Styrene	273 J/mol·K
Ethylene	90 J/mol·K	Ethylbenzene	299 J/mol·K
Benzene	201 J/mol·K	Hydrogen	30 J/mol·K
Toluene	249 J/mol·K	Steam	40 J/mol·K

$$\rho = 2137 \text{ kg/m}^3 \text{ of pellet}$$

$$\phi = 0.4$$

$$\Delta H_{\text{Rx1EB}} = 118{,}000 \text{ kJ/kmol ethylbenzene}$$

$$\Delta H_{\text{Rx2EB}} = 105{,}200 \text{ kJ/kmol ethylbenzene}$$

$$\Delta H_{\text{Rx3EB}} = -53{,}900 \text{ kJ/kmol ethylbenzene}$$

$$K_{p1} = \exp\left\{ b_1 + \frac{b_2}{T} + b_3 \ln(T) + [(b_4 T + b_5) T + b_6] T \right\} \text{ atm}$$

$b_1 = -17.34$	$b_4 = -2.314 \times 10^{-10}$
$b_2 = -1.302 \times 10^4$	$b_5 = 1.302 \times 10^{-6}$
$b_3 = 5.051$	$b_6 = -4.931 \times 10^{-3}$

The kinetic rate laws for the formation of styrene (St), benzene (B), and toluene (T), respectively, are as follows. ($_{\text{EB}}$ = ethylbenzene)

$$r_{1S} = \rho(1-\phi) \exp\left(-0.08539 - \frac{10{,}925}{T} \right)\left(P_{\text{EB}} - \frac{P_{St} P_{H_2}}{K_{p1}} \right) \quad (\text{kmol/m}^3 \cdot \text{s})$$

$$r_{2B} = \rho(1-\phi) \exp\left(13.2392 - \frac{25{,}000}{T} \right)(P_{\text{EB}}) \quad (\text{kmol/m}^3 \cdot \text{s})$$

$$r_{3T} = \rho(1-\phi) \exp\left(0.2961 - \frac{11{,}000}{T} \right)(P_{\text{EB}} P_{H_2}) \quad (\text{kmol/m}^3 \cdot \text{s})$$

The temperature T is in kelvin.

P8-31$_B$ The liquid-phase reactions

(1) $A + B \xrightarrow{\;k_1\;} D$ (desired reaction)

(2) $A + B \xrightarrow{\;k_2\;} U$ (undesired reaction)

are carried out in a perfectly insulated CSTR. The desired reaction is first order in A and zero order in B, while the undesired reaction is zero order in A and first order in B. The feed rate is equimolar in A and B. Species A enters the reactor at a temperature of 100°C and species B enters at a temperature of 50°C. The operating temperature of the reactor is 400 K. The molar flow rate of A entering the reactor is 60 mol/min: $C_{P_A} = 20$ cal/mol·K, $C_{P_B} = 30$ cal/mol·K, $C_{P_D} = 50$ cal/mol·K, and $C_{P_U} = 40$ cal/mol·K.

For reaction 1: $\Delta H_{Rx} = -3000$ cal/mol of A at 300 K

For reaction 2: $\Delta H_{Rx} = -5000$ cal/mol of A at 300 K

$$k_1 = 1000 \exp\left(-\frac{2000}{T}\right) \quad \text{min}^{-1}$$

$$(T \text{ is in kelvin})$$

$$k_2 = 2000 \exp\left(-\frac{3000}{T}\right) \quad \text{min}^{-1}$$

$$C_{A0} = 0.01 \text{ mol/dm}^3 \qquad 0.001 \frac{\text{mol}}{\text{dm}^3} < C_{B0} < 2 \text{ mol/dm}^3$$

(a) What will be the exit molar flow rates of U and D from the reactor?
(b) What is the CSTR reactor volume for the conditions specified above?
(c) Is there a more effective way to maximize D? Explain.
(*Hint:* Start with a mole balance on A. Outline your method before beginning any calculations.)

P8-32$_C$ The liquid phase reactions

$$A \xrightarrow{k_{1A}} B \xrightarrow{k_{2B}} C$$

$$B \xrightarrow{k_{3B}} D$$

can be carried out in a number of CSTRs. Currently, the following sizes are available, 4 dm³, 40 dm³, 400 dm³, 4000 dm³. You can use up to 4 CSTRs of any one size. The heat flow to each reactor is controled by adjusting the functional area $f(0 \leq f < 1.0)$.

$Q_r = 100\, f(T - T_a)$ (cal/min) where the ambient temperature can be varied between 0°C and 100°C.

$$k_{1A} = 0.3\, e^{7000\left(\frac{1}{300} - \frac{1}{T}\right)} \quad \text{min}^{-1}$$

$$k_{2B} = 0.03\, e^{7000\left(\frac{1}{300} - \frac{1}{T}\right)} \quad \text{min}^{-1}$$

$$k_{3B} = 0.1\, e^{4000\left(\frac{1}{300} - \frac{1}{T}\right)} \quad \text{dm}^3/\text{mol·min}$$

The concentration of pure A is 5 M. A can be diluted with solvent. The feed can be cooled to 0°C or heated to 100°C.

$$\Delta H_{Rx1A} = +10,000 \text{ cal/mol A}$$

$$\Delta H_{Rx2B} = -10,000 \text{ cal/mol B}$$

$$\Delta H_{Rx3B} = -100,000 \text{ cal/mol B}$$

Discuss how should the reaction be carried (i.e. $v_0 =$, $T_0 =$, etc.) out to produce 100 mol B/day?[No ans. given in solutions manual to this revised problem]

JOURNAL CRITIQUE PROBLEMS

P8C-1 Equation (8) in an article in *J. Chem. Technol. Biotechnol.*, *31*, 273 (1981) is the kinetic model of the system proposed by the authors. Starting from Equation (2), derive the equation that describes the system. Does it agree with Equation (8)? By using the data in Figure 1, determine the new reaction order. The data in Table 2 show the effect of temperature. Figure 2 illustrates this effect. Use Equation (8) and Table 2 to obtain Figure 2. Does it agree with the article's results? Now use Table 2 and your equation. How does the figure obtained compare with Figure 2? What is your new E_{ac} ?

P8C-2 The kinetics of the reaction of sodium hypochlorite and sodium sulfate were studied by the flow thermal method in *Ind. Eng. Chem. Fundam.*, *19*, 207 (1980). What is the flow thermal method? Can the energy balance developed in this article be applied to a plug-flow reactor, and if not, what is the proper energy balance? Under what conditions are the author's equations valid?

P8C-3 In an article on the kinetics of sucrose inversion by invertase with multiple steady states in a CSTR [*Chem. Eng. Commun.*, *6*, 151 (1980)], consider the following challenges: Are the equations for K_i and K_m correct? If not, what are the correct equations for these variables? Can an analysis be applied to this system to deduce regions of multiple steady states?

P8C-4 Review the article in *AIChE J.*, *16*, 415 (1970). How was the $G(T)$ curve generated? Is it valid for a CSTR? Should $G(T)$ change when the space time changes? Critique the article in light of these questions.

CD-ROM MATERIAL

- **Learning Resources**
  1. *Summary Notes for Lectures 13, 14, 15, 16, 17, and 35A*
  3. *Interactive Computer Modules*
     A. Heat Effects I
     B. Heat Effects II
  4. *Solved Problems*
     A. Example CD8–1 $\Delta H_{Rx}(T)$ for Heat Capacities Expressed as Quadratic Functions of Temperature
     B. Example CD8–2 Second Order Reaction Carried Out Adiabatically in a CSTR
  5. *PFR/PBR Solution Procedure for a Reversible Gas-Phase Reaction*
- **Living Example Problems**
  1. *Example 8–5 CSTR with a Cooling Coil*
  2. *Example 8–6 Liquid Phase Isomerization of Normal Butane*
  3. *Example 8–7 Production of Acetic Anhydride*
  4. *Example 8–10 Oxidation of SO_2*
  5. *Example 8–11 Parallel Reaction in a PFR with Heat Effects*
  6. *Example 8–12 Multiple Reactions in a CSTR*

- **Professional Reference Shelf**
   1. *Steady State Bifurcation Analysis*
      A. Fundamentals
      B. Example CD8–3 Determine the Parameters That Give Multiple Steady
                                States (MSS)
- **Additional Homework Problems**

CDP8-A$_B$ The exothermic reaction

$$A \; \rightleftharpoons \; 2B$$

is carried out in both a plug-flow reactor and a CSTR with heat exchange. You are requested to plot conversion as a function of reactor length for both adiabatic and nonadiabatic operation as well as to size a CSTR. [2nd Ed. P8-16]

CDP8-B$_B$ Use bifurcation theory (Section 8.6.5 on the CD-ROM) to determine the possible regions for multiple steady states for the gas reaction with the rate law

$$-r''_A = \frac{k_1 C_A}{(k_2 + k_3 C_A)^2}$$

[2nd Ed. P8-26]

CDP8-C$_B$ In this problem bifurcation theory (CD-ROM Section 8.6.5) is used to determine if multiple steady states are possible for each of three types of catalyst. [2nd Ed. P8-27$_B$]

CDP8-D$_B$ In this problem bifurcation theory (CD-ROM Section 8.6.5) is used to determine the regions of multiple steady states for the autocatalytic reaction

$$A + B \; \longrightarrow \; 2B$$

[2nd Ed. P8-28$_B$]

CDP8-E$_C$ This problem concerns the SO$_2$ reaction with heat losses. [2nd Ed. P8-33]

CDP8-F$_C$ This problem concerns the use of interstage cooling in SO$_2$ oxidation. [2nd Ed. P8-34a]

CDP8-G$_B$ This problem is a continuation of the SO$_2$ oxidation example problem. Reactor costs are considered in the analysis cooling. [2nd Ed. P8-34(b and c)]

CDP8-H$_B$ Parallel reactions take place in a CSTR with heat effects. [1st Ed. P9-21]

CDP8-I$_B$ This problem concerns multiple steady states for the second-order reversible liquid-phase reaction. [Old exam problem]

CDP8-J$_B$ Series reactions take place in a CSTR with heat effects. [1st Ed. P9-23]

CDP8-K$_B$ A drug intermediate is produced in a batch reactor with heat effects. The reaction sequence is

$$2A + B \; \longrightarrow \; C + D$$
$$C + A + B \; \longrightarrow \; E + D$$

The desired product is C.

CDP8-L$_B$ In the multiple steady state for

$$A \longrightarrow B$$

The phase plane of C_A vs. T shows a separatrix. [2nd Ed. P8-22]

CDP8-M$_B$ An acid-catalyzed second-order reaction is carried out adiabatically in a CSTR.

CDP8-N$_B$ A second-order reaction with multiple steady states is carried out in different solvents.

CDP8-O$_B$ Multiple reactions

$$A \longrightarrow 2B$$
$$2A + B \longrightarrow C$$

are carried out adiabatically in a PFR.

CDP8-P$_B$ Exothermic second-order reversible reaction carried out in a packed bed reactor.

SUPPLEMENTARY READING

1. An excellent development of the energy balance is presented in

ARIS, R., *Elementary Chemical Reactor Analysis*. Upper Saddle River, N.J.: Prentice Hall, 1969, Chaps. 3 and 6.

HIMMELBLAU, D. M., *Basic Principles and Calculations in Chemical Engineering*, 4th ed. Upper Saddle River, N.J.: Prentice Hall, 1982, Chaps. 4 and 6.

WESTERWERP, K. R., W. P. M. VAN SWAAIJ, and A. A. C. M. BEENACKERS, *Chemical Reactor Design and Operation*. New York: Wiley, 1984.

A number of example problems dealing with nonisothermal reactors can be found in

SMITH, J. M., *Chemical Engineering Kinetics*, 3rd ed. New York: McGraw-Hill, 1981, Chap. 5.

WALAS, S. M., *Reaction Kinetics for Chemical Engineers*. New York: McGraw-Hill, 1959, Chap. 3.

WALAS, S. M., *Chemical Reaction Engineering Handbook of Solved Problems*. Amsterdam: Gordon and Breach, 1995. See the following solved problems: Problem 4.10.1, page 444; Problem 4.10.08, page 450; Problem 4.10.09, page 451; Problem 4.10.13, page 454; Problem 4.11.02, page 456; Problem 4.11.09, page 462; Problem 4.11.03, page 459; Problem 4.10.11, page 463.

For a thorough discussion on the heat of reaction and equilibrium constant, one might also consult

DENBIGH, K. G., *Principles of Chemical Equilibrium*, 4th ed. Cambridge: Cambridge University Press, 1981.

2. A review of the multiplicity of the steady state is discussed by

LUSS, D., and V. BALAKOTAIAH, in *Chemical Reaction Engineering—Boston*, J. Wei and C. Georgakis, eds., ACS Symposium Series 196. Washington, D.C.: American Chemical Society.

PERLMUTTER, D. D., *Stability of Chemical Reactors*. Upper Saddle River, N.J.: Prentice Hall, 1972.

SCHMITZ, R. A., in *Chemical Reaction Engineering Reviews*, H. M. Hulburt, ed., Advances in Chemistry Series 148. Washington, D.C.: American Chemical Society, 1975, p. 156.

3. Partial differential equations describing axial and radial variations in temperature and concentration in chemical reactors are developed in

WALAS, S. M., *Reaction Kinetics for Chemical Engineers*. New York: McGraw-Hill, 1959, Chap. 8.

4. The heats of formation, $H_i(T)$, Gibbs free energies, $G_i(T_R)$, and the heat capacities of various compounds can be found in

PERRY, R. H., D. W. GREEN, and J. O. MALONEY, eds., *Chemical Engineers' Handbook*, 6th ed. New York: McGraw-Hill, 1984.

REID, R. C., J. M. PRAUSNITZ, and T. K. SHERWOOD, *The Properties of Gases and Liquids*, 3rd ed. New York: McGraw-Hill, 1977.

WEAST R. C., ed., *CRC Handbook of Chemistry and Physics*, 66th ed. Boca Raton, Fla.: CRC Press, 1985.

Unsteady-State Nonisothermal Reactor Design   9

> Chemical Engineers are not gentle people, they like high temperatures and high pressures.
>
> Steve LeBlanc

Up to now we have focused on the steady-state operation of nonisothermal reactors. In this section the unsteady-state energy balance will be developed and then applied to CSTRs, plug-flow reactors, and well-mixed batch and semibatch reactors.

We will then discuss reactor start-up, falling off the upper-steady state, the control of chemical reactors, and multiple reactions with heat effects.

9.1 The General Equation

We begin by recalling the unsteady-state form of the energy balance developed in Chapter 8.

$$\dot{Q} - \dot{W}_s + \sum_{i=1}^{n} F_i H_i \big|_{\text{in}} - \sum_{i=1}^{n} F_i H_i \big|_{\text{out}} = \left(\frac{\partial \hat{E}_{\text{sys}}}{\partial t} \right) \tag{8-9}$$

The total energy of the system is the sum of the products of specific energies, E_i, of the various species in the system volume and the number of moles of that species:

$$\hat{E}_{\text{sys}} = \sum_{i=1}^{n} N_i E_i \tag{9-1}$$

In evaluating $\hat{E}_{\text{sys}}$, we shall neglect changes in the potential and kinetic energies, and substitute for the internal energy U_i in terms of the enthalpy H_i:

$$\hat{E}_{\text{sys}} = \sum_{i=1}^{n} N_i E_i = \sum_{i=1}^{n} N_i U_i = \left[\sum_{i=1}^{n} N_i(H_i - PV_i) \right]_{\text{sys}} \tag{9-2}$$

Differentiating Equation (9-2) with respect to time and substituting into Equation (8-9) gives

Transient energy balance

$$\dot{Q} - \dot{W}_s + \sum_{i=1}^{n} F_i H_i \big|_{\text{in}} - \sum_{i=1}^{n} F_i H_i \big|_{\text{out}}$$

$$= \left[\sum_{i=1}^{n} N_i \frac{\partial H_i}{\partial t} + \sum_{i=1}^{n} H_i \frac{\partial N_i}{\partial t} - \frac{\partial \left(P \sum_{i=1}^{n} \overbrace{N_i V_i}^{V} \right)}{\partial t} \right]_{\text{sys}} \tag{9-3}$$

For brevity we shall write these sums as

$$\Sigma = \sum_{i=1}^{n}$$

unless otherwise stated.

9.2 Unsteady Operation of CSTRs and Semibatch Reactors

When no spatial variations are present in the system volume, and variations in the total pressure and volume are neglected, the energy balance, Equation (9-3), reduces to

$$\dot{Q} - \dot{W}_s + \Sigma F_{i0} H_{i0} - \Sigma F_i H_i = \Sigma N_i \frac{dH_i}{dt} + \Sigma H_i \frac{dN_i}{dt} \tag{9-4}$$

Recalling Equation (8-19),

$$H_i = H^\circ(T_R) + \int_{T_R}^{T} C_{pi} \, dT \tag{8-19}$$

and differentiating with respect to time, we obtain

$$\frac{dH_i}{dt} = C_{pi} \frac{dT}{dt} \tag{9-5}$$

Then substituting Equation (9-5) into (9-4) gives

$$\dot{Q} - \dot{W}_s + \Sigma F_{i0} H_{i0} - \Sigma F_i H_i = \Sigma N_i C_{pi} \frac{dT}{dt} + \Sigma H_i \frac{dN_i}{dt} \tag{9-6}$$

The mole balance on species i is

$$\frac{dN_i}{dt} = -v_i \, r_A V + F_{i0} - F_i \tag{9-7}$$

Using Equation (9-7) to substitute for dN_i/dt, Equation (9-6) becomes

$$\dot{Q} - \dot{W}_s + \Sigma F_{i0} H_{i0} - \Sigma F_i H_i$$

$$= \Sigma N_i C_{pi} \frac{dT}{dt} + \Sigma v_i H_i (-r_A V) + \Sigma F_{i0} H_i - \Sigma F_i H_i$$

Rearranging, we have

This form of the energy balance should be used when there is a phase change

$$\boxed{\frac{dT}{dt} = \frac{\dot{Q} - \dot{W}_s - \Sigma F_{i0}(H_i - H_{i0}) + (-\Delta H_{Rx})(-r_A V)}{\Sigma N_i C_{pi}}} \tag{9-8}$$

Substituting for H_i and F_{i0} for the case of no phase change gives us

Energy balance on a transient CSTR or semibatch reactor

$$\boxed{\frac{dT}{dt} = \frac{\dot{Q} - \dot{W}_s - \displaystyle\sum_{i=1}^{n} F_{i0} \tilde{C}_{pi} (T - T_{i0}) + [-\Delta H_{Rx}(T)](-r_A V)}{\displaystyle\sum_{i=1}^{n} N_i C_{pi}}} \tag{9-9}$$

Equation (9-9) applies to a semibatch reactor as well as unsteady operation of a CSTR.

For liquid-phase reactions where ΔC_p is small and can be neglected, the following approximation is often made:

$$\Sigma N_i C_{pi} \cong \Sigma N_{i0} C_{pi} = N_{A0} \Sigma \Theta_i C_{pi} = N_{A0} C_{ps}$$

where C_{ps} is the heat capacity of the solution. The units of $N_{A0} C_{ps}$ are (cal/K), or (Btu/°R) and

$$\Sigma F_{i0} C_{pi} = F_{A0} C_{ps}$$

where the units of $F_{A0} C_{ps}$ are (cal/s·K) or (Btu/h·R)[1]. With this approximation and assuming that every species enters the reactor at temperature T_0, we have

$$\frac{dT}{dt} = \frac{\dot{Q} - \dot{W}_s - F_{A0} C_{ps}(T - T_0) + [-\Delta H_{Rx}(T)](-r_A V)}{N_{A0} C_{ps}} \tag{9-10}$$

---

[1] We see that if heat capacity were given in terms of mass (i.e., $C_{psm} = $ cal/g·K) then both F_{A0} and N_{A0} would have to be converted to mass:

$$\dot{m}_{A0} C_{psm} = N_{A0} C_{ps}$$

and

$$\dot{m}_{A0} C_{psm} = F_{A0} C_{ps}$$

but the units of the products would still be the same (cal/K) and (cal/s·K), respectively.

9.2.1 Batch Reactors

A batch reactor is usually well mixed, so that we may neglect spatial variations in the temperature and species concentration. The energy balance on this reactor is found by setting F_{A0} equal to zero in Equation (9-9) yielding

$$\frac{dT}{dt} = \frac{\dot{Q} - W_s + (-\Delta H_{Rx})(-r_A V)}{\Sigma N_i C_{pi}} \qquad (9\text{-}11)$$

Equation (9-11) is the preferred form of the energy balance when the number of moles, N_i, is used in the mole balance rather than the conversion, X. The number of moles of species i at any X is

$$N_i = N_{A0}(\Theta_i + v_i X)$$

Consequently, in terms of conversion, the energy balance becomes

$$\frac{dT}{dt} = \frac{\dot{Q} - \dot{W}_s + (-\Delta H_{Rx})(-r_A V)}{N_{A0}(\Sigma\, \Theta_i C_{pi} + \Delta C_p X)} \qquad (9\text{-}12)$$

Batch reactor energy and mole balances

Equation (9-12) must be coupled with the mole balance

$$N_{A0}\frac{dX}{dt} = -r_A V \qquad (2\text{-}6)$$

and the rate law and then solved numerically.

9.2.2 Adiabatic Operation of a Batch Reactor

For adiabatic operation ($\dot{Q} = 0$) of a batch ($F_{i0} \equiv 0$) reactor and when the work done by the stirrer can be neglected ($\dot{W}_s \cong 0$), Equation (9-12) can be written as

$$-\Delta H_{Rx}(T)(-r_A V) = N_{A0}(C_{ps} + \Delta C_p X)\frac{dT}{dt} \qquad (9\text{-}13)$$

where as before

$$C_{ps} = \Sigma\, \Theta_i C_{pi} \qquad (9\text{-}14)$$

From the mole balance on a batch reactor we have

$$N_{A0}\frac{dX}{dt} = -r_A V \qquad (2\text{-}6)$$

We combine Equations (9-13) and (2-6) to obtain

$$-[\Delta H_{Rx} + \Delta C_p (T - T_R)]\frac{dX}{dt} = [C_{ps} + \Delta C_p X]\frac{dT}{dt} \qquad (9\text{-}15)$$

Canceling dt, separating variables, integrating, and rearranging (see CD-ROM) gives

$$X = \frac{C_{ps}(T - T_0)}{-\Delta H_{Rx}(T)} = \frac{\Sigma\,\Theta_i\,\tilde{C}_{pi}(T - T_0)}{-\Delta H_{Rx}(T)} \qquad (9\text{-}16)$$

$$T = T_0 + \frac{[-\Delta H_{Rx}(T_0)]X}{C_{ps} + X\,\Delta\hat{C}_p} = T_0 + \frac{[-\Delta H_{Rx}(T_0)]X}{\displaystyle\sum_{i=1}^{n}\Theta_i\,\tilde{C}_{pi} + X\,\Delta\hat{C}_p} \qquad (9\text{-}17)$$

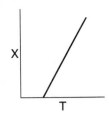

Temperature
conversion
relationship for
an adiabatic
batch reactor

We note that for adiabatic conditions the relationship between temperature and conversion is the same for batch reactors, CSTRs, PBRs, and PFRs. Once we have T as a function of X for a batch reactor, we can construct a table similar to Table E8-5.1 and use techniques analogous to those discussed in Section 8.3.2 to evaluate the design equation to determine the time necessary to achieve a specified conversion.

$$t = N_{A0}\int_0^x \frac{dX}{-r_A V} \qquad (2\text{-}9)$$

Example 9–1   Adiabatic Batch Reactor

Although you were hoping for a transfer to the Bahamas, you are still the engineer of the CSTR of Example 8-4, in charge of the production of propylene glycol. You are considering the installation of a new glass-lined 175-gal CSTR, and you decide to make a quick check of the reaction kinetics. You have an insulated instrumented 10-gal stirred batch reactor available. You charge this reactor with 1 gal of methanol and 5 gal of water containing 0.1 wt % H_2SO_4. The initial temperature of all these materials is 58°F.

How many minutes should it take the mixture inside the reactor to reach a conversion of 51.5% if the reaction rate law given in Example 8-4 is correct? What would be the temperature? Use the data presented in Example 8-4.

Solution

1. **Design equation**:

$$N_{A0}\frac{dX}{dt} = -r_A V \qquad (2\text{-}6)$$

Because there is a negligible change in density during the course of this reaction, the volume V is assumed to be constant.

2. **Rate law**:

$$-r_A = kC_A \qquad (E9\text{-}1.1)$$

3. **Stoichiometry**:

$$C_A = \left(\frac{N_{A0}}{V}\right)(1 - X) \qquad (E9\text{-}1.2)$$

4. **Combining** Equations (E9-1.1), (E9-1.2), and (2-6), we have

$$\frac{dX}{dt} = k(1 - X) \tag{E9-1.3}$$

From the data in Example 8-4,

$$k = (4.71 \times 10^9) \exp\left[\frac{-32,400}{(1.987)(T)}\right] \ \text{s}^{-1}$$

or

$$k = (2.73 \times 10^{-4}) \exp\left[\frac{32,400}{1.987}\left(\frac{1}{535} - \frac{1}{T}\right)\right] \ \text{s}^{-1} \tag{E9-1.4}$$

A table similar to that used in Example 8-6 can now be constructed.

5. **Energy balance**. Using Equation (9-17), the relationship between X and T for an adiabatic reaction is given by

$$T = T_0 + \frac{[-\Delta H_{\text{Rx}}(T_0)]X}{C_{ps} + \Delta \hat{C}_p X} \tag{E9-1.5}$$

6. **Evaluating the parameters** in the energy balance gives us the heat capacity of the solution:

$$
\begin{aligned}
C_{ps} &= \Sigma\, \Theta_i \tilde{C}_{pi} = \Theta_{\text{A}}\tilde{C}_{p\text{A}} + \Theta_{\text{B}}\tilde{C}_{p\text{B}} + \Theta_{\text{C}}\tilde{C}_{p\text{C}} + \Theta_{\text{I}}\tilde{C}_{p\text{I}} \\
&= (1)(35) + (18.65)(18) + 0 + (1.670)(19.5) \\
&= 403 \ \text{Btu/lb mol A} \cdot {}^\circ\text{F}
\end{aligned}
$$

From Example 8-4, $\Delta \hat{C}_p = -7$ Btu/lb mol·°F and consequently, the second term on the right-hand side of the expression for the heat of reaction,

$$
\begin{aligned}
\Delta H_{\text{Rx}}(T) &= \Delta H_{\text{Rx}}^\circ(T_R) + \Delta \tilde{C}_p(T - T_R) \\
&= -36,400 - 7(T - 528) \tag{E8-4.7}
\end{aligned}
$$

is very small compared with the first term [less than 2% at 51.5% conversion (from Example 8-4)].

Taking the heat of reaction at the initial temperature of 535°R,

$$\Delta H_{\text{Rx}}(T_0) = -36,400 - (7)(535 - 528)$$

$$= -36,450 \ \text{Btu/lb mol}$$

Because terms containing $\Delta \hat{C}_p$ are very small, it can be assumed that

$$\Delta \hat{C}_p \approx 0$$

In calculating the initial temperature, we must include the temperature rise from the heat of mixing the two solutions:

$$T_0 = (460 + 58) + 17$$

$$= 535°\text{R}$$

$$T = T_0 - \frac{X[\Delta H_{\text{Rx}}(T_0)]}{C_{ps}} = 535 - \frac{-36,450X}{403}$$

$$= 535 + 90.45\ X \tag{E9-1.6}$$

Hand Calculation

Integrating Equation (E9-1.3) gives

$$t = \int_0^{0.515} \frac{dX}{k(1-X)} \qquad \text{(E9-1.7)}$$

Again, the variation of Simpson's rule used in Example 8-6 is used. We now choose X, calculate T from Equation (E9-1.6), calculate k, and then calculate $(1/[k(1-X)])$ and tabulate it in Table E9-1.1.

TABLE E9-1.1

X	Temperature ($°R$)	k (s^{-1})	$\dfrac{1}{k(1-X)}$ (s)
0	535	2.73×10^{-4}	$3663 = f_0$
0.1288	547	5.33×10^{-4}	$2154 = f_1$
0.2575	558	9.59×10^{-4}	$1404 = f_2$
0.3863	570	17.7×10^{-4}	$921 = f_3$
0.5150	582	32.0×10^{-4}	$644 = f_4$

In evaluating Equation (E9-1.7) numerically, it was decided to use four equal intervals. Consequently, $\Delta X = h = 0.515/4 = 0.12875$. Using Simpson's rule, we have

$$t = \frac{h}{3}\,[f(x_0) + 4f(x_1) + 2f(x_2) + 4f(x_3) + f(x_4)]$$

$$= \frac{1}{3}\,(0.12875)[3663 + (4)(2154) + (2)(1404) + (4)(921) + 644]$$

$$= 833 \text{ s or } 13.9 \text{ min}$$

$$T = 582 \text{ R or } 122°F$$

Computer Solution

A software package (e.g., POLYMATH) was also used to combine Equations (E9-1.3), (E9-1.4), and (E9-1.6) to determine conversion and temperature as a function of time. Table E9-1.2 shows the program, and Figures E9-1.1 and E9-1.2 show the solution results.

TABLE E9-1.2.   POLYMATH PROGRAM

Equations:	Initial Values:
	0

```
d(x)/d(t)=k*(1-x)
t1=535+90.45*x
k=0.000273*exp(16306*((1/535)-(1/t1)))
```

$t_0 = 0, \qquad t_f = 1500$

CD Living

Example Problems

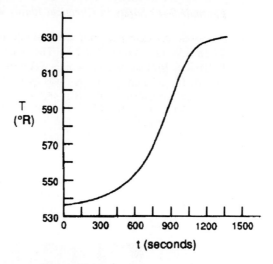

Figure E9-1.1   Temperature–time curve.

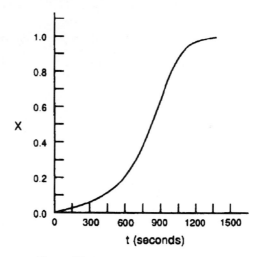

Figure E9-1.2   Conversion–time curve.

It is of interest to compare this residence time with the residence time in the 175-gal CSTR to attain the same conversion:

$$\tau = \frac{V}{v_0} = \frac{X}{k(1-X)} = \frac{0.515}{(0.0032)(0.485)} = 332 \text{ s}$$

This occasion is one when the increase in the reaction rate constant caused by the increase in temperature more than compensates for the decrease in rate caused by the decrease in concentration, so the residence time in the CSTR for this conversion is less than it would be in a batch or tubular plug-flow reactor.

Example 9–2   Safety in Chemical Plants with Exothermic Reactions [2]

A serious accident occurred at Monsanto plant in Sauget, Illinois, on August 8 at 12:18 A.M. (see Figure E9-2.1). The blast was heard as far as 10 miles away in Belleville, Illinois, where people were awakened from their sleep. The explosion occurred in a batch reactor that was used to produce nitroanaline from ammonia and o-nitrochlorobenzene (ONCB):

This reaction is normally carried out isothermally at 175°C and about 500 psi. The ambient temperature of the cooling water in the heat exchanger is 25°C. By adjusting the coolant rate the reactor temperature could be maintained at 175°C. At the maximum coolant rate the ambient temperature is 25°C throughout the heat exchange.

On the day of the accident, two changes in normal operation occurred.

1. The reactor was charged with 9.044 kmol of ONCB, 33.0 kmol of NH_3, and 103.7 kmol of H_2O. Normally, the reactor is charged with 3.17 kmol of ONCB, 103.6 kmol of H_2O, and 43 kmol of NH_3.

Was a Potential Problem Analysis carried out?

Figure E9-2.1 Aftermath of the explosion. (St. Louis Globe Democrat photo by Roy Cook. Courtesy of St. Louis Mercantile Library.)

---

[2] Adapted from the problem by Ronald Willey, *Seminar on a Nitroanaline Reactor Rupture.* Prepared for SACHE, Center for Chemical Process Safety, American Institute of Chemical Engineers, New York (1994). The values of ΔH_{Rx} and UA were estimated in the plant data of the temperature-time trajectory in the article by G. C. Vincent, *Loss Prevention*, **Vol. 5**, p. 46–52, AIChE, New York NY.

A management
decision was made
to triple production

2. The reaction is normally carried out isothermally at 175°C over a 24-h period. Approximately 45 min after the reaction was started, cooling to the reactor was halted, but only for 10 min. Cooling may have been halted for 10 min or so on previous occasions when the normal charge of 3.17 kmol of ONCB was used and no ill effects occurred.

   The reactor had a safety relief valve whose disk would rupture when the pressure exceeded approximately 700 psi. Once the disk ruptures, the water would vaporize, and the reaction would be cooled (quenched) by the release of the latent heat of vaporization.

   Plot the temperature–time trajectory up to a period of 120 min after the reactants were mixed and brought up to 175°C. Show that the following three conditions had to have been present for the explosion to occur: (1) increased ONCB charge, (2) reactor stopped for 10 min, and (3) relief system failure.

Additional Information: The rate law is

$$-r_{\text{ONCB}} = kC_{\text{ONCB}}C_{\text{NH}_3} \quad \text{with} \quad k = 0.00017 \, \frac{\text{m}^3}{\text{kmol} \cdot \text{min}} \text{ at } 188°C$$

The reaction volume for the charge of 9.044 kmol of ONCB:

$$V = 3.265 \text{ m}^3 \text{ ONCB/NH}_3 + 1.854 \text{ m}^3 \text{ H}_2\text{O} = 5.119 \text{ m}^3$$

The reaction volume for the charge of 3.17 kmol of ONCB:

$$V = 3.26 \text{ m}^3$$
$$\Delta H_{\text{Rx}} = -5.9 \times 10^5 \text{ kcal/kg mol}$$
$$E = 11{,}273 \text{ cal/mol}$$
$$C_{p\text{ONCB}} = C_{pA} = 40 \text{ cal/mol} \cdot \text{K}$$
$$C_{p\text{H}_2\text{O}} = C_{pW} = 18 \text{ cal/mol} \cdot \text{K} \qquad C_{\text{NH}_3} = C_{pB} = 8.38 \text{ cal/mol} \cdot \text{K}$$

Assume that $\Delta \tilde{C}_p \approx 0$:

$$UA = \frac{35.85 \text{ kcal}}{\text{min} \, °C} \text{ with } T_a = 298 \text{ K}$$

Solution

$$\text{A} + 2\text{B} \longrightarrow \text{C} + \text{D}$$

Mole balance:

$$\frac{dX}{dt} = -r_A \frac{V}{N_{A0}} \tag{E9-2.1}$$

Rate law:

$$-r_A = kC_AC_B \tag{E9-2.2}$$

Stoichiometry (liquid phase):

$$C_A = C_{A0}(1 - X) \tag{E9-2.3}$$

with
$$C_B = C_{A0}(\Theta_B - 2X) \qquad (E9\text{-}2.4)$$

$$\Theta_B = \frac{N_{B0}}{N_{A0}}$$

Combine:

$$-r_A = kC_{A0}^2(1-X)(\Theta_B - 2X) \qquad (E9\text{-}2.5)$$

$$k = 0.00017 \exp\left[\frac{11273}{1.987}\left(\frac{1}{461} - \frac{1}{T}\right)\right]$$

Energy Balance:

$$\frac{dT}{dt} = \frac{UA(T_a - T) + (r_A V)(\Delta H_{Rx})}{\sum N_i C_{pi}} \qquad (E9\text{-}2.6)$$

For $\Delta \hat{C}_p = 0$,

$$NC_p = \sum N_i C_{pi} = N_{A0}C_{pA} + N_{B0}C_{pB} + N_W C_{pW}$$

Let Q_g be the heat generated [i.e., $Q_g = (r_A V)(\Delta H_{Rx})$] and let Q_r be the heat removed [i.e., $Q_r = UA(T - T_a)$]:

$$\frac{dT}{dt} = \frac{\overbrace{UA(T_a - T)}^{-Q_r} + \overbrace{(r_A V)(\Delta H_{Rx})}^{Q_g}}{N_{A0}C_{pA} + N_{B0}C_{pB} + N_W C_{pW}} \qquad (E9\text{-}2.7)$$

$Q_g = (r_A V)(\Delta H_{Rx})$

$Q_r = UA(T - T_a)$

Then
$$\boxed{\frac{dT}{dt} = \frac{Q_g - Q_r}{NC_p}} \qquad (E9\text{-}2.8)$$

Parameter evaluation for day of explosion:

$$NC_p = \sum N_{A0}\Theta_i C_{pi} = (9.0448)(40) + (103.7)(18) + (33)(8.38)$$

$$\boxed{NC_p = 2504 \text{ kcal/K}}$$

A. Isothermal Operation Up to 45 Minutes

We will first carry out the reaction isothermally at 175°C up to the time the cooling was turned off at 45 min. Combining and canceling yields

$$\frac{dX}{dt} = kC_{A0}(1-X)(\Theta_B - 2X) \qquad (E9\text{-}2.9)$$

$$\Theta_B = \frac{33}{9.04} = 3.64$$

At 175°C = 448 K, $k = 0.0001167 \text{ m}^3/\text{kmol}\cdot\text{min}$. Integrating Equation (E9-2.9) gives us

$$t = \left[\frac{V}{kN_{A0}}\right]\left(\frac{1}{\Theta_B - 2}\right)\ln\frac{\Theta_B - 2X}{\Theta_B(1-X)} \qquad (E9\text{-}2.10)$$

Substituting the parameter values

$$45 \text{ min} = \left[\frac{5.119 \text{ m}^3}{0.0001167 \text{ m}^3/\text{kmol} \cdot \text{min} (9.044 \text{ kmol})} \right] \times \left(\frac{1}{1.64} \right) \ln \frac{3.64 - 2X}{3.64(1 - X)}$$

Solving for X, we find that at $t = 45$ min, then $X = 0.033$.

We will calculate the rate of generation Q_g at this temperature and conversion and compare it with the maximum rate of heat removal Q_R. The rate of generation Q_g is

$$Q_g = r_A V \, \Delta H_{Rx} = k \frac{N_{A0}(1 - X)N_{A0}[(N_{B0}/N_{A0}) - 2X]V(-\Delta H_{Rx})}{V^2} \tag{E9-2.11}$$

At this time (i.e., $t = 45$ min, $X = 0.033$, $T = 175°C$) we calculate k, then Q_r and Q_g. At 175°C, $k = 0.0001167$ m³/min·kmol.

$$Q_g = (0.0001167) \frac{(9.0448)^2 (1 - 0.033)}{5.119} \left[\frac{33}{(9.0448)} - 2(0.033) \right] 5.9 \times 10^5$$

$$= 3830 \text{ kcal/min}$$

The corresponding maximum cooling rate is

$$Q_r = UA(T - 298)$$
$$= 35.85(448 - 298) \tag{E9-2.12}$$
$$= 5378 \text{ kcal/min}$$

Therefore

Everything is OK

$$\boxed{Q_r > Q_g} \tag{E9-2.13}$$

The reaction can be controlled. There would have been no explosion had the cooling not been turned off.

B. Adiabatic Operation for 10 Minutes

The cooling was turned off for 45 to 55 min. We will now use the conditions at the end of the period of isothermal operation as our initial conditions for adiabatic operation period between 45 and 55 min:

$$t = 45 \text{ min} \quad X = 0.033 \quad T = 448$$

Between $t = 45$ and $t = 55$ min $UA = 0$. The POLYMATH program adiabatic operation is the same as that in Table E9.2-1 except that $Q_r = UA(T - 298)*(0)$, which is the same as setting $UA = 0$.

For the 45- to 55-min period without cooling, the temperature rose from 448 K to 468 K and the conversion from 0.033 to 0.0424. Using this temperature and conversion in Equation (E9-2.11), we calculate the rate of generation Q_g at 55 min as

$$Q_g = 6591 \text{ kcal/min}$$

The maximum rate of cooling at this reactor temperature is found from Equation (E9-2.12) to be

$$Q_r = 6093 \text{ kcal/min}$$

Here we see that

The point of no
return

$$\boxed{Q_g > Q_r} \tag{E9-2.14}$$

and the temperature will continue to increase. Therefore, the **point of no return** has been passed and the temperature will continue to increase, as will the rate of reaction until the explosion occurs.

C. Batch Operation with Heat Exchange

Return of the cooling occurs at 55 min. The values at the end of the period of adiabatic operation ($T = 468$ K, $X = 0.0423$) become the initial conditions for the period of operation with heat exchange. The cooling is turned on at its maximum capacity, $Q = UA(298 - T)$, at 55 min. Table E9-2.1 gives the POLYMATH program to determine the temperature–time trajectory.

TABLE E9-2.1.   POLYMATH PROGRAM

Equations:	Initial Values:
d(T)/d(t)=(Qg-Qr)/NCp	467.992
d(X)/d(t)=(-ra)*V/Nao	0.0423866
NCp=2504	
V=3.265+1.854	
Nao=9.0448	
UA=35.83	
dH=-590000	
Nbo=33	
k=.00017*exp(11273/(1.987)*(1/461-1/T))	
Qr=UA*(T-298)	
Theata=Nbo/Nao	
ra=-k*Nao**2*(1-X)*(Theata-2*X)/V**2	
rate=-ra	
Qg=ra*V*(dH)	
$t_0 = 55$, $t_f = 121$	

Interruptions in the
cooling system have
happened before
with no ill effects

Note that one can change N_{A0} and N_{B0} to 3.17 and 43 kmol in the program and show that if the cooling is shut off for 10 min, at the end of that 10 min Q_r will still be greater than Q_g and no explosion will occur.

The complete temperature–time trajectory is shown in Figure E9-2.2. One notes the long plateau after the cooling is turned back on. Using the values of Q_g and Q_r at 55 min and substituting into Equation (E9-2.8), we find that

$$\frac{dT}{dt} = \frac{(6591 \text{ kcal/min}) - (6093 \text{ kcal/min})}{2504 \text{ kcal/°C}} = 0.2\text{°C/min}$$

Consequently, even though dT/dt is positive, the temperature increases very slowly at first, 0.2°C/min. By 11:45, the temperature has reached 240°C and is beginning to increase more rapidly. One observes that 119 min after the batch was started the

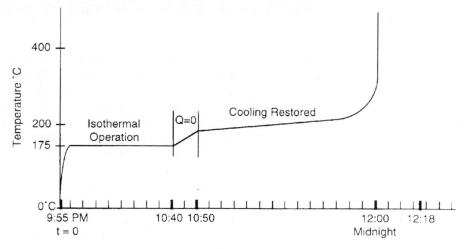

The explosion occurred shortly after midnight

Figure E9-2.2 Temperature–time trajection.

temperature increases sharply and the reactor explodes at approximately midnight. If the mass and heat capacity of the stirrer and reaction vessel had been included, the NC_p term would have increased by about 5% and extended the time until the explosion occurred by 15 or so minutes, which would predict the actual time the explosion occurred, at 12:18 A.M.

When the temperature reached 300°C, a secondary reaction, the decomposition of nitroaniline to noncondensable gases such CO, N_2, and NO_2, occurs, releasing even more energy. The total energy released was estimated to be 6.8×10^9 J, which is enough energy to lift the entire 2500-ton building 300 m (the length of three football fields) straight up.

D. Disk Rupture

We note that the pressure relief disk should have ruptured when the temperature reached 265°C (ca. 700 psi) but did not and the temperature continued to rise. If it had ruptured and all the water had vaporized, 10^6 kcal would have been drawn from the reacting solution, thereby lowering its temperature and quenching it.

If the disk had ruptured at 265°C (700 psi), the maximum mass flow rate, $\dot{m}$, out of the 2-in. orifice to the atmosphere (1 atm) would have been 830 kg/min at the time of rupture.

$$Q_r = \dot{m}_{vap}\,\Delta H_{vap} + UA(T - T_a)$$

$$= 830\,\frac{\text{kg}}{\text{min}} \times 540\,\frac{\text{kcal}}{\text{kg}} + 35.83\,\frac{\text{kcal}}{\text{K}}\,(538 - 298)\text{K}$$

$$= 4.48 \times 10^5\,\frac{\text{kcal}}{\text{min}} + 8604\,\frac{\text{kcal}}{\text{min}}$$

$$= 4.49 \times 10^5\,\frac{\text{kcal}}{\text{min}}$$

This value of Q_r is much greater than Q_g ($Q_g = 27{,}460$ kcal/min), so that the reaction could easily be quenched.

9.2.3 Transient CSTR, Batch, and Semibatch Reactors with Heat Exchanger—Ambient Temperature Not Spatially Uniform

In our past discussions of reactors with heat exchanges we assumed that the ambient temperature T_a was spatially uniform throughout the exchanger. This assumption is true if the system is a tubular reactor with the external pipe surface exposed to the atmosphere or if the system is a CSTR or batch where the coolant flow rate through exchanger is so rapid that the coolant temperatures entering and leaving the exchanger are virtually the same.

We now consider the case where the coolant temperature varies along the length of the exchanger while the temperature in the reactor is spatially uniform. The coolant enters the exchanger at a mass flow rate $\dot{m}_C$ at a temperature T_{a1} and leaves at a temperature T_{a2} (see Figure 9-1). As a first approximation, we assume a quasi-steady state for the coolant flow and neglect the accumulation term (i.e., $dT_a/dt = 0$). As a result, Equation (8-40) will give the rate of heat transfer *from* the exchanger *to* the reactor:

$$\dot{Q} = \dot{m}C_{pc}(T_{a1} - T)[1 - \exp(-UA/\dot{m}C_{pc})] \qquad (8\text{-}40)$$

Using Equation (8-40) to substitute for $\dot{Q}$ in Equation (9-9), we obtain

$$\boxed{\frac{dT}{dt} = \frac{\dot{m}_C C_{pc}(T_{a1} - T)[1 - \exp(-UA/\dot{m}_C C_{pc})] + (r_A V)(\Delta H_{Rx}) - \sum F_{i0}\tilde{C}_{pi}(T - T_0)}{\sum N_i C_{pi}}}$$

$$(9\text{-}18)$$

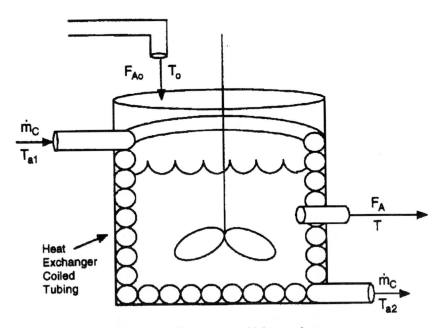

Figure 9-1   Tank reactor with heat exchanger.

At steady state ($dT/dt = 0$) Equation (9-18) can be solved for the conversion X as a function of reaction temperature by recalling that

$$F_{A0}X = -r_A V$$

and

$$\Sigma F_{i0} C_{pi}(T - T_0) = F_{A0} \Sigma \Theta_i C_{pi}(T - T_0)$$

and then rearranging Equation (9-18) to obtain

Steady-state energy balance

$$X = \frac{\dot{m}_C C_{pc}(T_{a1} - T)[1 - \exp(-UA/\dot{m}_C C_{pc})] - F_{A0} \Sigma \Theta_i \tilde{C}_{pi}(T - T_0)}{F_{A0}(-\Delta H_{Rx})}$$

(9-19)

Example 9–3   Heat Effects in a Semibatch Reactor

The second-order saponification of ethyl acetate is to be carried out in a semibatch reactor.

$$C_2H_5(CH_3COO)(aq) + NaOH(aq) \longleftrightarrow Na(CH_3COO)(aq) + C_2H_5OH(aq)$$

$$\text{A} \qquad\qquad \text{B} \qquad\qquad \text{C} \qquad\qquad \text{D}$$

Aqueous sodium hydroxide is to be fed at a concentration of 1 kmol/m³, a temperature of 300 K, and a rate of 0.004 m³/s to 0.2 m³ of a solution of water and ethyl acetate. The initial concentrations of ethyl acetate and water are 5 kmol/m³ and 30.7 kmol/m³, respectively. The reaction is exothermic and it is necessary to add a heat exchanger to keep its temperature below 315 K. A heat exchanger with $UA = 3000$ J/s·K and a coolant rate sufficiently high that the ambient coolant temperature is virtually constant at 290 K is available for use. Is this exchange adequate to keep the reactor temperature below 315 K? Plot temperature, C_A, C_B and C_C as a function of time.

Additional information: [3]

$$k = 0.39175 \exp\left[5472.7\left(\frac{1}{273} - \frac{1}{T}\right)\right] \text{ m}^3/\text{kmol} \cdot \text{s}$$

$$K_C = 10^{3885.44/T}$$

$$\Delta H_{Rx} = -79{,}076 \text{ kJ/kmol}$$

$$C_{p_A} = 170.7 \text{ J/mol/K}$$

$$C_{p_B} = C_{p_C} = C_{p_D} \cong C_{pW} = C_p = 75.24 \text{ J/mol} \cdot \text{K}$$

Feed: $C_{W0} = 55$ kmol/m³ $C_{B0} = 1.0$ kmol/m³

Initially: $C_{Wi} = 30.7$ kmol/m³ $C_{Ai} = 5$ kmol/m³ $C_{Bi} = 0$

---

[3] k from J. M. Smith, *Chemical Engineering Kinetics*, 3rd ed. (New York: McGraw-Hill, 1981, p. 205. ΔH_{Rx} and K_C calculated from values given in *Perry's Chemical Engineers' Handbook*, 6th ed. (New York: McGraw-Hill, 1984), pp. 3–147.

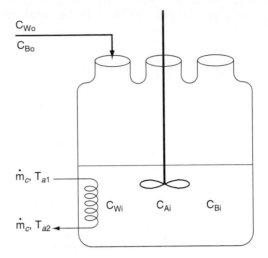

Solution

Mole balances:

$$\frac{dC_A}{dt} = r_A - \frac{v_0 C_A}{V} \tag{E9-3.1}$$

$$\frac{dC_B}{dt} = r_B + \frac{v_0 (C_{B0} - C_B)}{V} \tag{E9-3.2}$$

$$\frac{dC_C}{dt} = r_C - \frac{C_C v_0}{V} \tag{E9-3.3}$$

$$C_D = C_C$$

$$\frac{dN_W}{dt} = C_{Wi0} v_0 \tag{E9-3.4}$$

Initially,

$$N_{Wi} = V_i C_{Wi} = (0.2)(30.7) = 6.14 \text{ kmol}$$

Rate law:

$$-r_A = k\left(C_A C_B - \frac{C_C C_D}{K_C} \right) \tag{E9-3.5}$$

Stoichiometry:

$$-r_A = -r_B = r_C = r_D \tag{E9-3.6}$$

$$N_A = C_A V \tag{E9-3.7}$$

$$V = V_0 + v_0 t \tag{E9-3.8}$$

Energy balance: Next we replace $\sum_{i=1}^{n} F_{i0} \tilde{C}_{pi}$ in Equation (9-9). Because only B and water continually flow into the reactor

$$\sum_{i=1}^{n} F_{i0} \tilde{C}_{pi} = F_{B0} \tilde{C}_{pB} + F_{w0} \tilde{C}_W = F_{B0}\left(\tilde{C}_{pB} + \frac{F_{w0}}{F_{B0}} \tilde{C}_{pW} \right)$$

However, $\tilde{C}_{pB} = \tilde{C}_{pW}$:

$$\sum_{i=1}^{n} F_{i0}\tilde{C}_{pi} = F_{B0}\tilde{C}_{pB}(1 + \Theta_W)$$

where

$$\Theta_W = \frac{F_{W0}}{F_{B0}} = \frac{C_{W0}}{C_{B0}} = \frac{55}{1} = 55$$

$$\frac{dT}{dt} = \frac{UA(T_a - T) - F_{B0}C_{p_B}(1 + \Theta_W)(T - T_0) + (r_A V)\,\Delta H_{Rx}}{N_A C_{p_A} + N_B C_{p_B} + N_C C_{p_C} + N_D C_{p_D} + N_W C_{p_W}}$$

$$\frac{dT}{dt} = \frac{UA(T_a - T) - F_{B0}C_p(1 + \Theta_W)(T - T_0) + (r_A V)\,\Delta H_{Rx}}{C_p(N_B + N_C + N_D + N_W) + C_{p_A} N_A} \qquad \text{(E9-3.9)}$$

The POLYMATH program is given in Table E9-3.1. The solution results are shown in Figures E9-3.1 and E9-3.2.

TABLE E9-3.1.   POLYMATH PROGRAM FOR SEMIBATCH REACTOR

Equations	Initial Values
`d(Ca)/d(t)=ra-(v0*Ca)/V`	5
`d(Cb)/d(t)=rb+(v0*(Cb0-Cb)/V)`	0
`d(Cc)/d(t)=rc-(Cc*v0)/V`	0
`d(T)/d(t)=(UA*(Ta-T)-Fb0*cp*(1+55)*(T-T0)+ra*V*dh)/NCp`	300
`d(Nw)/d(t)=v0*Cw0`	6.14
`v0=0.004`	
`Cb0=1`	
`UA=3000`	
`Ta=290`	
`cp=75240`	
`T0=300`	
`dh=-7.9076e7`	
`Cw0=55`	
`k=0.39175*exp(5472.7*((1/273)-(1/T)))`	
`Cd=Cc`	
`Vi=0.2`	
`Kc=10**(3885.44/T)`	
`cpa=170700`	
`V=Vi+v0*t`	
`Fb0=Cb0*v0`	
`ra=-k*((Ca*Cb)-((Cc*Cd)/Kc))`	
`Na=V*Ca`	
`Nb=V*Cb`	
`Nc=V*Cc`	
`rb=ra`	
`rc=-ra`	
`Nd=V*Cd`	
`rate=-ra`	
`NCp=cp*(Nb+Nc+Nd+Nw)+cpa*Na`	
$t_0 = 0, \quad t_f = 360$	

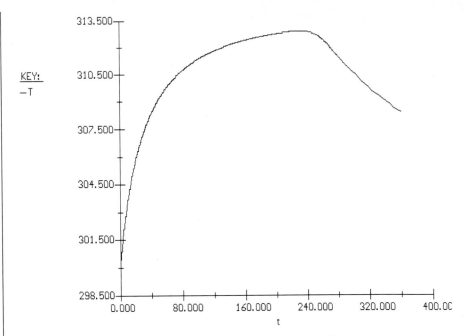

Figure E9-3.1   Temperature–time trajectory in a semibatch reactor.

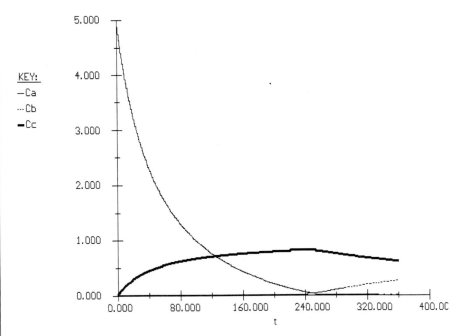

Figure E9-3.2   Concentration–time trajectories in a semibatch reactor.

9.3 Approach to the Steady State

In reactor startup it is often very important *how* temperature and concentrations approach their steady-state values. For example, a significant overshoot in temperature may cause a reactant or product to degrade, or the overshoot may be unacceptable for safe operation. If either case were to occur, we would say that the system exceeded its *practical stability limit*. Although we can solve the unsteady temperature-time and concentration–time equations numerically to see if such a limit is exceeded, it is often more insightful to study the approach to steady state by using the *temperature–concentration phase plane*. To illustrate these concepts we shall confine our analysis to a liquid-phase reaction carried out in a CSTR.

A qualitative discussion of how a CSTR approaches steady state is given in the CD-ROM. This analysis, summarized in Figure S-1 in the Summary, is developed to show the four different regions into which the phase plane is divided and how they allow one to sketch the approach to the steady state.

Example 9–4 Startup of a CSTR

Again we consider the production of propylene glycol (C) in a CSTR with a heat exchanger in Example 8-4. Initially there is only water at 75°F and 0.1 wt % H_2SO_4 in the 500-gallon reactor. The feed stream consists of 80 lb mol/h of propylene oxide (A), 1000 lb mol/h of water (B) containing 0.1 wt % H_2SO_4, and 100 lb mol/h of methanol (M). Plot the temperature and concentration of propylene oxide as a function of time, and a concentration vs. temperature graph for different entering temperatures and initial concentrations of A in the reactor.

The water coolant flows through the heat exchanger at a rate of 5 lb/s (1000 lb mol/h). The molar densities of pure propylene oxide (A), water (B), and methanol (M) are $\rho_{A0} = 0.932$ lb mol/ft³, $\rho_{B0} = 3.45$ lb mol/ft³, and $\rho_{M0} = 1.54$ lb mol/ft³, respectively.

$$UA = 16{,}000 \ \frac{\text{Btu}}{\text{h} \cdot °\text{F}} \qquad \text{with} \quad T_{a1} = 60°\text{F}$$

$$\dot{m}_C = 1000 \text{ lb mol/h} \qquad \text{with} \quad C_{p_C} = \frac{18 \text{ Btu}}{\text{lb mol} \cdot °\text{F}}$$

$$C_{p_A} = \frac{35 \text{ Btu}}{\text{lb mol} \cdot °\text{F}}, \quad C_{p_B} = \frac{18 \text{ Btu}}{\text{lb mol} \cdot °\text{F}}, \quad C_{p_C} = \frac{46 \text{ Btu}}{\text{lb mol} \cdot °\text{F}}, \quad C_{p_M} = \frac{19.5 \text{ Btu}}{\text{lb mol} \cdot °\text{F}}$$

Solution

$$A + B \longrightarrow C$$

Mole balances:

Initial Conditions

A: $\dfrac{dC_A}{dt} = r_A + \dfrac{(C_{A0} - C_A)v_0}{V}$ 0   (E9-4.1)

B: $\dfrac{dC_B}{dt} = r_B + \dfrac{(C_{B0} - C_B)v_0}{V}$ $C_{Bi} = 3.45 \dfrac{\text{lb mol}}{\text{ft}^3}$ (E9-4.2)

C: $\dfrac{dC_C}{dt} = r_C + \dfrac{-C_C v_0}{V}$ 0   (E9-4.3)

M: $\dfrac{dC_M}{dt} = \dfrac{v_0(C_{M0} - C_M)}{V}$ 0   (E9-4.4)

Rate law:

$$-r_A = kC_A \tag{E9-4.5}$$

Stoichiometry:

$$-r_A = -r_B = r_C \tag{E9-4.6}$$

Energy balance:

$$\frac{dT}{dt} = \frac{\dot{Q} - F_{A0} \sum \Theta_i C_{pi}(T - T_0) + (\Delta H_{Rx})(r_A V)}{\sum N_i C_{pi}} \tag{E9-4.7}$$

with

$$\dot{Q} = \dot{m}_C C_{P_C}(T_{a1} - T_{a2}) = \dot{m}_C C_{P_C}(T_{a1} - T)\left[1 - \exp\left(-\frac{UA}{\dot{m}_C C_{P_C}}\right)\right] \tag{E9-4.8}$$

and

$$T_{a2} = T - (T - T_{a1}) \exp\left(-\frac{UA}{\dot{m}_C C_{P_C}}\right)$$

Evaluation of parameters:

$$\sum N_i C_{pi} = C_{P_A} N_A + C_{P_B} N_B + C_{P_C} N_C + C_{P_D} N_D$$

$$= 35(C_A V) + 18(C_B V) + 46(C_C V) + 19.5(C_M V)$$

$$\sum \Theta_i C_{pi} = C_{P_A} + \frac{F_{B0}}{F_{A0}} C_{P_B} + \frac{F_{M0}}{F_{A0}} C_{P_M}$$

$$= 35 + 18\frac{F_{B0}}{F_{A0}} + 19.5\frac{F_{M0}}{F_{A0}}$$

$$v_0 = \frac{F_{A0}}{\rho_{A0}} + \frac{F_{B0}}{\rho_{B0}} + \frac{F_{M0}}{\rho_{M0}} = \left(\frac{F_{A0}}{0.923} + \frac{F_{B0}}{3.45} + \frac{F_{M0}}{1.54}\right)\frac{\text{ft}^3}{\text{hr}}$$

Neglecting $\Delta \tilde{C}_p$ because it changes the heat of reaction insignificantly over the temperature range of the reaction, the heat of reaction is assumed constant at:

$$\Delta H_{\text{Rx}} = -36,000 \ \frac{\text{Btu}}{\text{lb mol A}}$$

The POLYMATH program is shown in Table E9-4.1.

TABLE E9-4.1.   POLYMATH PROGRAM FOR CSTR STARTUP

Equations	Initial Values
`d(Ca)/d(t)=1/tau*(Ca0-Ca)+ra`	0
`d(Cb)/d(t)=1/tau*(Cb0-Cb)+rb`	3.45
`d(Cc)/d(t)=1/tau*(0-Cc)+rc`	0
`d(Cm)/d(t)=1/tau*(Cm0-Cm)`	0
`d(T)/d(t)=(Q-Fa0*ThetaCp*(T-T0)+(-36000)*ra*V)/NCp`	75
`Fa0=80`	
`T0=75`	
`V=(1/7.484)*500`	
`UA=16000`	
`Ta1=60`	
`k=16.96e12*exp(-32400/1.987/(T+460))`	
`Fb0=1000`	
`Fm0=100`	
`mc=1000`	
`ra=-k*Ca`	
`rb=-k*Ca`	
`rc=k*Ca`	
`Nm=Cm*V`	
`Na=Ca*V`	
`Nb=Cb*V`	
`Nc=Cc*V`	
`ThetaCp=35+Fb0/Fa0*18+Fm0/Fa0*19.5`	
`v0=Fa0/0.923+Fb0/3.45+Fm0/1.54`	
`Ta2=T-(T-Ta1)*exp(-UA/(18*mc))`	
`Ca0=Fa0/v0`	
`Cb0=Fb0/v0`	
`Cm0=Fm0/v0`	
`Q=mc*18*(Ta1-Ta2)`	
`tau=V/v0`	
`NCp=Na*35+Nb*18+Nc*46+Nm*19.5`	
$t_0 = 0, \qquad t_f = 4$	

Figures (E9-4.1) and (E9-4.2) show the reactor concentration and temperature of propylene oxide as a function of time, respectively, for an initial temperature of 75°F and only water in the tank (i.e., $C_{Ai} = 0$). One observes, both the temperature and concentration oscillate around their steady-state values ($T = 138°F$, $C_A = 0.039$ lb mol/ft³). Figure (E9-4.3) shows the phase plane of temperature and propylene oxide concentration for three different sets of initial conditions ($T_i = 75$, $C_{Ai} = 0$; $T_i = 150$, $C_{Ai} = 0$; and $T_i = 150$, $C_{Ai} = 0.14$).

Startup of a CSTR

KEY:
—Ca

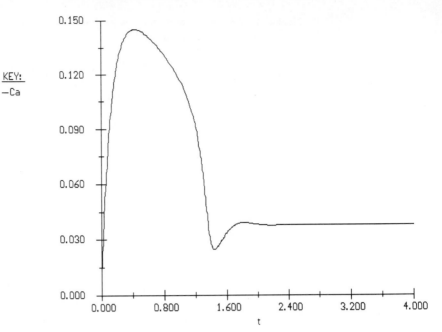

Figure E9-4.1   Propylene oxide concentration as a function of time.

KEY:
—T

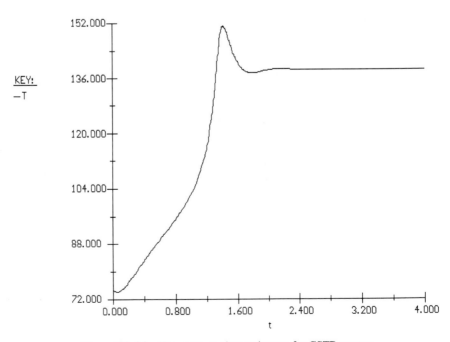

Figure E9-4.2   Temperature–time trajectory for CSTR startup.

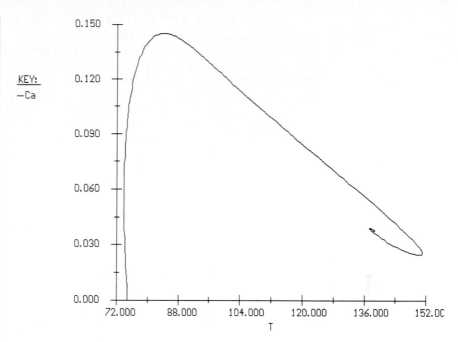

KEY:
—Ca

Figure E9-4.3   Concentration–temperature phase-plane trajectory.

Oops! The practical
stability limit was
exceeded.

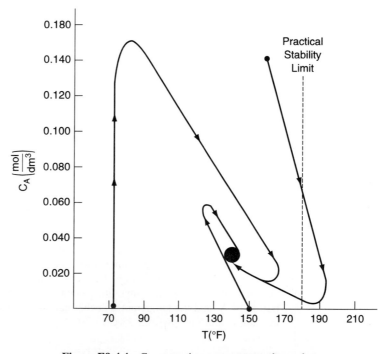

Figure E9-4.4   Concentration–temperature phase plane.

An upper limit of 180°F should not be exceeded in the tank. This temperature is the *practical stability limit*. The practical stability limit represent a temperature above which it is undesirable to operate because of unwanted side reactions, safety considerations, or damage to equipment. Consequently, we see if we started at an initial temperature of 160°F and an initial concentration of 0.14 mol/dm^3, the practical stability limit of 180°F would be exceeded as the reactor approached its steady-state temperature of 138°F. See concentration-temperature trajectory in Figure E9-4.4.

After about 1.6 h the reactor is operating at steady state with the following values:

$$C_A = 0.0379 \text{ lb mol/ft}^3 \quad C_C = 0.143 \text{ lb mol/ft}^3$$
$$C_B = 2.12 \text{ lb mol/ft}^3 \quad C_M = 0.2265 \text{ lb mol/ft}^3$$
$$T = 138.5°F$$

9.4 Control of Chemical Reactors

In any reaction system operating at steady state, there will always be upsets to the system—some large, some small—that may cause it to operate inefficiently, shut down, or perhaps explode. Examples of upsets that may occur are variations of the feed temperature, composition, and/or flow rate, the cooling jacket temperature, the reactor temperature, or some other variable. To correct for these upsets, a control system is usually added to make adjustments to the reaction system that will minimize or eliminate the effects of the upset. The material that follows gives at most a thumbnail sketch of how controllers help cause the reactor system to respond to unwanted upsets.

9.4.1 Falling Off the Steady State

We now consider what can happen to a CSTR operating at an upper steady state when an upset occurs in either the ambient temperature, the entering temperature, the flow rate, reactor temperature, or some other variable. To illustrate, let's reconsider the production of propylene glycol in a CSTR.

Example 9–5 Falling Off the Upper Steady State

In Example 9-4 we saw how a 500-gal CSTR used for the production of propylene glycol approached steady-state. For the flow rates and conditions (e.g., $T_0 = 75°F$, $T_{a1} = 60°F$), the steady-state temperature was 138°F and the corresponding conversion was 75.5%. Determine the steady-state temperature and conversion that would result if the entering temperature were to drop from 75°F to 70°F, assuming that all other conditions remain the same. First, sketch the steady state conversions calculated from the mole and energy balances as a function of temperature before and after the drop in entering temperature occurred. Next, plot the "conversion," concentration of A, and the temperature in the reactor as a function of time after the entering temperature drops from 75°C to 70°C.

Solution

The steady-state conversions can be calculated from the mole balance,

$$X_{MB} = \frac{\tau A e^{-E/RT}}{1 + \tau A e^{-E/RT}} \qquad \text{(E8-4.5)}$$

and from the energy balance,

$$X_{EB} = \frac{\sum \Theta_i \tilde{C}_{pi}(T - T_0) + [\dot{Q}/F_{A0}]}{-[\Delta H_{Rx}(T_R)]} \qquad \text{(E8-4.6)}$$

before ($T_0 = 75°F$) and after ($T_0 = 70°F$) the upset occurred. We shall use the parameter values given in Example 9-3 (e.g., $F_{A0} = 80$ lb mol/h, $UA = 16,000$ Btu/h·°F) to obtain a sketch of these conversions as a function of temperature, as shown in Figure E9-5.1.

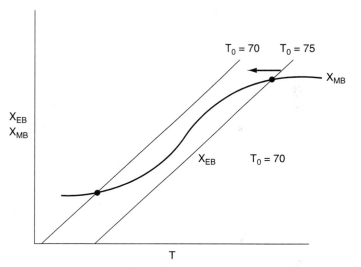

Figure E9-5.1   Conversion from mole and energy balances as a function of temperature.

We see that for $T_0 = 70°F$ the reactor has dropped below the extinction temperature and can no longer operate at the upper steady state. In Problem P9-16, we will see it is not always necessary for the temperature to drop below the extinction temperature in order to fall to the lower steady state. The equations describing the dynamic drop from the upper steady state to the lower steady state are identical to those given in Example 9-4; only the initial conditions and entering temperature are different. Consequently, the same POLYMATH and MATLAB programs can be used with these modifications. (See CD-ROM)

Initial conditions are taken from the final steady-state values given in Example 9-4.

$C_{Ai} = 0.039$ mol/ft³	$C_{Ci} = 0.143$ mol/ft³
$C_{Bi} = 2.12$ mol/ft³	$C_{Mi} = 0.226$ mol/ft³
$T_i = 138.5°F$	
Change T_0 to 70°F	

Because the system is not at steady state we cannot rigorously define a conversion in terms of the number of moles reacted because of the accumulation within the reactor. However, we can approximate the conversion by the equation $X = (1 - C_A/C_{A0})$. This equation is valid once the steady state is reached. Plots of the temperature and the conversion as a function of time are shown in Figures E9-5.2 and E9-5.3, respectively. The new steady-state temperature and conversion are $T = 83.6°F$ and $X = 0.19$.

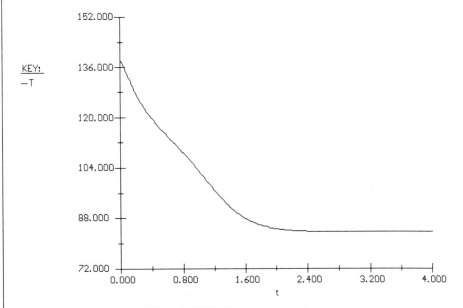

Figure E9-5.2   Temperature vs. time.

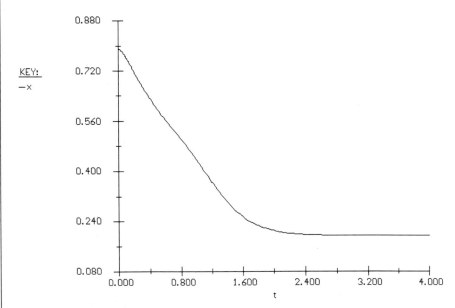

Figure E9-5.3   Conversion vs. time.

We will now see how we can make adjustments for upsets in the reactor operating conditions (such as we just saw in the drop in the entering temperature) so that we do not fall to the lower steady-state values. We can prevent this drop in conversion by adding a controller to the reactor.

9.4.2 Adding a Controller to a CSTR

Figure 9-2 shows a generic diagram for the control of a chemical process. The controller will function to minimize or correct for any unexpected disturbances that may upset the process. A control system will measure one of the output variables that must be controlled, Y (e.g., temperature, concentration), and compare it to a desired value Y_{sp}, called the set point. The difference, between the actual value, Y, and the desired value, Y_{sp}, is called the error signal, e. That is,

$$e = Y - Y_{sp} \tag{9-20}$$

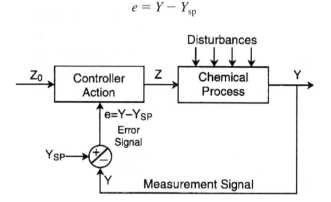

Figure 9-2   Control of a chemical process.

If the error signal is not equal to zero, a controller will make the appropriate changes in one of the system manipulated inputs, Z (e.g., flow rate, jacket temperature) to try to force the output variable, Y, to return to its set point, Y_{sp}.

For example, in the CSTR example just discussed, let's pick the effluent temperature in the reactor to be the output variable to be controlled (i.e., $Y = T$). Then define Y_{sp} to be the desired effluent temperature from the reactor. If the desired temperature is $T_{sp} = 180°C$ and the actual temperature in the reactor for some reason rises to 200°C (i.e., $T = 200°C$), then the error, e, would be

$$e = T - T_{sp} = 200 - 180 = 20$$

Because the error, e, is nonzero, this will cause the controller to react to reduce the error, e, to zero. The controller will send a signal that will manipulate one of the input variables, Z, such as the entering temperature, T_0, coolant flow rate, $\dot{m}_C$, or jacket temperature, T_a. For example, suppose that we choose the coolant flow rate, $\dot{m}_C$, as the manipulated variable to be changed; then the controller would increase $\dot{m}_C$ in an attempt to reduce Y (i.e., T) until it matches the set-point value $Y_{sp} = T_{sp} = 180$. A schematic of this process is shown in

Figure 9-3. Here we see that a fluctuation (e.g., rise) in the inlet temperature acts as a disturbance causing the reactor temperature to rise above the set-point temperature, producing an error signal. The error signal is acted upon by the controller to open the valve to increase the coolant flow rate, causing the reactor temperature to decrease and return to its set-point value.

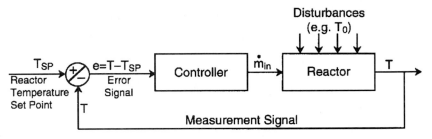

Figure 9-3 Control of a chemical reactor.

There are different types of controller actions that will make adjustments in the input, Z, based on the error between the output and set point. We shall consider four types: proportional, integral, derivative, and proportional-integral actions.

- *Proportional action (P):* The adjustment of the manipulated input variable Z, is proportional to the error, $e = Y_{sp} - Y$,

$$Z = Z_0 + k_C e \qquad (9\text{-}21)$$

The parameter Z_0 is called the controller bias and is the value of the manipulated input at the time the controller is turned on. The proportionality constant k_C is called the controller gain. The optimal value of k_C depends on the process. With proportional action the manipulated variable will continue to change to correct the offset as long as the measured variable keeps changing. Once the measured variable stops changing, the manipulated variable will stop changing whether or not the measured variable is at its set point. Since this controller uses the value of the error to adjust the input to the process, this type of controller can never fully return the output variable to its set point. This is a disadvantage of proportional action. Proportional action is seldom used by itself as a means of controlling the process.

- *Integral action (I):* The adjustment of the input variable is proportional to the integral of the error, $e = Y - Y_{sp}$:

$$Z = Z_0 + \frac{k_C}{\tau_I} \int_0^t e\, dt \qquad (9\text{-}22)$$

The parameter τ_I is called the integral time constant. For a reactor it is the order of magnitude at the space-time. Integral control action can also be expressed by the coupled set of equations

$$\frac{dI}{dt} = e \tag{9-23}$$

and

$$Z = Z_0 + \frac{k_C}{\tau_I} I \tag{9-24}$$

The optimal value of the proportionality constants, k_C and τ_I, are dependent on the process. With integral control the manipulated variables will continue to change to correct the upset as long as the integral of the error is nonzero. That is, the controller continues to manipulate the input variable until the error is zero. The main advantage of integral action is that over the long term the measured variable will always return to the set point. The disadvantage is that with this controller the response can become very oscillatory. Like proportional action, integral action is seldom used by itself.

- *Derivative action (D):*   The rate of adjustment of the input variable is proportional to the time rate of change of the error (i.e., de/dt).

$$Z = Z_0 + \tau_D \frac{de}{dt} \tag{9-25}$$

where τ_D is constant of proportionality call derivative time. The main problem with this type of action is that it can be very sensitive to noise and the error is normally filtered before entering the controller. Like proportional and integral action, derivative action is seldom used by itself.

- *Proportional-integral actions (PI):* The adjustment of the input variable is accomplished using both a proportional action and an integral action.

$$Z = Z_0 + k_C e + \frac{k_C}{\tau_I} \int_0^t e\, dt \tag{9-26}$$

The advantage of this controller is that it has quick response for large errors and does not have set-point offset. The measured variable can be returned to the set point without excessive oscillation. Methods to tune your controller for the best value τ_I, are discussed in Seborg, Edgar, and Mellichamp.[4] In addition, you can use the CD-ROM supplied with this book to vary these parameters in Example Problems E9-6 and E9-7 to learn their effects.

- *Proportional-integral derivative actions (PID):*   The adjustment of the input variable is accomplished using all three methods—proportional, integral, and derivative—as follows:

$$Z = Z_0 + k_C \left(e + \frac{1}{\tau_I} \int_0^t e\, dt + \tau_D \frac{de}{dt} \right) \tag{9-27}$$

---

[4] D. E. Seborg, T. F. Edgar, and D. A. Mellichamp, *Process Dynamics and Control* (New York: Wiley, 1989).

The advantage of PID action is that it gives rapid response. The disadvantage is that you have to tune three parameters

$$k_C, \tau_I, \text{ and } \tau_D.$$

Now let's apply the discussion above to control the exit temperature of a CSTR. The manipulated variable, Z, that will be used to control the temperature will be the coolant flow rate, $\dot{m}_C$, through the internal heat exchanger. A schematic diagram of this process is shown in Figure 9-4. To illustrate the response to a controller, we shall chose a controller with integral action to manipulate the coolant rate. For an integral action, the controller would activate the valve to change the flow rate using the following set of equations:

$$\dot{m}_C = \dot{m}_{C0} + \frac{k_C}{\tau_I} I \tag{9-28}$$

$$\frac{dI}{dt} = T - T_{sp} \tag{9-29}$$

The value $\dot{m}_{CO}$ is the coolant rate at the time the controller is turned on.

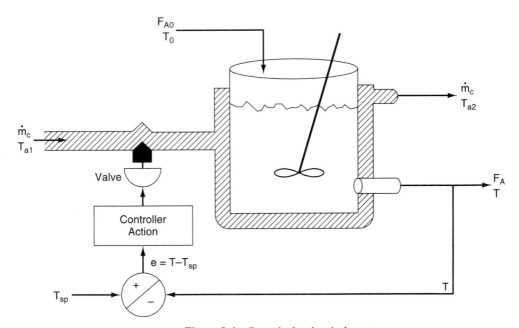

Figure 9-4   Control of a chemical reactor.

Example 9–6   Integral Control

We now reconsider Example 9-5 where a drop in the inlet temperature caused the reactor to operate at the lower steady state. However, we will now add an integral controller to the reaction system to learn what happens when the inlet temperature drops from 75°F to 70°F. The initial conditions at the time of the inlet temperature change are $C_A = 0.03789$, $C_B = 2.12$, $C_C = 0.143$, $C_M = 0.2266$, and $T = 138.5$°F.

Solution

For integral control we need to add two equations to our algorithm to obtain a relation between T and $\dot{m}_C$. For the CSTR reactor system at hand $Z = \dot{m}_{CO}$, $Y = T$ and Equations (9-23) and (9-24) become

$$\frac{dI}{dt} = T - T_{sp} \tag{E9-6.1}$$

$$\dot{m}_C = \dot{m}_{CO} + \frac{k_C}{\tau_I}(I) \tag{E9-6.2}$$

The adjustment to be made to the coolant rate is calculated from the integral of the error, e. The value of the proportionality constants, τ_I, and k_C, are chosen to maximize performance of the system without sacrificing stability. The coolant rate will need to increase in order to decrease the outlet temperature T and vice versa. In this example we will choose the controller gain as $k_C = 8.5$ and the integral time $\tau_I = V/v_0 = 0.165$ h. Methods of obtaining the best values of k_C and τ_I can be found in textbooks on chemical process control. Equations (E9-6.1) and (E9-6.2) are added to the POLYMATH to represent an integral controller along with the initial conditions. (Note that all equations are the same as in Examples 9-4 and 9-5 other than the perturbed variable, T_0, and the variable being controlled, which is the coolant rate, $\dot{m}_C$.) Figure E9-6.1 shows the reactor temperature as a function of time after the inlet temperature dropped to 70°F. The phase plane of reaction temperature and concentration is shown in Figure E9-6.2 as they return to their original values after the change in inlet temperature. See Problem P9-2(e) for further analysis of this example. (See CD-ROM)

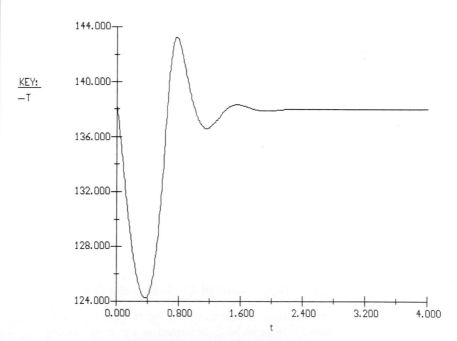

KEY:
—T

Figure E9-6.1 Temperature-time trajectory.

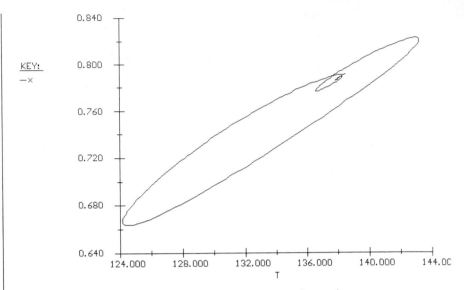

KEY:
—×

Figure E9-6.2   Conversion–temperature phase trajectory.

Example 9–7 Proportional-Integral (PI) Control

Repeat Example 9-6 using a proportional-integral controller instead of only an integral controller.

Solution

We note that while the output temperature returns to 138°F, the temperature oscillates over a 20°F range, which may be unacceptable in many situations. To overcome this we will add proportional to integral control. The equations here to be added to those in Table E9-4.1 (see the CD-ROM) to represent the PI controller are

$$\frac{dI}{dt} = T - T_{sp} \tag{E9-7.1}$$

$$\dot{m}_C = \dot{m}_{CO} + \frac{k_C}{\tau_I}(I) + k_C(T - T_{sp}) \tag{E9-7.2}$$

(Note that all equations are the same as before other than the perturbation variable and the variable being controlled.)

We will set $k_C = 8.5$ and $\tau_I = \tau_{av} = 0.165$ h. From the solution on the CD-ROM we see that the output outlet temperature does not oscillate as much as with just an integral controller. You can use the CD-ROM to try different controllers (differential, proportional, integral) and combinations thereof, along with different parameter values to study the behavior of CSTRs which have upsets to their system. Try $k_C = 55$ and $\tau_I = 0.1$.

9.5 Nonisothermal Multiple Reactions

For multiple reactions occurring in either a semibatch or batch reactor, Equation (9-18) can be generalized in the same manner as the steady-state energy balance, to give

$$\frac{dT}{dt} = \frac{\dot{m}_C C_{pc}(T_{a1} - T)[1 - \exp(-UA/\dot{m}_C C_{pc})] + \sum\limits_{i=1}^{q} r_{ij} V \Delta H_{Rxij}(T) - \sum F_{i0} \tilde{C}_{pi}(T - T_0)}{\sum N_i C_{pi}}$$

(9-30)

For large coolant rates Equation (9-30) becomes

$$\frac{dT}{dt} = \frac{UA(T_a - T) - \sum F_{i0} \tilde{C}_{pi}(T - T_0) + V \sum\limits_{i=1}^{q} r_{ij} \Delta H_{Rxij}}{\sum\limits_{i=1}^{n} N_i C_{pi}}$$

(9-31)

Example 9–8

The series reactions

$$2A \xrightarrow[\text{(1)}]{k_{1a}} B \xrightarrow[\text{(2)}]{k_{2b}} 3C$$

are catalyzed by H_2SO_4. All reactions are first order in the reactant concentration. The reaction is to be carried out in a semibatch reactor that has a heat exchanger inside with $UA = 35,000$ cal/h·K and an ambient temperature of 298 K. Pure A enters at a concentration of 4 mol/dm³, a volumetric flow rate of 240 dm³/hr, and a temperature of 305 K. Initially there is a total of 100 dm³ in the reactor, which contains 1.0 mol/dm³ of A and 1.0 mol/dm³ of the catalyst H_2SO_4. The reaction rate is independent of the catalyst concentration. The initial temperature of the reactor is 290 K.

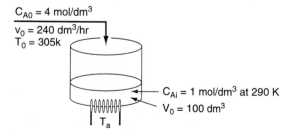

$C_{A0} = 4$ mol/dm³
$v_0 = 240$ dm³/hr
$T_0 = 305$k

$C_{Ai} = 1$ mol/dm³ at 290 K
$V_0 = 100$ dm³

T_a

Additional information:

$k_{1A} = 1.25/h^{-1}$ at 320 K with $E_{1A} = 9500$ cal/mol $C_{pA} = 30$ cal/mol·K
$k_{2B} = 0.08/h^{-1}$ at 300 K with $E_{2B} = 7000$ cal/mol $C_{pB} = 60$ cal/mol·K
$\Delta H_{Rx1A} = -6500$ cal/mol A $C_{pC} = 20$ cal/mol·K
$\Delta H_{Rx2B} = +8000$ cal/mol B $C_{pH_2SO_4} = 35$ cal/mol·K

Solution

Mole balances:

$$\frac{dC_A}{dt} = r_A + \frac{(C_{A0} - C_A)}{V} V_0 \qquad \text{(E9-8.1)}$$

$$\frac{dC_B}{dt} = r_B - \frac{C_B}{V} V_0 \qquad \text{(E9-8.2)}$$

$$\frac{dC_C}{dt} = r_C - \frac{C_C}{V} V_0 \qquad \text{(E9-8.3)}$$

Rate laws:

$$-r_{1A} = k_{1A} C_A \qquad \text{(E9-8.4)}$$

$$-r_{2B} = k_{2B} C_B \qquad \text{(E9-8.5)}$$

Stoichiometry (liquid phase): Use C_A, C_B, C_C

Relative rates:

$$-r_{2B} = \frac{r_{2C}}{3} \qquad \text{(E9-8.6)}$$

$$r_{2C} = -3\, r_{2B} \qquad \text{(E9-8.7)}$$

Net rates:

$$r_A = r_{1A} = -k_{1A} C_A \qquad \text{(E9-8.8)}$$

$$r_B = r_{1B} + r_{2B} = \frac{-r_{1A}}{2} + r_{2B} = \frac{k_{1A} C_A}{2} - k_{2B} C_B \qquad \text{(E9-8.9)}$$

$$r_C = 3\, k_{2B} C_B \qquad \text{(E9-8.10)}$$

$$N_i = C_i V \qquad \text{(E9-8.11)}$$

$$V = V_0 + v_0 t \qquad \text{(E9-8.12)}$$

$$N_{H_2 SO_2} = (C_{H_2 SO_{4,0}}) V_0 = \frac{1 \text{ mol}}{\text{dm}^3} \times 100 \text{ dm}^3 = 100 \text{ mol}$$

$$F_{A0} = \frac{4 \text{ mol}}{\text{dm}^3} \times 240 \frac{\text{dm}^3}{\text{h}} = 960 \frac{\text{mol}}{\text{h}}$$

Energy balance:

$$\frac{dT}{dt} = \frac{UA(T_a - T) - \sum F_{i0} C_{pi}(T - T_0) + \sum_{i=1}^{q} \Delta H_{Rxij} r_{ij} V}{\sum N_i C_{pi}} \qquad \text{(6-39)}$$

$$\frac{dT}{dt} = \frac{UA(T_a - T) - F_{A0} C_{pA}(T - T_0) + [(\Delta H_{Rx1A})(r_{1A}) + (\Delta H_{Rx2B})(r_{2B})] V}{[C_A C_{pA} + C_B C_{pB} + C_C C_{pC}] V + N_{H_2 SO_4} V_0 C_{pH_2 SO_4}}$$

$$\text{(E9-8.13)}$$

$$\frac{dT}{dt} = \frac{35,000\,(298 - T) - (4)(240)(30)(T - 300) + [(-6500)(-k_{1A}C_A) + (+8000)(-k_{2B}C_B)]\,V}{(30C_A + 60C_B + 20C_C)(100 + 240t) + (100)(35)}$$

(E9-8.14)

Equations (E9-8.1) through (E9-8.3) and (E9-8.8) through (E9-8.14) can be solved simultaneously with Equation (E9-8.14) using an ODE solver. The POLYMATH program is shown in Table E9-8.1 and the MATLAB program is on the CD-ROM. The time graphs are shown in Figures E9-8.1 and E9-8.2.

TABLE E9-8.1.   POLYMATH PROGRAM

Equations	Initial Values
d(Ca)/d(t)=ra+(Cao-Ca)*vo/V	1
d(Cb)/d(t)=rb-Cb*vo/V	0
d(Cc)/d(t)=rc-Cc*vo/V	0
d(T)/d(t)=(35000*(298-T)-Cao*vo*30*(T-305)+((-6500)*(-k1a*Ca)+(8000)*(-k2b*Cb))*V)/((Ca*30+Cb*60+Cc*20)*V+100*35)	290
Cao=4	
vo=240	
k1a=1.25*exp((9500/1.987)*(1/320-1/T))	
k2b=0.08*exp((7000/1.987)*(1/290-1/T))	
ra=-k1a*Ca	
V=100+vo*t	
rc=3*k2b*Cb	
rb=k1a*Ca/2-k2b*Cb	
$t_0 = 0,$ $t_f = 1.5$	

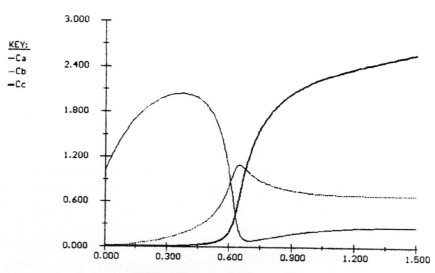

KEY:
—Ca
···Cb
—Cc

Figure E9-8.1   Concentration–time.

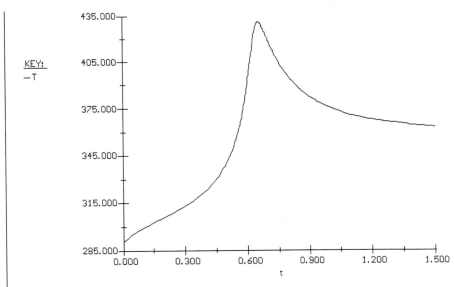

Figure E9-8.2 Temperature–time.

9.6 Unsteady Operation of Plug-Flow Reactors

In the CD-ROM, the unsteady energy balance is derived for a PFR. Neglecting changes in total pressure and shaft work, the following equation is derived:

Transient energy balance on a PFR

$$Ua(T_a - T) - \left(\sum_{i=1}^{n} F_i C_{p_i} \right) \frac{\partial T}{\partial V} + (-r_A)[-\Delta H_{Rx}(T)] = \sum_{i=1}^{n} C_i C_{p_i} \frac{\partial T}{\partial t} \quad (9\text{-}32)$$

This equation must be coupled with the mole balances:

Numerical solution required for these three coupled equations

$$\frac{-\partial F_i}{\partial V} + v_i(-r_A) = \frac{-\partial C_i}{\partial t} \quad (9\text{-}33)$$

and the rate law,

$$-r_A = k(T) \cdot \text{fn}(C_i) \quad (9\text{-}34)$$

and solved numerically. A variety of numerical techniques for solving equations of this type can be found in the book *Applied Numerical Methods*.[5]

---

[5] B. Carahan, H. A. Luther, and J. O. Wilkes, *Applied Numerical Methods* (New York: Wiley, 1969).

SUMMARY

1. Unsteady operation of CSTRs and semibatch reactors

$$\frac{dT}{dt} = \frac{\dot{Q} - \dot{W}_S - \sum\limits_{i=1}^{n} F_{i0}\tilde{C}_{P_i}(T - T_{i0}) + [-\Delta H_{Rx}(T)](-r_A V)}{\sum\limits_{i=1}^{n} N_i C_{p_i}} \qquad \text{(S9-1)}$$

For large heat-exchange coolant rates $(T_{a1} = T_{a2})$

$$\dot{Q} = UA(T_a - T) \qquad \text{(S9-2)}$$

For moderate to low coolant rates

$$\dot{Q} = \dot{m}_C C_{pC}(T - T_{a1})\left[1 - \exp\left(-\frac{UA}{\dot{m}_C C_p}\right)\right] \qquad \text{(S9-3)}$$

2. Batch reactors
   a. Nonadiabatic

$$\frac{dT}{dt} = \frac{\dot{Q} - \dot{W}_s + (-\Delta H_{Rx})(-r_A V)}{N_{A0}(\sum \Theta_i C_{pi} + \Delta C_p X)} \qquad \text{(S9-4)}$$

$$\frac{dT}{dt} = \frac{\dot{m}_C C_{pc}(T_{a1} - T)[1 - \exp(-UA/\dot{m}_C C_{pc})] + (r_A V)(\Delta H_{Rx}) - \sum F_{i0}\tilde{C}_{pi}(T - T_0)}{\sum N_i C_{pi}}$$

$$\text{(S9-5)}$$

   b. Adiabatic

$$\boxed{X = \frac{C_{ps}(T - T_0)}{-\Delta H_{Rx}(T)} = \frac{\sum \Theta_i \tilde{C}_{pi}(T - T_0)}{-\Delta H_{Rx}(T)}} \qquad \text{(S9-6)}$$

$$\boxed{T = T_0 + \frac{[-\Delta H_{Rx}(T_0)]X}{C_{ps} + X\,\Delta\hat{C}_p} = T_0 + \frac{[-\Delta H_{Rx}(T_0)]X}{\sum\limits_{i=1}^{n} \Theta_i \tilde{C}_{pi} + X\,\Delta\hat{C}_p}} \qquad \text{(S9-7)}$$

3. Startup of a CSTR (Figure S-1) and the approach to the steady state (CD-ROM). By mapping out regions of the concentration–temperature phase plane, one can view the approach to steady state and learn if the practical stability limit is exceeded.

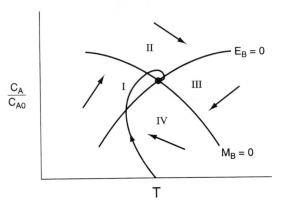

Figure S-1 Startup of a CSTR.

4. Control of chemical reactors
   - Proportional action ($e = Y_{sp} - Y$):

$$Z = Z_0 + k_C e \qquad (S9\text{-}8)$$

   - Integral action:

$$Z = Z_0 + \frac{k_C}{\tau_I} \int_0^t e\, dt \qquad (S9\text{-}9)$$

   - Proportional-integral actions:

$$Z = Z_0 + k_C e + \frac{k_c}{\tau_I} \int_0^t e\, dt \qquad (S9\text{-}10)$$

5. Multiple reactions (q reactions and n species)

$$\frac{dT}{dt} = \frac{\dot{m}_C C_{pc}(T_{a1} - T)[1 - \exp(-UA/\dot{m}_C C_{pc})] + \sum_{i=1}^{q} r_{ij} V \Delta H_{\text{Rx}ij}(T) - \sum F_{i0}\tilde{C}_{pi}(T - T_0)}{\sum N_i C_{pi}}$$

$$(S9\text{-}11)$$

QUESTIONS AND PROBLEMS

The subscript to each of the problem numbers indicates the level of difficulty: A, least difficult; D, most difficult.

A = ●   B = ■   C = ◆   D = ◆◆

In each of the questions and problems below, rather than just drawing a box around your answer, write a sentence or two describing how you solved the problem, the assumptions you made, the reasonableness of your answer, what you learned, and any other facts you want to include. You may wish to refer to W. Strunk and E. B. White, *The Elements of Style* (New York: Macmillian, 1979) and Joseph M. Williams, *Style: Ten Lessons in Clarity & Grace* (Glenview, Ill.: Scott, Foresman, 1989) to enhance the quality of your sentences.

P9-1$_A$ Read over the problems at the end of this chapter. Refer to the guidelines given in Problem 4-1, and make up an original problem that uses the concepts presented in this chapter. To obtain a solution:

(a) Make up your data and reaction.

(b) Use a real reaction and real data.

Also,

(c) Prepare a list of safety considerations for designing and operating chemical reactors.

See R. M. Felder, *Chem. Eng. Educ.*, *19*(4), 176 (1985). The August 1985 issue of *Chemical Engineering Progress* may be useful for part (c).

P9-2$_A$ Review the example problems in this chapter, choose one, and use a software package such as POLYMATH or MATLAB to carry out a parameter sensitivity analysis.

What if...

(a) Explore the ONCB explosion described in Example 9-2. Show that no explosion would have occurred if the cooling was not shut off for the 9.04-kmol charge of ONCB or if the cooling was shut off for 10 min after 45 min of operation for the 3.17-kmol ONCB charge. Show that if the cooling had been shut off for 10 min after 12 h of operation, no explosion would have occurred for the 9.04-kmol charge. Develop a set of guidelines as to when the reaction should be quenched should the cooling fail. Perhaps safe operation could be discussed using a plot of the time after the reaction began at which the cooling failed, t_0, versus the length of the cooling failure period, t_f, for the different charges of ONCB. Parameter values used in this example predict that the reactor will explode at midnight. What parameter values would predict the time the reactor would explode at the actual time of 18 min after midnight? Find a set of parameter values that would cause the explosion to occur at exactly 12:18 A.M. For example, include heat capacities of metal reactor and/or make a new estimate of *UA*. Finally, what if a 1/2-in. rupture disk rated at 800 psi had been installed and did indeed rupture at 800 psi (270°C)? The explosion still would have occurred. (*Note*: The mass flow rate $\dot{m}$ varies with the cross-sectional area of the disk. Consequently, for the conditions of the reaction the maximum mass flow rate out of the 1/2-in. disk can be found by comparing it with the mass flow rate of 830 lb/min of the 2-in. disk.

(b) Rework Example 9-4, *Startup of a CSTR*, for an entering temperature of 70°F, an initial reactor temperature of 160°F, and an initial concentration of polyethylene oxide of 0.1 *M*. Try other combinations of T_0, T_i, and C_{Ai} and report your results in terms of temperature–time trajectories and temperature–concentration phase planes.

(c) Rework Example 9-5, *Falling Off the Upper Steady State*. Try varying the entering temperature, T_0, to between 80 and 68°F and plot the steady-state conversion as a function of T_0. Vary the coolant rate between 10,000 and 400 mol/h. Plot conversion and reactor temperature as a function of coolant rate.

(d) Rework Example 9-6, concerning integral control of the CSTR, and vary the gain, k_C, between 0.1 and 500. Is there a lower value of k_C that will cause the reactor to fall to the lower steady state or an upper value to cause it to become unstable? What would happen if T_0 were to fall to 65°F or 60°F?

(e) Rework Example 9-7, concerning a CSTR with a PI controller, to learn the effects of the parameters k_C and τ_I. Which combination of parameter values generates the least and greatest oscillations in temperature? Which values of k_C and τ_I return the reaction to steady state the quickest?

P9-3$_B$ The following is an excerpt from *The Morning News*, Wilmington, Delaware (August 3, 1977): "Investigators sift through the debris from blast in quest for the cause [that destroyed the new nitrous oxide plant]. A company spokesman said it appears more likely that the [fatal] blast was caused by another gas— ammonium nitrate—used to produce nitrous oxide." An 83% (wt) ammonium nitrate and 17% water solution is fed at 200°F to the CSTR operated at a temperature of about 510°F. Molten ammonium nitrate decomposes directly to produce gaseous nitrous oxide and steam. It is believed that pressure fluctuations were observed in the system and as a result the molten ammonium nitrate feed to the reactor may have been shut off approximately 4 min prior to the explosion. Can you explain the cause of the blast? If the feed rate to the reactor just before shutoff was 310 lb of solution per hour, what was the exact temperature in the reactor just prior to shutdown? Using the data below, calculate the time it took to explode after the feed was shut off for the reactor. How would you start up or shut down and control such a reaction?

Assume that at the time the feed to the CSTR stopped, there was 500 lb of ammonium nitrate in the reactor at a temperature of 520°F. The conversion in the reactor is virtually complete at about 99.99%. Additional data for this problem are given in Problem 8-3. How would your answer change if 100 lb of solution were in the reactor? 310 lb? 800 lb?

P9-4$_B$ The first-order irreversible reaction

$$A(l) \longrightarrow B(g) + C(g)$$

is carried out adiabatically in a CSTR into which 100 mol/min of pure liquid A is fed at 400 K. The reaction goes virtually to completion (i.e., the feed rate into the reactor equals the reaction rate inside the reactor).

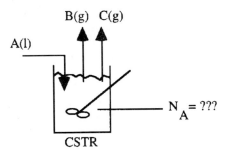

How many moles of liquid A are in the CSTR under steady-state conditions? Plot the temperature and moles of A in the reactor as a function of time after the feed to the reactor has been shut off.

Additional information:

Temperature (K)	400	800	1200
k (min^{-1})	0.19	0.32	2.5
H_A (kJ/mol)	−38	−30	−22
H_B (kJ/mol)	−26	−22	−18
H_C (kJ/mol)	−20	−16	−12

P9-5$_B$ The liquid-phase reaction in Problem P8-5 is to be carried out in a semibatch reactor. There is 500 mol of A initially in the reactor at 25°C. Species B is fed to the reactor at 50°C and a rate of 10 mol/min. The feed to the reactor is stopped after 500 mol of B has been fed.

(a) Plot the temperature and conversion as a function of time when the reaction is carried out adiabatically. Calculate to $t = 2$ h.

(b) Plot the conversion as a function of time when a heat exchanger ($UA = 100$ cal/min·K) is placed in the reactor and the ambient temperature is constant at 50°C. Calculate to $t = 3$ h.

(c) Repeat part (b) for the case where the reverse reaction cannot be neglected.

New parameter values:

$k = 0.01$ (dm^3/mol · min) at 300 K with $E = 10$ kcal/mol
$V_0 = 50$ dm^3, $v_0 = 1$ dm^3/min, $C_{A0} = C_{B0} = 10$ mol/dm^3
For the reverse reaction: $k_r = 10$ s^{-1} at 300 K with $E_r = 16$ kcal/mol.

P9-6$_B$ You are operating a batch reactor and the reaction is first-order, liquid-phase, and exothermic. An inert coolant is added to the reaction mixture to control the temperature. The temperature is kept constant by varying the flow rate of the coolant (see Figure P9-6).

$$A \longrightarrow B$$

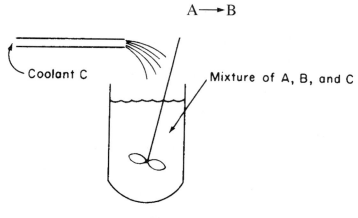

Coolant C   Mixture of A, B, and C

Figure P9-6

(a) Calculate the flow rate of the coolant 2 h after the start of the reaction. (*Ans.:* $F_C = 3.157$ lb/s.)

(b) It is proposed that rather than feeding a coolant to the reactor, a solvent be added that can be easily boiled off, even at moderate temperatures. The solvent has a heat of vaporization of 1000 Btu/lb and initially there are 25 lb mol of A placed in the tank. The initial volume of solvent and reactant is 300 ft^3. Determine the solvent evaporation rate as a function of time. What is the rate at the end of 2 h?

Additional information:

Temperature of reaction: 100°F
Value of k at 100°F: 1.2×10^{-4} s^{-1}

Temperature of coolant: 80°F

Heat capacity of all components: 0.5 Btu/lb·°F

Density of all components: 50 lb/ft³

ΔH°_{Rx}: $-25,000$ Btu/lb mol

Initially:

Vessel contains only A (no B or C present)

C_{A0}: 0.5 lb mol/ft³

Initial volume: 50 ft³

P9-7$_B$ The reaction

$$A + B \longrightarrow C$$

is carried out adiabatically in a constant-volume batch reactor. The rate law is

$$-r_A = k_1 C_A^{1/2} C_B^{1/2} - k_2 C_C$$

Plot the conversion and the concentrations of the reacting species as a function of time.

Additional information:

Entering Temperature = 100°C

k_1 (373 K) = 2 × 10⁻³ s⁻¹	E_1 = 100 kJ/mol
k_2 (373 K) = 3 × 10⁻⁵ s⁻¹	E_2 = 150 kJ/mol
C_{A0} = 0.1 mol/dm³	C_{pA} = 25 J/mol·K
C_{B0} = 0.125 mol/dm³	C_{pB} = 25 J/mol·K
ΔH°_{Rx}(298 K) = $-40,000$ J/mol A	C_{pC} = 40 J/mol·K

P9-8$_B$ Calculate the time necessary to achieve 90% adiabatic equilibrium conversion A in the reaction described in Problem CDP8-A when this reaction is carried out adiabatically in a batch reactor. The 50-ft³ reactor is charged (filled) with 900 lb mol which is 66.67% A and 33.33% inerts.

P9-9$_B$ The first order exothermic liquid-phase reaction

$$A \longrightarrow B$$

is carried out at 85°C in a jacketed 0.2-m³ CSTR. The coolant temperature in the reactor is 32°F. The heat-transfer coefficient is 120 W/m²·K. Determine the critical value of the heat-transfer area below which the reaction will run away and the reactor will explode [*Chem. Eng.*, *91*(10), 54 (1984)].

Additional information:

Specific reaction rate:

k = 1.1 min⁻¹ at 40°C

k = 3.4 min⁻¹ at 50°C

The heat capacity of the solution is 2 J/g·K. The solution density is 0.90 kg/dm³. The heat of reaction is -250 J/g. The feed temperature is 40°C and the feed rate is 90 kg/min. MW of A = 90 g/mol. C_{A0} = 2 *M*.

P9-10$_A$ Study the startup of a CSTR that initially does not contain any reactant. Plot concentration and temperature as a function of time using the data and reactions from the following problems:

(a) P8-12 with τ = 500 s.

(b) P8-5 with τ = 50 s.

P9-11$_B$ The elementary irreversible liquid-phase reaction

$$A + 2B \longrightarrow C$$

is to be carried out in a semibatch reactor in which B is fed to A. The volume of A in the reactor is 10 dm³, the initial concentration of A in the reactor is 5 mol/dm³, and the initial temperature in the reactor is 27°C. Species B is fed at a temperature of 52°C and a concentration of 4 *M*. It is desired to obtain at least 80% conversion of A in as short a time as possible, but at the same time the temperature of the reactor must not rise above 130°C. You should try to make approximately 120 mol of C in a 24 hr day allowing for 30 minutes to empty and fill the reactor between each batch. The coolant flowrate through the reactor is 2000 mol/min. There is a heat exchanger in the reactor.

(a) What volumetric feed rate (dm³/min) do you recommend?
(b) How would your answer or strategy change if the maximum coolant rate dropped to 200 mol/min?

Additional information:

$\Delta H_{Rx}^{\circ} = -55,000$ cal/mol A
$C_{pA} = 35$ cal/mol·K, $C_{pB} = 20$ cal/mol·K, $C_{pC} = 75$ cal/mol·K

$k = 0.0005 \dfrac{dm^6}{mol^2 \cdot min}$ at 27°C with $E = 8000$ cal/mol

$UA = 250 \dfrac{cal}{min \cdot K}$ with $T_a = 17°C$

C_p(coolant) = 18 cal/mol·K   [Old exam]

P9-12$_B$ Rework Example 9-6 using
(a) Only a proportional controller.
(b) Only an integral controller.
(c) A combined proportional and integral controller.

P9-13$_B$ Apply the different types of controllers to the reactions in Problem P9-10.

P9-14$_B$ **(a)** Rework Example 9-6 for the case of a 5°F decrease in the outlet temperature when the controlled input variable is the reactant feed rate.
(b) Consider a 5°F drop in the ambient temperature, T_a, when the controlled variable is the inlet temperature, 01.
(c) Use each of the controllers (P with $k_C = 10$, I with $y_J = 1$, D with $\tau_D = 0.1$) to keep the reactor temperature at the unstable steady state (i.e., $T = 112.5°F$ and $X = 0.3$).

P9-15$_B$ Rework Problem P9-3 for the case when a heat exchanger with $UA = 10,000$ Btu/h·ft² and a control system are added and the mass flow rate is increased to 310 lb of solution per hour.
(a) Plot temperature and mass of ammonium nitrate in the tank as a function of time when there is no control system on the reactor. Assume that all the ammonium nitrate reacts and show that the mass balance is

$$\frac{dM_A}{dt} = \dot{m}_{A0} - kM_A$$

There is 500 lb of A in the CSTR and the reactor temperature, T, is 516°F time at $t = 0$.

(b) Plot T and M_A as a function of time when a proportional controller is added to control T_a in order to keep the reactor temperature at 516°F. The controller gain, k_C, is -5 with T_{a0} set at 975°R.

(c) Plot T and M_A versus time when a PI controller is added with $\tau_I = 1$.

(d) Plot T and M_A versus time when two PI controllers are added to the reactor: one to control T and a second to control M by manipulating the feed rate $\dot{m}_{A0}$.

$$\dot{m}_{A0} = \dot{m}_{A00} + \frac{k_{C2}}{\tau_{12}} I_M + k_{C2}(M - M_{sp})$$

with

$$M_{sp} = 500 \text{ lb}$$

$$k_{C2} = 25 \text{ h}^{-1}$$

$$\tau_{12} = 1 \text{ h}$$

P9-16$_B$ The elementary liquid phase reaction

$$A \longrightarrow B$$

is carried out in a CSTR. Pure A is fed at a rate of 200 lb mol/hr at 530 R and a concentration of 0.5 lb mol/ft³. [M. Shacham, N. Brauner, and M. B. Cutlip, *Chem. Engr. Edu. 28*, No. 1, p. 30, Winter (1994).] The mass density of the solution is constant at 50 lb/ft³.

(a) Plot $G(T)$ and $R(T)$ as a function of temperature.

(b) What are the steady state concentrations and temperatures? [One answer $T = 628.645$ R, $C_A = 0.06843$ lb mol/ft³.] Which ones are stable? What is the extinction temperature?

(c) Apply the unsteady-state mole and energy balances to this system. Consider the upper steady state. Use the values you obtained in part (b) as your initial values to plot C_A and T versus time up to 6 hours and then to plot C_A vs. T. What did you find? Do you want to change any of your answers to part (b)?

(d) Expand your results for part (c) by varying T_o and T_a. [Hint: Try $T_o = 590$ R]. Describe what you find.

(e) What are the parameters in part (d) for the other steady states? Plot T and C_A as a function of time using the steady-state values as the initial conditions at the lower steady state by value of $T_o = 550$ R and $T_o = 560$ R. Compare your concentration and temperature versus time trajectories with those of $x1$ and $y1$ determined from linearized stability theory. Vary T_o.

(f) Explore this problem. Write a paragraph describing your results and what you learned from this problem.

(g) Carry out a linearized stability analysis. What are your values for τ, J, L, M, and N? What are the roots m_1 and m_2? See the CD-ROM for lecture notes on linearized stability.

(h) Normalize x and y by the steady-state values, $x1 = x/C_{As}$ and $y1 = y/T_s$, and plot $x1$ and $y1$ as a function of time and also $x1$ as a function of $y1$ [Hint: First try initial values of x and y of 0.02 and 2, respectively.] and use Equations 14 and 15 of the CD-ROM notes for Lecture 35.

Additional information:

$$v = 400 \text{ ft}^3/\text{hr} \qquad\qquad A = 250 \text{ ft}^2$$
$$C_{A0} = 0.50 \text{ mol/ft}^3 \qquad T_a = 530 \text{ R}$$
$$V = 48 \text{ ft}^3 \qquad\qquad\qquad T_o = 530 \text{ R}$$
$$A = 1.416 \times 10^{12} \text{ hr}^{-1} \qquad \Delta H_{\text{Rx}} = -30,000 \text{ BTU/lb mol}$$
$$E = 30,000 \text{ BTU/lb mol} \qquad C_p = 0.75 \text{ BTU/lbm-}°\text{R}$$
$$R = 1.987 \text{ BTU/lb mol }°\text{R} \qquad \rho = 50 \text{ lbm/ft}^3$$
$$U = 150 \text{ BTU/hr-ft}^2\text{-}°\text{R}$$

P9-17$_B$ The reactions in Example 8-12 are to be carried out in a batch reactor. Plot the temperature and the concentrations of A, B, and C as a function of time for the following cases:
 (a) Adiabatic operation.
 (b) Values of UA of 10,000, 40,000, and 100,000 J/min·K.
 (c) Use $UA = 40,000$ J/min·K and different initial reactor temperatures.

P9-18$_B$ The reaction in Problem P8-31 is to be carried out in a semibatch reactor.
 (a) How would you carry out this reaction (i.e., T_0, v_0, T_i)? The molar concentration of pure A and pure B are 5 and 4 mol/dm^3 respectively. Plot concentrations, temperatures, and the overall selectivity as a function of time for the conditions you chose.
 (b) Vary the reaction orders for each reaction and describe what you find.
 (c) Vary the heats of reaction and describe what you find.

CD-ROM MATERIAL

- **Learning Resources**
  1. *Summary Notes for Lectures 17, 18, and 35B*
  4. *Solved Problems*
     Example CD9–1 Startup of a CSTR
     Example CD9–2 Falling Off The Steady State
     Example CD9–3 Proportional Integral (PI) Control
- **Living Example Problems**
  1. *Example 9–1 Adiabatic Batch Reactor*
  2. *Example 9–2 Safety in Chemical Plants with Exothermic Reactions*
  3. *Example 9–3 Heat Effects in a Semibatch Reactor*
  4. *Example 9–4 Startup of a CSTR*
  5. *Example 9–5 Falling Off the Steady State*
  6. *Example 9–6 Integral Control of a CSTR*
  7. *Example 9–7 Proportion-Integral Control of a CSTR*
  8. *Example 9–8 Multiple Reactions in a Semibatch Reactor*
- **Professional Reference Shelf**
  1. *Intermediate Steps in Adiabatic Batch Reactor Derivation*
  2. *Approach to the Steady-State Phase-Plane Plots and Trajectories of Concentration versus Temperature*
  3. *Unsteady Operation of Plug Flow Reactors*
- **Additional Homework Problems**

CDP9-A$_B$ The production of propylene glycol discussed in Examples 8-4, 9-4, 9-5, 9-6, and 9-7 is carried out in a semibatch reactor.

CDP9-B$_C$ Reconsider Problem P9-14 when a PI controller is added to the coolant stream.

SUPPLEMENTARY READING

1. A number of solved problems for batch and semibatch reactors can be found in

WALAS, S. M., *Chemical Reaction Engineering Handbook*. Amsterdam: Gordon and Breach, 1995, pp. 386–392, 402, 460–462, and 469.

2. Basic control textbooks

SEBORG, D. E., T. F. EDGAR, and D. A. MELLICHAMP, *Process Dynamics and Control*. New York: Wiley, 1989.

OGUNNAIKE, B. A. and W. H. RAY, *Process Dynamics, Modeling and Control*. Oxford: Oxford University Press, 1994.

3. A nice historical perspective of process control is given in

EDGAR, T. F., "From the Classical to the Postmodern Era" *Chem. Eng. Educ.*, *31*, 12 (1997).

Catalysis and Catalytic Reactors **10**

> It isn't that they can't see the solution. It is that they can't see the problem.
>
> G. K. Chesterton

The objectives of this chapter are to develop an understanding of catalysts, reaction mechanisms, and catalytic reactor design. Specifically, after reading this chapter one should be able to (1) define a catalyst and describe its properties, (2) describe the steps in a catalytic reaction and apply the concept of a rate-limiting step to derive a rate law, (3) develop a rate law and determine the rate-law parameters from a set of gas–solid reaction rate data, (4) describe the different types of catalyst deactivation, determine an equation for catalytic activity from concentration–time data, define temperature–time trajectories to maintain a constant reaction rate, and (5) calculate the conversion or catalyst weight for packed (fixed) beds, moving beds, well-mixed (CSTR) and straight-through (STTR) fluid-bed reactors for both decaying and nondecaying catalysts. The various sections of this chapter roughly correspond to each of these objectives.

10.1 Catalysts

Catalysts have been used by humankind for over 2000 years.[1] The first observed uses of catalysts were in the making of wine, cheese, and bread. It was found that it was always necessary to add small amounts of the previous batch to make the current batch. However, it wasn't until 1835 that Berzelius began to tie together observations of earlier chemists by suggesting that small amounts of a foreign source could greatly affect the course of chemical reactions. This mysterious force attributed to the substance was called catalytic. In

---

[1] S. T. Oyama and G. A. Somorjai, *J. Chem. Educ., 65,* 765 (1986).

1894, Ostwald expanded Berzelius' explanation by stating that catalysts were substances that accelerate the rate of chemical reactions without being consumed. In over 150 years since Berzelius' work, catalysts have come to play a major economic role in the world market. In the United States alone, sales of process catalysts in 1996 were over $1 billion, the major uses being in petroleum refining and in chemical production.

10.1.1 Definitions

A *catalyst* is a substance that affects the rate of a reaction but emerges from the process unchanged. A catalyst usually changes a reaction rate by promoting a different molecular path ("mechanism") for the reaction. For example, gaseous hydrogen and oxygen are virtually inert at room temperature, but react rapidly when exposed to platinum. The reaction coordinate shown in Figure 10-1 is a measure of the progress along the reaction path as H_2 and O_2 approach each other and pass over the activation energy barrier to form H_2O. *Catalysis* is the occurrence, study, and use of catalysts and catalytic processes. Commercial chemical catalysts are immensely important. Approximately one-third of the material gross national product of the United States involves a catalytic process somewhere between raw material and finished product.[2] The development and use of catalysts is a major part of the constant search for new ways of increasing product yield and selectivity from chemical reactions. Because a catalyst makes it possible to obtain an end product by a different pathway (e.g. a lower energy barrier), it can affect both the yield and the selectivity.

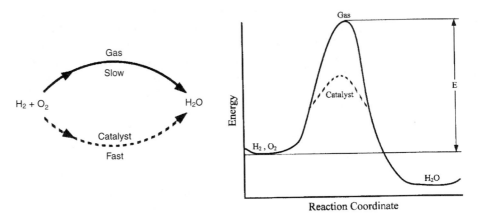

Figure 10-1   Different reaction paths.

Normally when we talk about a catalyst, we mean one that speeds up a reaction, although strictly speaking, a catalyst can either accelerate or slow the formation of a particular product species. *A catalyst changes only the rate of a reaction; it does not affect the equilibrium.*

---

[2] V. Haensel and R. L. Burwell, Jr., *Sci. Am.*, *225*(10), 46.

Homogeneous catalysis concerns processes in which a catalyst is in solution with at least one of the reactants. An example of homogeneous catalysis is the industrial Oxo process for manufacturing normal isobutylaldehyde. It has propylene, carbon monoxide, and hydrogen as the reactants and a liquid-phase cobalt complex as the catalyst.

$$CH_3\!-\!CH\!=\!CH_2 + CO + H_2 \xrightarrow{\ Co\ } \begin{cases} \underset{\displaystyle CH_3}{\overset{\displaystyle CHO}{CH_3\!-\!CH}} \\[2ex] CH_3\!-\!CH_2\!-\!CH_2\!-\!CHO \end{cases}$$

A *heterogeneous catalytic process* involves more than one phase; usually the catalyst is a solid and the reactants and products are in liquid or gaseous form. Much of the benzene produced in this country today is manufactured from the dehydrogenation of cyclohexane (obtained from the distillation of crude petroleum) using platinum-on-alumina as the catalyst:

cyclohexane   benzene   hydrogen

Sometimes the reacting mixture is in both the liquid and gaseous forms, as in the hydrodesulfurization of heavy petroleum fractions. Of these two types of catalysis, heterogeneous catalysis is the more common type. The simple and complete separation of the fluid product mixture from the solid catalyst makes heterogeneous catalysis economically attractive, especially because many catalysts are quite valuable and their reuse is demanded. Only heterogeneous catalysts will be considered in this chapter.

Examples of heterogeneous catalytic reactions

A heterogeneous catalytic reaction occurs at or very near the fluid–solid interface. The principles that govern heterogeneous catalytic reactions can be applied to both catalytic and noncatalytic fluid–solid reactions. These two other types of heterogeneous reactions involve gas–liquid and gas–liquid–solid systems. Reactions between gases and liquids are usually mass-transfer limited.

10.1.2 Catalyst Properties

Since a catalytic reaction occurs at the fluid–solid interface, a large interfacial area can be helpful or even essential in attaining a significant reaction rate. In many catalysts, this area is provided by a porous structure; the solid contains many fine pores, and the surface of these pores supplies the area needed for the high rate of reaction. The area possessed by some porous materials is surprisingly large. A typical silica-alumina cracking catalyst has a pore volume of 0.6 cm^3/g and an average pore radius of 4 nm. The corresponding surface area is 300 m^2/g.

Ten grams of this catalyst possess more surface area than a U.S. football field

A catalyst that has a large area resulting from pores is called a *porous catalyst*. Examples of these include the Raney nickel used in the hydrogenation of vegetable and animal oils, the platinum-on-alumina used in the reforming of petroleum naphthas to obtain higher octane ratings, and the promoted iron used in ammonia synthesis. Sometimes pores are so small that they will admit small molecules but prevent large ones from entering. Materials with this type of pore are called *molecular sieves*, and they may be derived from natural substances such as certain clays and zeolites, or be totally synthetic, such as some crystalline aluminosilicates (see Figure 10-2). These sieves can form the basis

Catalyst types:
· Porous
· Molecular sieves
· Monolithic
· Supported
· Unsupported

Typical
zeolite catalyst

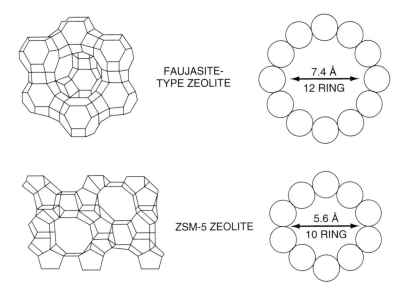

FAUJASITE-
TYPE ZEOLITE   7.4 Å
                            12 RING

ZSM-5 ZEOLITE   5.6 Å
                          10 RING

Figure 10-2 Framework structures and pore cross sections of two types of zeolites. (Top) Faujasite-type zeolite has a three-dimensional channel system with pores at least 7.4 Å in diameter. A pore is formed by 12 oxygen atoms in a ring. (Bottom) ZSM-5 zeolite has interconnected channels running in one direction, with pores 5.6 Å in diameter. ZSM-5 pores are formed by 10 oxygen atoms in a ring. [From N. Y. Chen and T. F. Degnan, *Chem. Eng. Prog.*, *84*(2), 33 (1988). Reproduced by permission of the American Institute of Chemical Engineers. Copyright © 1988 AIChE. All rights reserved.]

for quite selective catalysts; the pores can control the residence time of various molecules near the catalytically active surface to a degree that essentially allows *only* the desired molecules to react. For example, once inside the zeolite, the configuration of the reacting molecules may be able to be controlled by placement of the catalyst atoms at specific sites in the zeolite. This placement would facilitate cyclization reactions, such as orienting ethane molecules in a ring on the surface of the catalyst so that they form benzene:

$$\bigcirc \longrightarrow \bigcirc + 6H_2$$

Zeolites can also be used to form ethane molecules

$$CH_4 + \tfrac{1}{2}O_2 \longrightarrow CO + 2H_2 \longrightarrow CH_3OH \xrightarrow[\text{zeolite}]{\text{nonmetal}} CH_3CH_3$$

Not all catalysts need the extended surface provided by a porous structure, however. Some are sufficiently active so that the effort required to create a porous catalyst would be wasted. For such situations one type of catalyst is the monolithic catalyst. Monolithic catalysts are normally encountered in processes where pressure drop and heat removal are major considerations. Typical examples include the platinum gauze reactor used in the ammonia oxidation portion of nitric acid manufacture and catalytic converters used to oxidize pollutants in automobile exhaust. They can be porous (honeycomb) or non-porous (wire gauze). A photograph of a automotive catalytic converter is shown in Figure CD11-2. Platinum is a primary catalytic material in the monolith.

In some cases a catalyst consists of minute particles of an active material dispersed over a less active substance called a *support*. The active material is frequently a pure metal or metal alloy. Such catalysts are called *supported catalysts*, as distinguished from *unsupported catalysts*, whose active ingredients are major amounts of other substances called *promoters*, which increase the activity. Examples of supported catalysts are the automobile-muffler catalysts mentioned above, the platinum-on-alumina catalyst used in petroleum reforming, and the vanadium pentoxide on silica used to oxidize sulfur dioxide in manufacturing sulfuric acid. On the other hand, the platinum gauze for ammonia oxidation, the promoted iron for ammonia synthesis, and the silica–alumina dehydrogenation catalyst used in butadiene manufacture typify unsupported catalysts.

Most catalysts do not maintain their activities at the same levels for indefinite periods. They are subject to *deactivation*, which refers to the decline in a catalyst's activity as time progresses. Catalyst deactivation may be caused by an *aging* phenomenon, such as a gradual change in surface crystal structure, or by the deposit of a foreign material on active portions of the catalyst surface. The latter process is called *poisoning* or fouling of the catalyst. Deactivation may occur very fast, as in the catalytic cracking of petroleum naphthas, where the deposit of carbonaceous material (*coking*) on the catalyst requires that the catalyst be removed after only a couple of minutes in the reaction zone. In other processes poisoning might be very slow, as in automotive exhaust catalysts, which gradually accumulate minute amounts of lead even if unleaded gasoline is used because of residual lead in the gas station storage tanks.

For the moment, let us focus our attention on gas-phase reactions catalyzed by solid surfaces. For a catalytic reaction to occur, at least one and frequently all of the reactants must become attached to the surface. This attachment is known as *adsorption* and takes place by two different processes: physical adsorption and chemisorption. *Physical adsorption* is similar to condensation. The process is exothermic, and the heat of adsorption is relatively small, being on the order of 1 to 15 kcal/g mol. The forces of attraction between the gas molecules and the solid surface are weak. These van der Waals forces consist of interaction between permanent dipoles, between a permanent

Deactivation by:
· Aging
· Poisoning
· Coking

dipole and an induced dipole, and/or between neutral atoms and molecules. The amount of gas physically adsorbed decreases rapidly with increasing temperature, and above its critical temperature only very small amounts of a substance are physically adsorbed.

The type of adsorption that affects the rate of a chemical reaction is *chemisorption*. Here, the adsorbed atoms or molecules are held to the surface by valence forces of the same type as those that occur between bonded atoms in molecules. As a result the electronic structure of the chemisorbed molecule is perturbed significantly, causing it to be extremely reactive. Figure 10-3 shows the bonding from the adsorption of ethylene on a platinum surface to form chemisorbed ethylidyne. Like physical adsorption, chemisorption is an exothermic process, but the heats of adsorption are generally of the same magnitude as the heat of a chemical reaction (i.e., 10 to 100 kcal/g mol). If a catalytic reaction involves chemisorption, it must be carried out within the temperature range where chemisorption of the reactants is appreciable.

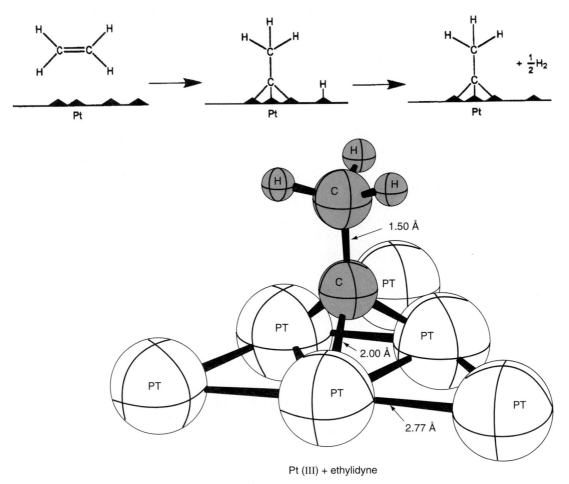

Pt (III) + ethylidyne

Figure 10-3 Ethylidyne as chemisorbed on platinum. (Adapted from G. A. Somorjai, *Introduction to Surface Chemistry and Catalysis*, Wiley, New York, 1994.)

Chemisorption on
active sites is what
catalyzes the
reaction
In a landmark contribution to catalytic theory, Taylor[3] suggested that a reaction is not catalyzed over the entire solid surface but only at certain *active sites* or centers. He visualized these sites as unsaturated atoms in the solids that resulted from surface irregularities, dislocations, edges of crystals, and cracks along grain boundaries. Other investigators have taken exception to this definition, pointing out that other properties of the solid surface are also important. The active sites can also be thought of as places where highly reactive intermediates (i.e., chemisorbed species) are stabilized long enough to react. However, for our purposes we will define an *active site* as *a point on the catalyst surface that can form strong chemical bonds with an adsorbed atom or molecule.*

One parameter used to quantify the activity of a catalyst is the *turnover frequency, N.* It is the number of molecules reacting per active site per second at the conditions of the experiment. When a metal catalyst such as platinum is deposited on a support, the metal atoms are considered active sites. The *dispersion, D*, of the catalyst is the fraction of the metal atoms deposited that are on the surface.

Example 10–1   Turnover Frequency in Fisher–Tropsch Synthesis

$$CO + 3H_2 \longrightarrow CH_4 + H_2O$$

The Fisher–Tropsch synthesis was studied using a commercial 0.5 wt % Ru on γ-Al_2O_3.[4] The catalyst dispersion percentage of atoms exposed, determined from hydrogen chemisorption, was found to be 49%. At a pressure of 988 kPa and a temperature of 475 K, a turnover frequency of 0.044 s^{-1} was reported for methane. What is the rate of formation of methane in mol/s·g of catalyst (metal plus support)?

Solution

$$-r'_A = N_{CH_4} D \left(\frac{1}{MW_{Ru}} \right) \frac{\% \, Ru}{100}$$

$$= \frac{0.044 \text{ molecules}}{(\text{surface atom Ru}) \cdot s} \times \frac{1 \text{ mol } CH_4}{6.02 \times 10^{23} \text{ molecules}}$$

$$\times \frac{0.49 \text{ surface atoms}}{\text{total atoms Ru}} \times \frac{6.02 \times 10^{23} \text{ atoms Ru}}{\text{g atom (mol) Ru}}$$

$$\times \frac{\text{g atoms Ru}}{101.1 \text{ g Ru}} \times \frac{0.005 \text{ g Ru}}{\text{g total}}$$

$$= 1.07 \times 10^{-6} \text{ mol/s} \cdot \text{g catalyst} \qquad\qquad (E10\text{-}1.1)$$

---

[3] H. S. Taylor, *Proc. R. Soc. London, A108,* 105 (1928).

[4] R. S. Dixit and L. L. Tavlarides, *Ind. Eng. Chem. Process Des. Dev.,* 22, 1 (1983).

Figure 10-4 shows the range of turnover frequencies (molecules/site·s) as a function of temperature and type of reaction. One notes that the turnover frequency in Example 10-1 is in the same range as the frequencies shown in the hydrogenation box.

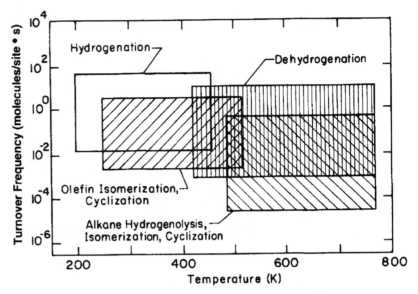

Figure 10-4 Range of turnover frequencies as a function for different reactions and temperatures. (Adapted from G. A. Somorjai, *Introduction to Surface Chemistry and Catalysis*, Wiley, New York, 1994.)

10.1.3 Classification of Catalysts

While platinum can be used for the reactions shown in Figure 10-4, we shall briefly discuss several classes of reactions and the catalysts used in each class.[5]

Alkylation and Dealkylation Reactions. *Alkylation* is the addition of an alkyl group to an organic compound. This type of reaction is commonly carried out in the presence of the Friedel–Crafts catalysts, $AlCl_3$ along with a trace of HCl. One such reaction is

$$C_4H_8 + i-C_4H_{10} \underset{\overset{AlCl_3}{\rightleftharpoons}}{} i-C_8H_{18}$$

A similar alkylation is the formation of ethyl benzene from benzene and ethylene:

$$C_6H_6 + C_2H_4 \longrightarrow C_6H_5C_2H_5$$

[5] J. H. Sinfelt, *Ind. Eng. Chem.*, *62*(2), 23 (1970); *62*(10), 66 (1970). Also, W. B. Innes, in P. H. Emmertt, Ed., *Catalysis*, Vol. 2, Reinhold, New York, 1955, p. 1.

The cracking of petroleum products is probably the most common dealkylation reaction. Silica-alumina, silica-magnesia, and a clay (montmorillonite) are common dealkylation catalysts.

Isomerization Reactions.   In petroleum production, the conversion of normal hydrocarbon chains to branched chains is important, since the latter has a higher octane number. Acid-promoted Al_2O_3 is a catalyst used in such isomerization reactions. Although this and other acid catalysts are used in isomerization reactions, it has been found that the conversion of normal paraffins to isoparaffins is easiest when both acid sites and hydrogenation sites are present, such as in the catalyst Pt on Al_2O_3.

Hydrogenation and Dehydrogenation Reactions.   The bonding strength between hydrogen and metal surfaces increases with an increase in vacant d-orbitals. Maximum catalytic activity will not be realized if the bonding is too strong and the products are not readily released. Consequently, this maximum in catalytic activity occurs when there is approximately one vacant d-orbital per atom. The most active metals for reactions involving hydrogen are generally Co, Ni, Rh, Ru, Os, Pd, Ir, and Pt. On the other hand, V, Cr, Cb, Mo, Ta, and W, each of which has a large number of vacant d-orbitals, are relatively inactive as a result of the strong adsorption for the reactants or the products or both. However, the oxides of Mo (MoO_2) and Cr (Cr_2O_3) are quite active for most reactions involving hydrogen. Dehydrogenation reactions are favored at high temperatures (at least 600°C), and hydrogenation reactions are favored at lower temperatures. Industrial butadiene, which has been used to produce synthetic rubber, can be obtained by the dehydrogenation of the butenes:

$$CH_3CH=CHCH_3 \xrightarrow{\text{catalyst}} CH_2=CHCH=CH_2 + H_2$$

(possible catalysts: calcium nickel phosphate, Cr_2O_3, etc.)

The same catalysts could also be used in the dehydrogenation of ethyl benzene to form styrene:

$$\phi CH_2CH_3 \xrightarrow{\text{catalyst}} H_2 + \phi CH=CH_2$$

An example of cyclization, which may be considered to be a special type of dehydrogenation, is the formation of cyclohexane from n-hexane.

Oxidation Reactions.   The transition group elements (group VIII) and subgroup I are used extensively in oxidation reactions. Ag, Cu, Pt, Fe, Ni, and their oxides are generally good oxidation catalysts. In addition, V_2O_5 and MnO_2 are frequently used for oxidation reactions. A few of the principal types of catalytic oxidation reactions are:

1. Oxygen addition:

$$2C_2H_4 + O_2 \xrightarrow{\text{Ag}} 2C_2H_4O$$
$$2SO_2 + O_2 \xrightarrow{V_2O_5} 2SO_3$$
$$2CO + O_2 \xrightarrow{\text{Cu}} 2CO_2$$

2. Oxygenolysis of carbon–hydrogen bonds:

$$2C_2H_5OH + O_2 \xrightarrow{\ Cu\ } 2CH_3CHO + 2H_2O$$

$$2CH_3OH + O_2 \xrightarrow{\ Ag\ } 2HCHO + 2H_2O$$

3. Oxygenation of nitrogen–hydrogen bonds:

$$5O_2 + 4NH_3 \xrightarrow{\ Pt\ } 4NO + 6H_2O$$

4. Complete combustion:

$$2C_2H_6 + 7O_2 \xrightarrow{\ Ni\ } 4CO_2 + 6H_2O$$

Platinum and nickel can be used for both oxidation reactions and hydrogenation reactions.

Hydration and Dehydration Reactions.   Hydration and dehydration catalysts have a strong affinity for water. One such catalyst is Al_2O_3, which is used in the dehydration of alcohols to form olefins. In addition to alumina, silica-alumina gels, clays, phosphoric acid, and phosphoric acid salts on inert carriers have also been used for hydration–dehydration reactions. An example of an industrial catalytic hydration reaction is the synthesis of ethanol from ethylene:

$$CH_2{=}CH_2 + H_2O \longrightarrow CH_3CH_2OH$$

Halogenation and Dehalogenation Reactions.   Usually, reactions of this type take place readily without utilizing catalysts. However, when selectivity of the desired product is low or it is necessary to run the reaction at a lower temperature, the use of a catalyst is desirable. Supported copper and silver halides can be used for the halogenation of hydrocarbons. Hydrochlorination reactions can be carried out with mercury copper or zinc halides.

Summary.   Table 10-1 gives a summary of the representative reactions and catalysts discussed above.

TABLE 10-1.   TYPES OF REACTIONS AND REPRESENTATIVE CATALYSTS

Reaction	Catalysts
1. Halogenation–dehalogenation	$CuCl_2$, $AgCl$, Pd
2. Hydration–dehydration	Al_2O_3, MgO
3. Alkylation–dealkylation	$AlCl_3$, Pd
4. Hydrogenation–dehydrogenation	Co, Pt, Cr_2O_3, Ni
5. Oxidation	Cu, Ag, Ni, V_2O_5
6. Isomerization	$AlCl_3$, Pt/Al_2O_3, Zeolites

If, for example, we were to form styrene from an equimolar mixture of ethylene and benzene, we could carry out an alkylation reaction to form ethyl benzene, which is then dehydrogenated to form styrene. We will need both an alkylation catalyst and a dehydrogenation catalyst:

$$C_2H_4 + C_6H_6 \xrightarrow[\text{trace HCl}]{AlCl_3} C_6H_5C_2H_5 \xrightarrow{\ Ni\ } C_6H_5CH{=}CH_2 + H_2$$

10.2 Steps in a Catalytic Reaction

A schematic diagram of a tubular reactor packed with catalytic pellets is shown in Figure 10-5a. The overall process by which heterogeneous catalytic reactions proceed can be broken down into the sequence of individual steps shown in Table 10-2 and pictured in Figure 10-6 for an isomerization.

The overall rate of reaction is equal to the rate of the slowest step in the mechanism. When the diffusion steps (1, 2, 6, and 7 in Table 10-2) are very fast compared with the reaction steps (3, 4, and 5), the concentrations in the immediate vicinity of the active sites are indistinguishable from those in the bulk fluid. In this situation, the transport or diffusion steps do not affect the overall rate of the reaction. In other situations, if the reaction steps are very fast

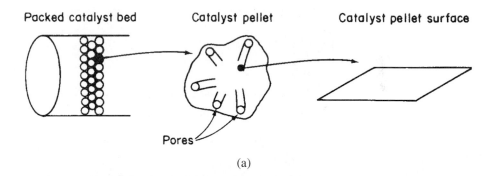

(a)

(b)

Figure 10-5   (a) Catalytic packed-bed reactor—schematic; (b) different shapes and sizes of catalyst. (Courtesy of the Engelhard Corporation.)

TABLE 10-2.   STEPS IN A CATALYTIC REACTION

1. Mass transfer (diffusion) of the reactant(s) (e.g., species A) from the bulk fluid to the external surface of the catalyst pellet
2. Diffusion of the reactant from the pore mouth through the catalyst pores to the immediate vicinity of the internal catalytic surface
3. Adsorption of reactant A onto the catalyst surface
4. Reaction on the surface of the catalyst (e.g., A ⟶ B)
5. Desorption of the products (e.g., B) from the surface
6. Diffusion of the products from the interior of the pellet to the pore mouth at the external surface
7. Mass transfer of the products from the external pellet surface to the bulk fluid

A reaction takes place *on* the surface, but the species involved in the reaction must get *to* and *from* the surface

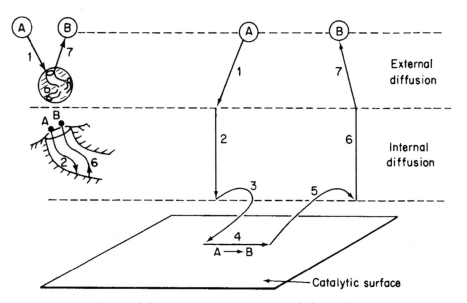

Figure 10-6   Steps in a heterogeneous catalytic reaction.

compared with the diffusion steps, mass transport does affect the reaction rate. In systems where diffusion from the bulk gas or liquid to the catalyst surface or to the mouths of catalyst pores affects the rate, changing the flow conditions past the catalyst should change the overall reaction rate. In porous catalysts, on the other hand, diffusion within the catalyst pores may limit the rate of reaction. Under these circumstances, the overall rate will be unaffected by external flow conditions even though diffusion affects the overall reaction rate.

There are many variations of the situation described in Table 10-2. Sometimes, of course, two reactants are necessary for a reaction to occur, and both of these may undergo the steps listed above. Other reactions between two substances have only one of them adsorbed.

In this chapter we focus on:
3. Adsorption
4. Surface reaction
5. Desorption

With this introduction, we are ready to treat individually the steps involved in catalytic reactions. In this chapter only the steps of adsorption, surface reaction, and desorption will be considered [i.e., it is assumed that the diffusion steps (1, 2, 6, and 7) are very fast, such that the overall reaction rate is not affected by mass transfer in any fashion]. Further treatment of the effects involving diffusion limitations is provided in Chapters 11 and 12.

Where Are We Heading?   As we saw in Chapter 5, one of the tasks of a chemical reaction engineer is to analyze rate data and to develop a rate law that can be used in reactor design. Rate laws in heterogeneous catalysis seldom follow power law models and hence are inherently more difficult to formulate from the data. In order to gain insight in developing rate laws from heterogeneous catalytic data, we are going to proceed in somewhat of a reverse manner than what is normally done in industry when one is asked to develop a rate law. That is, we will postulate catalytic mechanisms and *then* derive rate laws for the various mechanisms. The mechanism will typically have an adsorption step, a surface reaction step, and a desorption step, one of which is usually rate-limiting. Suggesting mechanisms and rate-limiting steps is not the first thing we normally do when presented with data. However, by deriving equations for different mechanisms we will observe the various forms of the rate law one can have in heterogeneous catalysis. Knowing the different forms that catalytic rate equations can take, it will be easier to view the trends in the data and deduce the appropriate rate law. This deduction is usually what is done first in industry before a mechanism is proposed. Knowing the form of the rate law, one can then numerically evaluate the rate law parameters and postulate a reaction mechanism and rate-limiting step that is consistent with the rate data. Finally, we use the rate law to design catalytic reactors. This procedure is shown in Figure 10-7. The dashed lines represent feedback to obtain new data in specific regions (e.g., concentrations, temperature) to evaluate the rate law parameters more precisely or to differentiate between reaction mechanisms.

An algorithm

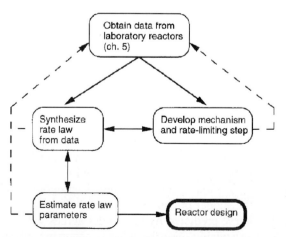

Figure 10-7   Collecting information for catalytic reactor design.

In the following analysis, adsorption will be discussed first, with emphasis on chemisorption. Adsorption can occur without reaction, and the mathematical treatment of nonreactive adsorption will lead naturally to the analysis of simultaneous adsorption, reaction, and desorption.

10.2.1 Adsorption Isotherms

Since chemisorption is usually a necessary part of a catalytic process, we shall discuss it before treating catalytic reaction rates. The letter S will represent an active site; alone it will denote a vacant site, with no atom, molecule, or complex adsorbed on it. The combination of S with another letter (e.g., $A \cdot S$) will mean that one unit of A will be adsorbed on the site S. Species A can be an atom, molecule, or some other atomic combination, depending on the circumstances. Consequently, the adsorption of A on a site S is represented by

$$A + S \rightleftharpoons A \cdot S$$

The total molar concentration of active sites per unit mass of catalyst is equal to the number of active sites per unit mass divided by Avogadro's number and will be labeled C_t (mol/g·cat.) The molar concentration of vacant sites, C_v, is the number of vacant sites per unit mass of catalyst divided by Avogadro's number. In the absence of catalyst deactivation we assume that the total concentration of active sites remains constant. Some further definitions include:

P_i partial pressure of species i in the gas phase, atm

$C_{i \cdot S}$ surface concentration of sites occupied by species i, g mol/g cat

A conceptual model depicting species A and B on two sites is shown below.

Vacant and occupied sites.

For the system shown, the total concentration of sites is

Site balance

$$C_t = C_v + C_{A \cdot S} + C_{B \cdot S} \tag{10-1}$$

This equation is referred to as a *site balance*.

Now consider the adsorption of a nonreacting gas onto the surface of a catalyst. Adsorption data are frequently reported in the form of adsorption *isotherms*. Isotherms portray the amount of a gas adsorbed on a solid at different pressures but at one temperature.

Postulate
models, then see
which one(s)
fit(s) the data

First, a model system is proposed and then the isotherm obtained from the model is compared with the experimental data shown on the curve. If the curve predicted by the model agrees with the experimental one, the model may reasonably describe what is occurring physically in the real system. If the predicted curve does not agree with that obtained experimentally, the model fails to match the physical situation in at least one important characteristic and perhaps more.

Two models will be postulated for the adsorption of carbon monoxide on metal—one in which CO is adsorbed as molecules, CO,

$$CO + S \rightleftarrows CO \cdot S$$

and the other in which carbon monoxide is adsorbed as oxygen and carbon atoms instead of molecules:

$$CO + 2S \rightleftarrows C \cdot S + O \cdot S$$

The former is called *molecular* or *nondissociated* (e.g., CO) *adsorption* and the latter is called *dissociative adsorption*. Whether a molecule adsorbs nondissociatively or dissociatively depends on the surface. For example, CO undergoes dissociative adsorption on iron and molecular adsorption on nickel.[6]

Two models:
1. Adsorption as CO
2. Adsorption as C and O

The adsorption of carbon monoxide molecules will be considered first. Since the carbon monoxide does not react further after being absorbed, we need only to consider the adsorption process:

$$CO + S \rightleftarrows CO \cdot S \tag{10-2}$$

In obtaining a rate law for the rate of adsorption, the reaction in Equation (10-2) can be treated as an *elementary reaction*. The rate of attachment of the carbon monoxide molecules to the surface is proportional to the number of collisions that these molecules make with the surface per second. In other words, a specific fraction of the molecules that strike the surface become adsorbed. The collision rate is, in turn, directly proportional to the carbon monoxide partial pressure, P_{CO}. Since carbon monoxide molecules adsorb only on vacant sites and not on sites already occupied by other carbon monoxide molecules, the rate of attachment is also directly proportional to the concentration of vacant sites, C_v. Combining these two facts means that the rate or attachment of carbon monoxide molecules to the surface is directly proportional to the product of the partial pressure of CO and the concentration of vacant sites; that is,

$$P_{CO} = C_{CO}RT$$

$$\text{rate of attachment} = k_A P_{CO} C_v$$

The rate of detachment of molecules from the surface can be a first-order process; that is, the detachment of carbon monoxide molecules from the surface is usually directly proportional to the concentration of sites occupied by the molecules e.g., $C_{CO \cdot S}$:

---

[6] R. L. Masel, *Principles of Adsorption and Reaction on Solid Surfaces*, Wiley, New York, 1996. http://www.uiuc.edu/ph/www/r-masel/

$$\text{rate of detachment} = k_{-A}C_{CO \cdot S}$$

The net rate of adsorption is equal to the rate of molecular attachment to the surface minus the rate of detachment from the surface. If k_A and k_{-A} are the constants of proportionality for the attachment and detachment processes, then

$$r_{AD} = k_A P_{CO} C_v - k_{-A} C_{CO \cdot S} \tag{10-3}$$

Adsorption

$$A + S \overset{}{\underset{}{\rightleftharpoons}} A \cdot S$$

$$r_{AD} =$$

$$k_A \times \left(P_A C_v - \frac{C_{A \cdot S}}{K_A} \right)$$

The ratio $K_A = k_A/k_{-A}$ is the *adsorption equilibrium constant*. Using it to rearrange Equation (10-3) gives

$$r_{AD} = k_A \left(P_{CO} C_v - \frac{C_{CO \cdot S}}{K_A} \right) \tag{10-4}$$

The adsorption rate constant k_A for molecular adsorption is virtually independent of temperature while the desorption constant k_{-A} increases exponentially with increasing temperature and equilibrium adsorption constant K_A decreases exponentially with increasing temperature. At a single temperature, in this case 25°C, they are, of course, constant in the absence of any catalyst deactivation.

$$r_{AD} = \left(\frac{mol}{gcat \cdot s} \right)$$

Since carbon monoxide is the only material adsorbed on the catalyst, the site balance gives

$$C_t = C_v + C_{CO \cdot S} \tag{10-5}$$

$$k_A = \left(\frac{1}{atm \cdot s} \right)$$

$$P_A = (atm)$$

At equilibrium, the net rate of adsorption equals zero. Setting the right-hand side of Equation (10-4) equal to zero and solving for the concentration of CO adsorbed on the surface, we get

$$C_v = \left(\frac{mol}{gcat} \right)$$

$$C_{CO \cdot S} = K_A C_v P_{CO} \tag{10-6}$$

$$K_A = \left(\frac{1}{atm} \right)$$

Using Equation (10-5) to give C_v in terms of $C_{CO \cdot S}$ and the total number of sites C_t, we can solve for $C_{CO \cdot S}$ in terms of constants and the pressure of carbon monoxide:

$$C_{A \cdot S} = \left(\frac{mol}{gcat} \right)$$

$$C_{CO \cdot S} = K_A C_v P_{CO} = K_A P_{CO}(C_t - C_{CO \cdot S})$$

Rearranging gives us

$$C_{CO \cdot S} = \frac{K_A P_{CO} C_t}{1 + K_A P_{CO}} \tag{10-7}$$

This equation thus gives $C_{CO \cdot S}$ as a function of the partial pressure of carbon monoxide, and is an equation for the adsorption isotherm. This particular type of isotherm equation is called a Langmuir *isotherm*.[7] Figure 10-8

---

[7] Named after Irving Langmuir (1881–1957), who first proposed it. He received the Nobel Prize in 1932 for his discoveries in surface chemistry.

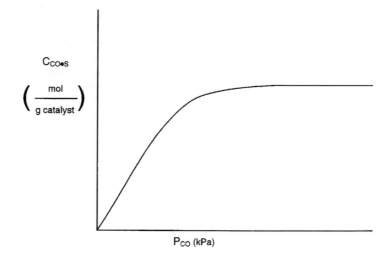

Figure 10-8   Langmuir isotherm.

shows a plot of the amount of CO adsorbed per unit mass of catalyst as a function of the partial pressure of CO.

One method of checking whether a model (e.g., molecular adsorption versus dissociative adsorption) predicts the behavior of the experimental data is to linearize the model's equation and then plot the indicated variables against one another. For example, Equation (10-7) may be arranged in the form

$$\frac{P_{CO}}{C_{CO \cdot S}} = \frac{1}{K_A C_t} + \frac{P_{CO}}{C_t} \tag{10-8}$$

and the linearity of a plot of $P_{CO}/C_{CO \cdot S}$ as a function of P_{CO} will determine if the data conform to a Langmuir single-site isotherm.

Next, the isotherm for carbon monoxide adsorbing as atoms is derived:

Dissociative adsorption

$$CO + 2S \; \rightleftharpoons \; C \cdot S + O \cdot S$$

When the carbon monoxide molecule dissociates upon adsorption, it is referred to as the *dissociative adsorption* of carbon monoxide. As in the case of molecular adsorption, the rate of adsorption here is proportional to the pressure of carbon monoxide in the system because this rate governs the number of gaseous collisions with the surface. For a molecule to dissociate as it adsorbs, however, two adjacent vacant active sites are required rather than the single site needed when a substance adsorbs in its molecular form. The probability of two vacant sites occurring adjacent to one another is proportional to the square of the concentration of vacant sites. These two observations mean that the rate of adsorption is proportional to the product of the carbon monoxide partial pressure and the square of the vacant-site concentration, $P_{CO} C_v^2$.

For desorption to occur, two occupied sites must be adjacent, meaning that the rate of desorption is proportional to the product of the occupied-site concentration, $(C \cdot S) \times (O \cdot S)$. The net rate of adsorption can then be expressed as

$$r_{AD} = k_A P_{CO} C_v^2 - k_{-A} C_{O \cdot S} C_{C \cdot S} \qquad (10\text{-}9)$$

Factoring out k_A, the equation for *dissociative adsorption* is

Rate of dissociative
adsorption

$$r_{AD} = k_A \left(P_{CO} C_v^2 - \frac{C_{C \cdot S} C_{O \cdot S}}{K_A} \right)$$

where

$$K_A = \frac{k_A}{k_{-A}}$$

For dissociative adsorption both k_A and k_{-A} increase exponentially with increasing temperature while K_A decreases with increasing temperature.

At equilibrium, $r_{AD} = 0$, and

$$k_A P_{CO} C_v^2 = k_{-A} C_{C \cdot S} C_{O \cdot S}$$

For $C_{C \cdot S} = C_{O \cdot S}$,

$$(K_A P_{CO})^{1/2} C_v = C_{O \cdot S} \qquad (10\text{-}10)$$

From Equation (10-1),

$$C_v = C_t - C_{C \cdot S} - C_{O \cdot S} = C_t - 2C_{O \cdot S}$$

This value may be substituted into Equation (10-10) to give an expression that can be solved for $C_{O \cdot S}$. The resulting isotherm equation is

Langmuir isotherm
for adsorption as
atomic carbon
monoxide

$$C_{O \cdot S} = \frac{(K_A P_{CO})^{1/2} C_t}{1 + 2(K_A P_{CO})^{1/2}} \qquad (10\text{-}11)$$

Taking the inverse of both sides of the equation, then multiplying through by $(P_{CO})^{1/2}$, yields

$$\frac{(P_{CO})^{1/2}}{C_{O \cdot S}} = \frac{1}{C_t (K_A)^{1/2}} + \frac{2(P_{CO})^{1/2}}{C_t} \qquad (10\text{-}12)$$

If dissociative adsorption is the correct model, a plot of $(P_{CO}^{1/2}/C_{O \cdot S})$ versus $P_{CO}^{1/2}$ should be linear.

When more than one substance is present, the adsorption isotherm equations are somewhat more complex. The principles are the same, though, and the isotherm equations are easily derived. It is left as an exercise to show that the adsorption isotherm of A in the presence of adsorbate B is given by the relationship

$$C_{A \cdot S} = \frac{K_A P_A C_t}{1 + K_A P_A + K_B P_B} \qquad (10\text{-}13)$$

When the adsorption of both A and B are first-order processes, the desorptions are also first-order, and both A and B are adsorbed as molecules. The derivations of other Langmuir isotherms are left an exercise.

Note assumptions
in the model
and check their
validity In obtaining the Langmuir isotherm equations, several aspects of the adsorption system were presupposed in the derivations. The most important of these, and the one that has been subject to the greatest doubt, is that a *uniform* surface is assumed. In other words, any active site has the same attraction for an impinging molecule as does any other site. Isotherms different from the Langmuir types may be derived based on various assumptions concerning the adsorption system, including different types of nonuniform surfaces.

10.2.2 Surface Reaction

The rate of adsorption of species A onto a solid surface,

$$A + S \rightleftharpoons A \cdot S$$

is given by

$$r_{AD} = k_A \left(P_A C_v - \frac{C_{A \cdot S}}{K_A} \right) \qquad (10\text{-}14)$$

Surface reaction
models
Once a reactant has been adsorbed onto the surface, it is capable of reacting in a number of ways to form the reaction product. Three of these ways are:

1. The surface reaction may be a single-site mechanism in which only the site on which the reactant is adsorbed is involved in the reaction. For example, an adsorbed molecule of A may isomerize (or perhaps decompose) directly on the site to which it is attached:

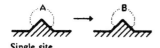
Single site

$$A \cdot S \rightleftharpoons B \cdot S$$

Because in each step the reaction mechanism is elementary, the rate law is

Single Site

$$k_S = \left(\frac{1}{s} \right)$$

$$r_S = k_S \left(C_{A \cdot S} - \frac{C_{B \cdot S}}{K_S} \right) \qquad (10\text{-}15)$$

where K_S is the surface reaction equilibrium constant $K_S = k_S / k_{-S}$

2. The surface reaction may be a dual-site mechanism in which the adsorbed reactant interacts with another site (either unoccupied or occupied) to form the product. For example, adsorbed A may react with an adjacent vacant site to yield a vacant site and a site on which the product is adsorbed:

Dual site

$$A \cdot S + S \rightleftharpoons B \cdot S + S$$

Dual Site

$$r_S = \left(\frac{\text{mol}}{\text{gcat} \cdot \text{s}}\right)$$

$$k_S = \left(\frac{\text{gcat}}{\text{mol} \cdot \text{s}}\right)$$

$$K_S = \text{(dimensionless)}$$

The corresponding rate law is

$$r_S = k_S\left(C_{A \cdot S}C_v - \frac{C_{B \cdot S}C_v}{K_S}\right) \tag{10-16}$$

Another example of a dual-site mechanism is the reaction between two adsorbed species:

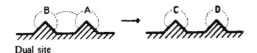

Dual site

$$A \cdot S + B \cdot S \; \rightleftharpoons \; C \cdot S + D \cdot S$$

with the rate law

$$r_S = k_S\left(C_{A \cdot S}C_{B \cdot S} - \frac{C_{C \cdot S}C_{D \cdot S}}{K_S}\right) \tag{10-17}$$

A third dual-site mechanism is the reaction of two species adsorbed on different types of sites S and S':

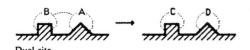

Dual site

$$A \cdot S + B \cdot S' \; \rightleftharpoons \; C \cdot S' + D \cdot S$$

with the rate law

$$r_S = k_S\left(C_{A \cdot S}C_{B \cdot S'} - \frac{C_{C \cdot S'}C_{D \cdot S}}{K_S}\right) \tag{10-18}$$

Langmuir-Hinshelwood kinetics

Reactions involving either single- or dual-site mechanisms described above are sometimes referred to as following *Langmuir–Hinshelwood kinetics.*

3. A third mechanism is the reaction between an adsorbed molecule and a molecule in the gas phase:

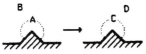

Eley–Rideal mechanism

$$A \cdot S + B(g) \; \rightleftharpoons \; C \cdot S + D(g)$$

with the rate law

Eley–Rideal mechanism

$$r_S = k_S\left(C_{A \cdot S}P_B - \frac{C_{C \cdot S}P_D}{K_S}\right) \tag{10-19}$$

This type of mechanism is referred to as an *Eley–Rideal mechanism.*

10.2.3 Desorption

In each of the cases above, the products of the surface reaction adsorbed on the surface are subsequently desorbed into the gas phase. For the rate of desorption of a species, e.g., C,

$$C \cdot S \rightleftharpoons C + S$$

$$k_D = \left(\frac{1}{s}\right)$$

$$K_{DC} = (atm)$$

$$K_C = \left(\frac{1}{atm}\right)$$

the desorption rate law is

$$r_D = k_D \left(C_{C \cdot S} - \frac{P_C C_v}{K_{DC}} \right) \tag{10-20}$$

where K_D is the desorption equilibrium constant. We note that the desorption step for C is just the reverse of the adsorption step for C and that the rate of desorption of C, r_D, is just opposite in sign to the rate of adsorption of C, r_{ADC}:

$$r_D = -r_{ADC}$$

In addition, we see that the desorption equilibrium constant K_{DC} is just the reciprocal of the adsorption equilibrium constant for C, K_C:

$$K_{DC} = \frac{1}{K_C}$$

10.2.4 The Rate-Limiting Step

When heterogeneous reactions are carried out at steady state, the rates of each of the three reaction steps in series (adsorption, surface reaction, and desorption) are equal to one another:

$$\boxed{-r_A' = r_{AD} = r_S = r_D}$$

However, one particular step in the series is usually found to be *rate-limiting* or *rate-controlling*. That is, if we could make this particular step go faster, the entire reaction would proceed at an accelerated rate. Consider the analogy to the electrical circuit shown in Figure 10-9. A given concentration of reactants is analogous to a given driving force or electromotive force (EMF). The current I (with units of C/s) is analogous to the rate of reaction, $-r_A'$ (mol/s·g cat.), and a resistance R_i is associated with each step in the series. Since the resistances are in series, the total resistance is just the sum of the individual resistances, so the current I is

$$I = \frac{E}{R_{tot}} = \frac{E}{R_{AD} + R_S + R_D} \tag{10-21}$$

The concept of a rate-limiting step

Since we observe only the total resistance, R_{tot}, it is our task to find which resistance is much larger (say, 100 Ω) than the other two (say, 0.1 Ω). Thus, if

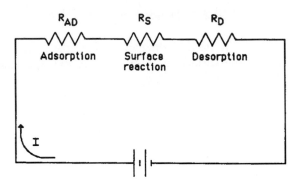

Figure 10-9   Electrical analog to heterogeneous reactions.

we could lower the largest resistance, the current I (e.g., $-r'_A$), would be larger for a given voltage, E. Analogously, we want to know which step in the adsorption–reaction–desorption series is limiting the overall rate of reaction.

The approach in determining catalytic and heterogeneous mechanisms is usually termed the *Langmuir–Hinshelwood approach*, since it is derived from ideas proposed by Hinshelwood[8] based on Langmuir's principles for adsorption. The Langmuir–Hinshelwood approach was popularized by Hougen and Watson[9] and occasionally includes their names. It consists of first assuming a sequence of steps in the reaction. In writing this sequence a choice must be made between such mechanisms as molecular or atomic adsorption, and single- or dual-site reaction. Next, rate laws are written for the individual steps as shown in the preceding section, assuming that all steps are reversible. Finally, a rate-limiting step is postulated and steps that are not rate-limiting are used to eliminate all coverage-dependent terms. The most questionable assumption in using this technique to obtain a rate law is the hypothesis that the activity of the surface toward adsorption, desorption, or surface reaction is independent of coverage; that is, the surface is essentially uniform as far as the various steps in the reaction are concerned.

An algorithm to determine the rate-limiting step

An example of an adsorption-limited reaction is the synthesis of ammonia; the reaction of carbon monoxide and nitric oxide is an example of a surface-limited reaction. The synthesis of ammonia from hydrogen and nitrogen,

$$3H_2 + N_2 \longrightarrow 2NH_3$$

over an iron catalyst proceeds by the following mechanism.[10]

---

[8] C. N. Hinshelwood, *The Kinetics of Chemical Change*, Clarendon Press, Oxford, 1940.

[9] O. A. Hougen and K. M. Watson, *Ind. Eng. Chem.*, *35*, 529 (1943).

[10] From the literature cited in G. A. Somorjai, *Introduction to Surface Chemistry and Catalysis*, Wiley, New York, 1994, p. 482.

$$H_2 + 2S \longrightarrow 2H \cdot S \qquad\qquad \text{Rapid}$$

Dissociative adsorption limits

$$\left.\begin{array}{l} N_2 + S \rightleftharpoons N_2 \cdot S \\ N_2 \cdot S + S \longrightarrow 2N \cdot S \end{array}\right\} \qquad \text{Rate-limiting}$$

$$\left.\begin{array}{l} N \cdot S + H \cdot S \rightleftharpoons HN \cdot S + S \\ NH \cdot S + H \cdot S \rightleftharpoons H_2N \cdot S + S \\ H_2N \cdot S + H \cdot S \rightleftharpoons NH_3 \cdot S + S \\ NH_3 \cdot S \rightleftharpoons NH_3 + S \end{array}\right\} \qquad \text{Rapid}$$

The rate-limiting step is believed to be the adsorption of the N_2 molecule as an N atom.

The reaction of two noxious automobile exhaust products, CO and NO,

$$CO + NO \longrightarrow CO_2 + \tfrac{1}{2} N_2$$

is carried out over copper catalyst to form environmentally acceptable products, N_2 and CO_2:

$$\left.\begin{array}{l} CO + S \rightleftharpoons CO \cdot S \\ NO + S \rightleftharpoons NO \cdot S \end{array}\right\} \qquad \text{Rapid}$$

Surface reaction limits

$$NO \cdot S + CO \cdot S \rightleftharpoons CO_2 + N \cdot S + S\} \qquad \text{Rate-limiting}$$

$$\left.\begin{array}{l} N \cdot S + N \cdot S \rightleftharpoons N_2 \cdot S \\ N_2 \cdot S \longrightarrow N_2 + S \end{array}\right\} \qquad \text{Rapid}$$

Analysis of the rate law suggests that CO_2 and N_2 are weakly adsorbed (see Problem P10-7).

10.3 Synthesizing a Rate Law, Mechanism, and Rate-Limiting Step

We now wish to develop rate laws for catalytic reactions that are not diffusion-limited. In developing the procedure to obtain a mechanism, a rate-limiting step, and a rate law consistent with experimental observation, we shall discuss a particular catalytic reaction, the decomposition of cumene to form benzene and propylene. The overall reaction is

$$C_6H_5CH(CH_3)_2 \longrightarrow C_6H_6 + C_3H_6$$

A conceptual model depicting the sequences of steps in this platinum-catalyzed reaction is shown in Figure 10-10. Figure 10-10 is only a schematic representation of the adsorption of cumene; a more realistic model is the formation of a complex of the π orbitals of benzene with the catalytic surface, as shown in Figure 10-11.

· Adsorption
· Surface
  reaction
· Desorption

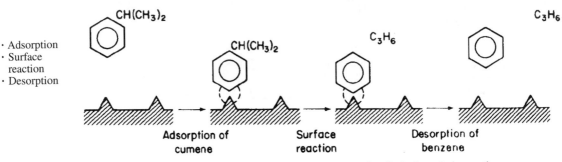

Adsorption of   Surface   Desorption of
cumene   reaction   benzene

Figure 10-10   Sequence of steps in reaction-limited catalytic reaction.

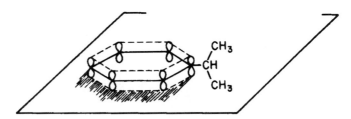

Figure 10-11 π-orbital complex on surface.

The following nomenclature will be used to denote the various species in this reaction: C = cumene, B = benzene, P = propylene, and I = inhibitor. The reaction sequence for this decomposition is

These three steps represent the mechanism for cumene decomposition

$$C + S \underset{k_{-A}}{\overset{k_A}{\rightleftarrows}} C \cdot S \qquad\qquad \text{adsorption of cumene on the surface} \qquad (10\text{-}22)$$

$$C \cdot S \underset{k_{-S}}{\overset{k_S}{\rightleftarrows}} B \cdot S + P \qquad \begin{array}{l}\text{surface reaction to form adsorbed}\\ \text{benzene and propylene in the}\\ \text{gas phase}\end{array} \qquad (10\text{-}23)$$

$$B \cdot S \underset{k_{-D}}{\overset{k_D}{\rightleftarrows}} B + S \qquad\qquad \text{desorption of benzene from surface} \qquad (10\text{-}24)$$

Equations (10-22) through (10-24) represent the mechanism proposed for this reaction.

When writing rate laws for these steps, we treat each step as an elementary reaction; the only difference is that the species concentrations in the gas phase are replaced by their respective partial pressures:

$$C_C \longrightarrow P_C$$

The rate expression for the adsorption of cumene as given in Equation (10-22) is

$C + S \underset{k_{-A}}{\overset{k_A}{\rightleftarrows}} C \cdot S$ $r_{AD} = k_A P_C C_v - k_{-A} C_{C \cdot S}$

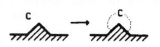

$$\text{Adsorption:} \qquad r_{AD} = k_A \left(P_C C_v - \frac{C_{C\cdot S}}{K_C} \right) \qquad (10\text{-}25)$$

If r_{AD} has units of (mol/g cat.·s) and $C_{C\cdot S}$ has units of (mol cumene adsorbed/g cat.) typical units of k_A, k_{-A}, and K_C would be

$$[k_A] \equiv (\text{kPa}\cdot\text{s})^{-1} \text{ or } (\text{atm}\cdot\text{h})^{-1}$$

$$[k_{-A}] \equiv \text{h}^{-1} \text{ or } \text{s}^{-1}$$

$$[K_C] \equiv \left[\frac{k_A}{k_{-A}} \right] \equiv \text{kPa}^{-1}$$

The rate law for the surface reaction step producing adsorbed benzene and propylene in the gas phase,

$$\text{C}\cdot\text{S} \underset{k_{-S}}{\overset{k_S}{\rightleftharpoons}} \text{B}\cdot\text{S} + \text{P}(g) \qquad (10\text{-}23)$$

is

$$r_S = k_S C_{C\cdot S} - k_{-S} P_P C_{B\cdot S}$$

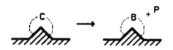

$$\text{Surface reaction:} \qquad r_S = k_S \left(C_{C\cdot S} - \frac{P_P C_{B\cdot S}}{K_S} \right) \qquad (10\text{-}26)$$

with the *surface reaction equilibrium constant* being

$$K_S = \frac{k_S}{k_{-S}}$$

Typical units for k_S and K_S are s^{-1} and atm, respectively.

Propylene is not adsorbed on the surface. Consequently, its concentration on the surface is zero.

$$C_{P\cdot S} = 0$$

The rate of benzene desorption [see Equation (10-24)] is

$$r_D = k_D C_{B\cdot S} - k_{-D} P_B C_v \qquad (10\text{-}27)$$

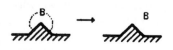

$$\text{Desorption:} \qquad r_D = k_D \left(C_{B\cdot S} - \frac{P_B C_v}{K_{DB}} \right) \qquad (10\text{-}28)$$

Typical units of k_D and K_{DB} are s^{-1} and kPa, respectively. By viewing the desorption of benzene,

$$\text{B}\cdot\text{S} \rightleftharpoons \text{B} + \text{S}$$

from right to left, we see that desorption is just the reverse of the adsorption of benzene. Consequently, it is easily shown that the benzene adsorption equilibrium constant K_B is just the reciprocal of the benzene desorption constant K_{DB}:

$$K_B = \frac{1}{K_{DB}}$$

and Equation (10-28) can be written as

$$\boxed{\text{Desorption:} \qquad r_D = k_D(C_{B \cdot S} - K_B P_B C_v)} \qquad (10\text{-}29)$$

Because there is no accumulation of reacting species on the surface the rates of each step in the sequence are all equal:

$$-r_C' = r_{AD} = r_S = r_D \qquad (10\text{-}30)$$

For the mechanism postulated in the sequence given by Equations (10-22) through (10-24), we wish to determine which step is rate-limiting. We first assume one of the steps to be rate-limiting (rate-controlling) and then formulate the reaction rate law in terms of the partial pressures of the species present. From this expression we can determine the variation of the initial reaction rate with the initial total pressure. If the predicted rate varies with pressure in the same manner as the rate observed experimentally, the implication is that the assumed mechanism and rate-limiting step are correct.

10.3.1   Is the Adsorption of Cumene Rate-Limiting?

To answer this question we shall assume that the adsorption of cumene is indeed rate-limiting, derive the corresponding rate law, and then check to see if it is consistent with experimental observation. By assuming that this (or any other) step is rate-limiting, we are considering that the reaction rate constant of this step (in this case k_A) is small with respect to the specific rates of the other steps (in this case k_S and k_D).[†] The rate of adsorption is

Need to express C_v
and $C_{C \cdot S}$ in terms
of P_C, P_B, and P_P

$$-r_C' = r_{AD} = k_A\left(P_C C_v - \frac{C_{C \cdot S}}{K_C}\right) \qquad (10\text{-}25)$$

Since we cannot measure either C_v or $C_{C \cdot S}$, we must replace these variables in the rate law with measurable quantities for the equation to be meaningful.

For steady-state operation we have

$$-r_C' = r_{AD} = r_S = r_D \qquad (10\text{-}30)$$

For adsorption-limited reactions, k_A is small and k_S and k_D are large. Consequently, the ratios r_S/k_S and r_D/k_D are very small (approximately zero), whereas the ratio r_{AD}/k_A is relatively large. The surface reaction rate expression is

$$r_S = k_S\left(C_{C \cdot S} - \frac{C_{B \cdot S} P_P}{K_S}\right) \qquad (10\text{-}31)$$

[†] Strictly speaking one should compare the product $k_A P_C$ with k_S and k_D. (See Summary Notes for Lecture 19 on CD-ROM.) The end result is the same however.

For adsorption-limited reactions the surface specific reaction rate k_S is large by comparison and we can set

$$\frac{r_S}{k_S} \simeq 0 \tag{10-32}$$

and solve Equation (10-31) for $C_{C \cdot S}$:

$$C_{C \cdot S} = \frac{C_{B \cdot S} P_P}{K_S} \tag{10-33}$$

To be able to express $C_{C \cdot S}$ solely in terms of the partial pressures of the species present, we must evaluate $C_{B \cdot S}$. The rate of desorption of benzene is

$$r_D = k_D (C_{B \cdot S} - K_B P_B C_v) \tag{10-29}$$

Using
$$\frac{r_S}{k_S} \simeq 0 \simeq \frac{r_D}{k_D}$$
to find $C_{B \cdot S}$ and $C_{C \cdot S}$ in terms of partial pressures

However, for adsorption-limited reactions, k_D is large by comparison and we can set

$$\frac{r_D}{k_D} \simeq 0 \tag{10-34}$$

and then solve Equation (10-29) for $C_{B \cdot S}$:

$$C_{B \cdot S} = K_B P_B C_v \tag{10-35}$$

After combining Equations (10-33) and (10-35), we have

$$C_{C \cdot S} = K_B \frac{P_B P_P}{K_S} C_v \tag{10-36}$$

Replacing $C_{C \cdot S}$ in the rate equation by Equation (10-36) and then factoring C_v, we obtain

$$r_{AD} = k_A \left(P_C - \frac{K_B P_B P_P}{K_S K_C} \right) C_v = k_A \left(P_C - \frac{P_B P_P}{K_P} \right) C_v \tag{10-37}$$

Observe that by setting $r_{AD} = 0$, the term $(K_S K_C / K_B)$ is simply the overall partial pressure equilibrium constant, K_P, for the reaction

$$C \rightleftharpoons B + P$$

$$\boxed{\frac{K_S K_C}{K_B} = K_P} \tag{10-38}$$

The equilibrium constant can be determined from thermodynamic data and is related to the change in the Gibbs free energy, ΔG°, by the equation (see Appendix C)

$$\boxed{RT \ln K = -\Delta G^\circ} \tag{10-39}$$

where R is the ideal gas constant and T is the absolute temperature.

The concentration of vacant sites, C_v, can now be eliminated from Equation (10-37) by utilizing the site balance to give the total concentration of sites, C_t, which is assumed constant:[11]

$$\boxed{\text{total sites} = \text{vacant sites} + \text{occupied sites}}$$

Since cumene and benzene are adsorbed on the surface, the concentration of occupied sites is $(C_{C \cdot S} + C_{B \cdot S})$, and the total concentration of sites is

Site balance

$$C_t = C_v + C_{C \cdot S} + C_{B \cdot S} \qquad (10\text{-}40)$$

Substituting Equations (10-35) and (10-36) into Equation (10-40), we have

$$C_t = C_v + \frac{K_B}{K_S} P_B P_P C_v + K_B P_B C_v$$

Solving for C_v, we have

$$C_v = \frac{C_t}{1 + P_B P_P K_B / K_S + K_B P_B} \qquad (10\text{-}41)$$

Combining Equations (10-41) and (10-37), we find that the rate law for the catalytic decompositon of cumene, assuming that the adsorption of cumene is the rate-limiting step, is

Cumene reaction rate law if adsorption were limiting step

$$\boxed{-r'_C = r_{AD} = \frac{C_t k_A (P_C - P_P P_B / K_P)}{1 + K_B P_P P_B / K_S + K_B P_B}} \qquad (10\text{-}42)$$

We now wish to sketch a plot of the initial rate as a function of the partial pressure of cumene, P_{C0}. Initially, no products are present; consequently, $P_P = P_B = 0$. The initial rate is given by

$$-r'_{C0} = C_t k_A P_{C0} = k P_{C0} \qquad (10\text{-}43)$$

If the cumene decompostion is adsorption limited, then the initial rate will be linear with the initial partial pressure of cumene as shown in Figure 10-12.

Before checking to see if Figure 10-12 is consistent with experimental observation, we shall derive the corresponding rate laws and initial rate plots when the surface reaction is rate-limiting and then when the desorption of benzene is rate-limiting.

---

[11]Some prefer to write the surface reaction rate in terms of the fraction of the surface of sites covered (i.e., f_A) rather than the number of sites $C_{A \cdot S}$ covered, the difference being the multiplication factor of the total site concentration, C_t. In any event, the final form of the rate law is the same because C_t, K_A, k_S, and so on, are all lumped into the specific reaction rate k.

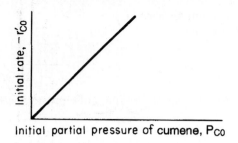

Figure 10-12 Uninhibited adsorption-limited reaction.

If adsorption were rate-limiting, the data should show $-r'_0$ increasing linearly with P_{CO}

10.3.2 Is the Surface Reaction Rate-Limiting?

The rate of surface reaction is

Single-site mechanism

$$r_S = k_S\left(C_{C\cdot S} - \frac{P_P C_{B\cdot S}}{K_S}\right) \tag{10-26}$$

Since we cannot readily measure the concentrations of the adsorbed species, we must utilize the adsorption and desorption steps to eliminate $C_{C\cdot S}$ and $C_{B\cdot S}$ from this equation.

From the adsorption rate expression in Equation (10-25) and the condition that k_A is large by comparison when surface reaction is controlling (i.e., $r_{AD}/k_A \simeq 0$), we obtain a relationship for the surface concentration for adsorbed cumene:

$$C_{C\cdot S} = K_C P_C C_v$$

In a similar manner, the surface concentration of adsorbed benzene can be evaluated from the desorption rate expression [Equation (10-29)] together with the approximation:

Using
$$\frac{r_{AD}}{k_A} \simeq 0 \simeq \frac{r_D}{k_D}$$
to find $C_{B\cdot S}$ and $C_{C\cdot S}$ in terms of partial pressures

$$\text{when } \frac{r_D}{k_D} \simeq 0 \qquad \text{then } C_{B\cdot S} = K_B P_B C_v$$

Substituting for $C_{B\cdot S}$ and $C_{C\cdot S}$ in Equation (10-26) gives us

$$r_S = k_S\left(P_C K_C - \frac{K_B P_B P_P}{K_S}\right)C_v = k_S K_C\left(P_C - \frac{P_B P_P}{K_P}\right)C_v$$

The only variable left to eliminate is C_v:

Site balance

$$C_t = C_v + C_{B\cdot S} + C_{C\cdot S}$$

Substituting for concentrations of the adsorbed species, $C_{B\cdot S}$, and $C_{C\cdot S}$ yields

$$C_v = \frac{C_t}{1 + K_B P_B + K_C P_C}$$

Cumene rate law for surface-reaction-limiting

$$-r'_C = r_S = \frac{\overbrace{k_S C_t K_C}^{k}\,(P_C - P_P P_B/K_P)}{1 + P_B K_B + K_C P_C} \tag{10-44}$$

The initial rate is

$$-r'_{\text{C0}} = \dfrac{\overbrace{k_{\text{S}} C_t K_{\text{C}}}^{k} P_{\text{C0}}}{1 + K_{\text{C}} P_{\text{C0}}} = \dfrac{k P_{\text{C0}}}{1 + K_{\text{C}} P_{\text{C0}}} \qquad (10\text{-}45)$$

At low partial pressures of cumene

$$1 \gg K_{\text{C}} P_{\text{C0}}$$

and we observe that the initial rate will increase linearly with the initial partial pressure of cumene:

$$-r'_{\text{C0}} \approx k P_{\text{C0}}$$

At high partial pressures

$$K_{\text{C}} P_{\text{C0}} \gg 1$$

and Equation (10-45) becomes

$$-r'_{\text{C0}} \cong \dfrac{k P_{\text{C0}}}{K_{\text{C}} P_{\text{C0}}} = \dfrac{k}{K_{\text{C}}}$$

and the rate is independent of the partial pressure of cumene. Figure 10-13 shows the initial rate of reaction as a function of initial partial pressure of cumene for the case of surface reaction controlling.

If surface reaction were rate-limiting, the data would show this behavior

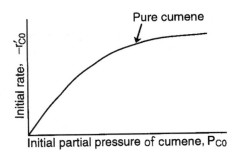

Figure 10-13   Surface-reaction-limited.

10.3.3 Is the Desorption of Benzene Rate-Limiting?

The rate expression for the desorption of benzene is

$$r_{\text{D}} = k_{\text{D}} (C_{\text{B} \cdot \text{S}} - K_{\text{B}} P_{\text{B}} C_v) \qquad (10\text{-}28)$$

For desorption-limited reactions, both k_{AD} and k_{S} are very large compared with k_{D}, which is small

From the rate expression for surface reaction, Equation (10-26), we set

$$\dfrac{r_{\text{S}}}{k_{\text{S}}} \simeq 0$$

to obtain

$$C_{B \cdot S} = K_S \left(\frac{C_{C \cdot S}}{P_P} \right) \tag{10-46}$$

Similarly, for the adsorption step, Equation (10-25), we set

$$\frac{r_{AD}}{k_A} \simeq 0$$

to obtain

$$C_{C \cdot S} = K_C P_C C_v$$

then substitute for $C_{C \cdot S}$ in Equation (10-46):

$$C_{B \cdot S} = \frac{K_C K_S P_C C_v}{P_P} \tag{10-47}$$

Combining Equations (10-28) and (10-47) gives us

$$r_D = k_D K_C K_S \left(\frac{P_C}{P_P} - \frac{P_B}{K_P} \right) C_v \tag{10-48}$$

where K_C is the cumene adsorption constant, K_S is the surface reaction equilibrium constant, and K_P is the gas-phase equilibrium constant for the reaction. To obtain an expression for C_v, we again perform a site balance:

Site balance: $C_t = C_{C \cdot S} + C_{B \cdot S} + C_v$

After substituting for the respective surface concentrations, we solve the site balance for C_v:

$$C_v = \frac{C_t}{1 + K_C K_S P_C / P_P + K_C P_C} \tag{10-49}$$

Replacing C_v in Equation (10-48) by Equation (10-49) and multiplying the numerator and denominator by P_P, we obtain the rate expression for desorption control:

Cumene
decomposition rate
law *if* desorption
were limiting

$$-r_C' = r_D = \frac{\overbrace{k_D C_t K_S K_C}^{k} (P_C - P_B P_P / K_P)}{P_P + P_C K_C K_S + K_C P_P P_C} \tag{10-50}$$

To determine the dependence of the initial rate on partial pressure of cumene, we again set $P_P = P_B = 0$; and the rate law reduces to

If desorption
controls, the initial
rate is independent
of partial pressure of
cumene

$$-r_{C0}' = k_D C_t$$

with the corresponding plot of $-r_{C0}'$ shown in Figure 10-14. If desorption were controlling, we would see that the initial rate would be independent of the initial partial pressure of cumene.

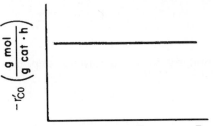

Initial partial pressure of cumene, P_{C0}

Figure 10-14   Desorption-limited reaction.

10.3.4 Summary of the Cumene Decomposition

Cumene
decomposition is
surface-reaction-
limited

The experimental observations of $-r'_{C0}$ as a function of P_{C0} are shown in Figure 10-15. From the plot in Figure 10-15 we can clearly see that neither adsorption nor desorption is rate-limiting. For the reaction and mechanism given by

$$C + S \rightleftharpoons C \cdot S \tag{10-22}$$

$$C \cdot S \rightleftharpoons B \cdot S + P \tag{10-23}$$

$$B \cdot S \rightleftharpoons B + S \tag{10-24}$$

the rate law derived by assuming that the surface reaction is rate-limiting agrees with the data.

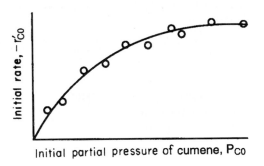

Initial partial pressure of cumene, P_{C0}

Figure 10-15   Actual initial rate as a function of partial pressure of cumene.

The rate law for the case of no inerts adsorbing on the surface is

$$-r'_C = \frac{k(P_C - P_B P_P / K_P)}{1 + K_B P_B + K_C P_C} \tag{10-44}$$

The forward cumene decomposition reaction is a single-site mechanism involving only adsorbed cumene while the reverse reaction of propylene in the gas phase reacting with adsorbed benzene is an Eley–Rideal mechanism.

If we were to have an adsorbing inert in the feed, the inert would not participate in the reaction but would occupy sites on the catalyst surface:

$$I + S \xrightleftharpoons{\hspace{1cm}} I \cdot S$$

Our site balance is now

$$C_t = C_v + C_{C \cdot S} + C_{B \cdot S} + C_{I \cdot S} \qquad (10\text{-}51)$$

Because the adsorption of the inert is at equilibrium, the concentration of sites occupied by the inert is

$$C_{I \cdot S} = K_I P_I C_v \qquad (10\text{-}52)$$

Substituting for the inert sites in the site balance, the rate law for surface reaction control when an adsorbing inert is present is

Adsorbing inerts

$$-r'_C = \frac{k(P_C - P_B P_P / K_P)}{1 + K_C P_C + K_B P_B + K_I P_I} \qquad (10\text{-}53)$$

Spark Plug

Octane Number and the Dual-Site Mechanism.   We now consider a dual-site mechanism which is a reforming reaction found in petroleum refining to upgrade the octane number of gasoline[t.] Fuels with low octane numbers can produce spontaneous combustion in the cylinder before the air/fuel mixture is compressed to its desired value and ignited by the spark plug. The margin figure shows the desired combustion wave front moving down from the spark plug and the unwanted spontaneous combustion wave in the lower right-hand corner. This spontaneous combustion produces detonation waves which constitute engine knock. The lower the octane number the greater the chance of engine knock.

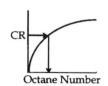

Calibration curve

CR

Octane Number

The octane number of a gasoline is determined from a calibration curve relating the compression ratio (CR) at which engine knock occurs to the octane number of the fuel. This curve is developed by carrying out a series of experiments on mixtures of *iso*-octane and heptane. The octane number is the percentage of *iso*-octane in this mixture. That is, pure *iso*-octane has an octane number of 100, 80% *iso*-octane/20% heptane has an octane number of 80, and so on. A fixed percentage of these two hydrocarbons is placed in a test engine and the compression ratio (CR) is increased continually until spontaneous combustion occurs, producing an engine knock. The compression ratio and the corresponding composition of the mixture are then recorded. The relative percentages of *iso*-octane and heptane are changed and the test repeated so that after a series of experiments a calibration curve is constructed. The gasoline to be calibrated is then used in the test engine, where the compression ratio is increased continually until the standard knock intensity is reached. Knowing the CR, the octane rating of the fuel is read off the calibration curve. A gasoline with an octane rating of 92 means that it matches the performance of a mixture of 92% *iso*-octane and 8% heptane.

---

[t]http://www.howstuffworks.com/question90.htm

The more compact the hydrocarbon molecule, the less likely it is to cause spontaneous combustion and engine knock. Consequently, it is desired to isomerize straight-chain hydrocarbon molecules into more compact molecules through a catalytic process called reforming. One such reforming catalyst is platinum on alumina. Platinum on alumina (Al_2O_3) is a bifunctional catalyst that can be prepared by exposing alumina pellets to a chloroplatinic acid solution, drying, and then heating in air 775 to 875 K for several hours. Next, the material is exposed to hydrogen at temperatures around 725 to 775 K to produce very small clusters of Pt on alumina. These clusters have sizes on the order of 10 Å, while the alumina pore sizes on which the Pt is deposited are on the order of 100 to 10,000 Å (i.e., 10 to 1000 nm).

Catalyst manufacture

As an example of catalytic reforming we shall consider the isomerization of *n*-pentane to *i*-pentane:

$$n\text{-pentane} \xrightarrow[\text{Al}_2\text{O}_3]{0.75 \text{ wt\% Pt}} i\text{-pentane}$$

Gasoline	
C_5	10%
C_6	10%
C_7	20%
C_8	25%
C_9	20%
C_{10}	10%
$C_{11\text{-}12}$	5%

Normal pentane has an octane number of 62, while *iso*-pentane has an octane number of 90! The *n*-pentane adsorbs onto the platinum, where it is dehydrogenated to form *n*-pentene. The *n*-pentene desorbs from the platinum and adsorbs onto the alumina, where it is isomerized to *i*-pentene, which then desorbs and subsequently adsorbs onto platinum, where it is hydrogenated to form *i*-pentane. That is,

$$n\text{-pentane} \underset{\text{Pt}}{\overset{-\text{H}_2}{\rightleftharpoons}} n\text{-pentene} \underset{}{\overset{\text{Al}_2\text{O}_3}{\rightleftharpoons}} i\text{-pentene} \underset{\text{Pt}}{\overset{+\text{H}_2}{\rightleftharpoons}} i\text{-pentane}$$

We shall focus on the isomerization step to develop the mechanism and the rate law:

$$n\text{-pentene} \overset{\text{Al}_2\text{O}_3}{\rightleftharpoons} i\text{-pentene}$$

$$\text{N} \rightleftharpoons \text{I}$$

The procedure for formulating a mechanism, rate-limiting step, and corresponding rate law is given in Table 10-3.

Table 10-4 gives the form of rate law for different reaction mechanisms that are irreversible and surface-reaction-limited.

We need a word of caution at this point. Just because the mechanism and rate-limiting step may fit the rate data does not imply that the mechanism is correct.[12] Usually, spectroscopic measurements are needed to confirm a mechanism absolutely. However, the development of various mechanisms and rate-limiting steps can provide insight into the best way to correlate the data and develop a rate law.

---

[12]R. I. Masel, *Principles of Adsorption and Reaction on Solid Surfaces*, Wiley, New York, 1996, p. 506. http://www.uiuc.edu/ph/www/r-masel/

TABLE 10-3.   ALGORITHM FOR DETERMINING REACTION MECHANISM AND RATE-LIMITING STEP

Isomerization of *n*-pentene (N) to *i*-pentene (I) over alumina

$$N \underset{\longleftarrow}{\overset{Al_2O_3}{\rightleftharpoons}} I$$

1. *Select a mechanism.* (Mechanism I)

    Adsorption: $N + S \rightleftharpoons N{\cdot}S$

    Surface reaction: $N{\cdot}S + S \rightleftharpoons I{\cdot}S + S$

    Desorption: $I{\cdot}S \rightleftharpoons I + S$

    Treat each reaction step as an elementary reaction when writing rate laws.

2. *Assume a rate-limiting step.* Choose the surface reaction first, since *more than 75% of all heterogeneous reactions that are not diffusion-limited are surface-reaction-limited.* The rate law for the surface reaction step is

$$-r'_N = r_S = k_S \left(C_v C_{N{\cdot}S} - \frac{C_{I{\cdot}S} C_v}{K_S} \right)$$

3. *Find the expression for concentration of the adsorbed species* $C_{i{\cdot}S}$. Use the other steps that are not limiting to solve for $C_{i{\cdot}S}$ (e.g., $C_{N{\cdot}S}$ and $C_{I{\cdot}S}$). For this reaction,

    From $\dfrac{r_{AD}}{k_{AD}} \simeq 0$: $C_{N{\cdot}S} = P_N K_N C_v$

    From $\dfrac{r_D}{k_D} \simeq 0$: $C_{I{\cdot}S} = \dfrac{P_I C_v}{K_D} = K_I P_I C_v$

4. *Write a site balance.*

$$C_t = C_v + C_{N{\cdot}S} + C_{I{\cdot}S}$$

5. *Derive the rate law.* Combine steps 2, 3, and 4 to arrive at the rate law:

$$-r'_N = r_S = \frac{\overbrace{k_S C_t^2 K_N}^{k} (P_N - P_I/K_P)}{(1 + K_N P_N + K_I P_I)^2}$$

6. *Compare with data.* Compare the rate law derived in step 5 with experimental data. If they agree, there is a good chance that you have found the correct mechanism and rate-limiting step. If your derived rate law (i.e., model) does not agree with the data:

   a. Assume a different rate-limiting step and repeat steps 2 through 6.

   b. If, after assuming that each step is rate-limiting, none of the derived rate laws agree with the experimental data, select a different mechanism (e.g., Mechanism II):

$$N + S \rightleftharpoons N{\cdot}S$$
$$N{\cdot}S \rightleftharpoons I{\cdot}S$$
$$I{\cdot}S \rightleftharpoons I + S$$

   and then proceed through steps 2 through 6.

   Mechanism II turns out to be the correct one. For mechanism II the rate law is

$$-r'_N = \frac{k(P_N - P_I/K_P)}{(1 + K_N P_N + K_I P_I)}$$

   c. If two or more models agree, the statistical tests discussed in Chapter 5 (e.g., comparison of residuals) should be used to discriminate between them (see the Supplementary Reading).

TABLE 10-4.   IRREVERSIBLE SURFACE-REACTION-LIMITED RATE LAWS

Single site

$$A \cdot S \longrightarrow B \cdot S \qquad\qquad -r'_A = \frac{kP_A}{1 + K_A P_A + K_B P_B}$$

Dual site

$$A \cdot S + S \longrightarrow B \cdot S + S \qquad\qquad -r'_A = \frac{kP_A}{(1 + K_A P_A + K_B P_B)^2}$$

$$A \cdot S + B \cdot S \longrightarrow C \cdot S + S \qquad\qquad -r'_A = \frac{kP_A P_B}{(1 + K_A P_A + K_B P_B + K_C P_C)^2}$$

Eley–Rideal

$$A \cdot S + B(g) \longrightarrow C \cdot S \qquad\qquad -r'_A = \frac{kP_A P_B}{1 + K_A P_A + K_C P_C}$$

10.3.5   Rate Laws Derived from the Pseudo-Steady-State Hypothesis

In Section 7.1 we discussed the PSSH where the net rate of formation of reactive intermediates was assumed to be zero. An alternative way to derive a catalytic rate law rather than setting

$$\frac{r_{AD}}{k_A} \cong 0$$

is to assume that each species adsorbed on the surface is a reactive intermediate. Consequently, the net rate of formation of species i adsorbed on the surface will be zero:

$$r^*_{i \cdot S} = 0 \qquad\qquad (10\text{-}54)$$

While this method works well for a single rate-limiting step, it also works well when two or more steps are rate-limiting (e.g., adsorption <u>and</u> surface reaction). To illustrate how the rate laws are derived using the PSSH, we shall consider the isomerization of normal pentene to *iso*-pentene by the following mechanism, shown as item 6 of Table 10-3:

$$N + S \underset{k_{-N}}{\overset{k_N}{\rightleftharpoons}} N \cdot S$$

$$N \cdot S \overset{k_S}{\longrightarrow} I \cdot S$$

$$I \cdot S \underset{k_{-I}}{\overset{k_I}{\rightleftharpoons}} I + S$$

The rate law for surface reaction is

$$-r'_N = r_S = k_S C_{N \cdot S} \qquad\qquad (10\text{-}55)$$

The net rate
of the adsorbed
species (i.e., active
intermediates)
is zero

The net rates of generation of $N \cdot S$ sites and $I \cdot S$ sites are

$$r^*_{N \cdot S} = k_N P_N C_v - k_{-N} C_{N \cdot S} - k_S C_{N \cdot S} = 0 \tag{10-56}$$

$$r^*_{I \cdot S} = k_S C_{N \cdot S} - k_I C_{I \cdot S} + k_{-I} P_I C_v = 0 \tag{10-57}$$

Solving for $C_{N \cdot S}$ and $C_{I \cdot S}$ gives

$$C_{N \cdot S} = \frac{k_N P_N C_v}{k_{-N} + k_S} \tag{10-58}$$

$$C_{I \cdot S} = \frac{k_S C_{N \cdot S} + k_{-I} P_I C_v}{k_I} = \left(\frac{k_S k_N P_N}{k_I (k_{-N} + k_S)} + \frac{k_{-I}}{k_I} P_I \right) C_v \tag{10-59}$$

and substituting for $C_{N \cdot S}$ in the surface reaction rate law gives

$$-r'_N = \frac{k_N k_S}{k_{-N} + k_S} P_N C_v \tag{10-60}$$

From a site balance we obtain

$$C_t = C_{N \cdot S} + C_{I \cdot S} + C_v \tag{10-61}$$

After substituting for $C_{N \cdot S}$ and $C_{I \cdot S}$, solving for C_v, which we then substitute in the rate law, we find

Use this method
when
· Some steps are
  irreversible
· Two or more
  steps are rate-
  limiting

$$-r'_N = \left(\frac{k_N k_S C_t}{k_{-N} + k_S} \right) \frac{P_N}{1 + \dfrac{k_N}{k_{-N} + k_S} \left(1 + \dfrac{k_S}{k_I} \right) P_N + \dfrac{k_{-I}}{k_I} P_I} \tag{10-62}$$

The adsorption constants are just the ratio of their respective rate constants:

$$K_I = \frac{k_{-I}}{k_I} \quad \text{and} \quad K_N = \frac{k_N}{k_{-N}}$$

$$-r'_N = \left(\overbrace{\frac{K_N k_S C_t}{1 + k_S/k_{-N}}}^{k} \right) \frac{P_N}{1 + \dfrac{1}{1 + k_S/k_{-N}} K_N P_N + K_I P_I}$$

We have taken the surface reaction step to be rate-limiting; therefore, the surface specific reaction constant, k_S, is much smaller than the rate constant for the desorption of normal pentene, k_{-N}; that is,

$$1 \gg \frac{k_S}{k_{-N}}$$

and the rate law given by Equation (10-62) becomes

$$-r'_A = \frac{kP_N}{1 + K_N P_N + K_I P_I} \tag{10-63}$$

This rate law [Equation (10-63)] is identical to the one derived assuming that $r_{AD}/k \approx 0$ and $r_D/k_{AD} \approx 0$. However, this technique is preferred if two or more steps are rate-limiting or if some of the steps are irreversible or if none of the steps are rate-limiting.

10.3.6 Temperature Dependence of the Rate Law

The specific reaction rate, k, will usually follow an Arrhenius temperature dependence and increase exponentially with temperature. However, the adsorption of all species on the surface is exothermic. Consequently, the higher the temperature, the smaller the adsorption equilibrium constant. Therefore, at high temperatures, the denominator of catalytic rate laws approaches 1. For example, for a surface-reaction-limited irreversible isomerization

$$A \longrightarrow B$$

in which both A and B are adsorbed on the surface, the rate law is

$$-r'_A = \frac{kP_A}{1 + K_A P_A + K_B P_B}$$

At high temperatures (low coverage)

$$1 \gg (P_A K_A + P_B K_B)$$

The rate law could then be approximated as

Neglecting the adsorbed species at high temperatures

$$-r'_A \simeq kP_A$$

or for a reversible isomerization we would have

$$-r'_A \simeq k\left(P_A - \frac{P_B}{K_P}\right) \tag{10-64}$$

The algorithm we can use as a start in postulating a reaction mechanism and rate-limiting step is shown in Table 10-3. We can never really prove a mechanism by comparing the derived rate law with experimental data. Independent spectroscopic or tracer experiments are usually needed to confirm the mechanism. We can, however, prove that a proposed mechanism is inconsistent with the experimental data by following the algorithm in Table 10-3. Rather than tak-

Available strategies for model building

ing all the experimental data and then trying to build a model from the data, Box et al.[13] describe techniques of sequential data taking and model building.

---

[13]G. E. P. Box, W. G. Hunter, and J. S. Hunter, *Statistics for Engineers*, Wiley, New York, 1978.

10.4 Design of Reactors for Gas–Solid Reactions

10.4.1 Basic Guidelines

Designing or analyzing a reactor in which a fluid reaction is promoted by a solid catalyst differs little from designing or analyzing a reactor for homogeneous reactions. The principles presented in the first four chapters of this book remain valid. The kinetic rate law is often more complex for a catalytic reaction than for a homogeneous one, and this complexity can make the fundamental design equation more difficult to solve analytically. Numerical solution of the reactor design equation thus is encountered rather frequently when designing reactors for catalytic reactions.

10.4.2 The Design Equations

All the design equations for ideal catalytic or fluid–solid reactors can be obtained from their homogeneous reactor analogs merely by substituting the catalyst or solid weight, W, for the reactor volume, V. The reactor volume is merely the catalyst weight W divided by the bulk density of the catalyst ρ_b. In the catalytic or fluid–solid reactor design equations, r_A', based on catalyst mass, must of course be used.

For an ideal batch reactor, the differential form of the design equation for a heterogeneous reaction is

Batch design equation

$$N_{A0} \frac{dX}{dt} = -r_A' W \qquad (10\text{-}65)$$

For a packed-bed reactor, the differential form of the design equation for a heterogeneous reaction is

Use differential form of design equation for catalyst decay and pressure drop

$$F_{A0} \frac{dX}{dW} = -r_A' \qquad (2\text{-}17)$$

The design equation for a perfectly mixed "fluidized" catalytic reactor can be replaced by that of a CSTR.

Tank reactors for solid-catalyzed gaseous or liquid reactions are seen much less frequently than tubular reactors because of the difficulty in separating the phases and in agitating a fluid phase in the presence of solid particles. One type of CSTR used to study catalytic reactions is the spinning basket reactor, which has the catalyst embedded in the blades of the spinning agitator.[14] Another is the Berty reactor, which uses an internal recycle stream to achieve perfectly mixed behavior.[15] These reactors (see Chapter 5) are frequently used in industry to evaluate reaction mechanisms and determine reaction kinetics.

---

[14]J. J. Carberry, *Ind. Eng. Chem.*, *56*(11), 39 (1964).

[15]J. Berty, *Chem. Eng. Prog.*, *70*(5), 78 (1974).

For the ideal CSTR, the design equation based on volume is

$$V = \frac{F_{A0}X}{-r_A} \tag{2-13}$$

and the equivalent equation for a catalytic or fluid–solid reactor (Figure 10-16) with the rate based on mass of solid is

"Fluidized" CSTR design equation

$$W = \frac{F_{A0}X}{-r_A'} \tag{10-66}$$

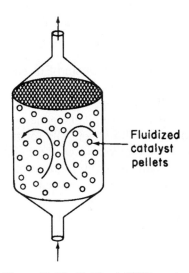

Fluidized catalyst pellets

Figure 10-16   Fluidized CSTR reactor.

Better models and design equations exist for fluidized-bed reactors and they should be used. One commonly used model is the bubbling bed model of Kunii and Levenspiel[16] and is discussed in Chapter 12.

The preceeding four sections conclude our discussion of elementary heterogeneous catalysis mechanisms and design equations. In the following section we shall work through an example problem using experimental data to (1) deduce a rate law, (2) determine a mechanism consistent with experimental data, (3) evaluate the rate law parameters, and (4) design a CSTR and packed-bed reactor.

Deduce
  Rate law
Find
  Mechanism
Evaluate
  Rate law
  parameters
Design
  PBR
  CSTR

10.5 Heterogeneous Data Analysis for Reactor Design

In this section we focus on four operations that reaction engineers need to be able to accomplish: (1) developing an algebraic rate law consistent with experimental observations, (2) analyzing the rate law in such a manner that the rate law parameters (e.g., k, K_A) can readily be extracted from the experimental data,

---

[16]D. Kunii and O. Levenspiel, *Fluidization Engineering*, R. E. Krieger, Huntington, N.Y., 1977.

(3) finding a mechanism and rate-limiting step consistent with the experimental data, and (4) designing a catalytic reactor to achieve a specified conversion. We shall use the hydrodemethylation of toluene to illustrate these four operations.

Hydrogen and toluene are reacted over a solid mineral catalyst containing clinoptilolite (a crystalline silica-alumina) to yield methane and benzene:[17]

$$C_6H_5CH_3 + H_2 \longrightarrow C_6H_6 + CH_4$$

We wish to design a packed-bed reactor and a fluidized CSTR to process a feed consisting of 30% toluene, 45% hydrogen, and 25% inerts. Toluene is fed at a rate of 50 mol/min at a temperature of 640°C and a pressure of 40 atm (4052 kPa). To design the PBR we must first determine the rate law from the differential reactor data presented in Table 10-5. In this table we find the rate of reaction of toluene as a function of the partial pressures of hydrogen (H), toluene (T), benzene (B), and methane (M). In the first two runs, methane was introduced into the feed together with hydrogen and toluene, while the other product, benzene, was fed to the reactor together with the reactants only in runs 3, 4, and 6. In runs 5 and 16 both methane and benzene were introduced in the feed. In the remaining runs, neither of the products were present in the feed-

TABLE 10-5.   DATA FROM A DIFFERENTIAL REACTOR

Run	$r_T' \times 10^{10}$ $\left(\dfrac{\text{g mol toluene}}{\text{g cat.}\cdot\text{s}}\right)$	Partial Pressure (atm)			
		Toluene, P_T	Hydrogen (H_2),[a] P_{H_2}	Methane, P_M	Benzene, P_B
Set A					
1	71.0	1	1	1	0
2	71.3	1	1	4	0
Set B					
3	41.6	1	1	0	1
4	19.7	1	1	0	4
5	42.0	1	1	1	1
6	17.1	1	1	0	5
Set C					
7	71.8	1	1	0	0
8	142.0	1	2	0	0
9	284.0	1	4	0	0
Set D					
10	47.0	0.5	1	0	0
11	71.3	1	1	0	0
12	117.0	5	1	0	0
13	127.0	10	1	0	0
14	131.0	15	1	0	0
15	133.0	20	1	0	0
16	41.8	1	1	1	1

[a] $P_H \equiv P_{H_2}$.

Unscramble the data to find the rate law

[17]J. Papp, D. Kallo, and G. Schay, *J. Catal.*, 23, 168 (1971).

stream; consequently, since the conversion was less than 1% in the differential reactor, the partial pressures of methane and benzene in these runs were essentially zero, and the reaction rates were equivalent to initial rates of reaction.

10.5.1 Deducing a Rate Law from the Experimental Data

Assuming that the reaction is essentially irreversible (which is reasonable after comparing runs 3 and 5), we ask what qualitative conclusions can be drawn from the data about the dependence of the rate of disappearance of toluene, $-r'_T$, on the partial pressures of toluene, hydrogen, methane, and benzene.

1. *Dependence on the product methane.* If the methane were adsorbed on the surface, the partial pressure of methane would appear in the denominator of the rate expression and the rate would vary inversely with methane concentration:

$$-r'_T \sim \frac{[\cdot]}{1 + K_M P_M + \cdots} \qquad (10\text{-}67)$$

However, from runs 1 and 2 we observe that a fourfold increase in the pressure of methane has little effect on $-r'_T$. Consequently, we assume that methane is either very weakly adsorbed (i.e., $K_M P_M \ll 1$) or goes directly into the gas phase in a manner similar to propylene in the cumene decomposition previously discussed.

2. *Dependence on the product benzene.* In runs 3 and 4, we observe that for fixed concentrations (partial pressures) of hydrogen and toluene the rate decreases with increasing concentration of benzene. A rate expression in which the benzene partial pressure appears in the denominator could explain this dependency:

If it is in the denominator, it is probably on the surface

$$-r'_T \sim \frac{1}{1 + K_B P_B + \cdots} \qquad (10\text{-}68)$$

The type of dependence of $-r'_T$ on P_B given by Equation (10-68) suggests that benzene is adsorbed on the clinoptilolite surface.

3. *Dependence on toluene.* At low concentrations of toluene (runs 10 and 11), the rate increases with increasing partial pressure of toluene, while at high toluene concentrations (runs 14 and 15), the rate is essentially independent of the toluene partial pressure. A form of the rate expression that would describe this behavior is

$$-r'_T \sim \frac{P_T}{1 + K_T P_T + \cdots} \qquad (10\text{-}69)$$

A combination of Equations (10-68) and (10-69) suggests that the rate law may be of the form

$$-r'_T \sim \frac{P_T}{1 + K_T P_T + K_B P_B + \cdots} \qquad (10\text{-}70)$$

4. *Dependence on hydrogen.* When we examine runs 7, 8, and 9 in Table 10-5, we see that the rate increases linearly with increasing hydrogen concentration and we conclude that the reaction is first-order in H_2. In light of this fact, hydrogen is either not adsorbed on the surface or it's coverage of the surface is extremely low $(1 \gg K_{H_2}P_{H_2})$ for the pressures used. If it were adsorbed, $-r'_T$ would have a dependence on P_{H_2} analogous to the dependence of $-r'_T$ on the partial pressure of toluene, P_T [see Equation (10-69)]. For first-order dependence on H_2,

$$-r'_T \sim P_{H_2} \tag{10-71}$$

Combining Equations (10-67) through (10-71), we find that the rate law

$$-r'_T = \frac{kP_{H_2}P_T}{1 + K_BP_B + K_TP_T}$$

is in qualitative agreement with the data shown in Table 10-5.

10.5.2 Finding a Mechanism Consistent with Experimental Observations

We now propose a mechanism for the hydrodemethylation of toluene. We assume that toluene is adsorbed on the surface and then reacts with hydrogen in the gas phase to produce benzene adsorbed on the surface and methane in the gas phase. Benzene is then desorbed from the surface. Since approximately 75% of all heterogeneous reaction mechanisms are surface-reaction-limited rather than adsorption- or desorption-limited, we begin by assuming the reaction between adsorbed toluene and gaseous hydrogen to be reaction-rate-limited. Symbolically, this mechanism and associated rate laws for each elementary step are:

Approximately 75% of all heterogeneous reaction mechanisms are surface-reaction-limited

Adsorption: $T(g) + S \rightleftharpoons T \cdot S$

$$r_{AD} = k_A \left(C_v P_T - \frac{C_{T \cdot S}}{K_T} \right) \tag{10-72}$$

Proposed mechanism

Surface reaction: $H_2(g) + T \cdot S \rightleftharpoons B \cdot S + M(g)$

$$r_S = k_S \left(P_{H_2}C_{T \cdot S} - \frac{C_{B \cdot S}P_M}{K_S} \right) \tag{10-73}$$

Desorption: $B \cdot S \rightleftharpoons B(g) + S$

$$r_D = k_D (C_{B \cdot S} - K_B P_B C_v) \tag{10-74}$$

For surface-reaction-limited mechanisms,

$$r_S = k_S \left(P_{H_2}C_{T \cdot S} - \frac{C_{B \cdot S}P_M}{K_S} \right) \tag{10-73}$$

we see that we need to replace $C_{\text{T}\cdot\text{S}}$ and $C_{\text{B}\cdot\text{S}}$ in Equation (10-73) by quantities that we can measure.

For surface-reaction-limited mechanisms, we use the adsorption rate Equation (10-72) to obtain $C_{\text{T}\cdot\text{S}}$:

$$\frac{r_{\text{AD}}}{k_{\text{A}}} \approx 0$$

Then

$$C_{\text{T}\cdot\text{S}} = K_{\text{T}}P_{\text{T}}C_v \tag{10-75}$$

and we use the desorption rate Equation (10-74) to obtain $C_{\text{B}\cdot\text{S}}$:

$$\frac{r_{\text{D}}}{k_{\text{D}}} \approx 0$$

Then

$$C_{\text{B}\cdot\text{S}} = K_{\text{B}}P_{\text{B}}C_v \tag{10-76}$$

The total concentration of sites is

Perform a site
balance to obtain C_v

$$\boxed{C_t = C_v + C_{\text{T}\cdot\text{S}} + C_{\text{B}\cdot\text{S}}} \tag{10-77}$$

Substituting Equations (10-75) and (10-76) into Equation (10-77) and re-arranging, we obtain

$$C_v = \frac{C_t}{1 + K_{\text{T}}P_{\text{T}} + K_{\text{B}}P_{\text{B}}} \tag{10-78}$$

Next, substitute for $C_{\text{T}\cdot\text{S}}$ and $C_{\text{B}\cdot\text{S}}$ and then substitute for C_v in Equation (10-73) to obtain the rate law for the case of surface-reaction control:

$$-r_{\text{T}}' = \frac{\overbrace{C_t k_{\text{S}} K_{\text{T}}}^{k}\, (P_{\text{H}_2}P_{\text{T}} - P_{\text{B}}P_{\text{M}}/K_P)}{1 + K_{\text{T}}P_{\text{T}} + K_{\text{B}}P_{\text{B}}} \tag{10-79}$$

Neglecting the reverse reaction we have

Rate law for
surface-reaction-
limited mechanism

$$\boxed{-r_{\text{T}}' = \frac{kP_{\text{H}_2}P_{\text{T}}}{1 + K_{\text{B}}P_{\text{B}} + K_{\text{T}}P_{\text{T}}}} \tag{10-80}$$

Again we note that the adsorption equilibrium constant of a given species is exactly the reciprocal of the desorption equilibrium constant of that species.

10.5.3 Evaluation of the Rate Law Parameters

In the original work on this reaction by Papp et al.,[18] over 25 models were tested against experimental data, and it was concluded that the mecha-

_____

[18]Ibid.

nism and rate-limiting step above (i.e., the surface reaction between adsorbed toluene and H_2 gas) is the correct one. Assuming that the reaction is essentially irreversible, the rate law for the reaction on clinoptilolite is

$$-r'_T = k\,\frac{P_{H_2}P_T}{1 + K_B P_B + K_T P_T}\qquad(10\text{-}80)$$

We now wish to determine how best to analyze the data to extract the rate law parameters, k, K_T, and K_B. This analysis is referred to as *parameter estimation*.[19] We now rearrange our rate law to obtain a linear relationship between our measured variables. For the rate law given by Equation (10-80), we see that if both sides of Equation (10-80) are divided by $P_{H_2}P_T$ and the equation is then inverted,

Linearize the rate equation to extract the rate law parameters

$$\boxed{\frac{P_{H_2}P_T}{-r'_T} = \frac{1}{k} + \frac{K_B P_B}{k} + \frac{K_T P_T}{k}}\qquad(10\text{-}81)$$

The multiple regression techniques described in Chapter 5 could be used to determine the rate law parameters by using the equation

A linear least squares analysis of the data shown in Table 10-5 is presented on the CD-ROM

$$\boxed{Y_j = a_0 + a_1 X_{1j} + a_2 X_{2j}}\qquad(5\text{-}30)$$

One can use the linearized least squares analysis to obtain initial estimates of the parameters k, K_T, K_B, in order to obtain convergence in nonlinear regression. However, in many cases it is possible to use a nonlinear regression analysis directly as described in Section 5.4 and in Example 5-6.

CD Professional

Reference Shelf

Example 10–2   Regression Analysis to Determine the Model Parameters k, K_B, and K_T

Solution

The data from Table 10-5 were entered into the POLYMATH nonlinear least squares program with the following modification. The rates of reaction in column 1 were multiplied by 10^{10}, so that each of the numbers in column 1 was entered directly (i.e., 71.0, 71.3, ...). The model equation was

$$\text{rate} = \frac{k P_T P_{H_2}}{1 + K_B P_B + K_T P_T}\qquad(\text{E}10\text{-}2.1)$$

The POLYMATH results are given in Table E10-2.1 and Figure E10-2.1. Converting the rate law to kilograms of catalyst and minutes,

$$-r'_T = \frac{1.45\times10^{-8}\,P_T P_{H_2}}{1 + 1.39 P_B + 1.038 P_T}\,\frac{\text{g mol T}}{\text{g cat.}\cdot\text{s}} \times \frac{1000\text{ g}}{1\text{ kg}} \times \frac{60\text{ s}}{\text{min}}\qquad(\text{E}10\text{-}2.2)$$

---

[19]See the Supplementary Reading for a variety of techniques for estimating the rate law parameters.

TABLE E10-2.1.   PARAMETER VALUES

Example 10-2

Matrix of correlation coefficients for the nonlinear correlation equation

	k	KB	KT
k	1	0.343228	0.95675
KB	0.343228	1	0.309403
KT	0.95675	0.309403	1

Param.	Converged Value	0.95 conf. interval	lower limit	upper limit
k	144.767	1.24044	143.527	146.008
KB	1.39053	0.0457982	1.34473	1.43632
KT	1.03841	0.0131607	1.02525	1.05157

Model: RATE=k*PT*PH2/(1+KB*PB+KT*PT)

k = 144.767   KT = 1.03841
KB = 1.39053
6 positive residuals, 10 negative residuals. Sum of squares = 3.26051

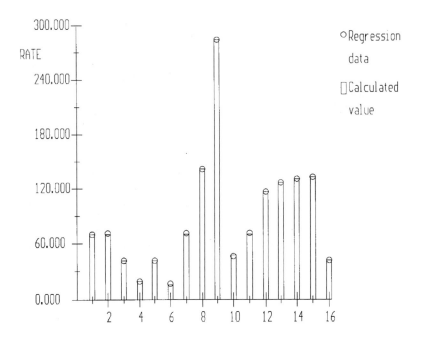

Model: RATE=k*PT*PH2/(1+KB*PB+KT*PT)

k = 144.767   KT = 1.03841

KB = 1.39053

6 positive residuals, 10 negative residuals. Sum of squares = 3.26051

Figure E10-2.1 Comparison of calculated and measured rates.

we have

$$-r'_T = \frac{8.7 \times 10^{-4} P_T P_{H_2}}{1 + 1.39 P_B + 1.038 P_T} \frac{\text{g mol T}}{\text{kg cat.} \cdot \text{min}} \qquad \text{(E10-2.3)}$$

Ratio of sites occupied by toluene to those occupied by benzene

Once we have the adsorption constants, K_T and K_B, we could calculate the ratio of sites occupied by the various adsorbed species. For example, the ratio of toluene sites to benzene sites at 40% conversion is

$$\frac{C_{T \cdot S}}{C_{B \cdot S}} = \frac{C_v K_T P_T}{C_v K_B P_B} = \frac{K_T P_T}{K_B P_B} = \frac{K_T P_{A0}(1 - X)}{K_B P_{A0} X}$$

$$= \frac{K_T (1 - X)}{K_B X} = \frac{1.038 (1 - 0.4)}{1.39 (0.4)} = 1.12$$

We see that at 40% conversion there are approximately 12% more sites occupied by toluene than by benzene.

10.5.4 Reactor Design

Our next step is to express the partial pressures P_T, P_B, and P_{H_2} as a function of X, combine the partial pressures with the rate law $-r'_A$ as a function of conversion, and carry out the integration of the packed-bed design equation

$$\frac{dX}{dW} = \frac{-r'_A}{F_{A0}} \qquad \text{(2-17)}$$

Example 10–3 Fixed (i.e., Packed)-Bed Reactor Design

The hydrodemethylation of toluene is to be carried out in a packed-bed reactor. Plot the conversion, pressure ratio, y, and the partial pressures of toluene, hydrogen, and benzene as a function of catalyst weight. The molar feed rate of toluene to the reactor is 50 mol/min and the reactor is operated at 40 atm and 640°C. The feed consists of 30% toluene, 45% hydrogen, and 25% inerts. Hydrogen is used in excess to help prevent coking. The pressure drop parameter is 9.8×10^{-5} kg^{-1}. Also determine the catalyst weight in a CSTR with a bulk density of 400 kg/m^3.

$$C_6H_5CH_3 + H_2 \longrightarrow C_6H_6 + CH_4$$

Solution

1. **Design equation:**

Balance on toluene (T)

$$\frac{dF_T}{dW} = -r'_T$$

$$\frac{dX}{dW} = \frac{-r'_T}{F_{T0}} \qquad \text{(E10-3.1)}$$

2. **Rate law.** From Equation (E10-2.3) we have

$$-r'_T = \frac{kP_{H_2}P_T}{1 + K_B P_B + K_T P_T} \qquad \text{(E10-3.2)}$$

with $k = 0.00087$ mol/atm^2/kg cat./min.

3. **Stoichiometry:**

$$P_T = C_T RT = C_{T0}RT_0 \left(\frac{1-X}{1+\varepsilon X}\right)y = P_{T0}\left(\frac{1-X}{1+\varepsilon X}\right)y$$

$$\varepsilon = y_{T0}\delta = 0.3(0) = 0$$

$$P_T = P_{T0}(1-X)y \qquad \text{(E10-3.3)}$$

Relating
Toluene (T)
Benzene (B)
Hydrogen (H₂)

$$P_{H_2} = P_{T0}(\Theta_{H_2} - X)y$$

$$\Theta_{H_2} = \frac{0.45}{0.30} = 1.5$$

$$P_{H_2} = P_{T0}(1.5 - X)y \qquad \text{(E10-3.4)}$$

$$P_B = P_{T0}Xy \qquad \text{(E10-3.5)}$$

Because $\varepsilon = 0$, we can use the integrated form of the pressure drop term.

P_0 = total pressure at the entrance

$$y = \frac{P}{P_0} = (1 - \alpha W)^{1/2} \qquad \text{(4-33)}$$

$$\alpha = 9.8 \times 10^{-5} \text{ kg}^{-1}$$

Note that P_{T0} designates the initial partial pressure of toluene. In this example the initial total pressure is designated P_0 to avoid any confusion. The initial mole fraction of toluene is 0.3 (i.e., $y_{T0} = 0.3$), so that the initial partial pressure of toluene is

$$P_{T0} = (0.3)(40) = 12 \text{ atm}$$

The maximum catalyst weight we can have and not fall below 1 atm is found from Equation (4-33) for an entering pressure of 40 atm.

$$\frac{1}{40} = (1 - 9.8 \times 10^{-5}W)^{1/2}$$

$$W = 10{,}197 \text{ kg}$$

The calculations for no ΔP are given on the CD-ROM

Consequently, we will set our final weight at 10,000 kg and determine the conversion as a function of catalyst weight up to this value. Equations (E10-3.1) through (E10-3.5) are shown below in the POLYMATH program in Table E10-3.1. The conversion is shown as a function of catalyst weight in Figure E10-3.1 and profiles of the partial pressures of toluene, hydrogen, and benzene are shown in Figure E10-3.2. We note that the pressure drop causes the partial pressure of benzene to go through a maximum as one traverses the reactor.

For the case of no pressure drop, the conversion that would have been achieved with 10,000 kg of catalyst weight would have been 79%, compared with

69% when there is pressure drop in the reactor. *For the feed rate given, eliminating or minimizing pressure drop would increase the production of benzene by up to 61 million pounds per year.*

CD Living

Example Problems

TABLE E10-3.1.   POLYMATH PROGRAM

Equations	Initial Values
d(x)/d(w)=-rt/ft0	0
ft0=50	
k=.0000000145*1000*60	
kt=1.038	
kb=1.39	
alpha=0.000098	
Po=40	
pt0=0.3*Po	
y=(1-alpha*w)**0.5	
ph=pt0*(1.5-x)*y	
pt=pt0*(1-x)*y	
pb=2*pt0*x*y	
rt=-k*kt*ph*pt/(1+kb*pb+kt*pt)	
rate=-rt	
$w_0 = 0$, $w_f = 10000$	

Conversion profile down the packed bed

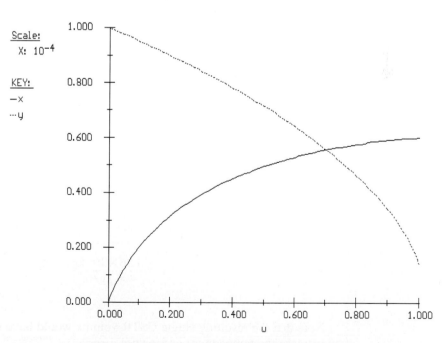

Scale:
X: 10^{-4}

KEY:
— x
··· y

Figure E10-3.1   Conversion and pressure ratio profiles.

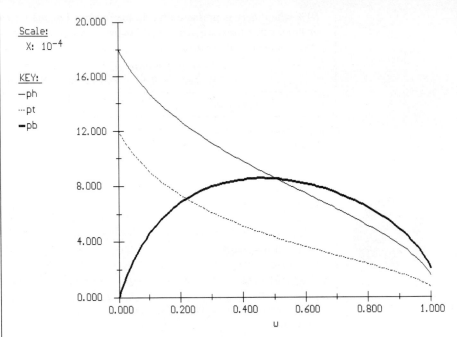

Note the partial
pressure of benzene
goes through a
maximum. Why?

Figure E10-3.2   Partial pressure profiles.

We will now calculate the fluidized CSTR catalyst weight necessary to achieve the same conversion as in the packed-bed reactor at the same operating conditions. The bulk density in the fluidized reactor is 0.4 g/cm^3. The design equation is

$$W = \frac{F_{T0}X}{-r_T'} \qquad (10\text{-}66)$$

At $X = 0.65$ we have

$$-r_T' = 2.39 \times 10^{-3} \ \frac{\text{g mol T}}{\text{kg cat.} \cdot \text{min}}$$

$$W = \frac{F_{T0}X}{-r_T'} = \frac{(50 \text{ g mol T/min})(0.65)}{2.39 \times 10^{-3} \text{ g mol T/kg cat.} \cdot \text{min}}$$

$$\boxed{W = 1.36 \times 10^4 \text{ kg of catalyst}}$$

$$V = \frac{1.36 \times 10^4 \text{ kg}}{400 \text{ kg/m}^3} = 34 \text{ m}^3$$

Note that the resulting single CSTR volume would have been too large—try to increase the temperature and/or use a series of CSTRs.

These values of the catalyst weight and reactor volume are quite high, especially for the low feed rates given. *Consequently, the temperature of the reacting mixture should be increased to reduce the catalyst weight, provided that side reactions do not become a problem at higher temperatures.*

How can the weight of catalyst be reduced?

Example 10-3 illustrated the major activities pertinent to catalytic reactor design described earlier in Figure 10-7. In this example the rate law was extracted directly from the data and then a mechanism was found that was consistent with experimental observation. However, developing a feasible mechanism may guide one in the synthesis of the rate law.

10.6 Chemical Vapor Deposition

We now extend the principle of the preceding sections to one of the emerging technologies in chemical engineering. Chemical engineers are now playing an important role in the electronics industry. Specifically, they are becoming more involved in the manufacture of electronic and photonic devices and recording materials. In the formation of microcircuits, electrically interconnected films are laid down by chemical reactions (see Section 12.8). One method by which these films are made is chemical vapor deposition (CVD).

The mechanisms by which CVD occur are very similar to those of heterogeneous catalysis discussed earlier in this chapter. The reactant(s) adsorb on the surface and then react on the surface to form a new surface. This process may be followed by a desorption step, depending on the particular reaction.

The growth of a germanium epitaxial film as an interlayer between a gallium arsenide layer and a silicon layer and as a contact layer is receiving increasing attention in the microelectronics industry.[20] Epitaxial germanium is also an important material in the fabrication of tandem solar cells. The growth of germanium films can be accomplished by CVD. A proposed mechanism is:

Gas-phase dissociation: $\quad GeCl_4(g) \; \rightleftharpoons \; GeCl_2(g) + Cl_2(g)$

Adsorption: $\qquad\qquad\qquad GeCl_2(g) + S \; \overset{k_A}{\rightleftharpoons} \; GeCl_2 \cdot S$

Adsorption: $\qquad\qquad\qquad H_2 + 2S \; \overset{k_H}{\rightleftharpoons} \; 2H \cdot S$

Surface reaction: $\qquad\quad GeCl_2 \cdot S + 2H \cdot S \; \overset{k_S}{\longrightarrow} \; Ge(s) + 2HCl(g) + 2S$

At first it may appear that a site has been lost when comparing the right- and left-hand sides of the surface reaction step. However, the newly formed germanium atom on the right-hand side is a site for the future adsorption of $H_2(g)$ or $GeCl_2(g)$ and there are three sites on both the right- and left-hand sides of the surface reaction step. These sites are shown schematically in Figure 10-17.

---

[20]H. Ishii and Y. Takahashi, *J. Electrochem. Soc.*, *135*, 1539 (1988).

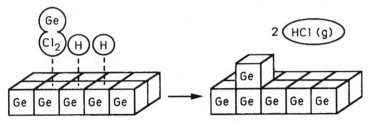

Figure 10-17

The surface reaction between adsorbed molecular hydrogen and germanium dichloride is believed to be rate-limiting:

$$r''_{\text{Dep}} = k_{\text{S}} f_{\text{GeCl}_2} f_{\text{H}}^2 \qquad (10\text{-}82)$$

where r''_{Dep} = deposition rate, nm/s
$\quad\quad\quad k_{\text{S}}$ = surface specific reaction rate, nm/s
$\quad\quad\quad f_{\text{GeCl}_2}$ = fraction on the surface occupied by germanium dichloride
$\quad\quad\quad f_{\text{H}}^2$ = fraction of the surface covered by molecular hydrogen

The deposition rate (film growth rate) is usually expressed in nanometers per second and is easily converted to a molar rate (mol/m²·s) by multiplying by the molar density of solid germanium.

The difference between developing CVD rate laws and rate laws for catalysis is that the site concentration (e.g., C_v) is replaced by the fractional surface area coverage (e.g., the fraction of the surface that is vacant, f_v). The total fraction of surface available for adsorption should, of course, add up to 1.0.

Area balance   Fractional area balance: $\boxed{f_v + f_{\text{GeCl}_2} + f_{\text{H}} = 1}$ (10-83)

We will first focus our attention on the adsorption of $GeCl_2$. The rate of jumping on to the surface is proportional to the partial pressure of $GeCl_2$. The net rate of $GeCl_2$ adsorption is

$$r_{\text{AD}} = k_{\text{A}} \left(f_v P_{\text{GeCl}_2} - \frac{f_{\text{GeCl}_2}}{K_{\text{A}}} \right) \qquad (10\text{-}84)$$

Since the surface reaction is rate-limiting, in a manner analogous to catalysis reactions, we have for the adsorption of $GeCl_2$

$$\frac{r_{\text{AD}}}{k_{\text{A}}} \approx 0$$

Solving Equation (10-84) for the fractional surface coverage of $GeCl_2$ gives

$$\boxed{f_{\text{GeCl}_2} = f_v K_{\text{A}} P_{\text{GeCl}_2}} \qquad (10\text{-}85)$$

For the dissociative adsorption of hydrogen on the Ge surface, the equation analogous to (10-84) is

$$r_{H_2} = k_H \left(P_{H_2} f_v^2 - \frac{f_H^2}{K_H} \right) \qquad (10\text{-}86)$$

Since the surface reaction is rate-limiting,

$$\frac{r_{H_2}}{k_H} \approx 0$$

Then

$$\boxed{f_H = f_v \sqrt{K_H P_{H_2}}} \qquad (10\text{-}87)$$

Recalling the rate of deposition of germanium, we substitute for f_{GeCl_2} and f_H in Equation (10-82) to obtain

$$r''_{Dep} = f_v^3 k_S K_A P_{GeCl_2} K_H P_{H_2} \qquad (10\text{-}88)$$

We solve for f_v in an identical manner to that for C_v in heterogeneous catalysis. Substituting Equations (10-85) and (10-87) into Equation (10-83) gives

$$f_v + f_v \sqrt{K_H P_{H_2}} + f_v K_A P_{GeCl_2} = 1$$

Rearranging yields

$$f_v = \frac{1}{1 + K_A P_{GeCl_2} + \sqrt{K_H P_{H_2}}} \qquad (10\text{-}89)$$

Finally, substituting for f_v in Equation (10-88), we find that

$$r''_{Dep} = \frac{k_S K_H K_A P_{GeCl_2} P_{H_2}}{(1 + K_A P_{GeCl_2} + \sqrt{K_H P_{H_2}})^3}$$

and lumping K_A, K_H, and k_S into a specific reaction rate k' yields

$$\boxed{r''_{Dep} = \frac{k' P_{GeCl_2} P_{H_2}}{(1 + K_A P_{GeCl_2} + \sqrt{K_H P_{H_2}})^3}} \qquad (10\text{-}90)$$

If we assume that the gas-phase reaction

$$GeCl_4(g) \rightleftharpoons GeCl_2(g) + Cl_2(g)$$

is in equilibrium with $P_{GeCl_2} = P_{Cl_2}$, we have

$$K_p = \frac{P_{GeCl_2}P_{Cl_2}}{P_{GeCl_4}} = \frac{P^2_{GeCl_2}}{P_{GeCl_4}}$$

$$P_{GeCl_2} \sim P^{1/2}_{GeCl_4}$$

and if hydrogen is weakly adsorbed, we obtain the rate of deposition as

$$r''_{Dep} = \frac{kP^{1/2}_{GeCl_4}P_{H_2}}{(1 + K_1 P^{1/2}_{GeCl_4})^3} \tag{10-91}$$

It should also be noted that it is possible that $GeCl_2$ may also be formed by the reaction of $GeCl_4$ and a Ge atom on the surface, in which case a different rate law would result.

10.7 Catalyst Deactivation

In designing fixed and ideal fluidized-bed catalytic reactors, we have assumed up to now that the activity of the catalyst remains constant throughout the catalyst's life. That is, the total concentration of active sites, C_t, accessible to the reaction does not change with time. Unfortunately, Mother Nature is not so kind as to allow this behavior to be the case in most industrially significant catalytic reactions. One of the most insidious problems in catalysis is the loss of catalytic activity that occurs as the reaction takes place on the catalyst. A wide variety of mechanisms have been proposed by Butt and Petersen,[21] to explain and model catalyst deactivation.

Catalytic deactivation adds another level of complexity to sorting out the reaction rate law parameters and pathways. In addition, we need to make adjustments for the decay of the catalysts in the design of catalytic reactors. This adjustment is usually made by a quantitative specification of the catalyst's activity, $a(t)$. In analyzing reactions over decaying catalysts we divide the reactions into two categories: *separable kinetics* and *nonseparable kinetics*. In separable kinetics, we separate the rate law and activity:[21]

Separable kinetics: $-r'_A = a(\text{past history}) \times -r'_A \text{ (fresh catalyst)}$.

When the kinetics and activity are separable, it is possible to study catalyst decay and reaction kinetics independently. However, nonseparability,

Nonseparable kinetics: $-r'_A = -r'_A \text{ (past history, fresh catalyst)}$

must be accounted for by assuming the existence of a nonideal surface or by describing deactivation by a mechanism composed of several elementary steps.[22]

[21]J. B. Butt and E. E. Petersen, *Activation, Deactivation and Poisoning of Catalysts*, Academic Press, New York, 1988. See also S. Szépe and O. Levenspiel, *Chem. Eng. Sci.*, 23, 881–894 (1968).

[22]D. T. Lynch and G. Emig, *Chem. Eng. Sci.*, 44(6), 1275–1280 (1989).

In this section we shall consider only separable kinetics and define the activity of the catalyst at time t, $a(t)$, as the ratio of the rate of reaction on a catalyst that has been used for a time t to the rate of reaction on a fresh catalyst:

$a(t)$: catalyst
activity

$$a(t) = \frac{-r'_A(t)}{-r'_A(t=0)} \tag{10-92}$$

Because of the catalyst decay, the activity decreases with time and a typical curve of the activity as a function of time is shown in Figure 10-18.

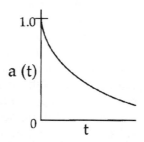

Figure 10-18   Activity as a function of time.

Combining Equations (10-92) and (3-1), the rate of disappearance of reactant A on a catalyst that has been utilized for a time t is

Reaction rate law
accounting for
catalyst activity

$$\boxed{-r'_A = a(t)\,k(T)\,\mathrm{fn}\,(C_A,\ C_B,\ \ldots,\ C_P)} \tag{10-93}$$

where $a(t)$ = catalytic activity, time-dependent
      $k(T)$ = specific reaction rate, temperature-dependent
      C_i = gas-phase concentration of reactants, products, or contaminant

The rate of catalyst decay, r_d, can be expressed in a rate law analogous to Equation (10-93):

Catalyst decay
rate law

$$r_d = -\frac{da}{dt} = p[a(t)]k_d(T)h(C_A, C_B, \ldots, C_P) \tag{10-94}$$

where k_d is the specific decay constant and $h(C_i)$ is the functionality of r_d on the reacting species concentrations. For the cases presented in this chapter this functionality will be either independent of concentration (i.e., $h = 1$), or be a linear function of species concentration (i.e., $h = C_i$).

The functionality of the activity term, $p[a(t)]$, in the decay law can, as we will soon see, take a variety of forms. For example, for a first-order decay,

$$p(a) = a \tag{10-95}$$

and for a second-order decay,

$$p(a) = a^2 \tag{10-96}$$

The particular function, $p(a)$, will vary with the gas catalytic system being used and the reason or mechanism for catalytic decay.

10.7.1 Types of Catalyst Deactivation

Sintering
Fouling
Poisoning
There are three categories into which the loss of catalytic activity can traditionally be divided: sintering or aging, fouling or coking, and poisoning.

Deactivation by Sintering (Aging).[23]   Sintering, also referred to as aging, is the loss of catalytic activity due to a loss of active surface area resulting from the prolonged exposure to high gas-phase temperatures. The active surface area may be lost either by crystal agglomeration and growth of the metals deposited on the support or by narrowing or closing of the pores inside the catalyst pellet. A change in the surface structure may also result from either surface recrystallization or from the formation or elimination of surface defects (active sites). The reforming of heptane over platinum on alumina is an example of catalyst deactivation as a result of sintering.

Figure 10-19 shows the loss of surface area resulting from the flow of the solid porous catalyst support at high temperatures to cause pore closure. Figure 10-20 shows the loss of surface by atomic migration and agglomeration of small metal sites deposited on the surface into a larger site where the interior atoms are not accessible to the reaction. Sintering is usually negligible at temperatures below 40% of the melting temperature of the solid.[24]

The catalyst support becomes soft and flows, resulting in pore closure

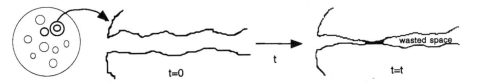

Figure 10-19   Decay by sintering: pore closure.

The atoms move along the surface and agglomerate

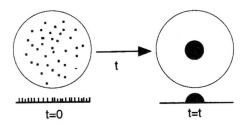

Figure 10-20   Decay by sintering: agglomeration of deposited metal sites.

Deactivation by sintering may in some cases be a function of the mainstream gas concentration. Although other forms of the sintering decay rate law exist, one of the most commonly used decay laws is second-order with respect to the present activity:

---

[23]See G. C. Kuczynski, Ed., *Sintering and Catalysis*, Vol. 10 of *Materials Science Research*, Plenum Press, New York, 1975.

[24]R. Hughes, *Deactivation of Catalysts*, Academic Press, San Diego, Calif., 1984.

$$r_d = k_d a^2 = -\frac{da}{dt} \tag{10-97}$$

Integrating, with $a = 1$ at time $t = 0$ yields

Sintering: second-
order decay

$$\boxed{a(t) = \frac{1}{1 + k_d t}} \tag{10-98}$$

The amount of sintering is usually measured in terms of the active surface area of the catalyst S_a:

$$S_a = \frac{S_{a0}}{1 + k_d t} \tag{10-99}$$

The sintering decay constant follows the Arrhenius equation

$$\boxed{k_d = k_d(T_0)\,\exp\left[\frac{E_d}{R}\left(\frac{1}{T_0} - \frac{1}{T}\right)\right]} \tag{10-100}$$

Minimizing
sintering

The decay activation energy, E_d, for the reforming of heptane on Pt/Al$_2$O$_3$ is on the order of 70 kcal/mol, which is rather high. Sintering can be reduced by keeping the temperature below 0.3 to 0.5 times the metal's melting point.

We will now stop and consider reactor design for a fluid–solid system with decaying catalyst. To analyze these reactors we only add one step to our algorithm, that is, determine the catalyst decay law. The sequence is shown below.

The algorithm

mole balance $\longrightarrow$ reaction rate law $\longrightarrow$ *decay rate law* $\longrightarrow$

stoichiometry $\longrightarrow$ combine and solve $\longrightarrow$ numerical techniques

Example 10–4   Calculating Conversion with Catalyst Decay in Batch Reactors

The first-order isomerization

$$A \longrightarrow B$$

is being carried out isothermally in a batch reactor on a catalyst that is decaying as a result of aging. Derive an equation for conversion as a function of time.

Solution

1. **Design equation:**

$$N_{A0}\frac{dX}{dt} = -r'_A W \tag{10-65}$$

2. **Reaction rate law:**

$$-r'_A = k'a(t)C_A \tag{E10-4.1}$$

One extra step
(number 3) is added
to the algorithm

3. **Decay law.** For second-order decay by sintering:

$$a(t) = \frac{1}{1 + k_d t} \tag{10-98}$$

4. **Stoichiometry:**

$$C_A = C_{A0}(1 - X) = \frac{N_{A0}}{V}(1 - X) \tag{E10-4.2}$$

5. **Combining** gives us

$$\frac{dX}{dt} = \frac{W}{V} k'a(t)(1 - X) \tag{E10-4.3}$$

Let $k = k'W/V$. Then, separating variables, we have

$$\frac{dX}{1 - X} = ka(t)\,dt \tag{E10-4.4}$$

Substituting for a and integrating yields

$$\int_0^X \frac{dX}{1 - X} = k\int_0^t \frac{dt}{1 + k_d t} \tag{E10-4.5}$$

$$\ln\frac{1}{1 - X} = \frac{k}{k_d}\ln(1 + k_d t) \tag{E10-4.6}$$

6. **Solving** for the conversion X at any time t, we find that

$$X = 1 - \frac{1}{(1 + k_d t)^{k/k_d}} \tag{E10-4.7}$$

This is the conversion that will be achieved in a batch reactor for a first-order reaction when the catalyst decay law is second-order. The purpose of this example was to demonstrate the algorithm for isothermal catalytic reactor design for a decaying catalyst.

Deactivation by Coking or Fouling. This mechanism of decay (see Figure 10-21) is common to reactions involving hydrocarbons. It results from a carbonaceous (coke) material being deposited on the surface of a catalyst.

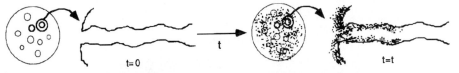

t=0

t

t=t

Figure 10-21   Schematic of decay by coking.

(a) Fresh catalyst   (b) Spent catalyst

Figure 10-22   Decay by coking. (Photos courtesy of Engelhard catalyst, copyright by Michael Gaffney Photographer, Mendham, N.J.)

The amount of coke on the surface after a time t has been found to obey the following empirical relationship:

$$C_C = At^n \qquad (10\text{-}101)$$

where C_C is the concentration of carbon on the surface and n and A are fouling parameters which can be functions of the feed rate. This expression was originally developed by Voorhies[25] and has been found to hold for a wide variety of catalysts and feedstreams. Representative values of A and n for the cracking of East Texas light gas oil[26] yield

$$\% \text{ coke} = 0.47 \sqrt{t(\text{min})}$$

Different functionalities between the activity and amount of coke on the surface have been observed. One commonly used form is

$$a = \frac{1}{C_C^p + 1} \qquad (10\text{-}102)$$

or, in terms of time,

$$a = \frac{1}{A^p t^{np} + 1} \qquad (10\text{-}103)$$

For light Texas gas oil being cracked at 750°F over a synthetic catalyst for short times, the decay law is

$$a = \frac{1}{1 + 7.6\, t^{1/2}} \qquad (10\text{-}104)$$

where t is in seconds.

[25]A. Voorhies, *Ind. Eng. Chem.*, *37*, 318 (1945).
[26]C. O. Prater and R. M. Lago, *Adv. Catal.*, *8*, 293 (1956).

Activity for
deactivation by
coking
Other commonly used forms are

$$a = e^{-\alpha_1 C_C} \tag{10-105}$$

and

$$a = \frac{1}{1 + \alpha_2 C_C} \tag{10-106}$$

A dimensionless fouling correlation has been developed by Pacheco and Petersen.[27]

Minimizing coking
When possible, coking can be reduced by running at elevated pressures (2000 to 3000 kPa) and hydrogen-rich streams. A number of other techniques for minimizing fouling are discussed by Bartholomew.[28] Catalysts deactivated by coking can usually be regenerated by burning off the carbon. The use of the shrinking core model to describe regeneration is discussed in Section 11.4.

Deactivation by Poisoning.   Deactivation by this mechanism occurs when the poisoning molecules become irreversibly chemisorbed to active sites, thereby reducing the number of sites available for the main reaction. The poisoning molecule, P, may be a reactant and/or a product in the main reaction, or it may be an impurity in the feedstream.

One of the most significant cases of catalyst poisoning occurred at the gasoline pump. For many years lead was used as an antiknock component in gasoline. While effective as an octane enhancer, it also poisoned the catalytic afterburner, which reduced NO_x, CO, and hydrocarbons in the exhaust. Consequently, lead had to be removed from gasoline.

Poison in the Feed.   Many petroleum feed stocks contain trace impurities such as sulfur, lead, and other components which are too costly to remove, yet poison the catalyst slowly over time. For the case of an impurity, P, in the feedstream, such as sulfur, for example, in the reaction sequence

Main
reaction:
$$\begin{cases} A + S \Leftrightarrow A \cdot S \\ A \cdot S \Leftrightarrow B \cdot S + C(g) \\ B \cdot S \Leftrightarrow B + S \end{cases} \quad -r'_A = a(t)\, \frac{k C_A}{1 + K_A C_A + K_B C_B}$$

Poisoning
reaction: $P + S \longrightarrow P \cdot S \qquad r_D = -\dfrac{da}{dt} = k'_d C_p^m a^q \tag{10-107}$

the surface sites would change with time as shown in Figure 10-23.

Progression of sites
being poisoned

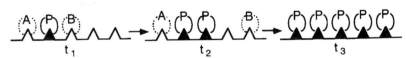

Figure 10-23   Decay by poisoning.

[27]M. A. Pacheco and E. E. Petersen, *J. Catal.*, 86, 75 (1984).
[28]C. Bartholomew, *Chem. Eng.*, Sept. 12, 1984, p. 96.

If we assume the rate of removal of the poison from the reactant gas stream onto the catalyst sites is proportional to the number of sites that are unpoisoned $(C_{t0} - C_{\text{P} \cdot \text{S}})$ and the concentration of poison in the gas phase, C_P:

$$r_{\text{P} \cdot \text{S}} = k(C_{t0} - C_{\text{P} \cdot \text{S}})C_\text{P}$$

where $C_{\text{P} \cdot \text{S}}$ is the concentration of poisoned sites and C_{t0} is the total number of sites initially available. Because every molecule that is adsorbed from the gas phase onto a site is assumed to poison the site, this rate is also equal to the rate of removal of sites from the surface:

$$\frac{dC_{\text{P} \cdot \text{S}}}{dt} = r_{\text{P} \cdot \text{S}} = k_d(C_{t0} - C_{\text{P} \cdot \text{S}})C_\text{P}$$

Dividing through by C_{t0} and letting f be the fraction of the total number of sites that have been poisoned yields

$$\frac{df}{dt} = k_d(1 - f)C_\text{P} \tag{10-108}$$

The fraction of sites available for adsorption $(1 - f)$ is essentially the activity $a(t)$. Consequently, Equation (10-108) becomes

$$-\frac{da}{dt} = a(t)k_d C_\text{P} \tag{10-109}$$

A number of examples of catalysts with their corresponding catalyst poisons are given by Bartholomew.[29]

Packed-Bed Reactors.   In packed-bed reactors where the poison is removed from the gas phase by being adsorbed on the specific catalytic sites, the deactivation process can move through the packed bed as a wave front. Here, at the start of the operation, only those sites near the entrance to the reactor will be deactivated because the poison (which is usually present in trace amounts) is removed from the gas phase by the adsorption, and consequently, the catalyst sites farther down the reactor will not be affected. However, as time continues, the sites near the entrance of the reactor become saturated and the poison must travel farther downstream before being adsorbed (removed) from the gas phase and attaching to a site to deactivate it. Figure 10-24 shows the corresponding activity profile for this type of poisoning process. We see in Figure 10-24 that by time t_4 the entire bed has become deactivated. The corresponding overall conversion at the exit of the reactor might vary with time as shown in Figure 10-25.

Poisoning by Either Reactants or Products.   For the case where the main reactant also acts as a poison, the rate laws are:

Main reaction: $A + S \longrightarrow B + S$ $-r'_A = k_A C_A^n$

Poisoning reaction: $A + S \longrightarrow A \cdot S$ $r_d = k'_d C_A^m a^q$

---

[29]Farrauto, R. J. and C. H. Bartholomew, "Fundamentals of Industrial Catalytic Processes," Blackie Academic and Professional, New York 1997.

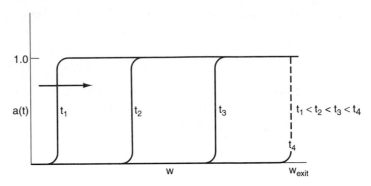

Figure 10-24   Movement of activity front in a packed bed.

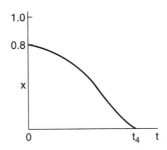

Figure 10-25   Exit conversion as a function of time.

An example where one of the reactants acting as a poison is in the reaction of CO and H_2 over ruthenium to form methane, with

$$-r_{CO} = ka(t)C_{CO}$$

$$-\frac{da}{dt} = r_d = k_d' a(t) C_{CO}$$

Similar rate laws can be written for the case when the product B acts as a poison.

For *separable deactivation kinetics* resulting from contacting a poison at a constant concentration C_{P_0}:

Separable
deactivation
kinetics
$$-\frac{da}{dt} = r_d = k_d' C_{P_0}^n a^q(t) = k_d a^q \tag{10-110}$$

The solution to this equation for the case of first-order decay

$$-\frac{da}{dt} = k_d a \tag{10-111}$$

is

$$a = e^{-k_d t} \tag{10-112}$$

Empirical Decay Laws.   Table 10-6 gives a number of empirical decay laws along with the reaction systems to which they apply.

TABLE 10-6.   DECAY RATE LAWS

Functional Form of Activity	Decay Reaction Order	Differential Form	Integral Form	Examples
Linear	0	$-\dfrac{da}{dt} = \beta_0$	$a = 1 - \beta_0 t$	Conversion of *para*-hydrogen on tungsten when poisoned with oxygen[a]
Exponential	1	$-\dfrac{da}{dt} = \beta_1 a$	$a = e^{-\beta_1 t}$	Ethylene hydrogenation on Cu poisoned with CO[b] Paraffin dehydrogenation on $Cr \cdot Al_2O_3$[c] Cracking of gas oil[d] Vinyl chloride monomer formation[e]
Hyperbolic	2	$-\dfrac{da}{dt} = \beta_2 a^2$	$\dfrac{1}{a} = 1 + \beta_2 t$	Vinyl chloride monomer formation[f] Cyclohexane dehydrogenation on Pt/Al_2O_3[g] Isobutylene hydrogenation on Ni[h]
Reciprocal power	$\dfrac{\beta_3 + 1}{\beta_3} = \gamma$	$-\dfrac{da}{dt} = \beta_3 a^n A_0^{1/5}$	$a = A_0 t^{-\beta_3}$	Cracking of gas oil and gasoline on clay[i]
	$\dfrac{\beta_4 + 1}{\beta_4} = n$	$-\dfrac{da}{dt} = \beta_4 a^n A_0^{1/5}$	$a = A_0 t^{-\beta_4}$	Cyclohexane aromatization on NiAl[j]

[a]D. D. Eley and E. J. Rideal, *Proc. R. Soc. London*, *A178*, 429 (1941).

[b]R. N. Pease and L. Y. Steward, *J. Am. Chem. Soc.*, *47*, 1235 (1925).

[c]E. F. K. Herington and E. J. Rideal, *Proc. R. Soc. London*, *A184*, 434 (1945).

[d]V. W. Weekman, *Ind. Eng. Chem. Process Des. Dev.*, *7*, 90 (1968).

[e]A. F. Ogunye and W. H. Ray, *Ind. Eng. Chem. Process Des. Dev.*, *9*, 619 (1970).

[f]A. F. Ogunye and W. H. Ray, *Ind. Eng. Chem. Process Des. Dev.*, *10*, 410 (1971).

[g]H. V. Maat and L. Moscou, *Proc. 3rd Int. Congr. Catal.*, North-Holland, Amsterdam, 1965, p. 1277.

[h]A. L. Pozzi and H. F. Rase, *Ind. Eng. Chem.*, *50*, 1075 (1958).

[i]A. Voorhies, Jr., *Ind. Eng. Chem.*, *37*, 318 (1945); E. B. Maxted, *Adv. Catal.*, *3*, 129 (1951).

[j]C. G. Ruderhausen and C. C. Watson, *Chem. Eng. Sci.*, *3*, 110 (1954).

Source: J. B. Butt, Chemical Reactor Engineering–Washington, *Advances in Chemistry Series 109* American Chemical Society, Washington, D.C., 1972, p. 259. Also see CES 23, 881(1968)

One should also see "Fundamentals of Industrial Catalytic Processes," by Farrauto, and Bartholomew, which contains rate laws similar to those in Table 10-6 , and also gives a comprehensive treatment of catalyst deactivation.

Example 10–5 Catalyst Decay in a Fluidized Bed Modeled as a CSTR

The gas-phase cracking reaction

$$\text{crude oil } (g) \longrightarrow \text{products } (g)$$

$$A \longrightarrow B + C$$

is carried out in a *fluidized* CSTR reactor. The feedstream contains 80% crude (A) and 20% inert I. The crude oil contains sulfur compounds which poison the catalyst. As a first approximation we will assume that the cracking reaction is first-order in the crude oil concentration. The rate of catalyst decay is first-order in the present activity, and first-order in the reactant concentration. Assuming that the bed can be modeled as a well-mixed CSTR, determine the reactant concentration, activity, and conversion as a function of time. The volumetric feed rate to the reactor is 5000 m³/h. There are 50,000 kg of catalyst in the reactor and the bulk density is 500 kg/m³.

Additional information:

$$C_{A0} = 0.8 \text{ mol/dm}^3 \qquad k = \rho_B k' = 45 \text{ h}^{-1}$$

$$C_{T0} = 1.0 \text{ mol/dm}^3 \qquad k_d = 9 \text{ dm}^3/\text{mol} \cdot \text{h}$$

Solution

1. **Mole balance** on reactant:

$$v_0 C_{A0} - v C_A + r'_A W = \frac{dN_A}{dt} \tag{E10-5.1}$$

Recalling $N_A = C_A V$ and $r_A V = r'_A W$, then for constant volume we have

$$v_0 C_{A0} - v C_A + r_A V = V \frac{dC_A}{dt} \tag{E10-5.2}$$

2. **Rate law:**

$$-r_A = k a C_A \tag{E10-5.3}$$

3. **Decay law:**

$$\boxed{-\frac{da}{dt} = k_d a C_A} \tag{E10-5.4}$$

4. **Stoichiometry** (gas-phase, $P = P_0$, $T = T_0$). From Equation (3-41) we have

$$v = v_0 \frac{F_T}{F_{T0}} = \frac{v_0 (F_A + F_B + F_C + F_{I0})}{F_{A0} + F_{I0}} \tag{E10-5.5}$$

For every mole of A that reacts, 1 mol each of C and B are formed:

$$F_B = F_C = F_{A0} - F_A$$

Substituting in Equation (E10-5.5) gives

$$\frac{v}{v_0} = \frac{F_{I0} + 2F_{A0} - F_A}{F_{T0}}$$

$$\frac{v}{v_0} = 1 + \frac{F_{A0}}{F_{T0}} - \frac{F_A}{F_{T0}} = 1 + y_{A0} - \frac{C_A v}{C_{T0} v_0}$$

Solving for v yields

$$v = v_0 \frac{1 + y_{A0}}{1 + C_A/C_{T0}} \qquad (E10\text{-}5.6)$$

$$y_{A0} = \frac{C_{A0}}{C_{T0}} \qquad (E10\text{-}5.7)$$

5. **Combining** gives us

$$v_0 C_{A0} - \frac{v_0(1 + y_{A0})}{1 + C_A/C_{T0}} C_A - kaC_A V = V \frac{dC_A}{dt} \qquad (E10\text{-}5.8)$$

Dividing both sides of Equation (E10-5.8) by the volume and writing the equation in terms of $\tau = V/v_0$, we obtain

$$\boxed{\frac{dC_A}{dt} = \frac{C_{A0}}{\tau} - \frac{(1 + y_{A0})/(1 + C_A/C_{T0}) + a\tau k}{\tau} C_A} \qquad (E10\text{-}5.9)$$

As an approximation we assume the conversion to be

$$\boxed{X = \frac{F_{A0} - F_A}{F_{A0}} = 1 - \frac{vC_A}{v_0 C_{A0}} = 1 - \left(\frac{1 + y_{A0}}{1 + C_A/C_{T0}}\right)\left(\frac{C_A}{C_{A0}}\right)} \qquad (E10\text{-}5.10)$$

Calculation of reactor volume and space time yields

$$V = \frac{W}{\rho_b} = \frac{50{,}000}{500 \text{ kg/m}^3} = 100 \text{ m}^3$$

$$\tau = \frac{V}{v_0} = \frac{100 \text{ m}^3}{5000 \text{ m}^3/\text{h}} = 0.02 \text{ h}$$

Equations (E10-5.4), (E10-5.9), and (E10-5.10) are solved using POLYMATH as the ODE solver. The POLYMATH program is shown in Table E10-5.1. The solution is shown in Figure E10-5.1.

The conversion variable X does not have much meaning in flow systems not at steady state, owing to the accumulation of reactant. However, here the space time is relatively short ($\tau = 0.02$ h) in comparison with the time of decay $t = 0.5$ h. Consequently, we can assume a quasi-steady state and consider the conversion as defined by Equation (E10-5.10) valid. Because the catalyst decays in less than an hour, a fluidized bed would not be a good choice to carry out this reaction.

TABLE E10-5.1.   POLYMATH PROGRAM

Equations	Initial Values
d(a)/d(t)=-kd*a*Ca	1
d(Ca)/d(t)=CaO/tau-((1+yao)/(1+Ca/CtO)+tau*a*k)*Ca/tau	0.8
kd=9	
CaO=.8	
tau=.02	
k=45	
CtO=1.	
yao=CaO/CtO	
X=1-(1+yao)/(1+Ca/CtO)*Ca/CaO	
t_0 = 0, t_f = 0.5	

C_A, X, and a time trajectories in a CSTR *not* at steady state

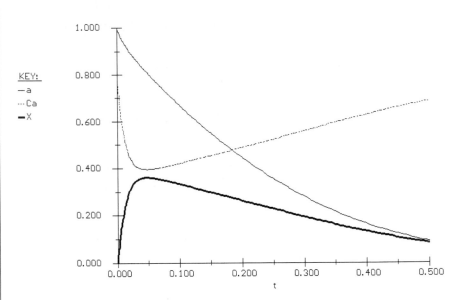

Figure E10-5.1   Variation of C_A, a, and X with time in a CSTR.

We will now consider three reaction systems that can be used to handle systems with decaying catalyst. We will classify these systems as those having slow, moderate, and rapid losses of catalytic activity. To offset the decline in chemical reactivity of decaying catalysts in continuous-flow reactors, the following three methods are commonly used:

Matching the reactor type with speed of catalyst decay

- Slow decay – *Temperature–Time Trajectories* (10.7.2)
- Moderate decay – *Moving-Bed Reactors* (10.7.3)
- Rapid decay – *Straight-Through Transport Reactors* (10.7.4)

10.7.2 Temperature–Time Trajectories

In many large-scale reactors, such as those used for hydrotreating, and reaction systems where deactivation by poisoning occurs, the catalyst decay is relatively slow. In these continuous-flow systems, constant conversion is usually necessary in order that subsequent processing steps (e.g., separation) are not upset. One way to maintain a constant conversion with a decaying catalyst in a packed or fluidized bed is to increase the reaction rate by steadily increasing the feed temperature to the reactor. (See Figure 10-26.)

Slow rate of catalyst decay

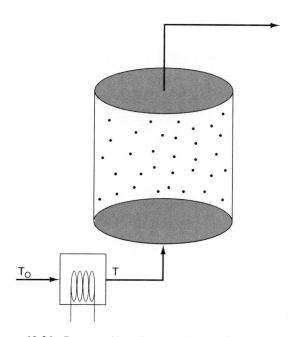

Figure 10-26   Reactor with preheater to increase feed temperature.

We are going to increase the feed temperature in such a manner that the reaction rate remains constant with time:

$$-r'_A (t = 0, T_0) = -r'_A (t, T) = a(t, T)[-r'_A (t = 0, T)]$$

For a first-order reaction we have

$$k(T_0)C_A = a(t, T)k(T)C_A$$

We will neglect any variations in concentration so that the product of the activity, a, and specific reaction rate, k, is constant, and equal to the specific reaction rate, k_0 at time $t = 0$ and temperature T_0, i.e.,

$$k(T)a(t, T) = k_0 \qquad (10\text{-}113)$$

Gradually raising the temperature can help offset effects of catalyst decay

The goal is to find *how* the temperature should be increased with time (i.e., the temperature–time trajectory) to maintain constant conversion. Substituting for k in terms of E_A gives

$$k_0 e^{(E_A/R)(1/T_0 - 1/T)} a = k_0 \qquad (10\text{-}114)$$

Solving for $1/T$ yields

$$\frac{1}{T} = \frac{R}{E_A} \ln a + \frac{1}{T_0} \qquad (10\text{-}115)$$

The decay law is

$$-\frac{da}{dt} = k_{d0} e^{(E_d/R)(1/T_0 - 1/T)} a^n \qquad (10\text{-}116)$$

Substituting Equation (10-115) into (10-116) and rearranging yields

$$-\frac{da}{dt} = k_{d0} \exp\left(-\frac{E_d}{E_A} \ln a\right) a^n = k_{d0} a^{(n - E_d/E_A)} \qquad (10\text{-}117)$$

Integrating with $a = 1$ at $t = 0$ for the case $n \neq (1 + E_d/E_A)$, we obtain

$$t = \frac{1 - a^{1 - n + E_d/E_A}}{k_{d0}(1 - n + E_d/E_A)} \qquad (10\text{-}118)$$

Solving Equation (10-114) for a and substituting in (10-118) gives

$$\boxed{t = \frac{1 - \exp\left[\dfrac{E_A - nE_A + E_d}{R}\left(\dfrac{1}{T} - \dfrac{1}{T_0}\right)\right]}{k_{d0}(1 - n + E_d/E_A)}} \qquad (10\text{-}119)$$

where k_{d0} = decay constant at temperature T_0, s^{-1}
$\qquad E_A$ = activation energy for the main reaction (e.g., A $\longrightarrow$ B), kJ/mol
$\qquad E_d$ = activation energy for catalyst decay, kJ/mol

Equation (10-119) tells us how the temperature of the catalytic reactor should be increased with time in order for the reaction rate to remain unchanged with time.

In many industrial reactions, the decay rate law changes as temperature increases. In hydrocracking, the temperature–time trajectories are divided into three regimes. Initially, there is fouling of the acidic sites of the catalyst followed by a linear regime due to slow coking, and finally, accelerated coking characterized by an exponential increase in temperature. The temperature–time trajectory for a deactivating hydrocracking catalyst is shown in Figure 10-27.

For a first-order decay, Krishnaswamy and Kittrell's expression [Equation (10-119)] for the temperature–time trajectory reduces to

$$t = \frac{E_A}{k_{d0} E_d} [1 - e^{(E_d/R)(1/T - 1/T_0)}] \qquad (10\text{-}120)$$

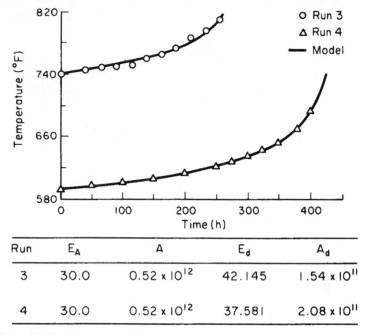

Comparing theory
and experiment

Run	E_A	A	E_d	A_d
3	30.0	0.52×10^{12}	42.145	1.54×10^{11}
4	30.0	0.52×10^{12}	37.581	2.08×10^{11}

Figure 10-27   Temperature–time trajectories for deactivating hydrocracking catalyst, runs 3 and 4. [Reprinted with permission from S. Krishnaswamy and J. R. Kittrell, *Ind. Eng. Chem. Process Des. Dev.*, *18*, 399 (1979). Copyright © 1979 American Chemical Society.]

10.7.3 Moving-Bed Reactors

Reaction systems with significant catalyst decay require the continual regeneration and/or replacement of the catalyst. Two types of reactors currently in commercial use that accommodate production with decaying catalysts are the moving-bed and straight-through transport reactor. A schematic diagram of a moving-bed reactor (used for catalytic cracking) is shown in Figure 10-28.

The freshly regenerated catalyst enters the top of the reactor and then moves through the reactor as a compact packed bed. The catalyst is coked continually as it moves through the reactor until it exits the reactor into the kiln, where air is used to burn off the carbon. The regenerated catalyst is lifted from the kiln by an airstream and then fed into a separator before it is returned to the reactor. The catalyst pellets are typically between $\frac{1}{8}$ and $\frac{1}{4}$ in. in diameter.

The reactant feedstream enters at the top of the reactor and flows rapidly through the reactor relative to the flow of the catalyst through the reactor (Figure 10-29). If the feed rates of the catalyst and the reactants do not vary with time, the reactor operates at steady state; that is, conditions at any point in the reactor do not change with time. The mole balance on reactant A over ΔW is

$$\begin{bmatrix} \text{flow} \\ \text{rate in} \end{bmatrix} - \begin{bmatrix} \text{flow} \\ \text{rate out} \end{bmatrix} + \begin{bmatrix} \text{rate of} \\ \text{generation} \end{bmatrix} = \begin{bmatrix} \text{rate of} \\ \text{accumulation} \end{bmatrix} \quad (10\text{-}121)$$

$$F_A(W) \quad - \quad F_A(W + \Delta W) \quad + \quad r'_A \Delta W \quad = \quad 0$$

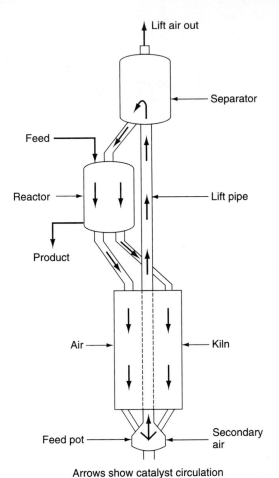

Used for reactions
with moderate rate
of catalyst decay.

Arrows show catalyst circulation

Figure 10-28 Thermofor catalytic cracking (TCC) unit. [From V. Weekman, *AIChE Monogr. Ser.*, *75*(11), 4 (1979). With permission of the AIChE. Copyright © 1979 AIChE. All rights reserved.]

Dividing by ΔW, letting ΔW approach zero, and expressing the flow rate in terms of conversion gives

$$F_{A0}\frac{dX}{dW} = -r'_A \qquad (2\text{-}17)$$

The rate of reaction at any time t is

$$-r'_A = a(t)[-r'_A(t=0)] = a(t)[k\,\mathrm{fn}(C_A, C_B, \ldots, C_P)] \qquad (10\text{-}91)$$

The activity, as before, is a function of the time the catalyst has been in contact with the reacting gas stream. The decay rate law is

$$-\frac{da}{dt} = k_d a^n \qquad (10\text{-}110)$$

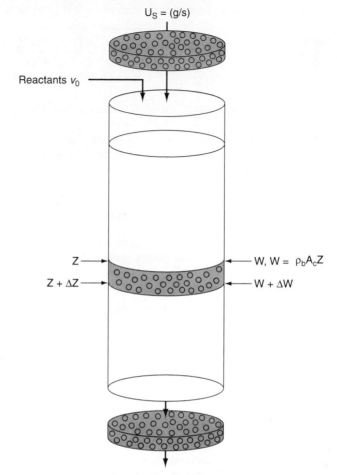

$$U_S = (g/s)$$

Reactants v_0

$Z \longrightarrow$ $\longleftarrow$ W, W $= \rho_b A_c Z$

$Z + \Delta Z \longrightarrow$ $\longleftarrow$ W $+ \Delta W$

Figure 10-29   Moving-bed reactor–schematic.

We now need to relate the contact time to the weight of the catalyst. Consider a point z in the reactor, where the reactant gas has passed cocurrently through a catalyst weight W. Since the solid catalyst is moving through the bed at a rate U_s (mass per unit time), the time t that the catalyst has been in contact with the gas when the catalyst reaches a point z is

$$t = \frac{W}{U_s} \tag{10-122}$$

If we now differentiate Equation (10-122)

$$dt = \frac{dW}{U_s} \tag{10-123}$$

and combine it with the decay rate law, we obtain

$$\boxed{-\frac{da}{dW} = \frac{k_d}{U_s} a^n} \tag{10-124}$$

The activity equation is combined with the mole balance:

The design equation
for moving-bed
reactors

$$\frac{dX}{dW} = \frac{a(W)[-r'_A(t=0)]}{F_{A0}}$$

(10-125)

Example 10–6   Catalytic Cracking in a Moving-Bed Reactor

The catalytic cracking of a gas oil charge, A, to form C_5+ (B) and to form coke and dry gas (C) is to be carried out in a screw-type conveyor moving-bed reactor at 900°F:

$$gas\ oil \begin{array}{c} \xrightarrow{k_B} C_{5^+} \\ \xrightarrow{k_C} dry\ gas,\ coke \end{array}$$

This reaction can also be written as

$$A \xrightarrow{k_1} products$$

While pure hydrocarbons are known to crack according to a first-order rate law, the fact that the gas oil exhibits a wide spectrum of cracking rates gives rise to the fact that the lumped cracking rate is well represented by a second-order rate law (see Problem P5-16) with the following specific reaction rate:[30]

$$-r'_A = 0.60 \frac{(dm)^6}{(g\ cat.)(mol)(min)} C_A^2$$

The catalytic deactivation is independent of gas-phase concentration and follows a first-order decay rate law, with a decay constant of 0.72 reciprocal minutes. The feedstream is diluted with nitrogen so that as a first approximation, volume changes can be neglected with reaction. The reactor contains 22 kg of catalyst that moves through the reactor at a rate of 10 kg/min. The gas oil is fed at a rate of 30 mol/min at a concentration of 0.075 mol/dm³. Determine the conversion that can be achieved in this reactor.

Solution

1. **Design equation:**

$$F_{A0} \frac{dX}{dW} = a(W)(-r'_A)$$

(E10-6.1)

2. **Rate law:**

$$-r'_A = kC_A^2$$

(E10-6.2)

3. **Decay law.** First-order decay

$$-\frac{da}{dt} = k_d a$$

---

[30]Estimated from V. W. Weekman and D. M. Nace, *AIChE J.*, *16*, 397 (1970).

Moving beds: moderate rate of catalyst decay

Using Equation (10-124), we obtain

$$-\frac{da}{dW} = \frac{k_d}{U_s} a \qquad\qquad \text{(E10-6.3)}$$

$$a = e^{-(k_d/U_s)W} \qquad\qquad \text{(E10-6.4)}$$

4. **Stoichiometry.** If $v \approx v_0$ **(See Problem P10-2g)** then

$$C_A = C_{A0}(1 - X) \qquad\qquad \text{(E10-6.5)}$$

5. **Combining**, we have

$$F_{A0} \frac{dX}{dW} = e^{-(k_d/U_s)W} k C_{A0}^2 (1 - X)^2 \qquad\qquad \text{(E10-6.6)}$$

6. **Separating and integrating** yields

$$\frac{F_{A0}}{k C_{A0}^2} \int_0^X \frac{dX}{(1 - X)^2} = \int_0^W e^{-(k_d/U_s)W} \, dW \qquad\qquad \text{(E10-6.7)}$$

$$\frac{X}{1 - X} = \frac{k C_{A0}^2 U_s}{F_{A0} k_d} \left(1 - e^{-k_d W/U_s}\right) \qquad\qquad \text{(E10-6.8)}$$

7. **Numerical evaluation:**

$$\frac{X}{1 - X} = \frac{0.6 \, \text{dm}^6}{\text{mol} \cdot \text{g cat.} \cdot \text{min}} \times \frac{(0.075 \, \text{mol/dm}^3)^2}{30 \, \text{mol/min}} \frac{10{,}000 \, \text{g cat./min}}{0.72 \, \text{min}^{-1}}$$

$$\times \left(1 - \exp\left[\frac{(-0.72 \, \text{min}^{-1})(22 \, \text{kg})}{10 \, \text{kg/min}}\right]\right)$$

$$= 1.24$$

$$X = 55\%$$

We will now rearrange Equation (E10-6.8) to a form more commonly found in the literature. Let λ be a dimensionless decay time:

$$\lambda = k_d t = \frac{k_d W}{U_s} \qquad\qquad \text{(10-126)}$$

and Da_2 be the Damköhler number for a second-order reaction (a reaction rate divided by a transport rate) for a packed-bed reactor:

$$Da_2 = \frac{(k C_{A0}^2)(W)}{F_{A0}} = \frac{k C_{A0} W}{v_0} \qquad\qquad \text{(10-127)}$$

Through a series of manipulations we arrive at the equation for the conversion in a *moving bed* for a second-order reaction:[31]

---

[31]Ibid.

$$X = \frac{\text{Da}_2(1 - e^{-\lambda})}{\lambda + \text{Da}_2(1 - e^{-\lambda})} \tag{10-128}$$

Similar equations are given or can easily be obtained for other reaction orders or decay laws.

Heat Effects in Moving Beds. We shall consider two cases for modeling the temperature profile in the moving-bed reactor. In one case the temperature of the solid catalyst and the temperature of the gas are different and in the other case they are the same.

Case 1 ($T \neq T_s$). The rate of heat transfer between the gas and solid catalyst particles is

$$Q_P = h\tilde{a}_P(T - T_S) \tag{10-129}$$

where h = heat transfer with coefficient, kJ/m$^2 \cdot$s$\cdot$K
$\tilde{a}_P$ = solid catalyst surface area per mass of catalyst in the bed, m^2/kg cat.
T_S = temperature of the solid, K

The energy balance on the gas phase is

$$\frac{dT}{dW} = \frac{U\tilde{a}_W(T_a - T) + h\tilde{a}_P(T_S - T) + (r'_A)(\Delta H_{\text{Rx}})}{\Sigma \, F_i \tilde{C}_{P_i}} \tag{10-130}$$

If D_P is the pipe diameter (m) and ρ_B is the bulk catalyst density (kg/m^3),

$$\tilde{a}_W = \frac{4}{D_P \rho_B} \tag{10-131}$$

is the heat exchange area of the pipe containing the mass of catalyst. The energy balance on the solid catalyst is

$$\frac{dT_S}{dW} = -\frac{h\tilde{a}_P(T_S - T)}{U_S C_{P_S}} \tag{10-132}$$

where C_{P_S} (J/kg$\cdot$K) is the heat capacity of the solids, U_s (kg/s) the catalyst loading, and $\tilde{a}_P$ is the external surface area of the catalyst pellet per unit mass of catalyst:

$$\tilde{a}_P = \frac{6}{d_P \rho_b} \tag{10-133}$$

where d_P is the pellet diameter.

Case 2 ($T_s = T$). If the product of the heat transfer coefficient and the surface is very large, we can assume that the solid and gas temperatures are identical. Under these circumstances the energy balance becomes

$$\frac{dT}{dW} = \frac{U\tilde{a}_W(T_a - T) + (r'_A)(\Delta H_{\text{Rx}})}{U_S C_{P_S} + \Sigma \, F_i C_{P_i}} \tag{10-134}$$

10.7.4 Straight-Through Transport Reactors

This reactor is used for reaction systems in which the catalyst deactivates very rapidly. Commercially, the STTR is used in the production of gasoline from the cracking of heavier petroleum fractions where coking of the catalyst pellets occurs very rapidly. In the STTR, the catalyst pellets and the reactant feed enter together and are transported very rapidly through the reactor. The bulk density of the catalyst particle in the STTR is significantly smaller than in moving-bed reactors, and often the particles are carried through at the same velocity as the gas velocity. In some places the STTR is also referred to as a circulating fluidized bed (CFB). A schematic diagram is shown in Figure 10-30.

A mole balance on the reactant A over the differential reactor volume

$$\Delta V = A_C \, \Delta z$$

STTR: Used when catalyst decay (usually coking) is very rapid

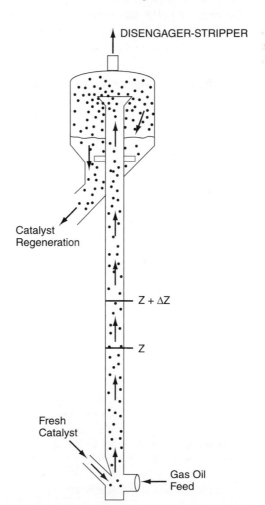

Figure 10-30 Straight-through transport reactor.

is

$$F_A\big|_z - F_A\big|_{z+\Delta z} + r_A A_C\, \Delta z = 0$$

Dividing by Δz and taking the limit as $\Delta z \to 0$ and recalling that $r_A = \rho_B r'_A$, we obtain

$$\frac{dF_A}{dz} = r_A A_C = r'_A \rho_B A_C \qquad (10\text{-}135)$$

In terms of conversion and catalyst activity

$$\frac{dX}{dz} = \left(\frac{\rho_B A_C}{F_{A0}}\right)[-r'_A(t=0)]\,a(t) \qquad (10\text{-}136)$$

For a catalyst particle traveling through the reactor with a velocity U_P, the time the catalyst pellet has been in the reactor when it reaches a height z is just

$$t = \frac{z}{U_P} \qquad (10\text{-}137)$$

Substituting for time t in terms of distance z [i.e., $a(t) = a(z/U_P)$], the mole balance now becomes

$$\frac{dX}{dz} = \frac{\rho_B A_C[-r'_A(t=0)]\,a(z/U_P)}{F_{A0}}$$

The molar flow rate, F_{A0}, can be expressed in terms of the gas velocity U_g, C_{A0}, and A_C:

$$F_{A0} = U_g A_C C_{A0}$$

Substituting for F_{A0}, we have

$$\boxed{\frac{dX}{dz} = \frac{\rho_B a\,(z/U_P)\,[-r'_A(t=0)]}{C_{A0} U_g}} \qquad (10\text{-}138)$$

Example 10–7 Decay in a Straight-Through Transport Reactor

The vapor-phase cracking of a gas oil is to be carried out in a straight-through transport reactor (STTR) that is 10 m high and 1.5 m in diameter. Gas oil is a mixture of normal and branched paraffins (C_{12}–C_{40}), naphthenes, and aromatics, all of which will be lumped as a single species, A. We shall lump the primary hydrocarbon products according to distillate temperature into two respective groups, dry gas (C–C_4) B and gasoline (C_5–C_{14}) C. The reaction

A typical cost of the catalyst in the system is $1 million

$$\text{gas oil } (g) \longrightarrow \text{ products } (g) + \text{coke}$$

can be written symbolically as

$$A \longrightarrow B + C + \text{coke}$$

Both B and C are adsorbed on the surface. The rate law for a gas-oil cracking reaction on fresh catalyst can be approximated by

$$-r'_A = \frac{k' P_A}{1 + K_A P_A + K_B P_B + K_C P_C}$$

with $k' = 0.0014$ kmol/kg cat.·s·atm, $K_A = 0.05$ atm^{-1}, $K_B = 0.15$ atm^{-1}, and $K_D = 0.1$ atm^{-1}. The catalyst decays by the deposition of coke, which is produced in most cracking reactions along with the reaction products. The decay law is

$$a = \frac{1}{1 + At^{1/2}} \quad \text{with } A = 7.6 \text{ s}^{-1/2}$$

Pure gas-oil enters at a pressure of 12 atm and a temperature of 400°C. The bulk density of catalyst in the STTR is 80 kg cat./m^3. Plot the activity and conversion of gas oil up the reactor for entering gas velocity $U_0 = 2.5$ m/s.

Solution

Mole balance:

$$F_{A0} \frac{dX}{dz} = -r_A A_C$$

$$F_{A0} = U_g A_C C_{A0}$$

$$\boxed{\frac{dX}{dz} = \frac{-r_A}{U_g C_{A0}}} \tag{E10-7.1}$$

Rate law:

$$-r_A = \rho_B (-r'_A) \tag{E10-7.2}$$

$$-r'_A = a[-r'_A(t=0)] \tag{E10-7.3}$$

On fresh catalyst

$$-r'_A(t=0) = k' \frac{P_A}{1 + K_A P_A + K_B P_B + K_C P_C} \tag{E10-7.4}$$

Combining Equations (E10-7.2) through (E10-7.4) gives

$$\boxed{-r_A = a \left(\rho_B k' \frac{P_A}{1 + K_A P_A + K_B P_B + K_C P_C} \right)} \tag{E10-7.5}$$

Decay law. Assuming that the catalyst particle and gas travel up the reactor at the velocity $U_P = U_g = U$, we obtain

$$t = \frac{z}{U} \tag{E10-7.6}$$

$$\boxed{a = \frac{1}{1 + A(z/U)^{1/2}}} \tag{E10-7.7}$$

where $U = v/A_C = v_0(1 + \varepsilon X)/A_C$ and $A_C = \pi D^2/4$.

Stoichiometry (gas phase isothermal and no pressure drop):

$$P_A = P_{A0} \frac{1-X}{1+\varepsilon X}$$ (E10-7.8)

$$P_B = \frac{P_{A0}X}{1+\varepsilon X}$$ (E10-7.9)

$$P_C = P_B$$ (E10-7.10)

Parameter evaluation:

$$\varepsilon = y_{A0}\delta = (1+1-1) = 1$$

$$U = U_0(1+\varepsilon X)$$

$$C_{A0} = \frac{P_{A0}}{RT_0} = \frac{12 \text{ atm}}{(0.082 \text{ m}^3 \cdot \text{atm}/\text{kmol} \cdot \text{K})/(673 \text{ K})} = 0.22 \frac{\text{kmol}}{\text{m}^3}$$

Equations (E10-7.1), (E10-7.5), (E10-7.7), and (E10-7.8) through (E10-7.10) are now combined and solved using an ODE solver. The POLYMATH program is shown in Table E10-7.1, and the computer output is shown in Table E10-7.2 and Figure E10-7.1.

TABLE E10-7.1.   EQUATIONS FOR THE STTR: LANGMUIR–HINSHELWOOD KINETICS

Equations	Initial Values
	0

```
d(X)/d(z)=-ra/U/Cao
Ka=0.05
Kb=.15
Pao=12
eps=1
A=7.6
R=0.082
T=400+273
Kc=0.1
rho=80
kprime=0.0014
D=1.5
Uo=2.5
U=Uo*(1+eps*X)
Pa=Pao*(1-X)/(1+eps*X)
Pb=Pao*X/(1+eps*X)
vo=Uo*3.1416*D*D/4
Cao=Pao/R/T
Kca=Ka*R*T
Pc=Pb
a=1/(1+A*(z/U)**0.5)
raprime=a*(-kprime*Pa/(1+Ka*Pa+Kb*Pb+Kc*Pc))


ra=rho*raprime
```

$z_0 = 0, \quad z_f = 10$

TABLE E10-7.2.   VALUES FOR THE STTR: LANGMUIR–HINSELWOOD KINETICS

Variables	Initial Values	Maximum Values	Minimum Values	Final Values
z	0	10	0	10
X	0	0.565647	0	0.565647
Ka	0.05	0.05	0.05	0.05
Kb	0.15	0.15	0.15	0.15
Pao	12	12	12	12
eps	1	1	1	1
A	7.6	7.6	7.6	7.6
R	0.082	0.082	0.082	0.082
T	673	673	673	673
Kc	0.1	0.1	0.1	0.1
rho	80	80	80	80
kprime	0.0014	0.0014	0.0014	0.0014
D	1.5	1.5	1.5	1.5
Uo	2.5	2.5	2.5	2.5
U	2.5	3.91412	2.5	3.91412
Pa	12	12	3.32913	3.32913
Pb	0	4.33544	0	4.33544
vo	4.41787	4.41787	4.41787	4.41787
Cao	0.217446	0.217446	0.217446	0.217446
Kca	2.7593	2.7593	2.7593	2.7593
Pc	0	4.33544	0	4.33544
a	1	1	0.0760585	0.0760585
raprime	-0.0105	-0.00015753	-0.0105	-0.00015753
ra	-0.84	-0.0126024	-0.84	-0.0126024

Summary of results

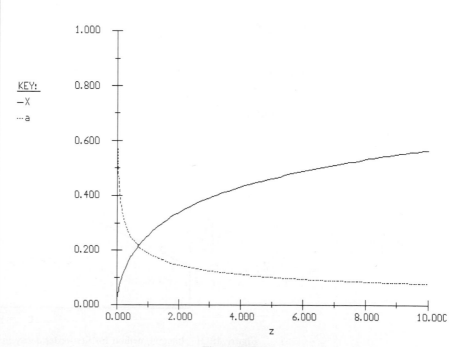

Figure E10-7.1

10.7.5 Determining the Order of Deactivation

Determination of
decay order for
known reaction
order, n
Integral Method   When the concentrations of reactants and products in the gas phase change very slowly with time, the pseudo-steady-state forms of the mole balances can be used. For the irreversible nth-order reaction carried out in a CSTR or in a differential reactor with catalyst weight W,

$$A \longrightarrow B$$

the mole balance is

$$F_{A0} - F_A = -r_A' a(t) W \tag{10-139}$$

Solving for the activity, $a(t)$, gives us

Using the pseudo-
steady-state
hypothesis
$$a(t) = \frac{v_0 C_{A0} - v_0 C_A}{W(-r_A')} = \frac{v_0}{W}\left(\frac{C_{A0} - C_A}{k C_A^n}\right) \tag{10-140}$$

where C_A is the effluent reactant concentration at time t.

In our efforts to determine the order of decay, we assume that the main reaction order is a known quantity. If it is not, a dual-model trial-and-error solution is necessary. *The idea is to find the simplest model that fits the data.* If we assume first-order decay,

$$a(t) = e^{-k_d t} = \frac{v_0}{Wk}\left(\frac{C_{A0} - C_A}{C_A^n}\right) \tag{10-141}$$

Substituting in $k_R = Wk/v_0$ and taking the log of both sides, we obtain

$$\boxed{k_d t = \ln k_R + \ln \frac{C_A^n}{C_{A0} - C_A}} \tag{10-142}$$

The plot of $\ln[C_A^n/(C_{A0} - C_A)]$ versus t should be a straight line whose slope is the deactivation rate constant (Figure 10-31). The specific reaction rate constant k can be calculated from the intercept.

First-order decay
in a CSTR

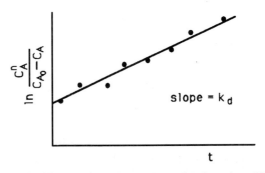

Figure 10-31   Integral method for first-order decay in a CSTR.

If first-order decay does not fit the data, one could try second-order decay:

$$a = \frac{1}{1 + k_d t} = \frac{C_{A0} - C_A}{k_R C_A^n}$$

$$\boxed{\frac{C_A^n}{C_{A0} - C_A} = \frac{1}{k_R} + \frac{k_d}{k_R} t} \qquad (10\text{-}143)$$

Consequently, both k_R and k_d can be found from a plot of $[C_A^n/(C_{A0} - C_A)]$ versus t, as shown in Figure 10-32. We can continue assuming decay orders in this manner until the decay rate law is found.

Second-order decay
in a CSTR

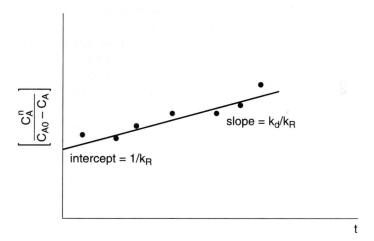

Figure 10-32   Integral method for second-order decay in a CSTR.

For a packed-bed reactor, the approach is quite similar to that described for a CSTR. For a first-order reaction, the combined mole balance and rate law is

$$v_0 \frac{dC_A}{dW} = -k a(t) C_A \qquad (10\text{-}144)$$

Solving for $a(t)$ for the case of uniform activity throughout the reactor yields

$$a(t) = \frac{v_0}{Wk} \ln \frac{C_{A0}}{C_A} \qquad (10\text{-}145)$$

For first-order decay, $a = e^{-k_d t}$ (see Figure 10-33), we take the log of both sides:

$$\boxed{-k_d t = \ln \frac{v_0}{Wk} + \ln \ln \frac{C_{A0}}{C_A}} \qquad (10\text{-}146)$$

If the concentration–time data do not fit the deactivation law, another decay law is chosen and the process is repeated.

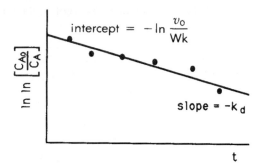

Figure 10-33   Integral method for first-order decay in a packed-bed reactor.

The general idea of the three previous cases (i.e., Figures 10-31 through 10-33) is to arrange the data in such a fashion as to arrive at functional groupings of measured variables that will be linear with time. The particular functional groups will vary with (1) type of reactor used to collect the data, (2) reaction order of the main reaction, and (3) the decay reaction order. For the three main

types of reactors, three main reaction rate laws, and three decay rates, *27 different types of plots could result.* We leave derivation of the equation for each of these plots to the reader and point out that only one additional step is needed in our solution algorithm. That step is the decay rate law:

mole balance $\longrightarrow$ reaction rate law $\longrightarrow$ *decay rate law* $\longrightarrow$

stoichiometry $\longrightarrow$ combine and solve $\longrightarrow$ numerical techniques

10.8 Reaction Engineering in Microelectronic Device Fabrication

Surface reactions play an important role in the manufacture of microelectronics devices. One of the single most important developments of this century was the invention of the integrated circuit. Advances in the development of integrated circuitry have led to the production of circuits that can be placed on a single semiconductor chip the size of a pinhead and perform a wide variety of tasks by controlling the electron flow through a vast network of channels. These channels, which are made from semiconductors such as silicon, gallium arsenide, indium phosphide and germanium, have led to the development of a multitude of novel microelectronic devices. Chemical reaction engineers play important roles in the manufacture of microelectronic and photonic devices and recording materials.

The manufacture of an integrated circuit requires the fabrication of a network of pathways for electrons. The principal reaction engineering steps of the fabrication process include depositing material on the surface of a material called a substrate (e.g., by chemical vapor deposition), changing the conductivity of regions of the surface (e.g., by boron doping or ion-implemention), and removing unwanted material (e.g., by etching). By applying these steps systematically, miniature electronic circuits can be fabricated on very small semiconductor chips. The fabrication of microelectronic devices may include as few as 30 or as many as 200 individual steps to produce chips with up to 10^9 elements per chip.

An abbreviated schematic of the steps involved in producing a typical MOSFET device is shown in Figure 10-34. Starting from the upper left, we see that single-crystal silicon ingots are grown in a Czochralski crystallizer, sliced into wafers, and chemically and physically polished. These polished wafers serve as starting materials for a variety of microelectronic devices. A typical fabrication sequence is shown for processing the wafer beginning with the formation of an SiO_2 layer on top of the silicon. The SiO_2 layer may be formed

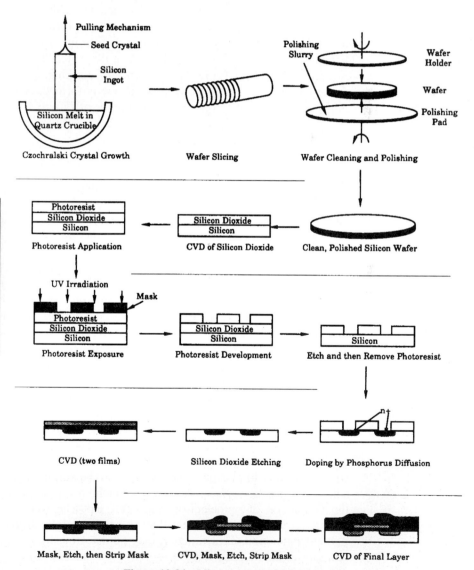

Figure 10-34 Microelectronic fabrication steps.

either by oxidizing a silicon layer or by laying down a SiO_2 layer by chemical vapor deposition (CVD). Next, the wafer is masked with a polymer photoresist, a template with the pattern to be etched onto the SiO_2 layer is placed over the photoresist, and the wafer is exposed to ultraviolet irradiation. If the mask is a positive photoresist, the light will cause the exposed areas of the polymer to dissolve when the wafer is placed in the developer. On the other hand, when a negative photoresist mask is exposed to ultraviolet irradiation, cross-linking of the polymer chains occurs and the *unexposed* areas dissolve in the developer. The undeveloped portion of the photoresist (in either case) will protect the covered areas from etching.

After the exposed areas of SiO_2 are etched to form trenches [either by wet etching (see Problem P5-12) or by plasma etching], the remaining photoresist is removed. Next, the wafer is placed in a furnace containing gas molecules of the desired dopant, which then diffuse into the exposed silicon. After diffusion of dopant to the desired depth in the wafer, the wafer is removed and covered with SiO_2 by CVD. The sequence of masking, etching, CVD, and metallization continues until the desired device is formed. A schematic of a final chip is shown in the lower right-hand corner of Figure 10-34. In Section 10.6 we discussed one of the key processing steps, CVD.

10.8.1 Etching

We have seen in Figure 10-34 that etching (i.e., the dissolution or physical or chemical removal of material) is also an important step in the fabrication process. Etching takes on a priority role in microelectronics manufacturing because of the need to create well-defined structures from an essentially homogeneous material. In integrated circuits, etching is necessary to remove unwanted material that could provide alternative pathways for electrons and thus hinder operation of the circuit. Etching is also of vital importance in the fabrication of micromechanical and optoelectronic devices. By selective etching of semiconductor surfaces it is possible to fabricate motors and valves, ultrasmall diaphragms that can sense differences in pressure, or cantilever beams that can sense acceleration. In each of these applications proper etching is crucial to remove material that would either short out a circuit or hinder movement of the micromechanical device.

There are two basic types of etching: wet etching and dry etching. The wet etching process, as described in Problem P5-12, uses liquids such as HF or KOH to dissolve the layered material that is unprotected by the photoresist mask. Wet etching is used primarily in the manufacture of micromechanical devices. Dry etching involves gas-phase reactions, which form highly reactive species, usually in plasmas, that impinge on the surface either to react with the surface, erode the surface, or both. Dry etching is used almost exclusively for the fabrication of optoelectronic devices. Optoelectronic devices differ from microelectronic devices in that they use light and electrons to carry out their particular function. That function may be detecting light, transmitting light, or emitting light. Etching is used to create the pathways or regions where light can travel and interact to produce the desired effects. Appliances using such

CD
Professional

Reference Shelf

devices include remote controls for TV sets, LED displays on clocks and microwave ovens, laser printers, and compact disc players. The material on the CD-ROM gives examples of both dry etching and wet etching. For dry etching, the reactive ion etching (RIE) of InP is described. Here the PSSH is used to arrive at a rate law for the rate of etching which is compared with experimental observation. In discussing wet etching the idea of dissolution catalysis is introduced and rate laws are derived and compared with experimental observation.

SUMMARY

1. **Types of adsorption:**
   a. Chemisorption
   b. Physical adsorption
2. The **Langmuir isotherm** relating the concentration of species A on the surface to the partial pressure of A in the gas phase is

$$C_{A \cdot S} = \frac{K_A C_t P_A}{1 + K_A P_A} \qquad (S10\text{-}1)$$

3. The sequence of steps for the solid-catalyzed isomerization

$$A \longrightarrow B \qquad (S10\text{-}2)$$

is:

   a. **Mass transfer of A** from the bulk fluid to the external surface of the pellet
   b. **Diffusion of A** into the interior of the pellet
   c. **Adsorption of A** onto the catalytic surface
   d. **Surface reaction of A** to form B
   e. **Desorption of B** from the surface
   f. **Diffusion of B** from the pellet interior to the external surface
   g. **Mass transfer of B** away from the solid surface to the bulk fluid

4. Assuming that mass transfer is not rate-limiting, the rate of adsorption is

$$r_{AD} = k_A \left(C_v P_A - \frac{C_{A \cdot S}}{K_A} \right) \qquad (S10\text{-}3)$$

The rate of surface reaction is

$$r_S = k_S \left(C_{A \cdot S} - \frac{C_{B \cdot S}}{K_S} \right) \qquad (S10\text{-}4)$$

The rate of desorption is

$$r_D = k_D \left(C_{B \cdot S} - K_B P_B C_v \right) \qquad (S10\text{-}5)$$

At steady state,

$$r_{AD} = r_S = r_D \tag{S10-6}$$

If there are no inhibitors present, the total concentration of sites is

$$C_t = C_v + C_{A \cdot S} + C_{B \cdot S} \tag{S10-7}$$

5. If we assume that the surface reaction is rate-limiting, we set

$$\frac{r_D}{k_D} \simeq 0 \qquad \frac{r_{AD}}{k_A} \simeq 0$$

and solve for $C_{A \cdot S}$ and $C_{B \cdot S}$ in terms of P_A and P_B. After substitution of these quantities in Equation (S10-4), the concentration of vacant sites is eliminated with the aid of Equation (S10-7):

$$-r_A' = r_S = \frac{C_t k_S K_A (P_A - P_B/K_P)}{1 + K_A P_A + K_B P_B} \tag{S10-8}$$

Recall that the equilibrium constant for desorption of species B is the reciprocal of the equilibrium constant for the adsorption of species B:

$$K_B = \frac{1}{K_{DB}} \tag{S10-9}$$

and

$$K_P = K_A K_S / K_B \tag{S10-10}$$

6. **Chemical vapor deposition:**

$$SiH_4 \rightleftharpoons SiH_2(g) + H_2(g) \tag{S10-11}$$

$$SiH_2 + S \longrightarrow SiH_2 \cdot S$$

$$SiH_2 \cdot S \longrightarrow Si(s) + H_2(g) \tag{S10-12}$$

$$r_{Dep} = \frac{k P_{SiH_4}}{P_{H_2} + K P_{SiH_4}}$$

7. **Catalyst deactivation.** The catalyst activity is defined as

$$a(t) = \frac{-r_A'(t)}{-r_A'(t = 0)} \tag{S10-13}$$

The rate of reaction at any time t is

$$-r_A' = a(t) k(T) \, fn(C_A, C_B, ..., C_P) \tag{S10-14}$$

The rate of catalyst decay is

$$r_d = -\frac{da}{dt} = p[a(t)]k_a'(T)g(C_A, C_B, \ldots, C_P) \quad \text{(S10-15)}$$

For first-order decay:

$$p(a) = a \quad \text{(S10-16)}$$

For second-order decay:

$$p(a) = a^2 \quad \text{(S10-17)}$$

8. For slow catalyst decay the idea of a temperature–time trajectory is to increase the temperature in such a way that the rate of reaction remains constant.

9. The coupled differential equations to be solved for a **moving-bed reactor** are

$$\frac{dX}{dW} = a(W)(-r_A') \quad \text{(S10-18)}$$

For nth-order activity decay and m order in a gas-phase concentration of species i,

$$-\frac{da}{dW} = \frac{k_D a^n C_i^m}{U_s} \quad \text{(S10-19)}$$

$$t = \frac{W}{U_s} \quad \text{(S10-20)}$$

For second-order decay ($n = 2$) and $m = 0$, the conversion for a second-order reaction in a moving-bed reactor with $v = v_0$ is

$$X = \frac{\text{Da}_2(1 - e^{-\lambda})}{\lambda + \text{Da}_2(1 - e^{-\lambda})} \quad \text{(S10-21)}$$

10. The coupled differential equations to be solved in a **straight-through transport reactor** (STTR) for the case when the particle and gas velocities, U, are identical are

$$\frac{dX}{dz} = \frac{a(t)[-r_A(t=0)]}{U} \quad \text{(S10-22)}$$

$$t = \frac{z}{U} \quad \text{(S10-23)}$$

For coking

$$a(t) = \frac{1}{1 + A't^{1/2}} \quad \text{(S10-24)}$$

ODE SOLVER ALGORITHM

The isomerization A → B is carried out over a decaying catalyst in a *moving-bed reactor*. Pure A enters the reactor and the catalyst flows through the reactor at a rate of 2.0 kg/s.

$$\frac{dX}{dW} = \frac{-r'_A}{F_{A0}}$$ $k = 0.1 \ mol/(kg\ cat. \cdot s \cdot atm)$

$$r'_A = \frac{-akP_A}{1 + K_A P_A}$$ $K_A = 1.5 \ atm^{-1}$

$$\frac{da}{dW} = \frac{-k_d a^2 P_B}{U_s}$$ $k_d = \dfrac{0.75}{s \cdot atm}$

$$P_A = P_{A0}(1 - X)y$$ $F_{A0} = 10 \ mol/s$

$$P_B = P_{A0}Xy$$ $P_{A0} = 2 \ atm$

$$\frac{dy}{dW} = -\frac{\alpha}{2y}$$ $U_s = 2.0 \ kg\ cat./s$

$\alpha = 0.004 \ kg^{-1}$ $W_f = 500 \ kg\ cat.$

QUESTIONS AND PROBLEMS

The subscript to each of the problem numbers indicates the level of difficulty: A, least difficult; D, most difficult.

$$A = \bullet \quad B = \blacksquare \quad C = \blacklozenge \quad D = \blacklozenge\blacklozenge$$

In each of the questions and problems below, rather than just drawing a box around your answer, write a sentence or two describing how you solved the problem, the assumptions you made, the reasonableness of your answer, what you learned, and any other facts you want to include. You may wish to refer to W. Strunk and E. B. White, *The Elements of Style* (New York: Macmillan, 1979) and Joseph M. Williams, *Style: Ten Lessons in Clarity & Grace* (Glenview, Ill.: Scott, Foresman, 1989) to enhance the quality of your sentences.

P10-1$_A$ Read over the problems at the end of this chapter. Make up an original problem that uses the concepts presented in this chapter. See Problem P4-1 for guidelines. To obtain a solution:
(a) Create your data and reaction.
(b) Use a real reaction and real data.
The journals listed at the end of Chapter 1 may be useful for part (b).

P10-2$_A$ **What if...**
(a) the entering pressure in Example 10-3 were increased to 80 atm or reduced to 2 atm?

(b) you were asked to sketch the temperature–time trajectories and to find the catalyst lifetimes for first- and for second-order decay when $E_A = 35$ kcal/mol, $E_d = 10$ kcal/mol, $k_{d0} = 0.01$ day^{-1}, and $T_0 = 400$ K? How would the trajectory of the catalyst lifetime change if $E_A = 10$ kcal/mol and $E_d = 35$ kcal/mol? At what values of k_{d0} and ratios of E_d to E_A would temperature–time trajectories not be effective? What would your temperature–time trajectory look like if $n = 1 + E_d/E_A$?

(c) the space-time in Example 10-5 were changed? How would the minimum reactant concentration change? Compare your results with the case when the reactor is full of inerts at time $t = 0$ instead of 80% reactant. Is your catalyst lifetime longer or shorter? What if the temperature were increased so that the specific rate constants increase to $k = 120$ and $k_d = 12$? Would your catalyst lifetime be longer or shorter than at the lower temperature?

(d) in Example 10-5 you were asked to describe how the minimum in reactant concentration changes as the space-time τ changes? What is the minimum if $\tau = 0.005$ h? If $\tau = 0.01$ h?

(e) the solids and reactants in Example 10-6 entered from opposite ends of the reactor? How would your answers change?

(f) the decay in the moving bed in Example 10-6 were second order? By how much must the catalyst charge, U_S, be increased to obtain the same conversion?

(g) What if in Example 6-6, $\varepsilon = 2$ instead of zero, how would the results be affected?

(h) you varied the parameters P_{A0}, U, A, and k' in the STTR in Example 10-7? What parameter has the greatest effect on either increasing or decreasing the conversion? Ask questions such as: What is the effect of varying the ratio of k to U or of k to A on the conversion? Make a plot of conversion versus distance as U is varied between 0.5 and 50 m/s. Sketch the activity and conversion profiles for $U = 0.025$ m/s, 0.25 m/s, 2.5 m/s, and 25 m/s. What generalizations can you make? Plot the exit conversion and activity as a function of gas velocity between velocities of 0.02 and 50 m/s. What gas velocity do you suggest operating at? What is the corresponding entering volumetric flow rate? What concerns do you have operating at the velocity you selected? Would you like to choose another velocity? If so, what is it?

P10-3$_A$ t-Butyl alcohol (TBA) is an important octane enhancer that is used to replace lead additives in gasoline [*Ind. Eng. Chem. Res., 27,* 2224 (1988)]. t-Butyl alcohol was produced by the liquid-phase hydration (W) of isobutene (I) over an Amberlyst-15 catalyst. The system is normally a multiphase mixture of hydrocarbon, water and solid catalysts. However, the use of cosolvents or excess TBA can achieve reasonable miscibility.
The reaction mechanism is believed to be

$$I + S \rightleftharpoons I \cdot S \tag{P10-3.1}$$

$$W + S \rightleftharpoons W \cdot S \tag{P10-3.2}$$

$$W \cdot S + I \cdot S \rightleftharpoons TBA \cdot S + S \tag{P10-3.3}$$

$$TBA \cdot S \rightleftharpoons TBA + S \tag{P10-3.4}$$

Derive a rate law assuming:
(a) The surface reaction is rate-limiting.
(b) The adsorption of isobutene is limiting.

(c) The reaction follows Eley–Rideal kinetics

$$I \cdot S + W \longrightarrow TBA \cdot S \qquad\qquad (P10\text{-}3.5)$$

and that the surface reaction is limiting.

(d) Isobutene (I) and water (W) are adsorbed on different sites

$$I + S_1 \underset{\longleftarrow}{\longrightarrow} I \cdot S_1 \qquad\qquad (P10\text{-}3.6)$$

$$W + S_2 \underset{\longleftarrow}{\longrightarrow} W \cdot S_2 \qquad\qquad (P10\text{-}3.7)$$

TBA is *not* on the surface, and the surface reaction is rate-limiting.

$$\left[Ans.:\ r'_{TBA} = -r'_I = \frac{k[C_I C_W - C_{TBA}/K_c]}{(1 + K_W C_W)(1 + K_I C_I)} \right]$$

(e) What generalization can you make by comparing the rate laws derived in parts (a) through (d)?
The process flow sheet for the commercial production of TBA is shown in Figure P10-3.

(f) What can you learn from this problem and the process flow sheet?

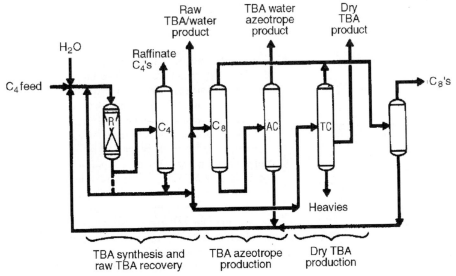

Figure P10-3   Hüls TBA synthesis process. R, reactor; C_4, C_4 column; C_8, C_8 column; AC, azeotrope column; TC, TBA column. (Adapted from R. E. Meyers, Ed., *Handbook of Chemicals Production Processes*, *Chemical Process Technology Handbook Series*, McGraw-Hill, New York, 1983, p. 1.19-3. ISBN 0-67-041 765-2.)

P10-4$_A$ The rate law for the hydrogenation (H) of ethylene (E) to form ethane (A) over a cobalt-molybdenum catalyst [*Collection Czech. Chem. Commun., 51,* 2760 (1988)] is

$$-r'_E = \frac{k P_E P_H}{1 + K_E P_E}$$

(a) Suggest a mechanism and rate-limiting step consistent with the rate law.
(b) What was the most difficult part in finding the mechanism?

P10-5$_B$ The dehydration of *n*-butyl alcohol (butanol) over an alumina-silica catalyst was investigated by J. F. Maurer (Ph.D. thesis, University of Michigan). The data in Figure P10-5 were obtained at 750°F in a modified differential reactor. The feed consisted of pure butanol.
(a) Suggest a mechanism and rate-controlling step that is consistent with the experimental data.
(b) Evaluate the rate law parameters.
(c) At the point where the initial rate is a maximum, what is the fraction of vacant sites? What is the fraction of occupied sites by both A and B?
(d) What generalizations can you make from studying this problem?

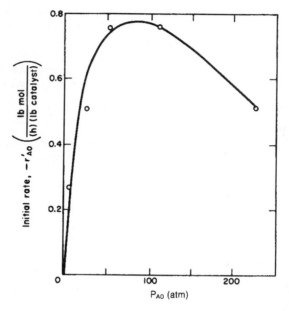

Figure P10-5

P10-6$_B$ The catalytic dehydration of methanol (ME) to form dimethyl ether (DME) and water was carried out over an ion exchange catalyst [K. Klusacek, *Collection Czech. Chem. Commun.*, *49*, 170(1984)]. The packed bed was initially filled with nitrogen and at $t = 0$ a feed of pure methanol vapor entered the reactor at 413 K, 100 kPa, and 0.2 cm³/s. The following partial pressures were recorded at the exit to the differential reactor containing 1.0 g of catalyst in 4.5 cm³ of reactor volume.

	t(s)						
	0	10	50	100	150	200	300
P_{N_2} (kPa)	100	50	10	2	0	0	0
P_{ME} (kPa)	0	2	15	23	25	26	26
P_{H_2O} (kPa)	0	10	15	30	35	37	37
P_{DME} (kPa)	0	38	60	45	40	37	37

Discuss the implications of these data.

P10-7$_B$ In 1981 the U.S. government put forth the following plan for automobile manufacturers to reduce emissions from automobiles over the next 23 years.

	Year		
	1981	1993	2004
Hydrocarbons	0.41	0.25	0.125
CO	3.4	3.4	1.7
NO	1.0	0.4	0.2

All values are in grams per mile. An automobile emitting 7.5 lb of CO and 2.2 lb of NO on a journey of 1000 miles would meet the current government requirements.

To remove oxides of nitrogen (assumed to be NO) from automobile exhaust, a scheme has been proposed that uses unburned carbon monoxide (CO) in the exhaust to reduce the NO over a solid catalyst, according to the reaction

$$CO + NO \longrightarrow products (N_2, CO_2)$$

Experimental data for a particular solid catalyst indicate that the reaction rate can be well represented over a large range of temperatures by

$$-r'_N = \frac{kP_N P_C}{(1 + K_1 P_N + K_2 P_C)^2}$$

where P_N = gas-phase partial pressure of NO
             P_C = gas-phase partial pressure of CO
k, K_1, K_2 = coefficients depending only on temperature

(a) Based on your experience with other such systems, you are asked to propose an adsorption-surface reaction-desorption mechanism that will explain the experimentally observed kinetics.

(b) A certain engineer thinks that it would be desirable to operate with a very large stoichiometric excess of CO to minimize catalytic reactor volume. Do you agree or disagree? Explain.

(c) When this reaction is carried out over a supported Rh catalyst [*J. Phys. Chem.*, 92, 389 (1988)], the reaction mechanism is believed to be

$$CO + S \rightleftharpoons CO \cdot S$$
$$NO + S \rightleftharpoons NO \cdot S$$
$$NO \cdot S + S \longrightarrow N \cdot S + O \cdot S$$
$$CO \cdot S + O \cdot S \longrightarrow CO_2 + 2S$$
$$N \cdot S + N \cdot S \longrightarrow N_2 + 2S$$

When the ratio of P_{CO}/P_{NO} is small, the rate law that is consistent with the experimental data is

$$-r'_{CO} = \frac{kP_{CO}}{(1 + K_{CO} P_{CO})^2}$$

What are the conditions for which the rate law and mechanism are consistent?

P10-8$_B$ Methyl ethyl ketone (MEK) is an important industrial solvent that can be produced from the dehydrogenation of butan-2-ol (Bu) over a zinc oxide catalyst [*Ind. Eng. Chem. Res.*, *27*, 2050 (1988)]:

$$Bu \rightarrow MEK + H_2$$

The following data giving the reaction rate for MEK were obtained in a differential reactor at 490°C.

P_{Bu} (atm)	2	0.1	0.5	1	2	1
P_{MEK} (atm)	5	0	2	1	0	0
P_{H_2} (atm)	0	0	1	1	0	10
r'_{MEK} (mol/h·g cat.)	0.044	0.040	0.069	0.060	0.043	0.059

(a) Suggest a rate law consistent with the experimental data.
(b) Suggest a reaction mechanism and rate-limiting step consistent with the rate law. (*Hint*: Some species might be weakly adsorbed.)
(c) What do you believe to be the point of this problem?
(d) Determine the catalyst weight necessary to achieve 90% conversion for an entering molar flow of pure butan-2-ol of 10 mol/min at a temperature of 490°C and an entering pressure of 10 atm. $W_{max} = 23$ kg.
(e) Plot the rate of reaction as a function of conversion.
(f) Repeat part (d) accounting for pressure drop and $\alpha = 0.03$ kg^{-1}. Plot P/P_0 and X as a function of catalyst weight down the reactor.

P10-9$_C$ The following data for the hydrogenation of *i*-octene to form *i*-octane were obtained using a differential reactor operated at 200°C.

Run	Rate (mol/g·h)	Partial Pressure (atm)		
		Hydrogen	*i*-Octene	*i*-Octane
1	0.0362	1	1	0
2	0.0239	1	1	1
3	0.0390	3	1	1
4	0.0351	1	3	1
5	0.0114	1	1	3
6	0.0534	10	1	0
7	0.0280	1	10	0
8	0.0033	1	1	10
9	0.0380	2	2	2
10	0.0090	1	1	4
11	0.0127	0.6	0.6	0.6
12	0.0566	5	5	5

(a) Develop a rate law and evaluate all the rate law parameters.
(b) Suggest a mechanism consistent with the experimental data.
      Hydrogen and *i*-octene are to be fed in stoichiometric proportions at a total rate of 5 mol/min at 200°C and 3 atm.
(c) Neglecting pressure drop, calculate the catalyst weight necessary to achieve 80% conversion of *i*-octene in a CSTR and in a plug-flow reactor.
(d) If pressure drop is taken into account and the $\frac{1}{8}$-in. catalyst pellets are packed in $1\frac{1}{2}$-in. schedule 80 pipes 35 ft long, what catalyst weight is necessary to achieve 80% conversion? The void fraction is 40% and the density of the catalyst is 2.6 g/cm^3.

P10-10$_B$ Cyclohexanol was passed over a catalyst to form water and cyclohexene:

$$\text{cyclohexanol} \longrightarrow \text{water} + \text{cyclohexene}$$

The following data were obtained.

Run	Reaction Rate (mol/dm$^3 \cdot$s) $\times$ 10^5	Partial Pressure of Cyclohexanol	Partial Pressure of Cyclohexene	Partial Pressure of Steam (H_2O)
1	3.3	1	1	1
2	1.05	5	1	1
3	0.565	10	1	1
4	1.826	2	5	1
5	1.49	2	10	1
6	1.36	3	0	5
7	1.08	3	0	10
8	0.862	1	10	10
9	0	0	5	8
10	1.37	3	3	3

It is suspected that the reaction may involve a dual-site mechanism, but it is not known for certain. It is believed that the adsorption equilibrium constant for cyclo-hexanol is around 1 and is roughly one or two orders of magnitude greater than the adsorption equilibrium constants for the other compounds. Using these data:

(a) Suggest a rate law and mechanism consistent with the data above.

(b) Determine the constants needed for the rate law.

[*Ind. Eng. Chem. Res.*, *32*, 2626–2632 (1993).]

(c) Why do you think estimates of the rate law parameters were given?

P10-11$_B$ The carbonation of allyl chloride (AC) [*Chem. Eng. Sci.*, *51*, 2069 (1996)] was carried out over a Pt catalyst:

$$CH_2CHCH_2Cl + CO + NaOH \xrightarrow{\text{Pd}} C_3H_5COOH + NaCl$$

The rate law is of the form

$$-r_{AC} = \frac{kC_{CO}C_{AC}C_{NaOH}}{(1 + K_{AC}C_{AC})^2}$$

The reaction mechanism is believed to be

$$Pd + CO \rightleftarrows Pd \cdot CO$$
$$Pd \cdot CO + NaOH \rightleftarrows (Pd \cdot CO \cdot NaOH)^*$$

Is there a rate-limiting step for which the rate law is consistent with the mechanism?

P10-12$_B$ A recent study of the *chemical vapor deposition* of silica from silane (SiH_4) is believed to proceed by the following irreversible two-step mechanism [*J. Electrochem. Soc., 139*(9), 2659 (1992)]:

$$SiH_4 + S \xrightarrow{k_1} SiH_2 \cdot S + H_2 \qquad (1)$$

$$SiH_2 \cdot S \xrightarrow{k_2} Si + H_2 \qquad (2)$$

This mechanism is somewhat different in that while SiH_2 is irreversibly adsorbed, it is highly reactive. In fact, adsorbed SiH_2 reacts as fast as it is formed [i.e., $r^*_{SiH_2 \cdot S} = 0$, i.e., PSSH (Chapter 7)], so that it can be assumed to behave as an active intermediate.

(a) Determine if this mechanism is consistent with the following data:

Deposition Rate (mm/min)	0.25	0.5	0.75	0.80
Silane Pressure (mtorr)	5	15	40	60

(b) At what partial pressures of silane would you take the next two data points?

P10-13$_A$ Vanadium oxides are of interest for various sensor applications, owing to the sharp metal–insulator transitions they undergo as a function of temperature, pressure, or stress. Vanadium triisopropoxide (VTIPO) was used to grow vanadium oxide films by *chemical vapor deposition* [*J. Electrochem. Soc., 136*, 897 (1989)]. The deposition rate as a function of VTIPO pressure is given below for two different temperatures:

$T = 120°C$:

Growth Rate (μm/h)	0.004	0.015	0.025	0.04	0.068	0.08	0.095	0.1
VTIPO Pressure (torr)	0.1	0.2	0.3	0.5	0.8	1.0	1.5	2.0

$T = 200°C$:

Growth Rate (μm/h)	0.028	0.45	1.8	2.8	7.2
VTIPO Pressure (torr)	0.05	0.2	0.4	0.5	0.8

In light of the material presented in this chapter, analyze the data and describe your results. Specify where additional data should be taken.

P10-14$_A$ Titanium dioxide is a wide-bandgap semiconductor that is showing promise as an insulating dielectric in VLSI capacitors and for use in solar cells. Thin films of TiO_2 are to be prepared by *chemical vapor deposition* from gaseous titanium tetraisopropoxide (TTIP). The overall reaction is

$$Ti(OC_3H_7)_4 \longrightarrow TiO_2 + 4C_3H_6 + 2H_2O$$

The reaction mechanism in a CVD reactor is believed to be [K. L. Siefering and G. L. Griffin, *J. Electrochem. Soc., 137*, 814 (1990)]

$$TTIP(g) + TTIP(g) \rightleftarrows I + P_1$$

$$I + S \rightleftarrows I \cdot S$$

$$I \cdot S \longrightarrow TiO_2 + P_2$$

where I is an active intermediate and P_1 is one set of reaction products and P_2 is another set. Assuming the homogeneous gas-phase reaction for TTIP is in equilibrium, derive a rate law for the deposition of TiO_2. The experimental results show that at 200°C the reaction is second-order at low partial pressures of TTIP and zero-order at high partial pressures, while at 300°C the reaction is second-order in TTIP over the entire pressure range. Discuss these results in light of the rate law you derived.

P10-15$_B$ In your plant, the reversible isomerization of compound A,

$$A \rightleftharpoons B,$$

is carried out over a *supported metal catalyst in an isothermal fixed-bed flow reactor*. A and B are liquid at the process conditions, and the volume change on reaction is negligible. The equilibrium constant for the reaction is 8.5 at 350°F and 6.0 at 400°F. The catalyzed reaction is pseudo-first-order in A, with an apparent Arrhenius activation energy of 26,900 Btu/lb mol.

Side reactions have a negligible effect on yield, but they slowly deactivate the catalyst. To increase the reaction rate and to partly offset deactivation, the temperature is raised on a schedule from 350°F to 400°F. At the end of an operating cycle, the catalyst is dumped and replaced. The nominal fresh feed of A is 300 gal/h, measured at 90°F. Unconverted A is separated from the reactor product and recycled to the feed. Total feed rate thus depends on catalyst activity.

The ACC Corporation offers an alternative, in which catalyst is to be *regenerated* in place by treatment with a solvent of proprietary composition. The cost is attractively low compared to catalyst replacement. ACC guarantees to achieve at least (1) 90% of fresh catalyst activity or (2) 2.25 times the activity of spent catalyst, whichever is less.

Following the use of the ACC regeneration procedure, a test run was made. Unfortunately, the feed rate was limited by trouble in another part of the plant. You are asked to review the following data and determine whether the guarantee was met. Support your conclusions by appropriate calculations.

Catalyst	Temperature (°F)	Fresh Feed Rate (gal/h)	Concentration of A in Reactor Product (%)
Fresh	350	310	25.0
Spent	400	300	32.2
Regenerated	350	220	19.0

(California Professional Engineers Exam)

P10-16$_B$ Rework Example 10-5 when
   (a) The reaction is carried out in a *moving-bed reactor* at a catalyst loading rate of 250,000 kg/h.
   (b) The reaction is carried out in a *packed-bed reactor* modeled as five CSTRs in series.
   (c) Repeat (a) when the catalyst and feed enter at opposite ends of the bed.
   (d) Determine the temperature–time trajectory to keep conversion constant in a CSTR if activation energies for reaction and decay are 30 kcal/mol and 10 kcal/mol, respectively.
   (e) How would your answer to part (d) change if the activation energies were reversed?

P10-17$_A$ Sketch *qualitatively* the reactant, product, and activity profiles as a function of length at various times for a *packed-bed reactor* for each of the following cases. In addition, sketch the effluent concentration of A as a function of time. The reaction is a simple isomerization:

$$A \longrightarrow B$$

(a) Rate law: $-r'_A = kaC_A$

　　Decay law: $r_d = k_d a C_A$

　　　　Case I: $k_d \ll k$

　　　　Case II: $k_d = k$

　　　　Case III: $k_d \gg k$

(b) $-r'_A = kaC_A$
　　$r_d = k_d a^2$

(c) $-r'_A = kaC_A$
　　$r_d = k_d a C_B$

(d) Sketch similar profiles for the rate laws in parts (a) and (c) in a *moving-bed reactor* with the solids entering at the same end of the reactor as the reactant.

(e) Repeat part (d) for the case where the solids and the reactant enter at opposite ends.

P10-18$_B$ The zero-order reaction A $\rightarrow$ B is carried out in a *moving-bed reactor* containing 1 kg of catalyst. The catalyst decay is also zero-order. The entering molar flow rate is pure A at 1 mol/min. Given the following information:

- The product sells for $160 per gram mole.
- The cost of operating the bed is $10 per kilogram of catalyst exiting the bed.

(a) What is the feed rate of solids (kg/min) that will give the maximum profit? (*Ans.: U_s = 4 kg/min.*)

(b) What are the activity and conversion exiting the reactor at this optimum? (*Note:* For the purpose of this calculation, ignore all other costs, such as the cost of the reactant, etc.)

(c) Redo parts (a) and (b) for the case when k_r = 5 mol/kg cat.·min and the reactant and catalyst are fed to opposite ends of the bed.

Additional information:

　　Specific reaction rate: k_r = 1.0 mol/kg cat.·min
　　Decay constant: k_d = 2.0 min^{-1}

P10-19$_B$ With the increasing demand for xylene in the petrochemical industry, the production of xylene from toluene disproportionation has gained attention in recent years [*Ind. Eng. Chem. Res.*, 26, 1854 (1987)]. This reaction,

$$2 \text{ toluene} \longrightarrow \text{benzene} + \text{xylene}$$

$$2T \xrightarrow{\text{catalyst}} B + X$$

was studied over a hydrogen mordenite catalyst that decays with time. As a first approximation, assume that the catalyst follows second-order decay,

$$r_d = k_d a^2$$

and the rate law for low conversions is

$$-r'_T = k_T P_T a$$

with $k_T = 20$ g mol/h·kg cat.·atm and $k_d = 1.6$ h^{-1} at 735 K.

(a) Compare the conversion time curves in a batch reactor containing 5 kg cat. at different initial partial pressures (1 atm, 10 atm, etc.). The reaction volume containing pure toluene initially is 1 dm^3 and the temperature is 735 K.

(b) What conversion can be achieved in a *moving-bed reactor* containing 50 kg of catalyst with a catalyst feed rate of 2 kg/h? Toluene is fed at a pressure of 2 atm and a rate of 10 mol/min.

(c) Explore the effect of catalyst feed rate on conversion.

(d) Suppose that $E_T = 25$ kcal/mol and $E_d = 10$ kcal/mol. What would the temperature–time trajectory look like for a CSTR? What if $E_T = 10$ kcal/mol and $E_d = 25$ kcal/mol?

(e) The decay law more closely follows the equation

$$r_d = k_d P_T^2 a^2$$

with $k_d = 0.2$ atm^{-2} h^{-1}. Redo parts (b) and (c) for these conditions.

P10-20$_B$ Catalytic reforming is carried out to make hydrocarbons more branched and in some cases cyclic, in order to increase the octane number of gasoline (*Recent Developments in Chemical Process and Plant Design*, Y. A. Liu, A. A. McGee, and W. R. Epperly, Eds., Wiley, New York, 1987, p. 33). A hydrocarbon feedstream is passed through a reforming catalytic reactor containing a platinum-on-alumina catalyst. The following data were obtained from measurements on the exit stream of the reactor at various times since fresh catalyst was placed in the reactor.

t (h)	0	35	100	200	300	500
Research Octane Number	106	102	99.7	98	96	94

Model the reactor as a fluidized-bed CSTR. The following temperature–time trajectory was implemented to offset the decay.

t (h)	0	175	250	500
T (°C)	500	515	520	530

(a) If possible, determine a decay law and a decay constant.

(b) What parameters, if any, can be estimated from the temperature–time trajectory?

(c) Why do you think the data were reported in terms of RON?

P10-21$_A$ The vapor-phase cracking of gas-oil in Example 10-7 is carried out over a different catalyst, for which the rate law is

$$-r'_A = k' P_A^2 \qquad \text{with } k' = 5 \times 10^{-5} \ \frac{\text{kmol}}{\text{kgcat.·s·atm}^2}$$

(a) Assuming that you can vary the entering pressure and gas velocity, what operating conditions would you recommend?

(b) What could go wrong with the conditions you chose?

Now assume the decay law is

$$-\frac{da}{dt} = k_D a \, C_{\text{coke}} \qquad \text{with } k_D = 100 \; \frac{\text{dm}^3}{\text{mol} \cdot \text{s}} \text{ at } 400°C$$

where the concentration, C_{coke}, in mol/dm^3 can be determined from a stoichiometric table.

(c) For a temperature of 400°C and a reactor height of 15 m, what gas velocity do you recommend? Explain. What is the corresponding conversion?

(d) The reaction is now to be carried in a STTR 15 m high and 1.5 m in diameter. The gas velocity is 2.5 m/s. You can operate in the temperature range between 100 and 500°C. What temperature do you choose, and what is the corresponding conversion?

(e) What would the temperature–time trajectory look like for a CSTR?

Additional information:

$E_R = 3000$ cal/mol
$E_D = 15,000$ cal/mol

P10-22$_\text{C}$ When the impurity cumene hydroperoxide is present in trace amounts in a cumene feedstream, it can deactivate the silica-alumina catalyst over which cumene is being cracked to form benzene and propylene. The following data were taken at 1 atm and 420°C in a differential reactor. The feed consists of cumene and a trace (0.08 mol %) of cumene hydroperoxide (CHP).

Benzene in Exit Stream (mol %)	2	1.62	1.31	1.06	0.85	0.56	0.37	0.24
t (s)	0	50	100	150	200	300	400	500

(a) Determine the order of decay and the decay constant. (*Ans.:* $k_d = 4.27 \times 10^{-3}$ s^{-1}.)

(b) As a first approximation (actually a rather good one), we shall neglect the denominator of the catalytic rate law and consider the reaction to be first-order in cumene. Given that the specific reaction rate with respect to cumene is $k = 3.8 \times 10^3$ mol/kg fresh cat.·s·atm, the molar flow rate of cumene (99.92% cumene, 0.08% CHP) is 200 mol/min, the entering concentration is 0.06 kmol/m^3, the catalyst weight is 100 kg, the velocity of solids is 1.0 kg/min, what conversion of cumene will be achieved in a *moving-bed reactor*?

P10-23$_\text{B}$ Benzene and propylene are produced from the cracking of cumene over a silica-alumina catalyst. Unfortunately, a trace amount of cumene hydroperoxide can deactivate the catalyst. The reaction is carried out in a batch reactor at a temperature sufficiently high that the adsorption constants of reactants and products are quite small. The reactor is charged with 0.1% cumene hydroperoxide, 90% N_2, and 9.9% cumene at a pressure of 20 atm. The following data were obtained in a well-mixed batch reactor with a catalyst concentration of 1 kg/m^3.

t(s)	0	10	20	30	40	60	80	100
% Conversion	0	3.7	7.1	10.2	13.0	18.0	22.3	25.9

t(s)	150	200	300	500	750	1000
% Conversion	33.0	37.9	44.1	49.3	51.2	51.7

Determine the reaction order and rate constant with respect to cumene. Determine the order of decay and the decay constant as well. (*Ans.:* First-order, $k = 6.63 \times 10^{-5} \text{ s}^{-1}$.)

P10-24$_C$ The reaction of cyclopentane to form *n*-pentane and coke was carried out over a palladium-alumina catalyst at 290°C [*J. Catal.*, *54*, 397 (1978)]. The following conversion–time data were obtained in a constant-volume batch reactor for a catalyst concentration of 0.01 kg/m³ and an initial reactant concentration of 0.03 kmol/m³.

t (min)	0	10	20	40	70	100	150	200	300	500	800	1200
% *X*	0	1.7	3.4	6.6	11	15	21.5	27.2	37.1	53.9	74.6	88.3

The following data were obtained in a differential reactor operated under the same conditions as the batch reactor above. The molar flow rate to the reactor was 1.5×10^{-5} kmol/min and the weight of catalyst in the reactor was 2 g.

t (min)	0	20	40	80	120	180	250	350	500	800	1200
% *X*	75	70.7	67	60.5	55.2	48.7	42	36	30	22	16.3

(a) Determine the reaction order with respect to cyclopentane and the specific reaction rate constant.

(b) Determine the order of decay and the decay constant.

(c) If this reaction is to be carried out in a *moving-bed reactor* in which pure cyclopentane is fed at a rate of 2 kmol/min and the catalyst at 1 kg/min, what weight of catalyst in the reactor is necessary to achieve 80% conversion of the cyclopentane?

(d) How would your answer to part (c) change if the catalyst feed rate were cut in half?

P10-25$_C$ The decomposition of spartanol to wulfrene and CO_2 is often carried out at high temperatures [*J. Theor. Exp.*, *15*, 15 (2014)]. Consequently, the denominator of the catalytic rate law is easily approximated as unity, and the reaction is first-order with an activation energy of 150 kJ/mol. Fortunately, the reaction is irreversible. Unfortunately, the catalyst over which the reaction occurs decays with time on stream. The following conversion–time data were obtained in a differential reactor:

For $T = 500$ K:

t (days)	0	20	40	60	80	120
X (%)	1	0.7	0.56	0.45	0.38	0.29

For $T = 550$ K:

t (days)	0	5	10	15	20	30	40
X (%)	2	1.2	0.89	0.69	0.57	0.42	0.33

(a) If the initial temperature of the catalyst is 480 K, determine the *temperature–time trajectory* to maintain a constant conversion.

(b) What is the catalyst lifetime?

P10-26$_B$ The hydrogenation of ethylbenzene to ethylcyclohexane over a nickel-mordenite catalyst is zero-order in both reactants up to an ethylbenzene conversion of 75% [*Ind. Eng. Chem. Res.*, 28(3), 260 (1989)]. At 553 K, $k = 5.8$ mol ethylbenzene/(dm^3 of catalyst·h). When a 100 ppm thiophene concentration entered the system, the ethylbenzene conversion began to drop.

Time (h)	0	1	2	4	6	8	12
Conversion	0.92	0.82	0.75	0.50	0.30	0.21	0.10

The reaction was carried out at 3 MPa and a molar ratio of H$_2$/ETB $= 10$. Discuss the catalyst decay. Be quantitative where possible.

P10-27   (Modified P8-8) The gas phase exothermic elementary reaction

$$A \xrightarrow{\ k\ } B + C$$

is carried out in a moving bed reactor.

$$k = 0.33 \exp\left[\frac{E_r}{R}\left(\frac{1}{450} - \frac{1}{T}\right)\right] \text{s}^{-1}$$

with

$$\frac{E_r}{R} = 3777 \text{ K}$$

Heat is removed by a heat exchanger jacketing the reactor.

$$\frac{Ua}{\rho_b} = \frac{0.8 \text{ J}}{\text{s·kg cat.·K}}$$

The flow rate of the coolant in the exchanger is sufficiently high that the ambient temperature is constant at 50°C. Pure A enters the reactor at a rate of 5.42 mol/s at a concentration of 0.27 mol/dm^3. Both the solid catalyst and the reactant enter the reactor at a temperature of 450 K and the heat transfer coefficient between the catalyst and gas is virtually infinite. The heat capacity of the solid catalyst is 100 J/kg cat./K.

The catalyst decay is first-order in activity with

$$k_d = 0.01 \exp\left[\frac{E_d}{R}\left(\frac{1}{450} - \frac{1}{T}\right)\right] \text{s}^{-1}$$

with

$$\left(\frac{E_d}{R}\right) = 7000 \text{ K}$$

There are 50 kg of catalyst in the bed.
(a) What catalyst charge rate (kg/s) will give the highest conversion?
(b) What is the corresponding conversion?
(c) Redo parts (a) and (b) for Case 1 $(T \neq T_S)$ of heat effects in moving beds. Use realistic values of the parameter values h and $\tilde{a}_p$ and $\tilde{a}_w$. Vary the entering temperatures of T_S and T.

Additional information:

$$C_{P_A} = 40 \text{ J/mol} \cdot \text{K}$$

$$C_{P_B} = 25 \text{ J/mol} \cdot \text{K}$$

$$C_{P_C} = 15 \text{ J/mol} \cdot \text{K}$$

$$\Delta H_{Rx} = -80 \text{ kJ/mol A}$$

JOURNAL CRITIQUE PROBLEMS

P10C-1 See "Catalytic decomposition of nitric oxide," *AIChE J.*, 7(4), 658 (1961). Determine if the following mechanism can also be used to explain the data in this paper.

$$NO + S \rightleftharpoons NO \cdot S$$

$$NO + NO \cdot S \rightleftharpoons N_2 + O_2 \cdot S$$

$$O_2 \cdot S \rightleftharpoons O_2 + S$$

P10C-2 In *J. Catal.*, *63*, 456 (1980), a rate expression is derived by assuming a reaction that is first-order with respect to the pressure of hydrogen and first-order with respect to the pressure of pyridine [Equation (10)]. Would another reaction order describe the data just as well? Explain and justify. Is the rate law expression derived by the authors correct?

P10C-3 See "The decomposition of nitrous oxide on neodymium oxide, dysprosium oxide and erbium oxide," *J. Catal.*, *28*, 428 (1973). Some investigators have reported the rate of this reaction to be independent of oxygen concentration and first-order in nitrous oxide concentration, while others have reported the reaction to be first-order in nitrous oxide concentration and negative one-half-order in oxygen concentration. Can you propose a mechanism that is consistent with both observations?

P10C-4 The kinetics of self-poisoning of Pd/Al_2O_3 catalysis in the hydrogenolysis of cyclopentane is discussed in *J. Catal.*, *54*, 397 (1978). Is the effective diffusivity used realistic? Is the decay homographic? The authors claim that the deactivation of the catalyst is independent of metal dispersion. If one were to determine the specific reaction rate as a function of percent dispersion, would this information support or reject the authors' hypotheses?

P10C-5 A packed-bed reactor was used to study the reduction of nitric oxide with ethylene on a copper-silica catalyst [*Ind. Eng. Chem. Process Des. Dev.*, *9*(3), 455 (1970)]. Develop the integral design equation in terms of the conversion and the initial pressure using the author's proposed rate law. If this equation is solved for conversion at various initial pressures and temperatures, is there a significant discrepancy between the experimental results shown in Figures 2 and 3 and the calculated results based on the proposed rate law? What is the possible source of this deviation?

P10C-6 The thermal degradation of rubber wastes was studied [*Int. J. Chem. Eng.*, *23*(4), 645 (1983)], and it was shown that a sigmoidal-shaped plot of conversion versus time would be obtained for the degradation reaction. Propose a model with physical significance that can explain this sigmoidal-shaped curve rather than merely curve fitting as the authors do. Also, what effect might the particle size distribution of the waste have on these curves? [*Hint:* See O. Levenspiel, *The Chemical Reactor Omnibook* (Corvallis, Ore.: Oregon State University Press, 1979) regarding gas-solid reactions.]

CD-ROM MATERIAL

- **Learning Resources**
    1. *Summary Notes for Lectures 19, 20, 21, 22, 23 and 24*
    3. *Interactive Computer Modules*
        A. Heterogeneous Catalysis
    4. *Solved Problems*
        Example CD10–1 Analysis of Heterogeneous Data [Class Problem, Winter 1997]
        Example CD10–2 Linear Least Squares to Determine the Rate Law Parameters
        Example CD10–3 Hydrodemethylation of Toluene in a PBR without Pressure Drop [2nd Ed. Example 6–3]
        Example CD10–4 Cracking of Texas Gas-Oil in a STTR [2nd Ed. Example 6–5]
- **Living Example Problems**
    1. *Example 10–2 Regression Analysis to Determine Model Parameters*
    2. *Example 10–3 Fixed-Bed Reactor Design*
    3. *Example 10–5 Catalyst Decay in a Fluidized Bed Modeled as a CSTR*
    4. *Example 10–7 Decay in a Straight Through Transport Reactor*
- **Professional Reference Shelf**
    1. *Hydrogen Adsorption*
        A. Molecular Adsorption
        B. Dissociative Adsorption
    2. *Catalyst Poisoning in a Constant Volume Batch Reactor*
    3. *Differential Method of Analysis to Determine the Decay Law*
    4. *Etching of Semiconductors*
        A. Dry Etching
        B. Wet Etching
        C. Dissolution Catalysis
- **Additional Homework Problems**

CDP10-A$_B$ Suggest a rate law and mechanism for the catalytic oxidation of ethanol over tantalum oxide when adsorption of ethanol and oxygen take place on different sites. [2nd ed. P6-17]

CDP10-B$_B$ Analyze the data for the vapor-phase esterification of acetic acid over a resin catalyst at 118°C.

CDP10-C$_B$ Silicon dioxide is grown by CVD according to the reaction

$$SiH_2Cl_2(g) + 2N_2O(g) \longrightarrow SiO_2(s) + 2N_2(g) + 2HCl(g)$$

Use the rate data to determine the rate law, reaction mechanism, and rate law parameters. [2nd ed. P6-13]

CDP10-D$_B$ The autocatalytic reaction A + B $\longrightarrow$ 2B is carried out in a moving-bed reactor. The decay law is first-order in B. Plot the activity and concentration of A and B as a function of catalyst weight.

CDP10-E$_B$ Determine the rate law and rate law parameters for the wet etching of an aluminum silicate.

CDP10-F$_B$ Titanium films are used in decorative coatings as well as wear-resistant tools because of their thermal stability and low electrical resistivity. TiN is produced by CVD from a mixture of TiCl$_4$ and NH$_3$TiN.

Develop a rate law, mechanism, and rate-limiting step and evaluate the rate law parameters.

CDP10-G$_C$ The decompositon of cumene is carried out over a LaY zeolite catalyst, and deactivation is found to occur by coking. Determine the decay law and rate law and use these to design a STTR.

CDP10-H$_B$ The dehydrogenation of ethylbenzene is carried out over a Shell catalyst. From the data provided, find the cost of the catalyst to produce a specified amount of styrene. [2nd ed. P6-20]

CDP10-I$_B$ A second-order reaction over a decaying catalyst takes place in a moving-bed reactor [Final Exam, Winter 1994]

CDP10-J$_B$ A first-order reaction A $\longrightarrow$ B + C takes place in a moving-bed reactor.

CDP10-K$_B$ For the cracking of normal paraffins (P$_n$), the rate has been found to increase with increasing temperature up to a carbon number of 15 (i.e., $n \leq 15$) and to decrease with increasing temperature for a carbon number greater than 16. [J. Wei, *Chem. Eng. Sci.*, *51*, 2995 (1996)]

CDP10-L$_B$ The formation of CH$_4$ from CO of H$_2$ is studied in a differential reactor.

CDP10-M$_B$ The reaction A + B $\longrightarrow$ C + D is carried out in a moving-bed reactor.

CDP10-N$_A$ Determine the rate law and mechanism for the reaction A + B $\longrightarrow$ C.

CDP10-O$_B$ Determine the rate law from data where the pressures are varied in such a way that the rate is constant. [2nd Ed. P6-18]

CDP10-P$_B$ Determine the rate law and mechanism for the vapor phase dehydration of ethanol. [2nd Ed. P6-21]

CDP10-Q$_A$ Second-order reaction and zero-order decay in a batch reactor.

CDP10-R$_B$ First-order decay in a moving bed reactor for the series reaction A $\longrightarrow$ B $\longrightarrow$ C.

SUPPLEMENTARY READING

1. A terrific discussion of heterogeneous catalytic mechanisms and rate-controlling steps may be found in

   BOUDART, M., and G. DJEGA-MARIADASSOU, *Kinetics of Heterogeneous Catalytic Reactors*. Princeton, N.J.: Princeton University Press, 1984.

   MASEL, R. I., *Principles of Adsorption and Reaction on Solid Surfaces*. New York: Wiley, 1996.

   SOMORJAI, G. A., *Introduction to Surface Chemistry and Catalysis*. New York: Wiley, 1994.

2. A truly excellent discussion of the types and rates of adsorption together with techniques used in measuring catalytic surface areas is presented in

   MASEL, R. I., *Principles of Adsorption and Reaction on Solid Surfaces*. New York: Wiley, 1996.

3. A discussion of the types of catalysis, methods of catalyst selection, methods of preparation, and classes of catalysts can be found in

   ENGELHARD CORPORATION, *Engelhard Catalysts and Precious Metal Chemicals Catalog*. Newark, N.J.: Engelhard Corp., 1985.

GATES, BRUCE C., *Catalytic Chemistry.* New York: Wiley, 1992.

GATES, B. C., J. R. KATZER, and G. C. A. SCHUIT, *Chemistry of Catalytic Processes.* New York: McGraw-Hill, 1979.

VAN SANTEN, R. A., and J. W. NIEMANTSVERDRIET, *Chemical Kinetics and Catalysis.* New York: Plenum Press, 1995.

4. Heterogeneous catalysis and catalytic reactors can be found in

SATERFIELD, C. N., *Heterogeneous Catalysis in Industrial Practice*, 2nd ed. New York: McGraw-Hill, 1991.

WHITE, M. G., *Heterogeneous Catalysis.* Upper Saddle River, N.J.: Prentice Hall, 1990.

and in the following journals: *Advances in Catalysis*, *Journal of Catalysis*, and *Catalysis Reviews.*

5. Techniques for discriminating between mechanisms and models can be found in

ATKINSON, A. C., and D. R. COX, "Planning experiments for discriminating between models," *J. R. Stat. Soc. Ser. B*, *36*, 321 (1974).

FROMENT, G. F., and K. B. BISCHOFF, *Chemical Reactor Analysis and Design.* New York: Wiley, 1979, Sec. 2.3.

6. Strategies for model building can be found in

BOX, G. E. P., W. G. HUNTER, and J. S. HUNTER, *Statistics for Experimenters.* New York: Wiley, 1978.

7. A reasonably complete listing of the different decay laws coupled with various types of reactors is given by

BUTT, J. B., and E. E. PETERSEN, *Activation, Deactivation, and Poisoning of Catalysts.* San Diego, Calif.: Academic Press, 1988.

FARRAUTO, R. J. and C. H. BARTHOLOMEW, "Fundamentals of Industrial Catalytic Processes," Blackie Academic and Professional, New York 1997.

HEGEDOUS, L. L., and R. MCCABE, *Catalyst Poisoning.* New York: Marcel Dekker, 1984.

HUGHES, R., *Deactivation of Catalysts.* San Diego, Calif.: Academic Press, 1984.

8. Examples of applications of catalytic principles to microelectronic manufacturing can be found in

HESS, D. W., and K. F. JENSEN, *Microelectronics Processing.* Washington, D.C.: American Chemical Society, 1989.

JENSEN, K. F., "Modeling of chemical vapor deposition reactors for the fabrication of microelectronic devices," in *Chemical and Catalytic Reactor Modeling.* Washington, D.C.: American Chemical Society, 1984.

LARRABEE, G. B., "Microelectronics," *Chem. Eng.*, *92*(12), 51 (1985).

LEE, H. H., *Fundamentals of Microelectronics Processing.* New York: McGraw-Hill, 1990.

External $\quad$ **11**
Diffusion Effects
on Heterogeneous
Reactions

Giving up is the ultimate tragedy.

Robert J. Donovan

or

The game's not over 'til it's over.

Yogi Berra

In many industrial reactions, the overall rate of reaction is limited by the rate of mass transfer of reactants and products between the bulk fluid and the catalytic surface. In the rate laws and catalytic reaction steps (i.e., diffusion, adsorption, surface reaction, desorption, and diffusion) presented in Chapter 10, we neglected the effects of mass transfer on the overall rate of reaction. In this chapter and the next we discuss the effects of diffusion (mass transfer) *resistance* on the overall reaction rate in processes that include both chemical reaction and mass transfer. The two types of diffusion resistance on which we focus

External vs. internal mass transfer resistance

attention are (1) *external resistance:* diffusion of the reactants or products between the bulk fluid and the external surface of the catalyst, and (2) *internal resistance:* diffusion of the reactants or products from the external pellet surface (pore mouth) to the interior of the pellet. In this chapter we focus on external resistance and in Chapter 12 we describe models for internal diffusional resistance with chemical reaction. After a brief presentation of the fundamentals of diffusion, including *Fick's first law*, we discuss representative correlations of mass transfer rates in terms of *mass transfer coefficients* for catalyst beds in which the external resistance is limiting. Qualitative observations will be made about the effects of fluid flow rate, pellet size, and pressure drop on reactor performance.

11.1 Mass Transfer Fundamentals

11.1.1 Definitions

Mass transfer usually refers to any process in which diffusion plays a role. *Diffusion* is the spontaneous intermingling or mixing of atoms or molecules by random thermal motion. It gives rise to motion of the species *relative* to motion of the mixture. In the absence of other gradients (such as temperature, electric potential, or gravitational potential), molecules of a given species within a single phase will always diffuse from regions of higher concentrations to regions of lower concentrations. This gradient results in a molar flux of the species (e.g., A), $\mathbf{W}_A$ (moles/area·time), in the direction of the concentration gradient. The flux of A, $\mathbf{W}_A$, is relative to a fixed coordinate (e.g., the lab bench) and is a vector quantity with typical units of mol/m²·s. In rectangular coordinates

$$\mathbf{W}_A = iW_{Ax} + jW_{Ay} + kW_{Az} \tag{11-1}$$

11.1.2 Molar Flux

The molar flux of A, $\mathbf{W}_A$, is the result of two contributions: $\mathbf{J}_A$, the molecular diffusion flux relative to the bulk motion of the fluid produced by a concentration gradient, and $\mathbf{B}_A$, the flux resulting from the bulk motion of the fluid:

Total flux = diffusion + bulk motion

$$\mathbf{W}_A = \mathbf{J}_A + \mathbf{B}_A \tag{11-2}$$

The bulk flow term for species A is the total flux of all molecules relative to a fixed coordinate times the mole fraction of A, y_A: $\mathbf{B}_A = y_A \Sigma \mathbf{W}_i$.

The bulk flow term $\mathbf{B}_A$ can also be expressed in terms of the concentration of A and the molar average velocity $\mathbf{V}$:

$$\mathbf{B}_A = C_A \mathbf{V} \tag{11-3}$$

$$\frac{\text{mol}}{\text{m}^2 \cdot \text{s}} = \frac{\text{mol}}{\text{m}^3} \cdot \frac{\text{m}}{\text{s}}$$

where the molar average velocity is

Molar average velocity

$$\mathbf{V} = \Sigma \, y_i \mathbf{V}_i$$

Here $\mathbf{V}_i$ is the particle velocity of species i and y_i the mole fraction of species i. By particle velocities, we mean the vector-average velocities of millions of A molecules at a point. For a binary mixture of species A and B, we let $\mathbf{V}_A$ and $\mathbf{V}_B$ be the particle velocities of species A and B, respectively. The flux of A with respect to a fixed coordinate system (e.g., the lab bench), $\mathbf{W}_A$, is just the product of the concentration of A and the particle velocity of A:

$$\mathbf{W}_A = C_A \mathbf{V}_A \tag{11-4}$$

$$\left(\frac{\text{mol}}{\text{dm}^2 \cdot \text{s}}\right) = \left(\frac{\text{mol}}{\text{dm}^3}\right)\left(\frac{\text{dm}}{\text{s}}\right)$$

The molar average velocity for a binary system is

$$\mathbf{V} = y_A \mathbf{V}_A + y_B \mathbf{V}_B \tag{11-5}$$

The total molar flux of A is given by Equation (11-1). $\mathbf{B}_A$ can be expressed either in terms of the concentration of A, in which case

$$\mathbf{W}_A = \mathbf{J}_A + C_A \mathbf{V} \tag{11-6}$$

or in terms of the mole fraction of A:

Binary system of
A and B

$$\mathbf{W}_A = \mathbf{J}_A + y_A(\mathbf{W}_A + \mathbf{W}_B) \tag{11-7}$$

11.1.3 Fick's First Law

Our discussion on diffusion will be restricted primarily to binary systems containing only species A and B. We now wish to determine *how* the molar diffusive flux of a species (i.e., $\mathbf{J}_A$) is related to its concentration gradient. As an aid in the discussion of the transport law that is ordinarily used to describe diffusion, recall similar laws from other transport processes. For example, in conductive heat transfer the constitutive equation relating the heat flux $\mathbf{q}$ and the temperature gradient is Fourier's law:

$$\mathbf{q} = -k_t \nabla T \tag{11-8}$$

where k_t is the thermal conductivity.

In rectangular coordinates, the gradient is in the form

Constitutive
equations in heat,
momentum, and
mass transfer

$$\nabla = i\,\frac{\partial}{\partial x} + j\,\frac{\partial}{\partial y} + k\,\frac{\partial}{\partial z} \tag{11-9}$$

The one-dimensional form of Equation (11-8) is

$$q_z = -k_t\,\frac{dT}{dz} \tag{11-10}$$

In momentum transfer, the constitutive relationship between shear stress, τ, and shear rate for simple planar shear flow is given by Newton's law of viscosity:

$$\tau = -\mu\,\frac{du}{dx} \tag{11-11}$$

The mass transfer flux law is analogous to the laws for heat and momentum transport. The constitutive equation for $\mathbf{J}_A$, the diffusional flux of **A** resulting from a concentration difference, is related to the concentration gradient by Fick's first law:

$$\mathbf{J}_A = -cD_{AB}\nabla y_A \tag{11-12}$$

where c is the total concentration (mol/dm³), D_{AB} is the diffusivity of A in B (dm²/s), and y_A is the mole fraction of A. Combining Equations (11-7) and (11-12), we obtain an expression for the molar flux of A:

Molar flux equation

$$\boxed{\mathbf{W}_A = -cD_{AB}\nabla y_A + y_A(\mathbf{W}_A + \mathbf{W}_B)} \tag{11-13}$$

11.2 Binary Diffusion

Although many systems involve more than two components, the diffusion of each species can be treated as if it were diffusing through another single species rather than through a mixture by defining an effective diffusivity. Methods and examples for calculating this effective diffusivity can be found in Hill.[1]

11.2.1 Evaluating the Molar Flux

The task is to now evaluate the bulk flow term

We now consider four typical conditions that arise in mass transfer problems and show how the molar flux is evaluated in each instance.

Equimolar Counterdiffusion. In equimolar counterdiffusion (EMCD), for every mole of A that diffuses in a given direction, one mole of B diffuses in the opposite direction. For example, consider a species A that is diffusing at steady state from the bulk fluid to a catalyst surface, where it isomerizes to form B. Species B then diffuses back into the bulk (see Figure 11-1). For every mole of A that diffuses to the surface, 1 mol of the isomer B diffuses away from the surface. The fluxes of A and B are equal in magnitude and flow counter to each other. Stated mathematically,

$$\mathbf{W}_A = -\mathbf{W}_B \tag{11-14}$$

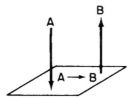

Figure 11-1 EMCD in isomerization reaction.

An expression for $\mathbf{W}_A$ in terms of the concentration of A, C_A, for the case of EMCD can be found by first substituting Equation (11-14) into Equation (11-7):

$$\mathbf{W}_A = \mathbf{J}_A + y_A[\mathbf{W}_A + (-\mathbf{W}_A)] = \mathbf{J}_A + 0$$
$$= \mathbf{J}_A = -cD_{AB}\nabla y_A \tag{11-15}$$

For constant total concentration

EMCD flux equation

$$\boxed{\mathbf{W}_A = \mathbf{J}_A = -D_{AB}\nabla C_A} \tag{11-16}$$

Dilute Concentrations. When the mole fraction of the diffusing solute and the bulk motion in the direction of the diffusion are small, the second term on the right-hand side of Equation (11-13) [i.e., $y_A(\mathbf{W}_A + \mathbf{W}_B)$] can usually be

[1] C. G. Hill, *Chemical Engineering Kinetics and Reactor Design*, Wiley, New York, 1977, p. 480.

neglected compared with the first term, $\mathbf{J}_A$. Under these conditions, together with the condition of constant total concentration, the flux of A is identical to that in Equation (11-16), that is,

Flux at dilute
concentrations

$$\boxed{\mathbf{W}_A \simeq \mathbf{J}_A = -D_{AB}\nabla C_A}$$

(11-17)

This approximation is almost always used for molecules diffusing within aqueous systems when the convective motion is small. For example, the mole fraction of a 1 M solution of a solute diffusing in water whose molar concentration, C_W, is

$$C_W = 55.6 \text{ mol/dm}^3$$

would be

$$y_A = \frac{C_A}{C_W + C_A} = \frac{1}{55.6 + 1} = 0.018$$

Consequently, in most liquid systems the concentration of the diffusing solute is small, and Equation (11-17) is used to relate $\mathbf{W}_A$ and the concentration gradient within the boundary layer.

Equation (11-13) also reduces to Equation (11-16) for porous catalyst systems in which the pore radii are very small. Diffusion under these conditions, known as *Knudsen diffusion*, occurs when the mean free path of the molecule is greater than the diameter of the catalyst pore. Here the reacting molecules collide more often with pore walls than with each other, and molecules of different species do not affect each other. The flux of species A for Knudsen diffusion (where bulk flow is neglected) is

$$\mathbf{W}_A = \mathbf{J}_A = -D_K\nabla C_A$$

(11-18)

where D_K is the Knudsen diffusivity.[2]

Diffusion Through a Stagnant Gas. The diffusion of a solute A through a stagnant gas B often occurs in systems in which two phases are present. Evaporation and gas absorption are typical processes in which this type of diffusion can be found. If gas B is stagnant, there is no net flux of B with respect to a fixed coordinate; that is,

$$\mathbf{W}_B = 0$$

Substituting into Equation (11-13) gives

$$\mathbf{W}_A = -cD_{AB}\nabla y_A + y_A\mathbf{W}_A$$

---

[2] C. N. Satterfield, *Mass Transfer in Heterogeneous Catalysis*, MIT Press, Cambridge, Mass., 1970, pp. 41–42, discusses Knudsen flow in catalysis and gives the expression for calculating D_K.

Rearranging yields

$$\mathbf{W}_A = \frac{-1}{1 - y_A} cD_{AB} \nabla y_A$$

or

Flux through a
stagnant gas

$$\boxed{\mathbf{W}_A = cD_{AB} \nabla \ln(1 - y_A) = cD_{AB} \nabla \ln y_B} \qquad (11\text{-}19)$$

Forced Convection.   In systems where the flux of A results primarily from forced convection, we assume that the diffusion in the direction of the flow (e.g., axial z direction), J_{Az}, *is small in comparison with the bulk flow* contribution in that direction, $B_{Az}(V_z \equiv U)$,

$$J_{Az} \simeq 0$$

Molar flux of
species A when
axial diffusion
effects are
negligible

$$W_{Az} \simeq B_{Az} = C_A V_z \equiv C_A U = \frac{v}{A_c} C_A \qquad (11\text{-}20)$$

where A_c is the cross-sectional area and v is the volumetric flow rate. Although the component of the diffusional flux vector of A in the direction of flow, J_{Az}, is neglected, the component of the flux of A in the x direction, J_{Ax}, which is normal to the direction of flow, may not necessarily be neglected (see Figure 11-2).

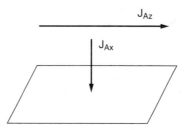

Figure 11-2   Forced axial convection with diffusion to surface.

In the previous material concerning the molar flow rate, F_A, diffusional effects were neglected, and F_A was written as the product of the volumetric flow rate and concentration:

Plug flow

$$F_A = vC_A$$

However, when accounting for diffusional effects, the molar flow rate of species A, F_A, in a specific direction z, is the product of molar flux in that direction, W_{Az}, and the cross-sectional area normal to the direction of flow, A_c:

$$F_{Az} = W_{Az}A_c = \left[-cD_{AB} \frac{dy_A}{dz} + y_A(W_z) \right] A_c \qquad (11\text{-}21)$$

where W_z is the total molar average flux in the z direction of all m species, that is,

$$W_z = \sum_{i=1}^{m} W_{iz} \tag{11-22}$$

11.2.2 Boundary Conditions

The most common boundary conditions are presented in Table 11-1.

TABLE 11-1. TYPES OF BOUNDARY CONDITIONS

1. Specify a concentration at a boundary (e.g., $z = 0$, $C_A = C_{A0}$). For an instantaneous reaction at a boundary, the concentration of the reactants at the boundary is taken to be zero (e.g., $C_{As} = 0$).
2. Specify a flux at a boundary.
   a. No mass transfer to a boundary,

$$W_A = 0 \tag{11-23}$$

   For example, at the wall of a nonreacting pipe,

$$\frac{dC_A}{dr} = 0 \qquad \text{at } r = R$$

   That is, because the diffusivity is finite, the only way the flux can be zero is if the concentration gradient is zero.
   b. Set the molar flux to the surface equal to the rate of reaction on the surface,

$$W_A(\text{surface}) = -r_A''(\text{surface}) \tag{11-24}$$

   c. Set the molar flux to the boundary equal to convective transport across a boundary layer,

$$W_A(\text{boundary}) = k_c(C_{Ab} - C_{As}) \tag{11-25}$$

   where k_c is the mass transfer coefficient and C_{As} and C_{Ab} are the surface and bulk concentrations, respectively.
3. Planes of symmetry. When the concentration profile is symmetrical about a plane, the concentration gradient is zero in that plane of symmetry. For example, in the case of radial diffusion in a pipe, at the center of the pipe

$$\frac{dC_A}{dr} = 0 \qquad \text{at } r = 0$$

11.2.3 Modeling Diffusion Without Reaction

In developing mathematical models for chemically reacting systems in which diffusional effects are important, the first steps are:

Step 1: Perform a differential mole balance on a particular species A.
Step 2: Substitute for F_{Az} in terms of W_{Az}.
Step 3: Replace W_{Az} by the appropriate expression for the concentration gradient.
Step 4: State the boundary conditions.
Step 5: Solve for the concentration profile.
Step 6: Solve for the molar flux.

Steps in modeling mass transfer

Example 11–1   Diffusion Through a Film to a Catalyst Particle

Species A, which is present in dilute concentrations, is diffusing at steady state from the bulk fluid through a stagnant film of B of thickness δ to the external surface of the catalyst (Figure E11-1.1). The concentration of A at the external boundary C_{Ab} and at the external catalyst surface is C_{As}, with $C_{Ab} > C_{As}$. Because the thickness of the "hypothetical stagnant film" next to the surface is small with regard to the diameter of the particle, we can neglect curvature and represent the diffusion in rectilinear coordinates as shown in Figure E11-1.2. Determine the concentration profile and the flux of A to the surface.

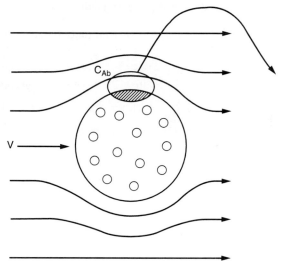

External Mass
Transfer

Figure E11-1.1   Transport to a sphere.

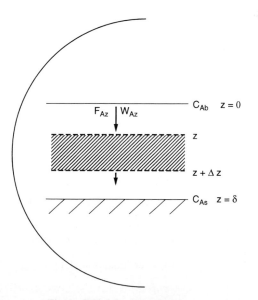

Figure E11-1.2   Boundary layer.

Additional information:

$$D_{AB} = 0.01 \text{ cm}^2/\text{s} = 10^{-6} \text{ m}^2/\text{s}$$

$$C_{Ab} = 0.01 \text{ mol/dm}^3$$

$$C_{As} = 0.002 \text{ mol/dm}^3$$

Solution

Our first step is to perform a mole balance on species A over a differential element of width Δz and cross-sectional area A_c and then arrive at a first-order differential equation in W_{Az} [i.e., Equation (E11-1.3)].

Step 1: The general **mole balance** equation is

An algorithm
- Mole balance
- Bulk flow = ?
- Differential equation
- Boundary conditions
- Concentration profile
- Molar flux

$$\begin{bmatrix} \text{rate} \\ \text{in} \end{bmatrix} - \begin{bmatrix} \text{rate} \\ \text{out} \end{bmatrix} + \begin{bmatrix} \text{rate of} \\ \text{generation} \end{bmatrix} = \begin{bmatrix} \text{rate of} \\ \text{accumulation} \end{bmatrix} \quad \text{(E11-1.1)}$$

$$F_{Az}|_z \quad - \quad F_{Az}|_{z+\Delta z} + \quad\quad 0 \quad\quad = \quad\quad 0$$

Dividing by $-\Delta z$ gives us

$$\frac{F_{Az}|_{z+\Delta z} - F_{Az}|_z}{\Delta z} = 0$$

and taking the limit as $\Delta z \to 0$, we obtain

$$\frac{dF_{Az}}{dz} = 0 \quad \text{(E11-1.2)}$$

Step 2: Next, **substitute** for F_A in terms of W_{Az} and A_c,

$$F_{Az} = W_{Az} A_c$$

Divide by A_c to get

$$\frac{dW_{Az}}{dz} = 0 \quad \text{(E11-1.3)}$$

Step 3: To **evaluate the bulk flow term**, we now must relate W_{Az} to the concentration gradient utilizing the specification of the problem statement. For diffusion of almost all solutes through a liquid, the concentration of the diffusing species is considered dilute. For a dilute concentration of the diffusion solute, we have for constant total concentration,

$$W_{Az} = -D_{AB} \frac{dC_A}{dz} \quad \text{(E11-1.4)}$$

Differentiating Equation (E11-1.4) for constant diffusivity yields

$$\frac{dW_{Az}}{dz} = -D_{AB} \frac{d^2C_A}{dz^2}$$

However, Equation (E11-1.3) yields

$$\frac{dW_{Az}}{dz} = 0$$

Therefore, the differential equation describing diffusion through a liquid film reduces to

$$\frac{d^2C_A}{dz^2} = 0 \qquad \text{(E11-1.5)}$$

Step 4: The **boundary** conditions are:

$$\text{When } z = 0, \qquad C_A = C_{Ab}$$

$$\text{When } z = \delta, \qquad C_A = C_{A\delta}$$

Step 5: **Solve for the concentration profile.** Equation (E11-1.5) is an elementary differential equation that can be solved directly by integrating twice with respect to z. The first integration yields

$$\frac{dC_A}{dz} = K_1$$

and the second yields

$$C_A = K_1 z + K_2 \qquad \text{(E11-1.6)}$$

where K_1 and K_2 are arbitrary constants of integration. We now use the boundary conditions to evaluate the constants K_1 and K_2.
At $z = 0$, $C_A = C_{Ab}$; therefore,

$$C_{Ab} = 0 + K_2$$

At $z = \delta$, $C_A = C_{As}$;

$$C_{As} = K_1\delta + K_2 = K_1\delta + C_{Ab}$$

Eliminating K_1 and rearranging gives the following concentration profile:

$$\boxed{\frac{C_A - C_{Ab}}{C_{As} - C_{Ab}} = \frac{z}{\delta}} \qquad \text{(E11-1.7)}$$

Rearranging (E11-1.7), we get the concentration profile shown in Figure E11-1.3.

$$C_A = C_{Ab} + (C_{As} - C_{Ab})\frac{z}{\delta} \qquad \text{(E11-1.8)}$$

Concentration
profile

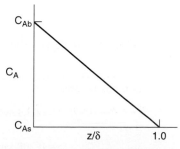

Figure E11-1.3 Concentration profile.

Step 6: The next step is to **determine the molar flux** of A diffusing through the stagnant film. For dilute solute concentrations and constant total concentration,

$$W_{Az} = -D_{AB}\left(\frac{dC_A}{dz}\right) \tag{E11-1.4}$$

To determine the flux, differentiate Equation (E11-1.8) with respect to z and then multiply by $-D_{AB}$:

$$W_{Az} = (-D_{AB})\frac{d}{dz}\left[C_{Ab} + (C_{As} - C_{Ab})\frac{z}{\delta}\right]$$

$$W_{Az} = -D_{AB}\left(\frac{C_{As} - C_{Ab}}{\delta}\right)$$

$$\boxed{W_{Az} = \frac{D_{AB}}{\delta}(C_{Ab} - C_{As})} \tag{E11-1.9}$$

In terms of mole fractions

$$W_{Az} = \frac{D_{AB}C_{T0}}{\delta}(y_{Ab} - y_{A\delta}) \tag{E11-1.10}$$

If $D_{AB} = 10^{-6}$ m²/s, $C_{T0} = 0.1$ mol/m³, and $\delta = 10^{-6}$ m, $y_{Ab} = 0.9$, and $y_{A\delta} = 0.2$, then

$$W_{Az} = \frac{(10^{-6}\ m^2/s)(0.1\ kmol/m^3)(0.9 - 0.2)}{10^{-6}\ m}$$

$$= 0.07\ kmol/m^2 \cdot s$$

EMCD or dilute concentration

Stagnant Film

If we had assumed diffusion through a stagnant film ($W_{bz} = 0$ and $B_{Az} = y_A W_A$) rather than dilute concentration or equal molar counter diffusion ($B_{Az} = 0$), we could use the solution procedure discussed above (see the CD-ROM), starting with

$$W_{Az} = -cD_{AB}\frac{dy_A}{dz} + y_A W_A \tag{E11-1.11}$$

to arrive at

Stagnant film

$$W_{Az} = \frac{cD_{AB}}{\delta}\ln\frac{1 - y_{A\delta}}{1 - y_{Ab}} \tag{E11-1.12}$$

The intermediate steps are given on the CD-ROM. For the same parameter values as before,

$$W_{Az} = \frac{(0.1\ kmol/m^3)(10^{-6}\ m^2/s)}{10^{-6}\ m}\ln\frac{1 - 0.02}{1 - 0.9}$$

$$= 0.208\ kmol/m^2 \cdot s$$

We see that for the case of diffusion through a stagnant film the flux is greater. Why is this?

For equal molar counterdiffusion or dilution, we can rearrange Equation (E11-1.8) to give the mole fraction as a function of z:

$$y_A = y_{Ab} - (y_{Ab} - y_{A\delta})\frac{z}{\delta} \qquad \text{(E11-1.13)}$$

For diffusion through a stagnant film, the mole fraction profile is shown on the CD-ROM to be y_{Ab}:

$$y_A = 1 - (1 - y_{Ab})\left(\frac{1 - y_{A\delta}}{1 - y_{Ab}}\right)^{z/\delta} \qquad \text{(E11-1.14)}$$

A comparison of the concentration profiles is shown in Figure E11-1.4.

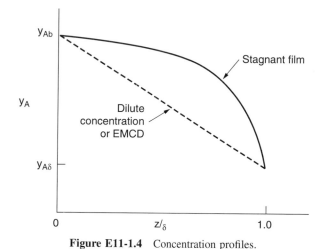

Figure E11-1.4   Concentration profiles.

11.2.4 Temperature and Pressure Dependence of D_{AB}

Before closing this brief discussion on mass transfer fundamentals, further mention should be made of the diffusion coefficient.[3] Equations for predicting gas diffusivities are given by Fuller[4] and are also given in Perry's *Handbook*.[5] The orders of magnitude of the diffusivities for gases, liquids,[6] and

---

[3] E. N. Fuller, P. D. Schettler, and J. C. Giddings, *Ind. Eng. Chem.*, *58*(5), 19 (1966). Several other equations for predicting diffusion coefficients can be found in R. C. Reid, J. M. Prausnitz, and T. K. Sherwood, *The Properties of Gases and Liquids*, 3rd ed., McGraw-Hill, New York, 1977, Chap. 11.

[4] For further discussion of mass transfer fundamentals, see R. B. Bird, W. E. Stewart, and E. N. Lightfoot, *Transport Phenomena*, Wiley, New York, 1960, Chaps. 16 and 17.

[5] R. H. Perry, D. W. Green, and J. O. Maloney, *Chemical Engineer's Handbook*, 6th ed., McGraw-Hill, New York, 1984.

[6] To estimate liquid diffusivities for binary systems, see K. A. Reddy and L. K. Doraiswamy, *Ind. Eng. Chem. Fund.*, *6*, 77 (1967).

solids and the manner in which they vary with temperature and pressure are given in Table 11-2. We note that the Knudsen, liquid, and solid diffusivities are independent of total pressure.

TABLE 11-2.   DIFFUSIVITY RELATIONSHIPS FOR GASES, LIQUIDS, AND SOLIDS

Phase	Order of Magnitude		Temperature and Pressure Dependences[a]
	cm²/s	m²/s	
Gas			
Bulk	10^{-1}	10^{-5}	$D_{AB}(T_2, P_2) = D_{AB}(T_1, P_1) \dfrac{P_1}{P_2}\left(\dfrac{T_2}{T_1}\right)^{1.75}$
Knudsen	10^{-2}	10^{-6}	$D_A(T_2) = D_A(T_1)\left(\dfrac{T_2}{T_1}\right)^{1/2}$
Liquid	10^{-5}	10^{-9}	$D_{AB}(T_2) = D_{AB}(T_1)\dfrac{\mu_1}{\mu_2}\left(\dfrac{T_2}{T_1}\right)$
Solid	10^{-9}	10^{-13}	$D_{AB}(T_2) = D_{AB}(T_1)\exp\left[\dfrac{E_D}{R}\left(\dfrac{T_2-T_1}{T_1 T_2}\right)\right]$

It is important to know magnitudes and T and P dependence of D_{AB} (margin note)

[a] μ_1, μ_2, liquid viscosities at temperatures T_1 and T_2, respectively; E_D, diffusion activation energy.

11.2.5 Modeling Diffusion with Chemical Reaction

The method used in solving diffusion problems similar to Example 11-3 is shown in Table 11-3. See also Cussler.[7]

CD Professional

Reference Shelf

TABLE 11-3.   STEPS IN MODELING CHEMICAL SYSTEMS WITH DIFFUSION AND REACTION

1. Define the problem and state the assumptions. **(See Problem Solving on the CD.)**
2. Define the system on which the balances are to be made.
3. Perform a differential mole balance on a particular species.
4. Obtain a differential equation in W_A by rearranging your balance equation properly and taking the limit as the volume of the element goes to zero.
5. Substitute the appropriate expression involving the concentration gradient for W_A from Section 10.2 to obtain a second-order differential equation for the concentration of A.[a]
6. Express the reaction rate r_A (if any) in terms of concentration and substitute into the differential equation.
7. State the appropriate boundary and initial conditions.
7a. Put the differential equations and boundary conditions in dimensionless form.
8. Solve the resulting differential equation for the concentration profile.
9. Differentiate this concentration profile to obtain an expression for the molar flux of A.
10. Substitute numerical values for symbols.

Expanding the previous six modeling steps just a bit (margin note)

[a] In some instances it may be easier to integrate the resulting differential equation in step 4 before substituting for W_A.

[7] E. L. Cussler, *Diffusion Mass Transfer in Fluid Systems*, 2nd ed., Cambridge University Press, New York, 1997.

The purpose of presenting algorithms (e.g., Table 11-3) to solve reaction engineering problems is to give the reader a starting point or framework with which to work if they were to get stuck. It is expected that once readers are familiar and comfortable using the algorithm/framework, they will be able to move in and out of the framework as they develop creative solutions to non-standard chemical reaction engineering problems.

Move in and out of the algorithm (Steps 1 → 6) to generate creative solutions

11.3 External Resistance to Mass Transfer

To begin our discussion on the diffusion of reactants from the bulk fluid to the external surface of a catalyst, we shall focus attention on the flow past a single catalyst pellet. Reaction takes place only on the catalyst and not in the fluid surrounding it. The fluid velocity in the vicinity of the spherical pellet will vary with position around the sphere. The hydrodynamic boundary layer is usually defined as the distance from a solid object to where the fluid velocity is 99% of the bulk velocity U_0. Similarly, the mass transfer boundary layer thickness, δ, is defined as the distance from a solid object to where the concentration of the diffusing species reaches 99% of the bulk concentration.

A reasonable representation of the concentration profile for a reactant A diffusing to the external surface is shown in Figure 11-3. As illustrated, the change in concentration of A from C_{Ab} to C_{As} takes place in a very narrow fluid layer next to the surface of the sphere. Nearly all of the resistance to mass transfer is found in this layer.

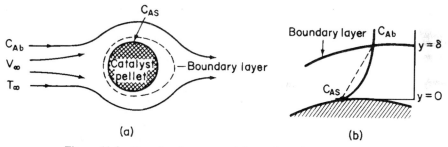

Figure 11-3   Boundary layer around the surface of a catalyst pellet.

11.3.1 Mass Transfer Coefficient

A useful way of modeling diffusive transport is to treat the fluid layer next to a solid boundary as a stagnant film of thickness δ. We say that *all* the resistance to mass transfer is found within this hypothetical stagnant film, and the properties (i.e., concentration, temperature) of the fluid at the outer edge of the film are identical to those of the bulk fluid. This model can readily be used to solve the differential equation for diffusion through a stagnant film. The dashed line in Figure 11-3b represents the concentration profile predicted by the hypothetical stagnant film model, while the solid line gives the actual profile. If the film thickness is much smaller than the radius of the pellet, curvature effects can be neglected. As a result, only the one-dimensional diffusion equation must be solved, as was shown in Section 11.1 (see also Figure 11-4).

The concept of a hypothetical stagnant film within which all the resistance to external mass transfer exists

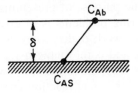

Figure 11-4 Concentration profile for EMCD in stagnant film model.

For either EMCD or dilute concentrations, the solution was shown to be in the form

$$W_{Ar}(P) = \frac{D_{AB}}{\tilde{\delta}}(C_{Ab} - C_{As}) \qquad (11\text{-}26)$$

where $W_{Ar}(P)$ is the flux at a specified position or point on the sphere.

The ratio of the diffusivity to the film thickness is the mass transfer coefficient, $\tilde{k}_c$. The tildas in the terms $\tilde{k}_c$ and $\tilde{\delta}$ denote that they are, respectively, the *local* transfer coefficient and the boundary layer thickness at a particular point P on the sphere,

$$\tilde{k}_c = \frac{D_{AB}}{\tilde{\delta}} \qquad (11\text{-}27)$$

The average mass transfer coefficient over the surface of area A is

Mass transfer coefficient

$$k_c = \frac{\displaystyle\int_A \tilde{k}_c \, dA}{A}$$

The average molar flux from the bulk fluid to the surface is

Molar flux of A to the surface

$$W_{Ar} = k_c(C_{Ab} - C_{As}) \qquad (11\text{-}28)$$

The mass transfer coefficient k_c is analogous to the heat transfer coefficient. The heat flux q from the bulk fluid at a temperature T_0 to a solid surface at T_s is

$$q_r = h(T_0 - T_s) \qquad (11\text{-}29)$$

For forced convection, the heat transfer coefficient is normally correlated in terms of three dimensionless groups: the Nusselt number, Nu, the Reynolds number, Re, and the Prandtl number, Pr. For the single spherical pellets discussed here, Nu and Re take the following forms:

$$\text{Nu} = \frac{hd_p}{k_t} \qquad (11\text{-}30)$$

$$\text{Re} = \frac{U\rho d_p}{\mu} \qquad (11\text{-}31)$$

The Prandtl number is not dependent on the geometry of the system.

The Nusselt, Prandtl, and Reynolds numbers are used in forced convection heat transfer correlations

$$Pr = \frac{\mu C_p}{k_t} = \frac{\mu}{\rho}\left(\frac{\rho C_p}{k_t}\right) = \frac{\nu}{\alpha_t} \qquad (11\text{-}32)$$

where α_t = thermal diffusivity, m²/s

$\nu = \dfrac{\mu}{\rho}$ = kinematic viscosity, m²/s

d_p = diameter of pellet, m
U = free-stream velocity, m/s
k_t = thermal conductivity, J/K·m·s
ρ = fluid density, kg/m³

The other symbols are as defined previously.

The heat transfer correlation relating the Nusselt number to the Prandtl and Reynolds numbers for flow around a sphere is[8]

$$Nu = 2 + 0.6Re^{1/2}Pr^{1/3} \qquad (11\text{-}33)$$

Although this correlation can be used over a wide range of Reynolds numbers, it can be shown theoretically that if a sphere is immersed in a stagnant fluid, then

$$Nu = 2 \qquad (11\text{-}34)$$

and that at higher Reynolds numbers in which the boundary layer remains laminar, the Nusselt number becomes

$$Nu \simeq 0.6Re^{1/2}Pr^{1/3} \qquad (11\text{-}35)$$

Although further discussion of heat transfer correlations is no doubt worthwhile, it will not help us to determine the mass transfer coefficient and the mass flux from the bulk fluid to the external pellet surface. However, the preceding discussion on heat transfer was not entirely futile, because, for similar geometries, *the heat and mass transfer correlations are analogous.* If a heat transfer correlation for the Nusselt number exists, the mass transfer coefficient can be estimated by replacing the Nusselt and Prandtl numbers in this correlation by the Sherwood and Schmidt numbers, respectively:

Converting a heat transfer correlation to a mass transfer correlation

$$Sh \longrightarrow Nu$$

$$Sc \longrightarrow Pr$$

The heat and mass transfer coefficients are analogous. The corresponding fluxes are

$$q_z = h(T - T_s) \qquad (11\text{-}36)$$

$$W_{Az} = k_c(C_A - C_{As}) \qquad (11\text{-}37)$$

The one-dimensional differential forms of the mass flux for EMCD and the heat flux are, respectively,

---

[8] W. E. Ranz and W. R. Marshall, Jr., *Chem. Eng. Prog.*, 48, 141–146, 173–180 (1952).

For EMCD the heat and molar flux equations are analogous

$$W_{Az} = -D_{AB} \frac{dC_A}{dz} \qquad (E11\text{-}1.4)$$

$$q_z = -k_t \frac{dT}{dz} \qquad (11\text{-}10)$$

If we replace h by k_c and k_t by D_{AB} in Equation (11-30), we obtain

$$\left.\begin{array}{c} h \longrightarrow k_c \\ k_t \longrightarrow D_{AB} \end{array}\right\} \text{Nu} \longrightarrow \text{Sh}$$

the mass transfer Nusselt number (i.e., the Sherwood number):

Sherwood number

$$\text{Sh} = \frac{k_c d_p}{D_{AB}} = \frac{(\text{m/s})(\text{m})}{\text{m}^2/\text{s}} \text{ dimensionless} \qquad (11\text{-}38)$$

The Prandtl number is the ratio of the kinematic viscosity (i.e., the momentum diffusivity) to the thermal diffusivity. Because the Schmidt number is analogous to the Prandtl number, one would expect that Sc is the ratio of the momentum diffusivity (i.e., the kinematic viscosity), ν, to the mass diffusivity D_{AB}. Indeed, this is true:

$$\alpha_t \longrightarrow D_{AB}$$

The Schmidt number is

Schmidt number

$$\text{Sc} = \frac{\nu}{D_{AB}} = \frac{\text{m}^2/\text{s}}{\text{m}^2/\text{s}} \text{ dimensionless} \qquad (11\text{-}39)$$

Consequently, the correlation for mass transfer for flow around a spherical pellet is analogous to that given for heat transfer [Equation (11-33)], that is,

$$\text{Sh} = 2 + 0.6\text{Re}^{1/2}\text{Sc}^{1/3} \qquad (11\text{-}40)$$

This relationship is often referred to as the *Frössling correlation*.[9]

11.3.2 Mass Transfer to a Single Particle

The Sherwood, Reynolds, and Schmidt numbers are used in forced convection mass transfer correlations

In this section we consider two limiting cases of diffusion and reaction on a catalyst particle.[10] In the first case the reaction is so rapid that the rate of diffusion of the reactant to the surface limits the reaction rate. In the second case, the reaction is so slow that virtually no concentration gradient exists in the gas phase (i.e., rapid diffusion with respect to surface reaction).

---

[9] N. Frössling, *Gerlands Beitr. Geophys.*, *52*, 170 (1938).

[10] A comprehensive list of correlations for mass transfer to particles is given by G. A. Hughmark, *Ind. Eng. Chem. Fund.*, *19*(2), 198 (1980).

Example 11–2 Rapid Reaction on a Catalyst Surface

Calculate the mass flux of reactant A to a single catalyst pellet 1 cm in diameter suspended in a large body of liquid. The reactant is present in dilute concentrations, and the reaction is considered to take place instantaneously at the external pellet surface (i.e., $C_{As} \simeq 0$). The bulk concentration of the reactant is 1.0 M, and the free-system liquid velocity is 0.1 m/s. The kinematic viscosity is 0.5 centistoke (cS; 1 centistoke $= 10^{-6}$ m²/s), and the liquid diffusivity of A is 10^{-10} m²/s.

Solution

For dilute concentrations of the solute the radial flux is

$$W_{Ar} = k_c(C_{Ab} - C_{As}) \tag{11-28}$$

Because reaction is assumed to occur instantaneously on the external surface of the pellet, $C_{As} = 0$. Also, C_{Ab} is given as 1 mol/dm³. The mass transfer coefficient for single spheres is calculated from the Frössling correlation:

$$\text{Sh} = \frac{k_c d_p}{D_{AB}} = 2 + 0.6\text{Re}^{1/2}\text{Sc}^{1/3} \tag{11-40}$$

<u>Liquid Phase</u>

Re = 2000

Sc = 5000

Sh = 460

$k_c = 4.6 \times 10^{-3}$ m/s

$$\text{Re} = \frac{\rho d_p U}{\mu} = \frac{d_p U}{\nu} = \frac{(0.01 \text{ m})(0.1 \text{ m/s})}{0.5 \times 10^{-6} \text{ m}^2/\text{s}} = 2000$$

$$\text{Sc} = \frac{\nu}{D_{AB}} = \frac{5 \times 10^{-7} \text{ m}^2/\text{s}}{10^{-10} \text{ m}^2/\text{s}} = 5000$$

Substituting the values above into Equation (11-40) gives us

$$\text{Sh} = 2 + 0.6(2000)^{0.5}(5000)^{1/3} = 460.7 \tag{E11-2.1}$$

$$k_c = \frac{D_{AB}}{d_p}\text{Sh} = \frac{10^{-10} \text{ m}^2/\text{s}}{0.01 \text{ m}} \times 460.7 = 4.61 \times 10^{-6} \text{ m/s} \tag{E11-2.2}$$

$$C_{Ab} = 1.0 \text{ mol/dm}^3 = 10^3 \text{ mol/m}^3$$

Substituting for k_c and C_{Ab} in Equation (10-28), the molar flux to the surface is

$$W_{Ar} = (4.61 \times 10^{-6}) \text{ m/s} (10^3 - 0) \text{ mol/m}^3 = 4.61 \times 10^{-3} \text{ mol/m}^2 \cdot \text{s}$$

Because $W_{Ar} = -r''_{As}$, this rate is also the rate of reaction per unit surface area of catalyst.

In Example 11-2, the surface reaction was extremely rapid and the rate of mass transfer to the surface dictated the overall rate of reaction. We now consider a more general case. The isomerization

$$A \longrightarrow B$$

is taking place on the surface of a solid sphere (Figure 11-5). The surface reaction follows a Langmuir–Hinshelwood single-site mechanism for which the rate law is

$$-r''_{As} = \frac{k_r C_{As}}{1 + K_A C_{As} + K_B C_{Bs}} \tag{11-41}$$

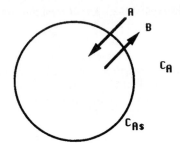

Figure 11-5 Diffusion to, and reaction on, external surface of pellet.

The temperature is sufficiently high that we need only consider the case of very weak adsorption (i.e., low surface coverage) of A and B; thus

$$(K_B C_{Bs} + K_A C_{As}) \ll 1$$

Therefore,

$$-r''_{As} = k_r C_{As} \tag{11-42}$$

Using boundary conditions 2b and 2c in Table 11-1, we obtain

$$W_A \big|_{surface} = -r''_{As} \tag{11-43}$$

$$W_A = k_c(C_A - C_{As}) = k_r C_{As} \tag{11-44}$$

The concentration C_{As} is not as easily measured as the bulk concentration. Consequently, we need to eliminate C_{As} from the equation for the flux and rate of reaction. Solving Equation (11-44) for C_{As} yields

$$C_{As} = \frac{k_c C_A}{k_r + k_c} \tag{11-45}$$

and the rate of reaction on the surface becomes

Molar flux of A to the surface is equal to the rate of consumption of A on the surface

$$\boxed{W_A = -r''_{As} = \frac{k_c k_r C_A}{k_r + k_c}} \tag{11-46}$$

One will often find the flux to or from the surface as written in terms of an *effective* transport coefficient k_{eff}:

$$W_A = -r''_{As} = k_{eff} C_A \tag{11-47}$$

where

$$k_{eff} = \frac{k_c k_r}{k_c + k_r}$$

Rapid Reaction.   We first consider how the overall rate of reaction may be increased when the rate of mass transfer to the surface limits the overall rate of reaction. Under these circumstances the specific reaction rate constant is much greater than the mass transfer coefficient

$$k_r \gg k_c$$

and

$$\frac{k_c}{k_r} \ll 1$$

$$-r_A'' = \frac{k_c C_A}{1 + k_c/k_r} \approx k_c C_A \qquad (11\text{-}48)$$

To increase the rate of reaction per unit surface area of solid sphere, one must increase C_A and/or k_c. In this gas-phase catalytic reaction example, and for most liquids, the Schmidt number is sufficiently large that the number 2 in Equation (11-40) is negligible with respect to the second term for Reynolds numbers greater than 25. As a result, Equation (11-40) gives

It is important to know how the mass transfer coefficient varies with fluid velocity, particle size, and physical properties

$$k_c = 0.6 \left(\frac{D_{AB}}{d_p} \right) \mathrm{Re}^{1/2} \mathrm{Sc}^{1/3}$$

$$= 0.6 \left(\frac{D_{AB}}{d_p} \right) \left(\frac{U d_p}{\nu} \right)^{1/2} \left(\frac{\nu}{D_{AB}} \right)^{1/3}$$

$$k_c = 0.6 \times \frac{D_{AB}^{2/3}}{\nu^{1/6}} \times \frac{U^{1/2}}{d_p^{1/2}} \qquad (11\text{-}49)$$

$$(\text{term 1}) \times (\text{term 2})$$

Term 1 is a function of temperature and pressure only. The diffusivity always increases with increasing temperature for both gas and liquid systems. However, the kinematic viscosity ν increases with temperature ($\nu \propto T^{3/2}$) for gases and decreases exponentially with temperature for liquids. Term 2 is a function of flow conditions and particle size. Consequently, to increase k_c and thus the overall rate of reaction per unit surface area, one may either decrease the particle size or increase the velocity of the fluid flowing past the particle. For this particular case of flow past a single sphere, we see that if the velocity is doubled, the mass transfer coefficient and consequently the rate of reaction is increased by a factor of

$$(U_2/U_1)^{0.5} = 2^{0.5} = 1.41 \text{ or } 41\%$$

Slow Reaction.   Here the specific reaction rate constant is small with respect to the mass transfer coefficient:

Mass transfer effects are not important when the reaction rate is limiting

$$k_r \ll k_c$$

$$-r_{As}'' = \frac{k_r C_A}{1 + k_r/k_c} \approx k_r C_A \qquad (11\text{-}50)$$

The specific reaction rate is independent of the velocity of fluid and for the solid sphere considered here, independent of particle size. *However*, for porous catalyst pellets, k_r may depend on particle size for certain situations, as shown in Chapter 12.

For the present case, Figure 11-6 shows the variation in reaction rate with particle size and velocity past the particle. At low velocities the mass transfer boundary layer thickness is large and diffusion limits the reaction. As the velocity past the sphere is increased, the boundary layer thickness decreases and the mass transfer across the boundary layer no longer limits the rate of reaction. One also notes that for a given velocity, reaction-limiting conditions can be achieved by using very small particles. However, the smaller the particle size, the greater the pressure drop in a packed bed. See Problem P4-21. When one is obtaining reaction rate data in the laboratory, one must operate at sufficiently high velocities or sufficiently small particle sizes to ensure that the reaction is not mass transfer–limited.

When collecting
rate law data,
operate in the
reaction-limited
region

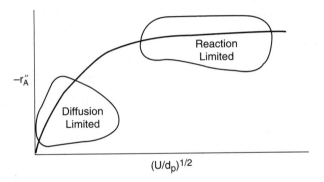

Figure 11-6 Regions of mass transfer–limited and reaction-limited reactions.

11.3.3 Mass Transfer–Limited Reactions in Packed Beds

A number of industrial reactions are potentially mass transfer–limited because they may be carried out at high temperatures without the occurrence of undesirable side reactions. In mass transfer–dominated reactions, the surface reaction is so rapid that the rate of transfer of reactant from the bulk gas or liquid phase to the surface limits the overall rate of reaction. Consequently, mass transfer–limited reactions respond quite differently to changes in temperature and flow conditions than do the rate-limited reactions discussed in previous chapters. In this section the basic equations describing the variation of conversion with the various reactor design parameters (catalyst weight, flow conditions) will be developed. To achieve this goal, we begin by carrying out a mole balance on the following mass transfer–limited reaction:

$$A + \frac{b}{a} B \longrightarrow \frac{c}{a} C + \frac{d}{a} D \qquad (2\text{-}2)$$

carried out in a packed-bed reactor (Figure 11-7). A steady-state mole balance on reactant A in the reactor segment between z and $z + \Delta z$ is

$$\begin{bmatrix} \text{molar} \\ \text{rate in} \end{bmatrix} - \begin{bmatrix} \text{molar} \\ \text{rate out} \end{bmatrix} + \begin{bmatrix} \text{molar rate of} \\ \text{generation} \end{bmatrix} = \begin{bmatrix} \text{molar rate of} \\ \text{accumulation} \end{bmatrix}$$

$$\left. F_{Az}\right|_z \quad - \quad \left. F_{Az}\right|_{z+\Delta z} \quad + \quad r''_A a_c (A_c \Delta z) \quad = \quad 0 \qquad (11\text{-}51)$$

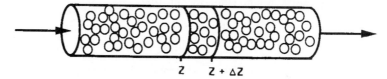

Figure 11-7   Packed-bed reactor.

where r''_A = rate of generation of A per unit catalytic surface area, mol/s·m²
$\quad a_c$ = external surface area of catalyst per volume of catalytic bed, m²/m³
$\quad\quad$ = $6(1 - \phi)/d_p$ for packed beds, m²/m³
$\quad \phi$ = porosity of the bed (i.e., porosity)[11]
$\quad d_p$ = particle diameter, m
$\quad A_c$ = cross-sectional area of tube containing the catalyst, m²

Dividing Equation (11-51) by $A_c \Delta z$ and taking the limit as $\Delta z \longrightarrow 0$, we have

$$-\frac{1}{A_c}\left(\frac{dF_{Az}}{dz}\right) + r''_A a_c = 0 \qquad (11\text{-}52)$$

We now need to express F_{Az} and r''_A in terms of concentration.

The molar flow rate of A in the axial direction is

$$F_{Az} = A_c W_{Az} = (J_{Az} + B_{Az})A_c \qquad (11\text{-}53)$$

Axial diffusion is neglected

In almost all situations involving flow in packed-bed reactors, the amount of material transported by diffusion or dispersion in the axial direction is negligible compared with that transported by convection (i.e., bulk flow):

$$J_{Az} \ll B_{Az}$$

(In Chapter 14 we consider the case where this assumption is not justified and dispersive effects must be taken into account.) From Equation (11-20), we have

$$F_{Az} = A_c W_{Az} = A_c B_{Az} = U C_A A_c \qquad (11\text{-}54)$$

where U is the superficial molar average velocity through the bed (m/s). Substituting for F_{Az} in Equation (11-52) gives us

$$-\frac{d(C_A U)}{dz} + r''_A a_c = 0 \qquad (11\text{-}55)$$

---

[11]In the nomenclature for Chapter 4, for Ergun Equation for pressure drop.

For the case of constant superficial velocity U,

$$-U\frac{dC_A}{dz} + r''_A a_c = 0 \qquad (11\text{-}56)$$

For reactions at steady state, the molar flux of A to the particle surface, W_{Ar} (mol/m²·s) (see Figure 11-8), is equal to the rate of disappearance of A on the surface $-r''_A$ (mol/m²·s); that is,

$$-r''_A = W_{Ar} \qquad (11\text{-}57)$$

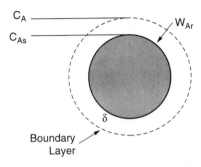

Figure 11-8   Diffusion across stagnant film surrounding catalyst pellet.

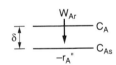

From Table 11-1, the boundary condition at the external surface is

$$-r''_A = W_{Ar} = k_c(C_A - C_{As}) \qquad (11\text{-}58)$$

where k_c = mass transfer coefficient = D_{AB}/δ
C_A = bulk concentration
C_{As} = concentration of A at the catalytic surface

Substituting for r''_A in Equation (11-56), we have

$$-U\frac{dC_A}{dz} - k_c a_c(C_A - C_{As}) = 0 \qquad (11\text{-}59)$$

In most mass transfer–limited reactions, the surface concentration is negligible with respect to the bulk concentration (i.e., $C_A \gg C_{As}$):

$$-U\frac{dC_A}{dz} = k_c a_c C_A \qquad (11\text{-}60)$$

Integrating with the limit, at $z = 0$, $C_A = C_{A0}$:

$$\frac{C_A}{C_{A0}} = \exp\left(-\frac{k_c a_c}{U} z\right) \qquad (11\text{-}61)$$

The corresponding variation of reaction rate along the length of the reactor is

$$-r''_A = k_c C_{A0} \exp\left(-\frac{k_c a_c}{U} z\right) \tag{11-62}$$

The concentration profile down a reactor of length L is shown in Figure 11-9.

Reactor concentration profile for a mass transfer–limited reaction

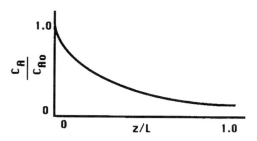

Figure 11-9   Axial concentration profile in a packed bed.

To determine the reactor length L necessary to achieve a conversion X, we combine the definition of conversion,

$$X = \frac{C_{A0} - C_{AL}}{C_{A0}} \tag{11-63}$$

with the evaluation of Equation (11-61) at $z = L$ to obtain

$$\boxed{\ln \frac{1}{1 - X} = \frac{k_c a_c}{U} L} \tag{11-64}$$

To learn the effect of flow rate and temperature on conversion, we need to know how these parameters affect the mass transfer coefficient. That is, we must determine the correlation for the mass transfer coefficient for the particular geometry and flow field. For flow through a packed bed, the correlation given by Thoenes and Kramers[12] for $0.25 < \phi < 0.5$, $40 < Re' < 4000$, and $1 < Sc < 4000$ is

$$Sh' = 1.0(Re')^{1/2} Sc^{1/3} \tag{11-65}$$

Thoenes–Kramers correlation for flow through packed beds

$$\left[\frac{k_c d_p}{D_{AB}}\left(\frac{\phi}{1 - \phi}\right)\frac{1}{\gamma}\right] = \left[\frac{U d_p \rho}{\mu (1 - \phi)\gamma}\right]^{1/2}\left(\frac{\mu}{\rho D_{AB}}\right)^{1/3} \tag{11-66}$$

where $Re' = \dfrac{Re}{(1 - \phi)\gamma}$

$\qquad Sh' = \dfrac{Sh\,\phi}{(1 - \phi)\gamma}$

---

[12]D. Thoenes, Jr. and H. Kramers, *Chem. Eng. Sci.*, **8**, 271 (1958).

d_p = particle diameter (equivalent diameter of sphere of the same volume), m

  = $[(6/\pi) \text{ (volume of pellet)}]^{1/3}$, m

ϕ = void fraction of packed bed

γ = shape factor (external surface area divided by πd_p^2)

U = superficial gas velocity through the bed, m/s

μ = viscosity, kg/m·s

ρ = fluid density, kg/m^3

$$\nu = \frac{\mu}{\rho} = \text{kinematic viscosity, m}^2/\text{s}$$

D_{AB} = gas-phase diffusivity, m^2/s

For constant fluid properties and particle diameter:

$$k_c \propto U^{1/2} \qquad (11\text{-}67)$$

We see that the mass transfer coefficient increases with the square root of the superficial velocity through the bed. Therefore, *for a fixed concentration, C_A,* such as that found in a differential reactor, the rate of reaction should vary with $U^{1/2}$:

For diffusion-limited reactions, reaction rate depends on particle size and fluid velocity

$$-r_A'' \propto k_c C_A \propto U^{1/2}$$

However, if the gas velocity is continually increased, a point is reached where the reaction becomes reaction rate–limited and consequently, is independent of the superficial gas velocity, as shown in Figure 11-8.

Most mass transfer correlations in the literature are reported in terms of the Colburn *J* factor as a function of the Reynolds number. The relationship between J_D and the numbers we have been discussing is

Colburn *J* factor

$$J_D = \frac{\text{Sh}}{\text{Sc}^{1/3}\,\text{Re}} \qquad (11\text{-}68)$$

Figure 11-10 shows data from a number of investigations for the *J* factor as a function of the Reynolds number for a wide range of particle shapes and gas flow conditions.

Dwidevi and Upadhyay[13] review a number of mass transfer correlations for both fixed and fluidized beds and arrive at the following correlation, which is valid for both gases (Re > 10) and liquids (Re > 0.01) in either fixed or fluidized beds:

A correlation for flow through packed beds in terms of the Colburn *J* factor

$$\phi J_D = \frac{0.765}{\text{Re}^{0.82}} + \frac{0.365}{\text{Re}^{0.386}} \qquad (11\text{-}69)$$

For nonspherical particles, the equivalent diameter used in the Reynolds and Sherwood numbers is $d_p = \sqrt{A_p/\pi} = 0.564\,\sqrt{A_p}$, where A_p is the external surface area of the pellet.

---

[13]P. N. Dwidevi and S. N. Upadhyay, *Ind. Eng. Chem. Process Des. Dev., 16,* 157 (1977).

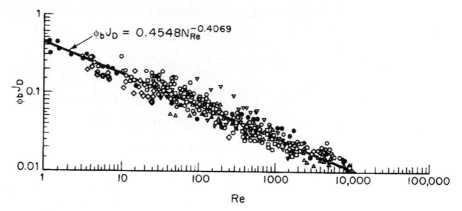

Figure 11-10   Mass transfer correlation for packed beds. [Reprinted with permission from P. N. Dwidevi and S. S. Upadhyay, *Ind. Eng. Chem. Process Des. Dev.*, *16*, 157 (1977). Copyright © 1977 American Chemical Society.]

To obtain correlations for mass transfer coefficients for a variety of systems and geometries, see either D. Kunii and O. Levenspiel, *Fluidization Engineering* (Huntington, N.Y.: Krieger, 1977), Chap. 7, or W. L. McCabe and J. C. Smith, *Unit Operations in Chemical Engineering*, 3rd ed. (New York: McGraw-Hill, 1976). For other correlations for packed beds with different packing arrangements, see I. Colquhoun-Lee and J. Stepanek, *Chemical Engineer*, February 1974, p. 108.

Example 11–3   Maneuvering a Space Satellite

Hydrazine has been studied extensively for use in monopropellant thrusters for space flights of long duration. Thrusters are used for altitude control of communication satellites. Here the decomposition of hydrazine over a packed bed of alumina-supported iridium catalyst is of interest.[14] In a proposed study, a 2% hydrazine in 98% helium mixture is to be passed over a packed bed of cylindrical particles 0.25 cm in diameter and 0.5 cm in length at a gas-phase velocity of 15 m/s and a temperature of 750 K. The kinematic viscosity of helium at this temperature is 4.5×10^{-4} m²/s. The hydrazine decomposition reaction is believed to be externally mass transfer–limited under these conditions. If the packed bed is 0.05 m in length, what conversion can be expected? Assume isothermal operation.

Additional information:

$D_{AB} = 0.69 \times 10^{-4}$ m²/s at 298 K
Bed porosity: 30%
Bed fluidicity: 95.7%

Solution

Rearranging Equation (11-64) gives us

$$X = 1 - e^{-(k_c a_c / U)L} \qquad \text{(E11-3.1)}$$

---

[14]O. I. Smith and W. C. Solomon, *Ind. Eng. Chem. Fund.*, *21*, 374 (1982).

A. Thoenes–Kramers correlation

1. First we find the volume-average particle diameter:

$$d_p = \left(\frac{6V}{\pi}\right)^{1/3} = \left(6\,\frac{\pi D^2}{4}\,\frac{L}{\pi}\right)^{1/3}$$

$$= [1.5(0.0025 \text{ m})^2 (0.005 \text{ m})]^{1/3}$$

$$= 3.61 \times 10^{-3} \text{ m}$$

(E11-3.2)

2. Surface area per volume of bed:

$$a_c = 6\left(\frac{1-\phi}{d_p}\right) = 6\left(\frac{1-0.3}{3.61 \times 10^{-3} \text{ m}}\right) = 1163 \text{ m}^2/\text{m}^3 \qquad \text{(E11-3.3)}$$

3. Mass transfer coefficient:

$$\text{Re} = \frac{d_p U}{\nu} = \frac{(3.61 \times 10^{-3} \text{ m})(15 \text{ m/s})}{4.5 \times 10^{-4} \text{ m}^2/\text{s}} = 120.3$$

$$\gamma = \frac{2\pi r L_p + 2\pi r^2}{\pi d_p^2} = \frac{(2)(0.0025/2)(0.005) + (2)(0.0025/2)^2}{(3.61 \times 10^{-3})^2} = 1.20 \quad \text{(E11-3.4)}$$

$$\text{Re}' = \frac{\text{Re}}{(1-\phi)\gamma} = \frac{120.3}{(0.7)(1.2)} = 143.2$$

Gas Phase

Re′ = 143

Sc = 1.3

Sh′ = 13.0

k_c = 3.5 m/s

Correcting the diffusivity to 750 K using Table 11-2 gives us

$$D_{AB}(750 \text{ K}) = D_{AB}(298 \text{ K}) \times \left(\frac{750}{298}\right)^{1.75} = (0.69 \times 10^{-4} \text{ m}^2/\text{s})(5.03)$$

$$= 3.47 \times 10^{-4} \text{ m}^2/\text{s} \qquad \text{(E11-3.5)}$$

$$\text{Sc} = \frac{\nu}{D_{AB}} = \frac{4.5 \times 10^{-4} \text{ m}^2/\text{s}}{3.47 \times 10^{-4} \text{ m}^2/\text{s}} = 1.30$$

Substituting Re′ and Sc into Equation (11-66) yields

$$\text{Sh}' = (143.2)^{1/2}(1.3)^{1/3} = (11.97)(1.09) = 13.05 \qquad \text{(E11-3.6)}$$

$$k_c = \frac{D_{AB}(1-\phi)}{d_p \phi}\,\gamma\,(\text{Sh}') = \left(\frac{3.47 \times 10^{-4} \text{ m}^2/\text{s}}{3.61 \times 10^{-3} \text{ m}}\right)\left(\frac{1-0.3}{0.3}\right)$$

$$\times (1.2)(13.05) = 3.52 \text{ m/s} \quad \text{(E11-3.7)}$$

The conversion is

$$X = 1 - \exp\left[-(3.52 \text{ m/s})\left(\frac{1163 \text{ m}^2/\text{m}^3}{15 \text{ m/s}}\right)(0.05 \text{ m})\right] \qquad \text{(E11-3.8)}$$

$$= 1 - 1.18 \times 10^{-6} \simeq 1.00 \qquad \text{virtually complete conversion}$$

B. **Colburn J_D factor.** Calculate the surface-area-average particle diameter. For cylindrical pellets the external surface area is

$$A = \pi dL_p + 2\pi \left(\frac{d^2}{4}\right) \tag{E11-3.9}$$

$$d_p = \sqrt{\frac{A}{\pi}} = \sqrt{\frac{\pi dL_p + 2\pi(d^2/4)}{\pi}} \tag{E11-3.10}$$

$$= \sqrt{(0.0025)(0.005) + \frac{(0.0025)^2}{2}} = 3.95 \times 10^{-3} \text{ m}$$

$$a = \frac{6(1-\phi)}{d_p} = 1063 \text{ m}^2/\text{m}^3$$

Gas Phase

Re = 130

$J_D = 0.23$

Sc = 130

Sh = 33

$k_c = 3$ m/s

$$\text{Re} = \frac{d_p U}{\nu} = \frac{(3.95 \times 10^{-3} \text{ m})(15 \text{ m/s})}{4.5 \times 10^{-4} \text{ m}^2/\text{s}}$$

$$= 131.6$$

$$\phi J_D = \frac{0.765}{\text{Re}^{0.82}} + \frac{0.365}{\text{Re}^{0.386}} \tag{11-69}$$

$$= \frac{0.765}{(131.6)^{0.82}} + \frac{0.365}{(131.6)^{0.386}} = 0.014 + 0.055 \tag{E11-3.11}$$

$$= 0.069$$

$$J_D = \frac{0.069}{0.3} = 0.23 \tag{E11-3.12}$$

$$\text{Sh} = \text{Sc}^{1/3}\text{Re}(J_D) \tag{E11-3.13}$$

$$= (1.3)^{1/3}(131.6)(0.23) = 33.0$$

$$k_c = \frac{D_{AB}}{d_p} \text{Sh} = \frac{3.47 \times 10^{-4}}{3.95 \times 10^{-3}} (33) = 2.9 \text{ m/s}$$

Then

$$X = 1 - \exp\left[-(2.9 \text{ m/s})\left(\frac{1063 \text{ m}^2/\text{m}^3}{15 \text{ m/s}}\right)(0.05 \text{ m})\right] \tag{E11-3.14}$$

$$= 1 - 0.0000345$$

$$\simeq 1 \qquad \text{again, virtually complete conversion}$$

If there were such a thing as the **bed fluidity**, given in the problem statement, it would be a useless piece of information. Make sure that you know what information you need to solve problems, and go after it. Do not let additional data confuse you or lead you astray with useless information or facts that represent someone else's bias, which are probably not well-founded. (See CD-ROM)

11.3.4 Mass Transfer–Limited Reaction on Metallic Surfaces

In this section we develop the design equations and give the mass transfer correlations for two common types of catalytic reactors: the wire screen or catalyst gauze reactor and the monolith reactor.

Catalyst Monolith.   The previous discussion in this chapter focused primarily on chemical reactions taking place in packed-bed reactors. However, when a gaseous feedstream contains significant amounts of particulate matter, dust tends to clog the catalyst bed. To process feedstreams of this type, parallel-plate reactors (monoliths) are commonly used. Figure 11-11 shows a schematic diagram of a monolith reactor. The reacting gas mixture flows between the parallel plates, and the reaction takes place on the surface of the plates.

Monoliths are used
as automobile
catalytic converters

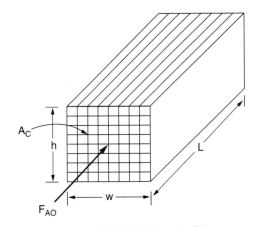

Figure 11-11   Catalyst monolith.

The rate of surface reaction is equal to mass flux to the surface. Taking the surface concentration equal to zero for mass transfer–limited reactions gives

$$-r''_A = k_c(C_A - 0) \tag{11-70}$$

Combining with the mole balance, we have

$$\frac{dF_A}{dV} = r''_A a_m = -k_c a_m C_A \tag{11-71}$$

where a_m is the catalytic surface area per unit volume. Correlations for a_m and the mass transfer coefficient, k_c, for monoliths can be found in the CD-ROM. Once these values are known, Equation (11-71) can be solved with a procedure similar to that in Example 10-3. One can then calculate the reactor volume necessary to achieve a specified conversion.

Wire Gauzes.   Wire gauzes are commonly used in the oxidation of ammonia and hydrocarbons. A gauze is a series of wire screens stacked one on top of another (Figure 11-12). The wire is typically made out of platinum or a platinum–rhodium alloy. The wire diameter ranges between 0.004 and 0.01 cm.

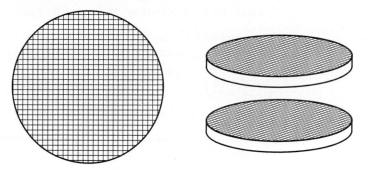

Figure 11-12   Wire gauzes.

As a first approximation, one can assume plug flow through the gauze, in which case the design equation is similar to that for monolith reactors,

Differential form of
the wire gauze
design equation

$$\frac{dF_A}{dV} = -k_c a_g C_A \qquad (11\text{-}72)$$

where a_g is the total screen surface area per total volume of one screen, m^2/m^3 or in^2/in^3. One can use the mass transfer correlations for wire gauzes found in the CD-ROM to solve Equation (11-72) to design a wire gauze reactor to achieve a specified conversion.

11.4 What If . . . ? (Parameter Sensitivity)

One of the most important skills of an engineer is to be able to predict the effects of changes of system variables on the operation of a process. The engineer needs to determine these effects quickly through approximate but reasonably close calculations, which are sometimes referred to as "back-of-the-envelope calculations." This type of calculation is used to answer such questions as "**What** will happen **if** I decrease the particle size?" "**What if** I triple the flow rate through the reactor?"

We will now proceed to show how this type of question can be answered using the packed-bed, mass transfer–limited reactors as a model or example system. Here we want to learn the effect of changes of the various parameters (e.g., temperature, particle size, superficial velocity) on the conversion. We begin with a rearrangement of the mass transfer correlation, Equation (11-49), to yield

$$k_c \propto \frac{D_{AB}^{2/3}}{\nu^{1/6}} \left(\frac{U^{1/2}}{d_p^{1/2}} \right) \qquad (11\text{-}73)$$

Find out how the
mass transfer
coefficient varies
with changes in
physical properties
and system
properties

The first term on the right-hand side is dependent on physical properties (temperature and pressure), whereas the second term is dependent on system properties (flow rate and particle size). One observes from this equation that the mass transfer coefficient increases as the particle size decreases. The use of sufficiently small particles offers another technique to escape from the mass transfer–limited regime into the reaction rate–limited regime.

Example 11–4   The Case of Divide and Be Conquered

A mass transfer–limited reaction is being carried out in two reactors of equal volume and packing, connected in series as shown in Figure E11-4.1. Currently, 86.5% conversion is being achieved with this arrangement. It is suggested that the reactors be separated and the flow rate be divided equally among each of the two reactors (Figure E11-4.2) to decrease the pressure drop and hence the pumping requirements. In terms of achieving a higher conversion, was this a good idea?

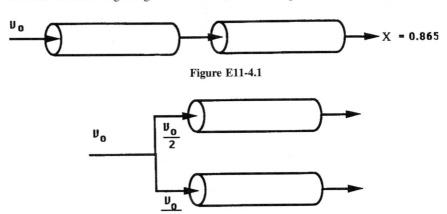

Figure E11-4.1

Reactors in series versus reactors in parallel

Figure E11-4.2   Different arrangements of 2 packed bed reactors.

Solution

As a first approximation, we neglect the effects of small changes in temperature and pressure on mass transfer. We recall Equation (11-64), which gives conversion as a function of reactor length. For a mass transfer–limited reaction

$$\ln \frac{1}{1-X} = \frac{k_c a_c}{U} L \tag{11-64}$$

For case 1, the undivided system:

$$\left(\ln \frac{1}{1-X_1}\right) = \frac{k_{c1} a_c}{U_1} L_1 \tag{E11-4.1}$$

$$X_1 = 0.865$$

For case 2, the divided system:

$$\left(\ln \frac{1}{1-X_2}\right) = \frac{k_{c2} a_c}{U_2} L_2 \tag{E11-4.2}$$

$$X_2 = ?$$

We now take the ratio of case 2 (divided system) to case 1 (undivided system):

$$\frac{\ln \dfrac{1}{1-X_2}}{\ln \dfrac{1}{1-X_1}} = \frac{k_{c2}}{k_{c1}} \left(\frac{L_2}{L_1}\right) \frac{U_1}{U_2} \tag{E11-4.3}$$

The surface area per unit volume a is the same for both systems.

From the conditions of the problem statement we know that

$$L_2 = \tfrac{1}{2}L_1$$

$$U_2 = \tfrac{1}{2}U_1$$

$$X_1 = 0.865$$

$$X_2 = ?$$

However, we must also consider the effect of the division on the mass transfer coefficient. From Equation (11-73) we know that

$$k_c \propto U^{1/2}$$

Then

$$\frac{k_{c2}}{k_{c1}} = \left(\frac{U_2}{U_1}\right)^{1/2} \tag{E11-4.4}$$

Multiplying by the ratio of superficial velocities yields

$$\frac{U_1}{U_2}\left(\frac{k_{c2}}{k_{c1}}\right) = \left(\frac{U_1}{U_2}\right)^{1/2} \tag{E11-4.5}$$

$$\ln\frac{1}{1-X_2} = \left(\ln\frac{1}{1-X_1}\right)\frac{L_2}{L_1}\left(\frac{U_1}{U_2}\right)^{1/2} \tag{E11-4.6}$$

$$= \left(\ln\frac{1}{1-0.865}\right)\left[\frac{\tfrac{1}{2}L_1}{L_1}\left(\frac{U_1}{\tfrac{1}{2}U_1}\right)^{1/2}\right]$$

$$= 2.00\left(\frac{1}{2}\right)\sqrt{2} = 1.414$$

Solving for X_2 gives us

$$X_2 = 0.76$$

Consequently, we see that although the divided arrangement will have the advantage of a smaller pressure drop across the bed, it is a bad idea in terms of product yield. Recall that if the reaction were reaction rate limited both arrangements would give the same conversion.

Example 11–5   The Case of the Overenthusiastic Engineers

The same reaction as that in Example 11-4 is being carried out in the same two reactors in series. A new engineer suggests that the rate of reaction could be increased by a factor of 2^{10} by increasing the reaction temperature from 400°C to 500°C, reasoning that the reaction rate doubles for every 10°C increase in temperature. Another engineer arrives on the scene and berates the new engineer with quotations from Chapter 3 concerning this rule of thumb. She points out that it is valid only for a specific activation energy within a specific temperature range. She then suggests that he go ahead with the proposed temperature increase but should only expect an increase on the order of 2^3 or 2^4. What do you think? Who is correct?

Solution

Because almost all surface reaction rates increase more rapidly with temperature than do diffusion rates, increasing the temperature will only increase the degree to which the reaction is mass transfer–limited.

We now consider the following two cases:

Case 1: $T = 400°C$ $X = 0.865$

Case 2: $T = 500°C$ $X = ?$

Taking the ratio of case 2 to case 1 and noting that the reactor length is the same for both cases ($L_1 = L_2$), we obtain

$$\frac{\ln \dfrac{1}{1 - X_2}}{\ln \dfrac{1}{1 - X_1}} = \frac{k_{c2}}{k_{c1}} \left(\frac{L_2}{L_1}\right)\frac{U_1}{U_2} = \frac{k_{c2}}{k_{c1}} \left(\frac{U_1}{U_2}\right) \tag{E11-5.1}$$

The molar feed rate F_{T0} remains unchanged:

$$F_{T0} = v_{01}\left(\frac{P_{01}}{RT_{01}}\right) = v_{02}\left(\frac{P_{02}}{RT_{02}}\right) \tag{E11-5.2}$$

Because $v = A_c U$, the superficial velocity at temperature T_2 is

$$U_2 = \frac{T_2}{T_1} U_1 \tag{E11-5.3}$$

We now wish to learn the dependence of the mass transfer coefficient on temperature:

$$k_c \propto \frac{U^{1/2}}{d_p^{1/2}}\left(\frac{D_{AB}^{2/3}}{v^{1/6}}\right) \tag{E11-5.4}$$

Taking the ratio of case 2 to case 1 and realizing that the particle diameter is the same for both cases gives us

$$\frac{k_{c2}}{k_{c1}} = \left(\frac{U_2}{U_1}\right)^{1/2}\left(\frac{D_{AB2}}{D_{AB1}}\right)^{2/3}\left(\frac{v_1}{v_2}\right)^{1/6} \tag{E11-5.5}$$

The temperature dependence of the gas-phase diffusivity is (from Table 11-2)

$$D_{AB} \propto T^{1.75} \tag{E11-5.6}$$

For most gases, viscosity increases with increasing temperature according to the relation

$$\mu \propto T^{1/2}$$

From the ideal gas law,

$$\rho \propto T^{-1}$$

Then

$$v = \frac{\mu}{\rho} \propto T^{3/2} \tag{E11-5.7}$$

$$\frac{\ln\dfrac{1}{1-X_2}}{\ln\dfrac{1}{1-X_1}} = \frac{U_1}{U_2}\left(\frac{k_{c2}}{k_{c1}}\right) = \left(\frac{U_1}{U_2}\right)^{1/2}\left(\frac{D_{AB2}}{D_{AB1}}\right)^{2/3}\left(\frac{\nu_1}{\nu_2}\right)^{1/6} \tag{E11-5.8}$$

$$= \left(\frac{T_1}{T_2}\right)^{1/2}\left[\left(\frac{T_2}{T_1}\right)^{1.75}\right]^{2/3}\left[\left(\frac{T_1}{T_2}\right)^{3/2}\right]^{1/6}$$

$$= \left(\frac{T_1}{T_2}\right)^{1/2}\left(\frac{T_2}{T_1}\right)^{7/6}\left(\frac{T_1}{T_2}\right)^{1/4} = \left(\frac{T_2}{T_1}\right)^{5/12} \tag{E11-5.9}$$

$$= \left(\frac{773}{673}\right)^{5/12} = 1.059$$

$$\ln\frac{1}{1-X_1} = \ln\frac{1}{1-0.865} = 2$$

$$\ln\frac{1}{1-X_2} = 1.059\left(\ln\frac{1}{1-X_1}\right) = 1.059(2) \tag{E11-5.10}$$

$$X_2 = 0.88$$

Consequently, we see that increasing the temperature from 400°C to 500°C increases the conversion by only 1.7%. Both engineers would have benefited from a more thorough study of this chapter.

For a packed catalyst bed, the temperature-dependence part of the mass transfer coefficient for a gas-phase reaction can be written as

$$k_c \propto U^{1/2}(D_{AB}^{2/3}/\nu^{1/6}) \tag{11-74}$$

$$k_c \propto U^{1/2}T^{11/12} \tag{11-75}$$

Important concept

Depending on how one fixes or changes the molar feed rate, F_{T0}, U may also depend on the feed temperature. *As an engineer, it is extremely important that you reason out the effects of changing conditions*, as illustrated in the preceding two examples.

11.5 The Shrinking Core Model

The shrinking core model is used to describe situations in which solid particles are being consumed either by dissolution or reaction and, as a result, the amount of the material being consumed is "shrinking." This model applies to areas ranging from pharmacokinetics (e.g., dissolution of pills in the stomach) to the formation of an ash layer around a burning coal particle, to catalyst regeneration. In this section we focus primarily on catalyst regeneration and leave other applications as exercises at the end of the chapter.

11.5.1 Catalyst Regeneration

Many situations arise in heterogeneous reactions where a gas-phase reactant reacts with a species contained in an inert solid matrix. One of the most common examples is the removal of carbon from catalyst particles that have been deactivated by fouling (see Section 10.7.1). Catalyst regeneration units used to reactivate the catalyst by burning off the carbon are shown in Figures 11-13 and 11-14. Figure 11-13 shows a schematic diagram of the removal of carbon from a single porous catalyst pellet as a function of time. Carbon is first removed from the outer edge of the pellet and then in the final stages of the regeneration from the center core of the pellet.

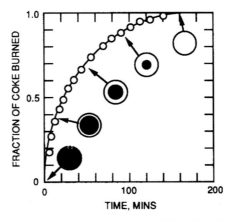

Progressive regeneration of fouled pellet

Figure 11-13 Shell progressive regeneration of fouled pellet. [Reprinted with permission from J. T. Richardson, *Ind. Eng. Chem. Process Des. Dev.*, *11*(1), 8 (1972); copyright American Chemical Society.]

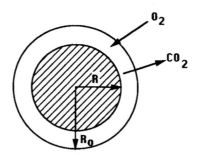

Figure 11-14 Partially regenerated catalyst pellet.

As the carbon continues to be removed from the porous catalyst pellet, the reactant gas must diffuse farther into the material as the reaction proceeds to reach the unreacted solid phase. Note that approximately 3 hours was required to remove all of the carbon from the pellets at these conditions. The regeneration time can be reduced by increasing the gas-phase oxygen concentration and temperature.

To illustrate the principles of the shrinking core model, we shall consider the removal of carbon from the catalyst particle just discussed. In Figure 11-15 a core of unreacted carbon is contained between $r = 0$ and $r = R$. Carbon has been removed from the porous matrix between $r = R$ and $r = R_o$. Oxygen diffuses from the outer radius R_o to the radius R, where it reacts with carbon to form carbon dioxide, which then diffuses out of the porous matrix. The reaction

Use steady-state profiles QSSA

$$C + O_2 \longrightarrow CO_2$$

at the solid surface is very rapid, so the rate of oxygen diffusion to the surface controls the rate of carbon removal from the core. Although the core of carbon is shrinking with time (an unsteady-state process), we assume the concentration profiles at any instant in time to be the steady-state profiles over the distance $(R_o - R)$. This assumption is referred to as the *quasi-steady state assumption* (QSSA).

Oxygen must diffuse through the porous pellet matrix until it reaches the unreacted carbon core

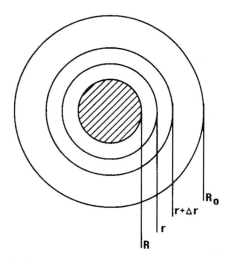

Figure 11-15   Sphere with unreacted carbon core of radius R.

To study how the radius of unreacted carbon changes with time we must first find the rate of diffusion of oxygen to the carbon surface. Next, we perform a mole balance on the elemental carbon and equate the rate of consumption of carbon to the rate of diffusion of oxygen to the gas carbon interface.

In applying a differential oxygen mole balance over the increment Δr located somewhere between R_o and R, we recognize that O_2 does not react in this region, and reacts only when it reaches the solid carbon interface located at $r = R$. We shall let species A represent O_2.

Step 1: The mole balance on O_2 (i.e., A) between r and $r + \Delta r$ is

$$\begin{bmatrix} \text{rate} \\ \text{in} \end{bmatrix} - \begin{bmatrix} \text{rate} \\ \text{out} \end{bmatrix} + \begin{bmatrix} \text{rate of} \\ \text{generation} \end{bmatrix} = \begin{bmatrix} \text{rate of} \\ \text{accumulation} \end{bmatrix}$$

$$W_{Ar} 4\pi r^2 \big|_r - W_{Ar} 4\pi r^2 \big|_{r+\Delta r} + \qquad 0 \qquad = \qquad 0$$

Dividing through by $-4\pi\Delta r$ and taking the limit gives

Mole balance on
oxygen

$$\lim_{\Delta r \to 0} \frac{W_{Ar}r^2|_{r+\Delta r} - W_{Ar}r^2|_r}{\Delta r} = \frac{d(W_{Ar}r^2)}{dr} = 0 \qquad (11\text{-}76)$$

Step 2: For every mole of O_2 that diffuses into the spherical pellet, 1 mol of CO_2 diffuses out ($W_{CO_2} = -W_{O_2}$), that is, EMCD. The constitutive equation for constant total concentration becomes

$$W_{Ar} = -D_e \frac{dC_A}{dr} \qquad (11\text{-}77)$$

where D_e is an *effective diffusivity* in the porous catalyst. In Chapter 12 we present an expanded discussion of effective diffusivities in a porous catalyst [cf. Equation (12-1)].

Step 3: Combining Equations (11-76) and (11-77) yields

$$\frac{d}{dr}\left(-D_e \frac{dC_A}{dr} r^2\right) = 0$$

Dividing by $-D_e$ gives

$$\boxed{\frac{d}{dr}\left(r^2 \frac{dC_A}{dr}\right) = 0} \qquad (11\text{-}78)$$

Step 4: The boundary conditions are:
At the outer surface of the particle, $r = R_o$: $C_A = C_{Ao}$
At the fresh carbon/gas interface, $r = R(t)$: $C_A = 0$
(rapid reaction)

Step 5: Integrating twice yields

$$r^2 \frac{dC_A}{dr} = K_1$$

$$C_A = \frac{-K_1}{r} + K_2$$

Using the boundary conditions to eliminate K_1 and K_2, the concentration profile is given by

Concentration
profile at a
given time, t
(i.e., radius, r)

$$\frac{C_A}{C_{Ao}} = \frac{1/R - 1/r}{1/R - 1/R_o} \qquad (11\text{-}79)$$

A schematic representation of the profile of O_2 is shown in Figure 11-16 at a time when the inner core is receded to a radius R. The zero on the r axis corresponds to the center of the sphere.

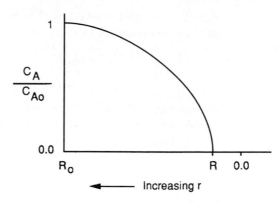

Figure 11-16   Oxygen concentration profile shown from the external radius of the pellet (R_o) to the pellet center. The gas–carbon interface is located at R.

Step 6: The molar flux of O_2 to the gas–carbon interface is

$$W_{Ar} = -D_e \frac{dC_A}{dr} = \frac{-D_e C_{Ao}}{(1/R - 1/R_o)\, r^2} \qquad (11\text{-}80)$$

Step 7: We now carry out an overall balance on elemental carbon. Elemental carbon does not enter or leave the particle.

$$\left[\begin{array}{c}\text{rate} \\ \text{in}\end{array}\right] - \left[\begin{array}{c}\text{rate} \\ \text{out}\end{array}\right] + \left[\begin{array}{c}\text{rate of} \\ \text{generation}\end{array}\right] = \left[\begin{array}{c}\text{rate of} \\ \text{accumulation}\end{array}\right]$$

Mole balance on shrinking core

$$0 \quad - \quad 0 \quad + \quad r_C'' 4\pi R^2 \quad = \quad \frac{d\left(\frac{4}{3}\pi R^3 \rho_C \phi_C\right)}{dt}$$

where ρ_C is the molar density of the carbon and ϕ_C is the volume fraction of carbon in the porous catalyst. Simplifying gives

$$\frac{dR}{dt} = \frac{r_C''}{\phi_C \rho_C} \qquad (11\text{-}81)$$

Step 8: The rate of disappearance of carbon is equal to the flux of O_2 to the gas–carbon interface:

$$-r_C'' = -W_{Ar}\bigg|_{r=R} = \frac{D_e C_{Ao}}{R - R^2/R_o} \qquad (11\text{-}82)$$

The minus sign arises with respect to W_{Ar} in Equation (11-82) because O_2 is diffusing in an inward direction [i.e., opposite to the increasing coordinate (r) direction]:

$$-\frac{dR}{dt} = \frac{D_e C_{Ao}}{\phi_C \rho_C}\left(\frac{1}{R - R^2/R_o}\right)$$

Step 9: Integrating with limits $R = R_0$ at $t = 0$, the time necessary for the solid carbon interface to recede inward to a radius R is

$$t = \frac{\rho_C R_0^2 \phi_C}{6 D_e C_{Ao}} \left[1 - 3 \left(\frac{R}{R_o} \right)^2 + 2 \left(\frac{R}{R_o} \right)^3 \right] \qquad (11\text{-}83)$$

We see that as the reaction proceeds, the reacting gas–solid moves closer to the center of the core. The corresponding oxygen concentration profiles at three different times are shown in Figure 11-17.

Concentration profiles at different times at inner core radii

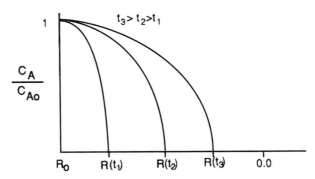

Figure 11-17   Oxygen concentration profile at various times. At t_1, the gas–carbon interface is located at $R(t_1)$; at t_2 it is located at $R(t_2)$.

The time necessary to consume all the carbon in the catalyst pellet is

Time to complete regeneration of the particle

$$t_c = \frac{\rho_C R_0^2 \phi_C}{6 D_e C_{Ao}} \qquad (11\text{-}84)$$

For a 1-cm diameter pellet with 4 volume fraction of carbon 0.04 the regeneration time is the order of 10s.

Variations on the simple system we have chosen here can be found on page 360 of Levenspiel[15] and in the problems at the end of this chapter.

11.5.2 Dissolution of Monodispersed Solid Particles

We now consider the case where the total particle is being completely consumed. We choose as an example the case where species A must diffuse to the surface to react with solid B at the liquid–solid interface. Reactions of this type are typically zero-order in B and first-order in A. The rate of mass transfer to the surface is equal to the rate of surface reaction.

$$W_{Ar} = k_c(C_A - C_{As}) = -r''_{As} = k_r C_{As}$$

$$\text{(diffusion)} \qquad\qquad \text{(surface reaction)}$$

---

[15]O. Levenspiel, *Chemical Reactor Engineering*, 2nd ed., Wiley, New York, 1972.

Eliminating C_{As}, we arrive at an equation identical to Equation (11-46):

$$W_{Ar} = -r''_{As} = \frac{k_c k_r}{k_c + k_r} C_A \qquad (11\text{-}46)$$

For the case of small particles and negligible shear stress at the fluid boundary, the Frössling equation, Equation (11-40), is approximated by

$$Sh = 2$$

or

$$k_c = \frac{2D_e}{D} \qquad (11\text{-}85)$$

where D is the diameter of the dissolving particle. Substituting Equation (11-85) into (11-46) and rearranging yields

Diameter at which mass transfer and reaction rate resistances are equal is D^*

$$-r''_{As} = \frac{k_r C_A}{1 + k_r/k_c} = \frac{k_r C_A}{1 + k_r D/2D_e} = \frac{k_r C_A}{1 + D/D^*} \qquad (11\text{-}86)$$

where $D^* = 2D_e/k_r$ is the diameter at which the resistances to mass transfer and reaction rate are equal.

$$\boxed{\begin{array}{ll} D > D^* & \text{mass transfer controls} \\ D < D^* & \text{reaction rate controls} \end{array}}$$

A mole balance on the solid particle yields

$$\text{in} - \text{out} + \text{generation} = \text{accumulation}$$

Balance on the dissolving solid B

$$0 - 0 + \quad r''_{Bs} \pi D^2 = \frac{d(\rho \pi D^3/6)}{dt}$$

where ρ is the molar density of species B. If 1 mol of A dissolves 1 mol of B, then $-r''_{As} = -r''_{Bs}$ and after differentiation and rearrangement we obtain

$$\frac{dD}{dt} = -\left[\frac{2(-r''_{As})}{\rho}\right] = -\frac{2k_r C_A}{\rho}\left(\frac{1}{1 + D/D^*}\right)$$

$$\boxed{\frac{dD}{dt} = -\frac{\alpha}{1 + D/D^*}} \qquad (11\text{-}87)$$

where

$$\alpha = \frac{2k_r C_A}{\rho}$$

At time $t = 0$, the initial diameter is $D = D_i$. Integrating Equation (11-87) for the case of excess concentration of reactant A, we obtain the following diameter–time relationship:

Excess A

$$D_i - D + \frac{1}{2D^*}(D_i^2 - D^2) = \alpha t \tag{11-88}$$

The time to complete dissolution of the solid particle is

$$t_c = \frac{1}{\alpha}\left(D_i + \frac{D_i^2}{2D^*}\right) \tag{11-89}$$

11.5.3 Flow and Dissolution in Porous Media

The concepts in the shrinking core model can also be applied to situations in which there is growth rather than shrinking. One example of this growth is the dissolution of pores in oil-bearing reservoirs to increase the oil flow out of the reservoir (recall Example 5-3). To model and understand this process, laboratory experiments are carried out by injecting HCl through calcium carbonate (i.e., limestone) cores. The carbonate core, the pore network, and the individual pores are shown in Figure 11-18. As acid flows through and dissolves the carbonate pore space, the pore radius increases so that the resistance to flow decreases, resulting in more acid flowing into the pore. Because there is a distribution of pore sizes, some pores will receive more acid than others. This uneven distribution results in even a greater dissolution rate of the pores receiving the most acid. This "autocatalytic-like" growth rate results in the emergence of a dominant channel, called a *wormhole*, that will be formed in the porous media.

Dissolution of limestone pore space by acid

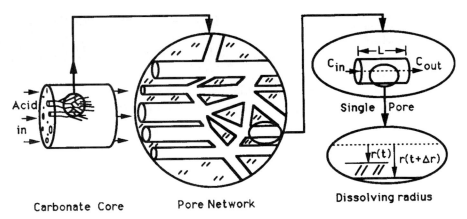

Figure 11-18   Network model of dissolving carbonate core.

The stoichiometric coefficient, $\vartheta_{c/a}$, times the moles of acid consumed in the pore in a time Δt is equal to the moles of material dissolved:

$$\vartheta_{c/a}[U\pi r^2(C_{Ain} - C_{Aout})\Delta t] = \rho_m 2\pi r L \Delta r \tag{11-90}$$

where $\vartheta_{c/a}$ is the moles of carbonate dissolved per moles of acid consumed, ρ_m is the carbonate molar density, and U is the fluid velocity.

The relationship between the inlet and outlet concentrations to the pore is analogous to Equation (11-61):

$$\frac{C_{Aout}}{C_{Ain}} = \exp\left(-\frac{2k_c}{rU}\,L\right) \tag{11-91}$$

We now combine Equations (11-90) and (11-91) with a mole balance on the acid to obtain the pore radius as a function of time.

Application of the mole balance to a specific pore in the network gives the radius of the pore as a function of time in increments of Δt:

Increase in pore radius

$$r_i(t + \Delta t) = r_i(t) + \frac{\vartheta_{c/a} Q_i C_{Ain}}{2\pi r_i L_i \rho_m}\left\{1 - \exp\left[-a\left(\frac{D_e L}{Q_i}\right)^{2/3}\right]\right\}\Delta t \tag{11-92}$$

where the subscript i refers to the ith pore, D_e is the effective diffusivity, $a = 1.75\pi$, and Q_i is the volumetric flow rate through the ith pore.

These equations can be coupled with the flow distribution through the porous media to yield the rate and pathway of the channel formation through the porous media. This channel can be visualized by filling the acidized carbonate pore space with molten woods metal, letting it solidify, and then imagining the etch channel by neutron radiography,[16] which is called a wormhole. A typical wormhole is shown in Figure 11-19a and the corresponding network simulation is shown in Figure 11-19b.

Wormholes etched by acid

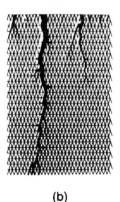

(a)   (b)

Figure 11-19   (a) Wood's alloy castings of a 0.127-m-length core showing the pathway that acid etched through the core. (b) Simulation results showing channeling in the network for the transport-limited case.

---

[16]H. S. Fogler and J. Jasti, *AIChE J.*, *36*(6), 827 (1990). See also M. L. Hoefner and H. S. Fogler, *AIChE J.*, *34*, 45 (1988), and S. D. Rege and H. S. Fogler, *AIChE J.*, *35*, 1177 (1989). C. Fredd and H. S. Fogler, *Soc. Petr. Engrg. J.*, *13*, p. 33 (1998). C. Fredd and H. S. Fogler, *AIChE J.* 44 p1933 (1998).

SUMMARY

1. The molar flux of A in a binary mixture of A and B is

$$\mathbf{W}_A = -cD_{AB}\nabla y_A + y_A(\mathbf{W}_A + \mathbf{W}_B) \qquad (S11\text{-}1)$$

   a. For equimolar counterdiffusion (EMCD) or for dilute concentration of the solute,

$$\mathbf{W}_A = -cD_{AB}\nabla y_A \qquad (S11\text{-}2)$$

   b. For diffusion through a stagnant gas,

$$\mathbf{W}_A = cD_{AB}\nabla \ln(1 - y_A) \qquad (S11\text{-}3)$$

   c. For negligible diffusion,

$$\mathbf{W}_A = y_A\mathbf{W} = C_A\mathbf{V} \qquad (S11\text{-}4)$$

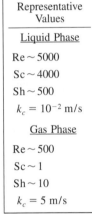

Representative
Values

Liquid Phase

$\text{Re} \sim 5000$

$\text{Sc} \sim 4000$

$\text{Sh} \sim 500$

$k_c = 10^{-2}\,\text{m/s}$

Gas Phase

$\text{Re} \sim 500$

$\text{Sc} \sim 1$

$\text{Sh} \sim 10$

$k_c = 5\,\text{m/s}$

2. The rate of mass transfer from the bulk fluid to a boundary at concentration C_{As} is

$$W_A = k_c(C_{Ab} - C_{As}) \qquad (S11\text{-}5)$$

   where k_c is the mass transfer coefficient.

3. The Sherwood and Schmidt numbers are, respectively,

$$\text{Sh} = \frac{k_c d_p}{D_{AB}} \qquad (S11\text{-}6)$$

$$\text{Sc} = \frac{\nu}{D_{AB}} \qquad (S11\text{-}7)$$

4. If a heat transfer correlation exists for a given system and geometry, the mass transfer correlation may be found by replacing the Nusselt number by the Sherwood number and the Prandtl number by the Schmidt number in the existing heat transfer correlation.

5. Increasing the gas-phase velocity and decreasing the particle size will increase the overall rate of reaction for reactions that are externally mass transfer–limited.

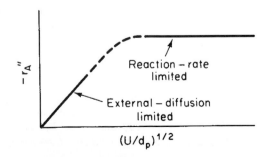

6. The conversion for externally mass transfer–limited reactions can be found from the equation

$$\ln \frac{1}{1-X} = \frac{k_c a}{U} L \qquad (S11\text{-}8)$$

7. Back-of-the-envelope calculations should be carried out to determine the magnitude and direction that changes in process variables will have on conversion. **What if . . .?**

8. The shrinking core model states that the time to regenerate a coked catalyst particle is

$$t_c = \frac{\rho_C R_0^2 \phi}{6 D_e C_{Ao}} \qquad (S11\text{-}9)$$

QUESTIONS AND PROBLEMS

The subscript to each of the problem numbers indicates the level of difficulty: A, least difficult; D, most difficult.

$$A = \bullet \quad B = \blacksquare \quad C = \blacklozenge \quad D = \blacklozenge\blacklozenge$$

In each of the questions and problems below, rather than just drawing a box around your answer, write a sentence or two describing how you solved the problem, the assumptions you made, the reasonableness of your answer, what you learned, and any other facts that you want to include. You may wish to refer to W. Strunk and E. B. White, *The Elements of Style* (New York: Macmillan, 1979) and Joseph M. Williams, *Style: Ten Lessons in Clarity & Grace* (Glenview, Ill.: Scott, Foresman, 1989) to enhance the quality of your sentences.

P11-1$_A$ Read over the problems at the end of this chapter. Make up an original problem that uses the concepts presented in this chapter. See Problem 4-1 for the guidelines. To obtain a solution:
   (a) Make up your data and reaction.
   (b) Use a real reaction and real data.
   The journals listed at the end of Chapter 1 may be useful for part (b).

P11-2$_A$ What if...
   (a) you were asked to rework Example 11-1 for the mass transfer–limited reaction

$$A \longrightarrow 2B \,?$$

   What would your concentration (mole fraction) profile look like? Using the same values for D_{AB}, and so on, in Example 11-1, what is the flux of A?
   (b) you were asked to rework Example 11-2 for the case when the reaction

$$A \longrightarrow 2B$$

   occurs at the surface? What would be the flux to the surface?
   (c) you were asked to rework Example 11-3 assuming a 50–50 mixture of hydrazine and helium? How would your answers change?
   (d) you were growing solid spheres rather than dissolving them? Could you modify Equations (11-80) and (11-83) so that they would predict the radius of the particle as a function of time?

(e) you were asked for representative values for Re, Sc, Sh, and k_c for both liquid- and gas-phase systems for a velocity of 10 cm/s and a pipe diameter of 5 cm (or a packed-bed diameter of 0.2 cm)? What numbers would you give?

P11-3$_B$ Pure oxygen is being absorbed by xylene in a catalyzed reaction in the experimental apparatus sketched in Figure P11-3. Under constant conditions of temperature and liquid composition the following data were obtained:

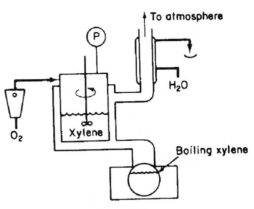

Figure P11-3

	Rate of Uptake of O_2 (mL/h) for System Pressure (absolute)			
Stirrer Speed (rpm)	1.2 atm	1.6 atm	2.0 atm	3.0 atm
400	15	31	75	152
800	20	59	102	205
1200	21	62	105	208
1600	21	61	106	207

No gaseous products were formed by the chemical reaction. What would you conclude about the relative importance of liquid-phase diffusion and about the order of the kinetics of this reaction? (**California Professional Engineers Exam**)

P11-4$_B$ The decomposition of cyclohexane to benzene and hydrogen is mass transfer–limited at high temperatures. The reaction is carried out in a 5-cm-ID pipe 20 m in length packed with cylindrical pellets 0.5 cm in diameter and 0.5 cm in length. The pellets are coated with the catalyst only on the outside. The bed porosity is 40%. The entering volumetric flow rate is 60 dm³/min.

(a) Calculate the number of pipes necessary to achieve 99.9% conversion of cyclohexane from an entering gas stream of 5% cyclohexane and 95% H_2 at 2 atm and 500°C.

(b) Plot conversion as a function of length.

(c) How much would your answer change if the pellet diameter and length were each cut in half?

(d) How would your answer to part (a) change if the feed were pure cyclohexane?

(e) What do you believe is the point of this problem?

P11-5$_B$ A plant is removing a trace of Cl_2 from a waste gas stream by passing it over a solid granular absorbent in a tubular packed bed (Figure P11-5). At present, 63.2% removal is being accomplished, but it is believed that greater removal could be achieved if the flow rate were increased by a factor of 4, the particle diameter were decreased by a factor of 3, and the packed tube length increased by 50%. What percentage of chlorine would be removed under the scheme proposed? (The chlorine transferring to the absorbent is removed completely by a virtually instantaneous chemical reaction.) (*Ans.: 98%.*) What guidelines (T, v, C_A) would you propose for efficient or optimum operation of this bed?

Cl_2 v_0 L

Figure P11-5

P11-6$_B$ In a certain chemical plant, a reversible fluid-phase isomerization

$$A \mathrel{\mathop{\rightleftharpoons}} B$$

is carried out over a solid catalyst in a tubular packed-bed reactor. If the reaction is so rapid that mass transfer between the catalyst surface and the bulk fluid is rate-limiting, show that the kinetics are described in terms of the bulk concentrations C_A and C_B by

$$-r_A'' = \frac{k_B[C_A - (1/K)C_B]}{1/K + k_B/k_A}$$

where $-r_A''$ = moles of A reacting per unit area catalyst per unit time
k_A, k_B = mass transfer coefficients for A and B
K = reaction equilibrium constant

It is desired to double the capacity of the existing plant by processing twice the feed of reactant A while maintaining the same fractional conversion of A to B in the reactor. How much larger a reactor, in terms of catalyst weight, would be required if all other operating variables are held constant? You may use the Thoenes–Kramers correlation for mass transfer coefficients in a packed bed. Describe the effects of the flow rate, temperature, particle size at conversion.

P11-7$_B$ The oxidation of ammonia is to be carried out over platinum gauze. The molar flow rate of ammonia is 10 mol/min at a temperature of 500 K and a pressure of 202.6 kPa. Is one 250-mesh screen 10 cm in diameter sufficient to achieve 60% conversion? The wire diameter is 0.044 mm. Assume 25% excess air, and ignore the effects of volume changes on the Reynolds number.

P11-8$_C$ In a diving-chamber experiment, a human subject breathed a mixture of O_2 and He while small areas of his skin were exposed to nitrogen gas. After awhile the exposed areas became blotchy, with small blisters forming on the skin. Model the skin as consisting of two adjacent layers, one of thickness δ_1 and the other of δ_2. If counterdiffusion of He out through the skin occurs at the same time as N_2 diffuses into the skin, at what point in the skin layers is the sum of the partial pressures a maximum? If the saturation partial pressure for the sum of the gases is 101 kPa, can the blisters be a result of the sum of the gas partial pressures exceeding the saturation partial pressure and the gas coming out of the solution (i.e., the skin)?

Before answering any of the questions above, derive the concentration profiles for N_2 and He in the skin layers.

Diffusivity of He and N_2 in the inner skin layer

$$= 5 \times 10^{-7} \text{ cm}^2/\text{s and } 1.5 \times 10^{-7} \text{ cm}^2/\text{s, respectively}$$

Diffusivity of He and N_2 in the outer skin layer

$$= 10^{-5} \text{ cm}^2/\text{s and } 3.3 \times 10^{-4} \text{ cm}^2/\text{s, respectively}$$

	External *Skin Boundary* *Partial Pressure*	*Internal* *Skin Boundary* *Partial Pressure*
N_2	101 kPa	0
He	0	81 kPa
δ_1	20 μm	Strotum corneum
δ_2	80 μm	Epidermis

P11-9$_A$ A spherical particle is dissolving in a liquid. The rate of dissolution is first-order in the solvent concentration, C. Assuming that the solvent is in excess, show that the following conversion time relationships hold.

Rate-Limiting Regime	Conversion Time Relationship
Surface reaction	$1 - (1 - X)^{1/3} = \dfrac{\alpha t}{D_i}$
Mass transfer	$\dfrac{D_i}{2D^*}[1 - (1 - X)^{2/3}] = \dfrac{\alpha t}{D_i}$
Mixed	$[1 - (1 - X)^{1/3}] + \dfrac{D_i}{2D^*}[1 - (1 - X)^{2/3}] = \dfrac{\alpha t}{D_i}$

P11-10$_C$ A powder is to be completely dissolved in an aqueous solution in a large, well-mixed tank. An acid must be added to the solution to render the spherical particle soluble. The particles are sufficiently small that they are unaffected by fluid velocity effects in the tank. For the case of excess acid, $C_0 = 2\ M$, derive an equation for the diameter of the particle as a function of time when

(a) Mass transfer limits the dissolution: $-\mathbf{W}_A = k_c C_{A0}$

(b) Reaction limits the dissolution: $-r''_A = k_r C_{A0}$

What is the time for complete dissolution in each case?

(c) Now assume that the acid is not in excess and that mass transfer is limiting the dissolution. One mole of acid is required to dissolve 1 mol of solid. The molar concentration of acid is 0.1 M, the tank is 100 L, and 9.8 mol of solid is added to the tank at time $t = 0$. Derive an expression for the radius of the particles as a function of time and calculate the time for the particles to dissolve completely.

(d) How could you make the powder dissolve faster? Slower?

Additional information:

$$D_e = 10^{-10} \text{ m}^2/\text{s}, \qquad k = 10^{-18}/\text{s}$$

$$\text{initial diameter} = 10^{-5} \text{ m}$$

P11-11$_B$ The irreversible gas-phase reaction

$$A \xrightarrow{\text{catalyst}} B$$

is carried out adiabatically over a packed bed of solid catalyst particles. The reaction is first-order in the concentration of A on the catalyst surface:

$$-r'_{As} = k'C_{As}$$

The feed consists of 50% (mole) A and 50% inerts and enters the bed at a temperature of 300 K. The entering volumetric flow rate is 10 dm³/s (i.e., 10,000 cm³/s). The relationship between the Sherwood number and the Reynolds number is

$$\text{Sh} = 100\ \text{Re}^{1/2}$$

As a first approximation, one may neglect pressure drop. The entering concentration of A is 1.0 *M*. Calculate the catalyst weight necessary to achieve 60% conversion of A for

(a) Isothermal operation.

(b) Adiabatic operation.

(c) What generalizations can you make after comparing parts **(a)** and **(b)**?

Additional information:

Kinematic viscosity: $\mu/\rho = 0.02$ cm²/s
Particle diameter: $d_p = 0.1$ cm
Superficial velocity: $U = 10$ cm/s
Catalyst surface area/mass of catalyst bed: $a = 60$ cm²/g cat.
Diffusivity of A: $D_e = 10^{-2}$ cm²/s
Heat of reaction: $\Delta H_{Rx} = -10,000$ cal/g mol A
Heat capacities:

$$C_{pA} = C_{pB} = 25 \text{ cal/g mol} \cdot \text{K}$$

$$C_{pS} \text{ (solvent)} = 75 \text{ cal/g mol} \cdot \text{K}$$

$$k' \text{ (300 K)} = 0.01 \text{ cm}^3/\text{s} \cdot \text{g cat with } E = 4000 \text{ cal/mol}$$

P11-12$_C$ (*Pills*) An antibiotic drug is contained in a solid inner core and is surrounded by an outer coating that makes it palatable. The outer coating and the drug are dissolved at different rates in the stomach, owing to their differences in equilibrium solubilities.

(a) If $D_2 = 4$ mm and $D_1 = 3$ mm, calculate the time necessary for the pill to dissolve completely.

(b) Assuming first-order kinetics ($k_A = 10$ h^{-1}) for the absorption of the dissolved drug (i.e., in solution in the stomach) into the bloodstream, plot the concentration in grams of the drug in the blood per gram of body weight as a function of time when the following three pills are taken simultaneously:

Pill 1: $D_2 = 5$ mm, $D_1 = 3$ mm

Pill 2: $D_2 = 4$ mm, $D_1 = 3$ mm

Pill 3: $D_2 = 3.5$ mm, $D_1 = 3$ mm

(c) Discuss how you would maintain the drug level in the blood at a constant level using different-size pills?

(d) How could you arrange a distribution of pill sizes so that the concentration in the blood was constant over a period (e.g., 3 hr) of time?

Additional information:

Amount of drug in inner core = 500 mg
Solubility of outer layer at stomach conditions = 1.0 kg/cm^3
Solubility of inner layer at stomach conditions = 0.4 kg/cm^3
Volume of fluid in stomach = 1.2 L
Typical body weight = 75 kg
Sh = 2., D_{AB} = 6×10^{-4} cm^2/min

P11-13$_B$ (*Seargeant Ambercromby*) Capt. Apollo is piloting a shuttle craft on his way to space station Klingon. Just as he is about to maneuver to dock his craft using the hydrazine system discussed in Example 11-3, the shuttle craft thrusters do not respond properly and it crashes into the station, killing Capt. Apollo. An investigation revealed that Lt. Darkside prepared the packed beds used to maneuver the shuttle and Lt. Data prepared the hydrazine–helium gas mixture. Foul play is suspected and Sgt. Ambercromby arrives on the scene to investigate.

(a) What are the first three questions he asks?

(b) Make a list of possible explanations for the crash, supporting each one by an equation or reason.

P11-14$_B$ If disposal of industrial liquid wastes by incineration is to be a feasible process, it is important that the toxic chemicals be completely decomposed into harmless substances. One study recently carried out concerned the atomization and burning of a liquid stream of "principal" organic hazardous constituents (POHCs) [*Environ. Prog.*, **8**, 152 (1989)]. The following data give the burning droplet diameter as a function of time (both diameter and time are given in arbitrary units):

Time	20	40	50	70	90	110
Diameter	9.7	8.8	8.4	7.1	5.6	4.0

What can you learn from these data?

P11-15$_B$ (*Estimating glacial ages*) The following oxygen-18 data were obtained from soil samples taken at different depths in Ontario, Canada. Assuming that all the ^{18}O was laid down during the last glacial age and that the transport of ^{18}O to the surface takes place by molecular diffusion, estimate the number of years since the last glacial age from the data below. Independent measurements give the diffusivity of ^{18}O in soil as 2.64×10^{-10} m^2/s.

Depth (m)	(surface) 0	3	6	9	12	18
^{18}O Conc. Ratio (C/C_0)	0	0.35	0.65	0.83	0.94	1.0

C_0 is the concentration of ^{18}O at 25 m.

JOURNAL ARTICLE PROBLEM

P11J-1   After reading the article "Designing gas-sparged vessels for mass transfer" [*Chem. Eng.*, *89*(24), p. 61 (1982)], design a gas-sparged vessel to saturate 0.6 m³/s of water up to an oxygen content of 4×10^{-3} kg/m³ at 20°C. A liquid holding time of 80 s is required.

JOURNAL CRITIQUE PROBLEMS

P11C-1   The decomposition of nitric oxide on a heated platinum wire is discussed in *Chem. Eng. Sci.*, *30*, 781 (1975). After making some assumptions about the density and the temperatures of the wire and atmosphere, and using a correlation for convective heat transfer, determine if mass transfer limitations are a problem in this reaction.

P11C-2   Given the proposed rate equation on page 296 of the article in *Ind. Eng. Chem. Process Des. Dev.*, *19*, 294 (1980), determine whether or not the concentration dependence on sulfur, C_s, is really second-order. Also, determine if the intrinsic kinetic rate constant, K_{2p}, is indeed only a function of temperature and partial pressure of oxygen and not of some other variables as well.

P11C-3   Read through the article on the oxidation kinetics of oil shale char in *Ind. Eng. Chem. Process Des. Dev.*, *18*, 661 (1979). Are the units for the mass transfer coefficient k_m in Equation (6) consistent with the rate law? Is the mass transfer coefficient dependent on sample size? Would the shrinking core model fit the authors' data as well as the model proposed by the authors?

CD-ROM MATERIAL

- **Learning Resources**
    1. Summary Notes for Lectures 27 and 28
    4. Solved Problems
        A. Example CD11–1 Calculating Steady State Fluxes
        B. Example CD11–2 Diffusion Through a Stagnant Gas
        C. Example CD11–3 Relating Fluxes W_A, B_A, and J_A

- **Professional Reference Shelf**
    1. Mass Transfer Limited Reactions on Metallic Surfaces
    - A. Catalyst Monoliths (Catalytic converter for autos)
    - B. Wire Gauze's
- **Additional Homework Problems**

CDP11-A$_A$ An isomerization reaction that follows Langmuir–Hinshelwood kinetics is carried out on monolith catalyst. [2nd Ed. P10-11]

CDP11-B$_B$ A parameter sensitivity analysis is required for this problem in which an isomerization is carried out over a 20-mesh gauze screen. [2nd Ed. P10-12]

CDP11-C$_C$ This problem examines the effect on temperature in a catalyst monolith. [2nd Ed. P10-13]

CDP11-D$_D$ A second-order catalytic reaction is carried out in a catalyst monolith. [2nd Ed. P10-14]

CDP11-E$_C$ Fracture acidizing is a technique to increase the productivity of oil wells. Here acid is injected at high pressures to fracture the rock and form a channel that extends out from the well bore. As the acid flows through the channel it etches the sides of the channel to make it larger and thus less resistant to the flow of oil. Derive an equation for the concentration profile of acid and the channel width as a function of distance from the well bore.

CDP11-F$_C$ The solid–gas reaction of silicon to form SiO_2 is an important process in microelectronics fabrication. The oxidation occurs at the Si–SiO_2 interface. Derive an equation for the thickness of the SiO_2 layer as a function of time. [2nd Ed. P10-17]

CDP11-G$_B$ Mass transfer limitations in CVD processing to product material with ferroelectric and piezoelectric properties. [2nd Ed. P10-17]

CDP11-H$_B$ Calculate multicomponent diffusivities. [2nd Ed. P10-9]

CDP11-I$_B$ Application of the shrinking core model to FeS_2 rock samples in acid mine drainage. [2nd Ed. P10-18]

SUPPLEMENTARY READING

1. The fundamentals of diffusional mass transfer can be found in

> BIRD, R. B., W. E. STEWART, and E. N. LIGHTFOOT, *Transport Phenomena*. New York: Wiley, 1960, Chaps. 16 and 17.

> CUSSLER, E. L., *Diffusion Mass Transfer in Fluid Systems*, 2nd ed. New York: Cambridge University Press, 1997.

> FAHIEN, R. W., *Fundamentals of Transport Phenomena*. New York: McGraw-Hill, 1983, Chap. 7.

> GEANKOPLIS, C. J., *Mass Transport Phenomena*. New York: Holt, Rinehart and Winston, 1972, Chaps. 1 and 2.

> HINES, A. L., and R. N. MADDOX, *Mass Transfer: Fundamentals and Applications*. Upper Saddle River, N.J.: Prentice Hall, 1984.

> LEVICH, V. G., *Physiochemical Hydrodynamics*. Upper Saddle River, N.J.: Prentice Hall, 1962, Chaps. 1 and 4.

2. Equations for predicting gas diffusivities are given in Appendix D. Experimental values of the diffusivity can be found in a number of sources, two of which are

PERRY, R. H., D. W. GREEN, and J. O. MALONEY, *Chemical Engineers' Handbook*, 6th ed. New York: McGraw-Hill, 1984.

SHERWOOD, T. K., R. L. PIGFORD, and C. R. WILKE, *Mass Transfer.* New York: McGraw-Hill, 1975.

3. A number of correlations for the mass transfer coefficient can be found in

LYDERSEN, A. L., *Mass Transfer in Engineering Practice.* New York: Wiley-Interscience, 1983, Chap. 1.

MCCABE, W. L., J. C. SMITH, and P. HARRIOTT, *Unit Operations of Chemical Engineering.* New York: McGraw-Hill, 1985, Chap. 21.

TREYBAL, R. E., *Mass Transfer Operations*, 3rd ed. New York: McGraw-Hill, 1980.

Diffusion 12
and Reaction
in Porous Catalysts

Research is to see what everybody else sees, and
to think what nobody else has thought.

Albert Szent-Gyorgyi

In our discussion of surface reactions in Chapter 11 we assumed that each point in the interior of the entire catalyst surface was *accessible* to the same reactant concentration. However, where the reactants diffuse into the pores within the catalyst pellet, the concentration at the pore mouth will be higher than that inside the pore, and we see that the entire catalytic surface is *not accessible* to the same concentration. To account for variations in concentration throughout the pellet, we introduce a parameter known as the effectiveness factor. In this chapter we will develop models for diffusion and reaction in two-phase systems, which include catalyst pellets and CVD reactors. The types of reactors discussed in this chapter will include packed beds, bubbling fluidized beds, slurry reactors, and trickle beds. After studying this chapter you will be able to describe diffusion and reaction in two- and three-phase systems, determine when internal pore diffusion limits the overall rate of reaction, describe how to go about eliminating this limitation, and develop models for systems in which both diffusion and reaction play a role (e.g., CVD).

> The concentration in the internal surface of the pellet is less than that of the external surface

In a heterogeneous reaction sequence, mass transfer of reactants first takes place from the bulk fluid to the external surface of the pellet. The reactants then diffuse from the external surface into and through the pores within the pellet, with reaction taking place only on the catalytic surface of the pores. A schematic representation of this two-step diffusion process is shown in Figures 10-3 and 12-1.

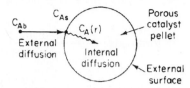

Figure 12-1   Mass transfer and reaction steps for a catalyst pellet.

12.1 Diffusion and Reaction in Spherical Catalyst Pellets

In this section we will develop the internal effectiveness factor for spherical catalyst pellets. The development of models that treat individual pores and pellets of different shapes is undertaken in the problems at the end of this chapter. We will first look at the internal mass transfer resistance to either the products or reactants that occurs between the external pellet surface and the interior of the pellet. To illustrate the salient principles of this model, we consider the irreversible isomerization

$$A \longrightarrow B$$

that occurs on the surface of the pore walls within the spherical pellet of radius R.

12.1.1 Effective Diffusivity

The pores in the pellet are not straight and cylindrical; rather, they are a series of tortuous, interconnecting paths of pore bodies and pore throats with varying cross-sectional areas. It would not be fruitful to describe diffusion within each and every one of the tortuous pathways individually; consequently, we shall define an effective diffusion coefficient so as to describe the average diffusion taking place at any position r in the pellet. We shall consider only radial variations in the concentration; the radial flux W_{Ar} will be based on the total area (voids and solid) normal to diffusion transport (i.e., $4\pi r^2$) rather than void area alone. This basis for W_{Ar} is made possible by proper definition of the effective diffusivity D_e.

The effective diffusivity accounts for the fact that:

1. Not all of the area normal to the direction of the flux is available (i.e., void) for the molecules to diffuse.
2. The paths are tortuous.
3. The pores are of varying cross-sectional areas.

An equation that relates D_e to either the bulk or the Knudsen diffusivity is

The effective diffusivity

$$D_e = \frac{D_A \phi_p \sigma}{\tilde{\tau}} \tag{12-1}$$

where

$$\tilde{\tau} = \text{tortuosity}[1] = \frac{\text{actual distance a molecule travels between two points}}{\text{shortest distance between those two points}}$$

$$\phi_p = \text{pellet porosity} = \frac{\text{volume of void space}}{\text{total volume (voids and solids)}}$$

$\sigma = $ constriction factor

The constriction factor accounts for the variation in the cross-sectional area that is normal to diffusion.[2] It is a function of the ratio of maximum to minimum pore areas (Figure 12-2a). When the two areas, A_1 and A_2, are equal, the constriction factor is unity, and when $\beta = 10$, the constriction factor is approximately 0.5.

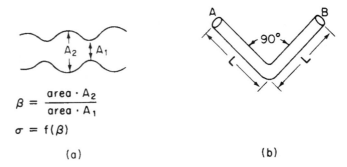

$$\beta = \frac{\text{area} \cdot A_2}{\text{area} \cdot A_1}$$

$$\sigma = f(\beta)$$

(a)   (b)

Figure 12-2   (a) Pore constriction; (b) pore tortuosity.

Example 12–1   Finding the Tortuosity

Calculate the tortuosity for the hypothetical pore of length, L (Figure 12-2b), from the definition of $\tilde{\tau}$.

Solution

$$\tilde{\tau} = \frac{\text{actual distance molecule travels from } A \text{ to } B}{\text{shortest distance between } A \text{ and } B}$$

The shortest distance between points A and B is $2L/\sqrt{2}$. The actual distance the molecule travels from A to B is $2L$.

---

[1] Some investigators lump constriction and tortuosity into one factor, called the tortuosity factor, and set it equal to $\tilde{\tau}/\sigma$. C. N. Satterfield, *Mass Transfer in Heterogeneous Catalysis* (Cambridge, Mass.: MIT Press, 1970), pp. 33–47, has an excellent discussion on this point.

[2] See E. E. Petersen, *Chemical Reaction Analysis*, Prentice Hall, Upper Saddle River, N.J., 1965, Chap. 3; C. N. Satterfield and T. K. Sherwood, *The Role of Diffusion in Catalysis*, Addison-Wesley, Reading, Mass., 1963, Chap. 1.

$$\tilde{\tau} = \frac{2L}{2L/\sqrt{2}} = \sqrt{2} = 1.414$$

Although this value is reasonable for $\tilde{\tau}$, values for $\tilde{\tau} = 6$ to 10 are not unknown. Typical values of the constriction factor, the tortuosity, and the pellet porosity are, respectively, $\sigma = 0.8$, $\tilde{\tau} = 3.0$, and $\phi_p = 0.40$.

12.1.2 Derivation of the Differential Equation Describing Diffusion and Reaction

We now perform a steady-state mole balance on species A as it enters, leaves, and reacts in a spherical shell of inner radius r and outer radius $r + \Delta r$ of the pellet (Figure 12-3). Note that even though A is diffusing inward toward the center of the pellet, the convention of our shell balance dictates that the flux be in the direction of increasing r. We choose the flux of A to be positive in the direction of increasing r (i.e., the outward direction). Because A is actually diffusing inward, the flux of A will have some negative value, such as $-10 \text{ mol/m}^2 \cdot \text{s}$, indicating that the flux is actually in the direction of decreasing r.

First we will derive the concentration profile of reactant A in the pellet

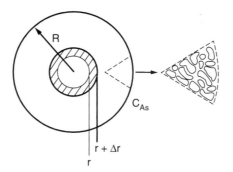

Figure 12-3   Shell balance on a catalyst pellet.

We now proceed to perform our shell balance on A. The area that appears in the balance equation is the total area (voids and solids) *normal* to the direction of the molar flux:

$$\text{rate of A in at } r = W_{Ar} \cdot \text{area} = W_{Ar} \times 4\pi r^2 \big|_r \tag{12-2}$$

$$\text{rate of A out at } (r + \Delta r) = W_{Ar} \cdot \text{area} = W_{Ar} \times 4\pi r^2 \big|_{r+\Delta r} \tag{12-3}$$

$$\begin{bmatrix} \text{rate of} \\ \text{generation} \\ \text{of A within a} \\ \text{shell thickness} \\ \text{of } \Delta r \end{bmatrix} = \begin{bmatrix} \text{rate of reaction} \\ \hline \text{mass of catalyst} \end{bmatrix} \times \begin{bmatrix} \text{mass catalyst} \\ \hline \text{volume} \end{bmatrix} \times \begin{bmatrix} \text{volume of shell} \end{bmatrix}$$

$$= \qquad r'_A \qquad \times \qquad \rho_c \qquad \times \qquad 4\pi r_m^2 \, \Delta r \tag{12-4}$$

Mole balance for diffusion and reaction inside the catalyst pellet

where r_m is some mean radius between r and $r + \Delta r$ that is used to approximate the volume ΔV of the shell.

The mole balance over the shell thickness Δr is

$$\text{(in at } r) \quad - \quad \text{(out at } r + \Delta r) \quad + \text{(generation within } \Delta r) = 0$$
$$(W_{Ar} \times 4\pi r^2 |_r) - (W_{Ar} \times 4\pi r^2 |_{r+\Delta r}) + (r_A' \times \rho_c \times 4\pi r_m^2 \Delta r) = 0 \qquad (12\text{-}5)$$

After dividing by $(-4\pi \Delta r)$ and taking the limit as $\Delta r \longrightarrow 0$, we obtain the following differential equation:

$$\frac{d(W_{Ar}r^2)}{dr} - r_A' \rho_c r^2 = 0 \qquad (12\text{-}6)$$

Because 1 mol of A reacts under conditions of constant temperature and pressure to form 1 mol of B, we have EMCD at constant total concentration (Section 11.2.1), or

The flux equation

$$W_{Ar} = -cD_e \frac{dy_A}{dr} = -D_e \frac{dC_A}{dr} \qquad (12\text{-}7)$$

where C_A is the number of moles of A per dm^3 of open pore volume (i.e., volume of gas) as opposed to (mol/vol of gas and solids). In systems where we do not have EMCD in catalyst pores, it may still be possible to use Equation (12-7) if the reactant gases are present in dilute concentrations.

After substituting Equation (12-7) into Equation (12-6) we arrive at the following differential equation describing diffusion with reaction in a catalyst pellet:

$$\frac{d[-D_e(dC_A/dr)r^2]}{dr} - r^2 \rho_c r_A' = 0 \qquad (12\text{-}8)$$

We now need to incorporate the rate law. In the past we have based the rate of reaction in terms of either per unit volume,

$$-r_A \; [=] \; (\text{mol/dm}^3 \cdot \text{s})$$

Inside the Pellet

$-r_A = \rho_c(-r_A')$

$-r_A' = S_a(-r_A'')$

$-r_A = \rho_c S_a(-r_A'')$

or per unit mass of catalyst,

$$-r_A' \; [=] \; (\text{mol/g cat.} \cdot \text{s})$$

When we study reactions on the internal surface area of catalysts, the rate of reaction and rate law are often based on per unit surface area,

$$-r_A'' \; [=] \; (\text{mol/m}^2 \cdot \text{s})$$

As a result, the surface area of the catalyst per unit mass of catalyst,

$$S_a \; [=] \; (\text{m}^2/\text{g cat.})$$

is an important property of the catalyst. The rate of reaction per unit mass of catalyst, $-r_A'$, and the rate of reaction per unit surface area of catalyst are related through the equation

S_a: 10 grams of catalyst may cover as much surface area as a football field

$$-r_A' = -r_A'' S_a$$

A typical value of S_a might be 150 m^2/g of catalyst.

The rate law

As mentioned previously, at high temperatures, the denominator of the catalytic rate law approaches 1. Consequently, for the moment, it is reasonable to assume that the surface reaction is of nth order in the gas-phase concentration of A within the pellet.

$$-r''_A = k_n C^n_A \qquad (12\text{-}9)$$

where

$$-r''_A: \qquad k_n [=] \left(\frac{m^3}{kmol}\right)^{n-1} \frac{m}{s}$$

Similarly,

$$-r'_A: \qquad S_a k_n [=] \left(\frac{m^3}{kmol}\right)^{n-1} \frac{m^3}{kg \cdot s}$$

$$-r_A: \qquad \rho_c S_a k_n [=] \left(\frac{m^3}{kmol}\right)^{n-1} \frac{1}{s}$$

Substituting the rate law equation (12-9) into Equation (12-8) gives

$$\frac{d[r^2(-D_e\, dC_A/dr)]}{dr} + r^2 k_n \rho_c S_a C^n_A = 0 \qquad (12\text{-}10)$$

By differentiating the first term and dividing through by $-r^2 D_e$, Equation (12-10) becomes

Differential
equation and
boundary conditions
describing diffusion
and reaction in a
catalyst pellet

$$\boxed{\frac{d^2 C_A}{dr^2} + \frac{2}{r}\left(\frac{dC_A}{dr}\right) - \frac{k_n \rho_c S_a}{D_e} C^n_A = 0} \qquad (12\text{-}11)$$

The boundary conditions are:

1. The concentration remains finite at the center of the pellet:

$$\boxed{C_A \text{ is finite} \qquad \text{at } r = 0}$$

2. At the external surface of the catalyst pellet, the concentration is C_{As}:

$$\boxed{C_A = C_{As} \qquad \text{at } r = R}$$

12.1.3 Writing the Equation in Dimensionless Form

We now introduce dimensionless variables φ and λ so that we may arrive at a parameter that is frequently discussed in catalytic reactions, the *Thiele modulus*. Let

$$\varphi = \frac{C_A}{C_{As}} \tag{12-12}$$

$$\lambda = \frac{r}{R} \tag{12-13}$$

With the transformation of variables, the boundary condition

$$C_A = C_{As} \qquad \text{at } r = R$$

becomes

$$\varphi = \frac{C_A}{C_{As}} = 1 \qquad \text{at } \lambda = 1$$

and the boundary condition

$$C_A \text{ is finite} \qquad \text{at } r = 0$$

becomes

$$\varphi \text{ is finite} \qquad \text{at } \lambda = 0$$

We now rewrite the differential equation for the molar flux in terms of our dimensionless variables. Starting with

$$W_{Ar} = -D_e \frac{dC_A}{dr} \tag{11-7}$$

we use the chain rule to write

$$\frac{dC_A}{dr} = \left(\frac{dC_A}{d\lambda}\right) \frac{d\lambda}{dr} = \frac{d\varphi}{d\lambda} \left(\frac{dC_A}{d\varphi}\right) \frac{d\lambda}{dr} \tag{12-14}$$

Then differentiate Equation (12-12) with respect to φ and Equation (12-13) with respect to r, and substitute the resulting expressions,

$$\frac{dC_A}{d\varphi} = C_{As} \qquad \text{and} \qquad \frac{d\lambda}{dr} = \frac{1}{R}$$

into the equation for the concentration gradient to obtain

$$\frac{dC_A}{dr} = \frac{d\varphi}{d\lambda} \frac{C_{As}}{R} \tag{12-15}$$

The flux of A in terms of the dimensionless variables, φ and λ, is

$$W_{Ar} = -D_e \frac{dC_A}{dr} = -\frac{D_e C_{As}}{R} \left(\frac{d\varphi}{d\lambda}\right) \tag{12-16}$$

At steady state, the net flow of species A that enters into the pellet at the external pellet surface reacts completely within the pellet. The overall rate of

The total rate of consumption of A inside the pellet, M_A (mol/s)

reaction is therefore equal to the total molar flow of A into the catalyst pellet. The overall rate of reaction, M_A, can be obtained by multiplying the molar flux at the outer surface by the external surface area of the pellet, $4\pi R^2$:

$$M_A = -4\pi R^2 W_{Ar}\big|_{r=R} = +4\pi R^2 D_e \frac{dC_A}{dr}\bigg|_{r=R} = 4\pi R D_e C_{As} \frac{d\varphi}{d\lambda}\bigg|_{\lambda=1} \quad (12\text{-}17)$$

Consequently, to determine the overall rate of reaction, which is given by Equation (12-17), we first solve Equation (12-11) for C_A, differentiate C_A with respect to r, and then substitute the resulting expression into Equation (12-17). Differentiating the concentration gradient, Equation (12-15), yields

All the reactant that diffuses into the pellet is consumed (a black hole)

$$\frac{d^2C_A}{dr^2} = \frac{d}{dr}\left(\frac{dC_A}{dr}\right) = \frac{d}{d\lambda}\left(\frac{d\varphi}{d\lambda}\frac{C_{As}}{R}\right)\frac{d\lambda}{dr} = \frac{d^2\varphi}{d\lambda^2}\left(\frac{C_{As}}{R^2}\right) \quad (12\text{-}18)$$

After dividing by C_{As}/R^2, the dimensionless form of Equation (12-11) is written as

$$\frac{d^2\varphi}{d\lambda^2} + \frac{2}{\lambda}\frac{d\varphi}{d\lambda} - \frac{k_n R^2 S_a \rho_c C_{As}^{n-1}}{D_e}\varphi^n = 0$$

Then

Dimensionless form of equations describing diffusion and reaction

$$\boxed{\frac{d^2\varphi}{d\lambda^2} + \frac{2}{\lambda}\left(\frac{d\varphi}{d\lambda}\right) - \phi_n^2\,\varphi^n = 0} \quad (12\text{-}19)$$

where

$$\boxed{\phi_n^2 = \frac{k_n R^2 S_a \rho_c C_{As}^{n-1}}{D_e}} \quad (12\text{-}20)$$

Thiele modulus

The square root of the coefficient of φ^n, i.e., ϕ_n, is called the Thiele modulus. The Thiele modulus, ϕ_n, will always contain a subscript (e.g., n) which will distinguish this symbol from symbol for porosity, ϕ, defined in Chapter 4, which has no subscript. The quantity ϕ_n^2 is a measure of the ratio of "a" surface reaction rate to "a" rate of diffusion through the catalyst pellet:

$$\boxed{\phi_n^2 = \frac{k_n C_{As}^{n-1}\rho_c S_a R^2}{D_e} = \frac{k_n C_{As}^n \rho_c S_a R}{D_e[(C_{As}-0)/R]} = \frac{\text{"a" surface reaction rate}}{\text{"a" diffusion rate}}} \quad (12\text{-}20)$$

When the Thiele modulus is large, internal diffusion usually limits the overall rate of reaction; when ϕ_n is small, the surface reaction is usually rate-limiting. If for the reaction

$$A \longrightarrow B$$

the surface reaction were rate-limiting with respect to the adsorption of A and the desorption of B, and if species A and B are weakly adsorbed (i.e. low coverage) and present in very dilute concentrations, we can write the apparent first-order rate law

$$-r_A'' \simeq k_1 C_A \tag{12-21}$$

The units of k_1 are $m^3/m^2 s$ (= m/s).

For a first-order reaction, Equation (12-19) becomes

$$\frac{d^2\varphi}{d\lambda^2} + \frac{2}{\lambda}\frac{d\varphi}{d\lambda} - \phi_1^2\varphi = 0 \tag{12-22}$$

where

$$\phi_1 = R\sqrt{\frac{k_1 \rho_c S_a}{D_e}}$$

$$[k_1 \rho_c S_a] = \left(\frac{m}{s} \cdot \frac{g}{m^3} \cdot \frac{m^2}{g}\right) = 1/s$$

$$\frac{k_1 \rho_c S_a}{D_e} = \left(\frac{1/s}{m^2/s}\right) = \frac{1}{m^2}$$

$$\phi_1 = R\sqrt{\frac{k_1 \rho_c S_a}{D_e}} = m\left(\frac{1}{m^2}\right)^{1/2} = \frac{1}{1} \qquad \text{(Dimensionless)}$$

The boundary conditions are

$$\text{B.C. 1: } \varphi = 1 \qquad \text{at } \lambda = 1 \tag{12-23}$$

$$\text{B.C. 2: } \varphi \text{ is finite} \qquad \text{at } \lambda = 0 \tag{12-24}$$

12.1.4 Solution to the Differential Equation for a First-Order Reaction

Differential equation (12-22) is readily solved with the aid of the transformation $y = \varphi\lambda$:

$$\frac{d\varphi}{d\lambda} = \frac{1}{\lambda}\left(\frac{dy}{d\lambda}\right) - \frac{y}{\lambda^2}$$

$$\frac{d^2\varphi}{d\lambda^2} = \frac{1}{\lambda}\left(\frac{d^2y}{d\lambda^2}\right) - \frac{2}{\lambda^2}\left(\frac{dy}{d\lambda}\right) + \frac{2y}{\lambda^3}$$

With these transformations, Equation (12-22) reduces to

$$\frac{d^2y}{d\lambda^2} - \phi_1^2 y = 0 \tag{12-25}$$

This differential equation has the following solution (Appendix A.4):

$$y = A_1 \cosh\phi_1\lambda + B_1 \sinh\phi_1\lambda$$

In terms of φ,

$$\varphi = \frac{A_1}{\lambda} \cosh\phi_1\lambda + \frac{B_1}{\lambda} \sinh\phi_1\lambda \qquad (12\text{-}26)$$

The arbitrary constants A_1 and B_1 can easily be evaluated with the aid of the boundary conditions. At $\lambda = 0$; $\cosh\phi_1\lambda \to 1$, $(1/\lambda) \to \infty$, and $\sinh\phi_1\lambda \to 0$. Because the second boundary condition requires φ to be finite at the center, A_1 must be zero.

The constant B_1 is evaluated from B.C. 1 (i.e., $\varphi = 1$, $\lambda = 1$) and the dimensionless concentration profile is

Concentration
profile

$$\boxed{\varphi = \frac{C_A}{C_{As}} = \frac{1}{\lambda}\left(\frac{\sinh\phi_1\lambda}{\sinh\phi_1}\right)} \qquad (12\text{-}27)$$

Figure 12-4 shows the concentration profile for three different values of the Thiele modulus, ϕ_1. Small values of the Thiele modulus indicate surface reaction controls and a significant amount of the reactant diffuses well into the pellet interior without reacting. Large values of the Thiele modulus indicate that the surface reaction is rapid and that the reactant is consumed very close to the external pellet surface and very little penetrates into the interior of the pellet. Consequently, if the porous pellet is to be plated with a precious metal catalyst (e.g., Pt), it should only be plated in the immediate vicinity of the external surface when large values of ϕ_n characterize the diffusion and reaction. That is, it would be a waste of the precious metal to plate the entire pellet when internal diffusion is limiting because the reacting gases are consumed near the outer surface. Consequently, the reacting gases would never contact the center portion of the pellet.

For large values of
the Thiele modulus,
internal diffusion
limits the rate of
reaction

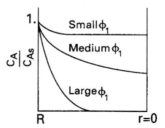

Figure 12-4   Concentration profile in a spherical catalyst pellet.

12.2 Internal Effectiveness Factor

The magnitude of the effectiveness factor (ranging from 0 to 1) indicates the relative importance of diffusion and reaction limitations. The internal effectiveness factor is defined as

η is a measure of
how far the
reactant diffuses
into the pellet
before reacting

$$\eta = \frac{\text{actual overall rate of reaction}}{\substack{\text{rate of reaction that would result if entire interior surface were}\\\text{exposed to the external pellet surface conditions } C_{As}, T_s}} \quad \text{(12-28)}$$

The overall rate, $-r'_A$, is also referred to as the observed rate of reaction $(-r_A(\text{obs}))$. In terms of symbols, the effectiveness factor is

$$\eta = \frac{-r_A}{-r_{As}} = \frac{-r'_A}{-r'_{As}} = \frac{-r''_A}{-r''_{As}}$$

To derive the effectiveness factor for a first-order reaction it is easiest to work in reaction rates of (moles per unit time), M_A, rather than in moles per unit time per unit mass of catalyst, i.e., $-r'_A$

$$\eta = \frac{-r'_A}{-r'_{As}} = \frac{-r'_A \times (\text{mass of catalyst})}{-r'_{As} \times (\text{mass of catalyst})} = \frac{M_A}{M_{As}}$$

First we shall consider the denominator, M_{As}. If the entire surface were exposed to the concentration at the external surface of the pellet, C_{As}, the rate for a first-order reaction would be

$$M_{As} = (\text{rate per unit surface area}) \times \left(\frac{\text{surface area}}{\text{mass of catalyst}}\right) \times (\text{mass of catalyst})$$

$$= \underbrace{(k_1 C_{As}) \times (S_a)}_{-r'_{As}} \times \left(\frac{4}{3}\pi R^3 \rho_c\right)$$

$$= -r'_{As} \times \left(\frac{4}{3}\pi R^3 \rho_c\right) \quad \text{(12-29)}$$

The subscript s indicates that the rate $-r'_{As}$ is evaluated at the conditions present at the **external** surface of the pellet.

The actual rate of reaction is the rate at which the reactant diffuses into the pellet at the outer surface. We recall Equation (12-17) for the actual rate of reaction,

The actual rate of
reaction

$$M_A = 4\pi R D_e C_{As} \left.\frac{d\varphi}{d\lambda}\right|_{\lambda=1} \quad \text{(12-17)}$$

Differentiating Equation (12-27) and then evaluating the result at $\lambda = 1$ yields

$$\left.\frac{d\varphi}{d\lambda}\right|_{\lambda=1} = \left(\frac{\phi_1 \cosh\lambda\phi_1}{\lambda \sinh\phi_1} - \frac{1}{\lambda^2}\frac{\sinh\lambda\phi_1}{\sinh\phi_1}\right)_{\lambda=1} = (\phi_1 \coth\phi_1 - 1) \quad \text{(12-30)}$$

Substituting Equation (12-30) into (12-17) gives us

$$M_A = 4\pi R D_e C_{As}(\phi_1 \coth\phi_1 - 1) \quad \text{(12-31)}$$

We now substitute Equations (12-29) and (12-31) into Equation (12-28) to obtain an expression for the effectiveness factor:

$$\eta = \frac{M_A}{M_{As}} = \frac{M_A}{(-r'_{As})\left(\frac{4}{3}\pi R^3 \rho_c\right)} = \frac{4\pi R D_e C_{As}}{k_1 C_{As} S_a \rho_c \frac{4}{3}\pi R^3}(\phi_1 \coth \phi_1 - 1)$$

$$= 3 \underbrace{\frac{1}{k_1 S_a \rho_c R^2/D_e}}_{\phi_1^2}(\phi_1 \coth \phi_1 - 1)$$

Internal effectiveness factor for a first-order reaction in a spherical catalyst pellet

$$\boxed{\eta = \frac{3}{\phi_1^2}(\phi_1 \coth \phi_1 - 1)} \tag{12-32}$$

A plot of the effectiveness factor as a function of the Thiele modulus is shown in Figure 12-5. Figure 12-5a shows η as a function of ϕ for a spherical catalyst pellet for reactions of zero-, first-, and second-order. Figure 12-5b corresponds to a first-order reaction occurring in three differently shaped pellets of volume V_p and external surface area A_p. When volume change accompanies a reaction, the corrections shown in Figure 12-6 apply to the effectiveness factor for a first-order reaction.

If $\phi_1 > 2$

then $\eta \approx \frac{3}{\phi_1^2}[\phi_1 - 1]$

If $\phi_1 > 20$

then $\eta \approx \frac{3}{\phi_1}$

We observe that as the particle diameter becomes very small, ϕ decreases, so that the effectiveness factor approaches 1 and the reaction is surface-reaction-limited. On the other hand, when ϕ is large (~ 30), the internal effectiveness factor η is small (i.e., $\eta \ll 1$), and the reaction is diffusion-limited within the pellet. Consequently, factors influencing the rate of external mass transport will have a negligible effect on the overall reaction rate. For large values of the Thiele modulus, the effectiveness factor can be written as

$$\eta \simeq \frac{3}{\phi_1} = \frac{3}{R}\sqrt{\frac{D_e}{k_1 \rho_c S_a}} \tag{12-33}$$

To express the overall rate of reaction in terms of the Thiele modulus, we rearrange Equation (12-28) and use the rate law for a first-order reaction in Equation (12-29)

$$-r'_A = \left(\frac{\text{Actual reaction rate}}{\text{Reaction rate at } C_{As}}\right) \times (\text{Reaction rate at } C_{As})$$

$$= \eta(-r'_{As})$$

$$= \eta(k_1 C_{As})(S_a) \tag{12-34}$$

Combining Equations (12-33) and (12-34), the overall rate of reaction for a first-order, internal-diffusion-limited reaction is

$$-r'_A = \frac{3}{R}\sqrt{\frac{D_e S_a k_1}{\rho_c}}C_{As}$$

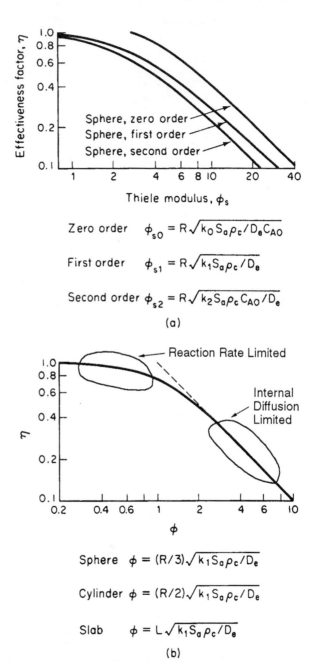

$$\text{Zero order} \quad \phi_{s0} = R\sqrt{k_0 S_a \rho_c / D_e C_{AO}}$$

$$\text{First order} \quad \phi_{s1} = R\sqrt{k_1 S_a \rho_c / D_e}$$

$$\text{Second order} \quad \phi_{s2} = R\sqrt{k_2 S_a \rho_c C_{AO} / D_e}$$

(a)

Internal effectiveness factor for different reaction orders and catalyst shapes

$$\text{Sphere} \quad \phi = (R/3)\sqrt{k_1 S_a \rho_c / D_e}$$

$$\text{Cylinder} \quad \phi = (R/2)\sqrt{k_1 S_a \rho_c / D_e}$$

$$\text{Slab} \quad \phi = L\sqrt{k_1 S_a \rho_c / D_e}$$

(b)

Figure 12-5 (a) Effectiveness factor plot for nth-order kinetics spherical catalyst particles (from *Mass Transfer in Heterogeneous Catalysis*, by C. N. Satterfield, 1970; reprint edition: Robert E. Krieger Publishing Co., 1981; reprinted by permission of the author). (b) First-order reaction in different pellet geometrics (from R. Aris, *Introduction to the Analysis of Chemical Reactors*, 1965, p. 131; reprinted by permission of Prentice-Hall, Englewood Cliffs, NJ)

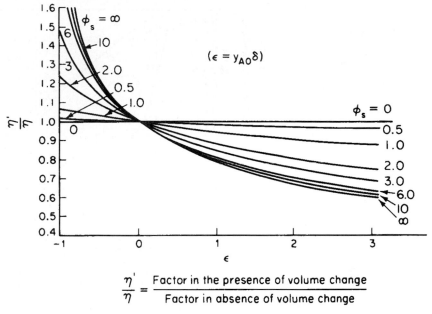

Correction for volume change with reaction (i.e., $\varepsilon \neq 0$)

$$\frac{\eta'}{\eta} = \frac{\text{Factor in the presence of volume change}}{\text{Factor in absence of volume change}}$$

Figure 12-6   Effectiveness factor ratios for first-order kinetics on spherical catalyst pellets for various values of the Thiele modulus, ϕ_s, for a sphere. [From V. W. Weekman and R. L. Goring, *J. Catal.*, *4*, 260 (1965).]

How can the rate of reaction be increased?

Therefore, to increase the overall rate of reaction, $-r'_A$: (1) decrease the radius R (make pellets smaller); (2) increase the temperature; (3) increase the concentration; and (4) increase the internal surface area. For reactions of order n, we have, from Equation (12-20),

$$\phi_n^2 = \frac{k_n R^2 S_a \rho_c C_{As}^{n-1}}{D_e} \tag{12-20}$$

For large values of the Thiele modulus, the effectiveness factor is

$$\eta = \left(\frac{2}{n+1}\right)^{1/2} \frac{3}{\phi_n} = \left(\frac{2}{n+1}\right)^{1/2} \frac{3}{R} \sqrt{\frac{D_e}{k_n S_a \rho_c}} C_{As}^{(1-n)/2} \tag{12-35}$$

Consequently, for reaction orders greater than 1, the effectiveness factor decreases with increasing concentration at the external pellet surface.

The above discussion of effectiveness factors is valid only for isothermal conditions. When a reaction is exothermic and nonisothermal, the effectiveness factor can be significantly greater than 1 as shown in Figure 12-7. Values of η greater than 1 occur because the external surface temperature of the pellet is less than the temperature inside the pellet where the exothermic reaction is taking place. Therefore, the rate of reaction inside the pellet is greater than the rate at the surface. Thus, because the effectiveness factor is the ratio of the actual reaction rate to the rate at surface conditions, the effectiveness factor

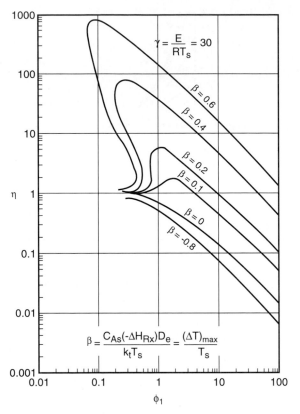

Figure 12-7 Nonisothermal effectiveness factor.

can be greater than 1, depending on the magnitude of the parameters β and γ. The parameter γ is sometimes referred to as the Arrhenius number, and the parameter β represents the maximum temperature difference that could exist in the pellet relative to the surface temperature T_s.

$$\gamma = \text{Arrhenius number} = \frac{E}{RT_s}$$

$$\beta = \frac{\Delta T_{max}}{T_s} = \frac{T_{max} - T_s}{T_s} = \frac{\Delta H_{Rx} D_e C_{As}}{k_t T_s}$$

(See Problem P12-15 for the derivation of β.) The Thiele modulus, ϕ_1, is evaluated at the external surface temperature. Typical values of γ for industrial processes range from a value of $\gamma = 6.5$ ($\beta = 0.025$, $\phi_1 = 0.22$) for the synthesis of vinyl chloride from HCl and acetone to a value of $\gamma = 29.4$ ($\beta = 6 \times 10^{-5}$, $\phi_1 = 1.2$) for the synthesis of ammonia.[3] The lower the thermal conductivity k_t and the higher the heat of reaction, the greater the temperature difference (see Problems P12-15 and P12-16). We observe from Figure 12-7 that multiple

[3] H. V. Hlavacek, N. Kubicek, and M. Marek, *J. Catal.*, *15*, 17 (1969).

steady states can exist for values of the Thiele modulus less than 1 and when β is greater than approximately 0.2. There will be no multiple steady states when the criterion developed by Luss[4] is fulfilled.

Criterion for no MSSs in the pellet

$$\boxed{4(1+\beta) > \beta\gamma}$$
(12-36)

12.3 Falsified Kinetics

You may not be measuring what you think you are

There are circumstances under which the measured reaction order and activation energy are not the true values. Consider the case in which we obtain reaction rate data in a differential reactor, where precautions are taken to virtually eliminate external mass transfer resistance. From these data we construct a log-log plot of the rate of reaction as a function of the bulk gas-phase concentration (Figure 12-8). The slope of this plot is the apparent reaction order n' and the rate law takes the form

$$-r_A' = k_n' C_{As}^{n'}$$
(12-37)

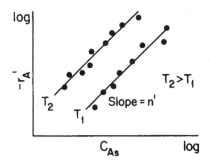

Measured rate with apparent reaction order n'

Figure 12-8   Determining the apparent reaction order.

We will now proceed to relate this measured reaction order n' to the true reaction order n. Using the definition of the effectiveness factor, note that the actual rate is the product of η and the rate of reaction, $-r_A'$, evaluated at the external surface, $k_n S_a C_{As}^n$:

$$-r_A' = \eta(-r_{As}') = \eta(k_n S_a C_{As}^n)$$
(12-38)

For large values of the Thiele modulus ϕ_n, we can use Equation (12-35) to obtain

$$-r_A' = \frac{3}{\phi_n}\sqrt{\frac{2}{n+1}}\, k_n S_a C_{As}^n = \frac{3}{R}\sqrt{\frac{2}{n+1}}\sqrt{\frac{D_e}{\rho_c S_a k_n}}\, k_n S_a C_{As}^n$$

$$= \frac{3}{R}\sqrt{\frac{2 D_e S_a}{(n+1)\rho_c}}\, k_n^{1/2} C_{As}^{(n+1)/2}$$
(12-39)

[4] D. Luss, *Chem. Eng. Sci.*, **23**, 1249 (1968).

We equate the true reaction rate, Equation (12-39), to the measured reaction rate, Equation (12-37), to get

$$-r'_A = \sqrt{\frac{2}{n+1}} \left(\frac{3}{R} \sqrt{\frac{D_e S_a}{\rho_c}} \; k_n^{1/2} C_{As}^{(n+1)/2} = k'_n C_{As}^{n'} \right) \tag{12-40}$$

Because the overall exponent of the concentration, C_{As}, must be the same for both the analytical and measured rates of reaction, the apparent reaction order n' is related to the true reaction order n by

The true and the apparent reaction order

$$\boxed{n' = \frac{1+n}{2}} \tag{12-41}$$

In addition to an apparent reaction order, there is also an apparent activation energy, E_{app}. This value is the activation energy we would calculate using the experimental data, from the slope of a plot of $\ln(-r'_A)$ as a function of $1/T$ at a fixed concentration of A. Substituting for the measured and true specific reaction rates in terms of the activation energy,

$$\underbrace{k'_n = A_{app} e^{-E_{app}/RT}}_{\text{measured}} \qquad \underbrace{k_n = A_T e^{-E_T/RT}}_{\text{true}}$$

into Equation (12-40), we find that

$$-r'_A = \frac{3}{R} \sqrt{\frac{2}{n+1} \frac{S_a D_e}{\rho_c}} A_T^{1/2} \left[\exp\left(\frac{-E_T}{RT} \right) \right]^{1/2} C_{As}^{(n+1)/2} = A_{app} \left[\exp\left(\frac{-E_{App}}{RT} \right) \right] C_{As}^{n'}$$

Taking the natural log of both sides gives us

$$\ln\left[\left(\frac{3}{R} \sqrt{\frac{2}{n+1} \frac{S_a D_e}{\rho_c}} \right) A_T^{1/2} C_{As}^{(n+1)/2} \right] - \frac{E_T}{2RT} = \ln(A_{app} C_{As}^{n'}) - \frac{E_{app}}{RT} \tag{12-42}$$

where E_T is the true activation energy.

Comparing the temperature-dependent terms on the right- and left-hand sides of Equation (12-42), we see that the true activation energy is equal to twice the apparent activation energy.

The true activation energy

$$\boxed{E_T = 2E_{app}} \tag{12-43}$$

This measurement of the apparent reaction order and activation energy results primarily when internal diffusion limitations are present and is referred to as *disguised* or *falsified kinetics*. Serious consequences could occur if the laboratory data were taken in the disguised regime and the reactor were operated in a different regime. For example, what if the particle size were reduced so that internal diffusion limitations became negligible? The higher activation energy, E_T, would cause the reaction to be much more temperature sensitive and there is the possibility for *runaway reaction conditions* to occur.

12.4 Overall Effectiveness Factor

For first-order reactions we can use an overall effectiveness factor to help us analyze diffusion, flow, and reaction in packed beds. We now consider a situation where external and internal resistance to mass transfer to and within the pellet are of the same order of magnitude (Figure 12-9). At steady state, the transport of the reactant(s) from the bulk fluid to the external surface of the catalyst is equal to the net rate of reaction of the reactant within and on the pellet.

Here, both internal
and external
diffusion are
important

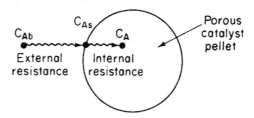

Figure 12-9   Mass transfer and reaction steps.

The molar rate of mass transfer from the bulk fluid to the external surface is

$$\text{molar rate} = (\text{molar flux}) \cdot (\text{external surface area})$$
$$M_A = W_{Ar} \cdot (\text{surface area/volume})(\text{reactor volume})$$
$$= W_{Ar} \cdot a_c \Delta V \tag{12-44}$$

where a_c is the external surface area per unit reactor volume (cf. Chapter 11) and ΔV is the volume.

This molar rate of mass transfer to the surface, M_A, is equal to the net (total) rate of reaction *on* **and** *within* the pellet:

$$M_A = -r''_A \cdot (\text{external area} + \text{internal area})$$

$$\text{External area} = \frac{\text{external area}}{\text{reactor volume}} \times \text{reactor volume}$$

$$= a_c \, \Delta V$$

$$\text{Internal area} = \frac{\text{internal area}}{\text{mass of catalyst}} \times \frac{\text{mass of catalyst}}{\text{volume of catalyst}} \times \frac{\text{volume of catalyst}}{\text{reactor volume}} \times \text{reactor volume}$$

$$= S_a \overbrace{\rho_c(1-\phi)}^{\rho_b} \Delta V$$

$$= S_a \rho_b \, \Delta V$$

$$M_A = -r''_A [a_c \, \Delta V + S_a \rho_b \, \Delta V] \tag{12-45}$$

Combining Equations (12-44) and (12-45) and canceling the volume ΔV, one obtains

$$W_A a_c = -r''_A \cdot (a_c + S_a \rho_b)$$

For most catalysts the internal surface area is much greater than the external surface area (i.e., $S_a \rho_b \gg a_c$), in which case we have

$$\boxed{W_A a_c = -r_A'' S_a \rho_b} \tag{12-46}$$

where $-r_A''$ is the overall rate of reaction within and on the pellet per unit surface area. The relationship for the rate of mass transport is

$$M_A = W_{Ar} a_c \, \Delta V = k_c(C_{Ab} - C_{As}) a_c \, \Delta V \tag{12-47}$$

Because internal diffusion resistance is also significant, not all of the interior surface of the pellet is accessible to the concentration at the external surface of the pellet, C_{As}. We have already learned that the effectiveness factor is a measure of this surface accessibility [see Equation (12-38)]:

$$(-r_A'' = -r_{As}'' \eta)$$

Assuming that the surface reaction is first-order with respect to A, we can utilize the internal effectiveness factor to write

$$-r_A'' = \eta k_1 C_{As} \tag{12-48}$$

Next we can use Equations (12-45) through (12-48) to eliminate C_{As} from Equation (12-47), so that the total molar transport of A from the bulk fluid to the external pellet surface can be expressed solely in terms of bulk concentration and other parameters of the system (e.g., the mass transfer coefficient, k_c, and the specific reaction rate, k_1).

The net rate of reaction given by Equation (12-46) is equal to the rate of mass transfer of A from the bulk fluid to the external pellet surface [Equation (12-47)]. We need to eliminate the surface concentration from any equation involving the rate of reaction or rate of mass transfer, because C_{As} cannot be measured by standard techniques. To accomplish this elimination, first substitute Equation (12-48) into Equation (12-46):

$$W_{Ar} a_c \, \Delta V = \eta k_1 S_a C_{As} \rho_b \, \Delta V$$

Canceling ΔV yields

$$k_c a_c(C_{Ab} - C_{As}) = \eta k_1 S_a \rho_b C_{As} \tag{12-49}$$

Solving for C_{As}, we obtain

Concentration at the pellet surface as a function of bulk gas concentration

$$C_{As} = \frac{k_c a_c C_{Ab}}{k_c a_c + \eta k_1 S_a \rho_b} \tag{12-50}$$

Substituting for C_{As} in Equation (12-48) gives

$$-r_A'' = \frac{\eta k_1 k_c a_c C_{Ab}}{k_c a_c + \eta k_1 S_a \rho_b} \tag{12-51}$$

In discussing the surface accessibility, we defined the internal effectiveness factor η with respect to the concentration at the external surface of the pellet, C_{As}:

$$\eta = \frac{\text{actual overall rate}}{\text{rate that would result if the entire surface were}} \qquad (12\text{-}28)$$
$$\text{exposed to the external surface concentration } C_{As}$$

We now define an overall effectiveness factor that is based on the bulk concentration:

Two different effectiveness factors

$$\Omega = \frac{\text{actual overall rate}}{\text{rate that would result if the entire surface were}} \qquad (12\text{-}52)$$
$$\text{exposed to the bulk concentration } C_{Ab}$$

Dividing the numerator and denominator of Equation (12-51) by $k_c a_c$, we obtain the net rate of reaction (total molar flow of A to the surface in terms of the bulk fluid concentration), which is a measurable quantity.

$$-r_A'' = \frac{\eta}{1 + \eta k_1 S_a \rho_b / k_c a_c} k C_{Ab} \qquad (12\text{-}53)$$

Consequently, the overall rate of reaction in terms of the bulk concentration C_{Ab} is

$$\boxed{-r_A'' = \Omega(-r_{Ab}'') = \Omega k_1 C_{Ab}} \qquad (12\text{-}54)$$

where

Overall effectiveness factor for a first-order reaction

$$\boxed{\Omega = \frac{\eta}{1 + \eta k_1 S_a \rho_b / k_c a_c}} \qquad (12\text{-}55)$$

The rates of reaction based on surface and bulk concentrations are related by

$$-r_A'' = \Omega(-r_{Ab}'') = \eta(-r_{As}'') \qquad (12\text{-}56)$$

where

$$-r_{As}'' = k_1 C_{As}$$
$$-r_{Ab}'' = k_1 C_{Ab}$$

The actual rate of reaction is related to the reaction rate evaluated at the bulk concentrations. The actual rate can be expressed in terms of the rate per unit volume, $-r_A$, the rate per unit mass, $-r_A'$, and the rate per unit surface area, $-r_A''$, which are related by the equation

$$-r_A = -r_A' \rho_b = -r_A'' S_a \rho_b$$

In terms of the overall effectiveness factor for a first-order reaction and the reactant concentration in the bulk

$$-r_A = r_{Ab} \Omega = r'_{Ab} \rho_b \Omega = -r''_{Ab} S_a \rho_b \Omega = k_1 C_{Ab} S_a \rho_b \Omega \qquad (12\text{-}57)$$

where again

$$\Omega = \frac{\eta}{1 + \eta k_1 S_a \rho_b / k_c a_c}$$

Recall that k_1 is given in terms of the catalyst surface area ($m^3/m^2 \cdot s$).

12.5 Estimation of Diffusion- and Reaction-Limited Regimes

In many instances it is of interest to obtain "quick and dirty" estimates to learn which is the rate-limiting step in a heterogeneous reaction.

12.5.1 Weisz–Prater Criterion for Internal Diffusion

The Weisz–Prater criterion uses measured values of the rate of reaction, $-r'_A$ (obs), to determine if internal diffusion is limiting the reaction. This criterion can be developed intuitively by first rearranging Equation (12-32) in the form

$$\eta \phi_1^2 = 3(\phi_1 \coth \phi_1 - 1) \qquad (12\text{-}58)$$

The left-hand side is the Weisz–Prater parameter.

$$C_{WP} = \eta \times \phi_1^2 \qquad (12\text{-}59)$$

$$= \frac{\text{observed (actual) reaction rate}}{\text{reaction rate evaluated at } C_{As}} \times \frac{\text{reaction rate evaluated at } C_{As}}{\text{a diffusion rate}}$$

$$= \frac{\text{actual reaction rate}}{\text{a diffusion rate}}$$

Substituting for

$$\eta = \frac{-r'_A (\text{obs})}{-r'_{As}} \qquad \text{and} \qquad \phi_1^2 = \frac{-r''_{As} S_a \rho_c R^2}{D_e C_{As}} = \frac{-r'_{As} \rho_c R^2}{D_e C_{As}}$$

in Equation (12-59) we have

$$C_{WP} = \frac{-r'_A (\text{obs})}{-r'_{As}} \left(\frac{-r'_{As} \rho_c R^2}{D_e C_{As}} \right) \qquad (12\text{-}60)$$

$$C_{WP} = \eta \phi_1^2 = \frac{-r'_A (\text{obs}) \rho_c R^2}{D_e C_{As}} \qquad (12\text{-}61)$$

Are there any internal diffusion limitations indicated from the Weisz–Prater criterion?

All the terms in Equation (12-61) are either measured or known. Consequently, we can calculate C_{WP}. If

$$\boxed{C_{WP} \ll 1}$$

there are no diffusion limitations and consequently no concentration gradient exists within the pellet. If

$$\boxed{C_{WP} \gg 1}$$

internal diffusion limits the reaction severely.

Example 12–2 Estimating Thiele Modulus and Effectiveness Factor

The first-order reaction

$$A \longrightarrow B$$

was carried out over two different-sized pellets. The pellets were contained in a spinning basket reactor that was operated at sufficiently high rotation speeds that external mass transfer resistance was negligible. The results of two experimental runs made under identical conditions are as given in Table E12-2.1. Estimate the Thiele modulus and effectiveness factor for each pellet. How small should the pellets be made to virtually eliminate all internal diffusion resistance?

These two experiments yield an enormous amount of information

TABLE E12-2.1

	Measured Rate (mol/g cat.·s) $\times 10^5$	Pellet Radius (m)
Run 1	3.0	0.01
Run 2	15.0	0.001

Solution

Combining Equations (12-58) and (12-61), we obtain

$$\frac{-r'_A(\text{obs})R^2\rho_c}{D_e C_{As}} = \eta\phi_1^2 = 3(\phi_1 \coth\phi_1 - 1) \tag{E12-2.1}$$

Letting the subscripts 1 and 2 refer to runs 1 and 2, we apply Equation (E12-2.1) to runs 1 and 2 and then take the ratio to obtain

$$\frac{-r'_{A2}R_2^2}{-r'_{A1}R_1^2} = \frac{\phi_{12}\coth\phi_{12} - 1}{\phi_{11}\coth\phi_{11} - 1} \tag{E12-2.2}$$

The terms ρ_c, D_e, and C_{As} cancel because the runs were carried out under identical conditions. The Thiele modulus is

$$\phi_1 = R \sqrt{\frac{-r'_{As}\rho_c}{D_e C_{As}}} \qquad \text{(E12-2.3)}$$

Taking the ratio of the Thiele moduli for runs 1 and 2, we obtain

$$\frac{\phi_{11}}{\phi_{12}} = \frac{R_1}{R_2} \qquad \text{(E12-2.4)}$$

or

$$\phi_{11} = \frac{R_1}{R_2}\phi_{12} = \frac{0.01 \text{ m}}{0.001 \text{ m}}\phi_{12} = 10\phi_{12} \qquad \text{(E12-2.5)}$$

Substituting for ϕ_{11} in Equation (E12-2.2) and evaluating $-r'_A$ and R gives us

$$\left(\frac{15\times 10^{-5}}{3\times 10^{-5}}\right)\frac{(0.001)^2}{(0.01)^2} = \frac{\phi_{12}\coth\phi_{12}-1}{10\phi_{12}\coth(10\phi_{12})-1} \qquad \text{(E12-2.6)}$$

$$0.05 = \frac{\phi_{12}\coth\phi_{12}-1}{10\phi_{12}\coth(10\phi_{12})-1} \qquad \text{(E12-2.7)}$$

We now have one equation and one unknown. Solving Equation (E12-2.7) we find that

$$\phi_{12} = 1.65 \qquad \text{for } R_2 = 0.001 \text{ m}$$

Then

$$\phi_{11} = 16.5 \qquad \text{for } R_1 = 0.01 \text{ m}$$

The corresponding effectiveness factors are

$$\text{For } R_2: \quad \eta_2 = \frac{3(\phi_{12}\coth\phi_{12}-1)}{\phi_{12}^2} = \frac{3(1.65\coth 1.65-1)}{(1.65)^2} = 0.856$$

$$\text{For } R_1: \quad \eta_1 = \frac{3(16.5\coth 16.5-1)}{(16.5)^2} \approx \frac{3}{16.5} = 0.182$$

Next we calculate the particle radius needed to virtually eliminate internal diffusion control (say, $\eta = 0.95$):

$$0.95 = \frac{3(\phi_{13}\coth\phi_{13}-1)}{\phi_{13}^2} \qquad \text{(E12-2.8)}$$

Solution to Equation (E12-2.8) yields $\phi_{13} = 0.9$:

$$R_3 = R_1\frac{\phi_{13}}{\phi_{11}} = (0.01)\left(\frac{0.9}{16.5}\right) = 5.5\times 10^{-4} \text{ m}$$

A particle size of 0.55 mm is necessary to virtually eliminate diffusion control (i.e., $\eta = 0.95$).

12.5.2 Mears' Criterion for External Diffusion

The Mears[5] criterion, like the Wiesz–Prater criterion, uses the measured rate of reaction, $-r'_A$, (kmol/kg cat.·s) to learn if mass transfer from the bulk gas phase to the catalyst surface can be neglected. Mears proposed that when

Is external diffusion limiting?

$$\frac{-r'_A \rho_b R n}{k_c C_{Ab}} < 0.15 \qquad (12\text{-}62)$$

where n = reaction order

R = catalyst particle radius, m

ρ_b = bulk density of catalyst bed, kg/m^3

  = $(1 - \phi)\rho_c$ (ϕ = porosity)

C_{Ab} = bulk concentration kmol/m^3

k_c = mass transfer coefficient, m/s

external mass transfer effects can be neglected.

The mass transfer coefficient can be calculated from the appropriate correlation, such as that of Thoenes–Kramers, for the flow conditions through the bed. When Equation (12-62) is satisfied, no concentration gradients exist between the bulk gas and external surface of the catalyst pellet.

Mears also proposed that the bulk fluid temperature, T, will be virtually the same as the temperature at the external surface of the pellet when

Is there a temperature gradient?

$$\left| \frac{-\Delta H_{Rx}(-r'_A) \rho_b R E}{h T^2 R_g} \right| < 0.15 \qquad (12\text{-}63)$$

where h = heat transfer coefficient, kJ/m^2·s·K

R_g = gas constant, kJ/mol·K

ΔH_{Rx} = heat of reaction, kJ/mol

E = activation energy, kJ/kmol

and the other symbols are as in Equation (12-62).

12.6 Mass Transfer and Reaction in a Packed Bed

We now consider the same isomerization taking place in a packed bed of catalyst pellets rather than on one single pellet (see Figure 12-10). The concentration C_A is the bulk gas-phase concentration of A at any point along the length of the bed.

---

[5] D. E. Mears, *Ind. Eng. Chem. Process Des. Dev., 10*, 541 (1971). Other interphase transport-limiting criteria can be found in *AIChE Symp. Ser. 143* (S. W. Weller, ed.), 70 (1974).

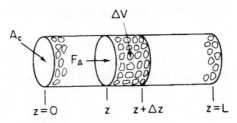

Figure 12-10   Packed-bed reactor.

We shall perform a balance on A over the volume element ΔV, neglecting any radial variations in concentration and assuming that the bed is operated at steady state. The following symbols will be used in developing our model:

$$A_c = \text{cross-sectional area of the tube, dm}^2$$

$$C_{Ab} = \text{bulk gas concentration of A, mol/dm}^3$$

$$\rho_b = \text{bulk density of the catalyst bed, g/dm}^3$$

$$v_0 = \text{volumetric flow rate, dm}^3/\text{s}$$

$$U = \text{superficial velocity} = v_0/A_c, \text{dm/s}$$

A mole balance on the volume element $(A_c \Delta z)$ yields

$$[\text{rate in}] - [\text{rate out}] + [\text{rate of formation of A}] = 0$$

$$A_c W_{Az}|_z - A_c W_{Az}|_{z+\Delta z} + r_A' \rho_b A_c \Delta z = 0$$

Now we will see how to use η and Ω to calculate conversion in a packed bed

Dividing by $A_c \, \Delta z$ and taking the limit as $\Delta z \longrightarrow 0$ yields

$$-\frac{dW_{Az}}{dz} + r_A' \rho_b = 0 \qquad (12\text{-}64)$$

Assuming that the total concentration c is constant, Equation (11-13) can be expressed as

$$W_{Az} = -D_{AB}\frac{dC_{Ab}}{dz} + y_{Ab}(W_{Az} + W_{Bz})$$

Also, writing the bulk flow term in the form

$$B_{Az} = y_{Ab}(W_{Az} + W_{Bz}) = y_{Ab}cU = UC_{Ab}$$

Equation (12-64) can be written in the form

$$D_{AB}\frac{d^2C_{Ab}}{dz^2} - U\frac{dC_{Ab}}{dz} + r_A' \rho_b = 0 \qquad (12\text{-}65)$$

The term $D_{AB}(d^2C_{Ab}/dz^2)$ is used to represent either diffusion and/or dispersion in the axial direction. Consequently, we shall use the symbol D_a for the dispersion coefficient to represent either or both of these cases. We will come back to this form of the diffusion equation when we discuss dispersion in

Chapter 14. The overall reaction rate within the pellet, r'_A, is the overall rate of reaction within and on the catalyst per unit mass of catalyst. It is a function of the reactant concentration within the catalyst. This overall rate can be related to the rate of reaction of A that would exist if the entire surface were exposed to the bulk concentration C_{Ab} through the overall effectiveness factor Ω:

$$-r'_A = -r'_{Ab} \times \Omega \tag{12-57}$$

For the first-order reaction considered here,

$$-r'_{Ab} = -r''_{Ab} S_a = k S_a C_{Ab} \tag{12-66}$$

Substituting Equation (12-66) into Equation (12-57), we obtain the overall rate of reaction per unit mass of catalyst in terms of the bulk concentration C_{Ab}:

$$-r'_A = \Omega k S_a C_A$$

Substituting the equation for $-r'_A$ above into Equation (12-65), we form the differential equation describing diffusion with a first-order reaction in a catalyst bed:

Flow and first-order reaction in a packed bed

$$D_a \frac{d^2 C_{Ab}}{dz^2} - U \frac{dC_{Ab}}{dz} - \Omega \rho_b k S_a C_{Ab} = 0 \tag{12-67}$$

As an example, we shall solve this equation for the case in which the flow rate through the bed is very large and the axial diffusion can be neglected. Finlayson[6] has shown that axial dispersion can be neglected when

Criterion for neglecting axial dispersion/diffusion

$$\left| \frac{-r'_A \rho_b d_p}{U_0 C_{Ab}} \right| \ll \left| \frac{U_0 d_p}{D_a} \right| \tag{12-68}$$

where U_0 is the superficial velocity, d_p the particle diameter, and D_a the effective axial dispersion coefficient. In Chapter 14 we will consider solutions to the complete form of Equation (12-67).

Neglecting axial dispersion with respect to forced axial convection,

$$\left| U \frac{dC_{Ab}}{dz} \right| \gg \left| D_a \frac{d^2 C_{Ab}}{dz^2} \right|$$

Equation (12-67) can be arranged in the form

$$\frac{dC_{Ab}}{dz} = -\left(\frac{\Omega \rho_b k S_a}{U} \right) C_{Ab} \tag{12-69}$$

With the aid of the boundary condition at the entrance of the reactor,

$$C_{Ab} = C_{Ab0} \quad \text{at } z = 0$$

[6] L. C. Young and B. A. Finlayson, *Ind. Eng. Chem. Fund.*, *12*, 412 (1973).

Equation (12-69) can be integrated to give

$$C_{Ab} = C_{Ab0}\, e^{-(\rho_b k S_a \Omega z)/U}$$

(12-70)

The conversion at the reactor's exit, $z = L$, is

Conversion in a
packed-bed reactor

$$\boxed{X = 1 - \frac{C_{Ab}}{C_{Ab0}} = 1 - e^{-(\rho_b k S_a \Omega L)/U}}$$

(12-71)

Example 12–3   Reducing Nitrous Oxides in a Plant Effluent

In Section 7.2.2 we saw the role that nitric oxide plays in smog formation and the incentive we would have for reducing its concentration in the atmosphere. It is proposed to reduce the concentration of NO in an effluent stream from a plant by passing it through a packed bed of spherical porous carbonaceous solid pellets. A 2% NO–98% air mixture flows at a rate of 1×10^{-6} m³/s (0.001 dm³/s) through a 2-in.-ID tube packed with porous solid at a temperature of 1173 K and a pressure of 101.3 kPa. The reaction

$$NO + C \longrightarrow CO + \tfrac{1}{2}N_2$$

is first-order in NO, that is,

$$-r'_{NO} = k S_a C_{NO}$$

and occurs primarily in the pores inside the pellet, where

$$S_a = \text{internal surface area} = 530 \text{ m}^2/\text{g}$$

$$k = 4.42 \times 10^{-10} \text{ m}^3/\text{m}^2 \cdot \text{s}$$

Calculate the weight of porous solid necessary to reduce the NO concentration to a level of 0.004%, which is below the Environmental Protection Agency limit.

Additional information:

At 1173 K, the fluid properties are

$$\nu = \text{kinematic viscosity} = 1.53 \times 10^{-8} \text{ m}^2/\text{s}$$

$$D_{AB} = \text{gas-phase diffusivity} = 2.0 \times 10^{-8} \text{ m}^2/\text{s}$$

$$D_e = \text{effective diffusivity} = 1.82 \times 10^{-8} \text{ m}^2/\text{s}$$

The properties of the catalyst and bed are

$$\rho_c = \text{density of catalyst particle} = 2.8 \text{ g/cm}^3 = 2.8 \times 10^6 \text{ g/m}^3$$

$$\phi = \text{bed porosity} = 0.5$$

$$\rho_b = \text{bulk density of bed} = \rho_c(1 - \phi) = 1.4 \times 10^6 \text{ g/m}^3$$

$$R = \text{pellet radius} = 3 \times 10^{-3} \text{ m}$$

$$\gamma = 1.0$$

Solution

It is desired to reduce the NO concentration from 2.0% to 0.004%. Neglecting any volume change at these low concentrations gives us

$$X = \frac{C_{Ab0} - C_{Ab}}{C_{Ab0}} = \frac{2 - 0.004}{2} = 0.998$$

where A represents NO.

The variation of NO down the length of the reactor is

$$\frac{dC_{Ab}}{dz} = -\frac{\Omega k S_a \rho_b C_{Ab}}{U} \tag{12-69}$$

Multiplying the numerator and denominator on the right-hand side of Equation (12-69) by the cross-sectional area A_c, and realizing that the weight of solids up to a point z in the bed is

$$W = \rho_b A_c z$$

the variation of NO concentration with solids is

$$\frac{dC_{Ab}}{dW} = -\frac{\Omega k S_a C_{Ab}}{v} \tag{E12-3.1}$$

Because NO is present in dilute concentrations, we shall take $\varepsilon \ll 1$ and set $v = v_0$. We integrate Equation (E12-3.1) using the boundary condition that when $W = 0$, then $C_{Ab} = C_{Ab0}$:

$$X = 1 - \frac{C_{Ab}}{C_{Ab0}} = 1 - \exp\left(-\frac{\Omega k S_a W}{v_0}\right) \tag{E12-3.2}$$

where

$$\Omega = \frac{\eta}{1 + \eta k S_a \rho_c / k_c a_c} \tag{12-55}$$

Rearranging, we have

$$\boxed{W = \frac{v_0}{\Omega k S_a} \ln\frac{1}{1 - X}} \tag{E12-3.3}$$

1. *Calculating the internal effectiveness factor* for spherical pellets in which a first-order reaction is occurring, we obtain

$$\eta = \frac{3}{\phi_1^2} (\phi_1 \coth\phi_1 - 1) \tag{12-32}$$

As a first approximation, we shall neglect any changes in the pellet resulting from the reactions of NO with the porous carbon. The Thiele modulus for this system is[7]

---

[7] L. K. Chan, A. F. Sarofim, and J. M. Beer, *Combust. Flame*, *52*, 37 (1983).

$$\phi_1 = R \sqrt{\frac{k_1 \rho_c S_a}{D_e}} \qquad (E12\text{-}3.4)$$

where

$$R = \text{pellet radius} = 3 \times 10^{-3} \text{ m}$$
$$D_e = \text{effective diffusivity} = 1.82 \times 10^{-8} \text{ m}^2/\text{s}$$
$$\rho_c = 2.8 \text{ g/cm}^3 = 2.8 \times 10^6 \text{ g/m}^3$$
$$k = \text{specific reaction rate} = 4.42 \times 10^{-10} \text{ m}^3/\text{m}^2 \cdot \text{s}$$

$$\phi_1 = 0.003 \text{ m} \sqrt{\frac{(4.42 \times 10^{-10} \text{ m/s})(530 \text{ m}^2/\text{g})(2.8 \times 10^6 \text{ g/m}^3)}{1.82 \times 10^{-8} \text{ m}^2/\text{s}}}$$

$$\phi_1 = 18$$

Because ϕ_1 is large,

$$\eta = \frac{3}{18} = 0.167$$

2. *To calculate the external mass transfer coefficient*, the Thoenes–Kramers correlation is used.

$$\text{Sh}' = (\text{Re}')^{1/2} \text{Sc}^{1/3} \qquad (11\text{-}65)$$

For a 2-in.-ID pipe, $A_c = 2.03 \times 10^{-3}$ m^2. The superficial velocity is

$$U = \frac{v_0}{A_c} = \frac{10^{-6} \text{ m}^3/\text{s}}{2.03 \times 10^{-3} \text{ m}^2} = 4.93 \times 10^{-4} \text{ m/s}$$

Calculate
Re'
Sc
Then
Sh'
Then
k_c

$$\text{Re}' = \frac{U d_p}{(1-\phi)\nu} = \frac{(4.93 \times 10^{-4} \text{ m/s})(6 \times 10^{-3} \text{ m})}{(1-0.5)(1.53 \times 10^{-8} \text{ m}^2/\text{s})} = 386.7$$

Nomenclature note: ϕ with subscript 1, ϕ_1 = Thiele modulus
ϕ without subscript, ϕ = porosity

$$\text{Sc} = \frac{\nu}{D_{AB}} = \frac{1.53 \times 10^{-8} \text{ m}^2/\text{s}}{2.0 \times 10^{-8} \text{ m}^2/\text{s}} = 0.765$$

$$\text{Sh}' = (386.7)^{1/2}(0.765)^{1/3} = (19.7)(0.915) = 18.0$$

$$k_c = \frac{1-\phi}{\phi}\left(\frac{D_{AB}}{d_p}\right)\text{Sh}' = \frac{0.5}{0.5}\left(\frac{2.0 \times 10^{-8} \text{ m}^2/\text{s}}{6.0 \times 10^{-3} \text{ m}}\right)(18.0)$$

$$k_c = 6 \times 10^{-5} \text{ m/s}$$

3. *Calculating the external area per mass of solids*, we obtain

$$a_c = \frac{6(1-\phi)}{d_p} = \frac{6(1-0.5)}{6 \times 10^{-3} \text{ m}}$$
$$= 500 \text{ m}^2/\text{m}^3$$

4. *Evaluating the overall effectiveness factor.* Substituting into Equation (12-55), we have

$$\Omega = \frac{\eta}{1 + \eta k_1 S_a \rho_b / k_c a_c}$$

$$\Omega = \frac{0.167}{1 + \dfrac{(0.167)(4.4 \times 10^{-10} \text{ m}^3/\text{m}^2 \cdot \text{s})(530 \text{ m}^2/\text{g})(1.4 \times 10^6 \text{ g}/\text{m}^3)}{(6 \times 10^{-5} \text{ m/s})(500 \text{ m}^2/\text{m}^3)}}$$

$$= \frac{0.167}{1 + 1.83} = 0.059$$

In this example we see that both the external and internal resistances to mass transfer are significant.

5. *Calculating the weight of solid necessary to achieve 99.8% conversion.* Substituting into Equation (E12-3.3), we obtain

$$W = \frac{1 \times 10^{-6} \text{ m}^3/\text{s}}{(0.059)(4.42 \times 10^{-10} \text{ m}^3/\text{m}^2 \cdot \text{s})(530 \text{ m}^2/\text{g})} \ln \frac{1}{1 - 0.998}$$

$$= 450 \text{ g}$$

6. *The reactor length is*

$$L = \frac{W}{A_c \rho_b} = \frac{450 \text{ g}}{(2.03 \times 10^{-3} \text{ m}^2)(1.4 \times 10^6 \text{ g}/\text{m}^3)}$$

$$= 0.16 \text{ m}$$

12.7 Determination of Limiting Situations from Reaction Data

For external mass transfer-limited reactions in packed beds, the rate of reaction at a point in the bed is

$$-r_A' = k_c a_c C_A \tag{12-72}$$

Variation of reaction rate with system variables

The correlation for the mass transfer coefficient, Equation (11-66), shows that k_c is directly proportional to the square root of the velocity and inversely proportional to the square root of the particle diameter:

$$k_c \propto \frac{U^{1/2}}{d_p^{1/2}} \tag{12-73}$$

We recall from Equation (12-55) that the variation of external surface area with catalyst particle size is

$$a_c \propto \frac{1}{d_p}$$

Consequently, for external mass transfer-limited reactions, the rate is inversely proportional to the particle diameter to the three-halves power:

$$-r_A' \propto \frac{1}{d_p^{3/2}} \tag{12-74}$$

From Equation (11-75) we see that for gas-phase external mass transfer-limited reactions, the rate increases approximately linearly with temperature.

When internal diffusion limits the rate of reaction, we observe from Equation (12-39) that the rate of reaction varies inversely with particle diameter, is independent of velocity, and exhibits an exponential temperature dependence which is not as strong as that for surface-reaction-controlling reactions. For surface-reaction-limited reactions the rate is independent of particle size and is a strong function of temperature (exponential). Table 12-1 summarizes the dependence of the rate of reaction on the velocity through the bed, particle diameter, and temperature for the three types of limitations that we have been discussing.

TABLE 12-1

Type of Limitation	Variation of Reaction Rate with:		
	Velocity	Particle Size	Temperature
External diffusion	$U^{1/2}$	$(d_p)^{-3/2}$	$\approx$ Linear
Internal diffusion	Independent	$(d_p)^{-1}$	Exponential
Surface reaction	Independent	Independent	Exponential

Many heterogeneous reactions are diffusion limited

The exponential temperature dependence for internal diffusion limitations is usually not as strong a function of temperature as is the dependence for surface reaction limitations. If we would calculate an activation energy between 8 and 24 kJ/mol, chances are that the reaction is strongly diffusion-limited. An activation energy of 200 kJ/mol, however, suggests that the reaction is reaction rate-limited.

12.8 Multiphase Reactors

Multiphase reactors are reactors in which two or more phases are necessary to carry out the reaction. The majority of multiphase reactors involve gas and liquid phases which contact a solid. In the case of the slurry and trickle bed reactors, the reaction between the gas and the liquid takes place on a solid catalyst surface (see Table 12-2). However, in some reactors the liquid phase is an inert medium for the gas to contact the solid catalyst. The latter situation arises when a large heat sink is required for highly exothermic reactions. In many cases the catalyst life is extended by these milder operating conditions.

The multiphase reactors discussed in this edition of the book are the slurry reactor, fluidized bed, and the trickle bed reactor. The trickle bed reactor which has reaction and transport steps similar to the slurry reactor is discussed in the first edition of the book and on the CD-ROM along with the bubbling fluidized bed. In slurry reactors, the catalyst is suspended in the liquid and gas is bubbled through the liquid. A slurry reactor may be operated in either a semibatch or continuous mode.

TABLE 12-2.   APPLICATIONS OF THREE-PHASE REACTORS

I. *Slurry reactor*
   A. Hydrogenation
      1. of fatty acids over a supported nickel catalyst
      2. of 2-butyne-1,4-diol over a Pd-CaCO$_3$ catalyst
      3. of glucose over a Raney nickel catalyst
   B. Oxidation
      1. of C$_2$H$_4$ in an inert liquid over a PdCl$_2$-carbon catalyst
      2. of SO$_2$ in inert water over an activated carbon catalyst
   C. Hydroformation
      of CO with high-molecular-weight olefins on either a cobalt or ruthenium complex bound to polymers
   D. Ethynylation
      Reaction of acetylene with formaldehyde over a CaCl$_2$-supported catalyst
II. *Trickle bed reactors*
   A. Hydrodesulfurization
      Removal of sulfur compounds from crude oil by reaction with hydrogen on Co-Mo on alumina
   B. Hydrogenation
      1. of aniline over a Ni-clay catalyst
      2. of 2-butyne-1,4-diol over a supported Cu-Ni catalyst
      3. of benzene, α-CH$_3$ styrene, and crotonaldehyde
      4. of aromatics in napthenic lube oil distilate
   C. Hydrodenitrogenation
      1. of lube oil distillate
      2. of cracked light furnace oil
   D. Oxidation
      1. of cumene over activated carbon
      2. of SO$_2$ over carbon

Source: C. N. Satterfield, *AIChE J.*, *21*, 209 (1975); P. A. Ramachandran and R. V. Chaudhari, *Chem. Eng.*, *87*(24), 74 (1980); R. V. Chaudhari and P. A. Ramachandran, *AIChE J.*, *26*, 177 (1980).

12.8.1 Slurry Reactors

Uses of a slurry reactor

In recent years there has been an increased emphasis on the study of slurry reactors in chemical reactor engineering. A slurry reactor is a multiphase flow reactor in which reactant gas is bubbled through a solution containing solid catalyst particles. The solution may be either a reactant, as in the case of the hydrogenation of methyl linoleate, a product as in the case of the production of hydrocarbon wax, or an inert, as in the Fischer–Tropsch synthesis of methane. Slurry reactors may be operated in a batch or continuous mode. One of the main advantages of slurry reactors is that temperature control and heat recovery are easily achieved. In addition, constant overall catalytic activity can be maintained by the addition of small amounts of catalyst with each reuse during batch operation or with constant feeding during continuous operation.

Example 12–4 Industrial Slurry Reactor

Describe the operation an industrial slurry reactor used to convert synthesis gas (CO and H$_2$) to a hydrocarbon wax by the Fischer–Tropsch synthesis.

Solution

The Fischer–Tropsch reactions were discussed in Example 1-4. A schematic of the Sasol slurry reactor, used to make wax, is shown in Figure E12-4.1. In the slurry reactor, a typical reaction stoichiometry might be

$$25CO + 51H_2 \longrightarrow C_{25}H_{52} + 25H_2O$$

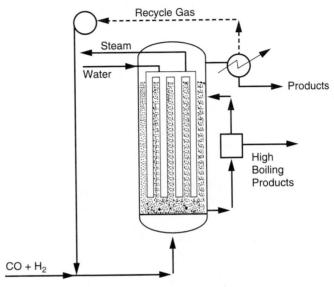

Figure E12-4.1   Sasol slurry reactor.

The reactor is 5 m in diameter and 22 m high and is operated at a temperature of 240°C and pressure of 20 atm. The synthesis gas is bubbled through a heavy oil which is usually a product of the Fischer–Tropsch reaction itself. The catalyst loading (i.e., density in the solution) is the order of 100 kg/m³ with a typical operating range of 1 to 20 wt % solids. The reactor is cooled by an internal heat exchanger through which the coolant stream enters as water exits as steam to maintain the reactor at 240°C.

Synthesis gas is fed at a rate of 150,000 m³/h (STP) and has a composition of 12% CH_4, 1% CO_2, 29% CO, and 58% H_2. The fresh gas is mixed with the recycled gas and the mixture enters the reactor at 120°C. The liquid wax product stream exits at flow rate of 4.5 m³/h and has a mixture of hydrocarbons with the general formula of C_nH_{2n}, with n varying between 20 and 50. The exit tail gas stream contains methane (38%), hydrogen (37%), CO_2 (14%) with CO, water, and light hydrocarbons C_2 to C_5 making up the remaining 11%. The Sasol reactor is actually modeled as three or four slurry reactors in series. The liquid-phase of each slurry reactor in series is modeled as being well mixed, while the gas phase is modeled as being in plug flow as it moves up the column.

A more detailed schematic diagram of a slurry reactor is shown in Figure 12-11. In modeling the slurry reactor we assume that the liquid phase is well mixed, the catalyst particles are uniformly distributed, and the gas phase is in plug flow. The reactants in the gas phase participate in five reaction steps:

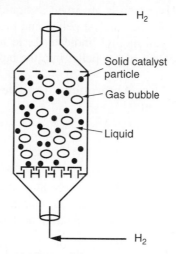

Figure 12-11   Slurry reactor for the hydrogenation of methyl linoleate.

Reaction steps in a
slurry reactor

1. Absorption from the gas phase into the liquid phase at the bubble surface
2. Diffusion in the liquid phase from the bubble surface to the bulk liquid
3. Diffusion from the bulk liquid to the external surface of the solid catalyst
4. Internal diffusion of the reactant in the porous catalyst
5. Reaction within the porous catalyst

The reaction products participate in the steps above but in reverse order (5 through 1). Each step may be thought of as a resistance to the overall rate of reaction R_A. These resistances are shown schematically in Figure 12-12. The concentration in the liquid phase is related to the gas-phase concentration through Henry's law:

Equilibrium at the
gas–liquid interface

$$C_i = P_i H' \tag{12-75}$$

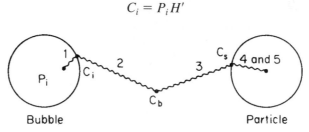

Figure 12-12   Steps in a slurry reactor.

One of the things we want to achieve in our analysis of slurry reactors is to learn how to detect which resistance is the largest (i.e., slowest step) and how we might operate the reactor to decrease the resistance of this step and thereby increase the efficiency of the reactor.

To illustrate the principles of slurry operation, we shall consider the hydrogenation of methyl linoleate, L, to form methyl oleate, O:

$$\text{methyl linoleate}(l) + \text{hydrogen}(g) \longrightarrow \text{methyl oleate}(l)$$
$$\text{L} \qquad + \quad \text{H}_2 \qquad \longrightarrow \qquad \text{O}$$

Hydrogen is absorbed in liquid methyl linoleate, diffuses to the external surface of the catalyst pellet, and then diffuses into the catalyst pellet, where it reacts with methyl linoleate, L, to form methyl oleate, O. Methyl oleate then diffuses out of the pellet into the bulk liquid.

Rate of Gas Absorption   The rate of absorption of H_2 per unit volume of linoleate oil is

$$R_A = k_b a_b (C_i - C_b) \qquad (12\text{-}76)$$

where k_b = mass transfer coefficient for gas absorption,[8] dm/s
$\quad\quad a_b$ = bubble surface area, dm²/(dm³ of solution)
$\quad\quad C_i$ = H_2 concentration in the oil at the oil–H_2 bubble interface, mol/dm³
$\quad\quad C_b$ = bulk concentration of H_2 in solution, mol/dm³

$$R_A [=] \frac{dm}{s} \left(\frac{dm^2}{dm^3 \text{ of solution}} \right) \frac{mol}{dm^3} = \frac{mol}{(dm^3 \text{ of solution}) \cdot s}$$

Equation (12-76) gives the rate of H_2 transport from the gas–liquid interface to the bulk liquid.

Transport to the Catalyst Pellet   The rate of mass transfer of H_2 from the bulk solution to the external surface of catalyst particles is

$$R_A = k_c a_p m (C_b - C_s) \qquad (12\text{-}77)$$

where k_c = mass transfer coefficient for particles, dm/s
$\quad\quad a_p$ = external surface area of particles, dm²/g of catalyst
$\quad\quad m$ = mass concentration of catalyst (g of catalyst/dm³ of solution); the
Catalyst loading $\quad\quad\quad\quad$ parameter m is also referred to as the *catalyst loading*
$\quad\quad C_s$ = concentration of H_2 at external surface of catalyst pellet, mol/dm³

$$R_A [=] \left(\frac{dm}{s} \right) \frac{dm^2}{g} \left(\frac{g}{dm^3 \text{ of solution}} \right) \frac{mol}{dm^3}$$

$$= \frac{mol}{(dm^3 \text{ of solution}) \cdot s}$$

Diffusion and Reaction in the Catalyst Pellet   In Section 12.2 we showed that the internal effectiveness factor was the ratio of the actual rate of reaction, $-r_A'$, to the rate r_{As}' that would exist if the entire interior of the pellet were exposed to the reactant concentration at the external surface, C_{As}. Consequently, the actual rate of reaction per unit mass of catalyst can be written

---

[8] Correlations for $k_b a_b$ for a wide variety of situations can be found in the review article "Design parameter estimations for bubble column reactors," by Y. T. Shah et al., *AIChE J.*, 28, 353 (1982).

$$\boxed{-r_A' = \eta(-r_{As}')}$$ (12-38)

Multiplying by the mass of catalyst per unit volume of solution, we obtain the rate of reaction per volume of solution:

$$R_A = m\eta(-r_{As}')$$

$$R_A [=] \frac{\text{g of catalyst}}{\text{dm}^3 \text{ of solution}} \left(\frac{1}{1}\right) \frac{\text{mol}}{\text{g cat.} \cdot \text{s}} = \frac{\text{mol}}{(\text{dm}^3 \text{ of solution}) \cdot \text{s}}$$ (12-78)

The Rate Law   The rate law is first-order in hydrogen and first-order in methyl linoleate. However, because the liquid phase is essentially all linoleate, it is in excess and its concentration, C_L, remains virtually constant at its initial concentration, C_{L0}, for small to moderate reaction times.

$$-r_A' = k'C_{L0}C = kC$$ (12-79)

The rate of reaction evaluated at the external pellet surface is

$$-r_{As}' = kC_s$$ (12-80)

where C_s = concentration of hydrogen at the external pellet surface, mol/dm³
k = specific reaction rate, dm³/g cat. · s

Determining the Limiting Step   Because at any point in the column the overall rate of transport is at steady state, the rate of transport from the bubble is equal to the rate of transport to the catalyst surface, which in turn is equal to the rate of reaction in the catalyst pellet. Consequently, for a reactor that is perfectly mixed, *or* one in which the catalyst, fluid, and bubbles all flow upward together in plug flow, we find that

$$R_A = k_b a_b (C_i - C_b) = k_c m a_p (C_b - C_s) = m\eta(-r_{As}')$$

Equations (12-76) through (12-80) can be arranged in the form

$$\frac{R_A}{k_b a_b} = C_i - C_b$$

$$\frac{R_A}{k_c a_p m} = C_b - C_s$$

$$\frac{R_A}{mk\eta} = C_s$$

Adding the equations above yields

$$R_A \left(\frac{1}{k_b a_b} + \frac{1}{k_c a_p m} + \frac{1}{k\eta m} \right) = C_i$$ (12-81)

Rearranging, we have

$$\boxed{\frac{C_i}{R_A} = \frac{1}{k_b a_b} + \frac{1}{m}\left(\frac{1}{k_c a_p} + \frac{1}{k\eta}\right)} \tag{12-82}$$

Each of the terms on the right-hand side can be thought of as a resistance to the overall rate of reaction such that

$$\frac{C_i}{R_A} = r_b + \frac{1}{m}(r_c + r_r) \tag{12-83}$$

or

$$\frac{C_i}{R_A} = r_b + \frac{1}{m}r_{cr} \tag{12-84}$$

where

$$r_b = \frac{1}{k_b a_b} = \text{resistance to gas absorption, s} \tag{12-85}$$

$$r_c = \frac{1}{k_c a_p} = \text{specific resistance to transport to surface of catalyst} \atop \text{pellet, gcat} \cdot \text{s/dm}^3 \tag{12-86}$$

$$r_r = \frac{1}{\eta k} = \text{specific esistance to diffusion and reaction within the} \atop \text{catalyst pellets, gcat} \cdot \text{s/dm}^3 \tag{12-87}$$

$$r_{cr} = r_r + r_c = \text{specific combined resistance to internal diffusion,} \atop \text{reaction, and external diffusion, gcat} \cdot \text{s/dm}^3 \tag{12-88}$$

For reactions other than first-order,

$$r_r = \frac{C_s}{\eta(-r'_{As})} \tag{12-89}$$

We see from Equation (12-84) that a plot of C_i/R_A as a function of the reciprocal of the catalyst loading $(1/m)$ should be a straight line. The slope will be equal to the specific combined resistance r_{cr} and the intercept will be equal to the gas absorption resistance r_b. Consequently, to learn the magnitude of the resistances, we would vary the concentration of catalyst (i.e., the catalyst loading, m) and measure the corresponding overall rate of reaction (Figure 12-13). The ratio of gas absorption resistance to diffusional resistance to and within the pellet at a particular catalyst loading m is

$$\frac{\text{absorption resistance}}{\text{diffusion resistance}} = \frac{r_b}{r_{cr}(1/m)} = \frac{\text{intercept} \times m}{\text{slope}}$$

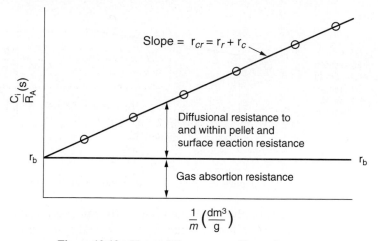

Finding the limiting resistance

Figure 12-13 Plot to delineate controlling resistances.

Suppose it is desired to change the catalyst pellet size (to make them smaller, for example). Because gas absorption is independent of catalyst particle size, the intercept will remain unchanged. Consequently, only one experiment is necessary to determine the combined diffusional and reaction resistances r_{cr}. As the particle size is decreased, both the effectiveness factor and mass transfer coefficient increase. As a result, the combined resistance, r_{cr}, decreases, as shown by the decreasing slope in Figure 12-14a. In Figure 12-14b we see that as the resistance to gas absorption increases, the intercept increases but the slope does not change. The two extremes of these controlling resistances are shown in Figure 12-15. Figure 12-15a shows a large intercept (r_b) and a small slope $(r_c + r_r)$, while Figure 12-15b shows a large slope $(r_c + r_r)$ and a small intercept. To decrease the gas absorption resistance, we might consider changing the sparger to produce more gas bubbles of smaller diameters.

Now that we have shown how we learn whether gas absorption r_b or diffusion-reaction $(r_c + r_r)$ is limiting by varying the catalyst loading, we will

If diffusion controls, decrease particle size, use more catalyst

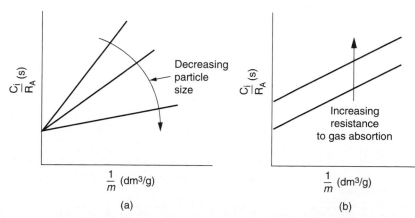

Figure 12-14 (a) Effect of particle size; (b) effect of gas absorption.

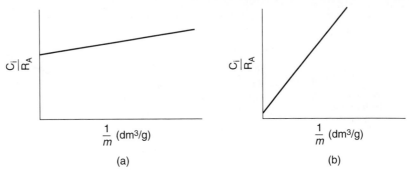

Figure 12-15   (a) Gas absorption controls; (b) diffusion and reaction control.

If gas absorption controls, could change the sparger to get smaller bubbles

focus on the case when both diffusion and reaction are limiting. The next step is to learn how we can separate r_c and r_r to learn whether

1. External diffusion is controlling,
2. Internal diffusion is controlling, or
3. Surface reaction is controlling.

To learn which of these steps controls, we must vary the particle size. After determining r_{cr} from the slope of C_i/R versus $1/m$ at each particle size, we can construct a plot of r_{cr} versus particle size, d_p.

Combined external and internal resistances

$$r_{cr} = \frac{1}{k_a a_p} + \frac{1}{\eta k} \tag{12-90}$$

a. **Small particles.** It has been shown earlier (see Figure 12-5) that as the particle diameter becomes small, the surface reaction controls and the effectiveness factor approaches 1.0. For small values of k (reaction control)

$$r_{cr} \simeq \frac{1}{k}$$

Surface reaction rate is independent of particle size

Consequently, r_{cr} and r_r are independent of particle size and a plot of $\ln r_{cr}$ as a function of $\ln d_p$ should yield a zero slope when surface reaction is limiting.

b. **Moderate-size particles.** For large values of the Thiele modulus we have shown [Equation (12-33)] that

$$\eta = \frac{3}{\phi} = \frac{6}{d_p}\left(\frac{D_e}{k\rho_c S_a}\right)^{1/2}$$

Then

If internal diffusion controls, resistance will vary linearly with particle size

$$r_r = \frac{1}{\eta k} = \alpha_1 d_p \tag{12-91}$$

We see that internal diffusion limits the reaction if a plot of r_{cr} versus d_p is linear. Under these conditions the overall rate of reaction can be increased by decreasing the particle size. However, the overall rate will be unaffected by the mixing conditions in the bulk liquid that would change the mass transfer boundary layer thickness next to the pellet surface.

c. **Moderate to large particles.** External resistance to diffusion was given by the equation

$$r_c = \frac{1}{k_c a_p} \tag{12-92}$$

The external surface area per mass of catalyst is

$$a_p = \frac{\text{external area}}{\text{mass of pellet}} = \frac{\pi d_p^2}{(\pi/6) d_p^3 \rho_c} = \frac{6}{d_p \rho_c} \tag{12-93}$$

Next we need to learn the variation of the mass transfer coefficient with particle size.

Case 1: No Shear Stress between Particles and Fluid   If the particles are sufficiently small, they move with the fluid motion such that there is no shear between the particle and the fluid. This situation is equivalent to diffusion to a particle in a stagnant fluid. Under these conditions the Sherwood number is 2:

$$\text{Sh} = \frac{k_c d_p}{D_{AB}} = 2 \tag{12-94}$$

Then

$$k_c = 2 \frac{D_{AB}}{d_p}$$

and

$$\boxed{r_c = \frac{\rho_c d_p^2}{12 D_{AB}} = \alpha_2 d_p^2} \tag{12-95}$$

Consequently, if external diffusion is controlling and there is no shear between the particle and fluid, the slope of a plot of $\ln r_{cr}$ versus $\ln d_p$ should be 2. Because the particle moves with the fluid, increasing the stirring would have no effect in increasing the overall rate of reaction.

Case 2: Shear between Particles and the Fluid   If the particles are sheared by the fluid motion, one can neglect the 2 in the Frössling correlation between the Sherwood number and Reynolds number, and

$$\text{Sh} = 2 + 0.6 \text{Re}^{1/2} \text{Sc}^{1/3} \tag{12-96}$$

becomes

$$\text{Sh} \propto \text{Re}^{1/2}$$

Then

$$\frac{k_c d_p}{D_{AB}} \propto \left(\frac{d_p U}{\nu}\right)^{1/2}$$

and

$$k_c \propto \frac{U^{1/2}}{d_p^{1/2}}$$

and

$$k_c a_c \propto \frac{U^{1/2}}{d_p^{1.5}}$$

> For external diffusion control resistance varies with square of particle size for smaller particles and $\frac{3}{2}$ power of particle size for larger particles

$$\boxed{r_c = \alpha_3 d_p^{1.5}} \tag{12-97}$$

Another correlation for mass transfer to spheres in a liquid moving at a low velocity[9] gives

$$\text{Sh}^2 = 4.0 + 1.21\,(\text{ReSc})^{2/3} \tag{12-98}$$

from which we can obtain, upon neglecting the first term on the right-hand side,

$$r_c = \alpha_4 d_p^{1.7}$$

If the combined resistance varies with d_p from the 1.5 to 1.7 power, then external resistance is controlling and the mixing (stirring speed) is important. Figure 12-16 shows a plot of the combined resistance r_{cr} as a function of particle diameter d_p on log-log paper for the various rate-limiting steps.

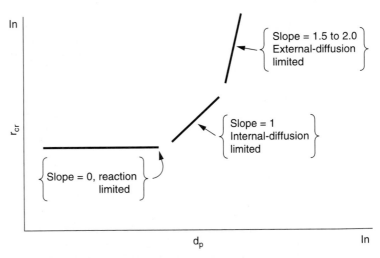

Figure 12-16   Effect of particle size on controlling resistance.

This figure is a schematic to show the various slopes of the resistance's and does not imply that internal diffusion limitations will always occur at particle sizes smaller than those shown for external diffusion limitations.

[9] C. N. Satterfield, *Mass Transfer in Heterogeneous Catalysis*, MIT Press, Cambridge, Mass., 1970, p. 114.

Given a set of reaction rate data, we can carry out the following procedure to determine which reaction step is limiting:

Algorithm to determine reaction-limiting step

1. Construct a series of plots of C_i/R as a function of $1/m$ for different pellet diameters.
2. Determine the combined resistance from the slopes of these plots for each corresponding particle diameter.
3. Plot r_{cr} as a function of d_p on log-log paper. From the slope of this plot determine which step is controlling. The slope should be 0, 1, 1.5, 1.7, or 2.
4. If the slope is between any of these values, say 0.5, this suggests that more than one resistance is limiting.

The variables that influence reactor operation under each of the limiting conditions just discussed are shown in Table 12-3.

TABLE 12-3.   VARIABLES AFFECTING OBSERVED REACTION RATE

Controlling Step	Variables with:		
	Major Influence	Minor Influence	Insignificant Influence
Gas–liquid mass transport	Stirring rate Reactor design (impeller, gas distributor, baffling, etc.) Concentration of reactant in gas phase	Temperature	Concentration of liquid-phase reactant Amount of catalyst Catalyst particle size Concentration of active component(s) on catalyst
Liquid–solid mass transport (gaseous reactant)	Amount of catalyst Catalyst particle size Concentration of reactant in gas phase	Temperature Stirring rate Reactor design Viscosity Relative densities	Concentration of liquid-phase reactant Concentration of active component(s) on catalyst
Liquid–solid mass transport (liquid reactant)	Amount of catalyst Catalyst particle size Concentration of reactant in liquid phase	Temperature Stirring rate Reactor design Viscosity Relative densities	Concentration of gas-phase reactant Concentration of active component(s) on catalyst
Chemical reaction (insignificant pore diffusion resistance)	Temperature Amount of catalyst Reactant concentrations Concentration of active component(s) on catalyst		Stirring rate Reactor design Catalyst particle size
Chemical reaction (significant pore diffusion resistance)	Amount of catalyst Reactant concentrations Temperature[a] Catalyst particle size Concentration of active component(s) on catalyst[a]	Pore structure	Stirring rate Reactor design

[a]These variables do not exert as strong an influence as when pore diffusion resistance is negligible.
Source: G. Roberts, in *Catalysis in Organic Synthesis*, P. N. Rylander and H. Greenfield, eds., San Diego, Calif., Academic Press, 1976. Reprinted with permission of Academic Press and Engelhard Industries, Edison, N.J.

Example 12–5   *Determining the Controlling Resistance*

The catalytic hydrogenation of methyl linoleate[10] was carried out in a laboratory-scale slurry reactor in which hydrogen gas was bubbled up through the liquid and catalyst. Unfortunately, the pilot-plant reactor did not live up to the laboratory reactor expectations. The catalyst particle size normally used was between 10 and 100 μm. In an effort to deduce the problem, the experiments listed in Table E12-5.1 were carried out on the pilot plant slurry reactor at 121°C.

TABLE E12-5.1.   RAW DATA

Run	Partial Pressure of H_2 (atm)	Solubility[a] of H_2 (kmol/m^3)	Size of Catalyst Particles (μm)	Catalyst Charge (kg/m^3)	H_2 Rate of Reaction, $-r_{H_2}$ (kmol/m$^3 \cdot$min)
1	3	0.007	40.0	5.0	0.0625
2	6	0.014	40.0	0.2	0.0178
3	6	0.014	80.0	0.16	0.0073

[a]Henry's law: $H'P_{H_2} = C_{H_2}$ with $H' = 0.00233$ mol H_2/atm$\cdot$dm^3.

    (a) What seems to be the problem (i.e., major resistance) with the pilot-plant reactor, and what steps should be taken to correct the problem? Support any recommendations with calculations.

    (b) For the 80-μm particle size, what are the various percentage resistances to absorption, diffusion, and so on, when the catalyst charge is 0.40 kg/m^3?

Solution

To determine the major resistance we need to plot $C_i/(-r_{H_2})$ as a function of $1/m$. First, compare the slope and intercept of the 40.0-μm particle size experiment to learn if gas absorption is the major resistance. From the data in Table E12-5.1, we develop Table E12-5.2. These data are plotted in Figure E12-5.1. For catalyst charges below 2.0 kg/m^3, diffusion is the major resistance to the overall reaction for the 80-μm particle.

TABLE E12-5.2.   PROCESSED DATA

Run	C_i/R_i (min)	$1/m$ (m^3/kg)
1	0.112	0.20
2	0.787	5.00
3	1.92	6.25

    (a) From the slope of the line corresponding to the 40-μm particle size, the combined external and internal diffusion and reaction resistance is

---

[10]W. A. Cordova and P. Harriott, *Chem. Eng. Sci.*, *30*, 1201 (1975).

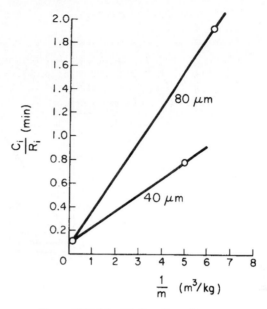

Figure E12-5.1   Finding the resistances.

$$r_{cr(40.0\ \mu m)} = \frac{0.787 - 0.112}{5.00 - 0.2} = 0.14 \ \frac{min \cdot kg}{m^3} \tag{E12-5.1}$$

For the 80.0 μm particle size, we obtain

$$r_{cr(80.0\ \mu m)} = \frac{1.92 - 1.00}{6.25 - 3.00} = 0.28 \ \frac{min \cdot kg}{m^3} \tag{E12-5.2}$$

Comparing Equations (E12-5.2) and (E12-5.1) gives us

$$\frac{r_{cr(80)}}{r_{cr(40)}} = \frac{0.28}{0.14} = 2.0$$

We see that when particle size is doubled, resistance is also doubled.

$$r_{cr} \propto d_p$$

Because the combined resistance is proportional to the particle diameter to the first power, internal diffusion is the controlling resistance of the three resistances. To decrease this resistance a smaller catalyst particle size should be used.

(b) For the 80.0-μm particle size at a catalyst charge of 0.4 kg/m³, the overall resistance at $1/m = 2.5$ is 0.84 min (Figure E12-5.2).

$$\text{percent gas absorption resistance} = \frac{0.08}{0.84} \times 100 = 9.5\%$$

$$\text{percent internal diffusion resistance} = \frac{0.84 - 0.08}{0.84} \times 100 = 90.5\%$$

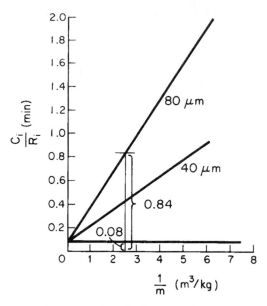

Figure E12-5.2 Finding the resistances.

Slurry Reactor Design In the previous material we discussed the transport and reaction steps and developed an equation for the overall resistance. A rearrangement of Equation (12-81) gives

$$R_A \equiv -r_A = \frac{C_i}{1/k_b a_b + 1/m(1/k_c a_c + 1/\eta k)} \tag{12-99}$$

To design slurry reactors we simply couple this rate law with the appropriate mole balance[11] (see Chapters 1 and 2).

Example 12–6 Slurry Reactor Design

Methyl linoleate is to be converted to methyl oleate in a 2-m³ slurry reactor. The molar feed rate of methyl linoleate to the reactor is 0.7 kmol/min. The partial pressure of H_2 is 6 atm and the reactor is considered to be well-mixed. Calculate the catalyst charge necessary to achieve 30% conversion for a 60-μm particle size. The reaction conditions are the same as those described in Example 12-5.

Solution

For a well-mixed reactor, the CSTR design equation is

$$V = \frac{F_{A0}X}{-r_A} \tag{E12-6.1}$$

---

[11]An interesting example of a slurry reactor modeled as a plug-flow reactor for Fischer–Tropsch synthesis is given by D. Stern, A. T. Bell, and H. Heinemann, *Chem. Eng. Sci.*, *38*, 597 (1983).

Multiplying by the solubility of hydrogen and rearranging gives us

$$\frac{VC_i}{F_{A0}X} = \frac{C_i}{-r_A} \tag{E12-6.2}$$

Equating Equations (E12-6.2) and (12-84), we have

$$\frac{VC_i}{F_{A0}X} = r_b + \frac{1}{m} r_{cr} \tag{E12-6.3}$$

From Example 12-5,

$$r_b = 0.08 \text{ min}$$

and for the 80-μm particle,

$$r_{cr} = 0.28 \text{ min} \cdot \text{kg/m}^3$$

Because internal diffusion controls,

$$r_{cr} = \alpha d_p$$

$$r_{cr(60 \text{ μm})} = \frac{60}{80}(0.28) = 0.21 \text{ min} \cdot \text{kg/m}^3$$

Substituting the parameter values into Equation (E12-6.3) gives us

$$\frac{(2 \text{ m}^3)(0.014 \text{ kmol/m}^3)}{(0.7 \text{ kmol/min})(0.3)} = 0.08 \text{ min} + \frac{1}{m}(0.21 \text{ kg} \cdot \text{min/m}^3)$$

which solves to a catalyst charge of

$$m = 3.9 \text{ kg/m}^3$$

Slurry reactors and ebulating bed reactors typically contain 1 to 5% catalyst by weight.

12.8.2 Trickle Bed Reactors

In a trickle bed reactor the gas and liquid flow (trickle) concurrently downward over a packed bed of catalyst particles. Industrial trickle beds are typically 3 to 6 m deep and up to 3 m in diameter and are filled with catalyst particles ranging from $\frac{1}{8}$ to $\frac{1}{32}$ in. in diameter. The pores of the catalyst are filled with liquid. In petroleum refining, pressures of 34 to 100 atm and temperatures of 350 to 425°C are not uncommon. A pilot-plant trickle bed reactor might be about 1 m deep and 4 cm in diameter. Trickle beds are used in such processes as the hydrodesulfurization of heavy oil stocks, the hydrotreating of lubricating oils, and reactions such as the production of butynediol from acetylene and aqueous formaldehyde over a copper acetylide catalyst. It is on this latter type of reaction,

$$A(g, l) + B(l) \longrightarrow C(l) \tag{12-100}$$

Characteristics and uses of a trickle bed reactor

that we focus in this section and on the CD-ROM. In a few cases, such as the Fischer–Tropsch synthesis, the liquid is inert and acts as a heat-transfer medium.

Fundamentals The basic reaction and transport steps in trickle bed reactors are similar to those in slurry reactors. The main differences are the correlations used to determine the mass transfer coefficients. In addition, if there is more than one component in the gas phase (e.g., liquid has a high vapor pressure or one of the entering gases is inert), there is one additional transport step in the gas phase. Figure 12-17 shows the various transport steps in trickle bed reactors. Following our analysis for slurry reactors we develop the equations for the rate of transport of each step. The steps involving **reactant A** in the gas phase are

Transport from bulk gas to gas–liquid interface to bulk liquid to solid–liquid interface

1. Transport from the bulk gas phase to the gas–liquid interface.
2. Equilibrium at the gas–liquid interface.
3. Transport from the interface to bulk liquid.
4. Transport from the bulk liquid to external catalyst surface.
5. Diffusion and reaction in the pellet.

Diffusion and reaction in catalyst pellet

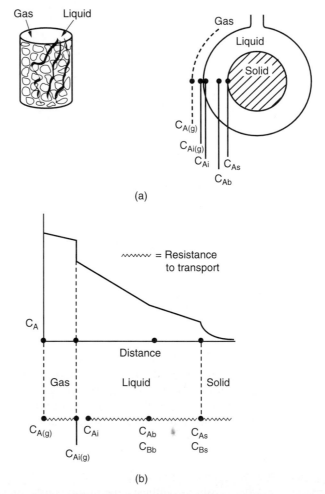

(a)

(b)

Figure 12-17 (a) Trickle bed reactor; (b) reactant concentration profile.

If we assume first-order reaction in dissolved gas A and in liquid B, we can combine all five steps as shown on the CD-ROM to arrive at

Overall rate
equation for A

$$-r_A' = \cfrac{1/H}{\underbrace{\cfrac{(1-\phi)\rho_c}{Hk_g a_i} + \cfrac{(1-\phi)\rho_c}{k_l a_i} + \cfrac{1}{k_c a_p} + \cfrac{1}{\eta k C_{Bs}}}_{k_{vg}}} C_A(g) \quad \frac{\text{mol}}{\text{g cat.} \cdot \text{s}} \quad (12\text{-}101)$$

that is,

$$-r_A' = k_{vg} C_A(g) \qquad (12\text{-}102)$$

where k_{vg} is the overall transfer coefficient for the gas into the pellet (m³ of gas/g cat.·s). H is Henry's law constant, k_g and k_l are the gas phase and liquid phase mass transfer coefficients, and a_i is the gas-liquid interfacial area per volume of bed, and other terms are as previously defined.

A mole balance on species A gives

Mole balance on A

$$\boxed{\frac{dF_A}{dW} = r_A' = -k_{vg} C_A(g)} \qquad (12\text{-}103)$$

We next consider the transport and reaction of the liquid-phase **reactant B**.

6. Transport of B from bulk liquid to solid catalyst interface.
7. Diffusion and reaction of B inside the catalyst pellet.

Overall rate
equation of B

$$-r_B' = \cfrac{1}{\underbrace{\cfrac{1}{k_c a_p} + \cfrac{1}{\eta k C_{As}}}_{k_{vl}}} C_B \qquad \frac{\text{mol}}{\text{g cat.} \cdot \text{s}} \qquad (12\text{-}104)$$

$$-r_B = k_{vl} C_B \qquad (12\text{-}105)$$

Mole balance on B

$$\boxed{\frac{dF_B}{dW} = v_l \frac{dC_B}{dW} = -r_B' = -k_{vl} C_B} \qquad (12\text{-}106)$$

One notes that the surface concentrations of A and B, C_{As} and C_{Bs}, appear in the denominator of the overall transport coefficients k_{vg} and k_{vl}. Consequently, Equations (12-101), (12-102), and (12-103) must be solved simultaneously. In some cases analytical solutions are available, but for complex rate laws, one resorts to numerical solutions.[12] However, we shall consider some limiting situations on the CD-ROM along with an example problem.

---

[12]A number of worked example problems for three-phase reactors can be found in an article by P. A. Ramachandran and R. Chaudhari, *Chem. Eng.*, *87*(24), 74 (1980).

12.9 Fluidized-Bed Reactors

The fluidized-bed reactor (FBR) has the ability to process large volumes of fluid. For the catalytic cracking of petroleum naphthas to form gasoline blends, as an example, the virtues of the fluidized-bed reactor drove its competitors from the market. Below is an outline of the FBR material found on the CD-ROM.

Fluidization occurs when small solid particles are suspended in an upward-flowing stream of fluid, as shown in Figure 12-18.

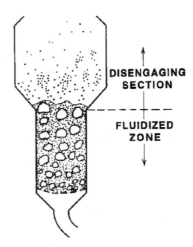

Figure 12-18   From Kunii and Levenspiel *Fluidization Engineering*, Copyright © 1969, Robert E. Krieger Pub. Co. Reproduced by permission of the publisher.

The fluid velocity is sufficient to suspend the particles, but not large enough to carry them out of the vessel. The solid particles swirl around the bed rapidly, creating excellent mixing among them. The material "fluidized" is almost always a solid. The "fluidizing medium" is either a liquid or gas, and the characteristics and behavior of a fluidized bed are strongly dependent on it. Nearly all the significant commercial applications of fluidized-bed technology concern gas-solid systems, so these will be treated in this chapter.

As shown in Figure 12-18 the containing vessel is usually cylindrical, though not necessarily so. At the bottom of the bed is a "distributor plate"— porous, pierced with holes, or perhaps containing bubble caps—which acts as a support for the bed and distributes the gas evenly over the vessel cross section. Above the bed is a space, termed the disengaging section, which allows the solids caught in the gas stream to fall back into the bed.

The material that follows is based upon what is seemingly the best model of the fluidized-bed reactor developed thus far—the bubbling bed model of Kunii and Levenspiel.[13]

---

[13]Kunii, D. and O. Levenspiel, *Fluidization Engineering 2nd Ed.* (Boston Butterworth, 1991).

12.10 An Over View

We are going to use the Kunii-Levenspiel bubbling bed model to describe reactions in fluidized beds. In this model, the reactant gas enters the bottom of the bed, and flows up the reactor in the form of bubbles.

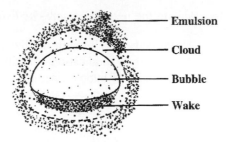

— Emulsion

— Cloud

— Bubble

— Wake

Figure 12-19   Schematic of bubble, cloud, and wake.

As the bubbles rise, mass transfer of the reactant gases takes place as they flow (diffuse) in and out of the bubble to contact the solid particles where the reaction product is formed. The product gases then flow back into a bubble and finally exit the bed when the bubble reaches the top of the bed. The rate at which the reactants and products transfer in and out of the bubble affects the conversion, as does the time it takes for the bubble to pass through the bed. Consequently, we need to describe the velocity at which the bubbles move through the column and the rate of transport of gases in and out of the bubbles. The bed is to be operated at a superficial velocity u_0. To calculate these parameters we need to determine a number of fluid-mechanics parameters associated with the fluidization process. Specifically, to determine the velocity of the bubble through the bed we need to first calculate:

Fluid Mechanics

Fluid mechanics:
Calculate

u_0
ε_{mf}
u_{mf}
d_{bm}
d_b
u_b

1) Porosity at minimum fluidization, ε_{mf}

$$\varepsilon_{mf} = 0.586\,\psi^{-0.72}\left(\frac{\mu^2}{\rho_g\,\eta\,d_p^3}\right)^{0.029}\left(\frac{\rho_g}{\rho_c}\right)^{0.021} \qquad (12\text{-}107)$$

2) Minimum fluidization velocity, u_{mf}

$$u_{mf} = \frac{(\psi d_p)^2}{150\mu}\underbrace{[g(\rho_c - \rho_g)]}_{\eta}\frac{\varepsilon_{mf}^3}{1 - \varepsilon_{mf}} \qquad (12\text{-}108)$$

3) Bubble size, d_b

$$\frac{d_{bm} - d_b}{d_{bm}} = e^{-0.3h/D_t} \qquad (12\text{-}109)$$

$$d_{bm} = 0.652\,[A_c\,(u_o - u_{mf})]^{0.4} \qquad \text{(12-110)}$$

cm   sq. cm.   cm/s

Mass Transport

To calculate the *mass transport coefficients* we must first calculate

Transport and
reaction:
Calculate

δ
γ_i
γ_c
$\gamma_b = 0.01$
K_{bc}
K_{ce}
K_R
$\downarrow$ X or W

1) Porosity at minimum fluidization, ε_{mf}
2) Minimum fluidization velocity, u_{mf}
3) Velocity of bubble rise, u_b

$$u_b = u_o - u_{mf} + (0.71)(gd_b)^{1/2} \qquad \text{(12-111)}$$

4) Bubble size, d_b
5) The transport coefficients K_{bc} and K_{ce}

$$K_{bc} = 4.5\left(\frac{u_{mf}}{d_b}\right) + 5.85\left(\frac{D^{1/2}g^{1/4}}{d_b^{5/4}}\right), \qquad \text{(12-112)}$$

$$K_{ce} = 6.78\left(\frac{\varepsilon_{mf} D u_b}{d_b^3}\right)^{1/2} \qquad \text{(12-113)}$$

where A_c = cross-sectional area, D_{AB} = diffusivity, d_p = catalyst particle
       diameter,
   g = gravitational constant, μ = viscosity, ρ_c = catalyst particle
       density,
   ρ_g = gas density, $\eta = g(\rho_c - \rho_g)$.

Reaction Rates

To determine the *reaction rate parameters* in the bed we need to first calculate

1) Fraction of the total bed occupied by bubbles, δ

$$\delta = \frac{u_o - u_{mf}}{u_b - u_{mf}(1 + \alpha)} \qquad \text{(12-114)}$$

2) Fraction of the bed consisting of wakes is $\alpha\delta$. α is the function of the
   particle size.
3) Volume of catalyst in the bubbles, γ_b, clouds, γ_c, and emulsion, γ_e.

$$\gamma_c = (1 - \varepsilon_{mf})\left[\frac{3(u_{mf}/\varepsilon_{mf})}{u_b - (u_{mf}/\varepsilon_{mf})} + \alpha\right] \qquad \text{(12-115)}$$

$$\gamma_e = (1 - \varepsilon_{mf})\left(\frac{1-\delta}{\delta}\right) - \gamma_c \qquad (12\text{-}116)$$

and $\gamma_b = 0.01$ to 0.001

Once each of these parameters is determined, one can calculate the catalyst weight necessary to achieve a given conversion using the equation

$$W = \frac{\rho_c A_c u_b (1 - \varepsilon_{mf})(1 - \delta)}{k_{cat} K_R} \ln \frac{1}{1 - X} \qquad (12\text{-}117)$$

where

$$K_R = \gamma_b + \cfrac{1}{\cfrac{k_{cat}}{K_{bc}} + \cfrac{1}{\gamma_c + \cfrac{1}{\cfrac{1}{\gamma_e} + \cfrac{k_{cat}}{K_{ce}}}}} \qquad (12\text{-}118)$$

and k_{cat} is the specific reaction rate determined from laboratory experiments (e.g., differential reactor).

The derivation of these equations along with two example problems are given on the CD-ROM.

CD Professional

Reference Shelf

12.11 Chemical Vapor Deposition Reactors

As discussed in Section 10.6, CVD is a very important process in the micro-electronics industry. The fabrication of microelectronic devices may include as few as 30 or as many as 200 individual steps to produce chips with up to 10^6 transducers per chip. An abbreviated schematic of the steps involved in producing a typical computer chip (MOSFET) is shown in Figure 10-32.

One of the key steps in the chip-making process is the deposition of different semiconductors and metals on the surface of the chip. This step can be achieved by CVD. CVD mechanisms were discussed in Chapter 10; consequently, this section will focus on CVD reactors. A number of CVD reactor types have been used, such as barrel reactors, boat reactors, and horizontal and vertical reactors. A description of these reactors and modeling equations are given by Jensen.[14]

One of the more common CVD reactors is the horizontal low-pressure CVD (LPCVD) reactor. This reactor operates at pressures of approximately 100 Pa. The main advantage of the LPCVD is its capability of processing a large number of wafers without detrimental effects to film uniformity. Owing

[14]K. F. Jensen, *Chem. Eng. Sci.*, *42*, 923 (1987).

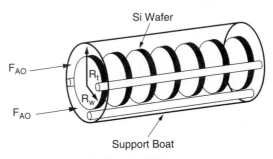

Figure 12-20   LPCVD boat reactor.

to the large increases in the diffusion coefficient at low pressures (recall Table 11-2), surface reactions are more likely to be controlling than mass transfer. A schematic of a LPCVD reactor is shown in Figure 12-20.

Modeling Concepts   We shall model the axial flow in the annular region as being laminar. This assumption is reasonable because a typical Reynolds number for flow in a LPCVD reactor is less than 1. As the reactant gases flow through the annulus, the reactants diffuse from the annulus radially inward between the wafers to coat them.[15]

The reacting gas flows through the annulus between the outer edges of the cylindrical wafers and the tube wall (Figure 12-20). Silicon is to be deposited on wafers in a LPCVD reactor. The reaction that is taking place is

$$SiH_2(g) \; \rightleftharpoons \; Si(s) + H_2(g)$$

Because SiH_2 is being consumed by CVD, the mole fraction of SiH_2 (i.e., the reactant) in the annulus, y_{AA}, decreases as the reactant flows down the length of the annulus.

The reacting gases diffuse out of the annular region into the space between the wafers, where the mole fraction is represented by y_A. As molecules diffuse radially *inward*, some of them are adsorbed and deposited on the wafer surface. The reaction products then diffuse radially *outward* into the gas stream axially flowing in the annulus (Figure 12-21). This system can be analyzed in a manner analogous to flow through a packed catalyst bed, where the reaction gases diffuse into the catalyst pellets. In this analysis we used an effectiveness factor to determine the overall rate of reaction per unit volume (or mass) of reactor bed. We can extend this idea to LPCVD reactors, where the reactants diffuse from the annular flow channel radially inward between the wafers. The concentration between the wafers is less than the concentration in the annulus. Consequently, the rate of deposition between the wafers will be less than the rate at conditions in the annulus. Fortunately, these two concentrations can be related by the effectiveness factor. We can determine the effectiveness factor once the concentration profile in the region between the wafers is obtained.

---

[15]K. F. Jensen, *J. Electrochem. Soc., 130*, 1450 (1983).

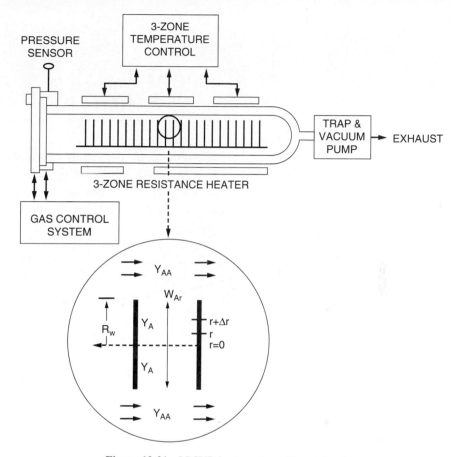

Figure 12-21   LPCVD boat reactor with peripherals.

As shown on the CD-ROM, one can solve the combined mole balance and rate law equations to obtain the concentration profile in between the wafers:

$$\frac{y_A}{y_{AA}} = \frac{I_o(\phi_1\lambda)}{I_o(\phi_1)}$$

(CDE12-7.9)

where I_o is the modified Bessel function, $\lambda = r/R_w$, and ϕ_1 = the Thiele modulus = $(2kR_w^2/D_{AB}\ell)^{1/2}$.

We now can obtain the effectiveness factor:

$$\eta = \frac{\text{actual rate of reaction}}{\begin{array}{c}\text{rate of reaction when entire wafer surface is exposed to}\\ \text{the concentration in the annulus, } C_{AA} \text{ (i.e., } y_{AA})\end{array}}$$

$$\eta = \frac{-W_{Ar}\big|_{r=R_w}(2\pi R_w\ell)}{2\pi R_w^2(-r''_{AA})}$$

(CDE12-7.10)

$$\eta = 2\,\frac{d\psi/d\lambda|_{\lambda=1}}{\phi_1^2}$$

$$\boxed{\eta = \frac{2I_1(\phi_1)}{\phi_1 I_o(\phi_1)}}$$ (CDE12-7.12)

where "I" is a modified Bessel function.

The concentration profile along the radius of the wafer disk as well as the wafer shape is shown in Figure 12-22 for different values of the Thiele modulus.

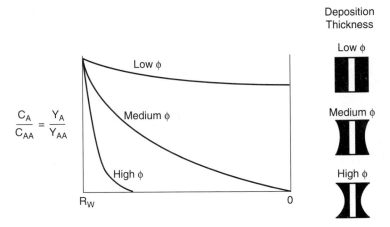

Figure 12-22   Radial concentration profile.

In addition to the possibility of a nonuniform wafer thickness in the radical direction, the thickness of the wafers can vary down the length of the boat reactor. We want to obtain an analytical solution of the silicon deposition rate and reactant concentration profile for the simplified version of the LPCVD reactor just discussed. Analytical solutions of this type are important in that an engineer can rapidly gain an understanding of the important parameters and their sensitivities without making a number of *runs* on the computer.

The thickness, T, of the deposit is obtained by integrating the deposition rate with respect to time:

$$T = \frac{kC_{A0}I_o(\phi_1 \cdot r/R_w)}{\rho I_o(\phi_1)}\{\exp[-\mathrm{Da}(z/L)]\}\,t$$

where Da is the Damköhler number for the reactor.

The reactant concentration profile and deposition thickness along the length of the reactor are shown schematically in Figure 12-23 for the case of small values of the Thiele modulus ($\eta \simeq 1$).

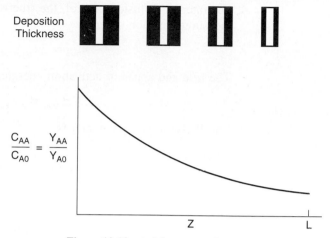

Deposition
Thickness

$$\frac{C_{AA}}{C_{A0}} = \frac{Y_{AA}}{Y_{A0}}$$

Z   L

Figure 12-23   Axial concentration profile.

SUMMARY

1. The concentration profile for a first-order reaction occurring in a spherical catalyst pellet is

$$\frac{C_A}{C_{As}} = \frac{R}{r}\left[\frac{\sinh(\phi_1 r/R)}{\sinh\phi_1}\right] \qquad (S12\text{-}1)$$

where ϕ_1 is the Thiele modulus. For a first-order reaction

$$\phi_1^2 = \frac{k\rho_c S_a}{D_e} R^2 \qquad (S12\text{-}2)$$

2. The effectiveness factors are

$$\begin{array}{c}\text{internal}\\ \text{effectiveness} = \eta =\\ \text{factor}\end{array} \frac{\text{actual rate of reaction}}{\begin{array}{c}\text{reaction rate if entire interior}\\ \text{surface exposed to concentration}\\ \text{at the external pellet surface}\end{array}}$$

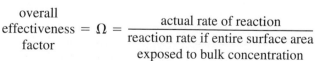

$$\begin{array}{c}\text{overall}\\ \text{effectiveness} = \Omega =\\ \text{factor}\end{array} \frac{\text{actual rate of reaction}}{\begin{array}{c}\text{reaction rate if entire surface area}\\ \text{exposed to bulk concentration}\end{array}}$$

3. For large values of the Thiele modulus,

$$\eta = \left(\frac{2}{n+1}\right)^{1/2}\frac{3}{\phi_n} \qquad (S12\text{-}3)$$

4. For internal diffusion control, the true reaction order is related to the measured reaction order by

$$n_{\text{true}} = 2n_{\text{apparent}} - 1 \qquad \text{(S12-4)}$$

The true and apparent activation energies are related by

$$E_{\text{true}} = 2E_{\text{app}} \qquad \text{(S12-5)}$$

5. The Weisz–Prater parameter is

$$C_{\text{WP}} = \phi_1^2 y = \frac{-r_A'(\text{observed})\rho_c R^2}{D_e C_{As}} \qquad \text{(S12-6)}$$

The Weisz–Prater criterion dictates that

$$\text{If } C_{\text{WP}} \ll 1 \qquad \text{no internal diffusion limitations present}$$

$$\text{If } C_{\text{WP}} \gg 1 \qquad \text{internal diffusion limitations present}$$

6. Slurry reactors

$$A(g,l) + B(l) \longrightarrow C(l)$$

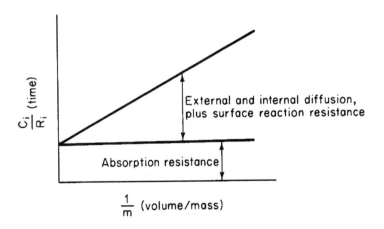

$$\frac{C_i}{R_A} = \frac{1}{k_b a_b} + \frac{1}{m}\left(\frac{1}{k_c a_c} + \frac{1}{\eta k}\right) \qquad \text{(S12-7)}$$

7. Trickle Bed Reactors

$$-r_A' = \frac{1/H}{\underbrace{\dfrac{(1-\phi_b)\rho_c}{H k_g a_i} + \dfrac{(1-\phi_b)\rho_c}{k_l a_i} + \dfrac{1}{k_c a_p} + \dfrac{1}{\eta k C_{Bs}}}_{k_{vg}}} C_A(g) \qquad \frac{\text{mol}}{\text{g cat.}\cdot\text{s}}$$

$$(12\text{-}101)$$

8. Fluidized Bed Reactors

$$W = \frac{\rho_c A_c u_b (1 - \varepsilon_{mf})(1 - \delta)}{k_{cat} K_R} \ln \frac{1}{1 - X} \qquad (12\text{-}117)$$

$$K_R = \gamma_b + \cfrac{1}{\cfrac{k_{cat}}{K_{bc}} + \cfrac{1}{\gamma_c + \cfrac{1}{\cfrac{1}{\gamma_e} + \cfrac{k_{cat}}{K_{ce}}}}} \qquad (12\text{-}118)$$

9. The deposition thickness on wafers in a boat reactor at a distance z down the reactor and at a wafer radius r is

$$T = \frac{k C_{A0} I_o (\phi_1 \cdot r / R_w)}{\rho I_o (\phi_1)} \{ \exp [-\text{Da}(z/L)] \} t \qquad (S12\text{-}8)$$

QUESTIONS AND PROBLEMS

The subscript to each of the problem numbers indicates the level of difficulty: A, least difficult; D, most difficult.

$$A = \bullet \quad B = \blacksquare \quad C = \blacklozenge \quad D = \blacklozenge\blacklozenge$$

In each of the questions and problems below, rather than just drawing a box around your answer, write a sentence or two describing how you solved the problem, the assumptions you made, the reasonableness of your answer, what you learned, and any other facts that you want to include. You may wish to refer to W. Strunk and E. B. White, *The Elements of Style* (New York: Macmillan, 1979) and Joseph M. Williams, *Style: Ten Lessons in Clarity & Grace* (Glenview, Ill.: Scott, Foresman, 1989) to enhance the quality of your sentences.

P12-1$_C$ Make up an original problem using the concepts presented in Section _____ (your instructor will specify the section). Extra credit will be given if you obtain and use real data from the literature. (See Problem P4-1 for the guidelines.)

P12-2$_B$ **What if...**

    (a) your internal surface area decreased with time because of sintering. How would your effectiveness factor change and the rate of reaction change with time if $k_d = 0.01$ hr^{-1} and $\eta = 0.01$ at $t = 0$? Explain.

    (b) someone had used the false kinetics (i.e., wrong E, wrong n)? Would their catalyst weight be overdesigned or underdesigned? What are other positive or negative effects that occur?

    (c) the gas velocity in Example 12-3 were increased? How would your reactor length change? Explain what other effects (e.g., temperature) would cause it to become larger and what would cause it to become smaller.

    (d) you were asked to compare the conditions (e.g., catalyst charge, conversions) and sizes of the reactors in Examples 12-4 and 12-6. What differences would you find? Are there any fundamental discrepancies between the two? If so, what are they, and what are some reasons for them?

(e) you were to assume the resistance to gas absorption in Example 12-4 were the same as in Example 12-6 and that the liquid phase reactor volume in Example 12-4 was 50% of the total, could you estimate the controlling resistance? If so, what is it? What other things could you calculate in Example 12-4 (e.g., selectivity, conversion, molar flow rates in and out)? Hint: Some of the other reactions that occur include

$$CO + 3H_2 \longrightarrow CH_4 + H_2O$$

$$H_2O + CO \longrightarrow CO_2 + H_2$$

(f) you applied the Mears and Weisz–Prater criteria to Examples 11-4 and 12-3? What would you find? What would you learn if $\Delta H_{Rx} = -25$ k cal/mol, $h = 100$ Btu/h·ft²·°F and $E = 20$ k cal/mol?

(g) the temperature in Example 12-5 were increased? How would the relative resistances in the slurry reactor change?

(h) significantly less liquid were exiting a trickle bed reactor. What are the first three questions you would ask?

(i) you were asked for all the things that could go wrong in the operation of a slurry reactor, what would you say?

(j) you were asked to give three important industrial reactions (other than those listed in the text)? How would you go about finding the answer, and what three reactions would you list?

(k) $\gamma = 30$, $\beta = 0.4$, and $\phi = 0.4$ in Figure 12-7? What would cause you to go from the upper steady state to the lower steady state and vise versa?

P12-3$_B$ The catalytic reaction

$$A \longrightarrow B$$

takes place within a fixed bed containing spherical porous catalyst X22. Figure P12-3 shows the overall rates of reaction at a point in the reactor as a function of temperature for various entering total molar flow rates, F_{T0}.

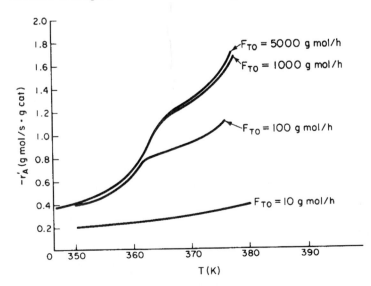

Figure P12-3

(a) Is the reaction limited by external diffusion?

(b) If your answer to part (a) was "yes," under what conditions [of those shown (i.e., T, F_{T0})] is the reaction limited by external diffusion?

(c) Is the reaction "reaction-rate-limited"?

(d) If your answer to part (c) was "yes," under what conditions [of those shown (i.e., T, F_{T0})] is the reaction limited by the rate of the surface reactions?

(e) Is the reaction limited by internal diffusion?

(f) If your answer to part (e) was "yes," under what conditions [of those shown (i.e., T, F_{T0})] is the reaction limited by the rate of internal diffusion?

(g) For a flow rate of 10 g mol/h, determine (if possible) the overall effectiveness factor, Ω, at 360 K.

(h) Estimate (if possible) the internal effectiveness factor, η, at 367 K.

(i) If the concentration at the external catalyst surface is 1 g mol/dm³, calculate (if possible) the concentration at $r = R/2$ inside the porous catalyst at 367 K. (Assume a first-order reaction.)

Additional information:

Gas properties:	**Bed properties:**
Diffusivity: 0.1 cm²/s	Tortuosity of pellet: 1.414
Density: 0.001 g/cm³	Bed permeability: 1 millidarcy
Viscosity: 0.0001 g/cm·s	Porosity = 0.3

P12-4$_B$ The reaction

$$A \longrightarrow B$$

is carried out in a differential packed-bed reactor at different temperatures, flow rates, and particle sizes. The results shown in Figure P12-4 were obtained.

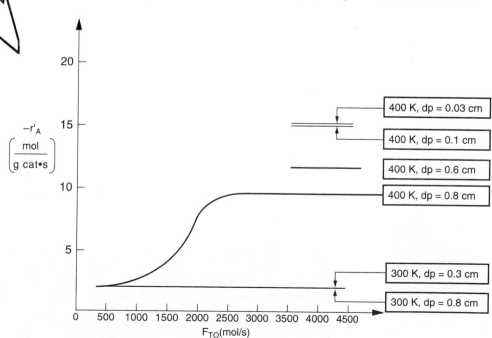

Figure P12-4

(a) What regions (i.e., conditions d_p, T, F_{T0}) are external mass transfer-limited?

(b) What regions are reaction-rate-limited?

(c) What region is internal-diffusion-controlled?

(d) What is the internal effectiveness factor at $T = 400$ and $d_p = 0.8$ cm?

P12-5$_A$ Curves A, B, and C in Figure P12-5 show the variations in reaction rate for three different reactions catalyzed by solid catalyst pellets. What can you say about each reaction?

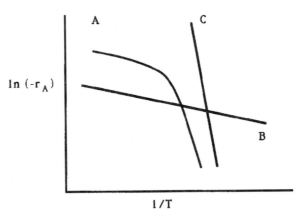

Figure P12-5

P12-6$_B$ A first-order heterogeneous irreversible reaction is taking place within a spherical catalyst pellet which is plated with platinum throughout the pellet (see Figure 12-3). The reactant concentration halfway between the external surface and the center of the pellet (i.e., $r = R/2$) is equal to one-tenth the concentration of the pellet's external surface. The concentration at the external surface is 0.001 g mol/dm³, the diameter ($2R$) is 2×10^{-3} cm, and the diffusion coefficient is 0.1 cm²/s.

$$A \longrightarrow B$$

(a) What is the concentration of reactant at a distance of 3×10^{-4} cm in from the external pellet surface? (*Ans.:* $C_A = 2.36 \times 10^{-4}$ mol/dm³.)

(b) To what diameter should the pellet be reduced if the effectiveness factor is to be 0.8? (*Ans.:* $D_2 = 6.8 \times 10^{-4}$ cm.)

(c) If the catalyst support were not yet plated with platinum, how would you suggest that the catalyst support be plated *after* it had been reduced by grinding?

P12-7$_B$ The swimming rate of a small animal [*J. Theoret. Biol.*, **26**, 11 (1970)] is related to the energy released by the hydrolysis of adenosine triphosphate (ATP) to adenosine diphosphate (ADP). The rate of hydrolysis is equal to the rate of diffusion of ATP from the midpiece to the tail (see Figure P12-7). The diffusion coefficient of ATP in the midpiece and tail is 3.6×10^{-6} cm²/s. ADP is converted to ATP in the midsection, where its concentration is 4.36×10^{-5} mol/cm³. The cross-sectional area of the tail is 3×10^{-10} cm².

(a) Derive an equation for diffusion and reaction in the tail.

(b) Derive an equation for the effectiveness factor in the tail.

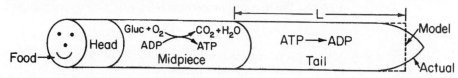

Figure P12-7

(c) Taking the reaction in the tail to be of zero-order, calculate the length of the tail. The rate of reaction in the tail is 23×10^{-18} mol/s.

(d) Compare your answer with the average tail length of 41 μm. What are possible sources of error?

P12-8$_B$ A first-order, heterogeneous, irreversible reaction is taking place within a catalyst pore which is plated with platinum entirely along the length of the pore (Figure P12-8). The reactant concentration at the plane of symmetry (i.e., equal distance from the pore mouths) of the pore is equal to one-tenth the concentration of the pore mouth. The concentration at the pore mouth is 0.001 g mol/dm^3, the pore length ($2L$) is 2×10^{-3} cm, and the diffusion coefficient is 0.1 cm^2/s.

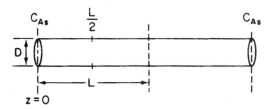

Figure P12-8

(a) Derive an equation for the effectiveness factor.

(b) What is the concentration of reactant at $L/2$?

(c) To what length should the pore length be reduced if the effectiveness factor is to be 0.8?

(d) If the catalyst support were not yet plated with platinum, how would you suggest the catalyst support be plated *after* the pore length, L, had been reduced by grinding?

P12-9$_A$ A first-order reaction is taking place inside a porous catalyst. Assume dilute concentrations and neglect any variations in the axial (x) direction.

(a) Derive an equation for both the internal and overall effectiveness factors for the rectangular porous slab shown in Figure P12-9.

(b) Repeat part (a) for a cylindrical catalyst pellet where the reactants diffuse inward in the radial direction.

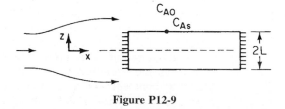

Figure P12-9

P12-10$_B$ The irreversible reaction

$$A \longrightarrow B$$

is taking place in the porous catalyst disk shown in Figure P12-9. The reaction is zero-order in A.

(a) Show that the concentration profile is

$$\frac{C_A}{C_{As}} = 1 + \frac{\phi_0^2}{2}\left[\left(\frac{z}{L}\right)^2 - 1\right] \tag{P12-9.1}$$

where

$$\phi_0 = \frac{kL^2}{D_e C_{As}}$$

(b) For a Thiele modulus of 1.414, at what point in the disk is the concentration zero? For $\phi_0 = 4$?

(c) The effectiveness factor can be written as

$$\eta = \frac{\int_0^L -r_A A_c \, dz}{-r_{As} A_c L} = \frac{\int_0^{z1} -r_A A_c \, dz + \int_{z1}^L -r_A A_c \, dz}{-r_{As} A_c L} \tag{P12-9.2}$$

where A_c is the cross-sectional area of the disk. Show for a zero-order reaction that

$$\eta = \begin{cases} 1 & \text{for } \phi_0 \le 1.414 \\ 1 - \left(1 - \frac{2}{\phi_0^2}\right)^{1/2} & \text{for } \phi_0 \ge 1.414 \end{cases}$$

(d) Make a sketch for η versus ϕ_0 similar to the one shown in Figure 12-5.
(e) Repeat parts (a) to (d) for a spherical catalyst pellet.
(f) Repeat parts (a) to (d) for a cylindrical catalyst pellet.
(g) What do you believe to be the point of this problem?

P12-11$_C$ The second-order decomposition reaction

$$A \longrightarrow B + 2C$$

is carried out in a tubular reactor packed with catalyst pellets 0.4 cm in diameter. The reaction is internal-diffusion-limited. Pure A enters the reactor at a superficial velocity of 3 m/s, a temperature of 250°C, and a pressure of 500 kPa. Experiments carried out on smaller pellets where surface reaction is limiting yielded a specific reaction rate of 0.05 m^6/mol·g cat.·s. Calculate the length of bed necessary to achieve 80% conversion. Critique the numerical answer.

Additional information:

Effective diffusivity: 2.66×10^{-8} m^2/s
Uneffective diffusivity: 0.00 m^2/s
Bed porosity: 0.4
Pellet density: 2×10^6 g/m^3
Internal surface area: 400 m^2/g

P12-12$_C$ Derive the concentration profile and effectiveness factor for cylindrical pellets 0.2 cm in diameter and 1.5 cm in length. Neglect diffusion through the ends of the pellet.

(a) Assume that the reaction is a first-order isomerization. (*Hint:* Look for a Bessel function.)

(b) Rework Problem P12-11 for these pellets.

P12-13$_B$ If you were to use Equation (12-35) for the effectiveness factor, would all the asymptotes of the lines of Figure 12-5a coalesce to one line?

P12-14$_B$ The reaction

$$A \longrightarrow 3B$$

takes place isothermally in a spherical catalyst pellet. Using the data in Problem P12-11, plot the concentration profile as a function of pellet radius. Compare this profile with the profile obtained for equal molar counter diffusion. What is the effectiveness factor in each case?

P12-15$_C$ Reconsider diffusion and reaction in a spherical catalyst pellet for the case where the reaction is not isothermal. Show that the energy balance can be written as

$$\frac{1}{r^2}\frac{d}{dr}\left(r^2 k_t \frac{dT}{dr}\right) + (-\Delta H_R)(-r_A) = 0 \qquad \text{(P12-15.1)}$$

where k_t is the effective thermal conductivity, cal/s·cm·K of the pellet with $dT/dr = 0$ at $r = 0$ and $T = T_s$ at $r = R$.

(a) Evaluate Equation (12-11) for a first-order reaction and combine with Equation (P12-15.1) to arrive at an equation giving the maximum temperature in the pellet.

$$T_{\max} = T_s + \frac{(-\Delta H_R)(D_e C_{As})}{k_t T_s} \qquad \text{(P12-15.2)}$$

Note: At $T_{\max}$, $C_A = 0$.

(b) Choose representative values of the parameters and use a software package to solve Equations (12-11) and (P12-15.1) simultaneously for $T(r)$ and $C_A(r)$ when the reaction is carried out adiabatically. Show that the resulting solution agrees qualitatively with Figure 12-7.

P12-16$_C$ Determine the effectiveness factor for a nonisothermal spherical catalyst pellet in which a first-order isomerization is taking place.

Additional information:

$A_i = 100 \text{ m}^2/\text{m}^3$
$\Delta H_R = -800{,}000 \text{ J/mol}$
$D_e = 8.0 \times 10^{-8} \text{ m}^2/\text{s}$
$C_{As} = 0.01 \text{ kmol/m}^3$
External surface temperature of pellet, $T_s = 400 \text{ K}$
$E = 120{,}000 \text{ J/mol}$
Thermal conductivity of pellet $= 0.004 \text{ J/m·s·K}$
$d_p = 0.005 \text{ m}$
Specific reaction rate $= 10^{-1} \text{ m/s at } 400 \text{ K}$
Density of calf's liver $= 1.1 \text{ g/dm}^3$

How would your answer change if the pellets were 10^{-2}, 10^{-4}, and 10^{-5} m in diameter? What are typical temperature gradients in catalyst pellets?

P12-17$_B$ The elementary isomerization reaction

$$A \longrightarrow B$$

is taking place on the walls of a cylindrical catalyst pore. In one run a catalyst poison P entered the reactor together with the reactant A. To estimate the effect of poisoning, we assume that the poison renders the catalyst pore walls near the pore mouth ineffective up to a distance z_1, so that no reaction takes place on the walls in this entry region.

(a) Show that before poisoning of the pore occurred, the effectiveness factor was given by

$$\eta = \frac{1}{\phi} \tanh \phi$$

where

$$\phi = L \sqrt{\frac{2k''}{rD_{AB}}}$$

with k'' = reaction rate constant (length/time)
         r = pore radius (length)
        D_{AB} = effective molecular diffusivity (area/time)

(b) Derive an expression for the concentration profile and also for the molar flux of A in the ineffective region $0 < z < z_1$, in terms of z_1, D_{AB}, C_{A1}, and C_{As}. Without solving any further differential equations, obtain the new effectiveness factor η' for the poisoned pore.

P12-18$_B$ (*Falsified kinetics*) The irreversible gas-phase dimerization

$$2A \longrightarrow A_2$$

is carried out at 8.2 atm in a stirred contained-solids reactor to which only pure A is fed. There is 40 g of catalyst in each of the four spinning baskets. The following runs were carried out at 227°C:

Total Molar Feed Rate, F_{T0} (g mol/min)	Mole Fraction A in Exit, y_A
1	0.21
2	0.33
4	0.40
6	0.57
11	0.70
20	0.81

The following experiment was carried out at 237°C:

$$F_{T0} = 9 \text{ g mol/min} \qquad y_A = 0.097$$

(a) What are the apparent reaction order and the apparent activation energy?
(b) Determine the true reaction order, specific reaction rate, and activation energy.
(c) Calculate the Thiele modulus and effectiveness factor.
(d) What diameter of pellets should be used to make the catalyst more effective?

(e) Calculate the rate of reaction on a rotating disk made of the catalytic material when the gas-phase reactant concentration is 0.01 g mol/L and the temperature is 527°C. The disk is flat, nonporous, and 5 cm in diameter.

Additional information:

Effective diffusivity: 0.23 cm²/s
Surface area of porous catalyst: 49 m²/g cat.
Density of catalyst pellets: 2.3 g/cm³
Radius of catalyst pellets: 1 cm
Color of pellets: blushing peach

P12-19$_A$ The following table was obtained from the data taken in a slurry reactor for the hydrogenation of methyl linoleate to form methyl oleate.

$$L + H_2 \longrightarrow O$$

$$S = \text{solubility of } H_2 \text{ in the liquid mixture, mol/dm}^3$$

$$m = \text{catalyst charge, g/dm}^3$$

$$-r_L' = \text{rate of reaction of methyl linoleate, mol/dm}^3 \cdot \text{min}$$

Catalyst Size	$S/-r_L'$ (min)	$1/m$ (dm³/g)
A	4.2	0.01
A	7.5	0.02
B	1.5	0.01
B	2.5	0.03
B	3.0	0.04

(a) Which catalyst size has the smaller effectiveness factor?
(b) If catalyst size A is to be used in the reactor at a concentration of 50 g/dm³, would a significant increase in the reaction be obtained if a more efficient gas sparger were used?
(c) If catalyst size B is to be used, what is the minimum catalyst charge that should be used to ensure that the combined diffusional resistances for the pellet are less than 50% of the total resistance?

P12-20$_B$ The catalytic hydrogenation of methyl linoleate to methyl oleate was carried out in a laboratory-scale slurry reactor in which hydrogen gas was bubbled up through the liquid containing spherical catalyst pellets. The pellet density is 2 g/cm³. The following experiments were carried out at 25°C:

Run	Partial Pressure of H_2 (atm)	Solubility of H_2 (g mol/dm³)	H_2 Rate of Reaction (g mol/dm³·min)	Catalyst Charge (g/dm³)	Catalyst Particle Size (μm)
1	3	0.007	0.014	3.0	12
2	18	0.042	0.014	0.5	50
3	3	0.007	0.007	1.5	50

(a) It has been suggested that the overall reaction rate can be enhanced by increasing the agitation, decreasing the particle size, and installing a more efficient sparger. With which, if any, of these recommendations do you agree? Are there other ways that the overall rate of reaction might be increased? Support your decisions with calculations.

(b) Is it possible to determine the effectiveness factor from the data above? If so, what is it?

(c) For economical reasons concerning the entrainment of the small solid catalyst particles in the liquid, it is proposed to use particles an order of magnitude larger. The following data were obtained from these particles at 25°C:

Run	Partial Pressure of H_2 (atm)	Solubility of H_2 (g mol/dm^3)	H_2 Rate of Reaction (g mol/dm$^3 \cdot$min)	Catalyst Charge (g/dm^3)	Catalyst Particle Size (μm)
4	3	0.007	0.00233	2.0	750

The Thiele modulus is 9.0 for the 750-μm particle size in run 4. Determine (if possible) the external mass transfer coefficient, k_c, and the percent (of the overall) of the external mass transfer resistance to the catalyst pellet.

P12-21$_B$ The hydrogenation of 2-butyne-1,4-diol to butenediol is to be carried out in a slurry reactor using a palladium-based catalyst. The reaction is first-order in hydrogen and in diol. The initial concentration of diol is 2.5 kmol/m^3. Pure hydrogen is bubbled through the reactor at a pressure of 35 atm at 35°C. The equilibrium hydrogen solubility at these conditions is 0.01 kmol/m^3, and the specific reaction rate is 0.048 m^6/kg·kmol·s. The catalyst charge is 0.1 kg/m^3 with a particle size of 0.01 cm and pellet density of 1500 kg/m^3.

(a) Calculate the percent of the overall resistance contributed by each of the transport steps.

(b) Plot conversion as a function of time up to 95%.

(c) How could the reaction time be reduced?

Additional information:

$$\text{diffusivity} = 10^{-9} \text{ m}^2/\text{s for } H_2 \text{ in organics}$$

$$k_b a_b = 0.3 \text{ s}^{-1}$$

$$k_c = 0.005 \text{ cm/s for } H_2 \text{ in organics}$$

$$k_c = 0.009 \text{ cm/s for 2-butyne-1,4-diol in butenediol}$$

Pellet density = 1.6 g/cm^3

Pellet porosity = 0.45

JOURNAL ARTICLE PROBLEMS

P12J-1 The article in *Trans. Int. Chem. Eng.*, *60*, 131 (1982) may be advantageous in answering the following questions.

(a) Describe the various types of gas–liquid–solid reactors.

     (b) Sketch the concentration profiles for gas absorption with:
- **(1)** An instantaneous reaction
- **(2)** A very slow reaction
- **(3)** An intermediate reaction rate

P12J-2 After reading the journal review by Y. T. Shah et al. [*AIChE J.*, *28*, 353 (1982)], design the following bubble column reactor. One percent carbon dioxide in air is to be removed by bubbling through a solution of sodium hydroxide. The reaction is mass transfer-limited. Calculate the reactor size (length and diameter) necessary to remove 99.9% of the CO_2. Also specify a type of sparger. The reactor is to operate in the bubbly flow regime and still process 0.5 m^3/s of gas. The liquid flow rate through the column is 10^{-3} m^3/s.

JOURNAL CRITIQUE PROBLEMS

P12C-1 Use the Weisz–Prater criterion to determine if the reaction discussed in *AIChE J.*, *10*, 568 (1964) is diffusion-rate-limited.

P12C-2 Use the references given in *Ind. Eng. Chem. Prod. Res. Dev.*, *14*, 226 (1975) to define the iodine value, saponification number, acid number, and experimental setup. Use the slurry reactor analysis to evaluate the effects of mass transfer and determine if there are any mass transfer limitations.

CD-ROM MATERIAL

- **Learning Resources**
  1. *Summary Notes for Lectures 27, 28, 29 and 30*
- **Professional Reference Shelf**
  1. *Trickle Bed Reactors*
     - A. Fundamentals
     - B. Limiting Situations
     - C. Evaluating the Transport Coefficients
  2. *Fluidized Bed Reactors*
     - A. Overview
     - B. Mechanics of Fluidized Beds
     - C. Descriptive Behavior of the Kunii–Levenspiel Bubbling Bed Model
     - D. Mass Transfer in Fluidized Beds
     - E. Reaction in a Fluidized Bed
     - F. Mole Balances on the (1) Bubble, (2) Cloud, and (3) Emulsion
     - G. Solution to the Balance Equations for a First Order Reaction
       - Example CD12-3 Catalytic Oxidation of Ammonia
     - H. Limiting Situations
       - Example CD12-4 Calculating the Resistance's
       - Example CD12-5 Effect of Particle Size of Catalyst Weight for a Slow Reaction
       - Example CD12-5 Effect of Catalyst Weight for a Rapid Reaction
     - I. Summary
  3. *CVD Boat Reactors*
     - A. Fundamentals
     - B. Examples
- **Additional Homework Problems**

CDP12-A$_B$ Determine the catalyst size that gives the highest conversion in a packed bed reactor.

CDP12-B$_D$ Determine the temperature profile to achieve a uniform thickness. [2nd ed. P11-18]

CDP12-C$_B$ Explain how varying a number of the parameters in the *CVD boat reactor* will affect the wafer shape. [2nd ed. P11-19]

CDP12-D$_B$ Determine the wafer shape in a CVD boat reactor for a series of operating conditions. [2nd ed. P11-20]

CDP12-E$_C$ Model the buildup of a silicon wafer on parallel sheets. [2nd ed. P11-21]

CDP12-F$_C$ Rework CVD boat reactor accounting for the reaction

$$SiH_4 \rightleftharpoons SiH_2 + H_2$$

[2nd ed. P11-22]

CDP12-G$_B$ Hydrogenation of an unsaturated organic is carried out in a *trickle bed reactor.* [2nd ed. P12-7]

CDP12-H$_B$ The oxidation of ethanol is carried out in a *trickle bed reactor.* [2nd ed. P12-9]

CDP12-J$_C$ Hydrogenation of aromatics in a *trickle bed reactor* [2nd Ed. P12-8$_B$]

CDP12-K$_C$ Open-ended fluidization problem that requires critical thinking to compare the two-phase fluid models with the three-phase bubbling bed model.

CDP12-L$_C$ Rework Example CD12-5 using the correlations of C. Chavorie and J. R. Grace [*I.E.C. Fundamentals*, **14**, p. 75 (1975)].

CDP12-M$_A$ This problem studies the effect of temperature on the minimum fluidization velocity u_{mf}.

CDP12-N$_A$ Calculate reaction rates at the top and the bottom of the bed for Example 12-5.

CDP12-O$_B$ Calculate the conversion for A → B in a bubbling fluidized bed.

CDP12-P$_B$ Calculate the effect of operating parameters on conversion for the reaction limited and transport limited operation.

CDP12-Q$_B$ Variation of Problem CD12-O$_B$.

CDP12-R$_B$ *Excellent Problem* Calculate all the parameters in Example CD12-3 for a different reaction and different bed.

CDP12-S$_B$ Plot conversion and concentration as a function of bed height in a bubbling fluidized bed.

CDP12-T$_B$ Use RTD studies to compare bubbling bed with a fluidized bed.

SUPPLEMENTARY READING

1. There are a number of books that discuss internal diffusion in catalyst pellets; however, one of the first books that should be consulted on this and other topics on heterogeneous catalysis is

LAPIDUS, L., and N. R. AMUNDSON, *Chemical Reactor Theory: A Review.* Upper Saddle River, N.J.: Prentice Hall, 1977.

In addition, see

ARIS, R., *Elementary Chemical Reactor Analysis.* Upper Saddle River, N.J.: Prentice Hall, 1969, Chap. 6. One should find the references listed at the end of this reading particularly useful.

LUSS, D., "Diffusion—Reaction Interactions in Catalyst Pellets," p. 239 in *Chemical Reaction and Reactor Engineering*, Marcel Dekker, New York, 1987.

The effects of mass transfer on reactor performance are also discussed in

DENBIGH, K., and J. C. R. TURNER, *Chemical Reactor Theory*, 2nd ed. Cambridge: Cambridge University Press, 1971, Chap. 7.

SATERFIELD, C. N., *Heterogeneous Catalysis in Industrial Practice*, 2nd Ed., McGraw-Hill, New York, 1991.

2. Diffusion with homogeneous reaction is discussed in

ASTARITA, G., *Mass Transfer with Chemical Reaction*. New York: Elsevier, 1967.

DANCKWERTS, P. V., *Gas–Liquid Reactions*. New York: McGraw-Hill, 1970.

Gas-liquid reactor design is also discussed in

CHARPENTIER, J. C., review article, *Trans. Inst. Chem. Eng.*, *60*, 131 (1982).

SHAH, Y. T., *Gas–Liquid–Solid Reactor Design*. New York: McGraw-Hill, 1979.

3. Modeling of CVD reactors is discussed in

HESS, D. W., K. F. JENSEN, and T. J. ANDERSON, "Chemical Vapor Deposition: A Chemical Engineering Perspective," *Rev. Chem. Eng.*, *3*, 97, 1985.

JENSEN, K. F., "Modeling of Chemical Vapor Deposition Reactors for the Fabrication of Microelectronic Devices," *Chemical and Catalytic Reactor Modeling*, ACS Symp. Ser. 237, M. P. Dudokovic, P. L. Mills, eds., Washington, D.C.: American Chemical Society, 1984, p. 197.

LEE, H. H., *Fundamentals of Microelectronics Processing*. New York: McGraw-Hill, 1990.

4. Multiphase reactors are discussed in

HERSKOWITZ, M., and J. M. SMITH, "Trickle Bed Reactors, A Review," *AIChE J.*, *29*, 1 (1983).

RAMACHANDRAN, P. A., and R. V. CHAUDHARI, *Three-Phase Catalytic Reactors*. New York: Gordon and Breach, 1983.

RODRIGUES, A. E., J. M. COLO, and N. H. SWEED, eds., *Multiphase Reactors*, Vol. 1: *Fundamentals*. Alphen aan den Rijn, The Netherlands: Sitjhoff and Noordhoff, 1981.

RODRIGUES, A. E., J. M. COLO, and N. H. SWEED, eds., *Multiphase Reactors*, Vol. 2: *Design Methods*. Alphen aan den Rijn, The Netherlands: Sitjhoff and Noordhoff, 1981.

SATTERFIELD, C. N., "Contacting Effectiveness in Trickle Bed Reactors," in *Chemical Reaction Engineering Reviews*, H. M. Hulburt, ed., ACS Adv. Chem. Ser. 148. Washington, D.C.: American Chemical Society, 1975, p. 50.

SATTERFIELD, C. N., "Trickle Bed Reactors" (journal review), *AIChE J.*, *21*, 209 (1975).

SHAH, Y. T., B. G. KELKAR, S. P. GODBOLE, and W. D. DECKWER, "Design Parameters Estimations for Bubble Column Reactors" (journal review), *AIChE J.*, *28*, 353 (1982).

TARHAM, M. O., *Catalytic Reactor Design*. New York: McGraw-Hill, 1983.

YATES, J. G., *Fundamentals of Fluidized-Bed Chemical Processes*, 3rd ed. London: Butterworth, 1983.

The following *Advances in Chemistry Series* volumes discuss a number of multiphase reactors:

FOGLER, H. S., ed., *Chemical Reactors*, ACS Symp. Ser. 168. Washington, D.C.: American Chemical Society, 1981, pp. 3–255.

WEI, J., and G. GEORGAKIS, eds., *Chemical Reactor Engineering—Boston*, ACS Symp. Ser. 196. Washington, D.C.: American Chemical Society, 1982, pp. 377–458.

5. Fluidization

In addition to Kunii and Levenspiel's book, many correlations can be found in

DAVIDSON, J. F., R. CLIFF, and D. HARRISON, *Fluidization*, 2nd Ed., Academic Press, Orlando, 1985.

A discussion of the different models can be found in

YATES, J. C., *Fundamentals of Fluidization*, Butterworth, London, 1983. Also see GELDART, D. ed. *Gas Fluidization Technology*, Wiley-Interscience, Chichester, 1986.

Distributions $\quad$ **13**
of Residence Times
for Chemical Reactors

Nothing in life is to be feared. It is only to be understood.

Marie Curie

13.1 General Characteristics

The reactors treated in the book thus far—the perfectly mixed batch, the plug-flow tubular, and the perfectly mixed continuous tank reactors—have been modeled as ideal reactors. Unfortunately, in the real world we often observe behavior very different from that expected from the exemplar; this behavior is true of students, engineers, college professors, and chemical reactors. Just as we must learn to work with people who are not perfect, so the reactor analyst must learn to diagnose and handle chemical reactors whose performance deviates from the ideal. Nonideal reactors and the principles behind their analysis form the subject of this chapter and the next.

We want to analyze and characterize nonideal reactor behavior

After studying this chapter the reader will be able to describe the cumulative $F(t)$, external age $E(t)$, and internal age $I(t)$ residence-time distribution functions and to recognize these functions for PFR, CSTR, and laminar flow reactions. The reader will also be able to apply these functions to calculate the conversion and concentrations exiting a reactor using the segregation model and the maximum mixedness model for both single and multiple reactions.

The basic ideas or concepts used to characterize and model nonideal reactors are really few in number. Before proceeding further, a few selected examples of nonideal mixing and modeling from the author's experiences will be presented.

System 1 In a gas–liquid continuous-stirred tank reactor (Figure 13-1), the gaseous reactant was bubbled into the reactor while the liquid reactant was fed through an inlet tube in the reactor's side. The reaction took place at the

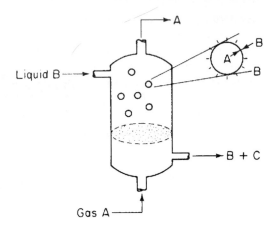

Figure 13-1   Gas–liquid reactor.

gas–liquid interface of the bubbles, and the product was a liquid. The continuous liquid phase could be regarded as perfectly mixed, and the reaction rate was proportional to the total bubble surface area. The surface area of a particular bubble depended on the time it had spent in the reactor. Because of their different sizes, some gas bubbles escaped from the reactor almost immediately, while others spent enough time in the reactor so that they were almost completely consumed. The time the bubble spends in the reactor is termed the *bubble residence time*. What was important in the analysis of this reactor was not the average residence time of the bubbles but rather the residence time of each bubble, i.e. the residence time distribution. The total reaction rate was found by summing over all the bubbles in the reactor. For this sum, the distribution of residence times of the bubbles leaving the reactor was required. An understanding of residence-time distributions (RTDs) and their effects on chemical reactor performance is thus one of the necessities of the technically competent reactor analyst.

Not all molecules are spending the same time in the reactor

System 2   A packed-bed reactor is shown in Figure 13-2. When a reactor is packed with catalyst, the reacting fluid usually does not flow through the reactor uniformly. Rather, there may be sections in the packed bed which offer little resistance to flow, and as a result a major portion of the fluid may channel through this pathway. Consequently, the molecules following this pathway do not spend as much time in the reactor as those flowing through the regions of high resistance to flow. We see that there is a distribution of time that molecules spend in the reactor in contact with the catalyst.

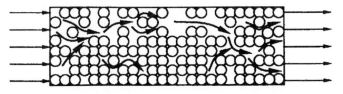

Figure 13-2   Packed-bed reactor.

System 3   In many continuous-stirred tank reactors, the inlet and outlet pipes are close together (Figure 13-3). In one operation it was desired to scale up pilot plant results to a much larger system. It was realized that some short-circuiting must occur, so the tanks were modeled as perfectly mixed CSTRs with a bypass stream. In addition to short circuiting, stagnant regions (dead zones) are often encountered. In these regions there is little or no exchange of material with the well-mixed regions, and consequently, virtually no reaction occurs there. Experiments were carried out to determine the amount of the material effectively bypassed and the volume of the dead zone. A simple modification of an ideal reactor successfully modeled the essential physical characteristics of the system and the equations were readily solvable.

We want to find ways of determining the dead volume and amount of bypassing

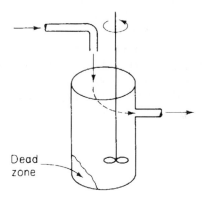

Dead zone

Figure 13-3   CSTR.

Three concepts used to describe nonideal reactors appear in the examples: *the distribution of residence times in the system, the quality of mixing,* and *the model used to describe the system.* All three of these concepts are considered when describing deviations from the mixing patterns assumed in ideal reactors. The three concepts can be regarded as characteristics of the mixing in nonideal reactors.

One way to order our thinking on nonideal reactors is to consider modeling the flow patterns in our reactors as either CSTRs or PFRs as a *first* approximation. In real reactors, however, nonideal flow patterns exist, resulting in ineffective contacting and lower conversions than in the case of ideal reactors. We must have a method of accounting for this nonideality, and to achieve this goal we use the next-higher level of approximation, which involves the use of *macromixing* information (RTD) (Sections 13.1 to 13.4). The next level uses information as a microscale (*micromixing*) to make predictions about the conversion of nonideal reactors. We address this third level of approximation in Sections 13.5 to 13.8 and in Chapter 14.

13.1.1 Residence-Time Distribution Function

The idea of using the distribution of residence times in the analysis of chemical reactor performance was apparently first proposed in a pioneering paper by MacMullin and Weber.[1] However, the concept did not appear to be used extensively until the early 1950s, when Danckwerts[2] gave organizational structure to the subject by defining most of the distributions of interest. The ever-increasing amount of literature on this topic since then has generally followed the nomenclature of Danckwerts, and this will be done here as well.

In an ideal plug-flow reactor, all the atoms of material leaving the reactor have been inside it for exactly the same amount of time. Similarly, in an ideal batch reactor, all the atoms of materials within the reactor have been inside it for an identical length of time. The time the atoms have spent in the reactor is called the *residence time* of the atoms in the reactor.

The idealized plug-flow and batch reactors are the only two classes of reactors in which all the atoms in the reactors have the same residence time. In all other reactor types, the various atoms in the feed spend different times inside the reactor; that is, there is a distribution of residence times of the material within the reactor. For example, consider the CSTR; the feed introduced into a CSTR at any given time becomes completely mixed with the material already in the reactor. In other words, some of the atoms entering the CSTR leave it almost immediately, because material is being continuously withdrawn from the reactor; other atoms remain in the reactor almost forever because all the material is never removed from the reactor at one time. Many of the atoms, of course, leave the reactor after spending a period of time somewhere in the vicinity of the mean residence time. In any reactor, the distribution of residence times can significantly affect its performance.

The *residence-time distribution* (RTD) of a reactor is a characteristic of the mixing that occurs in the chemical reactor. There is no axial mixing in a plug-flow reactor, and this omission is reflected in the RTD which is exhibited by this class of reactors. The CSTR is thoroughly mixed and possesses a far different kind of RTD than the plug-flow reactor. As will be illustrated later, not all RTDs are unique to a particular reactor type; markedly different reactors can display identical RTDs. Nevertheless, the RTD exhibited by a given reactor yields distinctive clues to the type of mixing occurring within it and is one of the most informative characterizations of the reactor.

> The "RTD": Some molecules leave quickly, others overstay their welcome

> We will use the RTD to characterize nonideal reactors

13.2 Measurement of the RTD

The RTD is determined experimentally by injecting an inert chemical, molecule, or atom, called a *tracer*, into the reactor at some time $t = 0$ and then measuring the tracer concentration, C, in the effluent stream as a function of time. In addition to being a nonreactive species that is easily detectable, the tracer should have physical properties similar to those of the reacting mixture and be completely soluble in the mixture. It also should not adsorb on the walls or other surfaces in the reactor. The latter requirements are needed so that the tracer's behavior will honestly reflect that of the material flowing

---

[1] R. B. MacMullin and M. Weber, Jr., *Trans. Am. Inst. Chem. Eng.*, *31*, 409 (1935).

[2] P. V. Danckwerts, *Chem. Eng. Sci.*, *2*, 1 (1953).

Use of tracers to
determine the RTD

through the reactor. Colored and radioactive materials are the two most common types of tracers. The two most used methods of injection are *pulse input* and *step input*.

13.2.1 Pulse Input

In a pulse input, an amount of tracer N_0 is suddenly injected in one shot into the feedstream entering the reactor in as short a time as possible. The outlet concentration is then measured as a function of time. Typical concentration–time curves at the inlet and outlet of an arbitrary reactor are shown in

The *C* curve

Figure 13-4. The effluent concentration–time curve is referred to as the *C* curve in RTD analysis. We shall analyze the injection of a tracer pulse for a single-input and single-output system in which *only flow* (i.e., no dispersion) carries the tracer material across system boundaries. First, we choose an increment of time Δt sufficiently small that the concentration of tracer, $C(t)$, exiting between time t and $t + \Delta t$ is essentially constant. The amount of tracer material, ΔN, leaving the reactor between time t and $t + \Delta t$ is then

$$\Delta N = C(t) v \Delta t \tag{13-1}$$

where v is the effluent volumetric flow rate. In other words, ΔN is the amount of material that has spent an amount of time between t and $t + \Delta t$ in the reac-

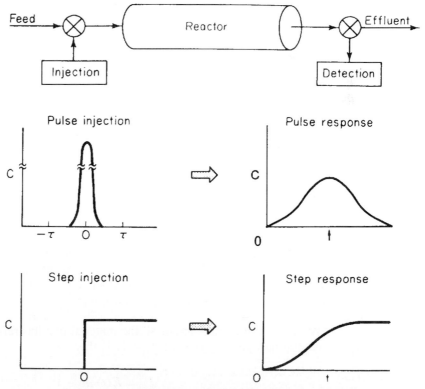

Figure 13-4 RTD measurements.

tor. If we now divide by the total amount of material that was injected into the reactor, N_0, we obtain

$$\frac{\Delta N}{N_0} = \frac{vC(t)}{N_0}\, \Delta t \tag{13-2}$$

which represents the fraction of material that has a residence time in the reactor between time t and $t + \Delta t$.

For pulse injection we define

$$E(t) = \frac{vC(t)}{N_0} \tag{13-3}$$

so that

$$\frac{\Delta N}{N_0} = E(t)\, \Delta t \tag{13-4}$$

The quantity **E(t)** is called the **residence-time distribution function**. It is the function that describes in a quantitative manner how much time different fluid elements have spent in the reactor.

If N_0 is not known directly, it can be obtained from the outlet concentration measurements by summing up all the amounts of materials, ΔN, between time equal to zero and infinity. Writing Equation (13-1) in differential form yields

$$dN = vC(t)\, dt \tag{13-5}$$

and then integrating, we obtain

$$N_0 = \int_0^\infty vC(t)\, dt \tag{13-6}$$

The volumetric flow rate v is usually constant, so we can define $E(t)$ as

We find the RTD function, **E(t)**, from the tracer concentration **C(t)**

$$\boxed{E(t) = \frac{C(t)}{\displaystyle\int_0^\infty C(t)\, dt}} \tag{13-7}$$

The integral in the denominator is the area under the C curve.

An alternative way of interpreting the residence-time function is in its integral form:

$$\begin{bmatrix} \text{fraction of material leaving the reactor} \\ \text{that has resided in the reactor} \\ \text{for times between } t_1 \text{ and } t_2 \end{bmatrix} = \int_{t_1}^{t_2} E(t)\, dt$$

We know that the fraction of all the material that has resided for a time t in the reactor between $t = 0$ and $t = \infty$ is 1; therefore,

Eventually all must leave

$$\int_0^\infty E(t)\, dt = 1 \tag{13-8}$$

Example 13–1   Constructing the C(t) and E(t) Curves

A sample of the tracer hytane at 320 K was injected as a pulse to a reactor and the effluent concentration measured as a function of time, resulting in the following data:

t (min)	0	1	2	3	4	5	6	7	8	9	10	12	14
C (g/m³)	0	1	5	8	10	8	6	4	3.0	2.2	1.5	0.6	0

The measurements represent the exact concentrations at the times listed and not average values between the various sampling tests. **(a)** Construct figures showing $C(t)$ and $E(t)$ as functions of time, **(b)** Determine both the fraction of material leaving the reactor that has spent between 3 and 6 min in the reactor and the fraction of material leaving that has spent between 7.75 and 8.25 min in the reactor, **(c)** 3 min or less.

Solution

(a) By plotting C as a function of time, the curve shown in Figure E13-1.1 is obtained. To obtain the $E(t)$ curve from the $C(t)$ curve, we just divide $C(t)$ by the integral

$$\int_0^\infty C(t)\, dt$$

The *C* curve

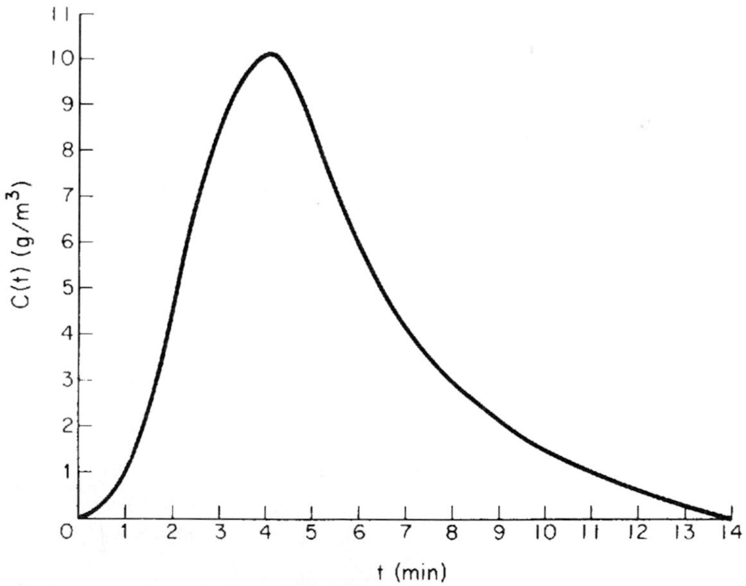

Figure E13-1.1

which is just the area under the *C* curve. This area can be found using the numerical integration formulas in Appendix A.4:

$$\int_0^\infty C(t)\,dt = \int_0^{10} C(t)\,dt + \int_{10}^{14} C(t)\,dt \qquad \text{(E13-1.1)}$$

$$\int_0^{10} C(t)\,dt = \tfrac{1}{3}[1(0) + 4(1) + 2(5) + 4(8)$$
$$+ 2(10) + 4(8) + 2(6)$$
$$+ 4(4) + 2(3.0) + 4(2.2) + 1(1.5)]$$
$$= 47.4 \text{ g} \cdot \text{min}/\text{m}^3$$

$$\int_{10}^{14} C(t)\,dt = \tfrac{2}{3}[1.5 + 4(0.6) + 0] = 2.6 \text{ g} \cdot \text{min}/\text{m}^3$$

$$\int_0^\infty C(t)\,dt = 50.0 \text{ g} \cdot \text{min}/\text{m}^3 \qquad \text{(E13-1.2)}$$

We now calculate

$$E(t) = \frac{C(t)}{\displaystyle\int_0^\infty C(t)\,dt} = \frac{C(t)}{50 \text{ g} \cdot \text{min}/\text{m}^3} \qquad \text{(E13-1.3)}$$

with the following results:

t (min)	1	2	3	4	5	6	7	8	9	10	12	14
$C(t)$ (g/m^3)	1	5	8	10	8	6	4	3	2.2	1.5	0.6	0
$E(t)$ (min^{-1})	0.02	0.1	0.16	0.2	0.16	0.12	0.08	0.06	0.044	0.03	0.012	0

(b) These data are plotted in Figure E13-1.2. The shaded area represents the fraction of material leaving the reactor that has resided in the reactor between 3 and 6 min.

The *E* curve

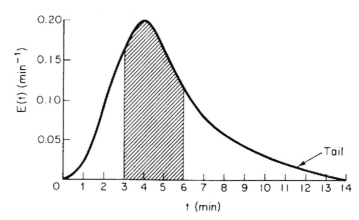

Figure E13-1.2

$$\int_3^6 E(t)\,dt = \text{shaded area}$$

$$= \tfrac{3}{8}\Delta t(f_1 + 3f_2 + 3f_3 + f_4)$$

$$= \tfrac{3}{8}(1)[0.16 + 3(0.2) + 3(0.16) + 0.12] = 0.51$$

Evaluating this area, we find that 51% of the material leaving the reactor spends between 3 and 6 min in the reactor.

Because the time between 7.75 and 8.25 min is very small relative to a time scale of 14 min we shall use an alternative technique to determine this fraction to reinforce the interpretation of the quantity $E(t)\,dt$. The average value of $E(t)$ between these times is 0.06 min^{-1}:

$$E(t)\,dt = (0.06\ \text{min}^{-1})(0.5\ \text{min}) = 0.03$$

The tail

Consequently, 3.0% of the fluid leaving the reactor has been in the reactor between 7.75 and 8.25 min. The long-time portion of the $E(t)$ curve is called the *tail*. In this example the tail is that portion of the curve between *say* 10 and 14 min.

(c) Finally, we shall consider the fraction of material that has been in the reactor for a time t or less, that is, the fraction that has spent between 0 and t minutes in the reactor. This fraction is just the shaded area under the curve up to $t = t$ minutes. This is shown in Figure E13-1.3 for $t = 3$ min. Calculating the area under the curve, we see that 20% of the material has spent 3 min *or less* in the reactor.

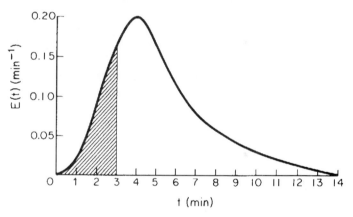

Figure E13-1.3

The principal difficulties with the pulse technique lie in the problems connected with obtaining a reasonable pulse at a reactor's entrance. The injection must take place over a period which is very short compared with residence times in various segments of the reactor or reactor system, and there must be a negligible amount of dispersion between the point of injection and the entrance to the reactor system. If these conditions can be fulfilled, this technique represents a simple and direct way of obtaining the RTD.

Drawbacks to the
pulse injection to
obtain the RTD

There are problems when the concentration–time curve has a long tail because the analysis can be subject to large inaccuracies. This problem principally affects the denominator of the right-hand side of Equation (13-7) [i.e., the integration of the $C(t)$ curve]. It is desirable to extrapolate the tail and analytically continue the calculation. The tail of the curve may sometimes be approximated as an exponential decay (see Problem 14-10). The inaccuracies introduced by this assumption are very likely to be much less than those resulting from either truncation or numerical imprecision in this region.

13.2.2 Step Tracer Experiment

Now that we have an understanding of the meaning of the RTD curve from a pulse input, we will formulate a more general relationship between a time-varying tracer injection and the corresponding concentration in the effluent. We shall state without development that the output concentration from a vessel is related to the input concentration by the convolution integral[3]:

$$C_{out}(t) = \int_0^t C_{in}(t - t')E(t')\, dt \qquad (13\text{-}9)$$

The inlet concentration most often takes the form of either perfect *pulse input* (Dirac delta function), *imperfect pulse injection* (see Figure 13-4), or a *step input*.

We will now analyze a *step input* in the tracer concentration for a system with a constant volumetric flow rate. Consider a constant rate of tracer addition to a feed that is initiated at time $t = 0$. Before this time no tracer was added to the feed. Stated symbolically, we have

$$C_0(t) = \begin{cases} 0 & t < 0 \\ (C_0)\ \text{constant} & t \geq 0 \end{cases}$$

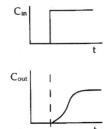

The concentration of tracer in the feed to the reactor is kept at this level until the concentration in the effluent is indistinguishable from that in the feed; the test may then be discontinued. A typical outlet concentration curve for this type of input is shown in Figure 13-4.

Because the inlet concentration is a constant with time, C_0, we can take it outside the integral sign, that is,

$$C_{out} = C_0 \int_0^t E(t')\, dt'$$

Dividing by C_0 yields

$$\left[\frac{C_{out}}{C_0} \right]_{step} = \int_0^t E(t')\, dt' = F(t) \qquad (13\text{-}10)$$

We differentiate this expression to obtain the RTD function $E(t)$:

$$E(t) = \frac{d}{dt}\left[\frac{C(t)}{C_0} \right]_{step} \qquad (13\text{-}11)$$

The positive step is usually easier to carry out experimentally than the pulse test, and it has the additional advantage that the total amount of tracer in the feed over the period of the test does not have to be known as it does in the

---

[3] A development can be found in O. Levenspiel, *Chemical Reaction Engineering*, 2nd ed., Wiley, New York, 1972, p. 263.

pulse test. One possible drawback in this technique is that it is sometimes difficult to maintain a constant tracer concentration in the feed. Obtaining the RTD from this test also involves differentiation of the data and presents an additional and probably more serious drawback to the technique, because differentiation of data can, on occasion, lead to large errors. A third problem lies with the large amount of tracer required for this test. If the tracer is very expensive, a pulse test is almost always used to minimize the cost.

Advantages and drawbacks to the step injection

Other tracer techniques exist, such as negative step (i.e., elution), frequency-response methods, and methods that use inputs other than steps or pulses. These methods are usually much more difficult to carry out than the ones presented and are not encountered as often. For this reason they will not be treated here, and the literature should be consulted for their virtues, defects, and the details of implementing them and analyzing the results. A good source for this information is Wen and Fan.[4]

13.3 Characteristics of the RTD

From $E(t)$ we can learn how long different molecules have been in the reactor

Sometimes $E(t)$ is called the *exit-age distribution function*. If we regard the "age" of an atom as the time it has resided in the reaction environment, then $E(t)$ concerns the age distribution of the effluent stream. It is the most used of the distribution functions connected with reactor analysis because it characterizes the lengths of time various atoms spend at reaction conditions.

Figure 13-5 illustrates typical RTDs resulting from different reactor situations. Figure 13-5(a) and (b) correspond to nearly ideal PFRs and CSTRs, respectively. In Figure 13-5(c) one observes that a principal peak occurs at a time smaller than the space time ($\tau = V/v$) (i.e., early exit of fluid) and also that fluid exits at a time greater than space-time τ. This curve is representative of the RTD for a packed-bed reactor with channeling and dead zones. One scenario by which this situation might occur is shown in Figure 13-5(d). Figure 13-5(e) shows the RTD for the CSTR in Figure 13-5(f), which has dead zones and bypassing. The dead zone serves to reduce the effective reactor volume, indicating that the active reactor volume is smaller than expected.

13.3.1 Integral Relationships

The fraction of the exit stream that has resided in the reactor for a period of time shorter than a given value t is equal to the sum over all times less than t of $E(t)\, \Delta t$, or expressed continuously,

The cumulative RTD function $F(t)$

$$\int_0^t E(t)\, dt = \begin{bmatrix} \text{fraction of effluent} \\ \text{which has been in reactor} \\ \text{for less than time } t \end{bmatrix} = F(t) \qquad (13\text{-}12)$$

---

[4] C. Y. Wen and L. T. Fan, *Models for Flow Systems and Chemical Reactors*, Marcel Dekker, New York, 1975.

RTDs that are
commonly
observed

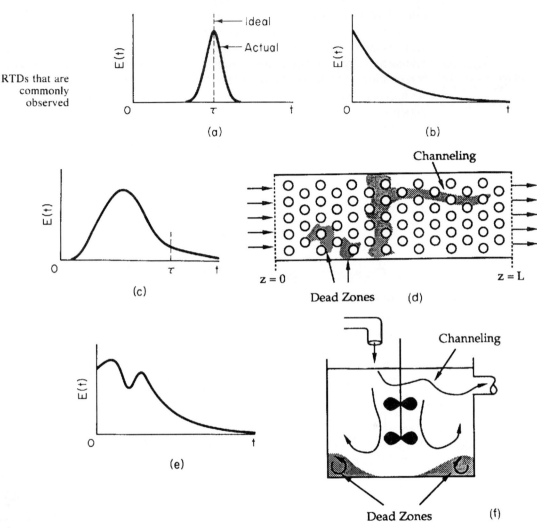

Figure 13-5 (a) RTD for near plug-flow reactor; (b) RTD for near perfectly mixed CSTR; (c) RTD for packed-bed reactor with dead zones and channeling; (d) packed-bed reactor; (e) tank reactor with short-circuiting flow (bypass); (f) CSTR with dead zone.

Analogously, we have

$$\int_{t}^{\infty} E(t)\, dt = \begin{bmatrix} \text{fraction of effluent} \\ \text{which has been in reactor} \\ \text{for longer than time } t \end{bmatrix} = 1 - F(t) \qquad (13\text{-}13)$$

Because t appears in the integration limits of these two expressions, Equations (13-12) and (13-13) are both functions of time. Danckwerts[5] defined Equation (13-12) as a *cumulative distribution function and called it F(t)*. We

---

[5] P. V. Danckwerts, *Chem. Eng. Sci.*, 2, 1 (1953).

can calculate $F(t)$ at various times t from the area under the curve of an $E(t)$ versus t plot. For example, in Figure E13-1.3 we saw that $F(t)$ at 3 min was 0.20, meaning that 20% of the molecules spent 3 min or less in the reactor. The shape of the $F(t)$ curve is shown for a tracer response to a step input in Figure 13-6. One notes from this curve that 80% [$F(t)$] of the molecules spend 40 min or less in the reactor, and 20% of the molecules [$1 - F(t)$] spend longer than 40 min in the reactor.

The F curve

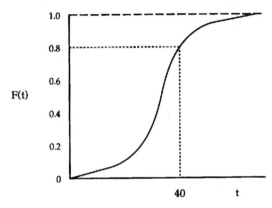

Figure 13-6   Cumulative distribution curve, $F(t)$.

The F curve is another function that has been defined as the normalized response to a particular input. Alternatively, Equation (13-12) has been used as a definition of $F(t)$, and it has been stated that as a result it can be obtained as the response to a positive-step tracer test. Sometimes the F curve is used in the same manner as the RTD in the modeling of chemical reactors. An excellent example is the study of Wolf and White,[6] who investigated the behavior of screw extruders in polymerization processes.

13.3.2 Mean Residence Time

In previous chapters treating ideal reactors, a parameter frequently used was the space-time or average residence time τ, which was defined as being equal to V/v. It will be shown that no matter what RTD exists for a particular reactor, ideal or nonideal, this nominal holding time, τ, is equal to the mean residence time, t_m.

As is the case with other variables described by distribution functions, the mean value of the variable is equal to the first moment of the RTD function, $E(t)$. Thus the mean residence time is

The first moment gives the average time the effluent molecules spent in the reactor

$$t_m = \frac{\displaystyle\int_0^\infty tE(t)\,dt}{\displaystyle\int_0^\infty E(t)\,dt} = \int_0^\infty tE(t)\,dt \qquad (13\text{-}14)$$

[6] D. Wolf and D. H. White, *AIChE J.*, 22, 122 (1976).

We now wish to show how we can determine the total reactor volume using the cumulative distribution function. Consider the following situation: We have a reactor completely filled with maize molecules. At time $t = 0$ we start to inject blue molecules to replace the maize molecules that currently fill the reactor. That is, the reactor volume V is equal to the volume occupied by the maize molecules. Now, in a time dt, the volume of molecules that will leave the reactor is $(v\,dt)$. The fraction of these molecules that have been in the reactor a time t or greater is $[1 - F(t)]$. Because only the maize molecules have been in the reactor a time t or greater, the volume of maize molecules, dV, leaving the reactor in a time dt is

$$dV = (v\,dt)[1 - F(t)] \tag{13-15}$$

If we now sum up all of the maize molecules that have left the reactor in time $0 < t < \infty$ we have

All we are doing here is proving that the space-time and mean residence time are equal

$$V = \int_0^\infty v\,[1 - F(t)]\,dt \tag{13-16}$$

Because the volumetric flow rate is constant,

$$V = v \int_0^\infty [1 - F(t)]\,dt \tag{13-17}$$

Using the integration-by-parts relationship gives

$$\int x\,dy = xy - \int y\,dx$$

and dividing by the volumetric flow rate gives

$$\frac{V}{v} = t[1 - F(t)]\Big|_0^\infty + \int_0^1 t\,dF \tag{13-18}$$

At $t = 0$, $F(t) = 0$; and as $t \to \infty$, then $[1 - F(t)] = 0$. The first term on the right-hand side is zero and the second term becomes

$$\frac{V}{v} = \tau = \int_0^1 t\,dF \tag{13-19}$$

However, $dF = E(t)\,dt$; therefore,

$$\tau = \int_0^\infty tE(t)\,dt \tag{13-20}$$

The right-hand side is just the mean residence time, and we see that the mean residence time is just the space time τ:

$\tau = t_m$, Q.E.D.

$$\boxed{\tau = t_m} \tag{13-21}$$

This result is true *only* for a *closed system* (i.e., no dispersion; see Chapter 14). The exact reactor volume is easily determined from the equation

$$\boxed{V = vt_m} \tag{13-22}$$

13.3.3 Other Moments of the RTD

It is very common to compare RTDs by using their moments instead of trying to compare their entire distributions (e.g., Wen and Fan[7]). For this purpose, three moments are normally used. The first is the mean residence time. The second moment commonly used is taken about the mean and is called the variance, or square of the standard deviation. It is defined by

The second moment about the mean is the variance

$$\sigma^2 = \int_0^\infty (t - t_m)^2 E(t)\, dt \qquad (13\text{-}23)$$

The magnitude of this moment is an indication of the "spread" of the distribution; the greater the value of this moment, the greater a distribution's spread.

The third moment is also taken about the mean and is related to the *skewness*. The skewness is defined by

The two parameters most commonly used to characterize the RTD are τ and σ

$$s^3 = \frac{1}{\sigma^{3/2}} \int_0^\infty (t - t_m)^3 E(t)\, dt \qquad (13\text{-}24)$$

The magnitude of this moment measures the extent that a distribution is skewed in one direction or another in reference to the mean.

Rigorously, for complete description of a distribution, all moments must be determined. Practically, these three are usually sufficient for a reasonable characterization of an RTD.

Example 13–2   Mean Residence Time and Variance Calculations

Calculate the mean residence time and the variance for the reactor characterized in Example 13-1 by the RTD obtained from a pulse input at 320 K.

Solution

First, the mean residence time will be calculated from Equation (13-14):

$$t_m = \int_0^\infty t E(t)\, dt \qquad (E13\text{-}2.1)$$

The area under the curve of a plot of $t E(t)$ as a function of t will yield t_m. Once the mean residence time is determined, the variance can be calculated from Equation (13-23):

$$\sigma^2 = \int_0^\infty (t - t_m)^2 E(t)\, dt \qquad (E13\text{-}2.2)$$

To calculate t_m and σ^2, Table E13-2.1 was constructed from the data given and interpreted in Example 13-1.

$$t_m = \int_0^\infty t E(t)\, dt = \int_0^{10} t E(t)\, dt + \int_{10}^\infty t E(t)\, dt$$

---

[7] Wen and Fan, *Models for Flow Systems and Chemical Reactors*, Chap. 11.

TABLE E13-2.1.   CALCULATING $E(t)$, t_m, AND σ^2

t	$C(t)$	$E(t)$	$tE(t)$	$t - t_m{}^a$	$(t - t_m)^2 E(t)^a$
0	0	0	0	−5.15	0
1	1	0.02	0.02	−4.15	0.34
2	5	0.10	0.20	−3.15	0.992
3	8	0.16	0.48	−2.15	0.74
4	10	0.20	0.80	−1.15	0.265
5	8	0.16	0.80	−0.15	0.004
6	6	0.12	0.72	0.85	0.087
7	4	0.08	0.56	1.85	0.274
8	3	0.06	0.48	2.85	0.487
9	2.2	0.044	0.40	3.85	0.652
10	1.5	0.03	0.30	4.85	0.706
12	0.6	0.012	0.14	6.85	0.563
14	0	0	0	8.85	0

^{a}These two columns are completed after the mean residence time (t_m) is found.

Again, using the numerical integration formulas in Appendix A.4, we have

$$t_m = \frac{h_1}{3}\,(f_1 + 4f_2 + 2f_3 + 4f_4 + \cdots + 4f_{n-1} + f_n)$$

$$+ \frac{h_2}{3}\,(f_{n+1} + 4f_{n+2} + f_{n+3})$$

Numerical
integration to find
the mean residence
time, t_m

$$t_m = \tfrac{1}{3}[1(0) + 4(0.02) + 2(0.2) + 4(0.48) + 2(0.8) + 4(0.8)$$
$$+ 2(0.72) + 4(0.56) + 2(0.48) + 4(0.40) + 1(0.3)]$$
$$+ \tfrac{2}{3}[0.3 + 4(0.14) + 0]$$
$$= 4.58 + 0.573 = 5.15 \text{ min}$$

Note: One could also use the spreadsheets in POLYMATH or Excel to formulate Table E13-2.1 and to calculate the mean residence time t_m and variance σ.

Calculating the
mean residence
time,

$\tau = t_m = \displaystyle\int_0^\infty tE(t)\,dt$

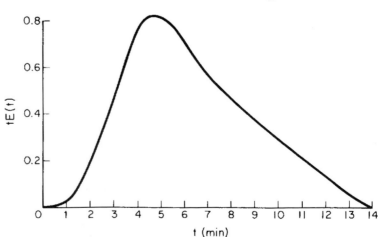

Figure E13-2.1   Calculating the mean residence time.

Plotting $tE(t)$ versus t we obtain Figure E13-2.1. The area under the curve is 5.15 min.

$$t_m = 5.15 \text{ min}$$

Now that the mean residence time has been determined, we can calculate the variance by calculating the area under the curve of a plot of $(t - t_m)^2 E(t)$ as a function of t (Figure E13-2.2). The area under the curve(s) is 6.10 min². Alternatively,

$$\sigma^2 = \int_0^\infty (t - t_m)^2 E(t)\, dt \tag{13-23}$$

Calculating the variance,

$\sigma^2 = \int_0^\infty (t - t_m)^2 E(t)\, dt$

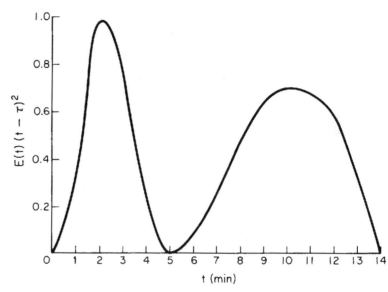

Figure E13-2.2   Calculating the variance.

Integrating between 0 and 10 min and 10 and 14 min, and then substituting values from Table E13-2.1, we have

$$\sigma^2 = \tfrac{1}{3}[1(0) + 4(0.34) + 2(0.992) + 4(0.74) + 2(0.265) + 4(0.004)$$

One could also use POLYMATH or Excel to make these calculations

$$+ 2(0.087) + 4(0.274) + 2(0.487) + 4(0.652) + 1(0.706)]$$

$$+ \tfrac{2}{3}[0.706 + 4(0.563) + 0]$$

$$= 4.136 + 1.972 = 6.1 \text{ min}^2$$

The square of the standard deviation is $\sigma^2 = 6.1$ min², so $\sigma = 2.4$ min.

13.3.4 Normalized RTD Function, $E(\Theta)$

Frequently, a normalized RTD is used instead of the function $E(t)$. If the parameter Θ is defined as

$$\Theta \equiv \frac{t}{\tau} \tag{13-25}$$

a dimensionless function $E(\Theta)$ can be defined as

$$E(\Theta) \equiv \tau E(t) \qquad (13\text{-}26)$$

and plotted as a function of Θ. The quantity Θ represents the number of reactor volumes of fluid based on entrance conditions that have flowed through the reactor in time t.

Why we use a normalized RTD

The purpose of creating this normalized distribution function is that the flow performance inside reactors of different sizes can be compared directly. For example, if the normalized function $E(\Theta)$ is used, *all* perfectly mixed CSTRs have numerically the same RTD. If the simple function $E(t)$ is used, numerical values of $E(t)$ can differ substantially for different CSTRs. As will be shown later, for a perfectly mixed CSTR,

$E(t)$ for a CSTR

$$E(t) = \frac{1}{\tau}\, e^{-t/\tau} \qquad (13\text{-}27)$$

and therefore

$$E(\Theta) = \tau E(t) = e^{-\Theta} \qquad (13\text{-}28)$$

From these equations it can be seen that the value of $E(t)$ at identical times can be quite different for two reactors of different average residence times. But for the same value of Θ, the value of $E(\Theta)$ is the same irrespective of the size of a perfectly mixed CSTR.

It is a relatively easy exercise to show that

$$\int_0^\infty E(\Theta)\, d\Theta = 1 \qquad (13\text{-}29)$$

and is recommended as a 93-s divertissement.

13.3.5 Internal-Age Distribution $I(\alpha)$

Tombstone jail

How long have you been here? $I(\alpha)$ When do you expect to get out?

Although this section is not a prerequisite to the remaining sections, the internal-age distribution is introduced here because of its close analogy to the external-age distribution. We shall let α represent the age of a molecule inside the reactor. The internal-age distribution function $I(\alpha)$ is a function such that $I(\alpha)\,\Delta\alpha$ is the fraction of material *inside the reactor* that has been inside the reactor for a period of time between α and $\alpha + \Delta\alpha$. It may be contrasted with $E(\alpha)\,\Delta\alpha$, which is used to represent the material *leaving the reactor* that has spent a time between α and $\alpha + \Delta\alpha$ in the reaction zone; $I(\alpha)$ characterizes the time the material has been (and still is) in the reactor at a particular time. The function $E(\alpha)$ is viewed outside the reactor and $I(\alpha)$ is viewed inside the reactor. In unsteady-state problems it can be important to know what the particular state of a reaction mixture is, and $I(\alpha)$ supplies this information. For example, in a catalytic reaction using a catalyst whose activity decays with time, the age distribution of the catalyst in the reactor is of importance and the internal-age distribution can be of use in modeling the reactor.

The relationship between the *internal-age* and *external-age distribution* can be demonstrated by analyzing a continuous reactor operating at steady state that is filled with material of volume V. Consider again that the volume of reactor is filled with maize-colored molecules, and at time $t = 0$ we start to inject blue molecules to replace the maize molecules. By definition of $I(\alpha)$, the volume of molecules inside the reactor that have been there between a time α and $\alpha + d\alpha$ is

$$dV = V[I(\alpha)]\, d\alpha \tag{13-30}$$

At $t = 0$ we will let $(v_0\, d\alpha)$ be the first volume of blue molecules that enter the reactor. We want to consider what has happened to the molecules in this volume at a time α after being injected. Some of the molecules will already have left the system at a time α, while others remain. The fraction of molecules that still remain in the system is $[1 - F(\alpha)]$. Consequently, the volume of molecules that entered the system between $t = 0$ and $t = d\alpha$ and are still in the system at a later time α is

Finding a relation between $E(\alpha)$ and $I(\alpha)$

$$dV = v_0\, d\alpha\, [1 - F(\alpha)] \tag{13-31}$$

This is the volume of molecules that have an age between α and $(\alpha + d\alpha)$. Equating Equations (13-30) and (13-31) and dividing by V and by $d\alpha$ gives

$$I(\alpha) = \frac{v}{V}[1 - F(\alpha)]$$

Then

Relating $I(\alpha)$ to $F(\alpha)$ and $E(\alpha)$

$$\boxed{I(\alpha) = \frac{1}{\tau}[1 - F(\alpha)] = \frac{1}{\tau}\left[1 - \int_0^\alpha E(\alpha)\, d\alpha\right]} \tag{13-32}$$

Differentiating Equation (13-32) and noting that

$$\frac{d[1 - F(\alpha)]}{d\alpha} = -E(\alpha)$$

gives

$$\boxed{E(\alpha) = -\frac{d}{d\alpha}[\tau I(\alpha)]} \tag{13-33}$$

As a brief exercise, the internal-age distribution of a perfectly mixed CSTR will be calculated. Equation (13-27) gives the RTD of the reactor, which upon substitution into Equation (13-32) gives

$$I(\alpha) = \frac{1}{\tau}\left(1 - \int_0^\alpha \frac{1}{\tau} e^{-\alpha/\tau}\, d\alpha\right)$$

$$= \frac{1}{\tau}\left(1 + e^{-\alpha/\tau}\Big|_0^\alpha\right)$$

$$= \frac{1}{\tau}e^{-\alpha/\tau} \tag{13-34}$$

True only for a
perfectly mixed
CSTR
Thus the internal-age distribution of a perfectly mixed CSTR is identical to the exit-age distribution, or RTD, because the composition of the effluent is identical to the composition of the material anywhere within the CSTR when it is perfectly mixed.

Example 13–3 CSTR with Fresh Catalyst Feed

When a catalyst is decaying, fresh catalyst must be fed to a reactor to keep a constant level of activity. The relation between catalyst weight, conversion, and catalyst activity is

$$W = \frac{F_{A0}X}{-r'_A} = \frac{F_{A0}X}{\bar{a}k_0 C_A^n} \qquad \text{(E13-3.1)}$$

where $\bar{a}$ is the mean activity in the reactor. Determine the mean activity for first-order decay in a CSTR.

Solution

Because there will be a distribution of times the various catalyst particles have spent in the reactor, there will be a distribution of activities. The mean activity is the integral of the product of the fraction of the particles that have been in the reactor (i.e., have ages) between time α and $\alpha + \Delta\alpha$, $I(\alpha)\, d\alpha$, and the activity at time α:

$$\bar{a} = \int_0^\infty a(\alpha)I(\alpha)\, d\alpha \qquad \text{(E13-3.2)}$$

For first-order decay,

$$a = e^{-k\alpha} \qquad \text{(E13-3.3)}$$

In a well-mixed CSTR,

$$I(\alpha) = \frac{1}{\tau} e^{-\alpha/\tau} \qquad \text{(13-34)}$$

Using $I(\alpha)$ and $a(\alpha)$
to find the mean
catalyst activity
$$\bar{a} = \int_0^\infty \frac{e^{-k_d\alpha} e^{-\alpha/\tau_c}}{\tau_c}\, d\alpha \qquad \text{(E13-3.4)}$$

where k_d is the decay constant and τ_c is the mean contact time, such that

$$\tau_c = \frac{W}{F_c} = \frac{\text{weight of catalyst (kg)}}{\text{feed rate of catalyst (kg/s)}} \qquad \text{(E13-3.5)}$$

Integrating yields

$$\boxed{\bar{a} = \frac{1}{\tau_c k_d + 1}} \qquad \text{(E13-3.6)}$$

We see that for a distribution of activities, each following first-order decay in an ideal CSTR, the form of the mean activity is identical to the integrated form for second-order catalyst decay. See Problem 13-2(a).

13.4 RTD in Ideal Reactors

13.4.1 RTDs in Batch and Plug-Flow Reactors

The RTDs in plug-flow reactors and ideal batch reactors are the simplest to consider. All the atoms leaving such reactors have spent precisely the same amount of time within the reactors. The distribution function in such a case is a spike of infinite height and zero width, whose area is equal to 1; the spike occurs at $t = V/v = \tau$, or $\Theta = 1$. Mathematically, this spike is represented by the Dirac delta function:

$E(t)$ for a plug-flow reactor

$$\boxed{E(t) = \delta(t - \tau)} \tag{13-35}$$

The Dirac delta function has the following properties:

$$\delta(x) = \begin{cases} 0 & \text{when } x \neq 0 \\ \infty & \text{when } x = 0 \end{cases} \tag{13-36}$$

$$\int_{-\infty}^{\infty} \delta(x)\, dx = 1$$
$$\int_{-\infty}^{\infty} g(x)\delta(x - \tau)\, dx = g(\tau) \tag{13-37}$$

This function is illustrated in Figure 13-7.

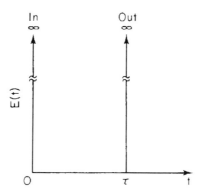

Figure 13-7   Ideal plug-flow response to a pulse tracer input.

13.4.2 Single-CSTR RTD

In an ideal CSTR the concentration of any substance in the effluent stream is identical to the concentration throughout the reactor. Consequently, it is possible to obtain the RTD from conceptual considerations in a fairly straightforward manner. A material balance on an inert tracer that has been injected as a pulse at time $t = 0$ into a CSTR yields for $t > 0$

From a tracer balance we can determine $E(t)$

$$\text{in} - \text{out} = \text{accumulation}$$

$$0 - vC = +V\frac{dC}{dt} \tag{13-38}$$

Because the reactor is perfectly mixed, C in this equation is the concentration of the tracer either in the effluent or within the reactor. Separating the variables and integrating with $C = C_0$ at $t = 0$ yields

$$C(t) = C_0 e^{-t/\tau} \tag{13-39}$$

This relationship gives the concentration of tracer in the effluent at any time t.

To find $E(t)$ for an ideal CSTR, we first recall Equation (13-7) and then substitute for $C(t)$ using Equation (13-39). That is,

$$E(t) = \frac{C(t)}{\displaystyle\int_0^\infty C(t)\, dt} = \frac{C_0 e^{-t/\tau}}{\displaystyle\int_0^\infty C_0 e^{-t/\tau}\, dt} \tag{13-40}$$

Evaluating the integral in the denominator completes the derivation of the RTD for an ideal CSTR given by Equations (13-27) and (13-28):

$E(t)$ for a CSTR

$$E(t) = \frac{e^{-t/\tau}}{\tau} \tag{13-27}$$

$$E(\Theta) = e^{-\Theta} \tag{13-28}$$

Recall $\Theta = t/\tau$.

This function is the RTD for a CSTR and is illustrated in Figure 13-8.

Response of an
ideal CSTR

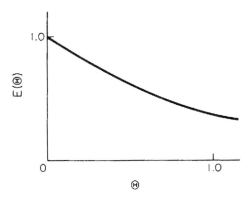

Figure 13-8   CSTR response to a pulse tracer input.

Earlier it was shown that generally the mean residence time in a reactor is equal to V/v, or τ. This relationship can be shown in a simpler fashion for the CSTR. Applying the definition of a mean residence time to the RTD for a CSTR, we obtain

$$t_m = \int_0^\infty tE(t)\, dt = \int_0^\infty \frac{t}{\tau} e^{-t/\tau}\, dt = \tau \tag{13-41}$$

Thus the nominal holding time (space time) $\tau = V/v$ is also the mean residence time that the material spends in the reactor.

The second moment about the mean is a measure of the spread of the distribution about the mean. The variance of residence times in a perfectly mixed tank reactor is (let $x = t/\tau$)

For a perfectly
mixed CSTR $\sigma = \tau$

$$\sigma^2 = \int_0^\infty \frac{(t-\tau)^2}{\tau} e^{-t/\tau} \, dt = \tau^2 \int_0^\infty (x-1)^2 e^{-x} \, dx = \tau^2 \qquad (13\text{-}42)$$

Then $\sigma = \tau$. The standard deviation is the square root of the variance. For a CSTR, the standard deviation of the residence-time distribution is as large as the mean itself.

13.4.3 Laminar Flow Reactor

Before proceeding to show how the RTD can be used to estimate conversion in a reactor, we shall derive $E(t)$ for a laminar flow reactor. For laminar flow in a tubular reactor, the velocity profile is parabolic, with the fluid in the center of the tube spending the shortest time in the reactor. A schematic diagram of the fluid movement after a time t is shown in Figure 13-9. The figure at the left shows how far down the reactor each concentric fluid element has traveled after a time t.

Molecules near the
center spend a
shorter time in the
reactor than those
close to the wall

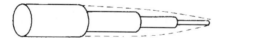

Figure 13-9   Schematic diagram of fluid elements in a laminar flow reactor.

The velocity profile in a pipe of outer radius R is

$$U = U_{\max}\left[1 - \left(\frac{r}{R}\right)^2\right] = 2U_{avg}\left[1 - \left(\frac{r}{R}\right)^2\right] = \frac{2v_0}{\pi R^2}\left[1 - \left(\frac{r}{R}\right)^2\right] \qquad (13\text{-}43)$$

where $U_{\max}$ is the centerline velocity and U_{avg} is the average velocity through the tube.

The time of passage of an element of fluid at a radius r is

$$t(r) = \frac{L}{U(r)} = \frac{\pi R^2 L}{v_0} \frac{1}{2[1 - (r/R)^2]}$$

$$= \frac{\tau}{2[1 - (r/R)^2]} \qquad (13\text{-}44)$$

The fraction of total fluid passing between r and $(r + dr)$ is dv/v_0:

$$\frac{dv}{v_0} = \frac{U(r)\,2\pi r \, dr}{v_0} \qquad (13\text{-}45)$$

We now need to relate the fluid fraction [Equation (13-45)] to the fraction of fluid spending between time t and $t + dt$ in the reactor. First we differentiate Equation (13-44):

$$dt = \frac{\tau}{2R^2} \frac{2r \, dr}{[1 - (r/R)^2]^2} = \frac{4}{\tau R^2} \left\{ \frac{\tau/2}{[1 - (r/R)^2]} \right\}^2 r \, dr$$

and then substitute for t using Equation (13-44) to yield

$$dt = \frac{4t^2}{\tau R^2} r \, dr \tag{13-46}$$

Combining Equations (13-45) and (13-46), we now have the fraction of fluid spending between time t and $t + dt$ in the reactor:

$$\frac{dv}{v_0} = \frac{L}{t} \left(\frac{2\pi r \, dr}{v_0} \right) = \frac{L}{t} \left(\frac{2\pi}{v_0} \right) \frac{\tau R^2}{4t^2} dt = \frac{\tau^2}{2t^3} dt$$

The minimum time the fluid may spend in the reactor is

$$t = \frac{L}{U_{\max}} = \frac{L}{2U_{\text{avg}}} \left(\frac{\pi R^2}{\pi R^2} \right) = \frac{V}{2v_0} = \frac{\tau}{2}$$

Consequently, the complete RTD function for a laminar flow reactor is

$E(t)$ for a laminar
flow reactor

$$E(t) = \begin{cases} 0 & t < \dfrac{\tau}{2} \\[2mm] \dfrac{\tau^2}{2t^3} & t \ge \dfrac{\tau}{2} \end{cases} \tag{13-47}$$

The cumulative distribution function for $t \ge \tau/2$ is

$$F(t) = \int_{\tau/2}^{t} \frac{\tau^2}{2t^3} dt = \frac{\tau^2}{2} \int_{\tau/2}^{t} \frac{dt}{t^3} = 1 - \frac{\tau^2}{4t^2} \tag{13-48}$$

The mean residence time t_m is

$$t_m = \int_{\tau/2}^{\infty} tE(t) \, dt = \frac{\tau^2}{2} \int_{\tau/2}^{\infty} \frac{dt}{t^2}$$

$$= \frac{\tau^2}{2} \left[-\frac{1}{t} \right]_{\tau/2}^{\infty} = \tau$$

This result was shown previously to be true for any reactor. The mean residence time is just the space-time τ.

The dimensionless form of the RTD function is

$$E(\Theta) = \begin{cases} 0 & \Theta < 0.5 \\ \dfrac{1}{2\Theta^3} & \Theta \geq 0.5 \end{cases}$$

(13-49)

and is plotted in Figure 13-10.

Normalized RTD
function for a
laminar flow
reactor

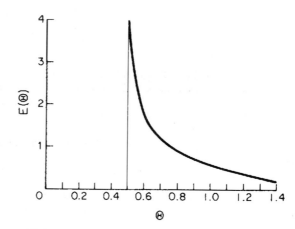

Figure 13-10   RTD curve for a laminar flow reactor.

13.4.4 PFR/CSTR Series RTD

Modeling the real
reactor as a CSTR
and a PFR in series

    In some stirred tank reactors, there is a highly agitated zone in the vicinity of the impeller which can be modeled as a perfectly mixed CSTR. Depending on the location of the inlet and outlet pipes, the reacting mixture may follow a somewhat tortuous path either before entering or after leaving the perfectly mixed zone—or even both. This tortuous path may be modeled as a plug-flow reactor. Thus this type of tank reactor may be modeled as a CSTR in series with a plug-flow reactor, and the PFR may either precede or follow the CSTR. In this section we develop the RTD for this type of tank reactor.

    First consider the CSTR followed by the PFR (Figure 13-11). The residence time in the CSTR will be denoted by τ_s and the residence time in the

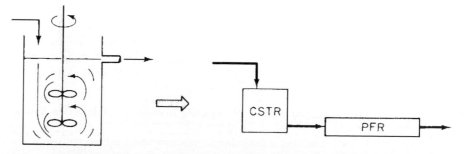

Figure 13-11   Real reactor modeled as a CSTR and PFR in series.

PFR by τ_p. If a pulse of tracer is injected into the entrance of the CSTR, the CSTR output concentration as a function of time will be

$$C = C_0 e^{-t/\tau_s}$$

This output will be delayed by a time τ_p at the outlet of the plug-flow section of the reactor system. Thus the RTD of the reactor system is

$$E(t) = \begin{cases} 0 & t < \tau_p \\ \dfrac{e^{-(t-\tau_p)/\tau_s}}{\tau_s} & t \ge \tau_p \end{cases} \tag{13-50}$$

See Figure 13-12.

The RTD is not unique to a particular reactor sequence

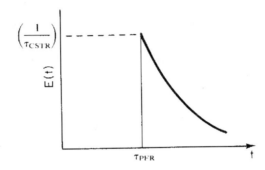

Figure 13-12   RTD curve for a CSTR and a PFR in series.

Next the reactor system in which the CSTR is preceded by the PFR will be treated. If the pulse of tracer is introduced into the entrance of the plug-flow section, then the same pulse will appear at the entrance of the perfectly mixed section τ_p seconds later, meaning that the RTD of the reactor system will be

$E(t)$ is the same no matter which reactor comes first

$$E(t) = \begin{cases} 0 & t < \tau_p \\ \dfrac{e^{-(t-\tau_p)/\tau_s}}{\tau_s} & t \ge \tau_p \end{cases} \tag{13-51}$$

which is *exactly* the same as when the CSTR was followed by the PFR.

It turns out that no matter where the CSTR occurs within the PFR/CSTR reactor sequence, the same RTD results as long as the sum of the residence times in the two sections is the same. Nevertheless, this is not the entire story.

Example 13–4   Comparing Second-Order Reaction Systems

Consider a second-order reaction being carried out in a *real* CSTR that can be modeled as two different reactor systems: In the first system an ideal CSTR is followed by an ideal PFR; in the second system the PFR precedes the CSTR. Let τ_s and τ_p

each equal 1 min, let the reaction rate constant equal 1.0 m³/kmol·min, and let the initial concentration of liquid reactant, C_{A0}, equal 1 kmol/m³. Find the conversion in each system.

Solution

Again, consider first the CSTR followed by the plug-flow section (Figure E13-4.1). A mole balance on the CSTR section gives

$$v_0(C_{A0} - C_{Ai}) = kC_{Ai}^2 V \qquad \text{(E13-4.1)}$$

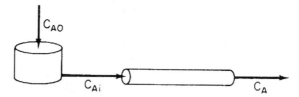

Figure E13-4.1   Early mixing scheme.

Rearranging, we have

$$\tau k C_{Ai}^2 + C_{Ai} - C_{A0} = 0$$

Then

$$C_{Ai} = \frac{\sqrt{1 + 4\tau_s k C_{A0}} - 1}{2\tau k} \qquad \text{(E13-4.2)}$$

Solving for C_{Ai} gives

$$C_{Ai} = \frac{-1 + \sqrt{1 + 4}}{2} = 0.618 \text{ kmol/m}^3 \qquad \text{(E13-4.3)}$$

and feeding this concentration into the plug-flow section gives

$$\frac{dF_A}{dV} = v_0 \frac{dC_A}{dV} = \frac{dC_A}{d\tau} = r_A = -kC_A^2 \qquad \text{(E13-4.4)}$$

$$\frac{1}{C_A} - \frac{1}{C_{Ai}} = \tau_p k \qquad \text{(E13-4.5)}$$

Substituting C_{Ai} = 0.618, τ = 1, and k = 1 in Equation (E13-4.5) yields

$$\frac{1}{C_A} - \frac{1}{0.618} = (1)(1)$$

Solving for C_A gives

$$C_A = 0.382 \text{ kmol/m}^3$$

as the concentration of reactant in the effluent from the reaction system. Thus, the conversion is 61.8%. (i.e. $X = ((1 - 0.382)/1) = 0.618$)

When the perfectly mixed section is preceded by the plug-flow section (Figure E13-4.2) the outlet of the PFR is the inlet to the CSTR, C_{Ai}:

$$\frac{1}{C_{Ai}} - \frac{1}{C_{A0}} = \tau_p k$$

$$\frac{1}{C_{Ai}} - \frac{1}{1} = (1)(1) \tag{E13-4.6}$$

$$C_{Ai} = 0.5 \text{ kmol/m}^3$$

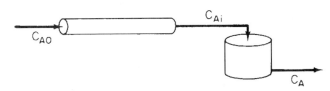

Figure E13-4.2 Late mixing scheme.

and a material balance on the perfectly mixed section (CSTR) gives

$$\tau_s k C_A^2 + C_A - C_{Ai} = 0$$

$$C_A = \frac{\sqrt{1 + 4\tau_s k C_{Ai}} - 1}{2\tau_s k} \tag{E13-4.7}$$

$$= \frac{-1 + \sqrt{1 + 2}}{2} = 0.366 \text{ kmol/m}^3 \tag{E13-4.8}$$

as the concentration of reactant in the effluent from the reaction system. The corresponding conversion is 63.4%.

In the one configuration, a conversion of 61.8% was obtained; in the other, 63.4%. While the difference in the conversions is small for the parameter values chosen, **the point is that there is a difference**.

The conclusion from this example is of extreme importance in reactor analysis: **The RTD is not a complete description of structure for a particular reactor or system of reactors**. The RTD is unique for a particular reactor. However, the reactor or reaction system is not unique for a particular RTD. When analyzing nonideal reactors, the RTD alone is not sufficient to determine its performance, and more information is needed. It will be shown that in addition to the RTD, an adequate model of the nonideal reactor flow pattern and knowledge of the quality of mixing or "degree of segregation" are both required to characterize a reactor properly.

13.5 Reactor Modeling with the RTD

There are many situations where the fluid in a reactor is neither well mixed nor approximates plug flow. Consequently, we must now ask the question:

How can we use the RTD to predict conversion in a real reactor?

To answer this question we will model a real reactor in a number of ways. We shall classify each model according to the number of adjustable parameters that are extracted from the RTD data (see Table 13-1). In this chapter we discuss only the segregation and maximum mixedness models. Other models are discussed in Chapter 14.

TABLE 13-1.   MODELS FOR PREDICTING CONVERSION FROM RTD DATA

1. Zero adjustable parameters

   a. Segregation model
   b. Maximum mixedness model

2. One adjustable parameter

   a. Tanks-in-series model
   b. Dispersion model

3. Two adjustable parameters: Real reactor modeled as combinations of ideal reactors

Ways we use the RTD data to predict conversion in nonideal reactors

The RTD tells us how long the various fluid elements have been in the reactor, but it does not tell us anything about the exchange of matter between the fluid elements (i.e., *the mixing*). The mixing of reacting species is one of the major factors controlling the behavior of chemical reactors. Fortunately for first-order reactions (referred to in some texts as linear reactions), knowledge of the length of time each molecule spends in the reactor is all that is needed to predict conversion. For first-order reactions the conversion is independent of concentration (recall Equation E9-1.3):

$$\frac{dX}{dt} = k(1 - X) \qquad \text{(E9-1.3)}$$

Consequently, mixing with the surrounding molecules is not important. Therefore, once the RTD is determined we can predict the conversion that will be achieved in the real reactor provided that the specific reaction rate for the first-order reaction is known. However, for reactions other than first-order, knowledge of the RTD is not sufficient to predict conversion. In these cases the degree of mixing of molecules must be known in addition to how long each molecule spends in the reactor. Consequently, we must develop models that account for the mixing of molecules inside the reactor.

The more complex models of nonideal reactors necessary to describe reactions other than first-order must contain information about *micromixing* in addition to that of *macromixing*. **Macromixing** produces a distribution of residence times *without*, however, specifying how molecules of different ages encounter one another in the reactor. **Micromixing**, on the other hand, describes how molecules of different ages encounter one another in the reactor. There are two extremes of *micromixing*: (1) all molecules of the same age group remain together as they travel through the reactor and are not mixed until they exit the reactor (i.e., complete segregation); (2) molecules of different age groups are completely mixed at the molecular level as soon as they enter the reactor (complete micromixing). For a given state of macromixing (i.e., a given RTD), these two extremes of micromixing will give the upper and lower limits on conversion in a nonideal reactor. However, there is a somewhat rare exception to the last statement and it is discussed on the **CD-ROM**. For reaction orders greater than 1, the segregation model will give the highest conversion, while for reaction order less than 1, the complete mixing model will give the highest conversion. A fluid in which the globules of a given age do not mix with other globules is called a *macrofluid*, while a fluid in which molecules are free to move everywhere is called a *microfluid*.[8] These two extremes of late and early mixing are shown in Figure 13-13(a) and (b), respectively. These extremes can also be seen by comparing Figures 13-15(a) and 13-16(a). The extremes of late and early mixing are referred to as *complete segregation* and *maximum mixedness*, respectively.

For a given RTD the bounds on conversion are:
Highest
   Segregation
Lowest
   Maximum
      Mixedness

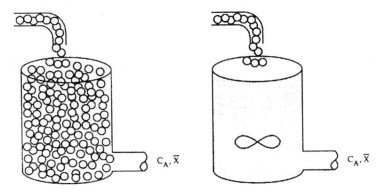

Figure 13-13   (a) Macrofluid; and (b) microfluid.

13.6 Zero-Parameter Models

13.6.1 Segregation Model

In a "perfectly mixed" CSTR, the entering fluid is assumed to be distributed immediately and evenly throughout the reacting mixture. This mixing is assumed to take place even on the micro scale, and elements of different ages mix together thoroughly to form a completely micromixed fluid. If fluid elements of different ages do not mix together at all, the elements remain segregated from each other and the fluid is termed *completely segregated*. The extremes of complete micromixing and complete segregation are the limits of the micromixing of a reacting mixture.

In developing the segregated mixing model, we first consider a CSTR because the application of the concepts of mixing quality are illustrated most easily using this reactor type. In the segregated flow model we visualize the flow through the reactor to consist of a continuous series of globules (Figure 13-14).

---

[8] J. Villermaux, *Chemical Reactor Design and Technology*, Martinus Nijhoff, Boston, 1986.

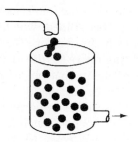

In the segregation
model globules
behave as batch
reactors operated
for different times

Figure 13-14   Little batch reactors (globules) inside a CSTR.

These globules retain their identity; that is, they do not interchange material with other globules in the fluid during their period of residence in the reaction environment. In addition, each globule spends a different amount of time
in the reactor. In essence, what we are doing is lumping all the molecules that
have the same residence time in the reactor into the same globule. The principles of reactor performance in the presence of completely segregated mixing
were first described by Danckwerts[9] and Zwietering.[10]

Another way of looking at the segregation model for a continuous-flow
system is the PFR shown in Figures 13-15(a) and (b). Because the fluid flows
down the reactor in plug flow, each exit stream corresponds to a specific residence time in the reactor. Batches of molecules are removed from the reactor
at different locations along the reactor in such a manner so as to duplicate the
RTD function, $E(t)$. The molecules removed near the entrance to the reactor

Minimum
Mixedness:
The segregation
model has mixing
at the latest
possible point

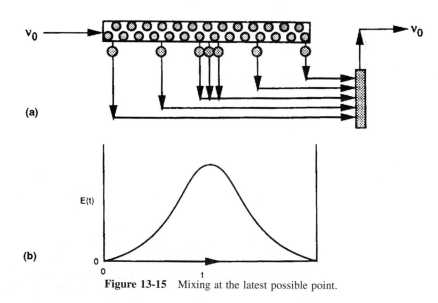

Figure 13-15   Mixing at the latest possible point.

[9] P. V. Danckwerts, *Chem. Eng. Sci.*, *8*, 93 (1958).
[10] T. N. Zwietering, *Chem. Eng. Sci.*, *11*, 1 (1959).

correspond to those molecules having short residence times in the reactor. Physically, this effluent would correspond to the molecules that channel rapidly through the reactor. The farther the molecules travel along the reactor before being removed, the longer their residence time. The points at which the various groups or batches of molecules are removed correspond to the RTD function for the reactor.

Because there is no molecular interchange between globules, each acts essentially as its own batch reactor. The reaction time in any one of these tiny batch reactors is equal to the time that the particular globule spends in the reaction environment. The distribution of residence times among the globules is given by the RTD of the particular reactor.

To determine the mean conversion in the effluent stream, we must average the conversions of various globules in the exit stream:

$$
\begin{bmatrix} \text{mean} \\ \text{conversion} \\ \text{of those globules} \\ \text{spending between} \\ \text{time } t \text{ and } t + dt \\ \text{in the reactor} \end{bmatrix} = \begin{bmatrix} \text{conversion} \\ \text{achieved after} \\ \text{spending a time } t \\ \text{in the reactor} \end{bmatrix} \times \begin{bmatrix} \text{fraction} \\ \text{of globules that} \\ \text{spend between } t \\ \text{and } t + dt \text{ in the} \\ \text{reactor} \end{bmatrix} \quad (13\text{-}52)
$$

$$
d\overline{X} = X(t) \times E(t)\,dt
$$

Summing over all globules, the mean conversion is

Mean conversion for the segregation model

$$
\overline{X} = \int_0^\infty X(t)E(t)\,dt \tag{13-53}
$$

Consequently, if we have the batch reactor equation for $X(t)$ and measure the RTD experimentally, we can find the mean conversion in the exit stream. *Thus, if we have the RTD and the reaction rate expression in a segregated flow situation, we have sufficient information to calculate the conversion.*

Consider the following first-order reaction:

$$
A \xrightarrow{\ k\ } \text{products}
$$

For a batch reactor we have

$$
-\frac{dN_A}{dt} = -r_A V
$$

For constant volume and with $N_A = N_{A0}(1 - X)$,

$$
N_{A0}\frac{dX}{dt} = -r_A V = kC_A V = kN_A = kN_{A0}(1 - X)
$$

$$
\boxed{\frac{dX}{dt} = k(1 - X)} \tag{13-54}
$$

Solving for $X(t)$, we have

$$X(t) = 1 - e^{-kt}$$

$$\overline{X} = \int_0^\infty (1 - e^{-kt})E(t)\,dt = \int_0^\infty E(t)\,dt - \int_0^\infty e^{-kt}E(t)\,dt \quad (13\text{-}55)$$

$$\overline{X} = 1 - \int_0^\infty e^{-kt}E(t)\,dt \qquad (13\text{-}56)$$

We will now determine the mean conversion predicted by the segregation model for an ideal PFR and a CSTR.

Example 13–5   Mean Conversion in a PFR and CSTR

Derive the equation of a first-order reaction using the segregation model when the RTD is equivalent to (a) an ideal PFR, and (b) an ideal CSTR. Compare these conversions with those obtained from the design equation.

Solution

(a) For the **PFR**, the RTD function is

$$E(t) = \delta(t - \tau) \qquad (13\text{-}35)$$

$$\overline{X} = \int_0^\infty X(t)E(t)\,dt = 1 - \int_0^\infty e^{-kt}E(t)\,dt \qquad (13\text{-}55)$$

Substituting for the RTD function for a PFR gives

$$\overline{X} = 1 - \int_0^\infty (e^{-kt})\delta(t - \tau)\,dt \qquad (\text{E}13\text{-}5.1)$$

Using the integral properties of the Dirac delta function, we obtain

$$\overline{X} = 1 - e^{-k\tau} \qquad (\text{E}13\text{-}5.2)$$

Recall that for a PFR after combining the mole balance, rate law, and stoichiometric relationships (cf. Chapter 4), we had

$$\frac{dX}{d\tau} = k(1 - X) \qquad (\text{E}13\text{-}5.3)$$

Integrating yields

$$X = 1 - e^{-k\tau} \qquad (\text{E}13\text{-}5.4)$$

which is identical to the conversion predicted by the segregation model.

(b) For the **CSTR**, the RTD function is

$$E(t) = \frac{1}{\tau} e^{-t/\tau} \qquad (13\text{-}27)$$

Recalling Equation (13-56), the mean conversion for a first-order reaction is

$$\overline{X} = 1 - \int_0^\infty e^{-kt} E(t) \, dt \qquad (13\text{-}56)$$

$$\overline{X} = 1 - \int_0^\infty \frac{e^{-(1/\tau + k)t}}{\tau} \, dt$$

$$\overline{X} = 1 + \frac{1}{k + 1/\tau} \frac{1}{\tau} e^{-(k+1/\tau)t} \bigg|_0^\infty$$

$$\overline{X} = \frac{\tau k}{1 + \tau k} \qquad (E13\text{-}5.5)$$

As expected, using the $E(t)$ for an ideal PRRs and CSTRs with segregation model gives a mean to conversion $\overline{X}$ identical to that obtained by using the algorithm in Ch. 4

Combining the CSTR mole balance, the rate law, and stoichiometry, we have

$$F_{A0}X = -r_A V$$

$$v_0 C_{A0} X = k C_{A0}(1 - X) V$$

$$X = \frac{\tau k}{1 + \tau k} \qquad (E13\text{-}5.6)$$

which is identical to the conversion predicted by the segregation model.

For a first-order reaction, knowledge of $E(t)$ is sufficient

We have just shown for a first-order reaction that whether you assume complete micromixing [Equation (E13-5.6)] or complete segregation [Equation (E13-5.5)] in a CSTR, the same conversion results. This phenomenon occurs because the rate of change of conversion for a first-order reaction does *not* depend on the concentration of the reacting molecules [Equation (13-54)]; it does not matter what kind of molecule is next to it or colliding with it. Thus the extent of micromixing does not affect a first-order reaction, so the segregated flow model can be used to calculate the conversion. As a result, *only the RTD is necessary to calculate the conversion for a first-order reaction in any type of reactor* (see Problem P13-3). Knowledge of neither the degree of micromixing nor the reactor flow pattern is necessary. We now proceed to calculate conversion in a real reactor using RTD data.

Example 13–6   Mean Conversion Calculations in a Real Reactor

Calculate the mean conversion in the reactor we have characterized by RTD measurements in Examples 13-1 and 13-2 for a first-order, liquid-phase, irreversible reaction in a completely segregated fluid:

$$A \longrightarrow \text{products}$$

The specific reaction rate is 0.1 min^{-1} at 320 K.

Solution

Because each globule acts as a batch reactor of constant volume, we use the batch reactor design equation to arrive at the equation giving conversion as a function of time:

$$X = 1 - e^{-kt} = 1 - e^{-0.1t} \qquad (E13\text{-}6.1)$$

To calculate the mean conversion we need to evaluate the integral:

$$\overline{X} = \int_0^\infty X(t)E(t)\,dt \tag{13-53}$$

The RTD function for this reactor was determined previously and given in Table E13-2.1 and is repeated in Table E13-6.1. To evaluate the integral we make a plot of $X(t)E(t)$ as a function of t as shown in Figure E13-6.1 and determine the area under the curve.

TABLE E13-6.1   PROCESSED DATA TO FIND THE MEAN CONVERSION $\overline{X}$

t (min)	$E(t)$ (min^{-1})	$X(t)$	$X(t)E(t)$ (min^{-1})
0	0.000	0	0
1	0.020	0.095	0.0019
2	0.100	0.181	0.0180
3	0.160	0.259	0.0414
4	0.200	0.330	0.0660
5	0.160	0.393	0.0629
6	0.120	0.451	0.0541
7	0.080	0.503	0.0402
8	0.060	0.551	0.0331
9	0.044	0.593	0.0261
10	0.030	0.632	0.01896
12	0.012	0.699	0.0084
14	0.000	0.75	0

These calculations are easily carried out with the aid of a spreadsheet such as Excel or POLYMATH

For a given RTD, the segregation model gives the upper bound on conversion

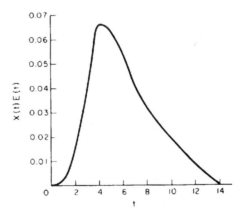

Figure E13-6.1   Plot of columns 1 and 4 from the data in Table E13-6.1.

$$\int_0^\infty X(t)E(t)\,dt = \int_0^{10} X(t)E(t)\,dt + \int_{10}^{14} X(t)E(t)\,dt$$

$$= \tfrac{1}{3}[0 + 4(0.0019) + 2(0.018) + 4(0.0414) + 2(0.066)$$

$$+ 4(0.0629) + 2(0.0541) + 4(0.0402) + 2(0.0331)$$

$$+ 4(0.0261) + 0.01896] + \tfrac{2}{3}[0.01896 + 4(0.0084) + 0]$$

$$= (0.350) + (0.035) = 0.385$$

$$\overline{X} = \text{area} = 0.385$$

The mean conversion is 38.5%. POLYMATH or Excel will give $\overline{X}$ directly after setting up columns 1 and 4 in Table E13-6.1. The area under the curve in Figure E13-6.1 is the mean conversion $\overline{X}$.

As discussed previously, because the reaction is *first-order*, the conversion calculated in Example 13-6 would be valid for a reactor with complete mixing, complete segregation, or any degree of mixing between the two. Although early or late mixing does not affect a first-order reaction, micromixing or complete segregation can modify the results of a second-order system significantly.

13.6.2 Maximum Mixedness

Segregation model mixing occurs at the latest possible point

In a reactor with a segregated fluid, mixing between particles of fluid does not occur until the fluid leaves the reactor. The reactor exit is, of course, the *latest* possible point that mixing can occur, and any effect of mixing is postponed until after all reaction has taken place. We can also think of completely segregated flow as being in a state of minimum mixedness. We now want to consider the other extreme, that of maximum mixedness consistent with a given residence-time distribution.

We return again to the plug-flow reactor with side entrances, only this time the fluid enters the reactor along its length (Figure 13-16). As soon as the fluid enters the reactor it is completely mixed radially (but not longitudinally) with the other fluid already in the reactor. The entering fluid is fed into the reactor through the side entrances in such a manner that the RTD of the plug-flow reactor with side entrances is identical to the RTD of the real reactor.

The globules at the far left of Figure 13-16 correspond to the molecules that spend a long time in the reactor while those at the far right correspond to the molecules that channel through the reactor. In the reactor with side

Maximum mixedness: mixing occurs at the earliest possible point

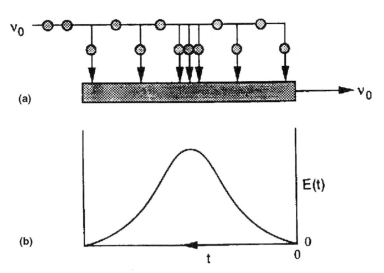

Figure 13-16   Mixing at the earliest possible point.

entrances, mixing occurs at the *earliest* possible moment consistent with the RTD. Thus the effect of mixing occurs as much as possible throughout the reactor, and this situation is termed the condition of *maximum mixedness*.[11] The approach to calculating conversion for a reactor in a condition of maximum mixedness will now be developed. In a reactor with side entrances, let λ be the time it takes for the fluid to move from a particular point to the end of the reactor. In other words, λ is the life expectancy of the fluid in the reactor at that point (Figure 13-17).

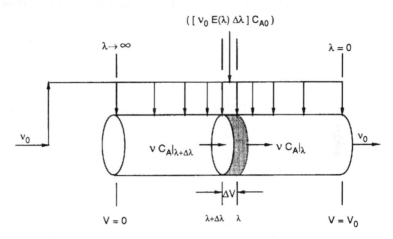

Figure 13-17   Modeling maximum mixedness by a plug-flow reactor with side entrances.

Moving down the reactor from left to right, λ decreases and becomes zero at the exit. At the left end of the reactor, λ approaches infinity or the maximum residence time if it is other than infinite. The volumetric flow rate of fluid fed into the side of the reactor in the interval corresponding to that between $\lambda + \Delta\lambda$ and λ is $v_0 E(\lambda)\,\Delta\lambda$. Because the flow rate is zero at the left end of the reactor, integrating the fluid entering reactor from the point corresponding to λ back to the left end of the reactor will give the volumetric flow rate in the reactor at the point corresponding to λ. That is, the volumetric flow rate inside the reactor at λ is

$$v(\lambda) = v_0 \int_{\lambda}^{\infty} E(\lambda)\,d\lambda = v_0[1 - F(\lambda)]$$

The volume of fluid with a life expectancy between λ and $\lambda + \Delta\lambda$ is

$$\Delta V = v_0[1 - F(\lambda)]\,\Delta\lambda \qquad (13\text{-}57)$$

The rate of generation of the substance A in this volume is

$$r_A \Delta V = r_A v_0[1 - F(\lambda)]\,\Delta\lambda \qquad (13\text{-}58)$$

[11]T. N. Zwietering, *Chem. Eng. Sci.*, *11*, 1 (1959).

We can now carry out a balance on substance A between λ and $\lambda + \Delta\lambda$:

$$\begin{bmatrix} \text{in} \\ \text{at } \lambda + \Delta\lambda \end{bmatrix} + \begin{bmatrix} \text{in} \\ \text{through side} \end{bmatrix} - \begin{bmatrix} \text{out} \\ \text{at } \lambda \end{bmatrix} + \begin{bmatrix} \text{generation} \\ \text{by reaction} \end{bmatrix} = 0$$

$$v_0[1 - F(\lambda)]C_A|_{\lambda+\Delta\lambda} + v_0 C_{A0} E(\lambda)\, \Delta\lambda$$

$$- v_0[1 - F(\lambda)]C_A|_\lambda + r_A v_0[1 - F(\lambda)]\, \Delta\lambda = 0 \qquad (13\text{-}59)$$

Dividing by $v_0\, \Delta\lambda$ and taking the limit as $\Delta\lambda \to 0$ gives

$$E(\lambda)C_{A0} + \frac{d\{[1 - F(\lambda)]C_A(\lambda)\}}{d\lambda} + r_A[1 - F(\lambda)] = 0 \qquad (13\text{-}60)$$

or

$$C_{A0}E(\lambda) + [1 - F(\lambda)]\frac{dC_A}{d\lambda} - C_A E(\lambda) + r_A[1 - F(\lambda)] = 0 \quad (13\text{-}61)$$

or

$$\boxed{\frac{dC_A}{d\lambda} = -r_A + (C_A - C_{A0})\frac{E(\lambda)}{1 - F(\lambda)}} \qquad (13\text{-}62)$$

We can rewrite Equation (13-62) in terms of conversion as

$$-C_{A0}\frac{dX}{d\lambda} = -r_A - C_{A0}X\frac{E(\lambda)}{1 - F(\lambda)} \qquad (13\text{-}63)$$

or

$$\boxed{\frac{dX}{d\lambda} = \frac{r_A}{C_{A0}} + \frac{E(\lambda)}{1 - F(\lambda)}(X)} \qquad (13\text{-}64)$$

The boundary condition is as $\lambda \to \infty$, then $C_A = C_{A0}$ for Equation (13-62) [or $X = 0$ for Equation (13-64)]. To obtain a solution, the equation is integrated backwards numerically, starting at a very large value of λ and ending with the final conversion at $\lambda = 0$. For a given RTD and reaction orders greater than one, the maximum mixedness model gives the lower bound on conversion.

MM gives the lower bound on X

Example 13–7 Conversion Bounds for a Nonideal Reactor

The liquid-phase, second-order dimerization

$$2A \longrightarrow B \qquad r_A = -kC_A^2$$

for which $k = 0.01$ dm^3/mol·min is carried out at a reaction temperature of 320 K. The feed is pure A with $C_{A0} = 8$ mol/dm^3. The reactor is nonideal and perhaps could be modeled as two CSTRs with interchange. The reactor volume is 1000 dm^3, and the feed rate for our dimerization is going to be 25 dm^3/min. We have run a tracer test on this reactor and the results are given in columns 1 and 2 of

Table E13-7.1. We wish to know the bounds on the conversion for different possible degrees of micromixing for the RTD of this reactor. What are these bounds?

Tracer test on tank reactor: $N_0 = 100$ g, $v = 25$ dm³/min.

TABLE E13-7.1

t (min)	C (mg/dm³)	$E(t)$ (min⁻¹)	$1 - F(t)$	$E(t)/[1 - F(t)]$ (min⁻¹)
0	112	0.0280	1.000	0.0280
5	95.8	0.0240	0.871	0.0276
10	82.2	0.0206	0.760	0.0271
15	70.6	0.0177	0.663	0.0267
20	60.9	0.0152	0.584	0.0260
30	45.6	0.0114	0.472	0.0242
40	34.5	0.00863	0.353	0.0244
50	26.3	0.00658	0.278	0.0237
70	15.7	0.00393	0.174	0.0226
100	7.67	0.00192	0.087	0.0221
150	2.55	0.000638	0.024	0.0266
200	0.90	0.000225	0.003	0.075
1	2	3	4	5

Columns 3 through 5 are calculated from columns 1 and 2

Solution

The bounds on the conversion are found by calculating conversions under conditions of complete segregation and maximum mixedness.

Conversion if fluid is completely segregated. The batch reactor equation for a second-order reaction of this type is

$$X = \frac{kC_{A0}t}{1 + kC_{A0}t}$$

The conversion for a completely segregated fluid in a reactor is

$$\overline{X} = \int_0^\infty X(t)E(t)\, dt$$

The calculations for this integration are carried out in Table E13-7.2. The numerical integration uses the simple trapezoid rule. The conversion for this system if the fluid were completely segregated is 0.61 or 61%.

TABLE E13-7.2.   SEGREGATION MODEL

t (min)	$X(t)$	$X(t)E(t)$ (min⁻¹)	$X(t)E(t)\,\Delta t$
0	0	0	0
5	0.286	0.00686	0.0172
10	0.444	0.00916	0.0400
15	0.545	0.00965	0.0470
20	0.615	0.00935	0.0475
30	0.706	0.00805	0.0870
40	0.762	0.00658	0.0732
50	0.800	0.00526	0.0592
70	0.848	0.00333	0.0859
100	0.889	0.00171	0.0756
150	0.923	0.000589	0.0575
200	0.941	0.000212	<u>0.0200</u>
			0.610

Spreadsheets work quite well here

Conversion for maximum mixedness. The Euler method will be used for numerical integration:

$$X_{i+1} = X_i + (\Delta\lambda)\left[\frac{E(\lambda_i)}{1 - F(\lambda_i)}X_i - kC_{A0}(1 - X_i)^2\right]$$

Integrating this equation presents some interesting results. If the equation is integrated from the exit side of the reaction, starting with $\lambda = 0$, the solution is unstable and soon approaches large negative or positive values, depending on what the starting value of X is. If integrated from the point where $\lambda \longrightarrow \infty$, oscillations occur but are damped out and the equation approaches the same final value no matter what initial value of X between 0 and 1 is used. We shall start the integration at $\lambda = 200$ and let $X = 0$ at this point. If we set $\Delta\lambda$ too large, the solution will blow up, so we will start out with $\Delta\lambda = -25$ and use the average of the measured values of $E(t)/[(1 - F(t)]$ where necessary.

$\lambda = 175$:

$$X = 0 + (-25)[(0.075)(0) - (0.01)(8)(1)^2] = 2$$

$\lambda = 150$:

$$X = 2 + (-25)\left[\left(\frac{0.076 + 0.0266}{2}\right)(2) - (0.01)(8)(1 - 2)^2\right] = 1.46$$

$\lambda = 125$:

$$X = 1.46 + (-25)[(0.0266)(1.46) - (0.01)(8)(1 - 1.46)^2] = 0.912$$

$\lambda = 100$:

$$X = 0.912 + (-25)\left[\left(\frac{0.0266 + 0.0221}{2}\right)(0.912) - (0.01)(8)(1 - 0.912)^2\right]$$

$$= 0.372$$

$\lambda = 70$:

$$X = 0.372 + (-30)[(0.0221)(0.372) - (0.01)(8)(1 - 0.372)^2] = 1.071$$

$\lambda = 50$:

$$X = 1.071 + (-20)[(0.0226)(1.071) - (0.01)(8)(1 - 1.071)^2] = 0.595$$

$\lambda = 40$:

$$X = 0.595 + (-10)[(0.0237)(0.595) - (0.01)(8)(1 - 0.595)^2] = 0.585$$

Running down the values of X along the right-hand side of the equation above shows that the oscillations have now damped out. Carrying out the remaining calculations down to the end of the reactor completes Table E13-7.3. The conversion for a condition of maximum mixedness in this reactor is 0.56 or 56%. It is interesting to note that there is little difference in the conversions for the two conditions of complete segregation (61%) and maximum mixedness (56%). With bounds this narrow, there would not be much point in modeling the reactor to improve the predictability of conversion.

Summary

PFR	76%
Segregation	61%
CSTR	58%
Max. mix	56%

TABLE E13-7.3.   MAXIMUM MIXEDNESS MODEL

λ (min)	X
200	0.0
175	2.0
150	1.46
125	0.912
100	0.372
70	1.071
50	0.595
40	0.585
30	0.580
20	0.581
10	0.576
5	0.567
0	0.564

CD Professional

Reference Shelf

For comparison it is left for the reader to show that the conversion for a PFR of this size would be 0.76, and the conversion in a perfectly mixed CSTR with complete micromixing would be 0.58.

The segregation model and maximum mixedness model are further compared on the CD-ROM for reaction orders between zero and one.

Equations (13-62) and (13-64) can be written in a slightly more compact form by making use of the **intensity function**.[12] The intensity function $\Lambda(\lambda)$ is the fraction of fluid in the vessel with age λ that will leave in a time λ to $\lambda + d\lambda$. We can relate $\Lambda(\lambda)$ to $I(\lambda)$ and $E(\lambda)$ in the following manner:

$$
\begin{bmatrix} \text{volume of} \\ \text{fluid leaving} \\ \text{between times} \\ \lambda \text{ and } \lambda + d\lambda \end{bmatrix} = \begin{bmatrix} \text{volume of fluid} \\ \text{remaining at} \\ \text{time } \lambda \end{bmatrix} \begin{bmatrix} \text{fraction of} \\ \text{the fluid with} \\ \text{age } \lambda \text{ that} \\ \text{will leave between} \\ \text{time } \lambda \text{ and } \lambda + d\lambda \end{bmatrix}
$$

$$[v_0 E(\lambda) \, d\lambda] = [V I(\lambda)][\Lambda(\lambda) \, d\lambda] \qquad (13\text{-}65)$$

Then

$$\Lambda(\lambda) = \frac{E(\lambda)}{\tau I(\lambda)} = -\frac{d \ln[\tau I(\lambda)]}{d\lambda} = \frac{E(\lambda)}{1 - F(\lambda)} \qquad (13\text{-}66)$$

Combining Equations (13-64) and (13-66) gives

$$\frac{dX}{d\lambda} = \frac{r_A(\lambda)}{C_{A0}} + \Lambda(\lambda)X(\lambda) \qquad (13\text{-}67)$$

---

[12]D. M. Himmelblau and K. B. Bischoff, *Process Analysis and Simulation*, Wiley, New York, 1968.

We also note that the exit age, t, is just the sum of the internal age, α, and the life expectance, λ:

$$t = \alpha + \lambda \qquad (13\text{-}68)$$

In addition to defining *maximum mixedness* discussed above, Zwietering[13] also generalized a measure of micromixing proposed by Dankwerts[14] and defined the **degree of segregation**, J, as

$$J = \frac{\text{variance of ages between fluid ``points''}}{\text{variance of ages of all molecules in system}} \qquad (13\text{-}69)$$

A fluid "point" contains many molecules but is small compared to the scale of mixing. The two extremes of the *degree of segregation* are

$$J = 1: \qquad \text{complete segregation}$$
$$J = 0: \qquad \text{maximum mixedness}$$

Equations for the variance and J for the intermediate cases can be found in Himmelblau and Bischoff and in Zwietering.

Cautions   The segregation model and the maximum mixedness model will not give the proper bounds on the conversion for certain rate laws and for nonisothermal operation. These situations arise, for example, if the rate of reaction goes through both a maximum and a minimum when plotted as a function of conversion and the initial rate is higher than the maximum (Figure 13-18). One example of a reaction whose rate law has this functionality is the ammonia synthesis at low temperatures. In these situations one must use the

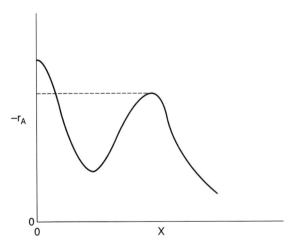

Figure 13-18   Functionality of rate laws that give exceptions to segregation/maximum mixedness bounds.

[13]T. N. Zwietering, *Chem. Eng. Sci.*, *11*, 1 (1959).

[14]P. V. Danckwerts, *Chem. Eng. Sci.*, *8*, 93 (1958).

attainable region analysis (ARA) of Glasser and Hildebrandt[15] to obtain the limits on conversion. The attainable region analysis, ARA, is discussed further on the CD-ROM.

In this section we have addressed the case where all we have is the RTD and no other knowledge about the flow pattern exists. Perhaps the flow pattern cannot be assumed because of a lack of information or other possible causes. Perhaps we wish to know the extent of possible error from assuming an incorrect flow pattern. We have shown how to obtain the conversion, using only the RTD, for two limiting mixing situations: the earliest possible mixing consistent with the RTD, or maximum mixedness, and mixing only at the reactor exit, or complete segregation. Calculating conversions for these two cases gives bounds on the conversions that might be expected for different flow paths consistent with the observed RTD.

13.6.3 Heat Effects

If tracer tests are carried out isothermally and then used to predict nonisothermal conditions, one must couple the segregation and maximum mixedness models with the energy balance to account for variations in the specific reaction rate. For adiabatic operation and $\Delta \hat{C}_P = 0$,

$$T = T_0 + \frac{(-\Delta H_{\text{Rx}})}{\sum \theta_i \hat{C}_{P_i}} X \qquad (\text{E8-6.9})$$

As before, the specific reaction rate is

$$k = k_1 \exp\left[\frac{E}{R}\left(\frac{1}{T_1} - \frac{1}{T}\right)\right] \qquad (\text{T8-2.3})$$

Assuming that $E(t)$ is unaffected by temperature variations in the reactor, one simply solves the segregation and maximum mixedness models, accounting for the variation of k with temperature [i.e., conversion; see Problem P13-2(h)].

13.7 Using Software Packages

Example 13-7 can be solved with an ODE solver such as POLYMATH by

1. **Fitting the $E(t)$ curve to a polynomial** and then using the
2. a. **Segregation model**
   Here we simply use the coupled set of differential equations for the mean or exit conversion, $\overline{X}$, and the conversion $X(t)$ inside a globule at any time, t.

$$\frac{d\overline{X}}{dt} = X(t)\,E(t) \qquad (13\text{-}70)$$

[15]D. Glasser, D. Hildebrandt, and S. Godorr, *Ind. Eng. Chem. Res.*, *33*, 1136 (1994). Also see http://www.engin.umich.edu/~cre/Chapters/ARpages/Intro/intro.htm **and** http://sunsite.wits.ac.za/wits/fac/engineering/procmat/ARHomepage/frame.htm

$$\frac{dX}{dt} = \frac{-r_A}{C_{A0}} \tag{13-71}$$

The rate of reaction is expressed as a function of conversion: for example,

$$-r_A = k_A C_{A0}^2 \frac{(1-X)^2}{(1+X)^2}$$

and the equations are then solved numerically with an ODE solver.

b. **Maximum mixedness model**

Because most software packages won't integrate backwards, we need to change the variable such that the integration proceeds forward as λ deviates from some large value to zero. We do this by forming a new variable, z, which is the difference between the longest time measured in the $E(t)$ curve, $\overline{T}$ and λ. In the case of Example 13-7, the longest time at which the tracer concentration was measured was 200 minutes (Table E13-7.1). Therefore we will set $\overline{T} = 200$.

$$z = \overline{T} - \lambda = 200 - \lambda$$

$$\lambda = \overline{T} - z = 200 - z$$

Then,

$$\frac{dX}{dz} = -\frac{r_A}{C_{A0}} - \frac{E(\overline{T} - z)}{1 - F(\overline{T} - z)} X \tag{13-72}$$

One now integrates between the limit $z = 0$ and $z = 200$ to find the exit conversion at $z = 200$ which corresponds to $\lambda = 0$.

In fitting $E(t)$ to a polynomial, one has to make sure that the polynomial does not become negative at large times. Another concern in the maximum mixedness calculations is that the term $1 - F(\lambda)$ does not go to zero. Setting the maximum value of $F(t)$ at 0.999 rather than 1.0 will eliminate this problem. It can also be circumvented by integrating the polynomial for $E(t)$ to get $F(t)$ and then setting the maximum value of $F(t)$ at 0.999.

Example 13-8   Using Software to Make Maximum Mixedness Model Calculations

Use an ODE solver to determine the conversion predicted by the **maximum mixedness model** for the $E(t)$ curve given in Example E13-7.

Solution

Because of the nature of the $E(t)$ curve, it is necessary to use two polynomials, a third order and a fourth order, each for a different part of the curve to express the

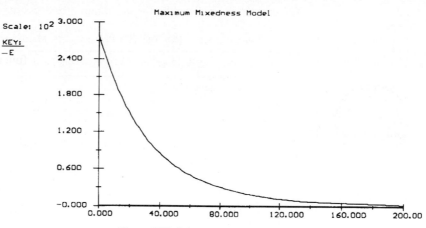

Figure E13-8.1 Polynomial fit of $E(t)$.

First fit $E(t)$

RTD, $E(t)$, as a function of time. The resulting $E(t)$ curve is shown in Figure E13-8.1.

To use POLYMATH to carry out the integration, we change our variable from λ to z using the largest time measurements that were taken from $E(t)$ in Table E13-7.1, which is 200 min:

$$z = 200 - \lambda$$

The equations to be solved are

$$\lambda = 200 - z \qquad\qquad \text{(E13-8.1)}$$

Maximum mixedness model

$$\frac{dX}{dz} = -\frac{r_A}{C_{A0}} - \frac{E(200-z)}{1-F(200-z)}X \qquad \text{(E13-8.2)}$$

For values of λ less than 70 we use the polynomial

$$E_1(\lambda) = 4.447e^{-10}\lambda^4 - 1.180e^{-7}\lambda^3 + 1.353e^{-5}\lambda^2 - 8.657e^{-4}\lambda + 0.028 \quad \text{(E13-8.3)}$$

For values of λ greater than 70 we use the polynomial

$$E_2(\lambda) = -2.640e^{-9}\lambda^3 + 1.3618e^{-6}\lambda^2 - 2.407e^{-4}\lambda + 0.015 \quad \text{(E13-8.4)}$$

$$\frac{dF}{d\lambda} = E(\lambda) \qquad\qquad \text{(E13-8.5)}$$

with $z = 0$ ($\lambda = 200$), $X = 0$, $F = 1$ [i.e., $F(\lambda) = 0.999$]. **Caution:** Because $1 - F(\lambda)$ cannot be zero at $X = 0$, we set the initial value of F at 0.999.

The POLYMATH equations are shown in Table E13-8.1. The solution is

$$\text{at } z = 200 \qquad X = 0.563$$

The conversion predicted by the maximum mixedness model is 56.3%.

TABLE E13-8.1.   POLYMATH PROGRAM FOR MAXIMUM MIXEDNESS MODEL

Equations:	Initial Values:

```
d(x)/d(z)=-(ra/cao+E/(1-F)*x)
k=.01
cao=8
lam=200-z
ca=cao*(1-x)
E1=4.44658e-10*lam^4-1.1802e-7*lam^3+1.35358e-5*lam^2-.00086
    5652*lam+.028004
E2=-2.64e-9*lam^3+1.3618e-6*lam^2-.00024069*lam+.015011
F1=4.44658e-10/5*lam^5-1.1802e-7/4*lam^4+1.35358e-5/3*lam^3-
    .000865652/2*lam^2+.028004*lam
F2=-(-9.30769e-8*lam^3+5.02846e-5*lam^2-.00941*lam+.61823-1)
ra=-k*ca^2
E=if(lam<=70)then(E1)else(E2)
F=if(lam<=70)then(F1)else(F2)
EF=E/(1-F)
```

$z_0 = 0,$ $z_f = 200$

Polynomials used to fit $E(t)$ and $F(t)$

13.8 RTD and Multiple Reactions

As discussed in Chapter 6, when multiple reactions occur in reacting systems, it is best to work in concentrations, moles, or molar flow rates rather than conversion.

13.8.1 Segregation Model

In the **segregation model** we consider each of the globules in the reactor to have different concentrations of reactants, C_A, and products, C_P. These globules are mixed together immediately upon exiting to yield the exit concentration of A, $\overline{C_A}$, which is the average of all the globules exiting:

$$\overline{C_A} = \int_0^\infty C_A(t)E(t)\,dt \qquad (13\text{-}73)$$

$$\overline{C_B} = \int_0^\infty C_B(t)E(t)\,dt \qquad (13\text{-}74)$$

The concentrations of the individual species, $C_A(t)$ and $C_B(t)$, in the different globules are determined from batch reactor calculations. For a constant-volume batch reactor, where q reactions are taking place, the coupled mole balance equations are

$$\frac{dC_A}{dt} = r_A = \sum_{i=1}^{i=q} r_{iA} \qquad (13\text{-}75)$$

$$\frac{dC_B}{dt} = r_B = \sum_{i=1}^{i=q} r_{iB} \qquad (13\text{-}76)$$

These equations are solved simultaneously with

$$\frac{d\overline{C_A}}{dt} = C_A(t)E(t) \qquad (13\text{-}77)$$

$$\frac{d\overline{C_B}}{dt} = C_B(t)E(t) \qquad (13\text{-}78)$$

to give the exit concentration. The RTD, $E(t)$, in Equations (13-77) and (13-78) are determined from experimental measurements and then fit to a polynomial.

13.8.2 Maximum Mixedness

For the **maximum mixedness model** we write Equation (13-62) for each species and replace r_A by the net rate of formation

$$\frac{dC_A}{d\lambda} = -\sum r_{iA} + (C_A - C_{A0}) \frac{E(\lambda)}{1 - F(\lambda)} \qquad (13\text{-}79)$$

$$\frac{dC_B}{d\lambda} = -\sum r_{iB} + (C_B - C_{B0}) \frac{E(\lambda)}{1 - F(\lambda)} \qquad (13\text{-}80)$$

After substitution for the rate laws for each reaction (e.g., $r_{1A} = k_1 C_A$) these equations are solved numerically by starting at a very large value of λ, say $\overline{T} = 200$, and integrating backwards to $\lambda = 0$ to yield the exit concentrations C_A, C_B, ...

We will now show how different RTDs with the *same* mean residence time can produce different product distributions for multiple reactions.

Example 13–9   RTD and Complex Reactions

Consider the following set of reactions:

$$A + B \xrightarrow{k_1} C$$

$$A \xrightarrow{k_2} D$$

$$B + D \xrightarrow{k_3} E$$

which are occurring in two different reactors with the same mean residence time $t_m = 1.26$ min. However, the RTD is very different for each of the reactors, as can be seen in Figures E13-9.1 and E13-9.2.

(a) Fit a polynomial to the RTDs.

(b) Determine the product distribution for
  1. The segregation model
  2. The maximum mixedness model

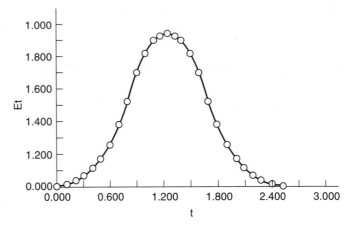

Figure E13-9.1 $E_1(t)$: asymmetric distribution.

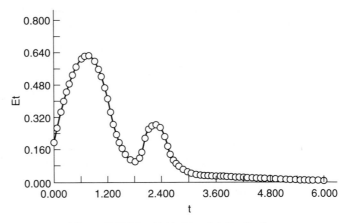

Figure E13-9.2 $E_2(t)$: bimodal distribution.

Solution

Segregation Model

Combining the mole balance and rate laws for a constant-volume batch reactor (i.e., globules), we have

$$\frac{dC_A}{dt} = r_A = r_{1A} + r_{2A} = -k_1 C_A C_B - k_2 C_A \qquad (E13\text{-}9.1)$$

$$\frac{dC_B}{dt} = r_B = r_{1B} + r_{3B} = -k_1 C_A C_B - k_3 C_B C_D \qquad (E13\text{-}9.2)$$

$$\frac{dC_C}{dt} = r_C = r_{1C} = k_1 C_A C_B \qquad (E13\text{-}9.3)$$

$$\frac{dC_D}{dt} = r_D = r_{2D} + r_{3D} = k_2 C_A - k_3 C_B C_D \qquad \text{(E13-9.4)}$$

$$\frac{dC_E}{dt} = r_E = r_{3E} = k_3 C_B C_D \qquad \text{(E13-9.5)}$$

and the concentration for each species exiting the reactor is found by integrating the equation

$$\frac{d\overline{C_i}}{dt} = C_i E(t) \qquad \text{(E13-9.6)}$$

over the life of the $E(t)$ curve. For this example the life of the $E_1(t)$ is 2.42 minutes and life of $E_2(t)$ is 6 minutes.

The initial conditions are $t = 0$, $C_A = C_B = 1$, and $C_C = C_D = C_E = 0$.

The POLYMATH program used to solve these equations is shown in Table E13-9.1 for the asymmetric RTD, $E_1(t)$.

TABLE E13-9.1.   POLYMATH PROGRAM
FOR SEGREGATION MODEL WITH ASYMMETRIC RTD (MULTIPLE REACTIONS)

Equations:	Initial Values:
d(ca)/d(t)=ra	1
d(cb)/d(t)=rb	1
d(cc)/d(t)=rc	0
d(cabar)/d(t)=ca*E	0
d(cbbar)/d(t)=cb*E	0
d(ccbar)/d(t)=cc*E	0
d(cd)/d(t)=rd	0
d(ce)/d(t)=re	0
d(cdbar)/d(t)=cd*E	0
d(cebar)/d(t)=ce*E	0
k1=1	
k2=1	
k3=1	
E1=-2.104*t^4+4.167*t^3-1.596*t^2+0.353*t-0.004	
E2=-2.104*t^4+17.037*t^3-50.247*t^2+62.964*t-27.402	
rc=k1*ca*cb	
re=k3*cb*cd	
ra=-k1*ca*cb-k2*ca	
rb=-k1*ca*cb-k3*cb*cd	
E=if(t<=1.26)then(E1)else(E2)	
rd=k2*ca-k3*cb*cd	
$t_0 = 0$, $t_f = 2.52$	

With the exception of the polynomial for $E_2(t)$, an identical program to that in Table E13-9.1 for the bimodal distribution is given on the CD-ROM. A comparison of the exit concentration and selectivities of the two RTD curves is shown in Table E13-9.2.

TABLE E13-9.2.   SEGREGATION MODEL RESULTS

Asymmetric Distribution		Bimodal Distribution	
The solution for $E_1(t)$ is:		The solution for $E_2(t)$ is:	
$\overline{C_A} = 0.151$	$\overline{C_E} = 0.178$	$\overline{C_A} = 0.245$	$\overline{C_E} = 0.162$
$\overline{C_B} = 0.434$	$\overline{X} = 84.9\%$	$\overline{C_B} = 0.510$	$\overline{X} = 75.5\%$
$\overline{C_C} = 0.357$	$S_{CD} = 1.18$	$\overline{C_C} = 0.321$	$S_{CD} = 1.21$
$\overline{C_D} = 0.303$	$S_{DE} = 1.70$	$\overline{C_D} = 0.265$	$S_{DE} = 1.63$

Maximum Mixedness Model

The equations for each species are

$$\frac{dC_A}{d\lambda} = -k_1 C_A C_B - k_2 C_A + (C_A - C_{A0}) \frac{E(\lambda)}{1 - F(\lambda)} \tag{E13-9.7}$$

$$\frac{dC_B}{d\lambda} = -k_1 C_A C_B - k_3 C_B C_D + (C_B - C_{B0}) \frac{E(\lambda)}{1 - F(\lambda)} \tag{E13-9.8}$$

$$\frac{dC_C}{d\lambda} = k_1 C_A C_B + (C_C - C_{C0}) \frac{E(\lambda)}{1 - F(\lambda)} \tag{E13-9.9}$$

$$\frac{dC_D}{d\lambda} = k_2 C_A - k_3 C_B C_D + (C_D - C_{D0}) \frac{E(\lambda)}{1 - F(\lambda)} \tag{E13-9.10}$$

$$\frac{dC_E}{d\lambda} = k_3 C_B C_D + (C_E - C_{E0}) \frac{E(\lambda)}{1 - F(\lambda)} \tag{E13-9.11}$$

The POLYMATH program for the bimodal distribution, $E_2(t)$, is shown in Table E13-9.3. The POLYMATH program for the asymmetric distribution is identical with the exception of the polynomial fit for $E_1(t)$ and is given on the CD-ROM. A comparison of the exit concentration and selectivities of the two RTD distributions is shown in Table E13-9.4.

TABLE E13-9.3.   POLYMATH PROGRAM
FOR MAXIMUM MIXEDNESS MODEL WITH BIMODAL DISTRIBUTION (MULTIPLE REACTIONS)

Equations:	Initial Values:
`d(ca)/d(z)=-(-ra+(ca-cao)*EF)`	1
`d(cb)/d(z)=-(-rb+(cb-cbo)*EF)`	1
`d(cc)/d(z)=-(-rc+(cc-cco)*EF)`	0
`d(F)/d(z)=-E`	0.99
`d(cd)/d(z)=-(-rd+(cd-cdo)*EF)`	0
`d(ce)/d(z)=-(-re+(ce-ceo)*EF)`	0
`cbo=1`	
`cao=1`	
`cco=0`	
`cdo=0`	
`ceo=0`	
`lam=6-z`	
`k2=1`	
`k1=1`	
`k3=1`	
`rc=k1*ca*cb`	
`re=k3*cb*cd`	
`E1=0.47219*lam^4-1.30733*lam^3+0.31723*lam^2+0.85688*lam+0.2`	
`   0909`	
`E2=3.83999*lam^6-58.16185*lam^5+366.2097*lam^4-1224.66963*la`	
`   m^3+2289.84857*lam^2-2265.62125*lam+925.46463`	
`E3=0.00410*lam^4-0.07593*lam^3+0.52276*lam^2-1.59457*lam+1.8`	
`   4445`	
`rb=-k1*ca*cb-k3*cb*cd`	
`ra=-k1*ca*cb-k2*ca`	
`rd=k2*ca-k3*cb*cd`	
`E=if(lam<=1.82)then(E1)else(if(lam<=2.8)then(E2)else(E3))`	
`EF=E/(1-F)`	
$z_0 = 0, \quad z_f = 6$	

TABLE E13-9.4.   MAXIMUM MIXEDNESS MODEL RESULTS

Asymmetric Distribution The solution for $E_1(t)$ (1) is:		Bimodal Distribution The solution for $E_2(t)$ (2) is:	
$C_A = 0.161$ $C_E = 0.192$		$C_A = 0.266$ $C_E = 0.190$	
$C_B = 0.467$ $X = 83.9\%$		$C_B = 0.535$ $X = 73.4\%$	
$C_C = 0.341$ $S_{CD} = 1.11$		$C_C = 0.275$ $S_{CD} = 1.02$	
$C_D = 0.306$ $S_{DE} = 1.59$		$C_D = 0.269$ $S_{DE} = 1.41$	

Calculations similar to those in Example 13-9 are given in an example on the CD-ROM for the series reaction

$$A \xrightarrow{k_1} B \xrightarrow{k_2} C$$

In addition, the effect of the variance of the RTD on the parallel reactions in Example 13-9 and on the series reaction in the CD-ROM is shown on the CD-ROM.

SUMMARY

1. The quantity $E(t)\,dt$ is the fraction of material that has spent between time t and $t + dt$ in the reactor.
2. The mean residence time

$$t_m = \int_0^\infty tE(t)\,dt = \tau \qquad \text{(S13-1)}$$

is equal to the space-time τ.

3. The variance about the mean residence time is

$$\sigma^2 = \int_0^\infty (t - t_m)^2 E(t)\,dt \qquad \text{(S13-2)}$$

4. The cumulative distribution function $F(t)$ gives the fraction of effluent material that has been in the reactor a time t or less:

$$F(t) = \int_0^t E(t)\,dt$$

$$1 - F(t) = \text{fraction of effluent material that has been in} \qquad \text{(S13-3)}$$
$$\text{the reactor a time } t \text{ or longer}$$

5. The RTD functions for an ideal reactor are

Plug-flow: $E(t) = \delta(t - \tau)$ \qquad (S13-4)

CSTR: $E(t) = \dfrac{e^{-t/\tau}}{\tau}$ \qquad (S13-5)

Laminar flow: $E(t) = 0 \qquad t < \dfrac{\tau}{2}$ \qquad (S13-6)

$E(t) = \dfrac{\tau^2}{2t^3} \qquad t \geq \dfrac{\tau}{2}$ \qquad (S13-7)

6. The dimensionless residence time is

$$\Theta = \frac{t}{\tau} \qquad \text{(S13-8)}$$

$$E(\Theta) = \tau E(t) \qquad \text{(S13-9)}$$

7. The internal-age distribution, $[I(\alpha)\,d\alpha]$, gives the fraction of material inside the reactor that has been inside between a time α and a time $(\alpha + d\alpha)$.

8. Segregation model: The conversion is

$$\overline{X} = \int_0^{\infty} X(t)E(t)\,dt \qquad\qquad \text{(S13-10)}$$

and for multiple reactions

$$\overline{C}_A = \int_0^{\infty} C_A(t)E(t)\,dt$$

9. Maximum mixedness: Conversion can be calculated by solving the following equations:

$$\frac{dX}{d\lambda} = \frac{r_A}{C_{A0}} + \frac{E(\lambda)}{1 - F(\lambda)}(X) \qquad\qquad \text{(S13-11)}$$

and for multiple reactions

$$\frac{dC_A}{d\lambda} = -r_{A_{net}} + (C_A - C_{A0})\frac{E(\lambda)}{1 - F(\lambda)} \qquad\qquad \text{(S13-12)}$$

$$\frac{dC_B}{d\lambda} = -r_{B_{net}} + (C_B - C_{B0})\frac{E(\lambda)}{1 - F(\lambda)} \qquad\qquad \text{(S13-13)}$$

from λ_{max} to $\lambda = 0$. To use an ODE solver let $z = \lambda_{max} - \lambda$.

QUESTIONS AND PROBLEMS

The subscript to each of the problem numbers indicates the level of difficulty: A, least difficult; D, most difficult.

$$\text{A} = \bullet \quad \text{B} = \blacksquare \quad \text{C} = \blacklozenge \quad \text{D} = \blacklozenge\blacklozenge$$

In each of the questions and problems below, rather than just drawing a box around your answer, write a sentence or two describing how you solved the problem, the assumptions you made, the reasonableness of your answer, what you learned, and any other facts that you want to include. You may wish to refer to W. Strunk and E. B. White, *The Elements of Style* (New York: Macmillan, 1979), and Joseph M. Williams, *Style: Ten Lessons in Clarity & Grace* (Glenview, Ill.: Scott, Foresman, 1989) to enhance the quality of your sentences.

P13-1$_A$ Read over the problems of this chapter. Make up an original problem that uses the concepts presented in this chapter. The guidelines are given in Problem P4-1. RTDs from real reactors can be found in *Ind. Eng. Chem.*, *49*, 1000 (1957); *Ind. Eng. Chem. Process Des. Dev.*, *3*, 381 (1964); *Can. J. Chem. Eng.*, *37*, 107 (1959); *Ind. Eng. Chem.*, *44*, 218 (1952); *Chem. Eng. Sci.*, *3*, 26 (1954); and *Ind. Eng. Chem.*, *53*, 381 (1961).

P13-2$_A$ **What if...**

(a) the catalyst decay law in Example 13-6 were 2nd order? 3rd order? What if the catalyst decay law followed that of West Texas crude in Example 10-7 with $t_m = 10$ s? What generalizations can you make?

(b) the cumulative RTD for two different packed-bed reactors (A and B) had the shapes shown in Figure P13-2? What could you say about each reactor?

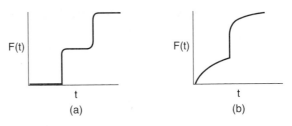

(a)   (b)

Figure P13-2

(c) you were asked to compare the results for the asymmetric and bimodal distributions in Tables E13-9.2 and E13-9.4. What similarities and differences do you observe? What generalizations can you make?

(d) the reaction in Examples 13-7 and 13-8 were third-order with $kC_{A0}^2 = 0.08$ min^{-1}? How would your answers change?

(e) the reaction in Examples 13-7 and 13-8 were half-order with $kC_{A0}^{1/2} = 0.08$ mol/dm$^3 \cdot$min? How would your answers change?

(f) you were asked to vary the specific reaction rates k_1 and k_2 in the series reaction A $\xrightarrow{k_1}$ B $\xrightarrow{k_2}$ C given on the CD-ROM? What would you find?

(g) you were asked to vary the temperature in Example 13-9 from 300 K, at which the rate constants are given, up to a temperature of 500 K? The activation energies in cal/mol are $E_1 = 5000$, $E_2 = 7000$, and $E_3 = 9000$. How would the selectivity change for each RTD curve?

(h) the reaction in Example 13-6 were exothermic and carried out adiabatically, the relationship between temperature and conversion [see Equation (T8-1.5)] were

$$T(K) = T_0 + \left(\frac{-\Delta H_{Rx}}{C_{P_A}}\right)X = 320 + 150X \qquad [P13\text{-}2(h).1]$$

and the activation energy were 30,000 J/mol? What would be the conversion? What generalizations can you make about the effect of temperature on the results (e.g. conversion) predicated from the RTD?

(i) the reaction in Example 13-7 were carried out adiabatically with the same parameters as those in Equation [P13-2(h).1]? How would your answers change?

Heat effects

(j) the reaction in Examples 13-8 and 13-6 were exothermic and carried out adiabatically with

$$T(K) = 320 - 100X \quad \text{and} \quad E = 45 \text{ kJ/mol?} \qquad [P13\text{-}2(j).1]$$

How would your answers change? What generalizations can you make about the effect of temperature on the results (e.g. conversion) predicated from the RTD?

(k) the reaction in Example 8-12 were carried out in the reactor described by the RTD in Example 13-9 with the exception that RTD is in seconds rather than minutes (i.e., $t_m = 1.26$ s)? How would your answers change?

P13-3$_C$ Show that for a first-order reaction

$$A \longrightarrow B$$

the exit concentration maximum mixedness equation

$$\frac{dC_A}{d\lambda} = kC_A + \frac{E(\lambda)}{1 - F(\lambda)}(C_A - C_{A0}) \qquad (P13\text{-}3.1)$$

is the same as the exit concentration given by the segregation model

$$C_A = C_{A0} \int_0^\infty E(t)e^{-kt}\, dt \qquad (P13\text{-}3.2)$$

[*Hint:* Verify

$$C_A = \frac{C_{A0}\, e^{k\lambda}}{1 - F(\lambda)} \int_\lambda^\infty E(t)e^{-kt}\, dt \qquad (P13\text{-}3.3)$$

is a solution to Equation (P13-3.1)].

P13-4$_C$ Consider the following power law rate law for the reaction

$$A \longrightarrow B$$

i.e.

$$-r_A = kC_A^n$$

The reaction is carried out isothermally in a

(a) For what values of n will a CSTR give a greater conversion than a PFR of equal volume?

(b) For what values of n will the maximum mixedness model give a greater conversion than the segregation model?

P13-5$_B$ The relative tracer concentrations obtained from pulse tracer tests on a commercial packed-bed desulfurization reactor are shown in Figure P13-5. After studying the RTD, what problems are occurring with the reactor during the period of poor operation (thin line)? The bed was repacked and the pulse tracer test again carried out with the results shown in Figure P13-5 (thick line). Calculate the conversion that could be achieved in the commercial desulfurization reactor during poor operation and during good operation (Figure P13-5) for the following reactions:

(a) A first-order isomerization with a specific reaction rate of 0.1 h^{-1}.

(b) A first-order isomerization with a specific reaction rate of 2.0 h^{-1}.
Also

(c) What do you conclude upon comparing the four conversions in parts (a) and (b)?

(d) How would your answers change if the reaction is carried out adiabatically with parameter values given by Equation [P13-2(h).1]?

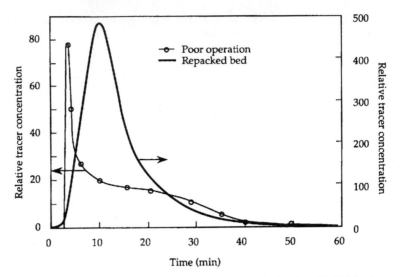

Figure P13-5 Pilot-plant RTD. [Reprinted with permission from E. V. Murphree, A. Voorhies, and F. Y. Mayer, *Ind. Eng. Chem. Process Des. Dev.*, *3*, 381 (1964). Copyright © 1964, American Chemical Society.]

P13-6   The irreversible liquid phase reaction

$$A \xrightarrow{k_1} B$$

is half order in A. The reaction is carried out in a non-ideal CSTR which can be modeled using the segregation model. RTD measurements on the reactor gave values of $\tau = 5$ min and $\sigma = 3$ min. For an entering concentration of pure A of 1.0 mol/dm^3 the mean exit conversion was 10%. Estimate the specific reaction rate constant, k_1.

P13-7$_B$ *(Distributions in a stirred tank)* Using a negative step tracer input, Cholette and Cloutier [*Can. J. Chem. Eng.*, *37*, 107 (1959)] studied the RTD in a tank for different stirring speeds. Their tank had a 30-in. diameter and a fluid depth of 30 in. inside the tank. The inlet and exit flow rates were 1.15 gal/min. Here are some of their tracer results for the relative concentration, C/C_0 (courtesy of the Canadian Society for Chemical Engineering):

	Impeller Speed (rpm)	
Time (min)	170	100
10	0.761	0.653
15	0.695	0.566
20	0.639	0.513
25	0.592	0.454
30	0.543	0.409
35	0.502	0.369
40	0.472	0.333
45	0.436	0.307
50	0.407	0.276
55	0.376	0.248
60	0.350	0.226
65	0.329	0.205

(a) Calculate and plot the cumulative exit-age distribution, the intensity function, and the internal-age distributions as a function of α for this stirred tank at the two impeller speeds. Can you tell anything about dead zones and bypassing at the different stirrer rates?

(b) What conversion can be expected for a second-order reaction with $kC_{A0} = 0.05/\text{min}$ at 320 K for the segregation model?

(c) Repeat part (b) for the maximum mixedness model.

(d) How would your answers to part (b) and (c) change if the reaction were carried out adiabatically with $T(K) = 320 + 80X$ and $E = 5000$ cal/mol?

P13-8$_B$ An isothermal pulse test on a piece of reaction equipment gave the following results: The output concentrations rose linearly from zero to 0.5 μmol/dm^3 in 5 min, then fell linearly to zero in 10 min after reaching the maximum value.

(a) Calculate in tabular form the values of $E(t)$ and $F(t)$ at 1-min intervals. Sketch these functions.

(b) What is the mean residence time? If the flow were 150 gal/min, what would be the total reactor volume? (*Ans.: $t_m = 6.67$ min, $V = 1000$ gal.*)

A second-order reaction with $kC_{A0} = 1.2$ min^{-1} at 325 K is carried out in the system.

(c) If the reactor were plug flow with the same flow and volume, what would be the conversion? (*Ans.: $X = 0.889$.*)

(d) If the reactor were a CSTR with the same flow and volume, what would be the conversion? (*Ans.: $X = 0.703$.*)

(e) If the flow were completely *segregated* with the $F(t)$ above, what would be the conversion? (*Ans.: $X = 0.86$.*)

(f) If the flow were in a state of maximum mixedness with the $E(t)$ above, what would be the conversion?

(g) How would your answers to part ____ (to be assigned) change if the reaction were carried out adiabatically with the parameter values given by Equation [P13-2(h).1]?

(h) What if the reaction were endothermic with $E = 10,000$ cal/mol and carried out adiabatically with

$$T(K) = 325 - 500X? \qquad\qquad [P13-7(h).1]$$

How would your answers change?

P13-9$_A$ Consider again the nonideal reactor characterized by the RTD data in Examples 13-1 and 13-2. The irreversible gas-phase nonelementary reaction

$$A + B \longrightarrow C + D$$

is first-order in A and second-order in B and is to be carried out isothermally. Calculate the conversion for:

(a) A PFR, a laminar flow reactor with complete segregation, and a CSTR.

(b) The cases of complete segregation and maximum mixedness.

Also

(c) Plot $I(\alpha)$ and $\Lambda(\lambda)$ as a function of time and then determine the mean age $\bar{\alpha}$ and the mean life expectancy $\bar{\lambda}$.

(d) How would your answers change if the reaction is carried out adiabatically with parameter values given by Equation [P13-2(h).1]?

Additional information:

$C_{A0} = C_{B0} = 0.0313$ mol/dm^3, $V = 1000$ dm^3,
$v_0 = 10$ dm^3/s, $k = 175$ dm^6/mol$^2 \cdot$s at 320 K.

P13-10$_B$ An irreversible first-order reaction takes place in a long cylindrical reactor. There is no change in volume, temperature, or viscosity. The use of the simplifying assumption that there is plug flow in the tube leads to an estimated degree of conversion of 86.5%. What would be the actually attained degree of conversion if the real state of flow is laminar, with negligible diffusion?

P13-11$_B$ For turbulent flow, the one-seventh-power-law model is usually used to describe the velocity profile, that is,

$$u = u_{max} \left(1 - \frac{r}{R} \right)^{1/7}$$

Derive the $E(t)$ function for turbulent flow in a tubular reactor. Calculate the conversion if an elementary isomerization occurs. The specific reaction rate is $k = 0.1 \text{ s}^{-1}$, the centerline velocity is 10 dm/s, and the reactor length is 100 dm. $R = 10$ dm.

P13-12$_B$ An RTD analysis was carried out on a liquid-phase reactor [*Chem. Eng. J.*, 1, 76 (1970)]. Analyze the following data:

t (s)	0	150	175	200	225	240	250	260
$C \times 10^3$ (g/dm^3)	0	0	1	3	7.4	9.4	9.7	9.4

t (s)	275	300	325	350	375	400	450
$C \times 10^3$ (g/dm^3)	8.2	5.0	2.5	1.2	0.5	0.2	0

(a) Plot the $E(t)$ curve for these data.
(b) What fraction of the material spends between 230 and 270 s in the reactor?
(c) Plot the $F(t)$ curve for these data.
(d) What fraction of the material spends less than 250 s in the reactor?
(e) What is the mean residence time?
(f) Plot $E(t)(t - t_m)^2$ as a function of time.
(g) What is the standard deviation?
(h) Suggest a model consistent with the experimental data.
(i) The hydrolysis of *t*-isobutyl chloride was carried out in this reactor. The specific reaction rate is 0.0115 s^{-1}. What conversions do the segregation model and the maximum mixedness models predict?
(j) The alkaline hydrolysis of ethyl acetate was also carried out in this reactor. The specific reaction rate is 10.55 dm^3/mol·min at 300 K. For an initial concentration of ethyl acetate of 0.01 M, use the segregation and maximum mixedness models to predict the limits of conversion in this reaction.
(k) How would your answers change if the reaction were carried out adiabatically with $T(K) = 300 + 150X$ and $E = 4000$ cal/mol?

P13-13$_B$ A decaying catalyst is fed to a fluidized CSTR at a rate of 20 kg/s. The 1-m^3 CSTR contains 200 kg of catalyst. The catalyst decay follows second-order kinetics. The elementary gas-phase reaction

$$A + B \longrightarrow C$$

takes place in the reactor. A and B are fed in stoichiometric proportions. The concentration of A is 0.04 mol/dm^3 and the total entering volumetric flow rate is 10 dm^3/s.

(a) What is the mean activity of the catalyst exiting the reactor?
(b) What is the mean activity of the catalyst inside the reactor?
(c) What conversion of A is predicted from the balance

$$W = \frac{F_{A0}X}{-r_{A0}\bar{a}}$$

(d) What conversion of A is predicted from the segregation model?
(e) What catalyst weight is necessary to achieve 75% conversion?
(f) What conversion would be achieved if the catalyst feed rate to reactor were increased by a factor of 5?
(g) How would your answers to parts (c), (d), and (e) change if the decay were first-order?

Additional information:

Decay constant k_d: 0.06 s^{-1}, void fraction: 0.7
Specific reaction rate for fresh catalyst: 0.03 dm^6/mol·s·g cat.

P13-14 The second-order liquid phase reaction

$$2A \xrightarrow{k_{1A}} B$$

is carried out in a non ideal CSTR. At 300 K the specific reaction rate is 0.5 dm^3/mol·min. In a tracer test, the tracer concentration rose linearly up to 1 mg/dm^3 at 1.0 minutes and then decreased linearly to zero at exactly 2.0 minutes. Pure A enters the reactor at a temperature of 300 K. The reaction is carried out adiabatically. The mean conversion exiting the reactor is 0.67.
(a) Assuming the segregation model best describes the mixing and reaction, estimate the activation energy for this reaction.
(b) Now consider the reaction to be carried out isothermally at 350 K where a second reaction now also takes place

$$A + B \xrightarrow{k_{2C}} C$$

So as not to impinge on the answer in part (a), we assume k_{1A} is 0.2 dm^3/mol·min at this temperature. If the selectivity of B to C is 2.38, estimate k_{2C}.

Additional information:

$$C_{P_A} = 50 \text{ J/mol·K}, \qquad C_{P_B} = 100 \text{ J/mol·K}$$

$$\Delta H_{Rx1A} = -7500 \text{ J/mol}$$

$$C_{A0} = 2 \text{ mol/dm}^3$$

P13-15$_B$ The reactions described in Problem 6-14 are to be carried out in the reactor whose RTD is described in Problem 13-13 ($\rho_b = 1$ kg/m^3.)
Determine the exit selectivities
(a) Using the segregation model.
(b) Using the maximum mixedness model.
(c) Compare the selectivities in parts (a) and (b) with those that would be found in an ideal PFR and ideal CSTR in which the space-time is equal to the mean residence time.

(d) What would your answers to parts (a) to (c) be if the reactor in Problem 13-6 were used?

P13-16$_B$ The reactions described in Example 6-8 are to be carried out in the reactor whose RTD is described in Example 13-7 with $C_{A0} = C_{B0} = 0.05$ mol/dm^3. Determine the exit selectivities

(a) Using the segregation model.

(b) Using the maximum mixedness model.

(c) Compare the selectivities in parts (a) and (b) with those that would be found in an ideal PFR and ideal CSTR in which the space-time is equal to the mean residence time.

(d) What would your answers to parts (a) to (c) be if the RTD curve rose from zero at $t = 0$ to a maximum of 50 mg/dm^3 after 10 min, and then fell linearly to zero at the end of 20 min?

P13-17$_B$ The reactions described in Problem 6-13 are to be carried out in the reactor whose RTD is described in Example 13-9. Determine the exit selectivities

(a) Using the segregation model.

(b) Using the maximum mixedness model.

(c) Compare the selectivities in parts (a) and (b) with those that would be found in an ideal PFR and ideal CSTR in which the space-time is equal to the mean residence time.

P13-18$_C$ The reactions described in Problem 6-11 are to be carried out in the reactor whose RTD is described in Problem 13-5 with $C_{A0} = 0.8$ mol/dm^3 and $C_{B0} = 0.6$ mol/dm^3. Determine the exit selectivities

(a) Using the segregation model.

(b) Using the maximum mixedness model.

(c) Compare the selectivities in parts (a) and (b) with those that would be found in an ideal PFR and ideal CSTR in which the space-time is equal to the mean residence time.

(d) How would your answer to parts (a) and (b) change if the reactor in Problem 13-14 were used?

P13-19$_B$ The 3rd order liquid phase reaction

$$A \xrightarrow{k_3} B$$

was carried out in a reactor that has the following RTD

E(t) = 0	for	t < 1 min
E(t) = 1.0	for	1 ≤ t ≤ 2 min
E(t) = 0	for	t > 2 min

The entering concentrations of A is 2 mol/dm^3.

(a) For isothermal operation, what is the conversion predicted by
1) the segregation model, X_{seg}
2) the maximum mixedness model, X_{MM}. Plot X vs. z (or λ) and explain why the curve looks the way it does.

(b) For isothermal operation, at what emperature is the discrepancy between X_{seg} and X_{MM} the greatest in the range $300 < T < 500$.

(c) Suppose the reaction is carried out adiabatically with an entering temperature of 305K.
1) Calculate X_{seg}

Additional Information

k = 0.3 min–1 at 300K	E/R = 20,000
ΔHrx = −40,000 cal/mol	$C_{PA} = C_{PB} = 25$ cal/mol/K

CD-ROM MATERIAL

- **Learning Resources**
  1. *Summary Notes for Lectures 31 and 32*
  2. *Web Modules*
     A. The Attainable Region Analysis
        (http://www.engin.umich.edu/~cre/Chapters/ARpages/Intro/intro.htm *and*
        http://www.wits.ac.za/fac/engineering/promat/aregion)
  4. *Solved Problems*
     A. Example CD13-1 Calculate the exit concentrations series reaction

     $$A \longrightarrow B \longrightarrow C$$

        using the segregation model and maximum mixedness
        model.
     B. Example CD13-2 Determination of the effect of variance on the exit
        concentrations for the series reaction

     $$A \longrightarrow B \longrightarrow C$$

- **Living Example Problems**
  1. *Example 13–8 Using Software to Make Maximum Mixedness Model Calculations*
  2. *Example 13–9 RTD and Complex Reactions*
- **Professional Reference Shelf**
  1. *Attainable Region Analysis*
  2. *Comparison of Conversion for Segregation and Maximum Mixedness Model for
     Conversion for Reaction Orders Between 0 and 1*
- **Additional Homework Problems**

CDP13-A$_C$ After showing that $E(t)$ for two CSTRs in series having different val-
              ues is

$$E(t) = \frac{1}{\tau(2m-1)}\left\{\exp\left(\frac{-t}{\tau - m}\right) - \exp\left[\frac{-t}{(\tau - m)\tau}\right]\right\}$$

              you are asked to make a number of calculations. [2nd ed. P13-11]

CDP13-B$_B$ Determine $E(t)$ and from data taken from a pulse test in which the
              pulse is not perfect and the inlet concentration varies with time. [2nd
              ed. P13-15]

CDP13-C$_B$ Derive the $E(t)$ curve for a Bingham plastic flowing in a cylindrical
              tube. [2nd ed. P13-16]

CDP13-D$_B$ The order of a CSTR and PFR in series is investigated for a third-order
              reaction. [2nd ed. P13-10]

CDP13-E$_B$ Review the Murphree pilot plant data when a second-order reaction
              occurs in the reactor. [1st ed. P13-15]

CDP13-F$_A$ Calculate the mean waiting time for gasoline at a service station and
              in a parking garage. [2nd Ed. P13-J]

CDP13-G$_B$ Apply the RTD given by

$$E(t) = \begin{cases} A - B(t_0 - t)^2 & \text{for } 0 \le t \le 2t_0 \\ 0 & \text{for } 0 > 2t_0 \end{cases}$$

to Examples 13-6 through 13-10. [2nd Ed. P13-2$_B$]

CDP13-H$_B$ The multiple reactions in Problem 6-16 are carried out in a reactor whose RTD is described in Example 13-7.

SUPPLEMENTARY READING

1. Discussions of the measurement and analysis of residence-time distribution can be found in

> CURL, R. L., and M. L. McMILLIN, "Accuracies in residence time measurements," *AIChE J.*, *12*, 819–822 (1966).

> HILL, C. G., *An Introduction to Chemical Engineering Kinetics and Reactor Design*. New York: Wiley, 1977, Chap. 11.

> LEVENSPIEL, O., *Chemical Reaction Engineering*, 2nd ed. New York: Wiley, 1972, Chaps. 9 and 10.

> SMITH, J. M., *Chemical Engineering Kinetics*, 3rd ed. New York: McGraw-Hill, 1981, Chap. 6.

2. An excellent discussion of segregation can be found in

> DOUGLAS, J. M., "The effect of mixing on reactor design," *AIChE Symp. Ser.*, *48*, Vol. 60, p. 1 (1964).

3. Also see

> CARBERRY, J., and A. VARMA, *Chemical Reaction and Reactor Engineering*. New York: Marcel Dekker, 1987.

> DUDUKOVIC, M., and R. FELDER, in *CHEMI Modules on Chemical Reaction Engineering*, Vol. 4, ed. B. Crynes and H. S. Fogler. New York: AIChE, 1985.

> HIMMELBLAU, D. M., and K. B. BISCHOFF, *Process Analysis and Stimulation: Deterministic Systems*. New York: Wiley, 1968.

> NAUMAN, E. B., "Residence time distributions and micromixing," *Chem. Eng. Commun.*, *8*, 53 (1981).

> NAUMAN, E. B., and B. A. BUFFHAM, *Mixing in Continuous Flow Systems*. New York: Wiley, 1983.

> ROBINSON, B. A., and J. W. TESTER, *Chem. Eng. Sci.*, *41*(3), 469–483 (1986).

> VILLERMAUX, J., "Mixing in chemical reactors," in *Chemical Reaction Engineering—Plenary Lectures*, ACS Symposium Series 226. Washington, D.C.: American Chemical Society, 1982.

> VILLERMAUX, J., *Chemical Reactor Design and Technology*. Boston: Martinus Nijhoff, 1986.

Models for Nonideal **14**
*Reactors**

14.1 Some Guidelines

Not all tank reactors are perfectly mixed nor do all tubular reactors exhibit plug-flow behavior. In these situations, some means must be used to allow for deviations from ideal behavior. Chapter 13 showed how the RTD was sufficient if the reaction was first-order *or* if the fluid was either in a state of complete segregation or maximum mixedness. We use the segregation and maximum mixedness models to bound the conversion when no adjustable parameters are used. For non-first-order reactions in a fluid with good micromixing, more than just the RTD is needed. These situations compose a great majority of reactor analysis problems and cannot be ignored. To predict conversions and product distribution for such systems, a model of reactor flow patterns is necessary. To model these patterns, we use combinations and/or modifications of ideal reactors to represent real reactors. With this technique we classify a model as being either a one-parameter model (e.g., tanks-in-series model or dispersion model) or a two-parameter model (e.g., reactor with bypassing and dead volume). The RTD is then used to evaluate the parameter(s) in the model. After completing this chapter the reader will be able to apply the tanks-in-series model and the dispersion model to tubular reactors. In addition, the reader will be able to suggest combinations of ideal reactors to model a real reactor.

Use the RTD to evaluate parameters

The choice of the particular model to be used depends largely on the engineering judgment of the person carrying out the analysis. It is this person's job to choose the model that best combines the conflicting goals of mathematical simplicity and physical realism. There is a certain amount of art in the development of a model for a particular reactor, and the examples presented below can only point toward a direction that an engineer's thinking might follow.

Conflicting goals

---

*Dr. Lee F. Brown of Los Alamos Scientific Laboratory contributed to this chapter.

For a given real reactor, it is not uncommon to use all the models discussed previously to predict conversion and then make a comparison. Usually, the real conversion will be *bounded* by the model calculations.

The following guidelines are suggested when developing modules for nonideal reactors:

A Model Must
  • Fit the data
  • Be able to extrapolate theory and experiment
  • Have realistic parameters

1. *The model must be mathematically tractable.* The equations used to describe a chemical reactor should be able to be solved without an inordinate expenditure of human or computer time.

2. *The model must realistically describe the characteristics of the nonideal reactor.* The phenomena occurring in the nonideal reactor must be reasonably described physically, chemically, and mathematically.

3. *The model must not have more than two adjustable parameters.* This constraint is used because an expression with more than two adjustable parameters can be fitted to a great variety of experimental data, and the modeling process in this circumstance is nothing more than an exercise in curve fitting. The statement "Give me four adjustable parameters and I can fit an elephant; give me five and I can include his tail!" is one that I have heard from many colleagues. Unless one is into modern art, a substantially larger number of adjustable parameters are necessary to draw a reasonable-looking elephant.[1] A one-parameter model is, of course, superior to a two-parameter model if the one-parameter model is sufficiently realistic. To be fair, however, in complex systems (e.g., internal diffusion and conduction, mass transfer limitations) where other parameters may be measured *independently*, then more than two parameters are quite acceptable.

14.2 One-Parameter Models

Here we use a single parameter to account for the nonideality of our reactor. This parameter is most always evaluated by analyzing the RTD determined from a tracer test. Examples of one-parameter models for a nonideal CSTR include the reactor dead volume V_D, where no reaction takes place, or the fraction f of fluid bypassing the reactor, thereby exiting unreacted. Examples of one-parameter models for tubular reactors include the tanks-in-series model and the dispersion model. For the tanks-in-series model, the parameter is the number of tanks, n, and for the dispersion model, it is the dispersion coefficient D_a. Knowing the parameter values, we then proceed to determine the conversion and/or effluent concentrations for the reactor.

Nonideal tubular reactors

We first consider nonideal tubular reactors. Tubular reactors may be empty, or they may be packed with some material that acts as a catalyst, heat-transfer medium, or means of promoting interphase contact. Until now when analyzing ideal tubular reactors, it usually has been assumed that the fluid moved through the reactor in piston-like flow (PFR), and every atom spends an identical length of time in the reaction environment. Here, the *velocity profile*

---

[1] J. Wei, *Chem. Technol.*, *5*, 128 (1975).

is flat and there is no axial mixing. Both of these assumptions are false to some extent in every tubular reactor; frequently, they are sufficiently false to warrant some modification. Most popular tubular reactor models need to have means to allow for failure of the plug-flow model and insignificant axial mixing assumptions; examples include the unpacked laminar flow tubular reactor, the unpacked turbulent flow, and packed-bed reactors. One of the two approaches is usually taken to compensate for failure of either or both of the ideal assumptions. One approach involves modeling the nonideal tubular reactor as a series of identically sized CSTRs. The other approach (the dispersion model) involves a modification of the ideal reactor by imposing axial dispersion on plug flow.

14.2.1 Tanks-in-Series Model

In this section we discuss the use of the tanks-in-series model to describe nonideal reactors and calculate conversion. We will analyze the RTD to determine the number of ideal tanks in series that will give approximately the same RTD as the nonideal reactor. Next we will apply the reaction engineering analysis developed in Chapters 1 through 4 to calculate conversion. We are first going to develop the RTD equation for three tanks in series (Figure 14.1) and then generalize to n reactors in series to derive an equation that gives the number of tanks in series that best fits the RTD data.

In Figure 2-9 we saw how tanks-in-series could approximate a PFR

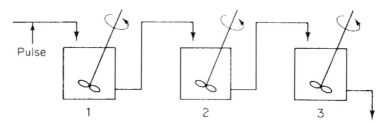

Figure 14-1   Tanks in series.

The RTD will be analyzed from a tracer pulse injected into the first reactor of three equally sized CSTRs in series. Using the definition of the RTD presented in Section 13.2, the fraction of material leaving the system of three reactors (i.e., leaving the third reactor) that has been in the system between time t and $t + \Delta t$ is

$$E(t)\,\Delta t = \frac{v\,C_3(t)\,\Delta t}{N_0} = \frac{C_3(t)}{\displaystyle\int_0^\infty C_3(t)\,dt}\,\Delta t$$

Then

$$E(t) = \frac{C_3(t)}{\displaystyle\int_0^\infty C_3(t)\,dt} \qquad (14\text{-}1)$$

In this expression, $C_3(t)$ is the concentration of tracer in the effluent from the third reactor and the other terms are as defined previously.

It is now necessary to obtain the outlet concentration of tracer, $C_3(t)$, as a function of time. As in a single CSTR, a material balance on the first reactor gives

$$V_1 \frac{dC_1}{dt} = -vC_1 \tag{14-2}$$

We perform a tracer balance on each reactor to obtain $C_3(t)$

Integrating gives the expression for the tracer concentration in the effluent from the first reactor:

$$C_1 = C_0 e^{-vt/V_1} = C_0 e^{-t/\tau_1} \tag{14-3}$$

$$C_0 = N_0/V_1 = \frac{v_0 \int_0^\infty C_3(t)\, dt}{V_1}$$

The volumetric flow rate is constant ($v = v_0$) and all the reactor volumes are identical ($V_1 = V_2 = V_i$); therefore, all the space-times of the individual reactors are identical ($\tau_1 = \tau_2 = \tau_i$). Because V_i is the volume of a single reactor in the series, τ_i here is the residence time in *one* of the reactors, *not* in the entire reactor system.

A material balance on the tracer in the second reactor gives

$$V_i \frac{dC_2}{dt} = vC_1 - vC_2$$

Using Equation (14-3) to substitute for C_1, we obtain the first-order ordinary differential equation

$$\frac{dC_2}{dt} + \frac{C_2}{\tau_i} = \frac{C_0}{\tau_i} e^{-t/\tau_i}$$

This equation is readily solved using an integrating factor e^{t/τ_i} along with the initial condition $C_2 = 0$ at $t = 0$, to give

$$C_2 = \frac{C_0 t}{\tau_i} e^{-t/\tau_i} \tag{14-4}$$

The same procedure used for the third reactor gives the expression for the concentration of tracer in the effluent from the third reactor (and therefore from the reactor system),

$$C_3 = \frac{C_0 t^2}{2\tau_i^2} e^{-t/\tau_i} \tag{14-5}$$

Substituting Equation (14-5) into Equation (14-1), we find that

$$E(t) = \frac{C_3(t)}{\displaystyle\int_0^\infty C_3(t)\, dt}$$

$$= \frac{t^2}{2\tau_i^3}\, e^{-t/\tau_i} \tag{14-6}$$

Generalizing this method to a series of n CSTRs gives the RTD for n CSTRs in series, $E(t)$:

RTD for equal-size tanks in series

$$\boxed{E(t) = \frac{t^{n-1}}{(n-1)!\,\tau_i^n}\, e^{-t/\tau_i}} \tag{14-7}$$

Because the total reactor volume is nV_i, then $\tau_i = \tau/n$, where τ represents the total reactor volume divided by the flow rate, υ:

$$E(\Theta) = \frac{n\,(n\Theta)^{n-1}}{(n-1)!}\, e^{-n\Theta} \tag{14-8}$$

where $\Theta = t/\tau$.

Figure 14-2 illustrates the RTDs of various numbers of CSTRs in series in a two-dimensional plot (a) and in a three-dimensional plot (b). As the number becomes very large, the behavior of the system approaches that of a plug-flow reactor.

We can determine the number of tanks in series by calculating the dimensionless variance σ_Θ^2 from a tracer experiment.

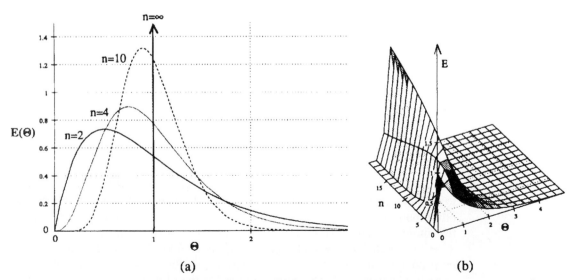

Figure 14-2   Tanks-in-series response to a pulse tracer input for different numbers of tanks.

$$\sigma_\Theta^2 = \frac{\sigma^2}{\tau^2} = \int_0^\infty (\Theta - 1)^2 E(\Theta)\, d\Theta$$

$$= \int_0^\infty \Theta^2 E(\Theta)\, d\Theta - 2 \int_0^\infty \Theta E(\Theta)\, d\Theta + \int_0^\infty E(\Theta)\, d\Theta \qquad (14\text{-}9)$$

$$\sigma_\Theta^2 = \int_0^\infty \Theta^2 E(\Theta)\, d\Theta - 1$$

$$= \int_0^\infty \Theta^2 \frac{n(n\Theta)^{n-1}}{(n-1)!} e^{-n\Theta}\, d\Theta - 1 \qquad (14\text{-}10)$$

$$\sigma_\Theta^2 = \frac{n^n}{(n-1)!} \int_0^\infty \Theta^{n+1} e^{-n\Theta}\, d\Theta - 1$$

$$= \frac{n^n}{(n-1)!} \left[\frac{(n+1)!}{n^{n+2}} \right] - 1$$

As the number of tanks increases, the variance decreases

$$= \frac{1}{n} \qquad (14\text{-}11)$$

The number of tanks in series is

$$\boxed{n = \frac{1}{\sigma_\Theta^2} = \frac{\tau^2}{\sigma^2}} \qquad (14\text{-}12)$$

This expression represents the number of tanks necessary to model the real reactor as n ideal tanks in series.

If the reaction is first-order, we can use Equation (4-11) to calculate the conversion,

$$X = 1 - \frac{1}{(1 + \tau_i k)^n} \qquad (4\text{-}11)$$

where

$$\tau_i = \frac{V}{v_0 n}$$

It is acceptable (and usual) for the value of n calculated from Equation (14-12) to be a noninteger in Equation (4-11) to calculate the conversion. For reactions other than first-order, an integer number of reactors must be used and sequential mole balances on each reactor must be carried out. For example, if $\tau^2/\sigma^2 = 2.8$ then one would round up to three tanks. The conversion and effluence concentrations would be solved sequentially using the algorithm developed in Chapter 4. That is, after solving for the effluent from the first tank, it would be used and the input to the second tank and so on.

14.2.2 Dispersion Model

The dispersion model is also used to describe nonideal tubular reactors. In this model, there is an axial dispersion of the material, which is governed by an analogy to Fick's law of diffusion, superimposed on the flow. So in addition to transport by bulk flow, UA_CC, every component in the mixture is transported through any cross section of the reactor at a rate equal to $[-D_aA_c(dC/dz)]$ resulting from molecular and convective diffusion. By convective diffusion we mean either Aris–Taylor dispersion in laminar flow reactors or turbulent diffusion resulting from turbulent eddies.

To illustrate how dispersion affects the concentration profile in a tubular reactor we consider the injection of a perfect tracer pulse. Figure 14-3 shows how dispersion causes the pulse to broaden as it moves down the reactor and becomes less concentrated.

Tracer pulse with
dispersion

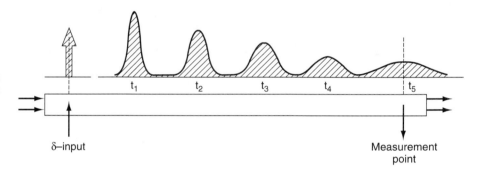

Figure 14-3   Dispersion in a tubular reactor. (From O. Levenspiel, *Chemical Reaction Engineering*, 2nd ed. Copyright © 1972 John Wiley & Sons, Inc. Reprinted by permission of John Wiley & Sons, Inc. All rights reserved.)

The molar flow rate of tracer (F_T) by both convection and dispersion is

$$F_T = -D_aA_c\,\frac{\partial C_T}{\partial z} + UA_cC_T$$

In this expression D_a is the effective dispersion coefficient (m²/s) and U (m/s) is the superficial velocity. Correlations for the dispersion coefficients in both liquid and gas systems may be found in Levenspiel.[2] Some of these correlations are given in Figures 14-5 through 14-7.

A mole balance on the inert tracer T gives

$$-\frac{\partial F_T}{\partial z} = A_c\,\frac{\partial C_T}{\partial t}$$

Substituting for F_T and dividing by the cross-sectional area A_c, we have

$$D_a\,\frac{\partial^2 C_T}{\partial z^2} - \frac{\partial(UC_T)}{\partial z} = \frac{\partial C_T}{\partial t} \qquad (14\text{-}13)$$

[2] O. Levenspiel, *Chemical Reaction Engineering*, Wiley, New York, 1962, pp. 290–293.

At first sight, this simple model appears to have the capability of accounting only for axial mixing effects. It will be shown, however, that this approach can compensate not only for problems caused by axial mixing, *but also for those caused by radial mixing and other nonflat velocity profiles*.[3] These fluctuations in concentration can result from different flow velocities and pathways and from molecular and turbulent diffusion.

Now that we have an intuitive feel for how dispersion affects the transport of molecules in a tubular reactor, we shall consider two types of dispersion, *laminar* and *turbulent*.

Dispersion in a Tubular Reactor with Laminar Flow. In a laminar flow reactor we know that the axial velocity varies in the radial direction according to the Hagen–Poiseuille equation:

$$u(r) = 2U \left[1 - \left(\frac{r}{R} \right)^2 \right]$$

where U is the average velocity. For laminar flow we saw that the RTD function $E(t)$ was given by

$$E(t) = \begin{cases} 0 & \text{for } t < \frac{\tau}{2} \left(\tau = \frac{L}{U} \right) \\ \\ \frac{\tau^2}{2t^3} & \text{for } t \geq \frac{\tau}{2} \end{cases} \qquad (13\text{-}47)$$

In arriving at this distribution $E(t)$ it was assumed that there was no transfer of molecules in the radial direction between streamlines. Consequently, with the aid of Equation (13-44), we know that the molecules on the center streamline ($r = 0$) exited the reactor at a time $t = \tau/2$, and molecules traveling on the streamline at $r = 3R/4$ exited the reactor at time

$$t = \frac{L}{u} = \frac{L}{2U[1 - (r/R)^2]} = \frac{\tau}{2\,[1 - (3/4)^2]}$$

$$= \frac{8}{7} \cdot \tau \qquad (13\text{-}44)$$

The question now arises: What would happen if some of the molecules traveling on the streamline at $r = 3R/4$ jumped (i.e., diffused) on to the streamline at $r = 0$? The answer is that they would exit sooner than if they had stayed on the streamline at $r = 3R/4$. Analogously, if some of the molecules from the faster streamline at $r = 0$ jumped (i.e., diffused) on to the streamline at $r = 3R/4$, they would take a longer time to exit (Figure 14-4). In addition to the molecules diffusing between streamlines, they can also move forward or backward relative to the average fluid velocity by molecular diffusion (Fick's law). With both axial and radial diffusion occurring, the question arises as to

---

[3] R. Aris, *Proc. R. Soc. (London), A235,* 67 (1956).

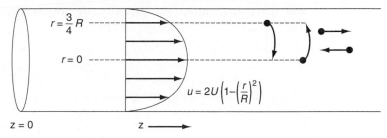

Figure 14-4   Radial diffusion in laminar flow.

Molecules diffusing between streamlines and back and forth along a streamline

what will be the distribution of residence times when molecules are transported between and along streamlines by diffusion. To answer this question we will derive an equation for the axial dispersion coefficient, D_a, that accounts for the axial and radial diffusion mechanisms. In deriving D_a, which is referred to as the Aris–Taylor dispersion coefficient, we closely follow the development given by Brenner and Edwards.[4]

The convective–diffusion equation for solute (e.g., tracer) transport in both the axial and radial direction is

$$\frac{\partial c}{\partial t} + u(r)\,\frac{\partial c}{\partial z} = D_{AB}\left\{ \frac{1}{r}\,\frac{\partial\,[r(\partial c/\partial r)]}{\partial r} + \frac{\partial^2 c}{\partial z^2}\right\} \tag{14-14}$$

where c is the solute concentration at a particular r, z, and t.

We are going to change the variable in the axial direction z to z^*, which corresponds to an observer moving with the fluid

$$z^* = z - Ut \tag{14-15}$$

A value of $z^* = 0$ corresponds to an observer moving with the fluid on the center streamline. Using the chain rule, we obtain

$$\left(\frac{\partial c}{\partial t}\right)_{z^*} + [u(r) - U]\,\frac{\partial c}{\partial z^*} = D_{AB}\left[\frac{1}{r}\,\frac{\partial}{\partial r}\left(r\,\frac{\partial c}{\partial r}\right) + \frac{\partial^2 c}{\partial z^{*2}}\right] \tag{14-16}$$

Because we want to know the concentrations and conversions at the exit to the reactor, we are really only interested in the average axial concentration $\overline{C}$, which is given by

$$\overline{C}(z, t) = \frac{1}{\pi R^2}\int_0^R c(r, z, t)\,2\pi r\,dr \tag{14-17}$$

Consequently, we are going to solve Equation (14-16) for the solution concentration as a function of r and then substitute the solution $c(r, z, t)$ into Equation (14-17) to find $\overline{C}(z, t)$. All the intermediate steps are given on the CD-ROM,

---

[4] H. Brenner and D. A. Edwards, *Macrotransport Processes*, Butterworth-Heinemann, Boston, 1993.

and the partial differential equation describing the variation of the average axial concentration with time and distance is

$$\frac{\partial \overline{C}}{\partial t} + U \frac{\partial \overline{C}}{\partial z} = D^* \frac{\partial^2 \overline{C}}{\partial z^{*2}}$$

(14-18)

where D^* is the Aris–Taylor dispersion coefficient:

Aris–Taylor
dispersion
coefficient

$$\boxed{D^* = D_{AB} + \frac{U^2 R^2}{48 D_{AB}}}$$

(14-19)

That is, for laminar flow in a pipe

$$D_a \equiv D^*$$

Figure 14-5 shows the dispersion coefficient D^* in terms of the ratio $D^*/U(2R) = D^*/Ud_1$ as a function of the product of the Reynolds and Schmidt numbers.

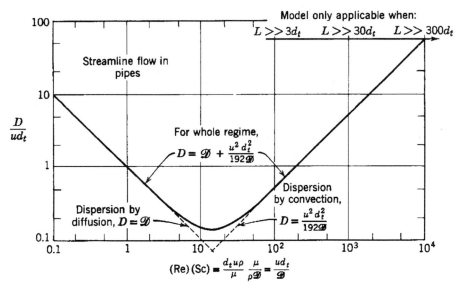

Figure 14-5 Correlation for dispersion for streamline flow in pipes. (From O. Levenspiel, *Chemical Reaction Engineering*, 2nd. ed. Copyright © 1972 John Wiley & Sons, Inc. Reprinted by permission of John Wiley & Sons, Inc. All rights reserved.) [*Note*: $D \equiv D_a$]

Dispersion for Turbulent Flow. When the flow is turbulent, an estimate of the dispersion coefficient, D_a, can be determined from Figure 14-6.

Dispersion in Packed Beds. For the case of gas–solid catalytic reactions which take place in packed-bed reactors, the dispersion coefficient, D_a, can be estimated by using Figure 14-7. Here d_p is the particle diameter and ε is the porosity.

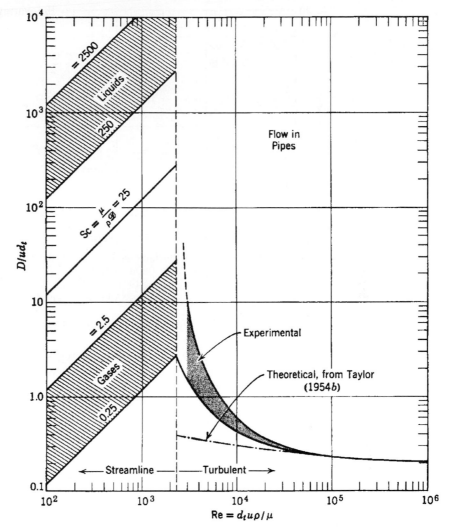

Once the Reynolds number is calculated, D_a can be found

Figure 14-6   Correlation for dispersion of fluids flowing in pipes. (From O. Levenspiel, *Chemical Reaction Engineering*, 2nd. ed. Copyright © 1972 John Wiley & Sons, Inc. Reprinted by permission of John Wiley & Sons, Inc. All rights reserved.) [*Note: $D \equiv D_a$*]

Experimental Determination of D_a.   The dispersion coefficient can be determined from a pulse tracer experiment. Here the effluent concentration of the reactor is measured as a function of time. From the effluent concentration data, the mean residence time and variance are calculated and these values are then used to determine D_a. To show how this is accomplished, we will write Equation (14-13),

$$D_a \frac{\partial^2 C_T}{\partial z^2} - \frac{\partial (U C_T)}{\partial z} = \frac{\partial C_T}{\partial t} \qquad (14\text{-}13)$$

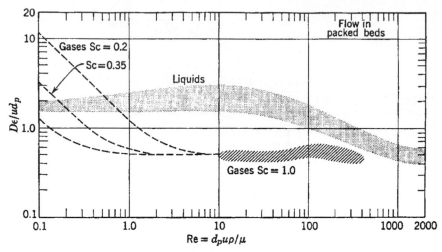

Figure 14-7 Experimental findings on dispersion of fluids flowing with mean axial velocity u in packed beds. (From O. Levenspiel, *Chemical Reaction Engineering*, 2nd. ed. Copyright © 1972 John Wiley & Sons, Inc. Reprinted by permission of John Wiley & Sons, Inc. All rights reserved.) [*Note*: $D \equiv D_a$]

Note: In these figures $D \equiv D_a$ and $\varepsilon \equiv \phi$ in the text nomenclature.

in dimensionless form, discuss the different types of boundary conditions at the reactor entrance and exit, solve for the exit concentration as a function of dimensionless time ($\Theta = t/\tau$), and then relate D_a, σ^2, and τ.

The first step is to put Equation (14-13) in dimensionless form to arrive at the dimensionless group(s) that characterize the process. Let

$$\Psi = \frac{C_T}{C_{T0}}, \quad \lambda = \frac{z}{L}, \quad \text{and} \quad \theta = \frac{tU}{L}$$

For a pulse input C_{T0} is defined as the mass of tracer injected, M, divided by the vessel volume, V. Then

$$\frac{D_a}{UL} \frac{\partial^2 \Psi}{\partial \lambda^2} - \frac{\partial \Psi}{\partial \lambda} = \frac{\partial \Psi}{\partial \theta} \tag{14-20}$$

The quantity UL/D_a is a form of the Peclet number, Pe. The Peclet number can be regarded as the ratio of

Peclet number

$$\text{Pe} = \frac{\text{rate of transport by convection}}{\text{rate of transport by diffusion or dispersion}} = \frac{Ul}{D_a}$$

in which l is the characteristic length term. There are two different types of Peclet numbers in common use. We can call Pe_r the reactor Peclet number; it uses the reactor length, L, for the characteristic length, so $\text{Pe}_r \equiv UL/D_a$. It is Pe_r that appears in Equation (14-20). The reactor Peclet number, Pe_r, for mass dispersion is often referred to in reacting systems as the Bodenstein number, Bo, rather than the Peclet number. The other type of Peclet number can be

For open tubes
$\text{Pe}_r \sim 10^6$,
$\text{Pe}_f \sim 10^4$

called the fluid Peclet number, Pe_f; it uses the characteristic length that determines the fluid's mechanical behavior. In a packed bed this length is the particle diameter d_p, and $\text{Pe}_f \equiv Ud_p/\phi D_a$. (The term U is the empty tube or superficial velocity. For packed beds we often wish to use the average interstitial velocity, and thus U/ϕ is commonly used for the packed-bed velocity term.) In an empty tube the fluid behavior is determined by the tube diameter d_t, and $\text{Pe}_f = Ud_t/D_a$. The fluid Peclet number, Pe_f, is given in all correlations relating the Peclet number to the Reynolds number because both are directly related to the fluid mechanical behavior. It is, of course, very simple to convert Pe_f to Pe_r: Multiply by the ratio L/d_p or L/d_t. The reciprocal of Pe_r, D_a/UL, is sometimes called the *vessel dispersion number.* Writing Equation (14-20) in terms of the Peclet number yields

For packed beds
$\text{Pe}_r \sim 10^3$,
$\text{Pe}_f \sim 10^1$

$$\text{Pe}_r = \frac{UL}{D_a}$$

$$\boxed{\frac{1}{\text{Pe}_r}\frac{\partial^2 \Psi}{\partial \lambda^2} - \frac{\partial \Psi}{\partial \lambda} = \frac{\partial \Psi}{\partial \theta}} \tag{14-21}$$

Boundary Conditions   There are two cases that we need to consider: boundary conditions for closed vessels and open vessels. In the case of closed-closed vessels we assume that there is no dispersion or radial variation in concentration either upstream (closed) or downstream (closed) of the reaction section, hence this is a closed-closed vessel. In an open vessel, dispersion occurs both upstream (open) and downstream (open) of the reaction section; hence this is an open-open vessel. These two cases are shown in Figure 14-8, where fluctuations in concentration due to dispersion are superimposed on the plug-flow velocity profile. A closed-open vessel boundary condition is one in which there is no dispersion in the entrance section but there is dispersion in the reaction and exit sections.

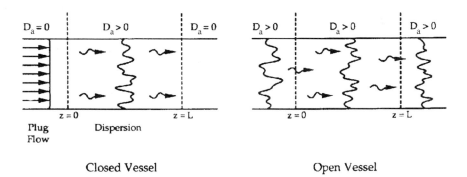

Figure 14-8   Types of boundary conditions.

$F_T \begin{array}{c} \rightarrow \\ \rightarrow \end{array} \overline{\left. 0^- \right| 0^+} \\ \overline{z=0}$

Closed–Closed Vessel Boundary Condition   For a closed-closed vessel, we have plug flow (no dispersion) to the immediate left of the entrance line ($z = 0^-$) (closed) and to the immediate right of the exit $z = L$ ($z = L^+$) (closed). However, between $z = 0^+$ and $z = L^-$ we have dispersion and reaction. The corresponding entrance boundary condition is

At $z = 0$: $\qquad\qquad\qquad F_T(0^-, t) = F_T(0^+, t)$

Substituting for F_T yields

$$UA_cC_T(0^-, t) = -A_cD_a\left(\frac{\partial C_T}{\partial z}\right)_{z=0^+} + UA_cC_T(0^+, t)$$

Solving for the entering concentration $C_T(0^-, t) = C_{T0}$:

Concentration boundary conditions at the entrance

$$\boxed{C_{T0} = \frac{-D_a}{U}\left(\frac{\partial C_T}{\partial z}\right)_{z=0^+} + C_T(0^+, t)}$$ (14-22)

At the exit to the reaction section the concentration is continuous and there is no gradient in tracer concentration:

Concentration boundary conditions at the exit

At $z = L$:

$$\boxed{\begin{array}{l} C_T(L^-) = C_T(L^+) \\[2mm] \dfrac{\partial C_T}{\partial z} = 0 \end{array}}$$ (14-23)

Danckwerts boundary conditions

These two boundary conditions, Equations (14-22) and (14-23), first stated by Danckwerts,[5] have become known as the famous *Danckwerts boundary conditions*. Bischoff[6] has given a rigorous derivation of them, solving the differential equations governing the dispersion of component A in the entrance and exit sections and taking the limit as D_a in entrance and exit sections approaches zero. From the solutions he obtained boundary conditions on the reaction section identical with those Danckwerts proposed. The initial condition is

Initial condition

$$\text{At } t = 0, \quad z > 0, \quad C_T(0^+, 0) = 0$$ (14-24)

The mass of tracer injected, M is

$$M = UA_c\int_0^\infty C_T(0^-, t)\, dt$$

In dimensionless form the Dankwerts boundary conditions are:

At $\lambda = 0$:

$$-\frac{1}{\text{Pe}_r}\frac{\partial \Psi}{\partial \lambda} + \Psi = \frac{C_T(0^-, t)}{C_{T0}} = 1$$ (14-25)

At $\lambda = 1$:

$$\frac{\partial \Psi}{\partial \lambda} = 0$$

Equation (14-21) has been solved numerically for a pulse injection and the resulting dimensionless effluent tracer concentration, Ψ, is shown as a function of the dimensionless time θ in Figure 14-9 for various Peclet numbers.

[5] P. V. Danckwerts, *Chem. Eng. Sci.*, 2, 1 (1953).
[6] K. B. Bischoff, *Chem. Eng. Sci.*, 16, 131 (1961).

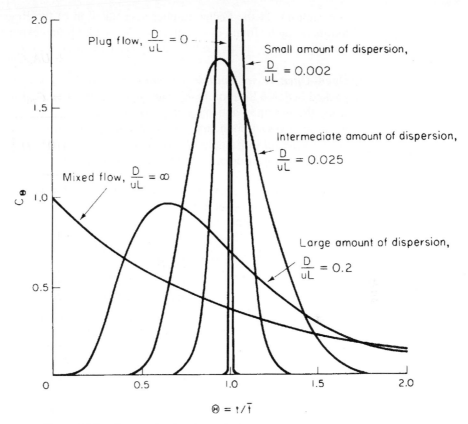

Figure 14-9 *C* curves in closed vessels for various extents of back-mixing as predicted by the dispersion model. (From O. Levenspiel, *Chemical Reaction Engineering*, 2nd ed. Copyright © 1972 John Wiley & Sons, Inc. Reprinted by permission of John Wiley & Sons, Inc. All rights reserved.) [*Note: D ≡ D_a*]

Although analytical solutions for ψ can be found, the result is an infinite series, the corresponding equations for the mean residence time, t_m, and the variance, σ^2, are[7]

$$\boxed{t_m = \tau} \tag{14-26}$$

and

Calculating Pe_r using t_m and σ^2 determined from RTD data

$$\frac{\sigma^2}{t_m^2} = \frac{1}{\tau^2} \int_0^\infty (t - \tau)^2 E(t)\, dt$$

$$\boxed{\frac{\sigma^2}{t_m^2} = \frac{2}{Pe_r} - \frac{2}{Pe_r^2}\left(1 - e^{-Pe_r}\right)} \tag{14-27}$$

---

[7] See K. Bischoff and O. Levenspiel, *Adv. Chem. Eng.*, **4**, 95 (1963).

Correlations for the Peclet number as a function of the Reynolds and Schmidt numbers can be found in Levenspiel.[8] Pe_r can be found experimentally by determining t_m and σ^2 from the RTD data and then solving Equation (14-27) for Pe_r.

Open–Open Vessel Boundary Conditions   When a tracer is injected into a packed bed at a location more than two or three particle diameters downstream from the entrance and measured some distance upstream from the exit, the open-open vessel boundary conditions apply. For an open-open system an analytical solution to Equation (14-21) can be obtained for a pulse tracer input.
    For an open-open system the boundary conditions at the entrance are

$$F_T(0^-, t) = F_T(0^+, t)$$

Then for the case when the dispersion coefficient is the same in the entrance and reaction sections:

$$\left. -D_a\left(\frac{\partial C_T}{\partial z}\right)\right|_{z=0^-} + UC_T(0^-, t) = \left. -D_a\left(\frac{\partial C_T}{\partial z}\right)\right|_{z=0^+} + UC_T(0^+, t) \Bigg\} \quad (14\text{-}28)$$

$$C_T(0^-, t) = C_T(0^+, t)$$

At the exit

$$C_T(L^-, t) = C_T(L^+, t)\} \quad (14\text{-}29)$$

Open at the exit

$$\left. -D_a\left(\frac{\partial C_T}{\partial z}\right)\right|_{z=L^-} + UC_T(L^-, t) = \left. -D_a\left(\frac{\partial C_T}{\partial z}\right)\right|_{z=L^+} + UC_T(L^+, t)$$

    There are a number of perturbations of these boundary conditions that can be applied. The dispersion coefficient can take on different values in each of the three regions ($z < 0; 0 \leq z \leq L$, and $z > 0$) and the tracer can also be injected at some point z_1 rather than at the boundary, $z = 0$. These cases and others can be found in the supplementary readings cited at the end of the chapter. We shall consider the case when there is no variation in the dispersion coefficient for all z and an impulse of tracer is injected at $z = 0$ at $t = 0$.
    Using the boundary conditions above, Equation (14-21) can be solved to determine the dimensionless effluent tracer concentration:

$$\Psi(1, \theta) = \frac{C_T(L, t)}{C_{T0}} = \frac{1}{2\sqrt{\pi\theta/Pe_r}} \exp\left[\frac{-(1-\theta)^2}{4\theta/Pe_r}\right] \quad (14\text{-}30)$$

The corresponding mean residence time is

Calculate τ for an
open–open system

$$\boxed{t_m = \left(1 + \frac{2}{Pe_r}\right)\tau} \quad (14\text{-}31)$$

---

[8] O. Levenspiel, *Chemical Reaction Engineering*, 2nd ed., Wiley, New York, 1972, pp. 282–284.

where τ is based on the volume between $z = 0$ and $z = L$ (i.e., reactor volume measured with a yardstick). We note that the mean residence time for an open system is greater than that for a closed system. The variance for an open system is

Calculate Pe_r for an open–open system

$$\boxed{\frac{\sigma^2}{t_m^2} = \frac{2}{Pe_r} + \frac{8}{Pe_r^2}} \qquad (14\text{-}32)$$

We now consider two cases for which we can use Equations (14-31) and (14-32) to determine the system parameters:

Case 1. The space-time τ is *known*. That is, V and v_0 are measured independently. Here we can determine the Peclet number by determining t_m and σ^2 from the concentration–time data and then using Equation (14-32) to calculate Pe_r. We can also calculate t_m and then use Equation (14-31) as a check, but this is usually less accurate.

Case 2. The space-time τ is *unknown*. This situation arises when there are dead or stagnant pockets that exist in the reactor along with the dispersion effects. To analyze this situation we first calculate t_m and σ^2 from the data as in case 1. Next we solve Equation (14-32) for Pe_r. Finally, we solve Equation (14-31) for τ and hence V. The dead volume is the difference between the measured volume (i.e., with a yardstick) and the volume calculated from the RTD.

Sloppy Tracer Inputs   It is not always possible to inject a tracer pulse cleanly as an input to a system, owing to the fact that it takes a finite time to inject the tracer. When the injection does not approach a perfect pulse input (Figure 14-10), the differences in the variances between the input and output tracer measurements are used to calculate the Peclet number:

$$\Delta\sigma^2 = \sigma_{in}^2 - \sigma_{out}^2$$

where σ_{in}^2 is the variance of the tracer measured at some point upstream (near the entrance) and σ_{out}^2 is the variance measured at some point downstream (near the exit).

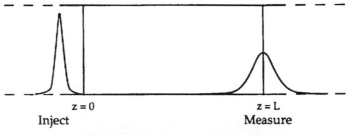

Figure 14-10   Imperfect tracer input.

For an open-open system, it has been shown[9] that the Peclet number can be calculated from the equation

$$\boxed{\frac{\Delta\sigma^2}{t_m^2} = \frac{2}{\mathrm{Pe}_r}}$$

(14-33)

Flow, Reaction, and Dispersion   Having discussed how to determine the dispersion coefficient we now return to the case where we have both dispersion and reaction in a tubular reactor. A mole balance is taken on a particular component of the mixture (say, species A) over a short length Δz of a tubular reactor in a manner identical to that in Chapter 1, to arrive at

$$-\frac{1}{A_c}\frac{dF_A}{dz} + r_A = 0$$

(14-34)

Combining Equation (14-34) and the equation for the molar flux F_A leads to

$$\frac{D_a}{U}\frac{d^2C_A}{dz^2} - \frac{dC_A}{dz} + \frac{r_A}{U} = 0$$

(14-35)

This equation is a second-order ordinary differential equation. It is nonlinear when r_A is other than zero- or first-order.

When the reaction rate r_A is first-order, the equation is linear:

Flow, reaction, and
dispersion

$$\frac{D_a}{U}\frac{d^2C_A}{dz^2} - \frac{dC_A}{dz} - \frac{kC_A}{U} = 0$$

(14-36)

and amenable to analytical solution. However, before obtaining a solution, we put our Equation (14-36) describing dispersion and reaction in dimensionless form by letting $\psi = C_A/C_{A0}$ and $\lambda = z/L$:

$$\frac{1}{\mathrm{Pe}_r}\frac{d^2\Psi}{d\lambda^2} - \frac{d\Psi}{d\lambda} - \mathrm{Da}\cdot\Psi = 0$$

(14-37)

The quantity kL/U appearing in Equation (14-37) is called the *Damköhler number* for convection, Da, and physically represents the ratio

Damköhler number

$$\mathrm{Da} = \frac{\text{rate of consumption of A by reaction}}{\text{rate of transport of A by convection}} = \frac{kC_{A0}^{n-1}L}{U} = kC_{A0}^{n-1}\tau \quad (14\text{-}38)$$

[For a first-order reaction, such as we have in Equation (14-36), Da $= kL/U$.] We shall consider the case of a closed-closed system, in which case we use the Danckwerts boundary conditions

$$-\frac{1}{\mathrm{Pe}_r}\frac{d\Psi}{d\lambda} + \Psi = 1 \qquad \text{at } \lambda = 0$$

---

[9] R. Aris, *Chem. Eng. Sci.*, 9, 266 (1959).

and

Closed-Closed
System

$$\frac{d\Psi}{d\lambda} = 0 \qquad \text{at } \lambda = 1$$

At the end of the reactor, where $\lambda = 1$, the solution is

Conversion for a
first-order reaction
in a tubular or
packed-bed reactor
with dispersion

$$
\boxed{
\begin{aligned}
\psi_L &= \frac{C_{AL}}{C_{A0}} = 1 - X \\[2mm]
&= \frac{4q \, \exp(\mathrm{Pe}_r/2)}{(1+q)^2 \, \exp(\mathrm{Pe}_r q/2) - (1-q)^2 \, \exp(-\mathrm{Pe}_r q/2)}
\end{aligned}
}
\qquad (14\text{-}39)
$$

where $q = \sqrt{1 + 4\mathrm{Da}/\mathrm{Pe}_r}$.

This solution was first obtained by Danckwerts[10] and has been published many places (e.g., Levenspiel[11]). With a slight rearrangement of Equation (14-39) we obtain the conversion as a function of Da and Pe_r.

$$
\boxed{
X = 1 - \frac{4q \, \exp(\mathrm{Pe}_r/2)}{(1+q)^2 \, \exp(\mathrm{Pe}_r q/2) - (1-q)^2 \, \exp(-\mathrm{Pe}_r q/2)}
}
\qquad (14\text{-}40)
$$

Outside the limited case of a first-order reaction, a numerical solution of the equation is required, and because this is a split-boundary-value problem, an iterative technique is required.

Example 14–1   Conversion Using Dispersion and Tanks-in-Series Models

The first-order reaction

$$A \longrightarrow B$$

is carried out in a 10-cm-diameter tubular reactor 6.36 m in length. The specific reaction rate is 0.25 min^{-1}. The results of a tracer test carried out on this reactor are shown in Table E14-1.1.

TABLE E14-1.1.   EFFLUENT TRACER CONCENTRATION AS A FUNCTION OF TIME

t (s)	0	1	2	3	4	5	6	7	8	9	10	12	14
C (mg/L)	0	1	5	8	10	8	6	4	3	2.2	1.5	0.6	0

Calculate conversion using **(a)** the closed vessel dispersion model, **(b)** PFR, **(c)** the tanks-in-series model, and **(d)** a single CSTR.

---

[10]P. V. Danckwerts, *Chem. Eng. Sci.*, 2, 1 (1953).

[11]Levenspiel, *Chemical Reaction Engineering*, 2nd ed.

Solution

(a) We will use Equation (14-40) to calculate the conversion

$$X = 1 - \frac{4q \ \exp(Pe_r/2)}{(1+q)^2 \ \exp(Pe_r q/2) - (1-q)^2 \ \exp(-Pe_r q/2)} \tag{14-40}$$

where $q = \sqrt{1 + 4Da/Pe_r}$, $Da = \tau k$, and $Pe_r = UL/D_a$. We calculate Pe_r from Equation (14-27):

$$\frac{\sigma^2}{\tau^2} = \frac{2}{Pe_r} - \frac{2}{Pe_r^2} (1 - e^{-Pe_r}) \tag{14-27}$$

First calculate t_m and
σ^2 from RTD data

However, we must find τ^2 and σ^2 from the tracer concentration data first.

$$\tau = \int_0^\infty tE(t)\, dt = \frac{V}{v} \tag{E14-1.1}$$

$$\sigma^2 = \int_0^\infty (t - \tau)^2 E(t)\, dt = \int_0^\infty t^2 E(t)\, dt - \tau^2 \tag{E14-1.2}$$

Consider the data listed in Table E14-1.2.

TABLE E14-1.2.   CALCULATIONS TO DETERMINE t_m AND σ^2

t	0	1	2	3	4	5	6	7	8	9	10	12	14
$C(t)$	0	1	5	8	10	8	6	4	3	2.2	1.5	0.6	0
$E(t)$	0	0.02	0.1	0.16	0.2	0.16	0.12	0.08	0.06	0.044	0.03	0.012	0
$tE(t)$	0	0.02	0.2	0.48	0.8	0.80	0.72	0.56	0.48	0.40	0.3	0.14	0
$t^2E(t)$	0	0.02	0.4	1.44	3.2	4.0	4.32	3.92	3.84	3.60	3.0	1.68	0

Here again
spreadsheets can be
used to calculate τ^2
and σ^2

$$\int_0^\infty C(t)\, dt = 50 \ \text{g} \cdot \text{min}$$

$$\tau = t_m = \int_0^\infty tE(t)\, dt = 5.15 \ \text{min}$$

Calculating the first term on the rhs of Equation (E14-1.2) we find

$$\int_0^\infty t^2 E(t)\, dt = (\tfrac{1}{3})[1(0) + 4(0.02) + 2(0.4) + 4(1.44) + 2(3.2) + 4(4.0)$$
$$+ 2(4.32) + 4(3.92) + 2(3.84) + 4(3.6) + 1(3.0)]$$
$$+ (\tfrac{2}{3})[3.0 + 4(1.68) + 0]$$
$$= 32.63 \ \text{min}^2$$

Substituting these values to Equation (E14-1.2) we obtain the variance, σ^2.

$$\sigma^2 = 32.63 - (5.15)^2 = 6.10 \ \text{min}^2$$

Most people, including the author, would use POLYMATH or Excel to form Table E14-1.2 and to calculate t_m and σ^2. Dispersion in a closed vessel is represented by

$$\frac{\sigma^2}{\tau^2} = \frac{2}{\text{Pe}_r^2}(\text{Pe}_r - 1 + e^{-\text{Pe}_r}) \qquad (14\text{-}27)$$

Calculate Pe$_r$ from t_m and σ^2

$$= \frac{6.1}{(5.15)^2} = 0.23 = \frac{2}{\text{Pe}_r^2}(\text{Pe}_r - 1 + e^{-\text{Pe}_r})$$

Solving for Pe$_r$ either by trial and error or using POLYMATH, we obtain

$$\text{Pe}_r = 7.5$$

Next, calculate D_a, q, and X

Next we calculate Da to be

$$\text{Da} = \tau k = (5.15 \text{ min})(0.25 \text{ min}^{-1}) = 1.29$$

Using the equations for q and X gives

$$q = \sqrt{1 + \frac{4\text{Da}}{\text{Pe}_r}} = \sqrt{1 + \frac{4(1.29)}{7.5}} = 1.30$$

Then

$$\frac{\text{Pe}_r q}{2} = \frac{(7.5)(1.3)}{2} = 4.87$$

Substitution into Equation (14-40) yields

Dispersion Model

$$X = 1 - \frac{4(1.30)\,e^{(7.5/2)}}{(2.3)^2 \exp(4.87) - (-0.3)^2 \exp(-4.87)}$$

$$= 0.68 \qquad 68\% \text{ conversion for the dispersion model}$$

When dispersion effects are present in this tubular reactor, 68% conversion is achieved.

(b) If the reactor were operating ideally as a plug-flow reactor, the conversion would be

PFR

$$X = 1 - e^{-\tau k} = 1 - e^{-\text{Da}} = 1 - e^{-1.29} = 0.725$$

That is, 72.5% conversion would be achieved in an ideal plug-flow reactor.

Tanks in series Model

(c) Conversion using the tanks-in-series model: We recall Equation (14-12) to calculate the number of tanks in series:

$$n = \frac{\tau^2}{\sigma^2} = \frac{(5.15)^2}{6.1} = 4.35$$

To calculate the conversion, we recall Equation (4-11). For a first-order reaction for n tanks in series, the conversion is

$$X = 1 - \frac{1}{(1 + \tau_i k)^n} = 1 - \frac{1}{[1 + (\tau/n)k]^n} = 1 - \frac{1}{(1 + 1.29/4.35)^{4.35}}$$

$$= \textbf{67.7\% for the tanks-in-series model}$$

(d) For a single CSTR,

CSTR

$$X = \frac{\tau k}{1 + \tau k} = \frac{1.29}{2.29} = 0.563$$

So 56.3% conversion would be achieved in a single ideal tank.

Summary:

Summary

> PFR: $X = 72.5\%$
>
> Dispersion: $X = 68.0\%$
>
> Tanks in series: $X = 67.7\%$
>
> CSTR: $X = 56.3\%$

In this example, correction for finite dispersion, whether by a dispersion model or a tanks-in-series model, is significant.

Tanks-in-Series Model Versus Dispersion Model. We have seen that we can apply both of these one-parameter models to tubular reactors using the variance of the RTD. For first-order reactions the two models can be applied with equal ease. However, the tanks-in-series model is mathematically easier to use to obtain the effluent concentration and conversion for reaction orders other than one and for multiple reactions. However, we need to ask what would be the accuracy of using the tanks-in-series model over the dispersion model. These two models are equivalent when the Peclet–Bodenstein number is related to the number of tanks in series, n, by the equation [12]

$$\text{Bo} = 2(n - 1) \tag{14-41}$$

Equivalency between models of tanks-in-series and dispersion

or

$$n = \frac{\text{Bo}}{2} + 1 \tag{14-42}$$

where $\text{Bo} = UL/D_a$, where U is the superficial velocity, L the reactor length, and D_a the dispersion coefficient.

For the conditions in Example 14-1, we see that the number of tanks calculated from the Bodenstein number, Bo (i.e., Pe_r), Equation (14-42), is 4.75, which is very close to the value of 4.35 calculated from Equation (14-12). Consequently, for reactions other then first-order, one would solve successively for the exit concentration and conversion from each tank in series for both a battery of four tanks in series and of five tanks in series in order to bound the expected values.

In addition to the one-parameter models of tanks-in-series and dispersion, many other one-parameter models exist when a combination of ideal reactors is to model the real reactor. For example, if the real reactor were modeled as a PFR and CSTR in series, the parameter would be the fraction, f, of the total reactor volume that behaves as a CSTR. Another one-parameter model would be the fraction of fluid that bypasses the ideal reactor. We can dream up many other situations which would alter the behavior of ideal reactors in a way that adequately describes a real reactor. However, it may be that one parameter is not sufficient to yield an adequate comparison between theory

---

[12]K. Elgeti, *Chem. Eng. Sci.*, *51*, 5077 (1996).

and practice. We explore these situations with combinations of ideal reactors in the section on two-parameter models.

14.3 Two-Parameter Models—Modeling Real Reactors with Combinations of Ideal Reactors

Creativity and engineering judgment are necessary for model formulation

It can be shown how a real reactor might be modeled by one of two different combinations of ideal reactors. These are but two of an almost unlimited number of combinations that could be made. However, if we limit the number of adjustable parameters to two (e.g., volume of the exchange reactor and exchange flow rate), the situation becomes much more tractable. Once a model has been chosen, what remains is to check to see whether it is a reasonable model and to determine the values of the model's parameters. Usually, the simplest means of obtaining the necessary data is some form of tracer test. These

A tracer experiment is used to evaluate the model parameters

tests have been described in Chapter 13, together with their uses in determining the RTD of a reactor system. Tracer tests can be used to determine the RTD, which can then be used in a similar manner to determine the suitability of the model and the value of its parameters.

In determining the suitability of a particular reactor model and the parameter values from tracer tests, it is usually not necessary to calculate the RTD function $E(t)$. The required information can be acquired directly from measurements of effluent concentration in a tracer test. The theoretical prediction of the particular tracer test in the chosen model system is compared with the tracer measurements from the real reactor. The parameters in the model are chosen so as to obtain the closest possible agreement between the model and experiment. If the agreement is then sufficiently close, the model is deemed reasonable. If not, another model must be chosen.

The quality of the agreement necessary to fulfill the criterion "sufficiently close" again depends creatively in developing the model and on engineering judgment. The most extreme demands are that the maximum error in the prediction not exceed the estimated error in the tracer test and that there be no observable trends with time in the difference between prediction (the model) and observation (the real reactor). Somewhat less stringent demands can usually be allowed with a reasonable model still resulting. To illustrate how the modeling is carried out, we will consider two different models for a CSTR.

14.3.1 Real CSTR Modeled Using Bypassing and Dead Space

A real CSTR is believed to be modeled as a combination of an ideal CSTR of volume V_s, a dead zone of volume V_d, and a bypass with a volumetric flow rate v_b (Figure 14-11). We have used a tracer experiment to evaluate the parameters of the model V_s and v_s. Because the total volume and volumetric flow rate are known, once V_s and v_s are found, v_b and V_d can readily be calculated.

The model system

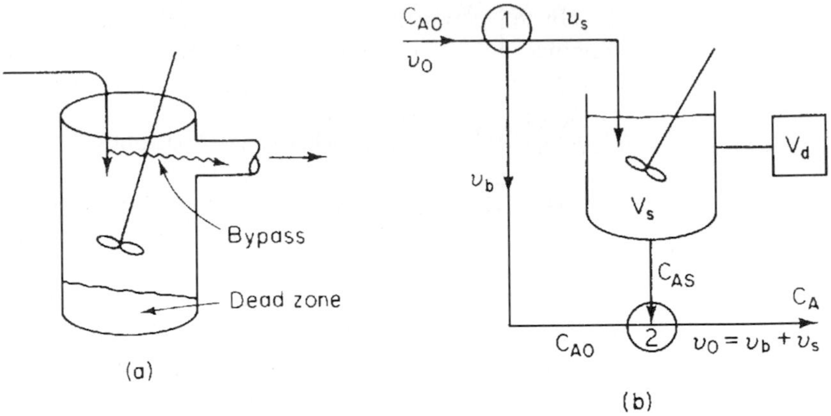

Figure 14-11   (a) Real system; (b) model system.

14.3.1A Solving the Model System for C_A and X

We shall calculate the conversion for this model for the first-order reaction

$$A \longrightarrow B$$

The bypass stream and effluent stream from the reaction volume are mixed at point 2. From a balance on species A around this point,

Balance at junction

$$\text{in} \qquad = \qquad \text{out}$$

$$C_{A0}v_b + C_{As}v_s = C_A(v_b + v_s) \tag{14-43}$$

We can solve for the concentration of A leaving the reactor,

$$C_A = \frac{v_b C_{A0} + C_{As}v_s}{v_b + v_s} = \frac{v_b C_{A0} + C_{As}v_s}{v_0}$$

Let $\alpha = V_s/V$ and $\beta = v_b/v_0$. Then

$$C_A = \beta C_{A0} + (1 - \beta)C_{As} \tag{14-44}$$

For a first-order reaction a mole balance on V_s gives

Mole balance on CSTR

$$v_s C_{A0} - v_s C_{As} - kC_{As}V_s = 0 \tag{14-45}$$

or, in terms of α and β,

$$C_{As} = \frac{C_{A0}(1 - \beta)v_0}{(1 - \beta)v_0 + \alpha V k} \tag{14-46}$$

Substituting Equation (14-46) into (14-44) gives the effluent concentration of species A:

Conversion as a function of model parameters

$$\boxed{\frac{C_A}{C_{A0}} = 1 - X = \beta + \frac{(1 - \beta)^2}{(1 - \beta) + \alpha \tau k}} \tag{14-47}$$

In the previous model we have attempted to model a real reactor with combinations of ideal reactors. The model had two parameters, α and β. If these parameters are known, we can readily predict the conversion. In the following section we shall see how we can use tracer experiments and RTD data to evaluate the model parameters.

14.3.1B Using a Tracer to Determine the Model Parameters in CSTR-with-Dead-Space-and-Bypass Model

In Section 14.3.1 we used the system shown in Figure 14-12, with bypass flow rate v_b and dead volume V_d, to model our real reactor system. We shall inject our tracer, T, as a positive-step input. The unsteady-state balance on the nonreacting tracer T in the reactor volume V_s is

Tracer balance for
step input

$$\text{in} - \text{out} \quad = \text{accumulation}$$

$$v_s C_{T0} - v_s C_{Ts} = \frac{dN_{Ts}}{dt} = V_s \frac{dC_{Ts}}{dt} \tag{14-48}$$

Model system

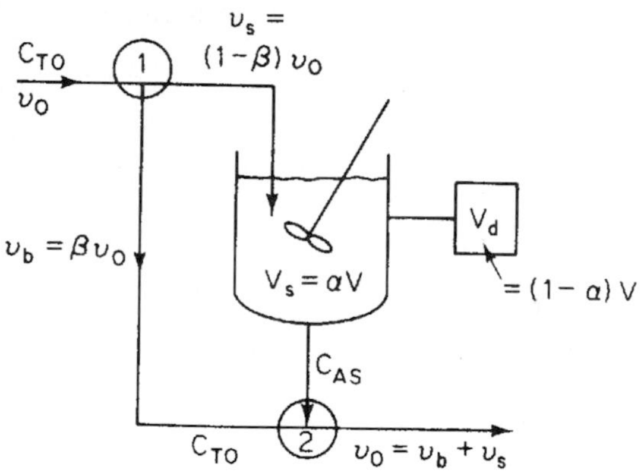

Figure 14-12   Model system: CSTR with dead volume and bypassing.

The conditions for the positive-step input are

$$\text{At } t < 0 \quad C_T = 0$$

$$\text{At } t \geq 0 \quad C_T = C_{T0}$$

A balance around junction point 2 gives

$$C_T = \frac{v_b C_{T0} + C_{Ts} v_s}{v_0} \tag{14-49}$$

As before,

$$V_s = \alpha V$$

$$v_b = \beta v_0$$

$$\tau = \frac{V}{v_0}$$

Integrating Equation (14-48) and substituting in terms of α and β gives

$$\frac{C_{Ts}}{C_{T0}} = 1 - \exp\left[-\frac{1-\beta}{\alpha}\left(\frac{t}{\tau}\right)\right] \tag{14-50}$$

Combining Equations (14-49) and (14-50), the effluent tracer concentration is

$$\frac{C_T}{C_{T0}} = 1 - (1-\beta)\exp\left[-\frac{1-\beta}{\alpha}\left(\frac{t}{\tau}\right)\right] \tag{14-51}$$

We now need to rearrange this equation to extract the model parameters, v_s and V_s, from the proper plot of the effluent tracer concentrations as a function of time. Rearranging yields

Evaluating the
model parameters

$$\ln\frac{C_{T0}}{C_{T0} - C_T} = \ln\frac{1}{1-\beta} + \left(\frac{1-\beta}{\alpha}\right)\frac{t}{\tau} \tag{14-52}$$

Consequently, we plot $\ln[C_{T0}/(C_{T0} - C_T)]$ as a function of t. If our model is correct, a straight line should result with a slope of $(1-\beta)/\tau\alpha$ and an intercept of $\ln[1/(1-\beta)]$.

Example 14–2   CSTR with Dead Space and Bypass

The elementary reaction

$$A + B \longrightarrow C + D$$

is to be carried out in the CSTR shown schematically in Figure 14-11. There is both bypassing and a stagnant region in this reactor. The tracer output for this reactor is shown in Table E14-2.1. The measured reactor volume is 1.0 m³ and the flow rate to the reactor is 0.1 m³/min. The reaction rate constant is 0.28 m³/kmol·min. The feed is equimolar in A and B with an entering concentration of A equal to 2.0 kmol/m³. Calculate the conversion that can be expected in this reactor (Figure E14-2.1).

TABLE E14-2.1   TRACER DATA

C_T (mg/dm³)	1000	1333	1500	1666	1750	1800
t (min)	4	8	10	14	16	18

Two-parameter
model

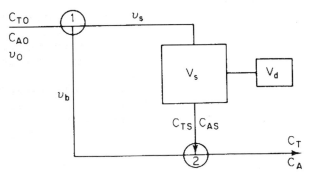

Figure E14-2.1   Schematic of real reactor modeled with dead space (V_d) and bypass (v_b).

Solution

Recalling Equation (14-52)

$$ln\frac{C_{T0}}{C_{T0}-C_T} = ln\frac{1}{1-\beta} + \frac{(1-\beta)}{\alpha}\frac{t}{\tau} \qquad (14\text{-}52)$$

Equation (14-52) suggests that we construct Table E14-2.2 from Table E14-2.1 and plot $C_{T0}/(C_{T0} - C_T)$ as a function of time on semilog paper. Using this table we get Figure E14-2.2.

TABLE E14-2.2   PROCESSED DATA

t (min)	4	8	10	14	16	18
$\dfrac{C_{T0}}{C_{T0}-C_T}$	2	3	4	6	8	10

Evaluating the parameters α and β

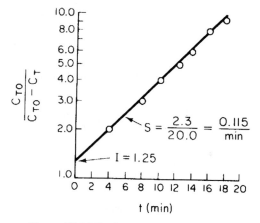

Figure E14-2.2   Response to a step input.

The volumetric flow rate to the well-mixed portion of the reactor, v_s, can be determined from the intercept, I:

$$\frac{1}{1-\beta} = I = 1.25$$

$$\beta = \frac{v_b}{v_0} = 0.2$$

The volume of the well-mixed region, V_s, can be calculated from the slope:

$$\frac{1-\beta}{\alpha\tau} = S = 0.115 \text{ min}^{-1}$$

$$\alpha\tau = \frac{1-0.2}{0.115} = 7 \text{ min}$$

We now proceed to determine the conversion corresponding to these model parameters.

1. **Balance on reactor volume V_s:**

$$\text{in} \quad - \quad \text{out} \quad + \text{ generation} = \text{ accumulation}$$

$$v_s C_{A0} - v_s C_{As} + \quad r_{As} V_s \quad = \quad 0 \qquad \text{(E14-2.1)}$$

2. **Rate law:**

Equimolar feed $\therefore C_{As} = C_{Bs}$

$$-r_A = k C_{As}^2$$

$$-r_{As} = k C_{As} C_{Bs} = k C_{As}^2 \qquad \text{(E14-2.2)}$$

3. **Combining** Equations (E14-2.1) and (E14-2.2) gives

$$v_s C_{A0} - v_s C_{As} - k C_{As}^2 V_s = 0 \qquad \text{(E14-2.3)}$$

Rearranging, we have

$$\tau_s k C_{As}^2 + C_{As} - C_{A0} = 0 \qquad \text{(E14-2.4)}$$

Solving for C_{As} yields

$$C_{As} = \frac{-1 + \sqrt{1 + 4\tau_s k C_{A0}}}{2\tau_s k} \qquad \text{(E14-2.5)}$$

4. **Balance around junction** point 2:

$$\text{in} \quad = \quad \text{out}$$

$$v_b C_{A0} + v_s C_{As} = v_0 C_A \qquad \text{(E14-2.6)}$$

Rearranging Equation (E14-2.6) gives us

$$C_A = \frac{v_0 - v_s}{v_0} C_{A0} + \frac{v_s}{v_0} C_{As} \qquad \text{(E14-2.7)}$$

5. **Parameter evaluation:**

$$v_s = 0.8 v_0 = (0.8)(0.1 \text{ m}^3/\text{min}) = 0.08 \text{ m}^3/\text{min}$$

$$V_s = (\alpha\tau) v_0 = (7.0 \text{ min})(0.1 \text{ m}^3/\text{min}) = 0.7 \text{ m}^3$$

$$\tau_s = \frac{V_s}{v_s} = 8.7 \text{ min}$$

$$C_{As} = \frac{\sqrt{1 + 4\tau_s k C_{A0}} - 1}{2\tau_s k} \qquad \text{(E14-2.8)}$$

$$= \frac{\sqrt{1 + (4)(8.7 \text{ min})(0.28 \text{ m}^3/\text{kmol}\cdot\text{min})(2 \text{ kmol}/\text{m}^3)} - 1}{(2)(8.7 \text{ min})(0.28 \text{ m}^3/\text{kmol}\cdot\text{min})}$$

$$= 0.724 \text{ kmol}/\text{m}^3$$

Substituting into Equation (E14-2.7) yields

$$C_A = \frac{0.1 - 0.08}{0.1} (2) + (0.8)(0.724) = 0.979$$

Finding the conversion

$$X = 1 - \frac{0.979}{2.0} = 0.51$$

If the real reactor were acting as an ideal CSTR, the conversion would be

$$C_A = \frac{\sqrt{1 + 4\tau k C_{A0}} - 1}{2\tau k} \qquad \text{(E14-2.9)}$$

$$\tau = \frac{V}{v_0} = \frac{1 \ m^3}{0.1 \ m^3/min} = 10 \ min$$

$$C_A = \frac{\sqrt{1 + 4(10)(0.28)(2)} - 1}{2(10)(0.28)} = 0.685$$

$X_{model} = 0.51$
$X_{Ideal} = 0.66$

$$X = 1 - \frac{C_A}{C_{A0}} = 1 - \frac{0.685}{2.0} = 0.66 \qquad \text{(E14-2.10)}$$

Other Models.   In Section 14.3.1 it was shown how we formulated a model consisting of ideal reactors to represent a real reactor. First we solved for the exit concentration and conversion for our model system in terms of two parameters α and β. We next evaluated these parameters from data of tracer concentration as a function of time. Finally, we substituted these parameter values into the mole balance, rate law, and stoichiometric equations to predict the conversion in our real reactor.

To reinforce this concept we will use one more example.

14.3.2 Real CSTR Modeled as Two CSTR Interchange

In this particular model there is a highly agitated region in the vicinity of the impeller; outside this region, there is a region with less agitation (Figure 14-13). There is considerable material transfer between the two regions. Both inlet and outlet flow channels connect to the highly agitated region. We shall model the highly agitated region as one CSTR, the quieter region as another CSTR, with material transfer between the two.

The model system

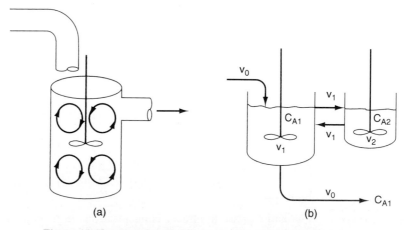

(a)   (b)

Figure 14-13   (a) Real reaction system; (b) model reaction system.

14.3.2A Solving the Model System for C_A and X

Let β represent that fraction of the total flow which is exchanged between reactors 1 and 2, that is,

$$v_1 = \beta v_0 \tag{14-53}$$

and let α represent that fraction of the total volume V occupied by the highly agitated region:

Two parameters:
α and β

$$V_1 = \alpha V \tag{14-54}$$

Then

$$V_2 = (1 - \alpha)V \tag{14-55}$$

The space-time is

$$\tau = \frac{V}{v_0}$$

As shown on the CD-ROM, for a first-order reaction, the exit concentration and conversion are

$$C_{A1} = \frac{C_{A0}}{1 + \beta + \alpha\tau k - \{\beta^2/[\beta + (1-\alpha)\tau k]\}} \tag{14-56}$$

and

Conversion for
two-CSTR model

$$\boxed{X = 1 - \frac{C_{A1}}{C_{A0}} = \frac{(\beta + \alpha\tau k)[\beta + (1-\alpha)\tau k] - \beta^2}{(1 + \beta + \alpha\tau k)[\beta + (1-\alpha)\tau k] - \beta^2}} \tag{14-57}$$

where C_{A1} is the reactor concentration exiting the first reactor and is analogous to C_{T1} in Figure 14-13(b).

14.3.2B Using a Tracer to Determine the Model Parameters in a CSTR with an Exchange Volume

The problem now is to evaluate the parameters α and β using the RTD data. A mole balance on a tracer pulse injected at $t = 0$ for each of the tanks is

$$\text{accumulation} = \text{rate in} - \text{rate out}$$

Unsteady-state
balance of inert
tracer

Reactor 1: $$V_1 \frac{dC_{T1}}{dt} = v_1 C_{T2} - (v_0 C_{T1} + v_1 C_{T1}) \tag{14-58}$$

Reactor 2: $$V_2 \frac{dC_{T2}}{dt} = v_1 C_{T1} - v_1 C_{T2} \tag{14-59}$$

C_{T1} and C_{T2} are the tracer concentrations in reactors 1 and 2, respectively, with $C_{T10} = N_{T0}/V_1$ and $C_{T20} = 0$.

Substituting in terms of α, β, and τ, we arrive at two coupled differential equations describing the unsteady behavior of the tracer that must be solved simultaneously.

Mass balance
on tracer

$$\tau\alpha\,\frac{dC_{T1}}{dt} = \beta C_{T2} - (1+\beta)C_{T1} \tag{14-60}$$

$$\tau(1-\alpha)\,\frac{dC_{T2}}{dt} = \beta C_{T1} - \beta C_{T2} \tag{14-61}$$

CD
Professional

Reference Shelf

Analytical solutions to Equations (14-60) and (14-61) are given in the CD-ROM. However, for more complicated systems, analytical solutions may not be possible to evaluate the system parameter.

$$\left(\frac{C_{T1}}{C_{T10}}\right)_{pulse} = \frac{(\alpha m_1 + \beta + 1)\,e^{m_2 t/\tau} - (\alpha m_2 + \beta + 1)\,e^{m_1 t/\tau}}{\alpha\,(m_1 - m_2)}$$

where $\hspace{13cm}$ (14-60)

$$m_1,\ m_2 = \left[\frac{1-\alpha+\beta}{2\alpha\,(1-\alpha)}\right]\left[-1 \pm \sqrt{1 - \frac{4\alpha\beta\,(1-\alpha)}{(1-\alpha+\beta^2)}}\right]$$

By an appropriate semilog plot of C_{T1}/C_{T0} vs. time, one can evaluate the model parameters α and β.

14.4 Use of Software Packages to Determine the Model Parameters

If analytical solutions are not available to obtain the model parameters from RTD data, one could use ODE solvers. Here, the RTD data would first be fit to a polynomial to the effluent concentration–time data.

Example 14–3

(a) Determine parameters α and β that can be used to model two CSTRs with interchange using the tracer concentration data listed in Table E14-3.1.

TABLE E14-3.1.   RTD DATA

t (min)	0.0	20	40	60	80	120	160	200	240
C_{Te}	2000	1050	520	280	160	61	29	16.4	10.0

(b) Determine the conversion of first-order reaction with $k = 0.03\ \text{min}^{-1}$ and $\tau = 40\ \text{min}$.

Solution

First we will use POLYMATH to fit the RTD to a polynomial. Because of the steepness of the curve, we shall use two polynomials. For $t \leq 80$ min,

$$C_{Te} = 2000 - 59.6t + 0.642t^2 - 0.00146t^3 - 1.04 \times 10^{-5}t^4 \quad \text{(E14-3.1)}$$

For $t > 80$,

$$C_{Te} = 921 - 17.3t + 0.129t^2 - 0.000438t^3 - 5.6 \times 10^{-7}t^4 \quad \text{(E14-3.2)}$$

Trial and error using software packages

where C_{Te} is the exit concentration of tracer determined experimentally. Next we would enter the tracer mole (mass) balances Equations (14-60) and (14-61) into an ODE solver. The POLYMATH program is shown in Table E14-3.2. Finally, we vary the parameters α and β and then compare the calculated effluent concentration C_{T1} with the experimental effluent tracer concentration C_{Te}. After a few trials we converge on the values $\alpha = 0.8$ and $\beta = 0.1$. We see from Figure E14-3.1 and Table E4-3.3 that the agreement between the RTD data and the calculated data are quite good, indicating the validity of our values of α and β. The graphical solution to this problem is given on the CD-ROM and in the 2nd Edition. We now substitute these values in Equation (14-57), and as shown in the CD-ROM, the corresponding conversion is 51% for the model system of two CSTRs with interchange:

$$\boxed{\frac{C_A}{C_{A0}} = 1 - X = \beta + \frac{(1-\beta)^2}{(1-\beta) + \alpha \tau k}}$$

$$\tau k = (40 \text{ min})(0.03 \text{ min}) = 1.2$$

$$X = \frac{[0.1 + (0.8)(1.2)][0.1 + (1 - 0.8)(1.2)] - (0.1)^2}{[1 + 0.1 + (0.8)(1.2)][0.1 + (1 - 0.8)(1.2) - (0.1)^2]}$$

$$X = 0.51$$

$$(X_{\text{model}} = 0.51) < (X_{\text{CSTR}} = 0.55) < (X_{\text{PFR}} = 0.7)$$

TABLE E14-3.2.   POLYMATH PROGRAM: TWO CSTRs WITH INTERCHANGE

Equations:	Initial Values:
`d(CT1)/d(t)=(beta*CT2-(1+beta)*CT1)/alpha/tau`	2000
`d(CT2)/d(t)=(beta*CT1-beta*CT2)/(1-alpha)/tau`	0
`CTe1=2000-59.6*t+0.64*t^2-0.00146*t^3-1.047*10^(-5)*t^4`	
`beta=0.1`	
`CTe2=921-17.3*t+0.129*t^2-0.000438*t^3+5.6*10^(-7)*t^4`	
`alpha=0.8`	
`tau=40`	
`CTe=if(t<80)then(CTe1)else(CTe2)`	
$t_0 = 0$, $t_f = 200$	

CD Professional

Reference Shelf

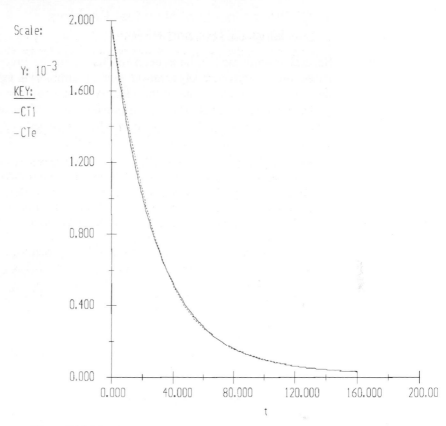

Figure E14-3.1   Comparison of model and experimental exit tracer concentrations.

TABLE E14-3.3.   TWO CSTRS WITH INTERCHANGE

t	CT1	CTe
0	2000	2000
10	1421.1968	1466.4353
20	1014.8151	1050.6448
30	728.96368	740.0993
40	527.42361	519.7568
50	384.9088	372.0625
60	283.76094	276.9488
70	211.6439	211.8353
80	159.93554	161.2816
90	122.6059	126.3396
100	95.434564	99
110	75.464723	77.9116
120	60.622197	61.8576
130	49.449622	49.7556
140	40.920928	40.6576
150	34.311835	33.75
160	29.109434	28.3536
170	24.949027	23.9236
180	21.570003	20.0496
190	18.78499	16.4556
200	16.458267	13

Agreement between
experiment and
model system

14.5 Other Models of Nonideal Reactors Using CSTRs and PFRs

Several reactor models have been discussed in the preceding pages. All are based on the physical observation that in almost all agitated tank reactors, there is a well-mixed zone in the vicinity of the agitator. This zone is usually represented by a CSTR. The region outside this well-mixed zone may then be modeled in various fashions. We have already considered the simplest models which have the main CSTR combined with a dead-space volume; if some short-circuiting of the feed to the outlet is suspected, a bypass stream can be added. The next step is to look at all possible combinations that we can use to model a nonideal reactor using only CSTRs, PFRs, dead volume, and bypassing. The rate of transfer between the two reactors is one of the model parameters. The positions of the inlet and outlet to the model reactor system depend on the physical layout of the real reactor.

Figure 14-14(a) describes a real PFR or PBR with channeling that is modeled as 2 PFRs/PBRs in series. The two parameters are the fraction of flow to the reactors (i.e. β and $(1 - \beta)$) and the fractional volume (i.e. α and $(1 - \alpha)$) of each reactor. Figure 14-14(b) describes a real PFR/PBR that has a backmix region and is modeled as a PFR/PBR in parallel with a CSTR. Figures 14-15(a) and (b) show a real CSTR modeled as two CSTRs with interchange. In one case the fluid exits from the top CSTR (a) and in the other case the fluid exits from the bottom CSTR. The parameter β represents the interchange volumetric flow rate and α the fractional volume of the top reactor, where the fluid exits the reaction system. We note that the reactor in model 14-15(b) was found to describe a real reactor used in production terephthalic acid extremely well. (Proc. Indian Inst. Chem. Eng. Golden Jubilee a congress Delhi p. 323 1997). A number of other combinations of ideal reactions can be found in Levenspiel.[13]

14.6 Using the RTD Versus Needing a Model

This chapter has discussed how to predict conversions in a nonideal reactor. Almost always, we would need to measure or estimate a reactor's RTD. If the reaction is first-order, the conversion can be predicted knowing only the RTD, as shown in Chapter 13. More extensive knowledge of flow pattern characteristics is not required for predicting conversions for first-order reactions.

If the reaction is not first-order, the RTD can be used to place bounds on the conversion using concepts associated with micromixing. One bound is obtained by treating the reactor fluid as if it were completely segregated; the other bound results from considering the fluid as being in a state of maximum mixedness.

---

[13]Levenspiel, *Chemical Reaction Engineering*, pp. 290–293.

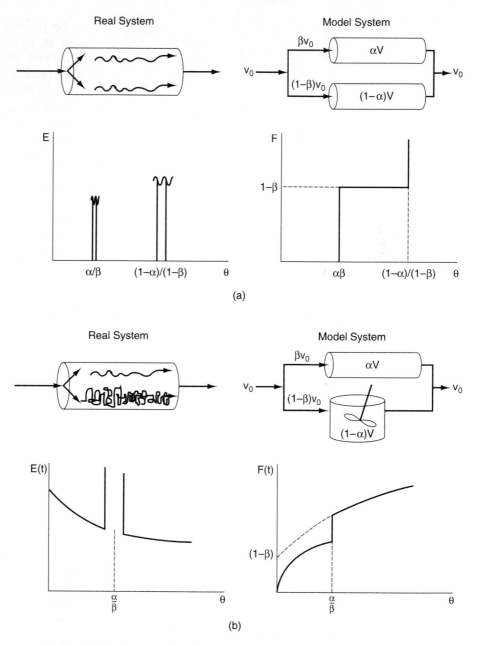

Figure 14-14 Combinations of ideal reactors used to model real PFRs. (a) two PFRs in parallel (b) PFR and CSTR in parallel.

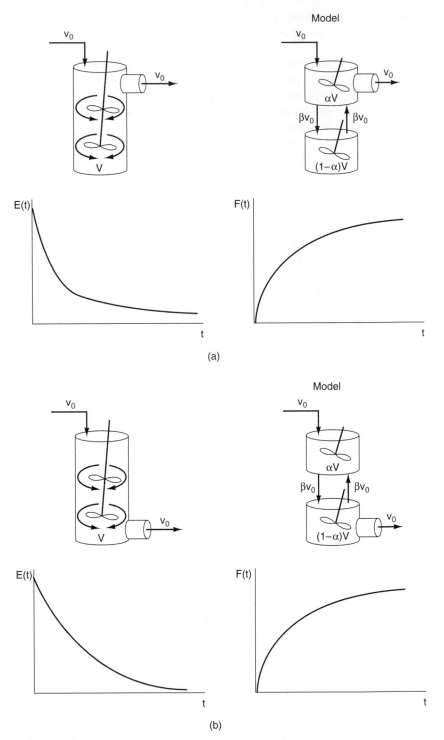

Figure 14-15 Combinations of ideal reactors to model a real CSTR. Two CSTRs with interchange (a) exit from the top CSTR (b) exit from the bottom CSTR.

Summary If the reaction is not first-order and a more precise estimate of reactor conversion is required than can be obtained from the bounds, a reactor model must be assumed. The choice of a proper model is almost pure art *requiring creativity and engineering judgment.* The flow pattern of the model must possess the most important characteristics of that in the real reactor. Standard models are available that have been used with some success, and these can be used as starting points. Models of tank reactors usually consist of combinations of PFRs, perfectly mixed CSTRs, and dead spaces in a configuration that matches as well as possible the flow pattern in the reactor. For tubular reactors, the simple dispersion model has proven most popular.

The parameters in the model, which with rare exception should not exceed two in number, are obtained from the RTD. Once the parameters are evaluated, the conversion in the model, and thus in the real reactor, can be calculated. For typical tank-reactor models, this is the conversion in a series–parallel reactor system. For the dispersion model, the second-order differential equation must be solved, usually numerically. Analytical solutions exist for the first-order situation, but as pointed out previously, no model has to be assumed for the first-order system if the RTD is available.

Correlations exist for the amount of dispersion that might be expected in common packed-bed reactors, so these systems can be designed using the dispersion model without obtaining or estimating the RTD. This situation is perhaps the only one where an RTD is not necessary for designing a nonideal reactor.

SUMMARY

1. The models for predicting conversion from RTD data are:
   a. Zero adjustable parameters
      (1) Segregation model
      (2) Maximum mixedness model
   b. One adjustable parameter
      (1) Tanks-in-series model
      (2) Dispersion model
   c. Two adjustable parameters: real reactor modeled as combinations of ideal reactors
2. Tanks-in-series model: Use RTD data to estimate the number of tanks in series,

$$n = \frac{\tau^2}{\sigma^2} \tag{S15-1}$$

For a first-order reaction

$$X = 1 - \frac{1}{(1 + \tau_i k)^n}$$

3. Dispersion model: For a first-order reaction, use the Danckwerts boundary conditions

$$X = 1 - \frac{4q \, \exp(\mathrm{Pe}_r/2)}{(1+q)^2 \, \exp(\mathrm{Pe}_r q/2) - (1-q)^2 \, \exp(-\mathrm{Pe}_r q/2)} \quad \text{(S15-2)}$$

where

$$q = \sqrt{1 + \frac{4\mathrm{Da}}{\mathrm{Pe}_r}} \quad \text{(S15-3)}$$

$$Da = \tau k \qquad \mathrm{Pe}_r = \frac{UL}{D_a} \qquad \mathrm{Pe}_f = \frac{Ud_p}{D_a \varepsilon} \quad \text{(S15-4)}$$

For laminar flow the dispersion coefficient is

$$D^* = D_{\mathrm{AB}} + \frac{U^2 R^2}{48 D_{\mathrm{AB}}} \quad \text{(S15-5)}$$

Use Equation (S15-6) to calculate Pe_r from the RTD data:

$$\frac{\sigma^2}{\tau^2} = \frac{2}{\mathrm{Pe}_r} - \frac{2}{\mathrm{Pe}_r^2}(1 - e^{-\mathrm{Pe}_r}) \quad \text{(S15-6)}$$

4. If a real reactor is modeled as a combination of ideal reactors, the model should have at most two parameters.

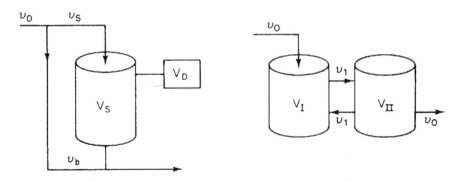

CSTR with bypass   Two CSTRs with

and dead volume   interchange

5. The RTD is used to extract model parameters.
6. Comparison of conversions for a PFR and CSTR with the zero-parameter and two-parameter models. X_{seg} symbolizes the conversion obtained from the segregation model and X_{mm} that from the maximum mixedness model for reaction orders greater than one.

$$\boxed{\begin{array}{c} X_{\mathrm{PFR}} > X_{\mathrm{seg}} > X_{\mathrm{mm}} > X_{\mathrm{CSTR}} \\[2mm] X_{\mathrm{PFR}} > X_{\mathrm{model}} \quad \text{with } X_{\mathrm{model}} < X_{\mathrm{CSTR}} \quad \text{or} \quad X_{\mathrm{model}} < X_{\mathrm{CSTR}} \end{array}}$$

Cautions: For rate laws with unusual concentration functionalities or for nonisothermal operation, these bounds may not be accurate and one must use attainable region analysis (ARA) discussed in Chapter 13.

QUESTIONS AND PROBLEMS

The subscript to each of the problem numbers indicates the level of difficulty: A, least difficult; D, most difficult.

A = ●   B = ■   C = ◆   D = ◆◆

In each of the questions and problems below, rather than just drawing a box around your answer, write a sentence or two describing how you solved the problem, the assumptions you made, the reasonableness of your answer, what you learned, and any other facts that you want to include. You may wish to refer to W. Strunk and E. B. White, *The Elements of Style* (New York: Macmillan, 1979), and Joseph M. Williams, *Style: Ten Lessons in Clarity & Grace* (Glenview, Ill.: Scott, Foresman, 1989) to enhance the quality of your sentences.

P14-1$_B$ Make up and solve an original problem. The guidelines are given in Problem P4-1. However, make up a problem in reverse by first choosing a model system such as a CSTR in parallel with a CSTR and PFR [with the PFR modeled as four small CSTRs in series; Figure P14-1(a)] or a CSTR with recycle and bypass [Figure P14-1(b)]. Write tracer mass balances and use an ODE solver to predict the effluent concentrations. In fact, you could build up an arsenal of tracer curves for different model systems to compare against real reactor RTD data. In this way you could deduce which model best describes the real reactor.

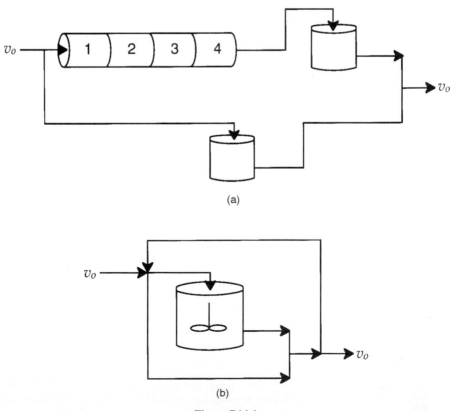

(a)

(b)

Figure P14-1

P14-2$_B$ **What if...**

(a) you were asked to design a tubular vessel that would minimize dispersion? What would be your guidelines? How would you maximize the dispersion? How would your design change for a packed bed?

(b) someone suggested you could use the solution to the flow-dispersion-reactor equation, Equation (14-40), for a second-order equation by linearizing the rate law by lettering $-r_A = k'C_A^2 \simeq (kC_{A0}/2)C_A = k'C_A$? Under what circumstances might this be a good approximation? Would you divide C_{A0} by something other than 2? What do you think of linearizing other non-first-order reactions and using Equation (14-40)? How could you test your results to learn if the approximation is justified?

(c) you were asked for the range of parameter values for which the tanks-in-series model and the dispersion model give similar results (e.g., conversion) and those values for which they give significantly different results?

(d) you varied D_a, k, U, and L in Example 14-1? To what parameters or groups of parameters (e.g., kL^2/D_a) would the conversion be most sensitive?

(e) the first-order reaction in Example 14-1 were carried out in tubular reactors of different diameters, but with the space-time, τ, remaining constant? The diameters would range from a diameter of 0.1 dm ($\mu = 0.01$ cm²/s, $U = 0.1$ cm/s, and $D_{AB} = 10^{-5}$ cm²/s) to a diameter of 1 m. How would your conversion change? Is there a diameter that would maximize or minimize conversion in this range?

(f) the reactor in Example 14-1 were an open vessel rather than a closed vessel? How would your answers change?

(g) the parameter values in Example 14-2 were $\alpha = 0.75$ and $\beta = 0.15$? How would your answers change?

(h) you were asked to explain physically the shapes of the curve in Figures 14-4(a) and (b) look the way they do, what would you say? What if the first pulse in Figure 14-4(a) broke through at $\theta = 0.5$ and the second pulse broke through at $\theta = 1.5$ in a tubular reactor in which a second-order liquid phase reaction

$$2A \longrightarrow B + C$$

was occurring. What would the conversion be if $\tau = 5$ min, $C_{A0} = 2$ mol/dm³, and $k = 0.1$ dm³/mol·min?

P14-3$_B$ The second-order liquid phase reaction

$$A \longrightarrow B + C$$

is to be carried out isothermally. The entering concentration of A is 1.0 mol/dm³. The specific reaction rate is 1.0 dm³/mol·min. A number of used reactors, (shown below) are available, each of which has been characterized by an RTD. There are two silver and white reactors and three maze and blue reactors available.

Reactor	σ (min)	τ (min)	Cost
Maize and Blue	2	2	$25,000
Green and White	4	4	50,000
Scarlet and Gray	3.05	4	50,000
Orange and Blue	2.31	4	50,000
Purple and White	5.17	4	50,000
Silver and Black	2.5	4	50,000
Crimson and White	2.5	2	25,000

(a) You have $50,000 available to spend. What is the greatest conversion you can achieve with the available money and reactors?

(b) How would your answer to (a) change if you had $75,000 available to spend?

(c) From which cities do you think the various used reactors came?

P14-4$_B$ The elementary liquid phase

$$A \xrightarrow{k_1} B, \qquad k_1 = 1.0 \text{ min}^{-1}$$

is carried out in a packed bed reactor in which dispersion is present. What is the conversion?

Additional information:

Porosity = 50%	Reactor length = 0.1 m.
Particle Size = 0.1 cm	Mean velocity = 1 cm/s
Kinematic viscosity = .01 cm^2/s	

P14-5$_A$ A gas-phase reaction is being carried out in a 5-cm-diameter tubular reactor that is 2 m in length. The velocity inside the pipe is 2 cm/s. As a very first approximation the gas properties can be taken as those of air (kinematic viscosity = 0.01 cm^2/s) and the diffusivities of the reacting species are approximately 0.005 cm^2/s.

(a) How many tanks in series would you suggest to model this reactor?

(b) If the second-order reaction A + B $\longrightarrow$ C + D is carried out for the case of equal molar feed and with C_{A0} = 0.01 mol/dm^3, what conversion can be expected at a temperature for which k = 25 dm^3/mol·s?

(c) How would your answers to parts (a) and (b) change if the fluid velocity were reduced to 0.1 cm/s? Increased to 1 m/s?

(d) How would your answers to parts (a) and (b) change if the superficial velocity was 4 cm/s through a packed bed of 0.2-cm-diameter spheres?

(e) How would your answers to parts (a) to (d) change if the fluid were a liquid with properties similar to water instead of a gas, and the diffusivity was 5 × 10^{-6} cm^2/s?

P14-6$_A$ Use the data in Example 13-2 to make the following determinations. (The volumetric feed rate to this reaction was 60 dm^3/min.)

(a) Calculate the Peclet numbers for both open and closed systems.

(b) For an open system, determine the space-time τ and then calculate the % dead volume in a reactor for which the manufacturer's specifications give a volume of 420 dm^3.

(c) Using the dispersion and tanks-in-series models, calculate the conversion for a closed vessel for the first-order isomerization

$$A \longrightarrow B$$

with k = 0.18 min^{-1}.

(d) Compare your results in part (c) with the conversion calculated from the tanks-in-series model, a PFR, and a CSTR.

P14-7$_A$ A tubular reactor has been sized to obtain 98% conversion and to process 0.03 m^3/s. The reaction is a first-order irreversible isomerization. The reactor is 3 m long, with a cross-sectional area of 25 cm^2. After being built, a pulse tracer test on the reactor gave the following data: t_m = 10 s and σ^2 = 65 s^2. What conversion can be expected in the real reactor?

P14-8$_B$ The degree of backmixing in a tall slurry reactor was analyzed by injecting a pulse of methyl orange into the column (presented at the AIChE Los Angeles meeting, November 1982). For a superficial gas velocity of 10 cm/s and a liquid velocity of 3 cm/s:

Tracer Concentration	0	0.2	0.5	0.7	0.85	1	1	0.95
t (min)	0.3	0.6	0.9	1.2	1.5	1.8	2.1	2.4

Tracer Concentration	0.85	0.8	0.65	0.5	0.35	0.25	0.125	0.0
t (min)	2.7	3.0	4.0	5.0	6.0	7.0	8.5	10.0

Note that no units are specified in the tracer concentration values.

(a) What is the mean residence time of a liquid molecule in the reactor?

(b) Develop a model that is consistent with the experimental data. Evaluate all model parameters.

(c) Using the data in Example 12-1, calculate the conversion of methyl linoleate that can be achieved under the same conditions as those of the tracer test. The reaction operating conditions are the same as those given in Example 12-2 (i.e., $m = 3.95$ kg/m^3, partial pressure of hydrogen $= 6$ atm).

P14-9$_C$ The following data were obtained from a step tracer input to a reactor:

t (min)	0	10	11	12	13	14	15	16	17	18	19	20
C (mg/dm^3)	0	0	1	4	7	8	8.5	8.9	9.2	9.5	10	11

t (min)	21	22	23	24	25	26	27	28	29	30
C (mg/dm^3)	15	20	22	23	23.5	24	24.3	24.5	24.5	24.5

Modeling nonideal reactors using combinations of ideal reactors

(a) Develop a model that is consistent with the experimental data.

(b) Evaluate all the model parameters. (*Ans.:* $V_{r1} = 0.26V_{tot}$, $F_{r1} = 0.37F_{tot}$.) For the second-order reaction

$$2A \longrightarrow B + C$$

with $kC_{A0} = 0.1$ min^{-1}, a reactor volume of 1 m^3, and a volumetric flow rate of 0.06 m^3/min, determine the conversion of A:

(c) Using the model developed in part (a). (*Ans.:* $X = 0.62$.)

(d) Using the segregation model.

(e) Using the maximum mixedness model.

(f) Using the dispersion model.

(g) Using the tanks-in-series model.

(h) Using both an ideal PFR and ideal CSTR.

P14-10$_B$ On August 19, 1973, a barge on the Mississippi River was damaged, spilling 140,000 gal of chloroform into the river. The following data were taken at a point 16.3 miles downstream from the spill:

t (h)	7	8	9	10	11	12	13	14	15	16	17
Concentration (ppb)	0	75	200	320	330	290	220	210	175	135	120

t (h)	18	19	20	21	22	24	26	28	36	52
Concentration (ppb)	120	120	120	85	75	75	75	75	70	65

Mississippi

Properties of the Mississippi River (averages):

$$\text{Volumetric flow: 268,000 ft}^3\text{/s} \qquad \text{Width: 4000 ft}$$

$$\text{Velocity: 1.26 mph} \qquad\qquad \text{Depth: 36.3 ft}$$

(*Hint:* Data are given only up to 52 hours. Don't forget the tail!)

(a) What is the mean time that a molecule spends traveling between the spill and the concentration monitoring site? (*Ans.:* 216 h.)

(b) Based on the flow rate, what is the shortest time until you would expect detection downstream?

(c) [Part (c) is a D-level problem.] Develop a model that fits the data. Evaluate the model parameters.

P14-11$_B$ A second-order irreversible reaction takes place in a nonideal, yet isothermal CSTR. The volume of the reactor is 1000 dm^3, and the flow rate of the reactant stream is 1 dm^3/s. At the temperature in the reactor, $k = 0.005$ dm^3/mol·s. The concentration of A in the feedstream is 10 mol/dm^3. The RTD is obtained from a tracer test on this reactor at the desired feed rate and reaction temperature. From the given RTD:

(a) Estimate the (1) maximum segregation and (2) minimum segregation (maximum mixedness) conversions that can be obtained from this reactor because of the different micromixing conditions that are possible.

(b) What conversions can be effected from the tanks-in-series model?

(c) What conversion can be effected by modeling this system as two CSTRs in parallel?

(d) Compare these conversions with those of a perfectly mixed CSTR and PFR of the same volume.

Following are the RTD data of the nonideal reactor:

t (s)	$E(t)$ (s^{-1})	$F(t)$
0	3.250×10^{-3}	
5	3.187×10^{-3}	
10	3.124×10^{-3}	0.032
25	2.945×10^{-3}	
40	2.776×10^{-3}	0.118
70	2.468×10^{-3}	
100	2.194×10^{-3}	0.265
175	1.637×10^{-3}	
250	1.224×10^{-3}	0.510
325	9.184×10^{-4}	
400	6.913×10^{-4}	0.648
700	2.366×10^{-4}	
1,000	9.755×10^{-5}	0.818
2,500	2.691×10^{-5}	
4,000	1.839×10^{-5}	0.928
7,000	8.689×10^{-6}	
10,000	4.104×10^{-6}	0.984
15,000	1.176×10^{-6}	
20,000	3.369×10^{-7}	1.00

P14-12$_C$ Consider a real tubular reactor in which dispersion is occurring.

    (a) For small deviations from plug flow, show that the conversion for a first-order reaction is given approximately as

$$X = 1 - \exp\left[-\tau k + \frac{(\tau k)^2}{Pe_r}\right] \qquad (P15\text{-}12.1)$$

    (b) Show that to achieve the same conversion, the relationship between the volume of a plug-flow reactor V_P and volume of a real reactor V in which dispersion occurs is

$$\frac{V}{V_P} = 1 + \frac{kD_e}{U^2} \qquad (P15\text{-}12.2)$$

    (c) For a Peclet number of 0.1 based on the PFR length, how much bigger than a PFR must the real reactor be to achieve the 99% conversion predicted by the PFR?

    (d) For an nth-order reaction, the ratio of exit concentration for reactors of the same length has been suggested as

$$\frac{C_A}{C_{A_{plug}}} = 1 + \frac{n}{Pe}\,(\tau k\, C_{A0}^{n-1})\,\ln\frac{C_{A0}}{C_{A_{plug}}} \qquad (P15\text{-}12.3)$$

    What do you think of this suggestion?

    (e) What is the effect of dispersion on zero-order reactions?

P14-13$_B$ The flow through a reactor is 10 dm³/min. A pulse test gave the following concentration measurements at the outlet:

t (min)	$c \times 10^5$	t (min)	$c \times 10^5$
0	0	15	238
0.4	329	20	136
1.0	622	25	77
2	812	30	44
3	831	35	25
4	785	40	14
5	720	45	8
6	650	50	5
8	523	60	1
10	418		

    (a) Plot the external age distribution $E(t)$ as a function of time.
    (b) Plot the external age cumulative distribution $F(t)$ as a function of time.
    (c) What is the mean residence time t_m?
    (d) What fraction of the material spends between 2 and 4 min in the reactor?
    (e) What fraction of the material spends longer than 6 min in the reactor?
    (f) What fraction of the material spends less than 3 min in the reactor?
    (g) Plot the normalized distributions $E(\Theta)$ and $F(\Theta)$ as a function of Θ.
    (h) What is the reactor volume?
    (i) Plot internal age distribution $I(t)$ as a function of time.
    (j) What is the mean internal age α_m?
    (k) Plot the intensity function, $\Lambda(t)$, as a function of time.

(l) The activity of a "fluidized" CSTR is maintained constant by feeding fresh catalyst and removing spent catalyst at constant rate. What is the mean catalytic activity if the catalyst decays according to the rate law

$$-\frac{da}{dt} = k_D a^2$$

with

$$k_D = 0.1 \text{ s}^{-1}?$$

(m) What conversion would be achieved in a PFR for a second-order reaction with $kC_{A0} = 0.1 \text{ min}^{-1}$?

(n) What conversion would be achieved in a laminar flow reactor?

(o) What conversion would be achieved in a CSTR?

(p) What would be the conversion for a second-order reaction with $kC_{A0} = 0.1 \text{ min}^{-1}$ using the segregation model?

(q) What would be the conversion for a second-order reaction with $kC_{A0} = 0.1 \text{ min}^{-1}$ using the maximum mixedness model?

(r) If the reactor is modeled as tanks in series, how many tanks are needed to represent this reactor? What is the conversion for a first-order reaction with $k = 0.1 \text{ min}^{-1}$?

(s) If the reactor is modeled by a dispersion model, what are the Peclet numbers for an open system and for a closed system? What is the conversion for a first-order reaction with $k = 0.1 \text{ min}^{-1}$ for each case?

(t) Use the dispersion model to estimate the conversion for a second-order reaction with $k = 0.1 \text{ dm}^3/\text{mol}\cdot\text{s}$ and $C_{A0} = 1 \text{ mol}/\text{dm}^3$.

(u) It is suspected that the reactor might be behaving as shown in Figure P14-13, with perhaps $V_1 = V_2$. What is the "backflow" from the second to the first vessel, as a multiple of v_0?

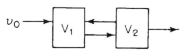

Figure P14-13

(v) If the model above is correct, what would be the conversion for a second-order reaction with $k = 0.1 \text{ dm}^3/\text{mol}\cdot\text{min}$ if $C_{A0} = 1.0 \text{ mol}/\text{dm}^3$?

(w) Prepare a table comparing the conversion predicted by each of the models described above.

(x) How would your answer to (P) change if the reaction were carried out adiabatically with the parameter values given in Problem P13-2(h)?

P14-14$_D$ It is proposed to use the elementary reactions

$$A + B \xrightarrow{k_1} C + D$$

$$C + B \xrightarrow{k_2} X + Y$$

to characterize mixing in a real reactor by monitoring the product distribution at different temperatures. The ratio of specific reaction rates (k_2/k_1) at temperatures T_1, T_2, T_3, and T_4 is 5.0, 2.0, 0.5, and 0.1, respectively. The corresponding values of $\tau k C_{A0}$ are 0.2, 2, 20, and 200.

(a) Calculate the product distribution for the CSTR and PFR in series described in Example 13-4 for $\tau_{CSTR} = \tau_{PFR} = 0.5\tau$.

(b) Compare the product distribution at two temperatures for RTD shown in Examples 13-1 and 13-2 for the complete segregation model and the maximum mixedness model.

(c) Explain how you could use the product distribution as a function of temperature (and perhaps flow rate) to characterize your reactor. For example, could you use the test reactions to determine whether the early mixing scheme or the late mixing scheme in Example 13-4 is more representative of a real reactor? Recall that both schemes have the same RTD.

(d) How should the reactions be carried out (i.e., at high or low temperatures) for the product distribution to best characterize the micromixing in the reactor?

P14-15$_D$ Choose one or more of the reaction schemes in Figure 14-14 and/or Examples 14-2 and 14-3. Use the reactions in one of the examples in Chapter 6 to apply to the these combinations of ideal reactors. Start with $\alpha = 0.5$ and $\beta = 0.5$ and then vary α and β.

CD-ROM MATERIAL

- **Learning Resources**
    1. *Summary Notes for Lectures 33 and 34*
    4. *Solved Problems*
        A. Example CD14-1 Two CSTRs with Interchange
- **Professional Reference Shelf**
    1. *Derivation of Equation for Taylor–Aris Dispersion*
    2. *Real Reactor Modeled in an Ideal CSTR with Exchange Volume*
- **Additional Homework Problems**

CDP14-A$_C$ A real reactor is modeled as

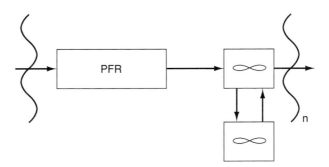

[2nd ed. P14-5]

CDP14-B$_B$ A batch reactor is modeled as

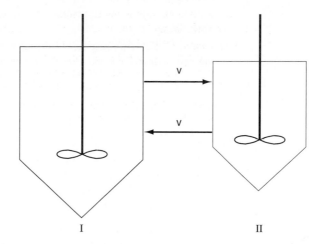

I   II

[2nd ed. P14-13]

CDP14-C$_B$ Develop a model for a real reactor for RTD obtained from a step input. [2nd ed. P14-10]

CDP14-D$_B$ Calculate D_a and X from sloppy tracer data. [2nd ed. P14-6$_A$]

CDP14-E$_B$ Use RTD data from Oak Ridge National Laboratory to calculate the conversion from the tanks-in-series and the dispersion models. [2nd ed. P14-7$_B$]

SUPPLEMENTARY READING

1. Excellent discussions of maximum mixedness can be found in

    DOUGLAS, J. M., "The effect of mixing on reactor design," *AIChE Symp. Ser. 48*, Vol. 60, p. 1 (1964).

    ZWIETERING TH. N., *Chem. Eng. Sci.*, *11*, 1 (1959).

2. Modeling real reactors with a combination of ideal reactors is discussed together with axial dispersion in

    LEVENSPIEL, O., *Chemical Reaction Engineering*, 2nd ed. New York: Wiley, 1972, Chaps. 9 and 10.

    SMITH, J. M., *Chemical Engineering Kinetics*, 3rd ed. New York: McGraw-Hill, 1981, Chap. 6.

    WEN, C. Y., and L. T. FAN, *Models for Flow Systems and Chemical Reactors*. New York: Marcel Dekker, 1975.

3. Mixing and its effects on chemical reactor design has been receiving increasingly sophisticated treatment. See, for example:

    BISCHOFF, K. B., "Mixing and contacting in chemical reactors," *Ind. Eng. Chem.*, *58*(11), 18 (1966).

    BRODKEY, R. S., "Fundamentals of turbulent motion, mixing, and kinetics," *Chem. Eng. Commun.*, *8*, 1 (1981).

NAUMAN, E. B., "Residence time distributions and micromixing," *Chem. Eng. Commun.*, *8*, 53 (1981).

NAUMAN, E. B., and B. A. BUFFHAM, *Mixing in Continuous Flow Systems*. New York: Wiley, 1983.

PATTERSON, G. K., "Applications of turbulence fundamentals to reactor modeling and scaleup," *Chem. Eng. Commun.*, *8*, 25 (1981).

4. See also

CARBERRY, J., and A. VARMA, *Chemical Reaction and Reactor Engineering*. New York: Marcel Dekker, 1987.

DUDUKOVIC, M., and R. FELDER, in *CHEMI Modules on Chemical Reaction Engineering*, Vol. 4, ed. B. Crynes and H. S. Fogler. New York: AIChE, 1985.

5. Dispersion. A discussion of the boundary conditions for closed-closed, open-open, closed-open, and open-closed vessels can be found in

ARIS, R., *Chem. Eng. Sci.*, *9*, 266 (1959).

DRANOFF, J. S., and HSU J. T., *Chem. Eng. Sci.*, *41*, 1930 (1986).

LEVENSPIEL, O., and BISCHOFF K. B., *Adv. in Chem. Eng.*, *4*, 95 (1963).

NAUMAN, E. B., *Chem. Eng. Commun.*, *8*, 53 (1981).

VANDERLAAN, E. Th., *Chem. Eng. Sci.*, *7*, 187 (1958).

WEHNER, J. F., and WILHELM R. H., *Chem. Eng. Sci.*, *6*, 89 (1956).

This is not the end.
It is not even the beginning of the end.
But it is the end of the beginning.

Winston Churchill

Appendix
CD-ROM

1. Appendix A   Example of Graphical Differentiation

2. Appendix D   Measurement of Slopes on Semilog Paper

3. Appendix E   Software Packages

4. Appendix H   Open-Ended Problems

5. Appendix I   How to Use the CD-ROM

6. Appendix J   Use of Computational Chemistry Software Packages

The following links give thermochemical data. (Heats of Formation, C_P, etc.)

1) http://www/uic.edu:80/~mansoori/Thermodynamic.Data.and.Property_html
2) http://webbook.nist.gov
3) http://funnelweb.utcc.utk.edu/~athas/databank/intro.html

Also see *Chem. Tech.* 28 No3 (March) p. 19 (1998).

Numerical   **A**
Techniques

A.1 Useful Integrals in Reactor Design

Also see http://www.integrals.com

$$\int_0^x \frac{dx}{1-x} = \ln\frac{1}{1-x} \tag{A-1}$$

$$\int_0^x \frac{dx}{(1-x)^2} = \frac{x}{1-x} \tag{A-2}$$

$$\int_0^x \frac{dx}{1+\varepsilon x} = \frac{1}{\varepsilon}\ln(1+\varepsilon x) \tag{A-3}$$

$$\int_0^x \frac{1+\varepsilon x}{1-x}\,dx = (1+\varepsilon)\ln\frac{1}{1-x} - \varepsilon x \tag{A-4}$$

$$\int_0^x \frac{1+\varepsilon x}{(1-x)^2}\,dx = \frac{(1-\varepsilon)x}{1-x} - \varepsilon\ln\frac{1}{1-x} \tag{A-5}$$

$$\int_0^x \frac{(1+\varepsilon x)^2}{(1-x)^2}\,dx = 2\varepsilon(1+\varepsilon)\ln(1-x) + \varepsilon^2 x + \frac{(1+\varepsilon)^2 x}{1-x} \tag{A-6}$$

$$\int_0^x \frac{dx}{(1-x)(\Theta_B - x)} = \frac{1}{\Theta_B - 1}\ln\frac{\Theta_B - x}{\Theta_B(1-x)} \qquad \Theta_B \neq 1 \tag{A-7}$$

$$\int_0^x \frac{dx}{ax^2 + bx + c} = \frac{-2}{2ax+b} + \frac{2}{b} \qquad \text{for } b^2 = 4ac \tag{A-8}$$

$$\int_0^x \frac{dx}{ax^2 + bx + c} = \frac{1}{a(p-q)}\ln\left(\frac{q}{p}\cdot\frac{x-p}{x-q}\right) \qquad \text{for } b^2 > 4ac \tag{A-9}$$

$$\int_0^W (1-\alpha W)^{1/2}\,dW = \frac{2}{3\alpha}[1 - (1-\alpha W)^{3/2}] \tag{A-10}$$

921

where p and q are the roots of the equation.

$$ax^2 + bx + c = 0 \qquad \text{i.e., } p, q = \frac{-b \mp \sqrt{b^2 - 4ac}}{2a}$$

$$\int_0^x \frac{a + bx}{c + gx} \, dx = \frac{bx}{g} + \frac{ag - bc}{g^2} \ln \frac{c + gx}{c} \qquad \text{(A-11)}$$

A.2 Equal-Area Graphical Differentiation

There are many ways of differentiating numerical and graphical data. We shall confine our discussions to the technique of equal-area differentiation. In the procedure delineated below we want to find the derivative of y with respect to x.

1. Tabulate the (y_i, x_i) observations as shown in Table A-1.
2. For each *interval*, calculate $\Delta x_n = x_n - x_{n-1}$ and $\Delta y_n = y_n - y_{n-1}$.

TABLE A-1

This method finds use in Chapter 5

x_i	y_i	Δx	Δy	$\dfrac{\Delta y}{\Delta x}$	$\dfrac{dy}{dx}$
x_1	y_1				$\left(\dfrac{dy}{dx}\right)_1$
		$x_2 - x_1$	$y_2 - y_1$	$\left(\dfrac{\Delta y}{\Delta x}\right)_2$	
x_2	y_2				$\left(\dfrac{dy}{dx}\right)_2$
		$x_3 - x_2$	$y_3 - y_2$	$\left(\dfrac{\Delta y}{\Delta x}\right)_3$	
x_3	y_3				$\left(\dfrac{dy}{dx}\right)_3$
		$x_4 - x_3$	$y_4 - y_3$	$\left(\dfrac{\Delta y}{\Delta x}\right)_4$	
x_4	y_4				$\left(\dfrac{dy}{dx}\right)_4$
		$x_5 - x_4$	$y_5 - y_4$	$\left(\dfrac{\Delta y}{\Delta x}\right)_5$	
x_5	y_5		etc.		

3. Calculate $\Delta y_n / \Delta x_n$ as an estimate of the *average* slope in an interval x_{n-1} to x_n.
4. Plot these values as a histogram versus x_i. The value between x_2 and x_3, for example, is $(y_3 - y_2)/(x_3 - x_2)$. Refer to Figure A-1.

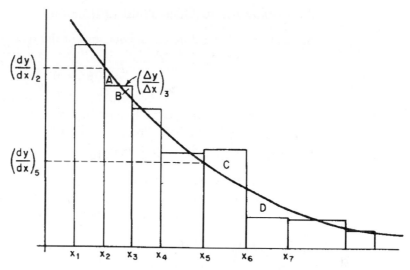

Figure A-1   Equal-area differentiation.

5. Next draw in the *smooth curve* that best approximates the *area* under the histogram. That is, attempt in each interval to balance areas such as those labeled A and B, but when this approximation is not possible, balance out over several intervals (as for the areas labeled C and D). From our definitions of Δx and Δy we know that

$$y_n - y_1 = \sum_{i=2}^{n} \frac{\Delta y}{\Delta x_i} \Delta x_i \qquad (\text{A-12})$$

The equal-area method attempts to estimate dy/dx so that

$$y_n - y_1 = \int_{x_1}^{x_n} \frac{dy}{dx}\, dx \qquad (\text{A-13})$$

that is, so that the area under $\Delta y/\Delta x$ is the same as that under dy/dx, *everywhere possible.*

6. Read estimates of dy/dx from this curve at the data points x_1, x_2, … and complete the table.

An example illustrating the technique is given on the CD-ROM.

    Differentiation is, at best, less accurate than integration. This method also *clearly indicates bad data* and allows for compensation of such data. Differentiation is only valid, however, when the data are presumed to differentiate *smoothly*, as in rate-data analysis and the interpretation of transient diffusion data.

A.3 Solutions to Differential Equations

Methods of solving differential equations of the type

$$\frac{d^2y}{dx^2} - \beta y = 0 \tag{A-14}$$

can be found in such texts as *Applied Differential Equations* by M. R. Spiegel (Upper Saddle River, N.J.: Prentice Hall, 1958, Chap. 4; a great book even though it's old) or in *Differential Equations* by F. Ayres (Schaum Outline Series, McGraw-Hill, New York, 1952). One method of solution is to determine the characteristic roots of

$$\left(\frac{d^2}{dx^2} - \beta\right)y = (m^2 - \beta)y \tag{A-15}$$

which are

$$m = \pm\sqrt{\beta} \tag{A-16}$$

> Solutions of this type are required in Chapter 12

The solution to the differential equation is

$$y = A_1 e^{-\sqrt{\beta}x} + B_1 e^{+\sqrt{\beta}x} \tag{A-17}$$

where A_1 and B_1 are arbitrary constants of integration. It can be verified that Equation (A-17) can be arranged in the form

$$y = A \sinh\sqrt{\beta}x + B \cosh\sqrt{\beta}x \tag{A-18}$$

Equation (A-18) is the more useful form of the solution when it comes to evaluating the constants A and B because $\sinh(0) = 0$ and $\cosh(0) = 1.0$. As an exercise you may want to verify that Equation (A-18) is indeed a solution to Equation (A-14).

A.4 Numerical Evaluation of Integrals

In this section we discuss techniques for numerically evaluating integrals for solving first-order differential equations.

1. *Trapezoidal rule* (two-point) (Figure A-2). This method is one of the simplest and most approximate, as it uses the integrand evaluated at the limits of integration to evaluate the integral:

$$\int_{X_0}^{X_1} f(X)\,dX = \frac{h}{2}[f(X_0) + f(X_1)] \tag{A-19}$$

when $h = X_1 - X_0$.

2. *Simpson's one-third rule* (three-point) (Figure A-3). A more accurate evaluation of the integral can be found with the application of Simpson's rule:

$$\int_{X_0}^{X_2} f(X)\, dX = \frac{h}{3}[f(X_0) + 4f(X_1) + f(X_2)]$$ (A-20)

where

$$h = \frac{X_2 - X_0}{2} \qquad X_1 = X_0 + h$$

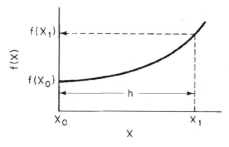

Figure A-2   Trapezoidal rule illustration.

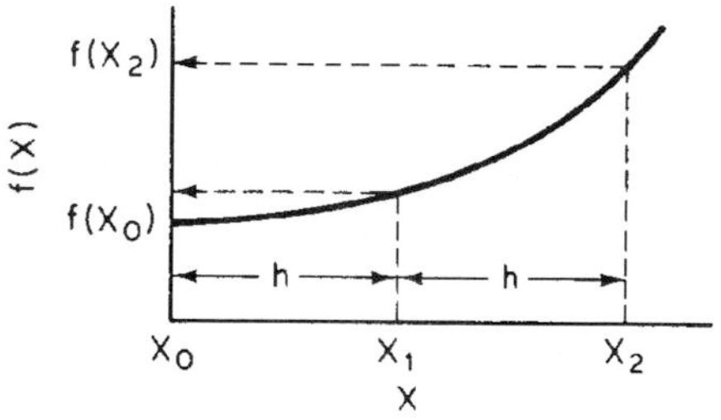

Figure A-3   Simpson's three-point rule illustration.

3. *Simpson's three-eighths rule* (four-point) (Figure A-4). An improved version of Simpson's one-third rule can be made by applying *Simpson's second rule*:

$$\int_{X_0}^{X_3} f(X)\, dX = \tfrac{3}{8}h[f(X_0) + 3f(X_1) + 3f(X_2) + f(X_3)]$$ (A-21)

where

$$h = \frac{X_3 - X_0}{3} \qquad X_1 = X_0 + h \qquad X_2 = X_0 + 2h$$

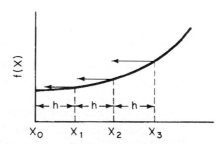

Figure A-4   Simpson's four-point rule illustration.

4. *Five-point quadrature formula:*

$$\int_{X_0}^{X_4} f(X)\, dX = \frac{h}{3}(f_0 + 4f_1 + 2f_2 + 4f_3 + f_4)$$ (A-22)

where

$$h = \frac{X_4 - X_0}{4}$$

5. For $N + 1$ points, where $(N/3)$ is an integer,

$$\int_{X_0}^{X_N} f(X)\,dX = \tfrac{3}{8}h\,[f_0 + 3f_1 + 3f_2 + 2f_3$$

$$+ 3f_4 + 3f_5 + 2f_6 + \cdots + 3f_{N-1} + f_N] \quad \text{(A-23)}$$

where

$$h = \frac{X_N - X_0}{N}$$

6. For $N + 1$ points, where N is even,

$$\int_{X_0}^{X_N} f(X)\,dX = \frac{h}{3}(f_0 + 4f_1 + 2f_2 + 4f_3 + 2f_4 + \cdots + 4f_{N-1} + f_N)$$

$$\text{(A-24)}$$

where

$$h = \frac{X_N - X_0}{N}$$

These formulas are useful in illustrating how the reaction engineering integrals and coupled ODEs (ordinary differential equation(s)) can be solved and also when there is an ODE solver power failure or some other malfunction.

A.5 Software Packages

Instructions on how to use POLYMATH, MatLab, and ASPEN can be found on the CD-ROM.

For the ordinary differential equation solver (ODE solver), contact:

POLYMATH
CACHE Corporation
P.O. Box 7939
Austin, TX   78713-7379

Aspen Technology, Inc.
10 Canal Park
Cambridge, Massachusetts
02141-2201 USA
E-mail: info@aspentech.com
Website: http://www.aspentech.com

Matlab
The Math Works, Inc.
20 North Main Street, Suite 250
Sherborn, MA   01770

Maple
Waterloo Maple Software
766884 Ontario, Inc.
160 Columbia Street West
Waterloo, Ontario,   Canada   N2L3L3

A critique of some of these software packages (and others) can be found in *Chemical Engineering Education*, Vol. XXV, Winter, p. 54 (1991).

Ideal Gas Constant and Conversion Factors   B

Ideal Gas Constant

$$R = \frac{8.314 \ \text{kPa} \cdot \text{dm}^3}{\text{mol} \cdot \text{K}}$$

$$R = \frac{1.987 \ \text{Btu}}{\text{lb mol} \cdot \degree\text{R}}$$

$$R = \frac{0.73 \ \text{ft}^3 \cdot \text{atm}}{\text{lb mol} \cdot \degree\text{R}}$$

$$R = \frac{8.3144 \ \text{J}}{\text{mol} \cdot \text{K}}$$

$$R = 0.082 \ \frac{\text{dm}^3 \cdot \text{atm}}{\text{mol} \cdot \text{K}} = \frac{0.082 \ \text{m}^3 \cdot \text{atm}}{\text{kmol} \cdot \text{K}}$$

$$R = \frac{1.987 \text{cal}}{\text{mol} \cdot \text{K}}$$

Volume of Ideal Gas

1 lb mol of an ideal gas at 32°F and 1 atm occupies 359 ft³.
1 g mol of an ideal gas at 0°C and 1 atm occupies 22.4 dm³.

$$C_A = \frac{P_A}{RT} = \frac{y_A P}{RT}$$

where C_A = concentration of A, mol/dm³
R = ideal gas constant, kPa·dm³/mol·K
T = temperature, K
P = pressure, kPa
y_A = mole fraction of A

Volume

1 cm³	= 0.001 dm³
1 in³	= 0.0164 dm
1 fluid oz	= 0.0296 dm³
1 ft³	= 28.32 dm³
1 m³	= 1000 dm³
1 U.S. gallon	= 3.785 dm³

Length

1 Å	= 10^{-8} cm
1 dm	= 10 cm
1 μm	= 10^{-4} cm
1 in.	= 2.54 cm
1 ft	= 30.48 cm
1 m	= 100 cm

$$\left(1 \ \text{ft}^3 = 28.32 \ \text{dm}^3 \times \frac{1 \ \text{gal}}{3.785 \ \text{dm}^3} = 7.482 \ \text{gal} \right)$$

Pressure

1 torr (1 mmHg) = 0.13333 kPa
1 in. H_2O = 0.24886 kPa
1 in. Hg   = 3.3843 kPa
1 atm   = 101.33 kPa
1 psi   = 6.8943 kPa
1 megadyne/cm^2 = 100 kPa

Energy (Work)

1 kg·m^2/s^2 = 1 J
1 Btu   = 1055.06 J
1 cal   = 4.1841 J
1 L·atm   = 101.34 J
1 hp·h   = 2.6806×10^6 J
1 kWh   = 3.6×10^6 J

Temperature

°F   = $1.8 \times$ °C + 32
°R   = °F + 459.69
K   = °C + 273.16
R   = $1.8 \times$ K
°Réamur = $1.25 \times$ °C

Mass

1 lb   = 454 g
1 kg   = 1000 g
1 grain   = 0.0648 g
1 oz (avoird.) = 28.35 g
1 ton   = 908,000 g

Viscosity

1 poise = 1 g/cm·s

Rate of change of energy with time

$$1 \text{ watt} = 1 \text{ J/s}$$

$$1 \text{ hp} = 746 \text{ J/s}$$

Force

$$1 \text{ dyne} = 1 \text{ g·cm/s}^2$$

$$1 \text{ Newton} = 1 \text{ kg·m/s}^2$$

$$1 \text{ Newton/m}^2 = 1 \text{ Pa}$$

Work

$$\text{Work} = \text{Force·Distance}$$

$$1 \text{ Joule} = 1 \text{ Newton·meter} = 1 \text{ kg m}^2/\text{s}^2 = 1 \text{ Pa·m}^3$$

Gravitational conversion factor

Gravitational constant

$$g = 32.2 \text{ ft/s}^2$$

American Engineering System

$$g_c = 32.174 \frac{(\text{ft})(\text{lb}_m)}{(\text{s}^2)(\text{lb}_f)}$$

SI/cgs System

$$g_c = 1 \text{ (Dimensionless)}$$

Thermodynamic C
Relationships
Involving
the Equilibrium
Constant[1]

For the gas-phase reaction

$$A + \frac{b}{a} B \rightleftharpoons \frac{c}{a} C + \frac{d}{a} D \qquad (2\text{-}2)$$

1. The true (dimensionless) equilibrium constant

$$RT \ln K = -\Delta G$$

$$\boxed{K = \frac{a_C^{c/a} \, a_D^{d/a}}{a_A a_B^{b/a}}}$$

where a_i is the activity of species i

$$a_i = \frac{f_i}{f_i^o}$$

where f_i = fugacity of species i
f_i^o = fugacity of the standard state. For gases the standard state is 1 atm.

$$a_i = \frac{f_i}{f_i^o} = \gamma_i P_i$$

---

[1] For the limitations and for further explanation of these relationships, see, for example, K. Denbigh, *The Principles of Chemical Equilibrium*, 3rd ed., Cambridge University Press, Cambridge, 1971, p. 138.

where γ_i is the activity coefficient

$$K = \frac{\gamma_C^{c/a}\,\gamma_D^{d/a}}{\underbrace{\gamma_A\,\gamma_B^{b/a}}_{K_\gamma}} \cdot \underbrace{\frac{P_C^{c/a}\,P_D^{d/a}}{P_A\,P_B^{b/a}}}_{K_p} = K_\gamma K_p$$

K_γ has units of $[\text{atm}]^{-\left(d/a+c/a-\frac{b}{a}-1\right)} = [\text{atm}]^{-\delta}$

K_p has units of $[\text{atm}]^{\left(d/a+c/a-\frac{b}{a}-1\right)} = [\text{atm}]^{\delta}$

For ideal gases $K_\gamma = 1.0$

It is important to be
able to relate
K, K_c, and K_p

2. The pressure equilibrium constant K_P is

$$K_P = \frac{P_C^{c/a}\,P_D^{d/a}}{P_A\,P_B^{b/a}} \tag{C-1}$$

P_i = partial pressure of species i, atm, kPa.

$$P_i = C_i\,RT$$

3. The concentration equilibrium constant K_C is

$$K_C = \frac{C_C^{c/a}\,C_D^{d/a}}{C_A\,C_B^{b/a}} \tag{C-2}$$

4. For ideal gases, K_C and K_P are related by

$$K_P = K_C(RT)^\delta \tag{C-3}$$

$$\delta = \frac{c}{a} + \frac{d}{a} - \frac{b}{a} - 1 \tag{C-4}$$

5. K_P is a function of temperature only, and the temperature dependence of K_P is given by van't Hoff's equation:

$$\boxed{\frac{d\ln K_P}{dT} = \frac{\Delta H_{\text{Rx}}(T)}{RT^2}} \tag{C-5}$$

$$\frac{d\ln K_P}{dT} = \frac{\Delta H_{\text{Rx}}^\circ(T_R) + \Delta\hat{C}_p(T - T_R)}{RT^2} \tag{C-6}$$

6. Integrating, we have

$$\ln\frac{K_P(T)}{K_P(T_1)} = \frac{\Delta H_{\text{Rx}}^\circ(T_R) - T_R\,\Delta\hat{C}_p}{R}\left(\frac{1}{T_1} - \frac{1}{T}\right) + \frac{\Delta\hat{C}_p}{R}\ln\frac{T}{T_1} \tag{C-7}$$

K_P and K_C are related by

$$K_C = \frac{K_P}{(RT)^\delta} \tag{C-8}$$

when

$$\delta = \left(\frac{d}{a} + \frac{c}{a} - \frac{b}{a} - 1\right) = 0$$

then

$$K_P = K_C$$

7. From Le Châtelier's principle we know that for exothermic reactions, the equilibrium shifts to the left (i.e., K and X_e decrease) as the temperature increases. Figures C-1 and C-2 show how the equilibrium constant varies with temperature for an exothermic reaction and for an endothermic reaction, respectively.

Variation of equilibrium constant with temperature

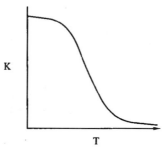

Figure C-1 Exothermic reaction.   **Figure C-2** Endothermic reaction.

8. The equilibrium constant at temperature T can be calculated from the change in the Gibbs free energy using

$$-RT \ln[K(T)] = \Delta G^{\circ}_{Rx}(T) \qquad (C\text{-}9)$$

$$\Delta G^{\circ}_{Rx} = \frac{c}{a} G^{\circ}_{C} + \frac{d}{a} G^{\circ}_{D} - \frac{b}{a} G^{\circ}_{B} - G^{\circ}_{A} \qquad (C\text{-}10)$$

9. Tables that list the standard Gibbs free energy of formation of a given species G°_i are available in the literature.
   1) http://www/uic.edu:80/~mansoori/Thermodynamic.Data.and. Property_html
   2) http://webbook.nist.gov
10. The relationship between the change in Gibbs free energy and enthalpy, H, and entropy, S, is

$$\Delta G = \Delta H - T \Delta S \qquad (C\text{-}11)$$

Example C–1   Water-Gas Shift Reaction

The water-gas shift reaction to produce hydrogen,

$$H_2O + CO \underset{}{\rightleftharpoons} CO_2 + H_2$$

is to be carried out at 1000 K and 10 atm. For an equal-molar mixture of water and carbon monoxide, calculate the equilibrium conversion and concentration of each species.

Data: At 1000 K and 10 atm the Gibbs free energies of formation are $G^{\circ}_{CO} = -47,860$ cal/mol; $G^{\circ}_{CO_2} = -94,630$ cal/mol; $G^{\circ}_{H_2O} = -46,040$ cal/mol; $G^{\circ}_{H_2} = 0$.

Solution

We first calculate the equilibrium constant. The first step in calculating K is to calculate the change in Gibbs free energy for the reaction. Applying Equation (C-10) gives us

$$\Delta G^{\circ}_{Rx} = G^{\circ}_{H_2} + G^{\circ}_{CO_2} - G^{\circ}_{CO} - G^{\circ}_{H_2O} \tag{EC-1.1}$$

$$= 0 + (-94{,}630) - (-47{,}860) - (-46{,}040)$$

$$= -730 \text{ cal/mol}$$

$$-RT \ln K = \Delta G^{\circ}_{Rx}(T) \tag{C-9}$$

$$\ln K = -\frac{\Delta G^{\circ}_{Rx}(T)}{RT} = \frac{-(-730 \text{ cal/mol})}{1.987 \text{ cal/mol} \cdot \text{K} \,(1000 \text{ K})} \tag{EC-1.2}$$

$$= 0.367$$

then

$$K = 1.44$$

Expressing the equilibrium constant first in terms of activities and then finally in terms of concentration, we have

$$K = \frac{a_{CO_2} a_{H_2}}{a_{CO} a_{H_2O}} = \frac{f_{CO_2} f_{H_2}}{f_{CO} f_{H_2O}} = \frac{\gamma_{CO_2} y_{CO_2} \gamma_{H_2} y_{H_2}}{\gamma_{CO} y_{CO} \gamma_{H_2O} y_{CO_2}} \tag{EC-1.3}$$

where a_i is the activity, f_i is the fugacity, γ_i is the activity coefficient (which we shall take to be 1.0 owing to high temperature and low pressure), and y_i is the mole fraction of species i.[2] Substituting for the mole fractions in terms of partial pressures gives

$$y_i = \frac{P_i}{P_T} = \frac{C_i RT}{P_T} \tag{EC-1.4}$$

$$K = \frac{P_{CO_2} P_{H_2}}{P_{CO} P_{H_2O}} = \frac{C_{CO_2} C_{H_2}}{C_{CO} C_{H_2O}} \tag{EC-1.5}$$

In terms of conversion for an equal molar feed we have

$$K = \frac{C_{CO,0} X_e C_{CO,0} X_e}{C_{CO,0} (1 - X_e) C_{CO,0} (1 - X_e)} \tag{EC-1.6}$$

$$= \frac{X_e^2}{(1 - X_e)^2} = 1.44 \tag{EC-1.7}$$

---

[2] See Chapter 9 in J. M. Smith, *Introduction to Chemical Engineering Thermodynamics*, 3rd ed., McGraw-Hill, New York, 1959, and Chapter 9 in S. I. Sandler, *Chemical and Engineering Thermodynamics*, Wiley, (1989) for a discussion of chemical equilibrium including nonideal effects.

From Figure EC-1.1 we read at 1000 K that log K_P = 0.15; therefore, K_P = 1.41, which is close to the calculated value. We note that there is no net change in the number of moles for this reaction; therefore,

$$K = K_p = K_c \text{ (dimensionless)}$$

Taking the square root of Equation (EC-1.7) yields

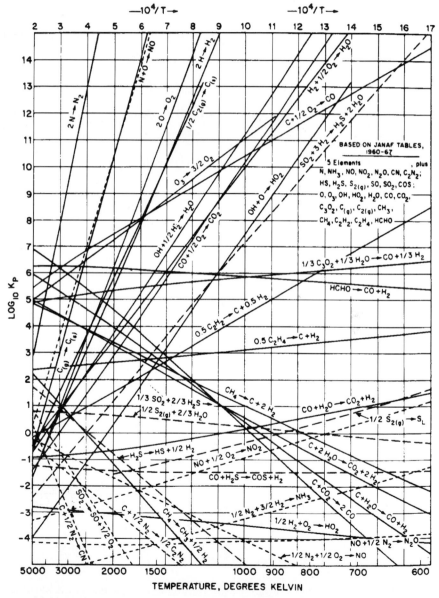

Figure EC-1.1   From M. Modell and R. Reid, *Thermodynamics and Its Applications*, © 1983. Reprinted by permission of Prentice Hall, Inc., Upper Saddle River, N.J.

$$\frac{X_e}{1 - X_e} = (1.44)^{1/2} = 1.2 \qquad \text{(EC-1.8)}$$

Solving for X_e, we obtain

$$X_e = \frac{1.2}{2.2} = 0.55$$

Then

$$C_{CO,0} = \frac{y_{CO,0} P_0}{RT_0}$$

$$= \frac{(0.5)(10 \text{ atm})}{(0.082 \text{ dm}^3 \cdot \text{atm}/\text{mol} \cdot \text{K})(1000 \text{ K})}$$

$$= 0.061 \text{ mol}/\text{dm}^3$$

$$C_{CO,e} = C_{CO,0}(1 - X_e) = (0.061)(1 - 0.55) = 0.0275 \text{ mol}/\text{dm}^3$$

$$C_{H_2O,e} = 0.0275 \text{ mol}/\text{dm}^3$$

$$C_{CO,e} = C_{H_2,e} = C_{CO,0}X_e = 0.0335 \text{ mol}/\text{dm}^3$$

Figure EC-1.1 gives the equilibrium constant as a function of temperature for a number of reactions. Reactions in which the lines increase from left to right are exothermic.

Measurement $\quad$ D

of Slopes

on Semilog Paper

By plotting data directly on the appropriate log-log or semilog graph paper, a great deal of time may be saved over computing the logs of the data and then plotting them on linear graph paper. In the CD-ROM we review the various techniques for plotting data and measuring slopes on semilog paper.

Software Packages E

A detailed explanation on how to use each of the following software packages along with examples can be found on the CD-ROM.

E.1 POLYMATH

POLYMATH is the most user friendly and is used throughout the text. All the following POLYMATH programs are used:

- Curve fitting
- Ordinary differential equation (ODE) solver
- Nonlinear algebraic equation solver
- Nonlinear regression

The most recent version of POLYMATH has both a normal and a stiff ODE solver along with a library to store home problems worked using POLYMATH. The example problems in the text that use POLYMATH are in the POLY-MATH library in the CD-ROM.

E.2 MATLAB

A number of schools use MATLAB as their basic software package. The disadvantage of the MATLAB ODE solver is that it is not particularly user friendly when trying to determine the variation of secondary parameter values. MATLAB will be used for the same four types of programs as POLYMATH.

E.3 ASPEN

ASPEN is a process simulator that is used primarily in senior design courses. It has the steepest learning curve of the software packages used in this text. It has a built-in database of the physical properties of reactants and products. Consequently, one has only to type in the chemicals and the rate law parameters. It is really too powerful to be used for the types of home problems given here. However, Example 8-3 is reworked in the CD-ROM using ASPEN. Perhaps one home assignment should be devoted to using ASPEN to solve a problem with heat effects in order to help familiarize the student with ASPEN.

Nomenclature F

A	Chemical species
A_c	Cross-sectional area (m^2)
A_p	Total external surface area of particle (m^2)
a_p	External surface area of catalyst per unit bed volume (m^2/m^3)
a	Area of heat exchange per unit volume of reactor (m^{-1})
a_c	External surface area per volume of catalyst pellets (m^2/m^3)
B	Chemical species
$\mathbf{B_A}$	Flux of A resulting from bulk flow (gmol/m$^2 \cdot$s)
Bo	Bodenstein number
C	Chemical species
C_i	Concentration of species i (gmol/dm^3)
$\tilde{C}_{pi}$	Heat capacity of species i at temperature T (cal/gmol$\cdot$K)
$\overline{C}_{pi}$	Mean heat capacity of species i between temperature T_0 and temperature T (cal/gmol$\cdot$K)
$\hat{C}_{pi}$	Mean heat capacity of species i between temperature T_R and temperature T (cal/gmol$\cdot$K)
c	Total concentration (gmol/dm^3) (Ch. 11)
D	Chemical species
D_{AB}	Binary diffusion coefficient of A in B (dm^2/s)
D_a	Dispersion coefficient (cm^2/s)
D_e	Effective diffusivity (dm^2/s)
D_K	Knudsen diffusivity (dm^2/s)
D^*	Taylor dispersion coefficient
E	Activation energy (cal/gmol)
(E)	Concentration of free (unbound) enzyme (gmol/dm^3)
F_i	Molar flow rate of species i (gmol/s)
F_{i0}	Entering molar flow of species i (gmol/s)
G	Superficial mass velocity (g/dm$^2 \cdot$s)
G_i	Rate of generation of species i (gmol/s)

$G_i^\circ(T)$	Gibbs free energy of species i at temperature T (cal/gmol·K)
$H_i(T)$	Enthalpy of species i at temperature T (cal/gmol i)
$H_{i0}(T)$	Enthalpy of species i at temperature T_0 (cal/gmol i)
H_i°	Enthalpy of formation of species i at temperature T_R (cal/gmol i)
h	Heat transfer coefficient (cal/m²·s·K)
$\mathbf{J_A}$	Molecular diffusive flux of species A (gmol/m²·s)
K_A	Adsorption equilibrium constant
K_c	Concentration equilibrium constant
K_e	Equilibrium constant (dimensionless)
K_P	Partial pressure equilibrium constant
k	Specific reaction rate
k_c	Mass transfer coefficient (m/s)
M_i	Molecular weight of species i (g/gmol)
m_i	Mass of species i (g)
N_i	Number of moles of species i (gmol)
n	Overall reaction order
Pe	Peclet number (Ch. 14)
P_i	Partial pressure of species i (atm)
Q	Heat flow from the surroundings to the system (cal/s)
R	Ideal gas constant
Re	Reynolds number
r	Radial distance (m)
r_A	Rate of generation of species A per unit volume (gmol A/s·dm³)
$-r_A$	Rate of disappearance of species A per unit volume (gmol A/s·dm³)
$-r_A'$	Rate of disappearance of species A per mass of catalyst (gmol A/g·s)
$-r_A''$	Rate of disappearance of A per unit area of catalytic surface (gmol A/m²·s)
S	An active site (Ch. 10)
(S)	Substrate concentration (gmol/dm³) (Ch. 7)
S_a	Surface area per unit mass of catalyst (m²/g)
S_{DU}	Selectivity parameter (instantaneous selectivety) (Ch. 6)
$\tilde{S}_{DU}$	Overall selectivety of D to U
Sc	Schmidt number (dimensionless) (Ch. 10)
Sh	Sherwood number (dimensionless) (Ch. 10)
SV	Space velocity (s⁻¹)
T	Temperature (K)
t	Time (s)
U	Overall heat transfer coefficient (cal/m²·s·K)
V	Volume of reactor (dm³)
V_0	Initial reactor volume (dm³)
v	Volumetric flow rate (dm³/s)
v_0	Entering volumetric flow rate (dm³/s)
W	Weight of catalyst (kg)
$\mathbf{W_A}$	Molar flux of species A (gmol/m²·s)

X	Conversion of key reactant, A
Y_i	Instantaneous yield of species i
$\tilde{Y}_i$	Overall yield of species i
y	Pressure ratio P/P_0
y_i	Mole fraction of species i
y_{i0}	Initial mole fraction of species i
Z	Compressibility factor
z	Linear distance (cm)

Subscripts

0	Entering or initial condition
b	Bed (bulk)
c	Catalyst
e	Equilibrium
p	Pellet

Greek symbols

α	Reaction order (Ch. 3)
α	Pressure drop parameter (Ch. 4)
α_i	Parameter in heat capacity (Ch. 8)
β_i	Parameter in heat capacity
β	Reaction order
γ_i	Parameter in heat capacity
δ	Change in the total number of moles per mole of A reacted
ε	Fraction change in volume per mole of A reacted resulting from the change in total number of moles
η	Internal effectiveness factor
Θ_i	Ratio of the number of moles of species i initially (entering) to the number of moles of A initially (entering)
λ	Dimensionless distance (z/L) (Ch. 12)
λ	Life expectancy (s) (Ch. 13)
μ	Viscosity (g/cm·s)
ρ	Density (g/cm^3)
ρ_c	Density of catalyst pellet (g/cm^3 of pellet)
ρ_b	Bulk density of catalyst (g/cm^3 of reactor bed)
τ	Space time (s)
ϕ	Void fraction
ϕ_n	Thiele modulus
ψ	Dimensionless concentration (C_A/C_{As})
Ω	External (overall) effectiveness factor

Molecular Dynamics **G**
of Chemical Reactions

In the past, theories attempting to predict the rate of reaction between two species A and B

$$A + B \longrightarrow C + D$$

have focused on the results of the average of a number of collisions. However, present theories to understand the kinetics of reactions focus on studying the kinetics of single collisions between molecules:

$$A(i, U_A) + B(j, U_B) \longrightarrow C(k, U_C) + D(l, U_D)$$

where molecule A has a velocity of U_A and is in internal state i, molecule B has a velocity U_B and is in internal state j, and C and D are defined in a similar manner. Part of the impetus for studying reactions of this type has resulted from the development of monoenergetic molecular beams which can produce pairs of molecules having exactly defined energies and directions. With this approach not only will one be able to study reaction mechanisms in more detail, but also use the information obtained from these monoenergetic studies to calculate the rate of reaction in gas mixtures with a distribution of energies by taking a weighted average over all the collisions taking place. In this appendix we discuss three methods currently being used and researched to predict the specific reaction rate from first principles: collision theory, transition state theory, and molecular dynamics.

G.1 Collision Theory

The origins of collision theory to predict the specific reaction rate lie in the kinetic theory of gases. The rate of reaction is calculated from the product of the frequency of collisions and the fraction of collisions that have enough energy to react. For the reaction involving species A and B to form species C and D,

$$A + B \longrightarrow C + D$$

we base the frequency of collisions (number per unit time), Z, on the average relative velocity of molecule A toward molecule B, U_R. With this basis, the collision theory can be used to predict the rate law as

$$-r_A = Ae^{-E/RT}C_A C_B \tag{G-1}$$

Both A and E may also be temperature dependent. We can estimate the activation energy from either potential energy surfaces or various empirical relationships and the frequency factor from either collision theory, transition state theory or from computational chemistry software (see Appendix J).

In the collision between molecules A and B it is classically assumed that these molecules behave as rigid spheres with radii σ_A and σ_B, respectively. Consequently, whenever molecule A "touches" molecule B, a collision is assumed to have taken place (Figure G-1). With this basis the collision cross-section, S, is

$$S = \pi(\sigma_A + \sigma_B)^2 = \pi\sigma_{AB}^2$$

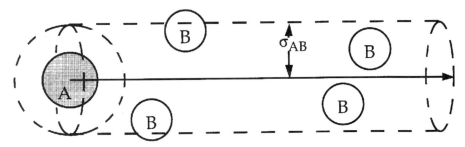

Figure G-1   Cylindrical collisional space swept out by molecule A. B molecules whose centers are within the cylinder would undergo collision.

By defining a relative velocity of A with respect to B, we analyze the B molecules as if they were stationary.

The distance molecule A travels with respect to molecule B is

$$\text{distance} = U_R \, \Delta t \tag{G-2}$$

In time Δt, molecule A sweeps out a volume

$$\Delta V = \pi(\sigma_A + \sigma_B)^2 U_R \, \Delta t = \pi\sigma_{AB}^2 U_R \, \Delta t \tag{G-3}$$

While sweeping out this volume, the A molecule will undergo collisions with the B molecules that are within this volume. The number of collisions that take place in time Δt will be equal to the number of B molecules in this volume, ΔV. That is, the number of collisions by a single A molecule per unit time is

$$Z_{1A \cdot B} = \pi\sigma_{AB}^2 U_R \tilde{C}_B \tag{G-4}$$

Then the collision frequency, Z_{AB}, of all A molecules per unit time per unit volume basis, is obtained by multiplying the collision frequency of one A molecule with all the B molecules, i.e. Z_{1A}, by the concentration of A, $\tilde{C}_A$ (molecules/dm³), giving

$$Z_{AB} = \pi \sigma_{AB}^2 U_R \tilde{C}_A \tilde{C}_B \tag{G-5}$$

From kinetic theory one may recall that the average velocity of a single molecule with mass m corresponding to molecular weight M is given by

$$U_R = \left(\frac{8kT}{\pi m}\right)^{1/2} = \left(\frac{8RT}{\pi M}\right)^{1/2} \tag{G-6}$$

For two water molecules at room temperature the average relative velocity is

$$U_R = \left[\frac{(8)(8314 \text{ g} \cdot \text{m}^2/\text{s}^2 \cdot \text{K})}{\pi \cdot 18 \text{ g}}(300) \text{ K}\right]^{1/2} = 594 \text{ m/s} = 2139 \text{ km/h}$$

For molecules of different molecular weights the relative velocity can be obtained by replacing M by the reduced mass:

$$U_R = \left(\frac{8RT}{\pi \mu_{AB}}\right)^{1/2} \qquad \mu_{AB} = \frac{M_A M_B}{M_A + M_B}$$

$$Z_{AB} = \left(\frac{8\pi RT}{\mu_{AB}}\right)^{1/2} \sigma_{AB}^2 \tilde{C}_A \tilde{C}_B \tag{G-7}$$

In collision theory we postulate that not all collisions are reactive; only those collisions with energy E or greater will react. Assume that the fraction of collisions which have enough energy to result in reaction is given by Maxwell's distribution function.

$$f = \exp\left(-\frac{E}{RT}\right) \tag{G-8}$$

The activation energy shown in Equation (G-8) is

$$E = \left[\begin{array}{c} \text{average energy} \\ \text{of molecules} \\ \text{that do react} \end{array}\right] - \left[\begin{array}{c} \text{average energy} \\ \text{of those molecules} \\ \text{that can react} \\ \text{(i.e., the reactants)} \end{array}\right]$$

The number of molecules reacting per unit volume per unit time is

$$-\tilde{r}_A = \left(\frac{8\pi RT}{\mu_{AB}}\right)^{1/2} \sigma_{AB}^2 e^{-E/RT} \tilde{C}_A \tilde{C}_B \tag{G-9}$$

Multiplying and dividing by Avogadro's number, N_{Av}, to convert the concentrations from number of molecules, $\tilde{C}_A$, to number of moles, C_A, we obtain

$$-r_A = \left[\left(\frac{8\pi RT}{\mu_{AB}}\right)^{1/2} \sigma_{AB}^2 N_{Av}\right] e^{-E/RT} C_A C_B \qquad \text{(G-10)}$$

$$-r_A = Ae^{-E/RT} C_A C_B \qquad \text{(G-1)}$$

If σ_{AB} is around 5 Å, at 400 K the frequency factor A is

$$A = \left[\frac{(8\pi)(8314 \text{ g} \cdot \text{m}^2/\text{s}^2 \cdot \text{K})}{18 \text{ g}} (400 \text{ K})\right]^{1/2} \left(\frac{0.5 \times 10^{-9} \text{ m}}{\text{molecule}}\right)^2 \left(6.023 \times 10^{23} \frac{\text{molecules}}{\text{mol}}\right)$$

$$= 3.2 \times 10^8 \text{ m}^3/\text{mol} \cdot \text{s}$$

The frequency factor calculated from collision theory is usually the upper bound and is usually multiplied by a steric factor. Because this number is an upper bound on the actual rate of reaction, a multiplication steric factor is used to adjust Z_{AB} empirically.

Extensions of Collision Theory.   One could also consider the case when the velocities of A and B are given by a Boltzmann distribution and the hard-sphere diameters are a function of the energy of approach. We now write Equation (G-10) in a little more general form,

$$-\tilde{r}_A = P_r \tilde{Z}_{AB} = \underbrace{P_r \pi (\sigma_A + \sigma_B)^2}_{S_r} U \tilde{C}_A \tilde{C}_B \qquad \text{(G-11)}$$

by defining the reaction cross section S_r as

$$S_r = P_r \pi_{AB}^2 \qquad \text{(G-12)}$$

where P_r is the probability of reaction and π_{AB}^2 is the hard-sphere cross section available for collision. The reaction cross section is a function of the relative velocity (U), the rotational (J) and vibrational (v) energy states of the molecule, and the impact parameter (b).

G.2 Transition State Theory

In transition state theory a transitory geometry is formed by the reactant(s), (A) and (B, C), as they proceed to products. First the molecules A and BC react to form an intermediate called an activated complex, $(ABC)^{\ddagger}$

$$A + BC \longrightarrow (ABC)^{\ddagger}$$

This complex can either decompose back to give the initial reactants

$$(ABC)^{\pm} \longrightarrow A + BC$$

or decompose into the reaction products

$$(ABC)^{\ddagger} \longrightarrow AB + C$$

Combining these steps, the reaction can be written

$$A + BC \underset{k_{-1}}{\overset{k_1}{\rightleftarrows}} (ABC)^{\pm} \overset{k_2}{\longrightarrow} AB + C$$

We can follow the pathway of this reaction by plotting the energy of interaction of molecules A and BC as a function distance between the species A and B (r_1) and B and C$_1$ (r_2) molecule

$$A \overset{r_1}{\longleftrightarrow} B \overset{r_2}{\longleftrightarrow} C \Rightarrow A \overset{r_1}{\longleftrightarrow} B \overset{r_2}{\longleftrightarrow} C \Rightarrow A \overset{r_1}{\longleftrightarrow} B \overset{r_2}{\longleftrightarrow} C$$

Perhaps this pathway is shown more clearly in three dimensions, where the distance between A and B decreases as the molecules come together and move over the potential energy barrier, and the distance between B and C increases as the molecules cross the barrier (Figure G-2). This distance is called the reaction coordinate and represents the progress along the reaction path.

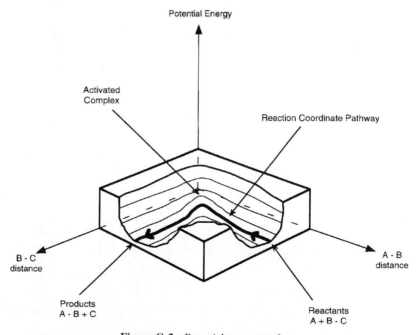

Figure G-2   Potential energy surface.

As the molecules come together, the interaction energy increases to some maximum value at the transition state. As the reaction proceeds past the transition state, the energy of interaction between molecules A and BC decreases as species A and B move close together and species B and C move farther apart. Figure G-3 shows this concept schematically.

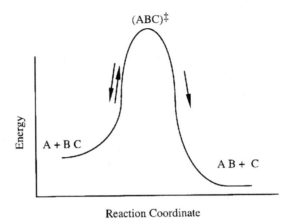

Figure G-3   Reaction path diagram.

The rate of reaction is the rate at which the complex $(ABC)^{\ddagger}$ crosses the energy barrier; that is, the rate of formation by A is

$$-r_A = k(ABC)^{\ddagger} \tag{G-13}$$

where

$$ABC^{\ddagger} \equiv C_{ABC^{\ddagger}}$$

Because the rate of crossing the barrier limits the overall rate of reaction, the reactants A and BC are assumed to be in equilibrium with the activated complex $(ABC)^{\ddagger}$; consequently, we have

$$\frac{(ABC)^{\ddagger}}{(A)(BC)} = K_c^{\ddagger} \tag{G-14}$$

Combining Equations (G-13) and (G-14) yields

$$-r_A = kK_c^{\ddagger}(A)(BC) \tag{G-15}$$

To determine the equilibrium constant, we again follow Laidler[1] and draw on statistical mechanics to obtain $K_c^{\ddagger}$

$$K_c^{\ddagger} = \frac{Q_{ABC}^{\ddagger}}{Q_A Q_B} e^{-E_0/RT} \tag{G-16}$$

---

[1] K. J. Laidler, *Chemical Kinetics*, 3rd ed., HarperCollins, New York, 1987.

where the Q's are the partition functions per unit volume and E_0 is the energy change going from reactants to products at absolute zero. The reacting molecules have electronic energy ε_e, vibrational energy ε_v, rotation energy ε_r, and translational energy ε_t, so that the total partition function Q is

$$Q = q_e q_v q_c q_t$$

with

$$q = \sum g_i e^{-\varepsilon_i/(kT)}$$

where k is Boltzmann's constant, ε_i is the energy of the ith state, and g_i is the degeneracy of the ith energy level. Examples of the partition functions are shown in Table G.1.

TABLE G-1.   PARTITION FUNCTIONS FOR DIFFERENT TYPES OF MOTION

Motion	Degrees of Freedom	Partition Function	Order of Magnitude
Translation	3	$\dfrac{(2\pi mkT)^{3/2}}{h^3}$ (per unit volume)	10^{31}–10^{32} m^{-3}
Rotation (linear molecule)	2	$\dfrac{8\pi^2 IkT}{\sigma h^2}$	10–10^2
Rotation (nonlinear molecule)	3	$\dfrac{8\pi^2 (8\pi^3 I_A I_B I_C)^{1/2} (kT)^{3/2}}{\sigma h^3}$	10^2–10^3
Vibration (per normal mode)	1	$\dfrac{1}{1 - e^{-h\nu/kT}}$	1–10
Free internal rotation	1	$\dfrac{(8\pi^2 I'kT)^{1/2}}{h}$	1–10

where m = mass of molecule
                    I = moment of inertia for linear molecule
I_A, I_B, and I_C = moments of inertia for a nonlinear molecule about three axes at right angles to one another
                   I' = moment of inertia for internal rotation
                    ν = normal-mode vibrational frequency
                    h = Planck's constant

It is useful to remember that the power to which h appears is equal to the number of degrees of freedom.

Source: K. J. Laidler, *Chemical Kinetics*, 3rd ed., HarperCollins, New York, 1987.

Combining Equations (G-15) and (G-16) gives

$$\boxed{\frac{(ABC^{\ddagger})}{[A][B]} = \frac{Q^{\ddagger}_{ABC}}{Q_A Q_B} e^{-E_0/RT}} \qquad (G\text{-}17)$$

    There are two approaches one can take at this point to derive the reaction rate. In one approach the rate of reaction is derived from the translational motion across the energy barrier. In the other, the rate is derived from the vibration of the molecules $(ABC)^{\ddagger}$, in which the vibration results in the separation of AB from $(ABC)^{\ddagger}$. We will choose the latter, where the frequency of crossing the barrier is just the vibrational frequency of the molecule, v, as the frequency approaches zero [i.e., the $(ABC)^{\ddagger}$ complex separates during vibration and therefore no longer vibrates].

Taking the limit of the vibrational partition function as the frequency approaches zero gives us

$$\lim_{v \to 0} \left(\frac{1}{1 - e^{-hv/kT}} \right) = \frac{1}{1 - (1 - hv/kT)} = \frac{kT}{hv} \tag{G-18}$$

We let $Q_{ABC^{\ddagger}}$ be the product of all the partition functions of $ABC^{\ddagger}$ except the vibration partition function. Then

$$\frac{[ABC]^{\#}}{[A][B]} = \frac{kT}{hv} \frac{Q_{ABC^{\ddagger}}}{Q_A Q_B} e^{-E_0/RT} \tag{G-19}$$

Rearranging, we have

$$[ABC]^{\ddagger} v = [A][B] \left[\left(\frac{kT}{h} \right) \frac{Q_{ABC^{\ddagger}}}{Q_A Q_B} e^{-E_0/RT} \right] \tag{G-20}$$

The product of the frequency of crossing the energy barrier, v, and the concentration of the activated complex is just the rate of reaction $[(\text{mol/dm}^3) \times (1/\text{s})]$

$$-r_{AB} = v(ABC)^{\ddagger} = kC_A C_B \tag{G-21}$$

where

$$\boxed{k = \left(\frac{kT}{h} \right) \frac{Q_{ABC^{\ddagger}}}{Q_A Q_B} e^{-E_0/kT}} \tag{G-22}$$

The term kT/h is the order of magnitude typically found for the frequency factor A. At 300 K, $kT/h = 6.25 \times 10^{12}$ s^{-1}. The task now is to evaluate the partition functions, and techniques for doing this evaluation can be found in Laidler[2] and are beyond the scope of this discussion.

G.3 Molecular Dynamics

The theory above does not fully take into consideration the energy states of the reacting molecules and the offset of the collision as measured by the impact parameter. To account for these parameters, Karplus[3] carried out a number of dynamic simulations to calculate collision trajectories.

---

[2] Laidler, *Chemical Kinetics*, p. 109.

[3] M. Karplus, R. N. Porter, and R. D. Sharma, *J. Chem. Phys.*, *43*(9), 3259 (1965).

Figure G-4 shows schematically the position of the molecules at the start and end of the calculation. One notes that for this trajectory calculation, no reaction occurred.

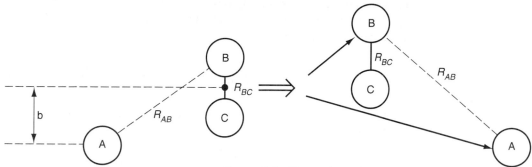

Figure G-4   Nonreactive trajectory.

Figure G-5 shows the distances from separation R_{AB}, R_{AC}, and R_{BC} as a function of time as A and BC approach each other and then separate (e.g. R_{AB} is the distance of separation between species A and species B). One also notes in this figure the vibration of the B—C molecule.

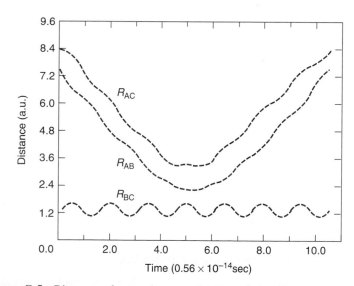

Figure G-5   Distances of separation as a function of time for a non reactive trajectory. [Courtesy of K. J. Laidler, *Chemical Kinetics*, 3rd ed., HarperCollins, New York, 1987. Redrawn from M. Karplus, R. N. Porter, and R. D. Sherma, *J. Chem. Phys.*, *43*(9), 3259 (1965).]

Figure G-6 shows schematically the position of the molecules at the start and the end of the calculation. Figure G-7 shows another trajectory calculation; however, this trajectory results in a reaction. We see from the figure that at a value

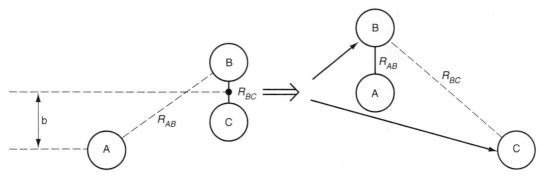

Figure G-6   Reactive trajectory.

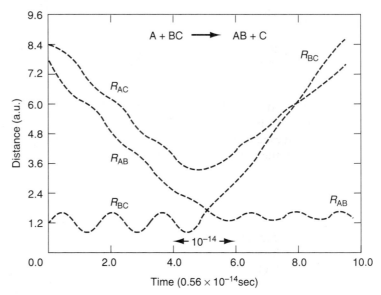

Figure G-7   Distances of separation as a function of time for a reactive trajectory. [Courtesy of K. J. Laidler, *Chemical Kinetics*, 3rd ed., HarperCollins, New York, 1987. Redrawn from M. Karplus, R. N. Porter, and R. D. Sherma, *J. Chem. Phys.*, *43*(9), 3259 (1965).]

(time) approximately 5 on the *x*-axis, the distance of A—B and B—C are equal. After that time, A and B combine and begin to vibrate as C is separated from B.

A large number of simulations are carried out and the trajectories determined. Next, one simply counts the number of trajectories that resulted in the reaction (e.g., Figure G-7). The reaction probability, P_r, is just the ratio of the number of trajectories resulting in reaction, N_r, to the total number of trajectories simulated, N, as N becomes large:

$$P_r = \lim_{N \to x} \left[\frac{N_r(U, v, J, b)}{N(U, v, J, b)} \right] \tag{G-23}$$

Figure G-8 shows the reaction probability as a function of the impact parameter for a particular U_R when B—C is in the ground state ($J = 0$, $v = 0$).

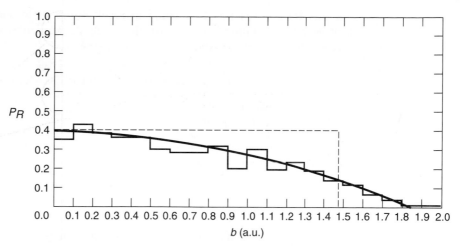

Figure G-8   Reaction probability as a function of the impact parameter. [Courtesy of K. J. Laidler, *Chemical Kinetics*, 3rd ed., HarperCollins, New York, 1987. Redrawn from M. Karplus, R. N. Porter, and R. D. Sherma, *J. Chem. Phys.*, *43*(9), 3259 (1965).]

We see that when the impact parameter is greater than $b_{max} = 1.85$ Å (a.u. in Figure G-8), the reaction probability is zero. As the relative velocity is increased, the solid curve shifts upward and b_{max} increases. For a given energy state the reaction cross section is

$$S_r(U, v, J) = \int_0^x 2P_r \pi b \, db \tag{G-24}$$

Figure G-9 shows the reaction cross section as a function of the relative velocity for the case of the vibrational and rotational ground states, i.e. $v = 0$, $J = 0$. One notes that there is a threshold velocity (i.e., approach energy) nec-

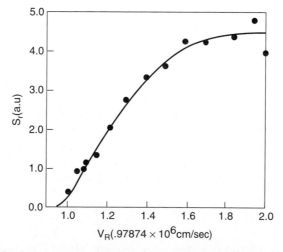

Figure G-9   Reaction cross section as a function of relative velocity for $J = 0$, $v = 0$.

essary for any reaction to occur. This result brings us to one of the key points of Karplus's findings, a comparison of the various energies involved in the reaction. This comparison is shown in Table G-2.

TABLE G-2.   ENERGIES IN H_2 + H EXCHANGE REACTION

Barrier height	9.13 kcal/mol
Vibration energy ($J = 0$, $v = 0$) $E = \frac{1}{2} h\nu$ = 6.2	kcal/mol
Threshold energy	5.89 kcal/mol
Calculated activation energy	7.44 kcal/mol
Measured activation energy	7.45 kcal/mol

We see that the minimum threshold energy plus the ground-state vibrational energy, 12.09 kcal/mol, is the total energy required to cross over the barrier (9.13 kcal/mol) in order for a reaction to occur. This result indicates that all the ground-state vibrational energy is not available for reaction. In addition, the activation energy E in the Arrhenius equation,

$$k = Ae^{-E/RT} \tag{G-25}$$

($E = 7.44$ kcal/mol) is not equal to the barrier height nor the minimum total energy (12.09 kcal/mol) necessary to react. The point is that all these energies (e.g. barrier height, threshold energy) are different from the activation energy. See Appendix J.

Open-Ended Problems H

The following describes open-ended problems on the CD-ROM.

H.1 Design of Reaction Engineering Experiment

The experiment is to be used in the undergraduate laboratory and cost less than $500 to build. The judging criteria are the same as the criteria for the National AIChE Student Chapter Competition. The design is to be displayed on a poster board and explained to a panel of judges. Guidelines for the poster board display are provided by Jack Fishman and are given on the CD-ROM.

H.2 Effective Lubricant Design

Lubricants used in car engines are formulated by blending a base oil with additives to yield a mixture with the desirable physical attributes. In this problem, students examine the degradation of lubricants by oxidation and design an improved lubricant system. The design should include the lubricant system's physical and chemical characteristics, as well as an explanation as to how it is applied to automobiles. Focus: automotive industry, petroleum industry.

H.3 Peach Bottom Nuclear Reactor

The radioactive effluent stream from a newly constructed nuclear power plant must be made to conform with Nuclear Regulatory Commission standards. Students use chemical reaction engineering and creative problem solving to propose solutions for the treatment of the reactor effluent. Focus: problem analysis, safety, ethics.

H.4 Underground Wet Oxidation

You work for a specialty chemicals company, which produces large amounts of aqueous waste. Your chief executive officer (CEO) read in a journal about an emerging technology for reducing hazardous waste, and you must evaluate the system and its feasibility. Focus: waste processing, environmental issues, ethics.

H.5 Hydrodesulfurization Reactor Design

Your supervisor at Kleen Petrochemical wishes to use a hydrodesulfurization reaction to produce ethylbenzene from a process waste stream. You have been assigned the task of designing a reactor for the hydrodesulfurization reaction. Focus: reactor design.

H.6 Continuous Bioprocessing

Most commercial bioreactions are carried out in batch reactors. The design of a continuous bioreactor is desired since it may prove to be more economically rewarding than batch processes. Most desirable is a reactor that can sustain cells that are suspended in the reactor while growth medium is fed in, without allowing the cells to exit the reactor. Focus: mixing modeling, separations, bioprocess kinetics, reactor design.

H.7 Methanol Synthesis

Kinetic models based on experimental data are being used increasingly in the chemical industry for the design of catalytic reactors. However, the modeling process itself can influence the final reactor design and its ultimate performance by incorporating different interpretations of experimental design into the basic kinetic models. In this problem, students are asked to develop kinetic modeling methods/approaches and apply them in the development of a model for the production of methanol from experimental data. Focus: kinetic modeling, reactor design.

H.8 Cajun Seafood Gumbo

Most gourmet foods are prepared by batch processes. In this problem, students are challenged to design a continuous process for the production of gourmet-quality Cajun seafood gumbo from an old family recipe. Focus: reactor design.

Most gourmet foods are prepared by a batch process (actually in a batch reactor). Some of the most difficult gourmet foods to prepare are Louisiana specialities, owing to the delicate balance between spices (hotness) and subtle flavors that must be achieved. In preparing Creole and Cajun food, certain flavors are released only by cooking some of the ingredients in hot oil for a period of time.

We shall focus on one specialty, Cajun seafood gumbo. Develop a continuous-flow reactor system that would produce 5 gal/h of a gourmet-quality seafood gumbo. Prepare a flow sheet of the entire operation. Outline certain experiments and areas of research that would be needed to ensure the success of your project. Discuss how you would begin to research these problems. Make a plan for any experiments to be carried out (see Section 5.7.2).

Following is an old family formula for Cajun seafood gumbo for batch operation (10 quarts, serves 40):

1 cup flour	4 bay leaves, crushed
$1\frac{1}{2}$ cups olive oil	$\frac{1}{2}$ cup chopped parsley
1 cup chopped celery	3 large Idaho potatoes (diced)
2 large red onions (diced)	1 tablespoon ground pepper
5 qt fish stock	1 tablespoon tomato paste
6 lb fish (combination of cod, red	5 cloves garlic (diced)
snapper, monk fish, and halibut)	$\frac{1}{2}$ tablespoon Tabasco sauce
12 oz crabmeat	1 bottle dry white wine
1 qt medium oysters	1 lb scallops
1 lb medium to large shrimp	

1. Make a roux (i.e., add 1 cup flour to 1 cup of boiling olive oil). Cook until dark brown. Add roux to fish stock.
2. Cook chopped celery and onion in boiling olive oil until onion is translucent. Drain and add to fish stock.
3. Add $\frac{1}{3}$ of the fish (2 lb) and $\frac{1}{3}$ of the crabmeat, liquor from oysters, bay leaves, parsley, potatoes, black pepper, tomato paste, garlic, Tabasco, and $\frac{1}{4}$ cup of the olive oil. Bring to a slow boil and cook 4 h, stirring intermittently.
4. Add 1 qt cold water, remove from the stove, and refrigerate (at least 12 h) until $2\frac{1}{2}$ h before serving.
5. Remove from refrigerator, add $\frac{1}{4}$ cup of the olive oil, wine, and scallops. Bring to a light boil, then simmer for 2 h. Add remaining fish (cut to bite size), crabmeat, and water to bring total volume to 10 qt. Simmer for 2 h, add shrimp, then 10 minutes later, add oysters and serve immediately.

How to Use the CD-ROM

The primary purpose of the CD-ROM is to serve as an enrichment resource. The benefits in using the CD are fourfold: (1) To provide you the option/opportunity for further study or clarification of a particular concept or topic through Summary Notes, additional examples, Interactive Computer Modules and Web Modules, (2) To provide the opportunity to practice critical thinking skills, creative thinking skills, and problem-solving skills through the use of **"What if..."** questions and "living example problems," (3) To provide additional technical material for the professional bookshelf on the CD-ROM, (4) To provide other tutorial information such as additional homework problems, thoughts on problem solving, information on how to use computational software in chemical reaction engineering, and a representative course structure.

I.1 Components of the CD-ROM

The following components are listed at the end of most chapters and can be accessed, by chapter, on the CD.

- **Learning Resources**
  1. *Summary Notes*
  2. *Web Modules*
  3. *Interactive Computer Modules*
  4. *Solved Problems*
- **Living Example Problems**
- **Professional Reference Shelf**
- **Frequently Asked Questions (FAQ)**
- **Additional Homework Problems**

The Table I-1 shows the various enrichment resources found in each chapter.

TABLE I-1.   CD-ROM ENRICHMENT RESOURCES

Chapter	1	2	3	4	5	6	7	8	9	10	11	12	13	14
Learning Resources														
Summary Notes	■	■	■	■	■	■	■	■	■	■	■	■	■	■
Web Modules	■			■		■								
Interactive Computer Modules	■	■	■	■	■			■		■				
Solved Problems	■	■	■	■	■	■	■	■						
Living Example Problems					■	■	■	■	■	■	■		■	■
Professional Reference Shelf	■			■	■	■	■	■		■	■	■	■	■
Additional Homework Problems	■	■	■	■	■	■	■	■	■	■	■	■	■	■
Frequently Asked Questions	■	■	■	■	■	■								

- **Other CD-ROM Material**
  - Software Toolbox
  - Representative Syllabi for a 3-Credit-Hour Course and a 4-Credit-Hour Course
  - Virtual Reality Module
  - Credits

Note. The Interactive Computer Modules(ICM) are a high memory use program. Due to this, there have been occasional intermitent problems (10-15% of the users) with the modules for chapters 2 and 10, namely Staging and Hetcat, respectively. The problem sometimes solves itself by trying the CD on another computer. In the ICM Heatfx 2 one can only do the first 3 reactors, and cannot continue on to part 2.

I.1 Usage

There are a number of ways you can use the CD-ROM in conjunction with the text. The CD provides you with *enrichment resources*. Pathways on how to use the materials to learn chemical reaction engineering are shown in Figures P-3 and P-4.

While the author recommends using the living examples before completing homework problems, they may be bypassed, as is the case with all the enrichment resources if time is not available. However, the enrichment resources aid in learning the material and they may also motivate you by the novel use of CRE principles.

I.2 Polymath

This version of POLYMATH is provided for execution directly from this CRE99 Cd-Rom that accompanies the textbook entitled Elements of Chemical Reaction Engineering by H. Scott Fogler. Printing is only supported when POLYMATH is installed directly on your personal computer.

From the windows desktop, go to start, run, and type in "polymath"

Double Click on the "Polymath" Icon that has a yellow thing in the middle.

When you are in the files **menu**, press "L" for load problem from file.

Type:

{cd}:\polymat4\[chname]\[filename]

For example, E9-2 from a computer where the cd is the d: drive one would type:

d:\polymat4\Ch9\E9-2

It is always going to be easier to run a file from the polymath program on the network. Installing the polymath program can be done, but there is no need for it unless it is a personal computer or a computer that doesn't have polymath already on it.

When loading up a file from Polymath, please use the list below to load up the correct example problem. CD Rom is drive D.

TABLE I-2.   POLYMATH EXAMPLE PROBLEMS

Chapter	Files	What to Type	Chapter	Files	What to Type
Ch 2	E2-3a	D:\polymat4\Ch2\E2-3a	Ch 9	E9-1	D:\polymat4\Ch9\E9-1
Ch 2	E2-3	D:\polymat4\Ch2\E2-3	Ch 9	E9-2	D:\polymat4\Ch9\E9-2
Ch 3	E3-8	D:\polymat4\Ch3\E3-8	Ch 9	E9-3	D:\polymat4\Ch9\E9-3
Ch 4	E4-10	D:\polymat4\Ch4\E4-10	Ch 9	E9-4	D:\polymat4\Ch9\E9-4
Ch 4	E4-11	D:\polymat4\Ch4\E4-11	Ch 9	E9-5	D:\polymat4\Ch9\E9-5
Ch 4	E4-7	D:\polymat4\Ch4\E4-7	Ch 9	E9-6	D:\polymat4\Ch9\E9-6
Ch 4	E4-8	D:\polymat4\Ch4\E4-8	Ch 9	E9-7	D:\polymat4\Ch9\E9-7
Ch 4	E4-9	D:\polymat4\Ch4\E4-9	Ch 9	E9-8	D:\polymat4\Ch9\E9-8
Ch 5	E5-6	D:\polymat4\Ch5\E5-6	Ch 10	E10-3	D:\polymat4\Ch10\E10-3
Ch 6	E6-6	D:\polymat4\Ch6\E6-6	Ch 10	E10-5	D:\polymat4\Ch10\E10-5
Ch 6	E6-7	D:\polymat4\Ch6\E6-7	Ch 10	E10-7	D:\polymat4\Ch10\E10-7
Ch 6	E6-8	D:\polymat4\Ch6\E6-8	Ch 10	E10-2	D:\polymat4\Ch10\E10-2
Ch 6	Cobra-1	D:\polymat4\Ch6\Cobra-1	Ch 13	Cde13-1	D:\polymat4\Ch13\Cde13-1
Ch 7	E7-2	D:\polymat4\Ch7\e7-2	Ch 13	cde13-1a	D:\polymat4\Ch13\cde13-1a
Ch 7	E7-3	D:\polymat4\Ch7\e7-3	Ch 13	cde13-1b	D:\polymat4\Ch13\cde13-1b
Ch 7	E7-9	D:\polymat4\Ch7\e7-9	Ch 13	cde13-1c	D:\polymat4\Ch13\cde13-1c
Ch 8	E8-10	D:\polymat4\ch8\e8-10	Ch 13	cde13-1d	D:\polymat4\Ch13\cde13-1d
Ch 8	E8-11	D:\polymat4\ch8\e8-11	Ch 13	cde13-1e	D:\polymat4\Ch13\cde13-1e
Ch 8	E8-12	D:\polymat4\ch8\e8-12	Ch 13	e13-2	D:\polymat4\Ch13\e13-2
Ch 8	E8-5	D:\polymat4\ch8\e8-5	Ch 13	e13-8	D:\polymat4\Ch13\e13-8
Ch 8	E8-6	D:\polymat4\ch8\e8-6	Ch 13	e13-9a	D:\polymat4\Ch13\e13-9a
Ch 8	E8-7	D:\polymat4\ch8\e8-7	Ch 13	e13-9b	D:\polymat4\Ch13\e13-9b
Ch 8	E8-7a	D:\polymat4\ch8\e8-7a	Ch 13	e13-9c	D:\polymat4\Ch13\e13-9c
Ch 9	Cde9-1	D:\polymat4\Ch9\Cde9-1	Ch 13	e13-9d	D:\polymat4\Ch13\e13-9d
Ch 9	Cde9-2	D:\polymat4\Ch9\Cde9-2	Ch 14	e14-3	D:\polymat4\Ch14\e14-3

Table I-2 by Mayur Valanju

Use of Computational Chemistry Software Packages $\quad$ **J**

J.1 Computational Chemical Engineering

As a prologue to the future, our profession is evolving to one of molecular chemical engineering. For chemical reaction engineers, computation chemistry and molecular modeling, this could well be our future. The 4th edition will reflect this change....

Thermodynamic properties of molecular species that are used in reactor design problems can be readily estimated from thermodynamic data tabulated in standard reference sources such as Perry's Handbook or the JANAF Tables. Thermochemical properties of molecular species not tabulated can usually be estimated using group contribution methods. Estimation of activation energies is, however, much more difficult due to the lack of reliable information on transition state structures, and the data required to carry out these calculations is not readily available.

Recent advances in computational chemistry and the advent of powerful easy-to-use software tools have made it possible to estimate important reaction rate quantities (such as activation energy) with sufficient accuracy to permit incorporation of these new methods into the reactor design process. Computational chemistry programs are based on theories and equations from quantum mechanics, which until recently, could only be solved for the simplest systems such as the hydrogen atom. With the advent of inexpensive high-speed desktop computers, the use of these programs in both engineering research and industrial practice is increasing rapidly. Molecular properties such as bond length, bond angle, net dipole moment, and electrostatic charge distribution can be calculated. Additionally, reaction energetics can be accurately determined by using quantum chemistry to estimate heats of formation of reactants, products, and also for transition state structures.

Examples of commercially available computational chemistry programs include SPARTAN developed by Wavefunction, Inc. (http://www. wavefun.com) and Cerius2 from Molecular Simulations, Inc. (http://www. msi.com). The following example utilizes SPARTAN 4.0 to estimate the activation energy for a nucleophilic substitution reaction (SN2). The following calculations were performed on an IBM 43-P RS-6000 UNIX workstation.

An example using SPARTAN to calculate the activation energy for the reaction

$$C_2H_5Cl + OH^- \longrightarrow C_2H_5OH + Cl^-$$

is given on the CD-ROM.

CDPApp.J-1$_A$ Redo Example Appendix J.1.

    (a) Choose different methods of calculation, such as using a value of 2.0Å to constrain the C—Cl and C—O bonds.

    (b) Choose different methods to calculate the potential energy surface. Compare the Ab Initio to the semi-empirical method.

    (c) Within the semi-empirical method, compare the AM1 and PM3 models.

Index

P R E N T I C E   H A L L

Professional Technical Reference
Tomorrow's Solutions for Today's Professionals.

Keep Up-to-Date with
PH PTR Online!

We strive to stay on the cutting-edge of what's happening in professional computer science and engineering. Here's a bit of what you'll find when you stop by **www.phptr.com**:

@ Special interest areas offering our latest books, book series, software, features of the month, related links and other useful information to help you get the job done.

Deals, deals, deals! Come to our promotions section for the latest bargains offered to you exclusively from our retailers.

$ Need to find a bookstore? Chances are, there's a bookseller near you that carries a broad selection of PTR titles. Locate a Magnet bookstore near you at www.phptr.com.

! What's New at PH PTR? We don't just publish books for the professional community, we're a part of it. Check out our convention schedule, join an author chat, get the latest reviews and press releases on topics of interest to you.

Subscribe Today! Join PH PTR's monthly email newsletter!

Want to be kept up-to-date on your area of interest? Choose a targeted category on our website, and we'll keep you informed of the latest PH PTR products, author events, reviews and conferences in your interest area.

Visit our mailroom to subscribe today! **http://www.phptr.com/mail_lists**

8.   **LIMITED WARRANTY AND DISCLAIMER OF WARRANTY:** The Company warrants that the SOFTWARE, when properly used in accordance with the Documentation, will operate in substantial conformity with the description of the SOFTWARE set forth in the Documentation. The Company does not warrant that the SOFTWARE will meet your requirements or that the operation of the SOFTWARE will be uninterrupted or error-free. The Company warrants that the media on which the SOFTWARE is delivered shall be free from defects in materials and workmanship under normal use for a period of thirty (30) days from the date of your purchase. Your only remedy and the Company's only obligation under these limited warranties is, at the Company's option, return of the warranted item for a refund of any amounts paid by you or replacement of the item. Any replacement of SOFTWARE or media under the warranties shall not extend the original warranty period. The limited warranty set forth above shall not apply to any SOFTWARE which the Company determines in good faith has been subject to misuse, neglect, improper installation, repair, alteration, or damage by you. EXCEPT FOR THE EXPRESSED WARRANTIES SET FORTH ABOVE, THE COMPANY DISCLAIMS ALL WARRANTIES, EXPRESS OR IMPLIED, INCLUDING WITHOUT LIMITATION, THE IMPLIED WARRANTIES OF MERCHANTABILITY AND FITNESS FOR A PARTICULAR PURPOSE. EXCEPT FOR THE EXPRESS WARRANTY SET FORTH ABOVE, THE COMPANY DOES NOT WARRANT, GUARANTEE, OR MAKE ANY REPRESENTATION REGARDING THE USE OR THE RESULTS OF THE USE OF THE SOFTWARE IN TERMS OF ITS CORRECTNESS, ACCURACY, RELIABILITY, CURRENTNESS, OR OTHERWISE.

IN NO EVENT, SHALL THE COMPANY OR ITS EMPLOYEES, AGENTS, SUPPLIERS, OR CONTRACTORS BE LIABLE FOR ANY INCIDENTAL, INDIRECT, SPECIAL, OR CONSEQUENTIAL DAMAGES ARISING OUT OF OR IN CONNECTION WITH THE LICENSE GRANTED UNDER THIS AGREEMENT, OR FOR LOSS OF USE, LOSS OF DATA, LOSS OF INCOME OR PROFIT, OR OTHER LOSSES, SUSTAINED AS A RESULT OF INJURY TO ANY PERSON, OR LOSS OF OR DAMAGE TO PROPERTY, OR CLAIMS OF THIRD PARTIES, EVEN IF THE COMPANY OR AN AUTHORIZED REPRESENTATIVE OF THE COMPANY HAS BEEN ADVISED OF THE POSSIBILITY OF SUCH DAMAGES. IN NO EVENT SHALL LIABILITY OF THE COMPANY FOR DAMAGES WITH RESPECT TO THE SOFTWARE EXCEED THE AMOUNTS ACTUALLY PAID BY YOU, IF ANY, FOR THE SOFTWARE.

SOME JURISDICTIONS DO NOT ALLOW THE LIMITATION OF IMPLIED WARRANTIES OR LIABILITY FOR INCIDENTAL, INDIRECT, SPECIAL, OR CONSEQUENTIAL DAMAGES, SO THE ABOVE LIMITATIONS MAY NOT ALWAYS APPLY. THE WARRANTIES IN THIS AGREEMENT GIVE YOU SPECIFIC LEGAL RIGHTS AND YOU MAY ALSO HAVE OTHER RIGHTS WHICH VARY IN ACCORDANCE WITH LOCAL LAW.

ACKNOWLEDGMENT

YOU ACKNOWLEDGE THAT YOU HAVE READ THIS AGREEMENT, UNDERSTAND IT, AND AGREE TO BE BOUND BY ITS TERMS AND CONDITIONS. YOU ALSO AGREE THAT THIS AGREEMENT IS THE COMPLETE AND EXCLUSIVE STATEMENT OF THE AGREEMENT BETWEEN YOU AND THE COMPANY AND SUPERSEDES ALL PROPOSALS OR PRIOR AGREEMENTS, ORAL, OR WRITTEN, AND ANY OTHER COMMUNICATIONS BETWEEN YOU AND THE COMPANY OR ANY REPRESENTATIVE OF THE COMPANY RELATING TO THE SUBJECT MATTER OF THIS AGREEMENT.

Should you have any questions concerning this Agreement or if you wish to contact the Company for any reason, please contact in writing at the address below.

Robin Short
Prentice Hall PTR
One Lake Street
Upper Saddle River, New Jersey 07458

About the CD-ROM

This CD-ROM is a companion to the book *Elements of Chemical Reaction Engineering, 3rd Edition,* by H. Scott Fogler. It contains web browser-driven content intended to supplement the information in the book.

We suggest that you use Netscape Navigator™ 3.x or greater or Netscape Communicator™ 4.x or greater to view the content of this CD. We have included the latest version of Netscape Navigator™ on this CD.

Also included on this CD are **executable software** intended to be installed and run directly from your computer. The software included are Windows® NT/95 and/or DOS 3.1-specific and do not include Macintosh versions. There are instructions and links to this software from the CD.

System Requirements:

Windows® 95/NT, or
Power Macintosh (cannot run executable modules)
CD-ROM drive, 4x or higher
Java-enabled, HTML browser (latest version of Netscape™ is included).

We recommend 36 MB of RAM and at least 16 MB of available space on your hard drive to install the software.

Prentice Hall does not offer technical support for this software. However, if there is a problem with the media, you may obtain a replacement copy by emailing us with your problem at discexchange@phptr.com.